S0-CFQ-284

A. A. STAGG HIGH SCHOOL
111th AND ROBERTS ROAD
PALOS HILLS, ILLINOIS 60465

Geometry Formulas

Square
Perimeter: $P = 4s$
Area: $A = s^2$

Rectangular Solid
Volume: $V = LWH$
Surface area: $A = 2HW + 2LW + 2LH$

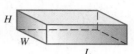

Rectangle
Perimeter: $P = 2L + 2W$
Area: $A = LW$

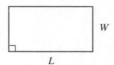

Cube
Volume: $V = e^3$
Surface area: $S = 6e^2$

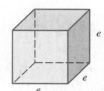

Triangle
Perimeter: $P = a + b + c$
Area: $A = \frac{1}{2}bh$

Right Circular Cylinder
Volume: $V = \pi r^2 h$
Surface area: $S = 2\pi rh + 2\pi r^2$
(Includes both circular bases)

Parallelogram
Perimeter: $P = 2a + 2b$
Area: $A = bh$

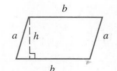

Cone
Volume: $V = \frac{1}{3}\pi r^2 h$
Surface area: $S = \pi r\sqrt{r^2 + h^2} + \pi r^2$
(Includes circular base)

Trapezoid
Perimeter: $P = a + b + c + B$
Area: $A = \frac{1}{2}h(b + B)$

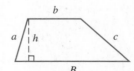

Right Pyramid
Volume: $V = \frac{1}{3}Bh$
B = area of the base

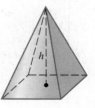

Circle
Diameter: $d = 2r$
Circumference: $C = 2\pi r = \pi d$
Area: $A = \pi r^2$

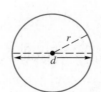

Sphere
Volume: $V = \frac{4}{3}\pi r^3$
Surface area: $S = 4\pi r^2$

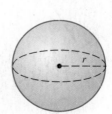

Algebra 2
Eighth Edition

Margaret L. Lial
American River College

John Hornsby
University of New Orleans

Terry McGinnis

PEARSON
Addison
Wesley

Boston • San Francisco • New York
London • Toronto • Sydney • Tokyo • Singapore • Madrid
Mexico City • Munich • Paris • Cape Town • Hong Kong • Montreal

Publisher	Greg Tobin
Editor in Chief	Maureen O'Connor
Project Editor	Lauren Morse
Editorial Assistant	Marcia Emerson
Managing Editor/Production Supervisor	Ron Hampton
Text and Cover Design	Dennis Schaefer
Supplements Production	Jason Miranda
Production Services	Elm Street Publishing Services, Inc.
Media Producers	Sharon Smith and Sara Anderson
Software Development	Math XL: John O'Brien; TestGen: Marty Wright
Marketing Manager	Jay Jenkins
Marketing Coordinator	Tracy Rabinowitz
Prepress Services Buyer	Caroline Fell
Technical Art Supervisor	Joseph K. Vetere
First Print Buyer	Hugh Crawford
Composition Services	Pre-Press Company, Inc.
Cover Photo	© Mike Dobel/Masterfile
Cover Image	Waterfall in Forillon National Park; Gaspe, Quebec, Canada

Photo Credits All photos from PhotoDisc, except the following: Photo Edit, p. 24 right; Beth Anderson, pp. 24 left, 165, 197, 327 right, 711, 725; BrandX Pictures RF, pp. 29, 647; Digital Vision, pp. 48, 87 left, 89, 202, 244, 365, 597 right, 639, 657, 688, 692, 720, 726; Michael Kim/Corbis, p. 49; The Kobal Collection, pp. 53, 75, 99 right, 308; Terry McGinnis, pp. 33, 69; Joe Skipper/Reuters/Corbis, p. 73 left; William Manning/Corbis, p. 73 right; Jeff Topping/ Reuters/Corbis, p. 80; Bettmann/Corbis, p. 87 bottom right; PAL, p. 99 left; Comstock RF, pp. 100, 113, 138, 228 left, 311, 327 left, 341, 597 left, 762; Dave Winter/Photo & Co./Corbis, p. 104; Jason Reed/Reuters/Corbis, p. 127; PictureQuest/Banana Stock RF, p. 190; NASA, p. 236, 534, 742, 752; Corbis RF, pp. 251, 304, 341, 364, 448, 578, 584, 610, 637, 651, 693, 721, 732, 736; Holt/Photo Researchers, p. 277; Ron Kuntz/ Reuters/Corbis, p. 279; AP/Mary Butkus, p. 286; Reuters/ Corbis, pp. 288, 477; Nigel Cattlin, Holt Studios Int., Photo Researchers, Inc., p. 289; Greg Fiume/New Sport/Corbis, p. 290 left; Duomo/Corbis, p. 290 right; AP Photo/Dennis Cook, p. 370; PAL/Getty Images, Inc./Stone Allstock, p. 371, 404; AP Photo/Middletown Journal, Pat Auckerman, p. 399; Rubberball RF, p. 461 left; Dana Fineman/Corbis/Sygma, p. 557, 603; Lee Snider/Corbis, p. 594, 609; Iowa State University, p. 645; AP Photo/The Grand Island Independence, p. 648; AP Wideworld Photo, p. 787; U.S. Mint, p. 793; Eyewire RF, p. 797.

Library of Congress Cataloging-in-Publication Data

Lial, Margaret L.

Algebra 2.—8th ed. / Margaret L. Lial, John Hornsby, Terry McGinnis.

p.   cm.

Includes index.

ISBN 0-321-27920-4 (College Edition)

ISBN 0-321-29223-5 (High School Edition)

1. Algebra. I. Hornsby, E. John. II. McGinnis, Terry. III. Title.

QA152.3.L534 2005

512.9—dc21   2004052580

Copyright © 2006 Pearson Education, Inc.

All rights reserved. No part of this publication may be reproduced, stored in a retrieval system, or transmitted, in any form or by any means, electronic, mechanical, photocopying, recording, or otherwise, without the prior written permission of the publisher. Printed in the United States of America. For information on obtaining permission for the use of material from this work, please submit a written request to Pearson Education, Inc., Rights and Contracts Department, 75 Arlington Street, Suite 300, Boston, MA 02116.

Contents

Index of Applications — v
Index of Focus on Real-Data Applications — ix
Preface — xi
Feature Walk-Through — xvii
An Introduction to Calculators — xxi
To the Student — xxvii
Diagnostic Pretest — xxix

CHAPTER 1 Review of the Real Number System — 1

1.1 Basic Concepts — 2
1.2 Operations on Real Numbers — 15
1.3 Exponents, Roots, and Order of Operations — 25
1.4 Properties of Real Numbers — 35
Summary — 43
Review Exercises — 47
Test — 51

CHAPTER 2 Linear Equations and Applications — 53

2.1 Linear Equations in One Variable — 54
2.2 Formulas — 65
2.3 Applications of Linear Equations — 77
2.4 Further Applications of Linear Equations — 91
Summary Exercises on Solving Applied Problems — 99
Summary — 101
Review Exercises — 105
Test — 109
Cumulative Review Exercises — 111

CHAPTER 3 Linear Inequalities and Absolute Value — 113

3.1 Linear Inequalities in One Variable — 114
3.2 Set Operations and Compound Inequalities — 129
3.3 Absolute Value Equations and Inequalities — 139
Summary Exercises on Solving Linear and Absolute Value Equations and Inequalities — 151
Summary — 153
Review Exercises — 157
Test — 161
Cumulative Review Exercises — 163

CHAPTER 4 Graphs, Linear Equations, and Functions — 165

4.1 The Rectangular Coordinate System — 166
4.2 Slope — 177
4.3 Linear Equations in Two Variables — 191
4.4 Linear Inequalities in Two Variables — 205
4.5 Introduction to Functions — 213
4.6 Variation — 229
Summary — 239
Review Exercises — 243
Test — 247
Cumulative Review Exercises — 249

CHAPTER 5 Systems of Linear Equations — 251

5.1 Systems of Linear Equations in Two Variables — 252
5.2 Systems of Linear Equations in Three Variables — 267
5.3 Applications of Systems of Linear Equations — 277
5.4 Solving Systems of Linear Equations by Matrix Methods — 291
Summary — 299
Review Exercises — 303
Test — 307
Cumulative Review Exercises — 309

CHAPTER 6 Exponents, Polynomials, and Polynomial Functions — 311

6.1 Integer Exponents and Scientific Notation — 312
6.2 Adding and Subtracting Polynomials — 329
6.3 Polynomial Functions — 335
6.4 Multiplying Polynomials — 343
6.5 Dividing Polynomials — 353
Summary — 359
Review Exercises — 363
Test — 367
Cumulative Review Exercises — 369

CHAPTER 7 Factoring — 371

7.1 Greatest Common Factors; Factoring by Grouping — 372
7.2 Factoring Trinomials — 379

7.3 Special Factoring 387
Summary Exercises on Factoring 393
7.4 Solving Equations by Factoring 395

Summary 405
Review Exercises 407
Test 409
Cumulative Review Exercises 411

CHAPTER 8 Rational Expressions and Functions 413

8.1 Rational Expressions and Functions;
Multiplying and Dividing 414
8.2 Adding and Subtracting Rational Expressions 425
8.3 Complex Fractions 435
8.4 Equations with Rational Expressions
and Graphs 441
*Summary Exercises on Rational Expressions
and Equations* 449
8.5 Applications of Rational Expressions 451

Summary 465
Review Exercises 469
Test 473
Cumulative Review Exercises 475

CHAPTER 9 Roots, Radicals, and Root Functions 479

9.1 Radical Expressions and Graphs 480
9.2 Rational Exponents 489
9.3 Simplifying Radical Expressions 499
9.4 Adding and Subtracting Radical Expressions 511
9.5 Multiplying and Dividing Radical Expressions 515
*Summary Exercises on Operations with Radicals
and Rational Exponents* 525
9.6 Solving Equations with Radicals 527
9.7 Complex Numbers 535

Summary 545
Review Exercises 549
Test 553
Cumulative Review Exercises 555

CHAPTER 10 Quadratic Equations, Inequalities, and Functions 557

10.1 The Square Root Property and
Completing the Square 558
10.2 The Quadratic Formula 569
10.3 Equations Quadratic in Form 577
Summary Exercises on Solving Quadratic Equations 589
10.4 Formulas and Further Applications 591

10.5 Graphs of Quadratic Functions 599
10.6 More About Parabolas; Applications 611
10.7 Quadratic and Rational Inequalities 623

Summary 633
Review Exercises 637
Test 643
Cumulative Review Exercises 647

CHAPTER 11 Exponential and Logarithmic Functions 651

11.1 Inverse Functions 652
11.2 Exponential Functions 661
11.3 Logarithmic Functions 669
11.4 Properties of Logarithms 679
11.5 Common and Natural Logarithms 687
11.6 Exponential and Logarithmic Equations;
Further Applications 695

Summary 707
Review Exercises 711
Test 715
Cumulative Review Exercises 717

CHAPTER 12 Nonlinear Functions, Conic Sections, and Nonlinear Systems 721

12.1 Additional Graphs of Functions; Composition 722
12.2 The Circle and the Ellipse 733
12.3 The Hyperbola, and Other Functions Defined
by Radicals 743
12.4 Nonlinear Systems of Equations 753
12.5 Second-Degree Inequalities and Systems
of Inequalities 763

Summary 771
Review Exercises 775
Test 779
Cumulative Review Exercises 783

Appendix A Strategies for Problem Solving 787

Appendix B Review of Fractions 801

Appendix C Determinants and Cramer's Rule 813

Appendix D Synthetic Division 823

Answers to Selected Exercises A-1

Index I-1

Index of Applications

Astronomy/Aerospace

Astronaut's weight, 233
Distance between the centers of the
  moon and Earth, 328
Dog's weight on the moon, 236
Light-year, 327
Meteorite fragments, 649
Orbits of planets, 742
Rocket traveling toward the sun, 321, 327
Rotational rate of a space station, 534
Satellite in orbit, 752, 776
Traveling from Venus to Mercury, 328

Automotive

Antifreeze, 84, 287
Collision impact of an automobile, 237
Gasoline and oil mixture, 68
Height of a truck, 742
Motor vehicle accidents, 461
Octane rating of gasoline, 84
Skidding car, 237, 597

Biology

Age of a female blue whale, 694
Dinosaur research, 598
Diversity of a species, 706
Growing algae, 305
Growth rate of a population, 474
Lengths of bones and height, 228
Population of mites, 673
Speed of a killer whale, 287
Tagging fish in a lake, 461, 474
Weight of a fish, 237

Business

Advertising slogans, 309, 720
Biggest U.S. companies, 285
Book buyers, 86
Break-even point, 762
CNBC profits, 246
Camera sales, 252, 266
Candy clerk, 106
Car sales, 47, 598
Company bankruptcy filings, 594, 609
Computer prices compared to Internet
  access prices, 762
Cost and revenue, 128, 162, 310
Daily newspapers, 112
Fortunes of frozen yogurt, 264
Franchised automobile tune-up
  shops, 783

Growth in on-line retail sales, 786
Hardware supplier, 290
Harlem Globetrotter Web site shop, 290
Local electronics store, 308
Manufacturing, 290, 555
Maximum revenue, 621
Merger of America Online and Time
  Warner, 649
Paper mill, 284
Production, 14, 265, 283–284
Restaurants, 461
Sales, 24, 184, 190, 202, 286, 648
Value of a business copy machine, 668
Wholesale drug prices, 604

Chemistry

Acid solution, 69, 74, 83, 84, 89, 90,
  107, 280, 286, 287, 306
Alcohol solution, 74, 89, 102–103, 112,
  280, 286, 308, 556
Chemical solution, 106
Classifying wetlands using
  pH values, 688, 692
Dye solution, 89
Half-life, 700, 706
Hydrogen peroxide solution, 305
Hydronium ion
  concentration, 688, 692, 713
Radioactive decay, 700, 706
Volume of a gas, 233

Construction

Building a box, 100
Building a wine rack, 458
Buildings in a sports complex, 752
Designing the Fleet Center arena in
  Boston, 189
Height of a structure, 87, 644, 742
Maximum load of a cylindrical
  column, 234
Pitch of a roof, 244
Stabilizing wall frames, 552
Tallest buildings in Kansas City,
  Missouri, 150
Upper deck at U.S. Cellular Field, 189

Consumer

Bread prices, 282–283
Cell phone plan, 202
Charge to customers for electricity, 230
Charter bus and flight fares, 621

Choosing the right air conditioner, 166
Comparing long-distance
  phone costs, 134, 197
Consumer price index, 88
Cost of airport parking, 731
Cost of gasoline, 166, 197, 236, 238
Cost of membership to an
  athletic club, 202, 244
Cost of parenthood, 76
Cost of pet food, 69
Currency exchange and best buys, 104
Discount, 99
Electricity use, 225, 609
Food or beverage costs, 277, 286,
  288, 289
Gallons of gasoline, 453
Mortgage shopping, 264
Natural gas consumption, 642, 678
Overnight delivery service charges, 725
Personal spending on
  medical care, 166, 173
Postage rates, 228, 555, 731
Price per hour for horseback rides, 236
Rental costs, 122, 128, 157, 202, 203,
  304, 713
Sale price, 726
Spending on home video games, 641
Taxicab charges, 228

Economics

Depreciation of a machine, 714
E-filing taxpayers, 190
Exports, 50, 285
Imports, 50
Inflation, 698–699, 713
Property taxes, 33
Sales tax, 88, 460
Social Security assets, 608, 622
Stock performance, 24
Stock splits, 10
Supply and demand, 597, 639, 762
Trade balance, 13
U.S. trade deficit, 608
U.S. trade with Mexico, 24

Education

ACT exam, 88
Average test score, 122, 127, 156,
  158, 161, 164
Bachelor's degrees in
  the U.S., 303, 762

Cedar Rapids schools' general
  reserve fund, 609
College expenses, 138
Grader's rate, 586
In-state tuition, 645
Number of girls vs. boys in a math
  class, 460
Research universities securing
  patents, 86
Teacher-to-student ratio, 461
Tuition and fees, 100, 198

Environment

Atmospheric carbon dioxide, 665–666
Atmospheric pressure, 666, 689, 716
Barometric pressure near the eye
  of a hurricane, 672–673
Carbon dioxide emissions, 694
Carbon monoxide emissions, 657
Duration of a storm, 498
Earthquake intensity, 322
Force of the wind, 249
Global warming, 668
Hazardous waste sites, 657
Insecticide solution, 89
Major Southern California
  earthquakes, 678
Moisture, 694
Natural gas providing energy, 237
Oil leak, 727, 732
Pollutant Standard Index, 653
Temperature, 23, 73, 106
Thermal inversion layer, 732
Windchill, 494

Finance

Average family income, 453
Bank debit cards, 341
Company borrowing money, 310
Compound interest, 698, 699, 705, 706,
  713, 714, 716, 785
Continuously compounded
  interest, 699, 705, 720, 785
Doubling time for an
  investment, 693, 699, 713
Finance charge on a loan, 75, 110
Interest earned, 237
Interest rate, 74, 106, 109, 639
Investment, 82, 88, 89, 90, 99, 107,
  108, 110, 112, 164, 234,
  286, 287, 461, 475, 622, 702
Mutual funds, 30
On-line bill paying, 336
Shares of common stock, 461
U.S. commercial bank failures, 609

Geometry

Angle measurement, 94, 97, 98, 100,
  110, 285, 289, 305,
  309, 369
Area of a circle, 231

Area of a rectangular region, 138
Area of a square, 352
Area of the Bermuda Triangle, 488
Area of the Vietnam Veterans'
  Memorial, 488
Base and height of a
  parallelogram, 398–399, 403
Base and height of a triangle, 403, 412
Circumference of a circle, 74
Circumference of the Roman
  Colosseum, 776
Diagonal of a box, 509
Diagonal of a television screen, 509
Diameter of a circle, 74
Dimensions of a rectangular
  solid, 100, 105
Dimensions of a rectangular-shaped
  object or region, 79, 87, 99, 106,
  157, 278, 285, 301, 304,
  370, 399, 403, 404, 408,
  410, 412, 596, 638, 761
Height of a triangle, 68, 475
Length of a side of a square, 74, 108,
  250, 403
Lengths of the sides of
  a triangle, 87, 100, 106, 108,
  289, 408, 596, 638
Maximum area, 615, 621, 641, 646
Measure of a strip or border around a
  rectangular region, 593, 596, 639, 644
Perimeter of a rectangular region, 138
Perimeter of a triangular region, 138, 552
Perimeter of an equilateral triangle, 95
Radius of a circle, 74, 106
Volume of a cylinder, 74, 236
Volume of a rectangular
  box, 234, 246, 358

Government

Electoral votes, 108
Federal spending on education, 34
Food stamp recipients, 189
Social Security, 24
U.S. budget, 327
U.S. Postal Service, 109, 203
U.S. Supreme Court cases, 341

Health/Life Sciences

AIDS cases, 105, 307
Average clotting time of blood, 149
Blood Alcohol Concentration (BAC), 34
Body Mass Index (BMI), 49, 112, 128,
  234, 237
Deaths caused by smoking, 448
Dieting, 164
Drug concentration in the
  bloodstream, 693
Drug dosage, 112, 462
Growth of outpatient surgery, 694
Health Maintenance Organizations
  (HMOs), 365

Health insurance, 340, 453
Higher-order multiple
  births, 603–604, 610
Medical doctors in the U.S., 367
Medication, 249
National health care expenditure, 321
Pharmacies paying wholesalers
  for drugs, 622
Recommended daily intake
  of calcium, 149
Target heart rate zone (THR), 128
Threshold weight, 487, 498
Thyroid, 129
Twin births, 370

Labor

Average hourly wages in Mexico, 476
Commission rate, 69, 74, 88
Company pension plan, 160
Major league baseball payrolls, 87
Median household income, 204, 244
Median weekly earnings, 159
Part-time job, 81
Racial composition of the U.S. work
  force, 106
Rates of employment change, 7
Wages, 230
Wal-Mart employees, 327
Women in math or computer science
  professions, 173
Women in the work force, 752
Women returning to work after
  having a baby, 701
Working alone to complete
  a job, 457, 464, 579–580,
  586, 638, 644
Working together to complete
  a job, 457, 464, 471,
  474, 478, 638, 784

Miscellaneous

Age, 98
Altitudes of mountains, 13, 48, 51
Average daily volume of first-class
  mail, 608
Bottles of perfume, 283
Caffeine amounts found in sodas, 711
Consecutive integers, 98, 100
Cutting yarn, 164
Depth of oceans, 13, 51
Dimensions of a cord of wood, 74
Distance to the horizon from an
  observer's point of view, 488, 510
Filling a tank, sink, or tub, 471, 472, 587
Height of a kite, 596
International times zones, 60, 70
Ladder leaning against a house, 592
Languages, 693
Long-distance area codes, 81
Long-distance phone calls, 237
m&m logos, 674

Mixing food or drinks,   83, 89, 280, 287, 288, 305, 308, 310, 718
Money denominations,   91, 95, 96, 112, 309, 556
Most important inventions,   369
Mowing a yard,   460
Occupations of stockholders,   110
Page numbers,   98
Powerball lottery,   327
Scrap value of a machine,   714
Search light movement,   639
Talking to automated teller machines (ATMs),   647
Toddler messing up a house,   464
Viewing distance of a camera,   246
Visitors to U.S. Presidents' homes,   87
Volume of water in a swimming pool,   225

Physics

Current in an electrical circuit,   236, 248
Decibel level,   693
Energy of an electric circuit,   488
Focal length of a lens,   451
Force of attraction on an object,   237
Freely falling objects,   399, 560, 566, 637
Frequency of a vibrating guitar string,   236
Height of a rocket,   399, 404
Hooke's law for an elastic spring,   230
Impedance of parallel resonant circuits,   510
Inlet and outlet pipes,   464
Law of tensions,   511
Light produced by a light source,   237
Object dropped,   236, 404
Object propelled upward,   404, 408, 410, 594, 597, 616, 621, 632, 639, 641
Ohm's law,   544
Resonant frequency of a circuit,   484, 487
Speed of light,   327
Strength of a contact lens,   472

Swing of a pendulum,   488, 549
Water emptied by a pipe,   236

Sports/Entertainment

Average sports ticket prices,   279
Base hits,   87
Batting average,   458
Boston marathon,   100
Concours d'elegance competition,   434
Daytona 500 race,   73, 109
Fan Cost Index (FCI),   286
Fresh Water Fun Run,   112
Highest-grossing domestic films,   133, 308
Home runs,   80, 305
Hurdle rates,   96
Indianapolis 500 race,   68, 73
Major league pitchers,   80
Olympic medals,   290, 306
Points scored in basketball,   477
Production of *The Music Man*,   96
Rugby,   752
Running,   96, 653
Skydiver's fall speed,   231
Television programs winning Emmy awards,   99
Television viewers,   76
Ticket sales,   96, 286, 288, 290, 622
Top-grossing American movies,   99
Top-grossing North American concert tour,   288
U.S. Olympic track and field trials,   127
Walking speed,   455–456
Win-loss record,   285, 290
Winning percentage of baseball teams,   75

Statistics/Demographics

Births to unmarried women,   341
Households owning at least one TV set,   75
Percent of high school students who smoke,   184, 610

Percent of working mothers of children under one year,   198
Population change,   13, 88
Population density,   366
Population growth,   265
Population of the United States,   327, 364
U.S. Hispanic population,   712

Technology

Average daily e-mail volume,   604
Cellular telephones,   215, 250
Computer network manager,   362
Computer speed,   327
Growth of e-mail boxes in North America,   640
Homes with multiple personal computers,   183
Internet access,   203

Transportation

Airports in the United States,   341
Average speed,   68
Distance apart,   93, 96, 249, 411
Distance between cities,   107, 460, 463, 657, 758
Distance,   93–94, 97, 202, 456, 463, 592, 596, 650
High-speed trains,   471
Location-finding system (LORAN),   777
Most expensive cities worldwide for business travelers,   286
Rate,   68, 95, 107, 110, 358, 458, 463, 474, 586, 644
Speed,   97, 103, 107, 164, 230, 281–282, 287, 288, 289, 305, 308, 454–455, 468, 471, 478, 556, 577–578, 579, 586, 638, 783
Time,   94, 95, 96, 97, 107, 108
Traffic intensity,   448
Traveling in opposite directions,   92, 93
U.S. travelers to foreign countries,   161, 718

Index of Focus on Real-Data Applications

SECTION	PAGE	TITLE	APPLICATION	OBJECTIVE
Chapter 1: Review of the Real Number System				
1.1	10	Stock Splits	Compute stock equity after a split.	Use fractions and decimals.
Chapter 2: Linear Equations and Applications				
2.1	60	International Time Zones	Find actual time to travel between cities.	Add and subtract integers.
2.2	70	Same (International) Time, Second Place	Use a 24-hour clock.	Add and subtract integers.
Summary	104	Currency Exchange and Best Buys	Compare different methods of changing currency.	Solve percent problems.
Chapter 3: Linear Inequalities and Absolute Value				
3.2	134	Comparing Long-Distance Costs	Compare cell phone rates.	Write and solve linear inequalities.
Summary	156	What Do I Have to Average on My Tests to Get the Grade I Want?	Compute final grades.	Write and solve linear inequalities.
Chapter 6: Exponents, Polynomials, and Polynomial Functions				
6.1	322	Earthquake Intensities Measured by the Richter Scale	Compare intensities of earthquakes.	Use scientific notation.
6.3	340	Comparing Mathematical Models	Estimate number of individuals covered by health insurance per year.	Use polynomial models to make predictions.
Summary	362	Reply All: The Computer Network Manager's Nightmare	Calculate number of e-mails.	Write and calculate exponential expressions.
Chapter 7: Factoring				
7.4	400	Idle Prime Time	Explore prime numbers.	Determine prime and composite numbers.
Chapter 8: Rational Expressions and Functions				
8.5	458	It Depends on What You Mean by "Average"	Find averages.	Set up and solve rational equations.
Chapter 9: Roots, Radicals, and Root Functions				
9.2	494	Windchill—A Radical Idea	Calculate windchill from wind speed and air temperature.	Evaluate a radical expression.
Chapter 10: Quadratic Equations, Inequalities, and Functions				
10.2	574	Almost, but Not Quite Right	Explore examples of incorrect use of the quadratic formula.	Learn the correct form of the quadratic formula.
10.3	584	Smile! You're on Golden Ratio!	Investigate the golden ratio and some of its applications.	Solve quadratic equations and work with radicals.

SECTION	PAGE	TITLE	APPLICATION	OBJECTIVE
Chapter 11: Exponential and Logarithmic Functions				
11.3	674	m&m's and Exponential Decay	Collect data using m&m's to derive a model.	Develop an exponential model.
11.6	702	Evaluating Investments: The Rule of 72	Compare growth from different interest rates.	Evaluate and interpret exponential and logarithmic expressions.
Chapter 12: Nonlinear Functions, Conic Sections, and Nonlinear Systems				
12.4	758	Who Arrived First?	Write equations to express distances after t hours.	Write and graph a system of equations.

Preface

The eighth edition of *Algebra 2* continues our ongoing commitment to provide the best possible text and supplements package to help teachers and students succeed. To that end, we have tried to address the diverse needs of today's students through an attractive design, updated figures and graphs, helpful features, careful explanations of topics, and a comprehensive package of supplements and study aids. We have taken special care to respond to the suggestions of users and reviewers and have added new examples and exercises based on their feedback. Students who have never studied algebra—as well as those who require further review of basic algebraic concepts before taking additional courses in mathematics, business, science, nursing, or other fields—will benefit from the text's student-oriented approach.

This text is part of a series that also includes the following books:

- *Essential Mathematics*, Second Edition, by Lial and Salzman
- *Basic College Mathematics*, Seventh Edition, by Lial, Salzman, and Hestwood
- *Prealgebra*, Third Edition, by Lial and Hestwood
- *Algebra 1*, Eighth Edition, by Lial, Hornsby, and McGinnis
- *Introductory and Intermediate Algebra*, Third Edition, by Lial, Hornsby, and McGinnis.

HALLMARK FEATURES

We believe students and teachers will welcome the following helpful features.

▶ *Chapter Openers* New and updated chapter openers feature real-world applications of mathematics that are relevant to students and tied to specific material within the chapters. Examples of topics include television, higher education costs, and gasoline prices. (See pp. 53, 113, and 165—Chapters 2, 3, and 4.)

▶ *Real-Life Applications* We are always on the lookout for interesting data to use in real-life applications. As a result, we have included new or updated examples and exercises from fields such as business, pop culture, sports, the life sciences, and technology that show the relevance of algebra to daily life. (See pp. 7, 190, and 282.) A comprehensive Index of Applications appears at the beginning of the text. (See pp. v–vii.)

▶ *Figures and Photos* Today's students are more visually oriented than ever. Thus, we have made a concerted effort to include mathematical figures, diagrams, tables, and graphs whenever possible. (See pp. 47, 77, and 173.) Many of the graphs use a style similar to that seen by students in today's print and electronic media. Photos have been incorporated to enhance applications in examples and exercises. (See pp. 87, 99, and 127.)

▶ *Emphasis on Problem Solving* Introduced in Chapter 2, our six-step problem-solving method is integrated throughout the text. The six steps, *Read, Assign a Variable, Write an Equation, Solve, State the Answer*, and *Check*, are emphasized in boldface type and repeated in examples and exercises to reinforce the problem-solving process for students. (See pp. 79, 83, and 278.) New **PROBLEM-SOLVING HINT** boxes provide students with helpful problem-solving tips and strategies. (See pp. 77, 92, and 456.)

Also new to this edition of the text is Appendix A, Strategies for Problem Solving. (See pp. 787–800.) This appendix provides examples of additional problem-solving techniques, such as working backward, using trial and error, and looking for patterns. A wide variety of applications are included.

▶ *Learning Objectives* Each section begins with clearly stated, numbered objectives, and the included material is directly keyed to these objectives so that students know exactly what is covered in each section. (See pp. 114, 252, and 312.)

▶ *Cautions and Notes* One of the most popular features of previous editions, **CAUTION** and **NOTE** boxes warn students about common errors and emphasize important ideas throughout the exposition. (See pp. 119, 142, and 197.) The text design makes them easy to spot: Cautions are highlighted in bright yellow and Notes are highlighted in purple.

▶ *Calculator Tips* These optional tips, marked with calculator icons, offer basic information and instruction for students using calculators in the course. (See pp. 253, 291, and 321.) An Introduction to Calculators is included at the beginning of the text. (See pp. xxi–xxv.)

▶ *Margin Problems* Margin problems, with answers immediately available at the bottom of the page, are found in every section of the text. (See pp. 30, 117, and 183.) This popular feature allows students to immediately practice the material covered in the examples in preparation for the exercise sets.

▶ *Ample and Varied Exercise Sets* The text contains a wealth of exercises to provide students with opportunities to practice, apply, connect, and extend the algebraic skills they are learning. Numerous illustrations, tables, graphs, and photos have been added to the exercise sets to help students visualize the problems they are solving. Problem types include writing, estimation, and calculator exercises as well as applications and multiple-choice, matching, true/false, and fill-in-the-blank problems. In the *Annotated Instructor's Edition* of the text, writing exercises are marked with ✍ icons so that teachers may assign these problems at their discretion. Exercises suitable for calculator work are marked in both the student and teacher editions with calculator icons ▦ . (See pp. 76, 96, and 112.)

▶ *Relating Concepts Exercises* These sets of exercises help students tie together topics and develop problem-solving skills as they compare and contrast ideas, identify and describe patterns, and extend concepts to new situations. (See pp. 97, 190, and 306.) These exercises make great collaborative activities for pairs or small groups of students.

▶ *Summary Exercises* Based on user feedback, we have added new sets of in-chapter summary exercises. These special exercise sets provide students with the all-important *mixed* review problems they need to master topics. Summaries of solution methods or additional examples are often included. (See pp. 99, 151, and 393.)

▶ *Study Skills Component* A desk-light icon at key points in the text directs students to a separate *Study Skills Workbook* containing activities correlated directly to the text. (See pp. 2, 43, and 161.) This unique workbook explains *how* the brain actually learns, so students understand *why* the study tips presented will help them succeed in the course. Students are introduced to the workbook in the To the Student section at the beginning of the text.

▶ *Focus on Real-Data Applications* These one-page activities present a relevant and in-depth look at how mathematics is used in the real world. Designed to help instructors answer the often-asked question, "When will I ever use this stuff?," these activities ask students to read and interpret data from newspaper articles, the Internet, and other familiar, real sources. (See pp. 10, 70, and 134.) The activities are well-suited to collaborative work and can also be completed by individuals or used for open-ended class discussions. A comprehensive Index of Focus on Real-Data Applications appears at the beginning of the text. (See pp. ix–x.) Teaching notes and extensions for the activities are provided in the *Printed Test Bank and Instructor's Resource Guide.*

▶ *Test Your Word Power* To help students understand and master mathematical vocabulary, this feature can be found in each chapter summary. Key terms from the chapter are presented along with four possible definitions in a multiple-choice format. Answers and examples illustrating each term are provided. (See pp. 240, 299, and 359.)

▶ *Ample Opportunity for Review* Each chapter concludes with a Chapter Summary that features Key Terms with definitions and helpful graphics, New Symbols, Test Your Word Power, and a Quick Review of each section's content with additional examples. A comprehensive set of Chapter Review Exercises, keyed to individual sections, is included, as are Mixed Review Exercises and a Chapter Test. Beginning with Chapter 2, each chapter concludes with a set of Cumulative Review Exercises that cover material going back to Chapter 1. (See pp. 153, 239, and 299.)

▶ *Diagnostic Pretest* A diagnostic pretest is included on pp. xxix–xxxii and covers material from the entire book, much like a sample final exam. This pretest can be used to facilitate student placement in the correct chapter according to skill level.

WHAT IS NEW IN THIS EDITION?

You will find many places in the text where we have polished individual presentations and added or updated examples, exercises, and applications based on reviewer feedback. Specific content changes you may notice include the following:

- Section 4.3, Linear Equations in Two Variables, includes increased emphasis on slope-intercept form, which is presented first and used in graphing lines and in modeling data.

- The presentation on functions in Section 4.5 has been rewritten.

- Former Chapter 6 has been split into two chapters. Chapter 6 now includes the sections on exponents, polynomial operations, and polynomial functions, while Chapter 7 covers factoring.

- All new sets of summary exercises appear in Chapters 9 and 10.

- Section 12.1 now includes graphing and applications of the greatest integer function and expanded coverage of composition of functions.

- Appendix A, Strategies for Problem Solving, is new to this edition. Synthetic division, formerly Section 6.5, is now covered in Appendix D.

WHAT SUPPLEMENTS ARE AVAILABLE?

For a comprehensive list of the supplements and study aids that accompany *Algebra 2*, Eighth Edition, see pp. xv and xvi.

ACKNOWLEDGMENTS

Previous editions of this text were published after thousands of hours of work, not only by the authors, but also by reviewers, teachers, students, answer checkers, and editors. To these individuals and all those who have worked in some way on this text over the years, we are most grateful for your contributions. We could not have done it without you. We especially wish to thank the following reviewers whose valuable contributions have helped to refine this edition of this text.

Mary Kay Abbey, *Montgomery College*
Randall Allbritton, *Daytona Beach College*
Sonya Armstrong, *West Virginia State College*
Linda Beller, *Brevard Community College*
Carla J. Bissell, *University of Nebraska at Omaha*
Vernon Bridges, *Durham Technical Community College*
Dawn Cox, *Cochise College*
Julie Dewan, *Mohawk Valley Community College*
Lucy Edwards, *Las Positas College*
Rob Farinelli, *Community College of Allegheny–Boyce Campus*
Anthony Hearn, *Community College of Philadelphia*

Jeffrey Kroll, *Brazosport College*
Barbara Krueger, *Cochise College*
Sandy Lofstock, *California Lutheran University*
Janice Rech, *University of Nebraska at Omaha*
Dwight Smith, *Prestonburg Community College*
Theresa Stalder, *University of Illinois–Chicago*
Mark Tom, *College of the Sequoias*

Over the years, we have come to rely on an extensive team of experienced professionals. Our sincere thanks go to these dedicated individuals at Addison-Wesley, who worked long and hard to make this revision a success: Greg Tobin, Maureen O'Connor, Jay Jenkins, Lauren Morse, Marcia Emerson, Sharon Smith, Sara Anderson, Tracy Rabinowitz, Ron Hampton, and Dennis Schaefer.

Thanks are due Gina Linko, Phyllis Crittenden, and Elm Street Publishing Services for their excellent production work. Sara Kueffer provided invaluable assistance updating the real data used in applications throughout the text. Jon Becker, Sara Kueffer, and Paul Lorczak did an outstanding job accuracy checking page proofs. Special thanks to Bernice Eisen who prepared the Index and Becky Troutman who compiled the Index of Applications.

As an author team, we are committed to the goal stated earlier in this Preface—to provide the best possible text and supplements package to help instructors and students succeed. We are most grateful to all those over the years who have aspired to this goal with us. As we continue to work toward it, we would welcome any comments or suggestions you might have via e-mail to math@awl.com.

Margaret L. Lial
John Hornsby
Terry McGinnis

Student Supplements

Student's Solutions Manual
- By Jeffery A. Cole, *Anoka-Ramsey Community College*
- Provides detailed solutions to the odd-numbered, section-level exercises and to all margin, Relating Concepts, Summary, Chapter Review, Chapter Test, and Cumulative Review Exercises
  ISBN: 0-321-28569-7

Study Skills Workbook
- By Diana Hestwood and Linda Russell
- Provides activities that teach students how to use the textbook effectively, plan their homework, take notes, make mind maps and study cards, manage study time, and prepare for and take tests
- Text desk-light icon at key points directs students to correlated activities in the workbook
  ISBN: 0-321-28563-8

Videotape Series
- Features an engaging team of lecturers
- Provides comprehensive coverage of each section and topic in the text
  ISBN: 0-321-28562-X

Digital Video Tutor
- Complete set of digitized videos on CD-ROM for student use at home or on campus
- Ideal for distance learning or supplemental instruction
  ISBN: 0-321-28570-0

New! Additional Skill and Drill Manual
- Provides additional practice and test preparation for students
  ISBN: 0-321-33166-4

MathXL® Tutorials on CD—*Optional*
- Provides algorithmically generated practice exercises that correlate at the objective level to the content of the text
- Includes an example and a guided solution to accompany every exercise and video clips for selected exercises
- Recognizes student errors and provides feedback; Generates printed summaries of students' progress
- **For purchase only on an annual basis**
  ISBN: 0-321-28565-4

Teacher Supplements

Annotated Instructor's Edition
- Provides answers to all text exercises in color next to the corresponding problems
- Includes icons to identify writing and calculator exercises
  ISBN: 0-321-28572-7

Instructor's Solutions Manual
- By Jeffery A. Cole, *Anoka-Ramsey Community College*
- Provides complete solutions to all even-numbered section-level exercises
  ISBN: 0-321-28568-9

Answer Book
- By Jeffery A. Cole, *Anoka-Ramsey Community College*
- Provides answers to all the exercises in the text
  ISBN: 0-321-28567-0

Printed Test Bank and Instructor's Resource Guide
- By James J. Ball, *Indiana State University*
- Contains two diagnostic pretests, six free-response and two multiple-choice test forms per chapter, and two final exams
- Includes teaching suggestions for each chapter, additional practice exercises for every objective of every section, a correlation guide from the seventh to the eighth edition, phonetic spellings for all key terms in the text, and teaching notes and extensions for the Focus on Real-Data Applications in the text
  ISBN: 0-321-28571-9

TestGen
- Enables teachers to build, edit, print, and administer tests
- Features a computerized bank of questions developed to cover all text objectives
- Available on a dual-platform Windows/Macintosh CD-ROM
  ISBN: 0-321-28566-2

InterAct Math Tutorial Web site: www.interactmath.com Get practice and tutorial help online! This interactive tutorial web site provides algorithmically generated practice exercises that correlate directly to the exercises in the textbook. Students can retry an exercise multiple times with new values each time for unlimited practice and mastery. Every exercise is accompanied by an interactive guided solution that provides helpful feedback for an incorrect answer. Students can also view a worked-out sample problem that steps them through an exercise similar to the one they're working on.

Math XL   **MathXL®—*Optional*** MathXL is a powerful online homework, tutorial, and assessment system that accompanies your Addison-Wesley textbook in mathematics or statistics. With MathXL, instructors can create, edit, and assign online homework and tests using algorithmically generated exercises correlated at the objective level to the textbook. All student work is tracked in MathXL's online gradebook. Students can take chapter tests in MathXL and receive personalized study plans based on their test results. The study plan diagnoses weaknesses and links students directly to tutorial exercises for the objectives they need to study and retest. Students can also access supplemental video clips and animations directly from selected exercises. MathXL is available to qualified adopters. For more information, visit our web site at www.mathxl.com, or contact your sales representative. **Available for purchase only on an annual basis.**

MyMathLab   **MyMathLab—*Optional*** MyMathLab is a series of text-specific, easily customizable online courses for Addison-Wesley textbooks in mathematics and statistics. MyMathLab is powered by CourseCompass—Pearson Education's online teaching and learning environment—and by MathXL—our online homework, tutorial, and assessment system. MyMathLab gives instructors the tools they need to deliver all or a portion of their course online, whether students are in a lab setting or working from home.

MyMathLab provides a rich and flexible set of course materials, featuring free-response exercises that are algorithmically generated for unlimited practice and mastery. Students can also use online tools, such as video lectures, animations, and a multimedia textbook, to independently improve their understanding and performance. Instructors can use MyMathLab's homework and test managers to select and assign online exercises correlated directly to the textbook, and they can import TestGen tests into MyMathLab for added flexibility. MyMathLab's online gradebook—designed specifically for mathematics and statistics—automatically tracks students' homework and test results and gives the instructor control over how to calculate final grades. Instructors can also add offline (paper-and-pencil) grades to the gradebook.

For more information, visit our Web site at www.mymathlab.com or contact your sales representative. **Available for purchase only on an annual basis.**

Feature Walk-Through

Chapter Openers Chapter openers feature real-world applications of mathematics that are relevant to students and tied to specific material within the chapters.

Linear Inequalities and Absolute Value
3

3.1 Linear Inequalities in One Variable

3.2 Set Operations and Compound Inequalities

3.3 Absolute Value Equations and Inequalities

Summary Exercises on Solving Linear and Absolute Value Equations and Inequalities

The cost of a college education has risen rapidly in the last decade. Average higher education tuition and fees for public institutions increased by 86% from the 1990–1991 school year to the 2001–2002 school year. (*Source:* National Center for Education Statistics, U.S. Department of Education.)

In Exercises 61–64 in Section 3.2, we apply the concepts of this chapter to college student expenses.

113

Section 3.1 Linear Inequalities in One Variable **127**

The weather forecast by time of day for one day of the U.S. Olympic Track and Field Trials, in Sacramento, California, is shown in the figure. Use this graph to work Exercises 46–49.

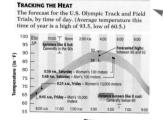

TRACKING THE HEAT
The forecast for the U.S. Olympic Track and Field Trials, by time of day. (Average temperature this time of year is a high of 93.5, low of 60.5.)

Source: Accuweather, Bee research.

46. Sprinters prefer Fahrenheit temperatures in the 90s. Using the upper boundary of the forecast, in what time period is the temperature expected to be at least 90°F?

47. Distance runners prefer cool temperatures. During what time period are temperatures predicted to be no more than 70°F? Use the lower forecast boundary.

48. What range of temperatures is predicted for the Women's 100-m event?

49. What range of temperatures is forecast for the Men's 10,000-m event?

Solve each problem. See Examples 8 and 9.

50. Amanda Pepper earned scores of 90 and 82 on her first two tests in English Literature. What score must

51. Finley Westmoreland scored 92 and 96 on his first two tests in Methods in Teaching Mathematics.

Figures and Photos Today's students are more visually oriented than ever. Thus, a concerted effort has been made to include mathematical figures, diagrams, tables, and graphs whenever possible. Many of the graphs use a style similar to that seen by students in today's print and electronic media. Photos have been incorporated to enhance applications in examples and exercises.

Relating Concepts   These sets of exercises help students tie together topics and develop problem-solving skills as they compare and contrast ideas, identify and describe patterns, and extend concepts to new situations. These exercises make great collaborative activities for pairs or small groups of students.

---

150   Chapter 3   Linear Inequalities and Absolute Value

RELATING CONCEPTS (EXERCISES 83–86)   For Individual or Group Work

The ten tallest buildings in Kansas City, Missouri, are listed along with their heights.

Building	Height (in feet)
KCTV Tower	1042
One Kansas City Place	632
Transamerica Tower	591
Hyatt Regency	504
Power and Light Building	476
City Hall	443
Federal Office Building	433
1201 Walnut Street	427
Commerce Tower	407
City Center Square	404

Source: Marshall Gerometta and Rick Bronson, www.skyscrapers.com; Council on Tall Buildings and Urban Habitat, Lehigh University.

Use this information to **work Exercises 83–86 in order.**

83. To find the average of a group of numbers, we add the numbers and then divide by the number of items added. Use a calculator to find the average of the heights.

84. Let k represent the average height of these buildings. If a height x satisfies the inequality

$$|x - k| < t,$$

then the height is said to be within t feet of the average. Using your result from Exercise 83, list the buildings that are within 50 ft of the average.

85. Repeat Exercise 84, but find the buildings that are within 75 ft of the average.

86. (a) Write an absolute value inequality that describes the height of a building that is *not* within 75 ft of the average.

(b) Solve the inequality you wrote in part (a).

(c) Use the result of part (b) to find the buildings that are not within 75 ft of the average.

Focus on *Real-Data Applications*

What Do I Have to Average on My Tests to Get the Grade I Want?

On the first day of class, you are typically given a syllabus that describes the course requirements. If the syllabus includes a grading scale for homework, tests, projects, and final exam, then you should be able to predict the points you need on the final exam to earn a specific grade.

One intermediate algebra teacher bases final grades on points earned for activities as given in the Graded Classwork table on the left. To determine final grades, the teacher strictly adheres to the point ranges given in the Grade Distribution table on the right.

GRADED CLASSWORK

Activity	Points Available
Homework and vocabulary	45
Daily activities (scaled)	55
Lab participation and completion	100
Major exams (3 at 100 points)	300
Final exam	150
Total points	650

GRADE DISTRIBUTION

Grade	Points Required
A	585–650
B	520–584
C	455–519
IP*	< 455 and active
F	< 455 and inactive

* In Progress

Notice that exams account for 450 of the possible 650 points. The remaining 200 points should be fairly easy to earn by keeping up with the day-to-day course requirements.

Assumption: You earn a "baseline" number of points based on the following criteria.

1. You earn *all* of the homework and vocabulary points.
2. You earn a minimum of 50 points based on daily activities.
3. You earn a minimum of 90 lab participation and completion points.

For Group Discussion

1. Assume that you earn the baseline number of points. Let x represent the test points to be earned. Write and solve linear inequalities to find the minimum number of points that you need in test scores to earn grades no lower than A, B, and C. What "test average" is each minimum score? Round *up* to the nearest whole percent.

2. To keep your scholarship, you must earn a B in the course. Write a compound inequality to find the range of points that you need in test scores to earn a B average. Solve the inequality. What range of "test averages" are those minimum scores? Round *up* to the nearest whole percent.

3. Mark does not like to do the homework or participate in labs. Assume that Mark earns only 15 points in homework and vocabulary, 40 points in daily activities, and 50 points in lab participation. Write and solve linear inequalities to find the minimum number of points that Mark needs in test scores to earn grades no lower than A, B, and C. What "test average" is each minimum score? Round *up* to the nearest whole percent.

156

Focus on Real-Data Applications   These one-page activities found throughout the text present even more relevant and in-depth looks at how mathematics is used in the real world. Designed to help teachers answer the often-asked question, "When will I ever use this stuff?," these activities ask students to read and interpret data from newspaper articles, the Internet, and other familiar, real sources. The activities are well suited to collaborative work and can also be completed by individuals or used for open-ended class discussions.

Calculator Tips   These optional tips, marked with calculator icons, offer basic information and instruction for students using calculators in the course.

Section 6.1   Integer Exponents and Scientific Notation   **321**

Calculator Tip   To enter numbers in scientific notation, you can use the (EE) or (EXP) key on a scientific calculator. For instance, to work Example 11 using a popular model calculator with an (EE) key, enter the following symbols.

1.92 (EE) 6 × 1.5 (EE) 3 (+/–) ÷ 1 3.2 (EE) 5 (+/–) × 4.5 (EE) 4 1 =

The (EXP) key is used in exactly the same way. Notice that the negative exponent −3 is entered by pressing 3, then (+/–). (*Keystrokes vary among different models of calculators*, so you should refer to your owner's manual if this sequence does not apply to your particular model.)

EXAMPLE 12   Using Scientific Notation to Solve Problems

In 1990, the national health care expenditure was $695.6 billion. By 2000, this figure had risen by a factor of 1.9; that is, it almost doubled in only 10 years. (*Source:* U.S. Centers for Medicare & Medicaid Services.)

12 The distance to the sun is 9.3×10^7 mi. How long would it take a rocket, traveling at 3.2×10^3 mph, to reach the sun? (*Hint:* $t = \frac{d}{r}$.)

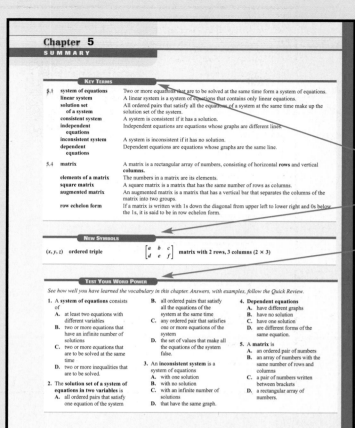

End-of-Chapter Material One of the most admired features of the Lial textbooks is the extensive and well-thought-out end-of-chapter material. At the end of each chapter, students will find a Summary that includes the following:

Key Terms are listed, defined, and referenced back to the appropriate section number.

New Symbols are listed for easy reference and study.

Test Your Word Power helps students understand and master mathematical vocabulary. Students are quizzed on Key Terms from the chapter in a multiple-choice format. Answers and examples illustrating each term are provided.

Quick Review sections give students not only the main concepts from the chapter (referenced back to the appropriate section), but also an adjacent example of each concept.

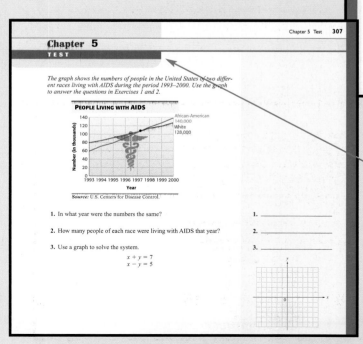

A Chapter Test helps students practice for the real thing.

Review Exercises are keyed to the appropriate sections so that students can refer to examples of that type of problem if they need help.

Chapter 5
REVIEW EXERCISES

[5.1] **1.** Solve the system by graphing.

$$x + 3y = 8$$
$$2x - y = 2$$

2. The graph shows the trends during the years 1974–2000 relating to bachelor's degrees awarded in the United States.

BACHELOR'S DEGREES IN THE U.S.

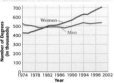

Source: U.S. National Center for Education Statistics, *Digest of Education Statistics,* annual.

(a) Between what years shown on the horizontal axis did the number of degrees for men and women reach equal numbers?

(b) When the number of degrees for men and women reached equal numbers, what was that number (approximately)?

Solve each system using the substitution method.

3. $3x + y = -4$
$x = \dfrac{2}{3}y$

4. $9x - y = -4$
$y = x + 4$

5. $-5x + 2y = -2$
$x + 6y = 26$

---

20. A plane flies 560 mi in 1.75 hr traveling with the wind. The return trip later against the same wind takes the plane 2 hr. Find the speed of the plane and the speed of the wind.

	r	t	d
With Wind	$x + y$	1.75	
Against Wind		2	

21. Sweet's Candy Store is offering a special mix for Valentine's Day. Ms. Sweet will mix some $2 per lb nuts with some $1 per lb chocolate candy to get 100 lb of mix, which she will sell at $1.30 per lb. How many pounds of each should she use?

	Price per Pound	Number of Pounds	Value
Nuts		x	
Chocolate		y	
Mixture		100	

22. A biologist wants to grow two types of algae, green and brown. She has 15 kg of nutrient X and 26 kg of nutrient Y. A vat of green algae needs 2 kg of nutrient X and 3 kg of nutrient Y, while a vat of brown algae needs 1 kg of nutrient X and 2 kg of nutrient Y. How many vats of each type of algae should she grow in order to use all the nutrients?

23. The sum of the measures of the angles of a triangle is 180°. The largest angle measures 10° less than the sum of the other two. The measure of the middle-sized angle is the average of the other two. Find the measures of the three angles.

24. How many liters each of 8%, 10%, and 20% hydrogen peroxide should be mixed together to get 8 L of 12.5% solution, if the amount of 8% solution used must be 2 L more than the amount of 20% solution used?

25. In the great baseball year of 1961, Yankee teammates Mickey Mantle, Roger Maris, and John Blanchard combined for 136 home runs. Mantle hit 7 fewer than Maris. Maris hit 40 more than Blanchard. What were the home run totals for each player? (*Source:* Neft, David S. and Richard M. Cohen, *The Sports Encyclopedia: Baseball 2003.*)

[5.4] *Solve each system using row operations.*

26. $2x + 5y = -4$
$4x - y = 14$

27. $6x + 3y = 9$
$-7x + 2y = 17$

28. $x + 2y - z = 1$
$3x + 4y + 2z = -2$
$-2x - y + z = -1$

MIXED REVIEW EXERCISES

Solve by any method.

29. $\dfrac{2}{3}x + \dfrac{1}{6}y = \dfrac{19}{2}$
$\dfrac{1}{3}x - \dfrac{2}{9}y = 2$

30. $2x - 5y = 8$
$3x + 4y = 10$

31. $x = 7y + 10$
$2x + 3y = 3$

Mixed Review Exercises require students to solve problems without the help of section references.

Cumulative Review Exercises
CHAPTERS 1–5

Evaluate.

1. $(-3)^4$

2. -3^4

3. $-(-3)^4$

4. $\sqrt{.49}$

5. $-\sqrt{.49}$

6. $\sqrt{-.49}$

Evaluate if $x = -4$, $y = 3$, and $z = 6$.

7. $|2x| + y^2 - z^3$

8. $-5(x^3 - y^3)$

9. $\dfrac{2x^2 - x + z}{y^2 - z}$

Solve each equation.

10. $7(2x + 3) - 4(2x + 1) = 2(x + 1)$

11. $.04x + .06(x - 1) = 1.04$

12. $ax + by = cx + d$ for x

13. $|6x - 8| = 4$

Solve each inequality.

14. $\dfrac{2}{3}x + \dfrac{5}{12}x \le 20$

15. $|3x + 2| \le 4$

16. $|12t + 7| \ge 0$

17. A recent survey measured public recognition of the most popular contemporary advertising slogans. Complete the results shown in the table if 2500 people were surveyed.

Slogan (product or company)	Percent Recognition (nearest tenth of a percent)	Actual Number Who Recognized Slogan (nearest whole number)
Please Don't Squeeze the . . . (Charmin)	80.4%	
The Breakfast of Champions (Wheaties)	72.5%	
The King of Beers (Budweiser)		1570
Like a Good Neighbor (State Farm)		1430

(Other slogans included "You're in Good Hands" (Allstate), "Snap, Crackle, Pop" (Rice Krispies), and "The Un-Cola" (7-Up).)
Source: Department of Integrated Marketing Communications, Northwestern University.

Solve each problem.

18. A jar contains only pennies, nickels, and dimes. The number of dimes is 1 more than the number of nickels, and the number of pennies is 6 more than the number of nickels. How many of each denomination can be

19. Two angles of a triangle have the same measure. The measure of the third angle is 4° less than twice the measure of each of the equal angles. Find the measures of the three angles.

Cumulative Review Exercises gather various types of exercises from preceding chapters to help students remember and retain what they are learning throughout the course.

An Introduction to Calculators

There is little doubt that the appearance of handheld calculators three decades ago and the later development of scientific and graphing calculators have changed the methods of learning and studying mathematics forever. For example, computations with tables of logarithms and slide rules made up an important part of mathematics courses prior to 1970. Today, with the widespread availability of calculators, these topics are studied only for their historical significance.

Calculators come in a large array of different types, sizes, and prices. *For the course for which this textbook is intended, the most appropriate type is the scientific calculator*, which costs $10–$20.

In this introduction, we explain some of the features of scientific and graphing calculators. However, remember that calculators vary among manufacturers and models, and that while the methods explained here apply to many of them, they may not apply to your specific calculator. *This introduction is only a guide and is not intended to take the place of your owner's manual.* Always refer to the manual whenever you need an explanation of how to perform a particular operation.

SCIENTIFIC CALCULATORS

Scientific calculators are capable of much more than the typical four-function calculator that you might use for balancing your checkbook. Most scientific calculators use *algebraic logic*. (Models sold by Texas Instruments, Sharp, Casio, and Radio Shack, for example, use algebraic logic.) A notable exception is Hewlett-Packard, a company whose calculators use *Reverse Polish Notation* (RPN). In this introduction, we explain the use of calculators with algebraic logic.

Arithmetic Operations To perform an operation of arithmetic, simply enter the first number, press the operation key (+, −, ×, or ÷), enter the second number, and then press the = key. For example, to add 4 and 3, use the following keystrokes.

Change Sign Key The key marked +/− allows you to change the sign of a display. This is particularly useful when you wish to enter a negative number. For example, to enter −3, use the following keystrokes.

Memory Key Scientific calculators can hold a number in memory for later use. The label of the memory key varies among models; two of these are M and STO . The M+ and M− keys allow you to add to or subtract from the value currently in memory. The memory recall key, labeled MR , RM , or RCL , allows you to retrieve the value stored in memory.

Suppose that you wish to store the number 5 in memory. Enter 5, then press the key for memory. You can then perform other calculations. When you need to retrieve the 5, press the key for memory recall.

If a calculator has a constant memory feature, the value in memory will be retained even after the power is turned off. Some advanced calculators have more than one memory. It is best to read the owner's manual for your model to see exactly how memory is activated.

Clearing/Clear Entry Keys The ⓒ or ⓒⒺ key allows you to clear the display or clear the last entry entered into the display. In some models, pressing the ⓒ key once will clear the last entry, while pressing it twice will clear the entire operation in progress.

Second Function Key This key, usually marked ②ⁿᵈ, is used in conjunction with another key to activate a function that is printed *above* an operation key (and not on the key itself). For example, suppose you wish to find the square of a number, and the squaring function (explained in more detail later) is printed above another key. You would need to press ②ⁿᵈ before the desired squaring function can be activated.

Square Root Key Pressing √ or √x will give the square root (or an approximation of the square root) of the number in the display. On some scientific calculators, the square root key is pressed *before* entering the number, while other calculators use the opposite order. Experiment with your calculator to see which method it uses. For example, to find the square root of 36, use the following keystrokes.

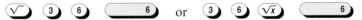

The square root of 2 is an example of an irrational number (**Chapter 9**). The calculator will give an approximation of its value, since the decimal for $\sqrt{2}$ never terminates and never repeats. The number of digits shown will vary among models. To find an approximation for $\sqrt{2}$, use the following keystrokes.

Squaring Key The x^2 key allows you to square the entry in the display. For example, to square 35.7, use the following keystrokes.

The squaring key and the square root key are often found on the same key, with one of them being a second function (that is, activated by the second function key previously described).

Reciprocal Key The key marked ¹/ₓ is the reciprocal key. (When two numbers have a product of 1, they are called *reciprocals*. See **Chapter 1**.) Suppose that you wish to find the reciprocal of 5. Use the following keystrokes.

Inverse Key Some calculators have an inverse key, marked INV. Inverse operations are operations that "undo" each other. For example, the operations of squaring and taking the square root are inverse operations. The use of the INV key varies among different models of calculators, so read your owner's manual carefully.

Exponential Key The key marked x^y or y^x allows you to raise a number to a power. For example, if you wish to raise 4 to the fifth power (that is, find 4^5, as explained in **Chapter 1**), use the following keystrokes.

Root Key Some calculators have this key specifically marked $\sqrt[x]{}$ or $\sqrt[y]{}$; with others, the operation of taking roots is accomplished by using the inverse key in conjunction with the exponential key. Suppose, for example, your calculator is of the latter type and you wish to find the fifth root of 1024. Use the following keystrokes.

Notice how this "undoes" the operation explained in the exponential key discussion.

Pi Key The number π is an important number in mathematics. It occurs, for example, in the area and circumference formulas for a circle. By pressing the π key, you can display the first few digits of π. (Because π is irrational, the display shows only an approximation.) One popular model gives the following display when the π key is pressed.

$$\boxed{3.1415927}$$ An approximation for π

Methods of Display When decimal approximations are shown on scientific calculators, they are either *truncated* or *rounded*. To see how a particular model is programmed, evaluate 1/18 as an example. If the display shows .0555555 (last digit 5), it truncates the display. If it shows .0555556 (last digit 6), it rounds the display.

When very large or very small numbers are obtained as answers, scientific calculators often express these numbers in scientific notation (**Chapter 6**). For example, if you multiply 6,265,804 by 8,980,591, the display might look like this:

$$\boxed{5.6270623\ 13}$$

The 13 at the far right means that the number on the left is multiplied by 10^{13}. This means that the decimal point must be moved 13 places to the right if the answer is to be expressed in its usual form. Even then, the value obtained will only be an approximation: 56,270,623,000,000.

GRAPHING CALCULATORS

While you are not expected to have a graphing calculator to study from this book, we include the following as background information and reference should your course or future courses require the use of graphing calculators.

Basic Features

In addition to the typical keys found on scientific calculators, graphing calculators have keys that can be used to create graphs, make tables, analyze data, and change settings. One of the major differences between graphing and scientific calculators is that a graphing calculator has a larger viewing screen with graphing capabilities. The screens below illustrate the graphs of $Y = X$ and $Y = X^2$.

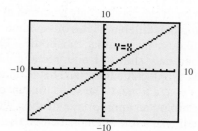
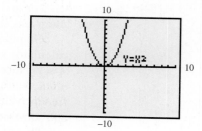

If you look closely at the screens, you will see that the graphs appear to be jagged rather than smooth, as they should be. The reason for this is that graphing calculators have much lower resolution than computer screens. Because of this, graphs generated by graphing calculators must be interpreted carefully.

Editing Input

The screen of a graphing calculator can display several lines of text at a time. This feature allows you to view both previous and current expressions. If an incorrect expression is entered, an error message is displayed. The erroneous expression can be viewed and corrected by using various editing keys, much like a word-processing program. You do not need to enter the entire expression again.

Many graphing calculators can also recall past expressions for editing or updating. The screen on the left below shows how two expressions are evaluated. The final line is entered incorrectly, and the resulting error message is shown in the screen on the right.

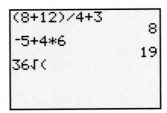

Order of Operations

Arithmetic operations on graphing calculators are usually entered as they are written in mathematical expressions. For example, to evaluate $\sqrt{36}$, you would first press the square root key, and then enter 36. See the screen on the left below. The order of operations on a graphing calculator is also important, and current models assist the user by inserting parentheses when typical errors might occur. The open parenthesis that follows the square root symbol is automatically entered by the calculator so that an expression such as $\sqrt{2 \times 8}$ will not be calculated incorrectly as $\sqrt{2} \times 8$. Compare the two entries and their results in the screen on the right.

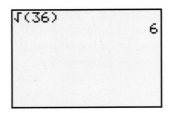
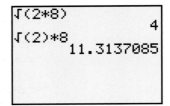

Viewing Windows

The viewing window for a graphing calculator is similar to the viewfinder in a camera. A camera usually cannot take a photograph of an entire view of a scene. The camera must be centered on some object and can capture only a portion of the available scenery. A camera with a zoom lens can photograph different views of the same scene by zooming in and out.

Graphing calculators have similar capabilities. The xy-coordinate plane is infinite. The calculator screen can only show a finite, rectangular region in the plane, and it must be specified before the graph can be drawn. This is done by setting both minimum and maximum values for the x- and y-axes. The scale (distance between tick marks) is usually specified as well. Determining an appropriate viewing window for a graph can be a challenge, and sometimes it will take a few attempts before a satisfactory window is found.

The screen on the left shows a standard viewing window, and the graph of $Y = 2X + 1$ is shown on the right. Using a different window would give a different view of the line.

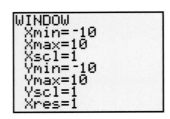
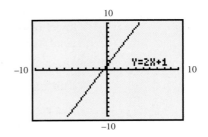

Locating Points on a Graph: Tracing and Tables

Graphing calculators allow you to trace along the graph of an equation and display the coordinates of points on the graph. See the screen on the left below, which indicates that the point (2, 5) lies on the graph of Y = 2X + 1. Tables for equations can also be displayed. The screen on the right shows a partial table for this same equation. Note the middle of the screen, which indicates that when X = 2, Y = 5.

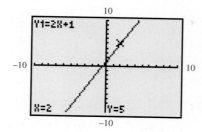

Additional Features

There are many features of graphing calculators that go far beyond the scope of this book. These calculators can be programmed, much like computers. Many of them can solve equations at the stroke of a key, analyze statistical data, and perform symbolic algebraic manipulations. Calculators also provide the opportunity to ask "What if . . . ?" more easily. Values in algebraic expressions can be altered and conjectures tested quickly.

Final Comments

Despite the power of today's calculators, they cannot replace human thought. *In the entire problem-solving process, your brain is the most important component.* Calculators are only tools and, like any tool, they must be used appropriately in order to enhance our ability to understand mathematics. Mathematical insight may often be the quickest and easiest way to solve a problem; a calculator may neither be needed nor appropriate. By applying mathematical concepts, you can make the decision whether or not to use a calculator.

To the Student: Success in Algebra

There are two main reasons students have difficulty with mathematics:

- Students start in a course for which they do not have the necessary background knowledge.

- Students don't know how to study mathematics effectively.

Your teacher can help you decide whether this is the right course for you. We can give you some study tips.

Studying mathematics *is* different from studying subjects like English and history. ***The key to success is regular practice.*** This should not be surprising. After all, can you learn to play the piano or ski well without a lot of regular practice? The same is true for learning mathematics. Working problems nearly every day is the key to becoming successful. Here is a list of things that will help you succeed in studying algebra.

1. **Attend class regularly.** Pay attention in class and take careful notes. In particular, note the problems your teacher works on the board and copy the complete solutions. Keep these notes separate from your homework.

2. **Ask questions.** Don't hesitate to ask questions in class. Other students may have the same questions but be reluctant to ask them, and everyone will benefit from the answers.

3. **Read your text carefully.** Many students go directly to the exercise sets without taking time to read the text and examples. Reading the *complete* section and working the margin problems will pay off when you tackle the homework problems.

4. **Reread your class notes.** Before starting your homework, rework the problems your teacher did in class. This will reinforce what you have learned. Teachers often hear the comment, *"I understand it perfectly when you do it, but I get stuck when I try to work the problem myself."*

5. **Practice by working problems.** Do your homework only *after* reading the text and reviewing your class notes. Check your work against the answer section or the *Student's Solutions Manual.* If you make an error and are unable to determine what went wrong, mark that problem and ask your teacher about it. Then work more problems of the same type to reinforce what you have learned.

6. **Work neatly.** Write symbols neatly. Skip lines between steps. Write large enough so that others can read your work. Use pencil. Make sure that problems are clearly separated from each other.

7. **Review the material.** After completing each section, look over the text again. Decide on the main objectives, and don't be content until you feel that you have mastered them. (In this book, objectives are clearly stated both at the beginning and within each section.) Write a summary of the section or make an outline for future reference.

8. **Prepare for tests.** The chapter summaries in the text are an excellent way for you to review key terms, new symbols, and important concepts from the chapter. After working through the chapter review exercises, use the chapter test as a practice test. Work the problems under test conditions, without looking at the text or answers until you are finished. Time yourself. When you have finished, check your answers against the answer section and rework any that you missed.

9. **Learn from your mistakes.** Keep all graded assignments, quizzes, and tests that are returned to you. Be sure to correct any errors on them and use them to study for future tests and the final exam.

10. **Be diligent and don't give up.** The authors of this text can tell you that they did not always understand a topic the first time they saw it. Don't worry if you also find this to be true. As you read more about a topic and work through the problems, you will gain understanding; the thrill of finally "getting it" is a great feeling. Listen to the words of the late Jim Valvano: *Never give up!*

NOTE
Reading a list of study tips is a good start, but you may need some help actually *applying* the tips to your work in this mathematics course.

Watch for this icon as you work in this textbook, particularly in the first few chapters. It will direct you to one of 12 activities in the *Study Skills Workbook*. Each activity helps you to actually *use* a study skills technique. These techniques will greatly improve your chances for success in this course.

- Find out *how your brain learns new material*. Then use that information to set up effective ways to learn mathematics.

- Find out *why short-term memory is so short* and what you can do to help your brain remember new material weeks and months later.

- Find out *what happens when you "blank out" on a test* and simple ways to prevent it from happening.

All the activities in the *Study Skills Workbook* are practical ways to enjoy and succeed at mathematics. Whether you need help with note taking, managing homework, taking tests, or preparing for a final exam, you'll find specific, clearly explained ideas that really work because they're based on research about how the brain learns and remembers.

Diagnostic Pretest

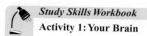
Study Skills Workbook
Activity 1: Your Brain

[Chapter 1]

1. Perform the indicated operations.

$$\frac{5 \cdot (-4)^2 - \sqrt{121}}{-4[-8 - (-3 - 9)] + (-7)}$$

1. _____

2. The table below shows the heights of some selected mountains and the depths of some selected trenches.

2. _____

Mountain	Height (in feet)	Trench	Depth (in feet)
Kennedy	16,286	Tonga	−35,433
Hood	11,239	Palau	−26,424
Washington	6,288	Vema	−21,004

Source: World Almanac and Book of Facts, 2004.

What is the difference between the height of Mt. Hood and the depth of the Vema Trench?

3. Evaluate $\dfrac{6m + 3n^3}{p + 4}$ if $m = 5$, $n = -1$, and $p = -7$.

3. _____

[Chapter 2]

4. Solve $.09x + .11(x + 8) = -.12$.

4. _____

5. Two trains leave from the same point at the same time, traveling in opposite directions. One travels 13 mph faster than the other. After 4 hr, they are 484 mi apart. Find the rate of each train.

5. _____

6. Find the measure of each angle.

6. _____

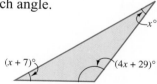

[Chapter 3]

Solve each inequality. Give the solution set in both interval and graph forms.

7. $3x - 2(x - 7) \geq 4(3x + 2) + x$

7. ────────┼────────▶

8. $-8t < 32$ and $3t - 5 \leq 10$

8. ────────┼────▶

9. Solve $|4 - 5y| = |3y + 9|$.

9. _____

[Chapter 4]

10. _____

10. Find the slope of the line through $(5, -12)$ and $(-9, 10)$.

11. x-intercept:_____

y-intercept:_____

11. Find the x- and y-intercepts and graph the equation $3x - 5y = -15$.

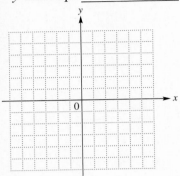

12. _____

12. Graph $4x + 3y < -12$.

13. The median ages for first marriage for women in the United States from 1998–2002 are shown in the table.

Year	1998	1999	2000	2001	2002
Marriage Age	25.0	25.1	25.1	25.1	25.3

Source: World Almanac and Book of Facts, 2004.

13. _____

Which set of ordered pairs from the table is a function? Explain.

A. (Marriage Age, Year)   **B.** (Year, Marriage Age)

[Chapter 5]

Solve each system.

14. _____

14. $-2x + 5y = 11$
$y = x + 7$

15. _____

15. $2x + 3y = 1$
$3x - 4y = 27$

16. _____

16. $3x + y - z = -2$
$x - 2y + z = 8$
$-2x + 3y + 3z = 1$

17. _____

17. A party mix is made by combining nuts that sell for $3.50 per lb with raisins that sell for $1.50 per lb. How much of each should be used to get 32 lb of a mix that will sell for $2.75 per lb?

[Chapter 6]

18. Simplify $(7s^0t^{-5})^2(t^{-3}s^2)^{-4}$ and write your answer with only positive exponents.

18. _____

19. Subtract $(11k^3 - 5k^2 + 6k - 2) - (8k^3 - 2k - 5)$.

19. _____

20. Find the product: $(4x - 9y)^2$.

20. _____

21. Divide $(2x^3 + 5x^2 - 13x + 7) \div (2x - 1)$.

21. _____

[Chapter 7]

22. Factor.

    (a) $6q^2 - 19pq - 20p^2$ **(b)** $y^3 + 3y^2 - 4y - 12$

22. (a) _____

    (b) _____

23. Solve $4x^2 + 17x - 15 = 0$.

23. _____

[Chapter 8]

24. Multiply $\dfrac{z^2 - 16}{z^2 - 4z - 5} \cdot \dfrac{z^2 - 10z + 25}{z^2 - 9z + 20}$.

24. _____

25. Subtract $\dfrac{5x}{x - 3} - \dfrac{4}{x + 3}$.

25. _____

26. Simplify the complex fraction: $\dfrac{16 - \dfrac{1}{x^2}}{\dfrac{4}{x} - \dfrac{1}{x^2}}$.

26. _____

[Chapter 9]

Simplify each expression. Assume that all variables represent positive real numbers.

27. $(-64m^{15}n^{-9})^{2/3}$

27. _____

28. $\sqrt[3]{250y^7z^{11}}$

28. _____

29. Solve $\sqrt{2x - 5} + 4 = x$.

29. _____

30. Multiply $(8 - 5i)(8 + 5i)$.

30. _____

[Chapter 10]

Solve each equation.

31. $(3x + 8)^2 = 49$

31. _____

32. $3x^2 + 5x - 1 = 0$

32. _____

33. Two cars left the same intersection at the same time, one heading due east and the other heading due south. Some time later, they were exactly 75 mi apart. The car headed east had gone 15 mi farther than the car headed south. How far had each car traveled?

33. _____

34. vertex: _____

domain: _____

range: _____

34. Graph $f(x) = -x^2 + 6x - 4$. Give the vertex, domain, and range.

35. _____

[Chapter 11]

35. Find $f^{-1}(x)$ for the one-to-one function defined by $f(x) = x^3 - 8$.

Solve.

36. _____

36. $5^{3x+2} = 25^{2x}$

37. _____

37. $\log_9 x = \dfrac{3}{2}$

38. _____

38. $\log_2 x + \log_2 (x - 6) = 4$

[Chapter 12]

39. (a) _____

(b) _____

(c) _____

(d) _____

39. Let $f(x) = x^2 + 3$ and $g(x) = 2x - 1$. Find each of the following.

(a) $(f \circ g)(2)$ **(b)** $(g \circ f)(2)$

(c) $(f \circ g)(x)$ **(d)** $(g \circ f)(x)$

Graph.

40.

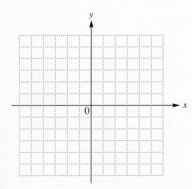

40. $25x^2 + 4y^2 = 100$

41.

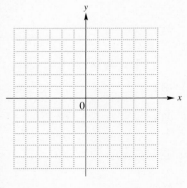

41. $4x^2 - 9y^2 = 36$

42. Solve the system.

$$y = x^2 - 4$$
$$2x - y = 1$$

42. _____

Review of the Real Number System

1.1 **Basic Concepts**

1.2 **Operations on Real Numbers**

1.3 **Exponents, Roots, and Order of Operations**

1.4 **Properties of Real Numbers**

Social Security is the largest source of income for elderly Americans. It is projected that in about 30 years, however, there will be twice as many older Americans as there are today, and the excess revenues now accumulating in Social Security's trust funds will be exhausted. (*Source:* Social Security Administration.)

To supplement their retirement incomes, more and more Americans have begun investing in mutual funds, pension plans, and other means of savings. In Section 1.3, we relate the concepts of this chapter to the percent of U.S. households investing in mutual funds.

1.1 Basic Concepts

OBJECTIVES

1 Write sets using set notation.

2 Use number lines.

3 Know the common sets of numbers.

4 Find additive inverses.

5 Use absolute value.

6 Use inequality symbols.

Study Skills Workbook
Activity 2: Your Textbook

1 Consider the set
$$\left\{ 0, 10, \frac{3}{10}, 52, 98.6 \right\}.$$

(a) Which elements of the set are natural numbers?

(b) Which elements of the set are whole numbers?

2 List the elements in each set.

(a) $\{x \mid x$ is a whole number less than 5$\}$

(b) $\{y \mid y$ is a whole number greater than 12$\}$

ANSWERS
1. **(a)** 10 and 52 **(b)** 0, 10, and 52
2. **(a)** $\{0, 1, 2, 3, 4\}$ **(b)** $\{13, 14, 15, \ldots\}$

In this chapter we review some of the basic symbols and rules of algebra.

OBJECTIVE 1 Write sets using set notation. A **set** is a collection of objects called the **elements** or **members** of the set. In algebra, the elements of a set are usually numbers. Set braces, { }, are used to enclose the elements. For example, 2 is an element of the set {1, 2, 3}. Since we can count the number of elements in the set {1, 2, 3}, it is a *finite set*.

In our study of algebra, we refer to certain sets of numbers by name. The set

$$N = \{1, 2, 3, 4, 5, 6, \ldots\}$$

is called the **natural numbers** or the **counting numbers.** The three dots show that the list continues in the same pattern indefinitely. We cannot list all of the elements of the set of natural numbers, so it is an *infinite set*.

When 0 is included with the set of natural numbers, we have the set of **whole numbers,** written

$$W = \{0, 1, 2, 3, 4, 5, 6, \ldots\}.$$

A set containing no elements, such as the set of whole numbers less than 0, is called the **empty set,** or **null set,** usually written $\emptyset$.

> **CAUTION**
> Do not write $\{\emptyset\}$ for the empty set; $\{\emptyset\}$ is a set with one element, $\emptyset$. Use only the notation $\emptyset$ for the empty set.

◀◀◀ Work Problem 1 at the Side.

In algebra, letters called **variables** are often used to represent numbers or to define sets of numbers. For example,

$$\{x \mid x \text{ is a natural number between 3 and 15}\}$$

(read "the set of all elements x such that x is a natural number between 3 and 15") defines the set

$$\{4, 5, 6, 7, \ldots, 14\}.$$

The notation $\{x \mid x$ is a natural number between 3 and 15$\}$ is an example of **set-builder notation.**

$$\{ x \mid x \text{ has property } P \}$$

the set of · all elements x · such that · x has a given property P

> **EXAMPLE 1 Listing the Elements in Sets**
>
> List the elements in each set.
>
> **(a)** $\{x \mid x$ is a natural number less than 4$\}$
> The natural numbers less than 4 are 1, 2, and 3. This set is $\{1, 2, 3\}$.
>
> **(b)** $\{y \mid y$ is one of the first five even natural numbers$\} = \{2, 4, 6, 8, 10\}$
>
> **(c)** $\{z \mid z$ is a natural number greater than or equal to 7$\}$
> The set of natural numbers greater than or equal to 7 is an infinite set, written with three dots as $\{7, 8, 9, 10, \ldots\}$.

◀◀◀ Work Problem 2 at the Side.

EXAMPLE 2 **Using Set-Builder Notation to Describe Sets**

Use set-builder notation to describe each set.

(a) {1, 3, 5, 7, 9}
There are often several ways to describe a set with set-builder notation. One way is {y | y is one of the first five odd natural numbers}.

(b) {5, 10, 15, . . . }
This set can be described as {x | x is a multiple of 5 greater than 0}.

Work Problem 3 at the Side.

OBJECTIVE 2 Use number lines. A good way to get a picture of a set of numbers is by using a **number line.** To construct a number line, choose any point on a horizontal line and label it 0. Next, choose a point to the right of 0 and label it 1. The distance from 0 to 1 establishes a scale that can be used to locate more points, with positive numbers to the right of 0 and negative numbers to the left of 0. The number 0 is neither positive nor negative. A number line is shown in Figure 1.

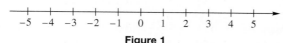

Figure 1

The set of numbers identified on the number line in Figure 1, including positive and negative numbers and 0, is part of the set of **integers,** written

$$I = \{\ldots, -3, -2, -1, 0, 1, 2, 3, \ldots\}.$$

Each number on a number line is called the **coordinate** of the point that it labels, while the point is the **graph** of the number. Figure 2 shows a number line with several selected points graphed on it.

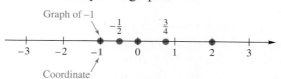

Figure 2

Work Problem 4 at the Side.

The fractions $-\frac{1}{2}$ and $\frac{3}{4}$, graphed on the number line in Figure 2, are examples of **rational numbers.** Rational numbers can be written in decimal form, either as terminating decimals such as $\frac{3}{5} = .6, \frac{1}{8} = .125,$ or $\frac{11}{4} = 2.75,$ or as repeating decimals such as $\frac{1}{3} = .33333\ldots$ or $\frac{3}{11} = .272727\ldots$. A repeating decimal is often written with a bar over the repeating digit(s). Using this notation, $.2727\ldots$ is written $.\overline{27}$.

Decimal numbers that neither terminate nor repeat are *not* rational, and thus are called **irrational numbers.** Many square roots are irrational numbers; for example, $\sqrt{2} = 1.4142136\ldots$ and $-\sqrt{7} = -2.6457513\ldots$ repeat indefinitely without pattern. (Some square roots *are* rational: $\sqrt{16} = 4, \sqrt{100} = 10,$ and so on.) Another irrational number is π, the ratio of the circumference of a circle to its diameter.

Some of the rational and irrational numbers discussed above are graphed on the number line in Figure 3 on the next page. The rational numbers together with the irrational numbers make up the set of **real numbers.** Every point on a number line corresponds to a real number, and every real number corresponds to a point on the number line.

3 Use set-builder notation to describe each set.

(a) {0, 1, 2, 3, 4, 5}

(b) {7, 14, 21, 28, . . . }

4 Graph the elements of each set.

(a) {−4, −2, 0, 2, 4, 6}

(b) $\left\{-1, 0, \frac{2}{3}, \frac{5}{2}\right\}$

(c) $\left\{5, \frac{16}{3}, 6, \frac{13}{2}, 7, \frac{29}{4}\right\}$

3. (a) One answer is {x | x is a whole number less than 6}. **(b)** One answer is {x | x is a multiple of 7 greater than 0}.

4. (a) −4 −2 0 2 4 6
(b) −2 −1 0 1 2 3
(c) 4 5 6 7 8

Real numbers

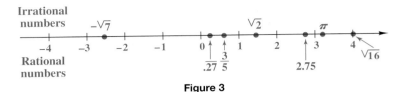

Figure 3

O B J E C T I V E **3** **Know the common sets of numbers.** The sets of numbers listed below will be used throughout the rest of this text.

Sets of Numbers

Natural numbers or counting numbers	$\{1, 2, 3, 4, 5, 6, \dots\}$
Whole numbers	$\{0, 1, 2, 3, 4, 5, 6, \dots\}$
Integers	$\{\dots, -3, -2, -1, 0, 1, 2, 3, \dots\}$
Rational numbers	$\left\{\dfrac{p}{q}\middle|\, p \text{ and } q \text{ are integers, } q \neq 0\right\}$
	Examples: $\frac{4}{1}, 1.3, -\frac{9}{2}, \frac{16}{8}$ or 2, $\sqrt{9}$ or 3, $.\overline{6}$
Irrational numbers	$\{x \mid x \text{ is a real number that is not rational}\}$
	Examples: $\sqrt{3}, -\sqrt{2}, \pi$
Real numbers	$\{x \mid x \text{ is represented by a point on a number line}\}*$

The relationships among these sets of numbers are shown in Figure 4; in particular, notice that the set of real numbers includes both the rational and irrational numbers. ***Every real number is either rational or irrational.*** Also, notice that the integers are elements of the set of rational numbers and that whole numbers and natural numbers are elements of the set of integers.

Real numbers

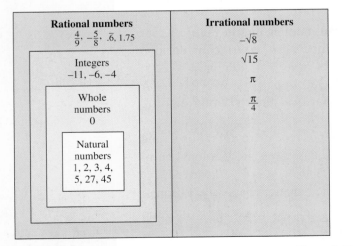

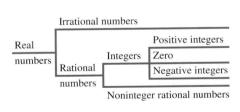

Figure 4 The Real Numbers

---

*An example of a number that is not a coordinate of a point on a number line is $\sqrt{-1}$. This number, called an *imaginary number,* is discussed in **Section 9.7.**

EXAMPLE 3 **Identifying Examples of Number Sets**

Which numbers in

$$\left\{ -8, -\sqrt{2}, -\frac{9}{64}, 0, .5, \frac{2}{3}, 1.\overline{12}, \sqrt{3}, 2 \right\}$$

are elements of each set?

(a) Integers $-8, 0,$ and 2 are integers.

(b) Rational numbers $-8, -\frac{9}{64}, 0, .5, \frac{2}{3}, 1.\overline{12},$ and 2 are rational numbers.

(c) Irrational numbers $-\sqrt{2}$ and $\sqrt{3}$ are irrational numbers.

(d) Real numbers
All the numbers in the given set are real numbers.

Work Problem 5 at the Side. ▶▶▶

EXAMPLE 4 **Determining Relationships between Sets of Numbers**

Decide whether each statement is *true* or *false*.

(a) All irrational numbers are real numbers.
This is true. As shown in Figure 4, the set of real numbers includes all irrational numbers.

(b) Every rational number is an integer.
This statement is false. Although some rational numbers are integers, other rational numbers, such as $\frac{2}{3}$ and $-\frac{1}{4}$, are not.

Work Problem 6 at the Side. ▶▶▶

OBJECTIVE 4 Find additive inverses.   Look again at the number line in Figure 1. For each positive number, there is a negative number on the opposite side of 0 that lies the same distance from 0. These pairs of numbers are called **additive inverses, negatives,** or **opposites** of each other. For example, 5 is the additive inverse of -5, and -5 is the additive inverse of 5.

Additive Inverse
For any real number a, the number $-a$ is the additive inverse of a.

Change the sign of a number to get its additive inverse. *The sum of a number and its additive inverse is always 0.*
The symbol "$-$" can be used to indicate any of the following:

1. a negative number, such as -9 or -15;

2. the additive inverse of a number, as in "-4 is the additive inverse of 4";

3. subtraction, as in $12 - 3$.

In the expression $-(-5)$, the symbol "$-$" is being used in two ways: the first $-$ indicates the additive inverse of -5, and the second indicates a negative number, -5. Since the additive inverse of -5 is 5, then $-(-5) = 5$. This example suggests the following property.

For any real number a, $-(-a) = a.$

5 Select all the sets from the following list that apply to each number.
Whole number
Rational number
Irrational number
Real number

(a) -6

(b) 12

(c) $.\overline{3}$

(d) $-\sqrt{15}$

(e) π

(f) $\frac{22}{7}$

(g) 3.14

6 Decide whether the statement is *true* or *false*. If *false*, tell why.

(a) All whole numbers are integers.

(b) Some integers are whole numbers.

(c) Every real number is irrational.

ANSWERS
5. **(a)** rational, real
   (b) whole, rational, real
   (c) rational, real
   (d) irrational, real
   (e) irrational, real
   (f) rational, real
   (g) rational, real
6. **(a)** true **(b)** true
   (c) false; Some real numbers are irrational, but others are rational numbers.

7 Give the additive inverse of each number.

(a) 9

Numbers written with positive or negative signs, such as $+4$, $+8$, -9, and -5, are called **signed numbers.** A positive number can be called a signed number even though the positive sign is usually left off. The following table shows the additive inverses of several signed numbers. Note that 0 is its own additive inverse.

Number	Additive Inverse
6	-6
-4	4
$\frac{2}{3}$	$-\frac{2}{3}$
-8.7	8.7
0	0

(b) -12

◀◀◀ Work Problem 7 at the Side.

OBJECTIVE 5 Use absolute value. Geometrically, the **absolute value** of a number a, written $|a|$, is the distance on the number line from 0 to a. For example, the absolute value of 5 is the same as the absolute value of -5 because each number lies five units from 0. See Figure 5. That is,

$$|5| = 5 \quad \text{and} \quad |-5| = 5.$$

(c) $-\dfrac{6}{5}$

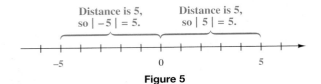

Distance is 5, so $|-5| = 5$. Distance is 5, so $|5| = 5$.

Figure 5

CAUTION
Because absolute value represents distance, and distance is always positive (or 0), *the absolute value of a number is always positive (or 0).*

(d) 0

The formal definition of absolute value follows.

Absolute Value

$$|a| = \begin{cases} a & \text{if } a \text{ is positive or } 0 \\ -a & \text{if } a \text{ is negative} \end{cases}$$

The second part of this definition, $|a| = -a$ if a is negative, requires careful thought. If a is a *negative* number, then $-a$, the additive inverse or opposite of a, is a positive number, so $|a|$ is positive. For example, if $a = -3$, then

(e) 1.5

$$|a| = |-3| = -(-3) = 3. \quad |a| = -a \text{ if } a \text{ is negative.}$$

EXAMPLE 5 Evaluating Absolute Value Expressions

Find the value of each expression.

(a) $|13| = 13$ **(b)** $|-2| = -(-2) = 2$

(c) $|0| = 0$

Continued on Next Page

ANSWERS
7. **(a)** -9 **(b)** 12   **(c)** $\frac{6}{5}$ **(d)** 0   **(e)** -1.5

(d) $-|8|$

Evaluate the absolute value first. Then find the additive inverse.

$$-|8| = -(8) = -8$$

(e) $-|-8|$

Work as in part (d): $|-8| = 8$, so

$$-|-8| = -(8) = -8.$$

(f) $|-2| + |5|$

Evaluate each absolute value first, and then add.

$$|-2| + |5| = 2 + 5 = 7$$

(g) $-|5 - 2| = -|3| = -3$

Work Problem 8 at the Side. ▶▶▶

Absolute value is useful when comparing size without regard to sign.

EXAMPLE 6 **Comparing Rates of Change in Industries**

The projected annual rates of employment change (in percent) in some of the fastest growing and most rapidly declining industries from 1994 through 2005 are shown in the table.

Industry (1994–2005)	Percent Rate of Change
Health services	5.7
Computer and data processing services	4.9
Child day care services	4.3
Footware, except rubber and plastic	−6.7
Household audio and video equipment	−4.2
Luggage, handbags, and leather products	−3.3

Source: U.S. Bureau of Labor Statistics.

What industry in the list is expected to see the greatest change? the least change?

We want the greatest *change*, without regard to whether the change is an increase or a decrease. Look for the number in the list with the largest absolute value. That number is found in footware, since $|-6.7| = 6.7$. Similarly, the least change is in the luggage, handbags, and leather products industry: $|-3.3| = 3.3$.

Work Problem 9 at the Side. ▶▶▶

OBJECTIVE 6 **Use inequality symbols.** The statement $4 + 2 = 6$ is an **equation;** it states that two quantities are equal. The statement $4 \neq 6$ (read "4 is not equal to 6") is an **inequality,** a statement that two quantities are *not* equal. When two numbers are not equal, one must be less than the other. The symbol $<$ means "is less than." For example,

$$8 < 9, \quad -6 < 15, \quad -6 < -1, \quad \text{and} \quad 0 < \frac{4}{3}.$$

The symbol $>$ means "is greater than." For example,

$$12 > 5, \quad 9 > -2, \quad -4 > -6, \quad \text{and} \quad \frac{6}{5} > 0.$$

In each case, ***the symbol "points" toward the smaller number.***

8 Find the value of each expression.

(a) $|6|$

(b) $|-3|$

(c) $-|5|$

(d) $-|-2|$

(e) $-|-7|$

(f) $|-6| + |-3|$

(g) $|-9| - |-4|$

(h) $-|9 - 4|$

9 Refer to the table in Example 6. Of the household audio/video equipment industry and computer/data processing services, which will show the greater change (without regard to sign)?

ANSWERS
8. (a) 6 **(b)** 3 **(c)** −5 **(d)** −2 **(e)** −7
(f) 9 **(g)** 5 **(h)** −5
9. Computer/data processing services

10 Insert < or > in each blank to make a true statement.

(a) 3 _____ 7

(b) 9 _____ 2

(c) −4 _____ −8

(d) −2 _____ −1

(e) 0 _____ −3

The number line in Figure 6 shows the numbers 4 and 9, and we know that $4 < 9$. On the graph, 4 is to the left of 9. The lesser of two numbers is always to the left of the other on a number line.

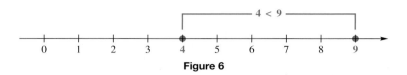

Figure 6

Inequalities on a Number Line

On a number line,

$a < b$ if a is to the left of b; $a > b$ if a is to the right of b.

We can use a number line to determine order. As shown on the number line in Figure 7, −6 is located to the left of 1. For this reason, $-6 < 1$. Also, $1 > -6$. From the same number line, $-5 < -2$, or $-2 > -5$.

Figure 7

CAUTION
Be careful when ordering negative numbers. Since −5 is to the left of −2 on the number line in Figure 7, $-5 < -2$, or $-2 > -5$. In each case, the symbol points to −5, the smaller number.

◀◀◀ **Work Problem 10 at the Side.**

The following table summarizes results about positive and negative numbers in both words and symbols.

Words	Symbols
Every negative number is less than 0.	If a is negative, then $a < 0$.
Every positive number is greater than 0.	If a is positive, then $a > 0$.
0 is neither positive nor negative.	

In addition to the symbols $\neq$, $<$, and $>$, the symbols $\leq$ and $\geq$ are often used.

INEQUALITY SYMBOLS

Symbol	Meaning	Example
$\neq$	is not equal to	$3 \neq 7$
$<$	is less than	$-4 < -1$
$>$	is greater than	$3 > -2$
$\leq$	is less than or equal to	$6 \leq 6$
$\geq$	is greater than or equal to	$-8 \geq -10$

ANSWERS
10. (a) < **(b)** > **(c)** > **(d)** < **(e)** >

The following table shows several inequalities and why each is true.

Inequality	Why It Is True
$6 \leq 8$	$6 < 8$
$-2 \leq -2$	$-2 = -2$
$-9 \geq -12$	$-9 > -12$
$-3 \geq -3$	$-3 = -3$
$6 \cdot 4 \leq 5(5)$	$24 < 25$

Notice the reason why $-2 \leq -2$ is true. With the symbol $\leq$, if *either* the $<$ part *or* the $=$ part is true, then the inequality is true. This is also the case with the $\geq$ symbol.

In the last line, recall that the dot in $6 \cdot 4$ indicates the product 6×4, or 24, and **5(5)** means 5×5, or 25. Thus, the inequality $6 \cdot 4 \leq 5(5)$ becomes $24 \leq 25$, which is true.

Work Problem 11 at the Side. ▶▶▶

11 Answer *true* or *false*.

(a) $-2 \leq -3$

(b) $8 \leq 8$

(c) $-9 \geq -1$

(d) $5 \cdot 8 \leq 7 \cdot 7$

(e) $3(4) > 2(6)$

ANSWERS
11. (a) false (b) true (c) false
(d) true (e) false

Focus on Real-Data Applications

Stock Splits

Buying stock in a corporation makes you part owner, and you can earn **dividends** from the profits. You can track the status of your holdings by reading stock exchange reports in daily papers such as the *Wall Street Journal*. Stock prices are also reported on Internet sites such as *Yahoo! Finance* or *Morningside.com*. In fact, many reputable Internet sites have on-line "education centers" to help you learn about investing in stocks, bonds, and mutual funds.

- Your **equity,** or the value of your stock holdings, is the product of the price per share times the number of shares. For example, if you owned 25.400 shares of Yahoo! Inc. stock, worth $46.22 per share on January 29, 2004, then your equity to the nearest cent would be

$$\$46.22 \times 25.400 = \$1173.99.$$

- The **number of shares** owned is typically given to three-decimal-place accuracy (nearest thousandth).
- The **price per share** is rounded to the nearest cent.

Occasionally, corporations announce a **stock split,** resulting in an increase in the number of shares, a decrease in the price per share, and a constant equity. For example, in a **2 for 1 split** the number of shares that you own doubles, the price per share is halved, and the equity remains the same. Proposed stock splits are published in advance.

Suppose that your 25.400 shares of Yahoo! Inc., priced at $46.22, split 3 for 2 on January 29, 2004.

1. The increased number of shares becomes $\frac{3}{2} \times 25.400 = 38.100$.

2. The decreased price per share becomes $\frac{2}{3} \times \$46.22 = \$30.813 = \$30.81$ (nearest cent).

Note: Because the price per share was rounded down, the number of shares would be adjusted slightly to maintain constant equity.

For Group Discussion

1. For the purpose of discussion only, assume that the stock prices in the table are those quoted on the day of a split. For each of the announced stock splits, calculate the *new* number of shares, price per share, and value of equity. To check your work, show that the equity is the same before and after the stock split.

Stock	Before Split		Split Ratio	After Split		Equity
	Number of Shares	Price per Share		Number of Shares	Price per Share	
(a) Chico's FAS, Inc.	25	$34.26	3–2			
(b) Ivax Corp	100	$29.50	5–4			
(c) State Street	52.406	$114.68	2–1			

2. What are some reasons why a corporation institutes a stock split?

1.1 Exercises

FOR EXTRA HELP Addison-Wesley Math Tutor Center MathXL Digital Video Tutor CD 1 Videotape 1 Student's Solutions Manual MyMathLab MyMathLab Interactmath.com

Study Skills Workbook
Activity 3: Homework

Write each set by listing its elements. See Example 1.

1. $\{x \mid x$ is a natural number less than 6$\}$

2. $\{m \mid m$ is a natural number less than 9$\}$

3. $\{z \mid z$ is an integer greater than 4$\}$

4. $\{y \mid y$ is an integer greater than 8$\}$

5. $\{a \mid a$ is an even integer greater than 8$\}$

6. $\{k \mid k$ is an odd integer less than 1$\}$

7. $\{x \mid x$ is an irrational number that is also rational$\}$

8. $\{r \mid r$ is a number that is both positive and negative$\}$

9. $\{p \mid p$ is a number whose absolute value is 4$\}$

10. $\{w \mid w$ is a number whose absolute value is 7$\}$

Write each set using set-builder notation. See Example 2. (More than one description is possible.)

11. $\{2, 4, 6, 8\}$

12. $\{11, 12, 13, 14\}$

13. $\{4, 8, 12, 16, \dots\}$

14. $\{\dots, -6, -3, 0, 3, 6, \dots\}$

Graph the elements of each set on a number line.

15. $\{-3, -1, 0, 4, 6\}$

16. $\{-4, -2, 0, 3, 5\}$

17. $\left\{-\dfrac{2}{3}, 0, \dfrac{4}{5}, \dfrac{12}{5}, \dfrac{9}{2}, 4.8\right\}$

18. $\left\{-\dfrac{6}{5}, -\dfrac{1}{4}, 0, \dfrac{5}{6}, \dfrac{13}{4}, 5.2, \dfrac{11}{2}\right\}$

Which elements of each set are (a) natural numbers, (b) whole numbers, (c) integers, (d) rational numbers, (e) irrational numbers, (f) real numbers? See Example 3.

19. $\left\{-8, -\sqrt{5}, -.6, 0, \dfrac{3}{4}, \sqrt{3}, \pi, 5, \dfrac{13}{2}, 17, \dfrac{40}{2}\right\}$

20. $\left\{-9, -\sqrt{6}, -.7, 0, \dfrac{6}{7}, \sqrt{7}, 4.\overline{6}, 8, \dfrac{21}{2}, 13, \dfrac{75}{5}\right\}$

Decide whether each statement is true *or* false. *If* false, *tell why. See Example 4.*

21. Every rational number is an integer.

22. Every natural number is an integer.

23. Every irrational number is an integer.

24. Every integer is a rational number.

25. Every natural number is a whole number.

26. Some rational numbers are irrational.

27. Some rational numbers are whole numbers.

28. Some real numbers are integers.

29. The absolute value of any number is the same as the absolute value of its additive inverse.

30. The absolute value of any nonzero number is positive.

*Give **(a)** the additive inverse and **(b)** the absolute value of each number. See the discussion of additive inverses and Example 5.*

31. 6

32. 8

33. -12

34. -15

35. $\dfrac{6}{5}$

36. .13

Find the value of each expression. See Example 5.

37. $|-8|$

38. $|-11|$

39. $\left|\dfrac{3}{2}\right|$

40. $\left|\dfrac{7}{4}\right|$

41. $-|5|$

42. $-|17|$

43. $-|-2|$

44. $-|-8|$

45. $-|4.5|$

46. $-|12.6|$

47. $|-2| + |3|$

48. $|-16| + |12|$

49. $|-9| - |-3|$

50. $|-10| - |-5|$

51. $|-1| + |-2| - |-3|$

52. $|-6| + |-4| - |-10|$

Solve each problem. See Example 6.

53. The table shows the percent change in population from 1995 through 2000 for selected cities in the world.

City	Percent Change
Mexico City, Mexico	1.52
Los Angeles, U.S.	1.28
Shanghai, China	−.34
Osaka, Japan	−.05
Beijing, China	.02
Mumbai (Bombay), India	2.80

Source: World Almanac and Book of Facts, 2004.

(a) Which city had the greatest change in population? What was this change? Was it an increase or a decline?

(b) Which city had the smallest change in population? What was this change? Was it an increase or a decline?

54. The table gives the net trade balance, in millions of dollars, for selected U.S. trade partners for April 2004.

Country	Trade Balance (in millions of dollars)
Germany	−2815
China	−7552
Netherlands	823
France	−951
Turkey	96

Source: U.S. Bureau of the Census.

A negative balance means that imports exceeded exports, while a positive balance means that exports exceeded imports.

(a) Which country had the greatest discrepancy between exports and imports? Explain.

(b) Which country had the smallest discrepancy between exports and imports? Explain.

Sea level refers to the surface of the ocean. The depth of a body of water such as an ocean or sea can be expressed as a negative number, representing average depth in feet below sea level. On the other hand, the altitude of a mountain can be expressed as a positive number, indicating its height in feet above sea level. The table gives selected depths and heights.

Body of Water	Average Depth in Feet (as a negative number)	Mountain	Altitude in Feet (as a positive number)
Pacific Ocean	−12,925	McKinley	20,320
South China Sea	−4,802	Point Success	14,158
Gulf of California	−2,375	Matlalcueyetl	14,636
Caribbean Sea	−8,448	Rainier	14,410
Indian Ocean	−12,598	Steele	16,644

Source: World Almanac and Book of Facts, 2004.

55. List the bodies of water in order, starting with the deepest and ending with the shallowest.

56. List the mountains in order, starting with the shortest and ending with the tallest.

57. *True* or *false:* The absolute value of the depth of the Pacific Ocean is greater than the absolute value of the depth of the Indian Ocean.

58. *True* or *false:* The absolute value of the depth of the Gulf of California is greater than the absolute value of the depth of the Caribbean Sea.

Use the number line to answer true *or* false *to each statement.*

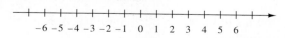

59. $-6 < -2$

60. $-4 < -3$

61. $-4 > -3$

62. $-2 > -1$

63. $3 > -2$

64. $5 > -3$

65. $-3 \geq -3$

66. $-4 \leq -4$

Use an inequality symbol to write each statement.

67. 7 is greater than y.

68. -4 is less than 12.

69. 5 is greater than or equal to 5.

70. -3 is less than or equal to -3.

71. $3t - 4$ is less than or equal to 10.

72. $5x + 4$ is greater than or equal to 19.

73. $5x + 3$ is not equal to 0.

74. $6x + 7$ is not equal to -3.

First simplify each side of the inequality. Then tell whether the resulting statement is true *or* false.

75. $-6 < 7 + 3$

76. $-7 < 4 + 2$

77. $2 \cdot 5 \geq 4 + 6$

78. $8 + 7 \leq 3 \cdot 5$

79. $-|-3| \geq -3$

80. $-|-5| \leq -5$

81. $-8 > -|-6|$

82. $-9 > -|-4|$

The graph shows egg production in millions of eggs in selected states for 2001 and 2002. Use this graph to work Exercises 83–87.

83. In 2001, was egg production in Iowa (IA) less than or greater than egg production in California (CA)?

84. In 2002, which states had production greater than 6500 million eggs?

85. In which states was the 2002 egg production greater than 2001 production?

86. If x represents 2002 egg production for Texas (TX) and y represents 2002 egg production for Ohio (OH), which is true: $x < y$ or $x > y$?

87. If x represents 2002 egg production for Indiana (IN) and y represents 2002 egg production for Minnesota (MN), which is true: $x < y$ or $x > y$?

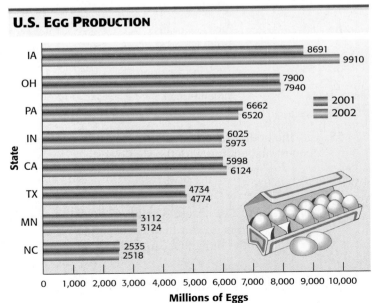

U.S. EGG PRODUCTION

Source: U.S. Department of Agriculture.

1.2 Operations on Real Numbers

In this section we review the rules for adding, subtracting, multiplying, and dividing real numbers.

OBJECTIVE **1** **Add real numbers.**   Number lines can be used to illustrate addition and subtraction of real numbers. To add two real numbers on a number line, start at 0. Move right (the *positive* direction) to add a positive number or left (the *negative* direction) to add a negative number. See Figure 8.

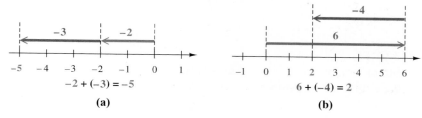

Figure 8

This procedure for adding real numbers can be generalized in the following rules.

Adding Real Numbers

Like signs   To add two numbers with the *same* sign, add their absolute values. The sign of the answer (either $+$ or $-$) is the same as the sign of the two numbers.

Unlike signs   To add two numbers with *different* signs, subtract the smaller absolute value from the larger. The sign of the answer is the same as the sign of the number with the larger absolute value.

Recall that the answer to an addition problem is called the **sum.**

EXAMPLE 1 **Adding Two Negative Numbers**

Find each sum.

(a) $-12 + (-8)$
First find the absolute values.
$$|-12| = 12 \quad \text{and} \quad |-8| = 8$$
Because -12 and -8 have the *same* sign, add their absolute values. Both numbers are negative, so the answer is negative.
$$-12 + (-8) = -(12 + 8) = -(20) = -20$$

(b) $-6 + (-3) = -(|-6| + |-3|) = -(6 + 3) = -9$

(c) $-1.2 + (-.4) = -(1.2 + .4) = -1.6$

(d) $-\dfrac{5}{6} + \left(-\dfrac{1}{3}\right) = -\left(\dfrac{5}{6} + \dfrac{1}{3}\right) = -\left(\dfrac{5}{6} + \dfrac{2}{6}\right) = -\dfrac{7}{6}$

Work Problem 1 at the Side.

OBJECTIVES

1 Add real numbers.

2 Subtract real numbers.

3 Multiply real numbers.

4 Find the reciprocal of a number.

5 Divide real numbers.

1 Find each sum.

(a) $-2 + (-7)$

(b) $-15 + (-6)$

(c) $-1.1 + (-1.2)$

(d) $-\dfrac{3}{4} + \left(-\dfrac{1}{2}\right)$

ANSWERS

1. (a) -9 (b) -21 (c) -2.3 (d) $-\dfrac{5}{4}$

2 Find each sum.

(a) $12 + (-1)$

(b) $3 + (-7)$

(c) $-17 + 5$

(d) $-\dfrac{3}{4} + \dfrac{1}{2}$

(e) $-1.5 + 3.2$

EXAMPLE 2 **Adding Numbers with Different Signs**

Find each sum.

(a) $-17 + 11$

First find the absolute values.

$$|-17| = 17 \quad \text{and} \quad |11| = 11$$

Because -17 and 11 have *different* signs, subtract their absolute values.

$$17 - 11 = 6$$

The number -17 has a larger absolute value than 11, so the answer is negative.

$$-17 + 11 = -6$$

Negative because $|-17| > |11|$

(b) $4 + (-1)$

Subtract the absolute values, 4 and 1. Because 4 has the larger absolute value, the sum must be positive.

$$4 + (-1) = 4 - 1 = 3$$

Positive because $|4| > |-1|$

(c) $-9 + 17 = 17 - 9 = 8$

(d) $-16 + 12$

The absolute values are 16 and 12. Subtract the absolute values. The negative number has the larger absolute value, so the answer is negative.

$$-16 + 12 = -(16 - 12) = -4$$

(e) $-\dfrac{4}{5} + \dfrac{2}{3}$

Write each number with a common denominator.

$$\frac{4}{5} = \frac{4 \cdot 3}{5 \cdot 3} = \frac{12}{15} \quad \text{and} \quad \frac{2}{3} = \frac{2 \cdot 5}{3 \cdot 5} = \frac{10}{15}$$

$$-\frac{4}{5} + \frac{2}{3} = -\frac{12}{15} + \frac{10}{15}$$

$$= -\left(\frac{12}{15} - \frac{10}{15} \right) \qquad -\frac{12}{15} \text{ has the larger absolute value.}$$

$$= -\frac{2}{15} \qquad \text{Subtract.}$$

(f) $-2.3 + 5.6 = 3.3$

Work Problem 2 at the Side.

OBJECTIVE 2 Subtract real numbers. Recall that the answer to a subtraction problem is called the **difference.** Thus, the difference between 6 and 4 is 2. To see how subtraction should be defined, compare the following two statements.

$$6 - 4 = 2$$
$$6 + (-4) = 2$$

The second statement is pictured on the number line in Figure 8(b) at the beginning of this section. Similarly, $9 - 3 = 6$ and $9 + (-3) = 6$ so that $9 - 3 = 9 + (-3)$. These examples suggest the following rule for subtraction.

ANSWERS

2. (a) 11 (b) -4 (c) -12 (d) $-\dfrac{1}{4}$

(e) 1.7

Subtraction

For all real numbers a and b,

$$a - b = a + (-b).$$

In words, change the sign of the second number (subtrahend) and add.

EXAMPLE 3 **Subtracting Real Numbers**

Find each difference.

Change to addition.
Change sign of second number (subtrahend).

(a) $6 - 8 = 6 + (-8) = -2$

Changed
Sign changed

(b) $-12 - 4 = -12 + (-4) = -16$

(c) $-10 - (-7) = -10 + [-(-7)]$ This step is often omitted.

$$= -10 + 7$$

$$= -3$$

(d) $-2.4 - (-8.1) = -2.4 + 8.1 = 5.7$

(e) $\dfrac{8}{3} - \left(-\dfrac{5}{3}\right) = \dfrac{8}{3} + \dfrac{5}{3} = \dfrac{13}{3}$

Work Problem 3 at the Side. ▶▶▶

When working a problem that involves both addition and subtraction, add and subtract in order from left to right. Work inside brackets or parentheses first.

EXAMPLE 4 **Adding and Subtracting Real Numbers**

Perform the indicated operations.

(a) $-8 + 5 - 6 = (-8 + 5) - 6$ Work from left to right.

$$= -3 - 6$$

$$= -3 + (-6)$$

$$= -9$$

(b) $15 - (-3) - 5 - 12 = (15 + 3) - 5 - 12$

$$= 18 - 5 - 12$$

$$= 13 - 12$$

$$= 1$$

(c) $-4 - (-6) + 7 - 1 = (-4 + 6) + 7 - 1$

$$= 2 + 7 - 1$$

$$= 9 - 1$$

$$= 8$$

Continued on Next Page

❸ Find each difference.

(a) $9 - 12$

(b) $-7 - 2$

(c) $-8 - (-2)$

(d) $-6.3 - (-11.5)$

(e) $12 - (-5)$

ANSWERS
3. **(a)** -3 **(b)** -9 **(c)** -6 **(d)** 5.2
   (e) 17

4 Perform the indicated operations.

(a) $-6 + 9 - 2$

(b) $12 - (-4) + 8$

(c) $-6 - (-2) - 8 - 1$

(d) $-3 - [(-7) + 15] + 6$

5 Find each product.

(a) $-7(-5)$

(b) $-.9(-15)$

(c) $-\dfrac{4}{7}\left(-\dfrac{14}{3}\right)$

(d) $7(-2)$

(e) $-.8(.006)$

(f) $\dfrac{5}{8}(-16)$

(g) $-\dfrac{2}{3}(12)$

ANSWERS
4. (a) 1 (b) 24 (c) -13 (d) -5
5. (a) 35 (b) 13.5 (c) $\dfrac{8}{3}$ (d) -14
   (e) $-.0048$ (f) -10 (g) -8

(d) $-9 - [-8 - (-4)] + 6 = -9 - [-8 + 4] + 6$ Work inside the brackets.
$$= -9 - [-4] + 6$$
$$= -9 + 4 + 6$$
$$= -5 + 6$$
$$= 1$$

>>> **Work Problem 4 at the Side.**

OBJECTIVE 3 Multiply real numbers. The answer to a multiplication problem is called the **product.** For example, 24 is the product of 8 and 3. The rules for finding signs of products of real numbers are given below.

Multiplying Real Numbers

Like signs   The product of two numbers with the *same* sign is positive.

Unlike signs   The product of two numbers with *different* signs is negative.

EXAMPLE 5   **Multiplying Real Numbers**

Find each product.

(a) $-3(-9) = 27$ Same sign; product is positive.

(b) $-.5(-.4) = .2$

(c) $-\dfrac{3}{4}\left(-\dfrac{5}{3}\right) = \dfrac{5}{4}$

(d) $6(-9) = -54$ Different signs; product is negative.

(e) $-.05(.3) = -.015$

(f) $\dfrac{2}{3}(-3) = -2$

(g) $-\dfrac{5}{8}\left(\dfrac{12}{13}\right) = -\dfrac{15}{26}$

>>> **Work Problem 5 at the Side.**

OBJECTIVE 4 Find the reciprocal of a number. Earlier, subtraction was defined in terms of addition. Now, division is defined in terms of multiplication. The definition of division depends on the idea of a **multiplicative inverse** or *reciprocal;* two numbers are *reciprocals* if they have a product of 1.

Reciprocal

The **reciprocal** of a nonzero number a is $\dfrac{1}{a}$.

Calculator Tip   Reciprocals (in decimal form) can be found with a calculator that has a key labeled $\boxed{1/x}$ or $\boxed{x^{-1}}$. For example, a calculator shows that the reciprocal of 25 is .04.

The table gives several numbers and their reciprocals.

Number	Reciprocal
$-\frac{2}{5}$	$-\frac{5}{2}$
-6	$-\frac{1}{6}$
$\frac{7}{11}$	$\frac{11}{7}$
$.05$	20
0	None

$-\frac{2}{5}\left(-\frac{5}{2}\right) = 1$

$-6\left(-\frac{1}{6}\right) = 1$

$\frac{7}{11}\left(\frac{11}{7}\right) = 1$

$.05(20) = 1$

There is no reciprocal for 0 because there is no number that can be multiplied by 0 to give a product of 1.

CAUTION
A number and its additive inverse have *opposite* signs; however, a number and its reciprocal always have the *same* sign.

Work Problem 6 at the Side. ▶▶▶

OBJECTIVE 5 Divide real numbers. The result of dividing one number by another is called the **quotient.** For example, when 45 is divided by 3, the quotient is 15. To define division of real numbers, we first write the quotient of 45 and 3 as $\frac{45}{3}$, which equals 15. The same answer will be obtained if 45 and $\frac{1}{3}$ are multiplied, as follows.

$$45 \div 3 = \frac{45}{3} = 45 \cdot \frac{1}{3} = 15$$

This suggests the following definition of division of real numbers.

Division
For all real numbers a and b (where $b \neq 0$),

$$a \div b = \frac{a}{b} = a \cdot \frac{1}{b}.$$

In words, multiply the first number by the reciprocal of the second number.

There is no reciprocal for the number 0, so *division by 0 is undefined.* For example, $\frac{15}{0}$ is undefined and $-\frac{1}{0}$ is undefined.

CAUTION
Division by 0 is undefined. However, dividing 0 by a nonzero number gives the quotient 0. For example,

$$\frac{6}{0} \text{ is undefined,} \quad \text{but} \quad \frac{0}{6} = 0 \quad (\text{since } 0 \cdot 6 = 0).$$

Be careful when 0 is involved in a division problem.

Work Problem 7 at the Side. ▶▶▶

6 Give the reciprocal of each number.

(a) 15

(b) -7

(c) $\dfrac{8}{9}$

(d) $-\dfrac{1}{3}$

(e) .125

7 Divide where possible.

(a) $\dfrac{9}{0}$

(b) $\dfrac{0}{9}$

(c) $\dfrac{-9}{0}$

(d) $\dfrac{0}{-9}$

ANSWERS
6. (a) $\dfrac{1}{15}$ (b) $-\dfrac{1}{7}$ (c) $\dfrac{9}{8}$ (d) -3 (e) 8
7. (a) undefined (b) 0 (c) undefined
   (d) 0

8 Find each quotient.

(a) $\dfrac{-16}{4}$

(b) $\dfrac{8}{-2}$

(c) $\dfrac{-15}{-3}$

(d) $\dfrac{\frac{3}{8}}{\frac{11}{16}}$

Since division is defined as multiplication by the reciprocal, the rules for signs of quotients are the same as those for signs of products.

Dividing Real Numbers

Like signs The quotient of two nonzero real numbers with the *same* sign is positive.

Unlike signs The quotient of two nonzero real numbers with *different* signs is negative.

EXAMPLE 6 Dividing Real Numbers

Find each quotient.

(a) $\dfrac{-12}{4} = -12 \cdot \dfrac{1}{4} = -3 \qquad \frac{a}{b} = a \cdot \frac{1}{b}$

(b) $\dfrac{6}{-3} = 6\left(-\dfrac{1}{3}\right) = -2 \qquad$ The reciprocal of -3 is $-\frac{1}{3}$.

(c) $\dfrac{-30}{-2} = -30\left(-\dfrac{1}{2}\right) = 15$

(d) $\dfrac{\frac{2}{3}}{\frac{5}{9}} = \dfrac{2}{3} \cdot \dfrac{9}{5} = \dfrac{6}{5} \qquad$ The reciprocal of the denominator $\frac{5}{9}$ is $\frac{9}{5}$.

This is a *complex fraction* **(Section 8.3)**, a fraction that has a fraction in the numerator, the denominator, or both.

> **Work Problem 8 at the Side.**

The rules for multiplication and division suggest the following results.

Equivalent Forms of a Fraction

The fractions $\dfrac{-x}{y}$, $-\dfrac{x}{y}$, and $\dfrac{x}{-y}$ are equal. (Assume $y \neq 0$.)

Example: $\dfrac{-4}{7} = -\dfrac{4}{7} = \dfrac{4}{-7}$.

The fractions $\dfrac{x}{y}$ and $\dfrac{-x}{-y}$ are equal.

Example: $\dfrac{4}{7} = \dfrac{-4}{-7}$.

9 Which of the following fractions are equal to $\frac{-3}{5}$?

A. $\dfrac{3}{5}$

B. $\dfrac{3}{-5}$

C. $-\dfrac{3}{5}$

D. $\dfrac{-3}{-5}$

The forms $\frac{x}{-y}$ and $\frac{-x}{-y}$ are not used very often.

Every fraction has three signs: the sign of the numerator, the sign of the denominator, and the sign of the fraction itself. Changing any two of these three signs does not change the value of the fraction. Changing only one sign, or changing all three, *does* change the value.

> **Work Problem 9 at the Side.**

ANSWERS

8. (a) -4 (b) -4 (c) 5 (d) $\dfrac{6}{11}$

9. B, C

1.2 Exercises

FOR EXTRA HELP

Tutor Center Addison-Wesley Math Tutor Center

MathXL MathXL

Digital Video Tutor CD 1 Videotape 1

Student's Solutions Manual

MyMathLab MyMathLab

Interactmath.com

Complete each statement and give an example.

1. The sum of a positive number and a negative number is 0 if _____.

2. The sum of two positive numbers is a _____ number.

3. The sum of two negative numbers is a _____ number.

4. The sum of a positive number and a negative number is negative if _____.

5. The sum of a positive number and a negative number is positive if _____.

6. The difference between two positive numbers is negative if _____.

7. The difference between two negative numbers is negative if _____.

8. The product of two numbers with like signs is _____.

9. The product of two numbers with unlike signs is _____.

10. The quotient formed by any nonzero number divided by 0 is _____, and the quotient formed by 0 divided by any nonzero number is _____.

Add or subtract as indicated. See Examples 1–3.

11. $13 + (-4)$

12. $19 + (-13)$

13. $-6 + (-13)$

14. $-8 + (-15)$

15. $-\dfrac{7}{3} + \dfrac{3}{4}$

16. $-\dfrac{5}{6} + \dfrac{3}{8}$

17. $-2.3 + .45$

18. $-.238 + 4.55$

19. $-6 - 5$

20. $-8 - 13$

21. $8 - (-13)$

22. $13 - (-22)$

23. $-16 - (-3)$

24. $-21 - (-8)$

25. $-12.31 - (-2.13)$

26. $-15.88 - (-9.22)$

27. $\dfrac{9}{10} - \left(-\dfrac{4}{3}\right)$

28. $\dfrac{3}{14} - \left(-\dfrac{1}{4}\right)$

29. $-2 - |-4|$

30. $9 - |-13|$

Multiply. See Example 5.

31. $5(-7)$

32. $6(-6)$

33. $-8(-5)$

34. $-10(-4)$

35. $-10\left(-\dfrac{1}{5}\right)$

36. $-\dfrac{1}{2}(-12)$

37. $\dfrac{3}{4}(-16)$

38. $\dfrac{4}{5}(-35)$

39. $-\dfrac{5}{2}\left(-\dfrac{12}{25}\right)$

40. $-\dfrac{9}{7}\left(-\dfrac{35}{36}\right)$

41. $-\dfrac{3}{8}\left(-\dfrac{24}{9}\right)$

42. $-\dfrac{2}{11}\left(-\dfrac{99}{4}\right)$

43. $-2.4(-2.45)$

44. $-3.45(-2.14)$

45. $3.4(-3.14)$

46. $5.66(-2.1)$

Give the reciprocal of each number.

47. 6

48. 8

49. -7

50. -11

51. $-\dfrac{2}{3}$

52. $-\dfrac{7}{8}$

53. $\dfrac{1}{5}$

54. $\dfrac{1}{4}$

55. $.02$

56. $.45$

57. $-.001$

58. $-.0003$

Divide where possible. See Example 6.

59. $\dfrac{-14}{2}$ **60.** $\dfrac{-26}{13}$ **61.** $\dfrac{-24}{-4}$ **62.** $\dfrac{-36}{-9}$ **63.** $\dfrac{100}{-25}$

64. $\dfrac{300}{-60}$ **65.** $\dfrac{0}{-8}$ **66.** $\dfrac{0}{-10}$ **67.** $\dfrac{5}{0}$ **68.** $\dfrac{12}{0}$

69. $-\dfrac{10}{17} \div \left(-\dfrac{12}{5}\right)$ **70.** $-\dfrac{22}{23} \div \left(-\dfrac{33}{4}\right)$ **71.** $\dfrac{\frac{12}{13}}{-\frac{4}{3}}$ **72.** $\dfrac{\frac{5}{6}}{-\frac{1}{30}}$

73. $-\dfrac{27.72}{13.2}$ **74.** $\dfrac{-126.7}{36.2}$ **75.** $\dfrac{-100}{-.01}$ **76.** $\dfrac{-50}{-.05}$

Perform the indicated operations. Work inside parentheses or brackets first. Remember to add and subtract in order, working from left to right. See Example 4.

77. $-7 + 5 - 9$ **78.** $-12 + 13 - 19$ **79.** $6 - (-2) + 8$

80. $7 - (-3) + 12$ **81.** $-9 - 4 - (-3) + 6$ **82.** $-10 - 5 - (-12) + 8$

83. $-4 - [(-4 - 6) + 12] - 13$ **84.** $-10 - [(-2 + 3) - 4] - 17$

Solve each problem.

85. The highest temperature ever recorded in Juneau, Alaska, was 90°F. The lowest temperature ever recorded there was −22°F. What is the difference between these two temperatures? (*Source: World Almanac and Book of Facts*, 2004.)

86. On August 10, 1936, a temperature of 120°F was recorded in Arkansas. On February 13, 1905, Arkansas recorded a temperature of −29°F. What is the difference between these two temperatures? (*Source: World Almanac and Book of Facts*, 2004.)

87. The low-carb diet craze was responsible for a first-quarter loss in Krispy Kreme doughnut sales in 2004. The company reported a loss of $24.4 million. One year earlier the company had reported a profit of $13.1 million. Express the difference between these two figures as a negative amount. (*Source:* Krispy Kreme Doughnuts.)

88. The Standard and Poor's 500, an index measuring the performance of 500 leading stocks, had an annual return of 37.58% in 1995. For 2000, its annual return was -9.10%. Find the difference between these two percents. (*Source:* Legg Mason Wood Walker, Inc.)

The table shows Social Security finances (in billions of dollars). Use this table to work Exercises 89 and 90.

Year	Tax Revenue	Cost of Benefits
2000	538	409
2010*	916	710
2020*	1479	1405
2030*	2041	2542

* Projected
Source: Social Security Board of Trustees.

89. Find the difference between Social Security tax revenue and cost of benefits for each year shown in the table.

90. Interpret your answer for the year 2030.

Use the graph of U.S. trade with Mexico to work Exercises 91–94.

91. What is the difference between the 1995 and 1994 trade balances?

92. What is the difference between the trade balances in 1994 and 1993?

93. Which of the following is the best estimate of the difference in millions of dollars between the 1997 and 1996 trade balances?

   A. 2000   **B.** 2500   **C.** 3000   **D.** 3500

94. Which of the following is the best estimate of the difference in millions of dollars between the 2002 and 2001 trade balances?

   A. 11,000   **B.** 8000   **C.** $-10,000$ **D.** -7000

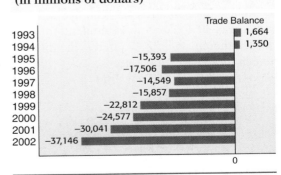

U.S. TRADE WITH MEXICO
(in millions of dollars)

Trade Balance

Year	Trade Balance
1993	1,664
1994	1,350
1995	−15,393
1996	−17,506
1997	−14,549
1998	−15,857
1999	−22,812
2000	−24,577
2001	−30,041
2002	−37,146

Source: Office of Trade and Economic Analysis, U.S. Department of Commerce.

1.3 Exponents, Roots, and Order of Operations

Two or more numbers whose product is a third number are **factors** of that third number. For example, 2 and 6 are factors of 12 since $2 \cdot 6 = 12$. Other factors of 12 are 1, 3, 4, 12, -1, -2, -3, -4, -6, and -12.

OBJECTIVE 1 Use exponents.   In algebra, we use *exponents* as a way of writing products of repeated factors. For example, the product $2 \cdot 2 \cdot 2 \cdot 2 \cdot 2$ is written

$$\underbrace{2 \cdot 2 \cdot 2 \cdot 2 \cdot 2}_{5 \text{ factors of } 2} = 2^5.$$

The number 5 shows that 2 is used as a factor 5 times. The number 5 is the **exponent**, and 2 is the **base**.

$$2^5 \;\leftarrow \text{Exponent}$$
$$\;\uparrow\!\!\text{—— Base}$$

Read 2^5 as "2 to the fifth power" or simply "2 to the fifth." Multiplying out the five 2s gives

$$2^5 = 2 \cdot 2 \cdot 2 \cdot 2 \cdot 2 = 32.$$

Exponential Expression

If a is a real number and n is a natural number,

$$a^n = \underbrace{a \cdot a \cdot a \cdots a,}_{n \text{ factors of } a}$$

where n is the **exponent**, a is the **base**, and a^n is an **exponential expression**. Exponents are also called **powers**.

EXAMPLE 1 Using Exponential Notation

Write each expression using exponents.

(a) $4 \cdot 4 \cdot 4$

Here, 4 is used as a factor 3 times, so

$$\underbrace{4 \cdot 4 \cdot 4}_{3 \text{ factors of } 4} = 4^3.$$

Read 4^3 as "4 cubed."

(b) $\dfrac{3}{5} \cdot \dfrac{3}{5} = \left(\dfrac{3}{5}\right)^2$ 2 factors of $\frac{3}{5}$

Read $\left(\frac{3}{5}\right)^2$ as "$\frac{3}{5}$ squared."

(c) $(-6)(-6)(-6)(-6) = (-6)^4$

(d) $(.3)(.3)(.3)(.3)(.3) = (.3)^5$

(e) $x \cdot x \cdot x \cdot x \cdot x \cdot x = x^6$

Work Problem 1 at the Side. ▶▶▶

OBJECTIVES

1 Use exponents.
2 Identify exponents and bases.
3 Find square roots.
4 Use the order of operations.
5 Evaluate expressions for given values of variables.

1 Write each expression using exponents.

(a) $3 \cdot 3 \cdot 3 \cdot 3 \cdot 3$

(b) $\dfrac{2}{7} \cdot \dfrac{2}{7} \cdot \dfrac{2}{7} \cdot \dfrac{2}{7}$

(c) $(-10)(-10)(-10)$

(d) $(.5)(.5)$

(e) $y \cdot y \cdot y \cdot y \cdot y \cdot y \cdot y \cdot y$

ANSWERS

1. (a) 3^5 (b) $\left(\dfrac{2}{7}\right)^4$ (c) $(-10)^3$
   (d) $(.5)^2$ (e) y^8

2 Write each expression without exponents.

(a) 5^3

(b) 3^4

(c) $(-4)^5$

(d) $(-3)^4$

(e) $(.75)^3$

(f) $\left(\dfrac{2}{5}\right)^4$

In parts (a) and (b) of Example 1, we used the terms *squared* and *cubed* to refer to powers of 2 and 3, respectively. The term *squared* comes from the figure of a square, which has the same measure for both length and width, as shown in Figure 9(a). Similarly, the term *cubed* comes from the figure of a cube. As shown in Figure 9(b), the length, width, and height of a cube have the same measure.

 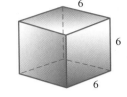

(a) $3 \cdot 3 = 3$ squared, or 3^2 **(b)** $6 \cdot 6 \cdot 6 = 6$ cubed, or 6^3

Figure 9

EXAMPLE 2 **Evaluating Exponential Expressions**

Write each expression without exponents.

(a) $5^2 = 5 \cdot 5 = 25$ 5 is used as a factor 2 times.

(b) $\left(\dfrac{2}{3}\right)^3 = \dfrac{2}{3} \cdot \dfrac{2}{3} \cdot \dfrac{2}{3} = \dfrac{8}{27}$ $\frac{2}{3}$ is used as a factor 3 times.

(c) $2^6 = 2 \cdot 2 \cdot 2 \cdot 2 \cdot 2 \cdot 2 = 64$

(d) $(-2)^4 = (-2)(-2)(-2)(-2) = 16$

(e) $(-3)^5 = (-3)(-3)(-3)(-3)(-3) = -243$

Parts (d) and (e) of Example 2 suggest the following generalization.

The product of an *even* number of negative factors is positive.
The product of an *odd* number of negative factors is negative.

Work Problem 2 at the Side.

Calculator Tip   Most calculators have a key labeled x^y or y^x that can be used to raise a number to a power. See "An Introduction to Calculators" at the beginning of this book for more information.

OBJECTIVE 2 Identify exponents and bases.

EXAMPLE 3 **Identifying Exponents and Bases**

Identify the exponent and the base. Then evaluate each expression.

(a) 3^6 The exponent is 6, and the base is 3.

$$3^6 = 3 \cdot 3 \cdot 3 \cdot 3 \cdot 3 \cdot 3 = 729$$

(b) 5^4 The exponent is 4, and the base is 5.

$$5^4 = 5 \cdot 5 \cdot 5 \cdot 5 = 625$$

Continued on Next Page

ANSWERS
2. **(a)** 125 **(b)** 81 **(c)** -1024 **(d)** 81
(e) .421875 **(f)** $\dfrac{16}{625}$

(c) $(-2)^6$

The exponent 6 applies to the number -2, so the base is -2.

$$(-2)^6 = (-2)(-2)(-2)(-2)(-2)(-2) = 64 \qquad \text{The base is } -2.$$

(d) -2^6

Since there are no parentheses, the exponent 6 applies *only* to the number 2, not to -2; the base is 2.

$$-2^6 = -(2 \cdot 2 \cdot 2 \cdot 2 \cdot 2 \cdot 2) = -64 \qquad \text{The base is 2.}$$

❸ Identify the exponent and the base. Then evaluate each expression.

(a) 7^3

CAUTION

As shown in Examples 3(c) and (d), it is important to distinguish between $-a^n$ and $(-a)^n$.

$$-a^n = -1 \underbrace{(a \cdot a \cdot a \cdots a)}_{n \text{ factors of } a} \qquad \text{The base is } a.$$

$$(-a)^n = \underbrace{(-a)(-a) \cdots (-a)}_{n \text{ factors of } -a} \qquad \text{The base is } -a.$$

(b) $(-5)^4$

▶▶▶ **Work Problem 3 at the Side.** ▶▶▶

OBJECTIVE ❸ Find square roots. As we saw in Example 2(a), $5^2 = 5 \cdot 5 = 25$, so 5 squared is 25. The opposite of squaring a number is called taking its **square root.** For example, a square root of 25 is 5. Another square root of 25 is -5 since $(-5)^2 = 25$; thus, 25 has two square roots, 5 and -5.

We write the positive or *principal* square root of a number with the symbol $\sqrt{}$, called a **radical sign.** For example, the positive or principal square root of 25 is written $\sqrt{25} = 5$. The negative square root of 25 is written $-\sqrt{25} = -5$. Since the square of any nonzero real number is positive, *the square root of a negative number, such as $\sqrt{-25}$, is not a real number.*

(c) -5^4

(EXAMPLE 4) Finding Square Roots

Find each square root that is a real number.

(a) $\sqrt{36} = 6$ since 6 is positive and $6^2 = 36$.

(b) $\sqrt{0} = 0$ since $0^2 = 0$.   **(c)** $\sqrt{\dfrac{9}{16}} = \dfrac{3}{4}$ since $\left(\dfrac{3}{4}\right)^2 = \dfrac{9}{16}$.

(d) $\sqrt{.16} = .4$ since $(.4)^2 = .16$.   **(e)** $\sqrt{100} = 10$ since $10^2 = 100$.

(d) $-(.9)^5$

(f) $-\sqrt{100} = -10$ since the negative sign is outside the radical sign.

(g) $\sqrt{-100}$ is not a real number since the negative sign is inside the radical sign. No *real number* squared equals -100.

Notice the difference among the expressions in parts (e), (f), and (g). Part (e) is the positive or principal square root of 100, part (f) is the negative square root of 100, and part (g) is the square root of -100, which is not a real number.

ANSWERS
3. **(a)** 3; 7; 343   **(b)** 4; -5; 625
    (c) 4; 5; -625 **(d)** 5; .9; $-.59049$

4 Find each square root that is a real number.

(a) $\sqrt{9}$

(b) $\sqrt{49}$

(c) $-\sqrt{81}$

(d) $\sqrt{\dfrac{121}{81}}$

(e) $\sqrt{.25}$

(f) $\sqrt{-9}$

(g) $-\sqrt{-169}$

5 Simplify.

(a) $5 \cdot 9 + 2 \cdot 4$

(b) $4 - 12 \div 4 \cdot 2$

CAUTION
The symbol $\sqrt{}$ is used only for the *positive* square root, except that $\sqrt{0} = 0$. The symbol $-\sqrt{}$ is used for the negative square root.

▶▶▶ **Work Problem 4 at the Side.**

⊞ **Calculator Tip** Most calculators have a square root key, usually labeled $\boxed{\sqrt{x}}$, that allows us to find the square root of a number. On some models, the square root key must be used in conjunction with the key marked $\boxed{\text{INV}}$ or $\boxed{\text{2nd}}$.

OBJECTIVE 4 Use the order of operations. To simplify an expression such as $5 + 2 \cdot 3$, what should we do first—add 5 and 2, or multiply 2 and 3? When an expression involves more than one operation symbol, we use the following **order of operations.**

Order of Operations
1. Work separately above and below any **fraction bar.**
2. If **grouping symbols** such as **parentheses ( )**, **square brackets [ ]**, or **absolute value bars | |** are present, start with the innermost set and work outward.
3. Evaluate all **powers, roots,** and **absolute values.**
4. Do any **multiplications** or **divisions** in order, working from left to right.
5. Do any **additions** or **subtractions** in order, working from left to right.

┌ **EXAMPLE 5** Using the Order of Operations

Simplify.

(a) $5 + 2 \cdot 3 = 5 + 6$ Multiply.

$\qquad\qquad\quad = 11$ Add.

(b) $24 \div 3 \cdot 2 + 6$
Multiplications and divisions are done *in the order in which they appear from left to right,* so divide first.

$$24 \div 3 \cdot 2 + 6 = 8 \cdot 2 + 6 \quad \text{Divide.}$$
$$= 16 + 6 \quad \text{Multiply.}$$
$$= 22 \quad \text{Add.}$$

▶▶▶ **Work Problem 5 at the Side.**

┌ **EXAMPLE 6** Using the Order of Operations

Simplify.

(a) $4 \cdot 3^2 + 7 - (2 + 8)$
Work inside the parentheses first.

── **Continued on Next Page**

ANSWERS

4. (a) 3 (b) 7 (c) -9 (d) $\dfrac{11}{9}$ (e) .5
   (f) not a real number
   (g) not a real number
5. (a) 53 (b) -2

$$4 \cdot 3^2 + 7 - (2 + 8) = 4 \cdot 3^2 + 7 - 10 \qquad \text{Add inside parentheses.}$$
$$= 4 \cdot 9 + 7 - 10 \qquad \text{Evaluate powers.}$$
$$= 36 + 7 - 10 \qquad \text{Multiply.}$$
$$= 43 - 10 \qquad \text{Add.}$$
$$= 33 \qquad \text{Subtract.}$$

(b) $\dfrac{1}{2} \cdot 4 + (6 \div 3 - 7)$

Work inside the parentheses, dividing before subtracting.

$$\frac{1}{2} \cdot 4 + (6 \div 3 - 7) = \frac{1}{2} \cdot 4 + (2 - 7) \qquad \text{Divide inside parentheses.}$$
$$= \frac{1}{2} \cdot 4 + (-5) \qquad \text{Subtract inside parentheses.}$$
$$= 2 + (-5) \qquad \text{Multiply.}$$
$$= -3 \qquad \text{Add.}$$

Work Problem 6 at the Side. ▶▶▶

EXAMPLE 7 Using the Order of Operations

Simplify $\dfrac{5 + 2^4}{6\sqrt{9} - 9 \cdot 2}$.

$$\frac{5 + 2^4}{6\sqrt{9} - 9 \cdot 2} = \frac{5 + 16}{6 \cdot 3 - 9 \cdot 2} \qquad \text{Evaluate powers and roots.}$$
$$= \frac{5 + 16}{18 - 18} \qquad \text{Multiply.}$$
$$= \frac{21}{0} \qquad \text{Add and subtract.}$$

Because division by 0 is undefined, the given expression is undefined.

Work Problem 7 at the Side. ▶▶▶

⌨ **Calculator Tip** Most calculators follow the order of operations given in this section. You may want to try some of the examples to see whether your calculator gives the same answers. Use the parentheses keys to insert parentheses where they are needed. To work Example 7 with a calculator, put parentheses around the numerator and the denominator.

OBJECTIVE 5 Evaluate expressions for given values of variables. Any collection of numbers, variables, operation symbols, and grouping symbols, such as

$$6ab, \quad 5m - 9n, \quad \text{and} \quad -2(x^2 + 4y), \qquad \text{Algebraic expressions}$$

is called an **algebraic expression.** Algebraic expressions have different numerical values for different values of the variables. We can evaluate such expressions by *substituting* given values for the variables.

Algebraic expressions are used in problem solving. For example, if movie tickets cost $8 each, the amount in dollars you pay for x tickets can be represented by the algebraic expression $8x$. We can substitute different numbers of tickets to get the costs to purchase those tickets.

---

6 Simplify.

(a) $(4 + 2) - 3^2 - (8 - 3)$

(b) $6 + \dfrac{2}{3}(-9) - \dfrac{5}{8} \cdot 16$

7 Simplify.

(a) $\dfrac{10 - 6 + 2\sqrt{9}}{11 \cdot 2 - 3(2)^2}$

(b) $\dfrac{-4(8) + 6(3)}{3\sqrt{49} - \dfrac{1}{2}(42)}$

ANSWERS
6. **(a)** -8 **(b)** -10
7. **(a)** 1 **(b)** undefined

8 Evaluate each expression if $w = 4$, $x = -12$, $y = 64$, and $z = -3$.

(a) $5x - 2w$

(b) $-6(x - \sqrt{y})$

(c) $\dfrac{5x - 3 \cdot \sqrt{y}}{x - 1}$

(d) $w^2 + 2z^3$

9 Use the expression in Example 9 to approximate the percent of U.S. households investing in mutual funds in 1990 and 2000. Round answers to the nearest tenth.

ANSWERS

8. (a) -68 (b) 120 (c) $\dfrac{84}{13}$ (d) -38

9. 1990: 26.0%; 2000: 48.0%

EXAMPLE 8 **Evaluating Expressions**

Evaluate each expression if $m = -4$, $n = 5$, $p = -6$, and $q = 25$.

(a) $5m - 9n = 5(-4) - 9(5) = -20 - 45 = -65$ Replace m with -4 and n with 5.

(b) $\dfrac{m + 2n}{4p} = \dfrac{-4 + 2(5)}{4(-6)} = \dfrac{-4 + 10}{-24} = \dfrac{6}{-24} = -\dfrac{1}{4}$

(c) $-3m^3 - n^2(\sqrt{q}) = -3(-4)^3 - (5)^2(\sqrt{25})$ Substitute; $m = -4$, $n = 5$, and $q = 25$.

$\qquad = -3(-64) - 25(5)$ Evaluate powers and roots.

$\qquad = 192 - 125$ Multiply.

$\qquad = 67$ Subtract.

CAUTION
To avoid errors when evaluating expressions, *use parentheses around any negative numbers that are substituted for variables.*

◀◀◀ **Work Problem 8 at the Side.**

EXAMPLE 9 **Evaluating an Expression to Approximate Mutual Fund Investors**

An approximation of the percent of U.S. households investing in mutual funds during the years 1980 through 2000 can be obtained by substituting a given year for x in the expression

$$2.2023x - 4356.6$$

and then evaluating. (*Source:* Investment Company Institute.)

(a) Approximate the percent of U.S. households investing in mutual funds in 1980. Round to the nearest tenth.

$2.2023x - 4356.6 = 2.2023(1980) - 4356.6$ Let $x = 1980$.

$\qquad\qquad\qquad \approx 4.0$ Use a calculator.

Recall that the symbol $\approx$ means "is approximately equal to." In 1980, about 4.0% of U.S. households invested in mutual funds.

◀◀◀ **Work Problem 9 at the Side.**

(b) Give the results found above and in Problem 9 at the side in a table. How has the percent of households investing in mutual funds changed during these years?

The table follows. The percent of U.S. households investing in mutual funds increased dramatically during these years.

Year	Percent of U.S. Households Investing in Mutual Funds
1980	4.0
1990	26.0
2000	48

← Percent in 2000 is twelve times percent in 1980.

1.3 Exercises

FOR EXTRA HELP   Addison-Wesley Math Tutor Center   MathXL   Digital Video Tutor CD 1 Videotape 1   Student's Solutions Manual   MyMathLab   Interactmath.com

Decide whether each statement is true *or* false. *If* false, *correct the statement so it is* true.

1. $-4^6 = (-4)^6$

2. $-4^7 = (-4)^7$

3. $\sqrt{16}$ is a positive number.

4. $3 + 5 \cdot 6 = 3 + (5 \cdot 6)$

5. $(-2)^7$ is a negative number.

6. $(-2)^8$ is a positive number.

7. The product of 8 positive factors and 8 negative factors is positive.

8. The product of 3 positive factors and 3 negative factors is positive.

9. In the exponential expression -3^5, -3 is the base.

10. $\sqrt{a}$ is positive for all positive numbers a.

11. Evaluate each exponential expression.
   (a) 8^2 **(b)** -8^2 **(c)** $(-8)^2$ **(d)** $-(-8)^2$

12. Evaluate each exponential expression.
   (a) 4^3 **(b)** -4^3 **(c)** $(-4)^3$ **(d)** $-(-4)^3$

Write each expression using exponents. See Example 1.

13. $8 \cdot 8 \cdot 8$

14. $10 \cdot 10 \cdot 10 \cdot 10$

15. $\dfrac{1}{2} \cdot \dfrac{1}{2}$

16. $\dfrac{3}{4} \cdot \dfrac{3}{4} \cdot \dfrac{3}{4} \cdot \dfrac{3}{4} \cdot \dfrac{3}{4}$

17. $(-4)(-4)(-4)(-4)$

18. $(-9)(-9)(-9)$

19. $z \cdot z \cdot z \cdot z \cdot z \cdot z \cdot z$

20. $a \cdot a \cdot a \cdot a \cdot a \cdot a$

Evaluate each expression. See Examples 2 and 3.

21. 4^2

22. 2^4

23. $.28^3$

24. $.91^3$

25. $\left(\dfrac{1}{5}\right)^3$

26. $\left(\dfrac{1}{6}\right)^4$

27. $\left(\dfrac{7}{10}\right)^3$

28. $\left(\dfrac{4}{5}\right)^4$

29. $(-5)^3$

30. $(-3)^5$

31. $(-2)^8$

32. $(-3)^6$

33. -3^6

34. -4^6

35. -8^4

36. -10^3

Identify the exponent and the base in each expression. Do not evaluate. See Example 3.

37. $(-4.1)^7$

38. $(-3.4)^9$

39. -4.1^7

40. -3.4^9

Find each square root. If it is not a real number, say so. See Example 4.

41. $\sqrt{81}$

42. $\sqrt{64}$

43. $\sqrt{169}$

44. $\sqrt{225}$

45. $-\sqrt{400}$

46. $-\sqrt{900}$

47. $\sqrt{\dfrac{100}{121}}$

48. $\sqrt{\dfrac{225}{169}}$

49. $-\sqrt{.49}$

50. $-\sqrt{.64}$

51. $\sqrt{-36}$

52. $\sqrt{-121}$

53. Match each square root with the appropriate value or description.

(a) $\sqrt{144}$ **A.** -12

(b) $\sqrt{-144}$ **B.** 12

(c) $-\sqrt{144}$ **C.** Not a real number

54. Explain why $\sqrt{-900}$ is not a real number.

55. If a is a positive number, is $-\sqrt{-a}$ positive, negative, or not a real number?

56. If a is a positive number, is $-\sqrt{a}$ positive, negative, or not a real number?

Simplify each expression. See Examples 5–7.

57. $12 + 3 \cdot 4$

58. $15 + 5 \cdot 2$

59. $2[-5 - (-7)]$

60. $3[-8 - (-2)]$

61. $-12\left(-\dfrac{3}{4}\right) - (-5)$

62. $-7\left(-\dfrac{2}{14}\right) - (-8)$

63. $6 \cdot 3 - 12 \div 4$

64. $9 \cdot 4 - 8 \div 2$

65. $10 + 30 \div 2 \cdot 3$ **66.** $12 + 24 \div 3 \cdot 2$ **67.** $-3(5)^2 - (-2)(-8)$ **68.** $-9(2)^2 - (-3)(-2)$

69. $5 - 7 \cdot 3 - (-2)^3$ **70.** $-4 - 3 \cdot 5 + 6^2$ **71.** $-7(\sqrt{36}) - (-2)(-3)$ **72.** $-8(\sqrt{64}) - (-3)(-7)$

73. $\dfrac{-8 + (-16)}{-3}$ **74.** $\dfrac{-9 + (-11)}{-2}$ **75.** $\dfrac{(-5 + \sqrt{4})(-2^2)}{-5 - 1}$

76. $\dfrac{(-9 + \sqrt{16})(-3^2)}{-4 - 1}$ **77.** $6|4 - 5| - 24 \div 3$
(*Hint:* Start inside the absolute value bars.)   **78.** $-4|2 - 4| + 8 \cdot 2$

79. $\dfrac{2(-5) + (-3)(-2)}{-8 + 3^2 - 1}$ **80.** $\dfrac{3(-4) + (-5)(-8)}{2^3 - 2 - 6}$

Evaluate each expression if $a = -3$, $b = 64$, and $c = 6$. See Example 8.

81. $3a + \sqrt{b}$ **82.** $-2a - \sqrt{b}$ **83.** $\sqrt{b} + c - a$ **84.** $\sqrt{b} - c + a$

85. $4a^3 + 2c$ **86.** $-3a^4 - 3c$ **87.** $\dfrac{2c + a^3}{4b + 6a}$ **88.** $\dfrac{3c + a^2}{2b - 6c}$

 Solve each problem. See Example 9.

Residents of Linn County, Iowa, in the Cedar Rapids Community School District can use the expression

$$(v \times .5485 - 4850) \div 1000 \times 31.44$$

to determine their property taxes, where v is home value. (*Source: The Gazette,* August 19, 2000.) Use the expression to calculate the amount of property taxes to the nearest dollar that the owner of a home with each of the following values would pay. Follow the order of operations.

89. $100,000   **90.** $150,000   **91.** $200,000

The Blood Alcohol Concentration (BAC) of a person who has been drinking is given by the expression

number of oz $\times$ % alcohol $\times$ .075 $\div$ body weight in lb $-$ hr of drinking $\times$ .015.

(*Source:* Lawlor, J., *Auto Math Handbook: Mathematical Calculations, Theory, and Formulas for Automotive Enthusiasts,* HP Books, 1991.)

92. Suppose a policeman stops a 190-lb man who, in 2 hr, has ingested four 12-oz beers (48 oz), each having a 3.2% alcohol content.

(a) Substitute the values in the formula, and write the expression for the man's BAC.

(b) Calculate the man's BAC to the nearest thousandth. Follow the order of operations.

93. Find the BAC to the nearest thousandth for a 135-lb woman who, in 3 hr, has drunk three 12-oz beers (36 oz), each having a 4.0% alcohol content.

94. Calculate the BACs in Exercises 92 and 93 if each person weighs 25 lb more and the rest of the variables stay the same. How does increased weight affect a person's BAC?

95. Predict how decreased weight would affect the BAC of each person in Exercises 92 and 93. Calculate the BACs if each person weighs 25 lb less and the rest of the variables stay the same.

96. An approximation of federal spending on education in billions of dollars from 1996 through 2001 can be obtained using the expression

$$3.31714x - 6597.86,$$

where x represents the year. (*Source:* U.S. Department of Education.)

(a) Use this expression to complete the following table. Round answers to the nearest tenth.

Year	Education Spending (in billions of dollars)
1997	26.5
1998	29.8
1999	
2000	
2001	

(b) Describe the trend in the amount of federal spending on education during these years.

1.4 Properties of Real Numbers

The study of any object is simplified when we know the properties of the object. For example, a property of water is that it freezes when cooled to 0°C. Knowing this helps us to predict the behavior of water.

The study of numbers is no different. The basic properties of real numbers reflect results that occur consistently in work with numbers, so they have been generalized to apply to expressions with variables as well.

OBJECTIVES

1 Use the distributive property.

2 Use the inverse properties.

3 Use the identity properties.

4 Use the commutative and associative properties.

5 Use the multiplication property of 0.

OBJECTIVE 1 Use the distributive property. Notice that

$$2(3 + 5) = 2 \cdot 8 = 16$$

and

$$2 \cdot 3 + 2 \cdot 5 = 6 + 10 = 16,$$

so

$$2(3 + 5) = 2 \cdot 3 + 2 \cdot 5.$$

This idea is illustrated by the divided rectangle in Figure 10.

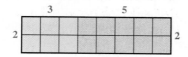

Area of left part is $2 \cdot 3 = 6$.
Area of right part is $2 \cdot 5 = 10$.
Area of total rectangle is $2(3 + 5) = 16$.

Figure 10

Similarly,

$$-4[5 + (-3)] = -4(2) = -8$$

and

$$-4(5) + (-4)(-3) = -20 + 12 = -8,$$

so

$$-4[5 + (-3)] = -4(5) + (-4)(-3).$$

These arithmetic examples are generalized to *all* real numbers as the **distributive property of multiplication with respect to addition,** or simply the **distributive property.**

Distributive Property

For any real numbers a, b, and c,

$$a(b + c) = ab + ac \quad \text{and} \quad (b + c)a = ba + ca.$$

The distributive property can also be written

$$ab + ac = a(b + c) \quad \text{and} \quad ba + ca = (b + c)a.$$

It can be extended to more than two numbers as well.

$$a(b + c + d) = ab + ac + ad$$

This property is important because it provides a way to rewrite a *product* $a(b + c)$ as a *sum* $ab + ac$, or a *sum* as a *product*.

1 Use the distributive property to rewrite each expression.

(a) $8(m + n)$

(b) $-4(p - 5)$

(c) $3k + 6k$

(d) $-6m + 2m$

(e) $2r + 3s$

(f) $5(4p - 2q + r)$

2 Use the distributive property to calculate each expression.

(a) $14 \cdot 5 + 14 \cdot 85$

(b) $78 \cdot 33 + 22 \cdot 33$

NOTE
When we rewrite $a(b + c)$ as $ab + ac$, we sometimes refer to the process as "removing" or "clearing" parentheses.

EXAMPLE 1 Using the Distributive Property

Use the distributive property to rewrite each expression.

(a) $3(x + y)$
Use the first form of the property.

$$3(x + y) = 3x + 3y$$

(b) $-2(5 + k) = -2(5) + (-2)(k)$
$$= -10 - 2k$$

(c) $4x + 8x$
Use the second form of the property.

$$4x + 8x = (4 + 8)x = 12x$$

(d) $3r - 7r = 3r + (-7r)$ Definition of subtraction
$$= [3 + (-7)]r$$ Distributive property
$$= -4r$$

(e) $5p + 7q$
Because there is no common number or variable here, we cannot use the distributive property to rewrite the expression.

(f) $6(x + 2y - 3z) = 6x + 6(2y) + 6(-3z)$
$$= 6x + 12y - 18z$$

As illustrated in Example 1(d), the distributive property can also be used for subtraction, so
$$a(b - c) = ab - ac.$$

Work Problem 1 at the Side.

The distributive property can be used to mentally perform calculations.

EXAMPLE 2 Using the Distributive Property for Calculation

Calculate $38 \cdot 17 + 38 \cdot 3$.
$$38 \cdot 17 + 38 \cdot 3 = 38(17 + 3)$$ Distributive property
$$= 38(20)$$
$$= 760$$

Work Problem 2 at the Side.

OBJECTIVE 2 Use the inverse properties. In **Section 1.1** we saw that the additive inverse of a number a is $-a$ and that the sum of a number and its additive inverse is 0. For example, 3 and -3 are additive inverses, as are -8 and 8. The number 0 is its own additive inverse.

ANSWERS
1. (a) $8m + 8n$ (b) $-4p + 20$ (c) $9k$
   (d) $-4m$ (e) cannot be rewritten
   (f) $20p - 10q + 5r$
2. (a) 1260 (b) 3300

In **Section 1.2,** we saw that two numbers with a product of 1 are recipro-cals. As mentioned there, another name for reciprocal is *multiplicative inverse.* This is similar to the idea of an additive inverse. Thus, 4 and $\frac{1}{4}$ are multiplica-tive inverses, as are $-\frac{2}{3}$ and $-\frac{3}{2}$. (Recall that reciprocals have the same sign.) We can extend these properties of arithmetic, the **inverse properties** of addi-tion and multiplication, to the real numbers of algebra.

Inverse Properties

For any real number a, there is a single real number $-a$ such that

$$a + (-a) = 0 \quad \text{and} \quad -a + a = 0.$$

The inverse "undoes" addition with the result 0.

For any *nonzero* real number a, there is a single real number $\frac{1}{a}$ such that

$$a \cdot \frac{1}{a} = 1 \quad \text{and} \quad \frac{1}{a} \cdot a = 1.$$

The inverse "undoes" multiplication with the result 1.

> **Work Problem 3 at the Side.** ▶▶▶

OBJECTIVE ▓3▓ **Use the identity properties.** The number 0 can be added to any number to get that number. That is, adding 0 leaves the identity of a number unchanged. Thus, 0 is the **identity element for addition** or the **additive identity.** Similarly, multiplying by 1 leaves the identity of any num-ber unchanged, so 1 is the **identity element for multiplication** or the **multiplicative identity.** The following **identity properties** summarize this discussion and extend these properties from arithmetic to algebra.

Identity Properties

For any real number a, $\quad a + 0 = 0 + a = a.$

Start with a number a; add 0. The answer is "identical" to a.

Also, $\qquad\qquad a \cdot 1 = 1 \cdot a = a.$

Start with a number a; multiply by 1. The answer is "identical" to a.

EXAMPLE 3 **Using the Identity Property $1 \cdot a = a$**

Simplify each expression.

(a) $12m + m = 12m + 1m \qquad$ Identity property

$\qquad\qquad = (12 + 1)m \qquad$ Distributive property

$\qquad\qquad = 13m \qquad\qquad$ Add inside parentheses.

(b) $y + y = 1y + 1y \qquad$ Identity property

$\qquad\quad = (1 + 1)y \qquad$ Distributive property

$\qquad\quad = 2y \qquad\qquad$ Add inside parentheses.

(c) $-(m - 5n) = -1(m - 5n) \qquad$ Identity property

$\qquad\qquad = -1(m) + (-1)(-5n) \qquad$ Distributive property

$\qquad\qquad = -m + 5n \qquad$ Multiply.

> **Work Problem 4 at the Side.** ▶▶▶

---

❸ Complete each statement.

(a) $4 + \underline{\qquad} = 0$

(b) $-7.1 + \underline{\qquad} = 0$

(c) $-9 + 9 = \underline{\qquad}$

(d) $5 \cdot \underline{\qquad} = 1$

(e) $-\dfrac{3}{4} \cdot \underline{\qquad} = 1$

(f) $7 \cdot \dfrac{1}{7} = \underline{\qquad}$

❹ Simplify each expression.

(a) $p - 3p$

(b) $r + r + r$

(c) $-(3 + 4p)$

(d) $-(k - 2)$

ANSWERS

3. (a) -4 (b) 7.1 (c) 0 (d) $\dfrac{1}{5}$
   (e) $-\dfrac{4}{3}$ (f) 1
4. (a) $-2p$ (b) $3r$ (c) $-3 - 4p$
   (d) $-k + 2$

Expressions such as $12m$ and $5n$ from Example 3 are examples of *terms*. A **term** is a number or the product of a number and one or more variables. Terms with exactly the same variables raised to exactly the same powers are called **like terms.** Some examples of like terms are

$$5p \text{ and } -21p \qquad -6x^2 \text{ and } 9x^2. \qquad \text{Like terms}$$

Some examples of unlike terms are

$$3m \text{ and } 16x \qquad 7y^3 \text{ and } -3y^2. \qquad \text{Unlike terms}$$

The numerical factor in a term is called the **numerical coefficient,** or just the **coefficient.** For example, in the term $9x^2$, the coefficient is 9.

OBJECTIVE 4 Use the commutative and associative properties. Simplifying expressions as in parts (a) and (b) of Example 3 is called **combining like terms.** Only like terms may be combined. To combine like terms in an expression such as

$$-2m + 5m + 3 - 6m + 8,$$

we need two more properties. From arithmetic, we know that

$$3 + 9 = 12 \quad \text{and} \quad 9 + 3 = 12.$$

Also,

$$3 \cdot 9 = 27 \quad \text{and} \quad 9 \cdot 3 = 27.$$

Furthermore, notice that

$$(5 + 7) + (-2) = 12 + (-2) = 10$$

and

$$5 + [7 + (-2)] = 5 + 5 = 10.$$

Also,

$$(5 \cdot 7)(-2) = 35(-2) = -70$$

and

$$(5)[7 \cdot (-2)] = 5(-14) = -70.$$

These arithmetic examples can now be extended to algebra.

Commutative and Associative Properties

For any real numbers a, b, and c,

$$a + b = b + a$$

and

$$ab = ba. \qquad \Big\} \text{ Commutative properties}$$

Interchange the order of the two terms or factors.

Also, $\quad a + (b + c) = (a + b) + c$

and $\quad\quad a(bc) = (ab)c. \qquad \Big\} \text{ Associative properties}$

Shift parentheses among the three terms or factors; order stays the same.

The commutative properties are used to change the *order* of the terms or factors in an expression. Think of *commuting* from home to work and then from work to home. The associative properties are used to *regroup* the terms or factors of an expression. Remember, to *associate* is to be part of a group.

EXAMPLE 4 **Using the Commutative and Associative Properties**

Simplify $-2m + 5m + 3 - 6m + 8$.

$$-2m + 5m + 3 - 6m + 8$$

$$= (-2m + 5m) + 3 - 6m + 8 \qquad \text{Order of operations}$$

$$= (-2 + 5)m + 3 - 6m + 8 \qquad \text{Distributive property}$$

$$= 3m + 3 - 6m + 8$$

By the order of operations, the next step would be to add $3m$ and 3, but they are unlike terms. To get $3m$ and $-6m$ together, use the associative and commutative properties. Begin by inserting parentheses and brackets according to the order of operations.

$$[(3m + 3) - 6m] + 8$$

$$= [3m + (3 - 6m)] + 8 \qquad \text{Associative property}$$

$$= [3m + (-6m + 3)] + 8 \qquad \text{Commutative property}$$

$$= [(3m + [-6m]) + 3] + 8 \qquad \text{Associative property}$$

$$= (-3m + 3) + 8 \qquad \text{Combine like terms.}$$

$$= -3m + (3 + 8) \qquad \text{Associative property}$$

$$= -3m + 11 \qquad \text{Add.}$$

In practice, many of the steps are not written down, but you should realize that the commutative and associative properties are used whenever the terms in an expression are rearranged and regrouped to combine like terms.

EXAMPLE 5 **Using the Properties of Real Numbers**

Simplify each expression.

(a) $5y - 8y - 6y + 11y$

$$= (5 - 8 - 6 + 11)y \qquad \text{Distributive property}$$

$$= 2y \qquad \text{Combine like terms.}$$

(b) $3x + 4 - 5(x + 1) - 8$

$$= 3x + 4 - 5x - 5 - 8 \qquad \text{Distributive property}$$

$$= 3x - 5x + 4 - 5 - 8 \qquad \text{Commutative property}$$

$$= -2x - 9 \qquad \text{Combine like terms.}$$

(c) $8 - (3m + 2) = 8 - 1(3m + 2) \qquad \text{Identity property}$

$$= 8 - 3m - 2 \qquad \text{Distributive property}$$

$$= 6 - 3m \qquad \text{Combine like terms.}$$

(d) $(3x)(5)(y) = [(3x)(5)]y \qquad \text{Order of operations}$

$$= [3(x \cdot 5)]y \qquad \text{Associative property}$$

$$= [3(5x)]y \qquad \text{Commutative property}$$

$$= [(3 \cdot 5)x]y \qquad \text{Associative property}$$

$$= (15x)y \qquad \text{Multiply.}$$

$$= 15(xy) \qquad \text{Associative property}$$

$$= 15xy$$

As previously mentioned, many of these steps are not usually written out.

Work Problem 5 at the Side. ▶▶▶

⑤ Simplify each expression.

(a) $12b - 9b + 4b - 7b + b$

(b) $-3w + 7 - 8w - 2$

(c) $-3(6 + 2t)$

(d) $9 - 2(a - 3) + 4 - a$

(e) $(4m)(2n)$

ANSWERS

5. (a) b **(b)** $-11w + 5$ **(c)** $-18 - 6t$
   (d) $19 - 3a$ **(e)** $8mn$

6 Complete each statement.

(a) $197 \cdot 0 =$ _____

CAUTION
Be careful. Notice that the distributive property does not apply in Example 5(d), because there is no addition involved.

$$(3x)(5)(y) \neq (3x)(5) \cdot (3x)(y)$$

OBJECTIVE 5 Use the multiplication property of 0. The additive identity property gives a special property of 0, namely that $a + 0 = a$ for any real number a. The **multiplication property of 0** gives a special property of 0 that involves multiplication: The product of any real number and 0 is 0.

Multiplication Property of 0
For any real number a,

$$a \cdot 0 = 0 \quad \text{and} \quad 0 \cdot a = 0.$$

(b) $0\left(-\dfrac{8}{9}\right) =$ _____

◀◀◀ **Work Problem 6 at the Side.**

(c) $0 \cdot$ _____ $= 0$

ANSWERS
6. (a) 0 **(b)** 0 **(c)** any real number

1.4 Exercises

FOR EXTRA HELP Addison-Wesley Math Tutor Center MathXL Digital Video Tutor CD 1 Videotape 1 Student's Solutions Manual MyMathLab Interactmath.com

Choose the correct response in Exercises 1–4.

1. The identity element for addition is

 A. $-a$ **B.** 0 **C.** 1 **D.** $\dfrac{1}{a}$.

2. The identity element for multiplication is

 A. $-a$ **B.** 0 **C.** 1 **D.** $\dfrac{1}{a}$.

3. The additive inverse of a is

 A. $-a$ **B.** 0 **C.** 1 **D.** $\dfrac{1}{a}$.

4. The multiplicative inverse of a, where $a \neq 0$, is

 A. $-a$ **B.** 0 **C.** 1 **D.** $\dfrac{1}{a}$.

Complete each statement.

5. The multiplication property of 0 says that the _____ of 0 and any real number is _____.

6. The commutative property is used to change the _____ of two terms or factors.

7. The associative property is used to change the _____ of three terms or factors.

8. Like terms are terms with the _____ variables raised to the _____ powers.

9. When simplifying an expression, only _____ terms can be combined.

10. The coefficient in the term $-8yz^2$ is _____.

Use the properties of real numbers to simplify each expression. See Examples 1 and 3.

11. $5k + 3k$

12. $6a + 5a$

13. $-9r + 7r$

14. $-4n + 6n$

15. $-8z + 4w$

16. $-12k + 3r$

17. $-a + 7a$

18. $-s + 9s$

19. $2(m + p)$

20. $3(a + b)$

21. $-5(2d - f)$

22. $-2(3m - n)$

Use the distributive property to calculate each value mentally. See Example 2.

23. $96 \cdot 19 + 4 \cdot 19$

24. $27 \cdot 60 + 27 \cdot 40$

25. $58 \cdot \dfrac{3}{2} - 8 \cdot \dfrac{3}{2}$

26. $\dfrac{8}{5} \cdot 17 + \dfrac{8}{5} \cdot 13$

27. $4.31(69) + 4.31(31)$

28. $\dfrac{4}{5}(17) + \dfrac{4}{5}(23)$

Simplify each expression. See Examples 1 and 3–5.

29. $-12y + 4y + 3 + 2y$

30. $-5r - 9r + 8r - 5$

31. $-6p + 11p - 4p + 6 + 5$

32. $-8x - 5x + 3x - 12 + 9$

33. $3(k + 2) - 5k + 6 + 3$

34. $5(r - 3) + 6r - 2r + 4$

35. $-2(m + 1) + 3(m - 4)$

36. $6(a - 5) - 4(a + 6)$

37. $.25(8 + 4p) - .5(6 + 2p)$

38. $.4(10 - 5x) - .8(5 + 10x)$

39. $-(2p + 5) + 3(2p + 4) - 2p$

40. $-(7m - 12) - 2(4m + 7) - 8m$

41. $2 + 3(2z - 5) - 3(4z + 6) - 8$

42. $-4 + 4(4k - 3) - 6(2k + 8) + 7$

Complete each statement so that the indicated property is illustrated. Simplify each answer, if possible.

43. $5x + 8x = $ _____
(distributive property)

44. $9y - 6y = $ _____
(distributive property)

45. $5(9r) = $ _____
(associative property)

46. $-4 + (12 + 8) = $ _____
(associative property)

47. $5x + 9y = $ _____
(commutative property)

48. $-5(7) = $ _____
(commutative property)

49. $1 \cdot 7 = $ _____
(identity property)

50. $-12x + 0 = $ _____
(identity property)

51. $8(-4 + x) = $ _____
(distributive property)

52. $3(x - y + z) = $ _____
(distributive property)

53. Give an "everyday" example of a commutative operation.

54. Give an "everyday" example of inverse operations.

RELATING CONCEPTS (EXERCISES 55–60) For Individual or Group Work

While it may seem that simplifying the expression $3x + 4 + 2x + 7$ to $5x + 11$ is fairly easy, there are several important steps that require mathematical justification. These steps are usually done mentally. **Work Exercises 55–60 in order,** *providing the property that justifies each statement in the given simplification. (These steps could be done in other orders.)*

55. $3x + 4 + 2x + 7 = (3x + 4) + (2x + 7)$ _____

56. $\qquad = 3x + (4 + 2x) + 7$ _____

57. $\qquad = 3x + (2x + 4) + 7$ _____

58. $\qquad = (3x + 2x) + (4 + 7)$ _____

59. $\qquad = (3 + 2)x + (4 + 7)$ _____

60. $\qquad = 5x + 11$ _____

Chapter 1

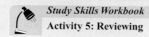

Study Skills Workbook
Activity 5: Reviewing

KEY TERMS

1.1	**set**	A set is a collection of objects.
	elements	The elements (**members**) of a set are the numbers or objects that make up the set.
	empty set	The set with no elements is called the empty (null) set.
	variable	A variable is a letter used to represent a number or a set of numbers.
	set-builder notation	Set-builder notation is used to describe a set of numbers without listing them.
	number line	A number line is a line with a scale to indicate the set of real numbers.
	coordinate	The number that corresponds to a point on the number line is its coordinate.
	graph	The point on the number line that corresponds to a number is its graph.
	additive inverse	The additive inverse (**negative, opposite**) of a number a is $-a$.
	signed numbers	Positive and negative numbers are signed numbers.
	absolute value	The absolute value of a number is its distance from 0 on a number line.
	equation	An equation is a mathematical statement that two quantities are equal.
	inequality	An inequality is a mathematical statement that two quantities are not equal.

Graph of -1

$$-3 \; -2 \; -1 \; 0 \; 1 \; 2 \; 3$$

Coordinate

1.2	**sum**	The answer to an addition problem is called the sum.	$2 + 3 = 5 \leftarrow$ Sum
	difference	The answer to a subtraction problem is called the difference.	$5 - 4 = 1 \leftarrow$ Difference
	product	The answer to a multiplication problem is called the product.	$2 \cdot 3 = 6 \leftarrow$ Product
	reciprocals	Two numbers whose product is 1 are reciprocals (**multiplicative inverses**).	
	quotient	The answer to a division problem is called the quotient.	$20 \div 4 = 5 \leftarrow$ Quotient

1.3	**factors**	Two (or more) numbers whose product is a third number are factors of that third number.
	exponent	An exponent (**power**) is a number that shows how many times a factor is repeated in a product.
	base	The base is the number that is a repeated factor in a product.
	exponential expression	A base with an exponent is called an exponential expression.
	square root	A square root of a number r is a number that can be squared to obtain r.
	algebraic expression	Any collection of numbers, variables, operation symbols, and grouping symbols is an algebraic expression.

$2^5 \leftarrow$ Exponent

Base

1.4	**term**	A term is a number or the product of a number and one or more variables.
	like terms	Like terms are terms with the same variables raised to the same powers.
	coefficient	A coefficient (**numerical coefficient**) is the numerical factor of a term.
	combining like terms	Combining like terms is a method of adding or subtracting like terms by using the properties of real numbers.

NEW SYMBOLS

$\{a, b\}$	set containing the elements a and b
$\emptyset$	empty (null) set
$\{x \mid x$ has property $P\}$	set-builder notation
$\lvert x \rvert$	absolute value of x
$\neq$	is not equal to
$<$	is less than
$\leq$	is less than or equal to
$>$	is greater than
$\geq$	is greater than or equal to
a^m	m factors of a
$\sqrt{}$	radical sign
$\sqrt{a}$	the positive (or principal) square root of a
$\approx$	is approximately equal to

TEST YOUR WORD POWER

See how well you have learned the vocabulary in this chapter. Answers, with examples, follow the Quick Review.

1. The **empty set** is a set
   A. with 0 as its only element
   B. with an infinite number of elements
   C. with no elements
   D. of ideas.

2. A **variable** is
   A. a symbol used to represent an unknown number
   B. a value that makes an equation true
   C. a solution of an equation
   D. the answer in a division problem.

3. The **absolute value** of a number is
   A. the graph of the number
   B. the reciprocal of the number
   C. the opposite of the number
   D. the distance between 0 and the number on a number line.

4. The **reciprocal** of a nonzero number a is
   A. a
   B. $\frac{1}{a}$
   C. $-a$
   D. 1.

5. A **factor** is
   A. the answer in an addition problem
   B. the answer in a multiplication problem
   C. one of two or more numbers that are added to get another number
   D. any number that divides evenly into a given number.

6. An **exponential expression** is
   A. a number that is a repeated factor in a product
   B. a number or a variable written with an exponent
   C. a number that shows how many times a factor is repeated in a product
   D. an expression that involves addition.

7. A **term** is
   A. a numerical factor
   B. a number or a product of numbers and variables raised to powers
   C. one of several variables with the same exponents
   D. a sum of numbers and variables raised to powers.

8. A **numerical coefficient** is
   A. the numerical factor in a term
   B. the number of terms in an expression
   C. a variable raised to a power
   D. the variable factor in a term.

9. The **identity element** for multiplication is
   A. 0
   B. a
   C. 1
   D. $\frac{1}{a}$.

QUICK REVIEW

Concepts	Examples

1.1 Basic Concepts

Sets of Numbers

Natural Numbers $\{1, 2, 3, 4, \dots\}$

Whole Numbers $\{0, 1, 2, 3, 4, \dots\}$

Integers $\{\dots, -2, -1, 0, 1, 2, \dots\}$

Rational Numbers

$\left\{ \dfrac{p}{q} \,\middle|\, p \text{ and } q \text{ are integers, } q \neq 0 \right\}$

(all terminating or repeating decimals)

Irrational Numbers

$\{x \mid x \text{ is a real number that is not rational}\}$

(all nonterminating, nonrepeating decimals)

Real Numbers

$\{x \mid x \text{ is represented by a point on a number line}\}$

(all rational and irrational numbers)

Absolute Value $|a| = \begin{cases} a & \text{if } a \text{ is positive or } 0 \\ -a & \text{if } a \text{ is negative} \end{cases}$

Examples:

$10, 25, 143$

$0, 8, 47$

$-22, -7, 0, 4, 9$

$-\dfrac{2}{3}, -.14, 0, 6, \dfrac{5}{8}, .33333\dots$

$\pi, .125469\dots, \sqrt{3}, -\sqrt{22}$

$-3, .7, \pi, -\dfrac{2}{3}$

$|12| = 12$

$|-12| = 12$

1.2 Operations on Real Numbers

Addition

Like Signs: To add two numbers with the same sign, add the absolute values. The answer has the same sign as the two numbers.

Unlike Signs: To add two numbers with different signs, subtract the smaller absolute value from the larger. The answer has the sign of the number with the larger absolute value.

Subtraction

Change the sign of the second number (subtrahend) and add.

Multiplication and Division

Like Signs: The answer is positive when multiplying or dividing two numbers with the same sign.

Unlike Signs: The answer is negative when multiplying or dividing two numbers with different signs.

Examples:

$-2 + (-7) = -(2 + 7) = -9$

$-5 + 8 = 8 - 5 = 3$
$-12 + 4 = -(12 - 4) = -8$

$-5 - (-3) = -5 + 3 = -2$

$-3(-8) = 24 \qquad \dfrac{-15}{-5} = 3$

$-7(5) = -35 \qquad \dfrac{-24}{12} = -2$

1.3 Exponents, Roots, and Order of Operations

The product of an even number of negative factors is positive.
The product of an odd number of negative factors is negative.

Order of Operations

1. Work separately above and below any fraction bar.

2. If parentheses, brackets, or absolute value bars are present, start with the innermost set and work outward.
3. Evaluate all exponents, roots, and absolute values.

4. Multiply or divide in order from left to right.
5. Add or subtract in order from left to right.

Examples:

$(-5)^2$ is positive: $(-5)^2 = (-5)(-5) = 25$
$(-5)^3$ is negative: $(-5)^3 = (-5)(-5)(-5) = -125$

$\dfrac{12 + 3}{5 \cdot 2} = \dfrac{15}{10} = \dfrac{3}{2}$

$(-6)[2^2 - (3 + 4)] + 3 = (-6)[2^2 - 7] + 3$

$= (-6)[4 - 7] + 3$

$= (-6)[-3] + 3$

$= 18 + 3$

$= 21$

Concepts	Examples

1.4 *Properties of Real Numbers*

Distributive Property

$a(b + c) = ab + ac$ (Remove parentheses.)

$$12(4 + 2) = 12 \cdot 4 + 12 \cdot 2$$

Inverse Properties

$a + (-a) = 0$ and $-a + a = 0$

(The additive inverse "undoes" addition to give 0.)

$a \cdot \dfrac{1}{a} = 1$ and $\dfrac{1}{a} \cdot a = 1$

(The multiplicative inverse "undoes" multiplication to give 1.)

$$5 + (-5) = 0 \qquad -12 + 12 = 0$$

$$5 \cdot \frac{1}{5} = 1 \qquad -\frac{1}{3}(-3) = 1$$

Identity Properties

$a + 0 = 0 + a = a$

(Start with a number a, add 0; the answer is identical to a.)

$a \cdot 1 = 1 \cdot a = a$

(Start with a number a, multiply by 1; the answer is identical to a.)

$$-32 + 0 = -32$$

$$17.5 \cdot 1 = 17.5$$

Commutative Properties

$a + b = b + a$ and $ab = ba$

(Two terms or factors; interchange the order.)

$$9 + (-3) = -3 + 9$$
$$6(-4) = (-4)6$$

Associative Properties

$a + (b + c) = (a + b) + c$ and $a(bc) = (ab)c$

(Three terms or factors; same order, parentheses shifted.)

$$7 + (5 + 3) = (7 + 5) + 3$$
$$-4(6 \cdot 3) = (-4 \cdot 6)3$$

Multiplication Property of 0

$a \cdot 0 = 0$ and $0 \cdot a = 0$

(Multiplying any number by 0 gives 0.)

$$4 \cdot 0 = 0 \qquad 0(-3) = 0$$

ANSWERS TO TEST YOUR WORD POWER

1. C; *Example:* The set of whole numbers less than 0 is the empty set, written $\emptyset$.
2. A; *Examples:* a, b, c
3. D; *Examples:* $|2| = 2$ and $|-2| = 2$
4. B; *Examples:* 3 is the reciprocal of $\frac{1}{3}$; $-\frac{5}{2}$ is the reciprocal of $-\frac{2}{5}$.
5. D; *Examples:* 2 and 5 are factors of 10 since both divide evenly (without remainder) into 10; other factors of 10 are $-10, -5, -2, -1, 1,$ and 10.
6. B; *Examples:* 3^4 and x^{10}
7. B; *Examples:* $6, \frac{x}{2}, -4ab^2$
8. A; *Examples:* The term $8z$ has numerical coefficient 8, and $-10x^3y$ has numerical coefficient -10.
9. C; *Example:* $1 \cdot 5 = 5 \cdot 1 = 5$

Chapter 1
REVIEW EXERCISES

If you need help with any of these Review Exercises, look in the section indicated in brackets.

[1.1] *Graph the elements of each set on a number line.*

1. $\left\{ -4, -1, 2, \dfrac{9}{4}, 4 \right\}$

2. $\left\{ -5, -\dfrac{11}{4}, -.5, 0, 3, \dfrac{13}{3} \right\}$

Find the value of each expression.

3. $|-16|$

4. $|23|$

5. $-|-4|$

6. $|-8| - |-3|$

Let set $S = \{-9, -\frac{4}{3}, -\sqrt{4}, -.25, 0, .\overline{35}, \frac{5}{3}, \sqrt{7}, \sqrt{-9}, \frac{12}{3}\}$. *Simplify the elements of S as necessary, and then list the elements that belong to the specified set.*

7. Whole numbers

8. Integers

9. Rational numbers

10. Real numbers

Write each set by listing its elements.

11. $\{x \mid x$ is a natural number between 3 and 9$\}$

12. $\{y \mid y$ is a whole number less than 4$\}$

Write true *or* false *for each inequality.*

13. $4 \cdot 2 \le |12 - 4|$

14. $2 + |-2| > 4$

15. $4(3 + 7) > -|40|$

The graph shows the percent change in car sales from January 2000 to January 2001 for various automakers. Use this graph to work Exercises 16–19.

16. Which automaker had the greatest change in sales? What was that change?

17. Which automaker had the least change in sales? What was that change?

18. *True* or *false:* The absolute value of the percent change for Honda was greater than the absolute value of the percent change for Toyota.

19. *True* or *false:* The percent change for Hyundai was more than four times greater than the percent change for Honda.

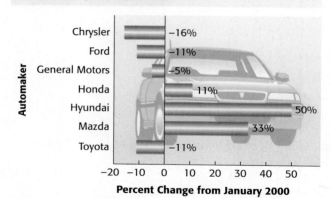

CAR SALES, JANUARY 2001

Source: Automakers.

[1.2] *Add or subtract as indicated.*

20. $-\dfrac{5}{8} - \left(-\dfrac{7}{3}\right)$

21. $-\dfrac{4}{5} - \left(-\dfrac{3}{10}\right)$

22. $-5 + (-11) + 20 - 7$

23. $-9.42 + 1.83 - 7.6 - 1.9$

24. $-15 + (-13) + (-11)$

25. $-1 - 3 - (-10) + (-7)$

26. $\dfrac{3}{4} - \left(\dfrac{1}{2} - \dfrac{9}{10}\right)$

27. $-\dfrac{2}{3} - \left(\dfrac{1}{6} - \dfrac{5}{9}\right)$

28. $-|-12| - |-9| + (-4) - |10|$

29. Telescope Peak, altitude 11,049 ft, is next to Death Valley, 282 ft below sea level. Find the difference between these altitudes. (*Source: World Almanac and Book of Facts*, 2004.)

Find each product or quotient.

30. $2(-5)(-3)(-3)$

31. $-\dfrac{3}{7}\left(-\dfrac{14}{9}\right)$

32. $-4.6(2.48)$

33. $\dfrac{75}{-5}$

34. $\dfrac{-2.3754}{-.74}$

35. Which one of the following is undefined: $\dfrac{5}{7-7}$ or $\dfrac{7-7}{5}$?

[1.3] *Evaluate each expression.*

36. 10^4

37. $\left(\dfrac{3}{7}\right)^3$

38. $(-5)^3$

39. -5^3

40. $(1.7)^2$

Find each square root. If it is not a real number, say so.

41. $\sqrt{400}$

42. $-\sqrt{196}$

43. $\sqrt{\dfrac{64}{121}}$

44. $-\sqrt{.81}$

45. $\sqrt{-64}$

Use the order of operations to simplify each expression.

46. $-14\left(\dfrac{3}{7}\right) + 6 \div 3$

47. $-\dfrac{2}{3}[5(-2) + 8 - 4^3]$

48. $\dfrac{-5(3^2) + 9(\sqrt{4}) - 5}{6 - 5(-2)}$

Evaluate each expression if $k = -4$, $m = 2$, and $n = 16$.

49. $4k - 7m$

50. $-3\sqrt{n} + m + 5k$

51. $\dfrac{4m^3 - 3n}{7k^2 - 10}$

52. The following expression for *body mass index* (BMI) can help determine ideal body weight.

$$704 \times (\text{weight in pounds}) \div (\text{height in inches})^2$$

A BMI of 19 to 25 corresponds to a healthy weight. (*Source: Washington Post.*)

(a) Carlos Beltran is 6 ft 1 in. tall and weighs 190 lb. (*Source: Street & Smith's Baseball 2004 Yearbook.*) Find his BMI (to the nearest whole number).

(b) Calculate your BMI.

[1.4] *Use the properties of real numbers to simplify each expression.*

53. $2q + 19q$

54. $13z - 17z$

55. $-m + 6m$

56. $5p - p$

57. $-2(k + 3)$

58. $6(r + 3)$

59. $9(2m + 3n)$

60. $-(-p + 6q) - (2p - 3q)$

61. $-3y + 6 - 5 + 4y$

62. $2a + 3 - a - 1 - a - 2$

63. $-3(4m - 2) + 2(3m - 1) - 4(3m + 1)$

Complete each statement so that the indicated property is illustrated. Simplify each answer, if possible.

64. $2x + 3x =$ _____
(distributive property)

65. $-4 \cdot 1 =$ _____
(identity property)

66. $2(4x) =$ _____
(associative property)

67. $-3 + 13 =$ _____
(commutative property)

68. $-3 + 3 =$ _____
(inverse property)

69. $5(x + z) =$ _____
(distributive property)

70. $0 + 7 =$ _____
(identity property)

71. $8 \cdot \dfrac{1}{8} =$ _____
(inverse property)

72. $3a + 5a + 6a =$ _____
(distributive property)

73. $\dfrac{9}{28} \cdot 0 =$ _____
(multiplication property of 0)

MIXED REVIEW EXERCISES*

The table gives U.S. exports and imports with Canada, in millions of dollars, for three recent years.

Year	Exports	Imports
2000	178,941	230,838
2001	163,424	216,268
2002	160,923	209,088

Source: Office of Trade and Economic Analysis, U.S. Department of Commerce.

Determine the absolute value of the difference between imports and exports for each year. Is the balance of trade (exports minus imports) in each year positive or negative?

74. 2000

75. 2001

76. 2002

Perform the indicated operations.

77. $\left(-\dfrac{4}{5}\right)^4$

78. $-\dfrac{5}{8}(-40)$

79. $-25\left(-\dfrac{4}{5}\right) + 3^3 - 32 \div \sqrt{4}$

80. $-8 + |-14| + |-3|$

81. $\dfrac{6 \cdot \sqrt{4} - 3 \cdot \sqrt{16}}{-2 \cdot 5 + 7(-3) - 10}$

82. $-\sqrt{25}$

83. $-\dfrac{10}{21} \div \left(-\dfrac{5}{14}\right)$

84. $.8 - 4.9 - 3.2 + 1.14$

85. -3^2

86. $\dfrac{-38}{-19}$

87. $-2(k - 1) + 3k - k$

88. $-\sqrt{-100}$

89. Evaluate $-m(3k^2 + 5m)$ if $k = -4$ and $m = 2$.

90. To evaluate $(3 + 2)^2$, should you work within the parentheses first, or should you square 3 and square 2 and then add?

---

* The order of exercises in this final group does not correspond to the order in which topics occur in the chapter. This random ordering should help you prepare for the chapter test in yet another way.

Chapter 1

CHAPTER 1 TEST

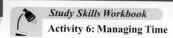
Study Skills Workbook
Activity 6: Managing Time

1. Graph $\{-3, .75, \frac{5}{3}, 5, 6.3\}$ on the number line.

1. +++++++++++→

Let $A = \{-\sqrt{6}, -1, -.5, 0, 3, \sqrt{25}, 7.5, \frac{24}{2}, \sqrt{-4}\}$. *First simplify each element as needed, and then list the elements from A that belong to each set.*

2. Whole numbers

2. _____

3. Integers

3. _____

4. Rational numbers

4. _____

5. Real numbers

5. _____

Perform the indicated operations.

6. $-6 + 14 + (-11) - (-3)$

6. _____

7. $10 - 4 \cdot 3 + 6(-4)$

7. _____

8. $7 - 4^2 + 2(6) + (-4)^2$

8. _____

9. $\dfrac{10 - 24 + (-6)}{\sqrt{16}(-5)}$

9. _____

10. $\dfrac{-2[3 - (-1 - 2) + 2]}{\sqrt{9}(-3) - (-2)}$

10. _____

11. $\dfrac{8 \cdot 4 - 3^2 \cdot 5 - 2(-1)}{-3 \cdot 2^3 + 1}$

11. _____

The table shows the heights in feet of some selected mountains and the depths in feet (as negative numbers) of some selected ocean trenches.

Mountain	Height	Trench	Depth
Foraker	17,400	Philippine	−32,995
Wilson	14,246	Cayman	−24,721
Pikes Peak	14,110	Java	−23,376

Source: World Almanac and Book of Facts, 2004.

12. What is the difference between the height of Mt. Foraker and the depth of the Philippine Trench?

12. _____

13. What is the difference between the height of Pikes Peak and the depth of the Java Trench?

13. _____

14. How much deeper is the Cayman Trench than the Java Trench?

14. _____

15. _____

16. _____

17. _____

18. (a) _____

(b) _____

(c) _____

19. _____

20. _____

21. _____

22. _____

23. _____

24. _____

25. _____

26. _____

27. _____

28. _____

29. _____

30. _____

Find each square root. If it is not a real number, say so.

15. $\sqrt{196}$ **16.** $-\sqrt{225}$

17. $\sqrt{-16}$

18. For the expression $\sqrt{a}$, under what conditions will its value be
   (a) positive,   (b) not real,   (c) 0?

Evaluate each expression if $k = -3$, $m = -3$, and $r = 25$.

19. $\sqrt{r} + 2k - m$ **20.** $\dfrac{8k + 2m^2}{r - 2}$

21. Use the properties of real numbers to simplify
$$-3(2k - 4) + 4(3k - 5) - 2 + 4k.$$

22. How does the subtraction sign affect the terms $-4r$ and 6 when simplifying $(3r + 8) - (-4r + 6)$? What is the simplified form?

Match each statement in Column I with the appropriate property in Column II. Answers may be used more than once.

I	II
23. $6 + (-6) = 0$	**A.** Distributive property
24. $4 + 5 = 5 + 4$	**B.** Inverse property
25. $-2 + (3 + 6) = (-2 + 3) + 6$	**C.** Identity property
26. $5x + 15x = (5 + 15)x$	**D.** Associative property
27. $13 \cdot 0 = 0$	**E.** Commutative property
28. $-9 + 0 = -9$	**F.** Multiplication property of 0
29. $4 \cdot 1 = 4$	
30. $(a + b) + c = (b + a) + c$	

Linear Equations and Applications

2

2.1 **Linear Equations in One Variable**

2.2 **Formulas**

2.3 **Applications of Linear Equations**

2.4 **Further Applications of Linear Equations**

Summary Exercises on Solving Applied Problems

Television, first operational in the 1940s, has become the most widespread form of communication in the world. In 2003, 106.7 million homes, 98% of all U.S. households, owned at least one TV set, and average viewing time among all viewers exceeded 30 hours per week. Favorite prime-time television programs were *CSI* and *Friends*, which concluded a highly successful 10-year run with a finale episode on May 6, 2004. (*Source:* Nielsen Media Research; *Microsoft Encarta Encyclopedia 2002*.)

In Section 2.2 we discuss the concept of *percent*—one of the most common everyday applications of mathematics—and use it in Exercises 45–50 to determine additional information about televisions in U.S. households.

2.1 Linear Equations in One Variable

OBJECTIVES

1. Decide whether a number is a solution of a linear equation.

2. Solve linear equations using the addition and multiplication properties of equality.

3. Solve linear equations using the distributive property.

4. Solve linear equations with fractions or decimals.

5. Identify conditional equations, contradictions, and identities.

Study Skills Workbook
Activity 2: Your Textbook

1 Are the given numbers solutions of the given equations?

(a) $3k = 15$; 5

(b) $r + 5 = 4$; 1

(c) $-8m = 12; \dfrac{3}{2}$

In the previous chapter we began to use *algebraic expressions*. Some examples of algebraic expressions are

$$8x + 9, \quad y - 4, \quad \text{and} \quad \frac{x^3 y^8}{z}. \quad \text{Algebraic expressions}$$

Equations and inequalities compare algebraic expressions, just as a balance scale compares the weights of two quantities. Many applications of mathematics lead to *equations,* statements that two algebraic expressions are equal. A *linear equation in one variable* involves only real numbers and one variable raised to the first power. Examples are

$$x + 1 = -2, \quad x - 3 = 5, \quad \text{and} \quad 2k + 5 = 10. \quad \text{Linear equations}$$

It is important to be able to distinguish between algebraic expressions and equations. ***An equation always contains an equals sign, while an expression does not.***

Linear Equation in One Variable

A **linear equation in one variable** can be written in the form
$$Ax + B = C,$$
where A, B, and C are real numbers, with $A \neq 0$.

A linear equation is also called a **first-degree equation** since the greatest power on the variable is one. Some examples of equations that are not linear (that is, *nonlinear*) are

$$x^2 + 3y = 5, \quad \frac{8}{x} = -22, \quad \text{and} \quad \sqrt{x} = 6. \quad \text{Nonlinear equations}$$

OBJECTIVE 1 Decide whether a number is a solution of a linear equation. If the variable in an equation can be replaced by a real number that makes the statement true, then that number is a **solution** of the equation. For example, 8 is a solution of the equation $x - 3 = 5$, since replacing x with 8 gives a true statement, $8 - 3 = 5$. An equation is *solved* by finding its **solution set,** the set of all solutions. The solution set of the equation $x - 3 = 5$ is $\{8\}$.

◄◄◄ Work Problem 1 at the Side.

Equivalent equations are equations that have the same solution set. To solve an equation, we usually start with the given equation and replace it with a series of simpler equivalent equations. For example,

$$5x + 2 = 17, \quad 5x = 15, \quad \text{and} \quad x = 3 \quad \text{Equivalent equations}$$

are all equivalent since each has the solution set $\{3\}$.

OBJECTIVE 2 Solve linear equations using the addition and multiplication properties of equality. Two important properties that are used in producing equivalent equations are the **addition property of equality** and the **multiplication property of equality.**

ANSWERS
1. (a) yes (b) no (c) no

Addition and Multiplication Properties of Equality

Addition Property of Equality

For all real numbers A, B, and C, the equations

$$A = B \quad \text{and} \quad A + C = B + C$$

are equivalent.

In words, the same number may be added to each side of an equation without changing the solution set.

Multiplication Property of Equality

For all real numbers A and B, and for $C \neq 0$, the equations

$$A = B \quad \text{and} \quad AC = BC$$

are equivalent.

In words, each side of an equation may be multiplied by the same nonzero number without changing the solution set.

Because subtraction and division are defined in terms of addition and multiplication, respectively, these properties can be extended:

The same number may be subtracted from each side of an equation, and each side of an equation may be divided by the same nonzero number, without changing the solution set.

EXAMPLE 1 **Using the Addition and Multiplication Properties to Solve a Linear Equation**

Solve $4x - 2x - 5 = 4 + 6x + 3$.

The goal is to get x alone on one side of the equation. First, combine like terms on each side of the equation to get

$$2x - 5 = 7 + 6x.$$

Next, use the addition property to get the terms with x on the same side of the equation and the remaining terms (the numbers) on the other side. One way to do this is to first subtract $6x$ from each side.

$2x - 5 - 6x = 7 + 6x - 6x$	Subtract $6x$.
$-4x - 5 = 7$	Combine like terms.
$-4x - 5 + 5 = 7 + 5$	Add 5.
$-4x = 12$	Combine like terms.
$\dfrac{-4x}{-4} = \dfrac{12}{-4}$	Divide by -4.
$x = -3$	Proposed solution

Check by substituting -3 for x in the *original* equation.

Check:	$4x - 2x - 5 = 4 + 6x + 3$		Original equation
	$4(-3) - 2(-3) - 5 = 4 + 6(-3) + 3$	?	Let $x = -3$.
	$-12 + 6 - 5 = 4 - 18 + 3$	?	Multiply.
	$-11 = -11$		True

The true statement indicates that $\{-3\}$ is the solution set.

Work Problem 2 at the Side. ▶▶▶

2 Solve and check.

(a) $3p + 2p + 1 = -24$

(b) $3p = 2p + 4p + 5$

(c) $4x + 8x = 17x - 9 - 1$

(d) $-7 + 3t - 9t = 12t - 5$

ANSWERS

2. (a) $\{-5\}$ (b) $\left\{-\dfrac{5}{3}\right\}$ (c) $\{2\}$

(d) $\left\{-\dfrac{1}{9}\right\}$

The steps to solve a linear equation in one variable are as follows.

3 Solve and check.

(a) $5p + 4(3 - 2p)$
$= 2 + p - 10$

Solving a Linear Equation in One Variable

Step 1 **Clear fractions.** Eliminate any fractions by multiplying each side by the least common denominator.

Step 2 **Simplify each side separately.** Use the distributive property to clear parentheses and combine like terms as needed.

Step 3 **Isolate the variable terms on one side.** Use the addition property to get all terms with variables on one side of the equation and all numbers on the other.

Step 4 **Isolate the variable.** Use the multiplication property to get an equation with just the variable (with coefficient 1) on one side.

Step 5 **Check.** Substitute the proposed solution into the original equation.

(b) $3(z - 2) + 5z = 2$

OBJECTIVE 3 **Solve linear equations using the distributive property.** In Example 1 we did not use Step 1 or the distributive property in Step 2 as given in the box. Many equations require one or both of these steps.

EXAMPLE 2 **Using the Distributive Property to Solve a Linear Equation**

Solve $2(k - 5) + 3k = k + 6$.

Step 1 Since there are no fractions in this equation, Step 1 does not apply.

(c) $-2 + 3(x + 4) = 8x$

Step 2 Use the distributive property to simplify and combine terms on the left side of the equation.

$$2(k - 5) + 3k = k + 6$$
$$2k - 10 + 3k = k + 6 \qquad 2(k - 5) = 2(k) - 2(5) = 2k - 10$$
$$5k - 10 = k + 6 \qquad \text{Combine like terms.}$$

Step 3 Next, use the addition property of equality.

$$5k - 10 - k = k + 6 - k \qquad \text{Subtract } k.$$
$$4k - 10 = 6 \qquad \text{Combine like terms.}$$
$$4k - 10 + 10 = 6 + 10 \qquad \text{Add 10.}$$
$$4k = 16 \qquad \text{Combine like terms.}$$

(d) $6 - (4 + m)$
$= 8m - 2(3m + 5)$

Step 4 Use the multiplication property of equality to get just k on the left.

$$\frac{4k}{4} = \frac{16}{4} \qquad \text{Divide by 4.}$$
$$k = 4$$

Step 5 Check by substituting 4 for k in the original equation.

$$\text{Check:} \quad 2(k - 5) + 3k = k + 6 \qquad \text{Original equation}$$
$$2(4 - 5) + 3(4) = 4 + 6 \quad ? \quad \text{Let } k = 4.$$
$$2(-1) + 12 = 10 \qquad ?$$
$$10 = 10 \qquad \text{True}$$

The solution checks, so the solution set is {4}.

ANSWERS
3. (a) {5} (b) {1} (c) {2} (d) {4}

◄◄◄ Work Problem 3 at the Side.

NOTE
Notice in Examples 1 and 2 that the equals signs are aligned in columns. Do not use more than one equals sign in a horizontal line of work when solving an equation.

OBJECTIVE 4 Solve linear equations with fractions or decimals. When fractions or decimals appear as coefficients in equations, our work can be made easier if we multiply each side of the equation by the least common denominator (LCD) of all the fractions. This is an application of the multiplication property of equality, and it produces an equivalent equation with integer coefficients.

EXAMPLE 3 Solving a Linear Equation with Fractions

Solve $\dfrac{x + 7}{6} + \dfrac{2x - 8}{2} = -4$.

Step 1 Start by eliminating the fractions. Multiply each side by the LCD, 6.

$$6\left(\frac{x + 7}{6} + \frac{2x - 8}{2}\right) = 6(-4)$$

Step 2 $\quad 6\left(\dfrac{x + 7}{6}\right) + 6\left(\dfrac{2x - 8}{2}\right) = 6(-4)$ Distributive property

$\qquad\qquad (x + 7) + 3(2x - 8) = -24$ Multiply.

$\qquad\qquad x + 7 + 3(2x) - 3(8) = -24$ Distributive property

$\qquad\qquad x + 7 + 6x - 24 = -24$ Multiply.

$\qquad\qquad 7x - 17 = -24$ Combine like terms.

Step 3 $\qquad\qquad 7x - 17 + 17 = -24 + 17$ Add 17.

$\qquad\qquad 7x = -7$ Combine like terms.

Step 4 $\qquad\qquad \dfrac{7x}{7} = \dfrac{-7}{7}$ Divide by 7.

$\qquad\qquad x = -1$

Step 5 Check by substituting -1 for x in the original equation.

$$\frac{x + 7}{6} + \frac{2x - 8}{2} = -4 \qquad \text{Original equation}$$

$$\frac{-1 + 7}{6} + \frac{2(-1) - 8}{2} = -4 \quad ? \quad \text{Let } x = -1.$$

$$\frac{6}{6} + \frac{-10}{2} = -4 \quad ?$$

$$1 - 5 = -4 \quad ?$$

$$-4 = -4 \qquad \text{True}$$

The solution checks, so the solution set is $\{-1\}$.

Work Problem 4 at the Side. ▶▶▶

4 Solve and check.

(a) $\dfrac{2p}{7} - \dfrac{p}{2} = -3$

(b) $\dfrac{k + 1}{2} + \dfrac{k + 3}{4} = \dfrac{1}{2}$

ANSWERS
4. (a) $\{14\}$ (b) $\{-1\}$

5 Solve and check.

(a) $.04x + .06(20 - x)$
     $= .05(50)$

In **Sections 2.2** and **2.3** we solve problems involving interest rates and concentrations of solutions. These problems involve percents that are converted to decimals. The equations that are used to solve such problems involve decimal coefficients. We can clear these decimals by multiplying by a power of 10, such as $10^1 = 10$, $10^2 = 100$, and so on, that will allow us to obtain integer coefficients.

EXAMPLE 4 Solving a Linear Equation with Decimals

Solve $.06x + .09(15 - x) = .07(15)$.

Because each decimal number is given in hundredths, multiply each side of the equation by 100. A number can be multiplied by 100 by moving the decimal point two places to the right.

$$.06x + .09(15 - x) = .07(15)$$

$$.06x + .09(15 - x) = .07(15) \qquad \text{Multiply by 100.}$$

$$6x + 9(15 - x) = 7(15)$$

$$6x + 9(15) - 9(x) = 7(15) \qquad \text{Distributive property}$$

$$6x + 135 - 9x = 105 \qquad \text{Multiply.}$$

$$-3x + 135 = 105 \qquad \text{Combine like terms.}$$

$$-3x + 135 - 135 = 105 - 135 \qquad \text{Subtract 135.}$$

$$-3x = -30 \qquad \text{Combine like terms.}$$

$$\frac{-3x}{-3} = \frac{-30}{-3} \qquad \text{Divide by } -3.$$

$$x = 10$$

Check by substituting 10 for x in the original equation.

$$\begin{aligned}
\textit{Check:} \qquad .06x + .09(15 - x) &= .07(15) \qquad &\text{Original equation} \\
.06(10) + .09(15 - 10) &= .07(15) \qquad &? \quad \text{Let } x = 10. \\
.06(10) + .09(5) &= .07(15) \qquad &? \\
.6 + .45 &= 1.05 \qquad &? \\
1.05 &= 1.05 \qquad &\text{True}
\end{aligned}$$

(b) $.10(x - 6) + .05x$
     $= .06(50)$

The solution set is $\{10\}$.

Work Problem 5 at the Side.

NOTE
Because of space limitations, we will not always show the check when solving an equation. To be sure that your solution is correct, *you should always check your work.*

OBJECTIVE 5 Identify conditional equations, contradictions, and identities. All of the preceding equations had solution sets containing one element; for example, $2(k - 5) + 3k = k + 6$ has solution set $\{4\}$. Some linear equations, however, have no solutions, while others have an infinite number of solutions. The table on the next page gives the names of these types of equations.

ANSWERS
5. (a) $\{-65\}$ **(b)** $\{24\}$

Type of Linear Equation	Number of Solutions	Indication When Solving
Conditional	One	Final line is $x =$ a number. (See Example 5(a).)
Contradiction	None; solution set $\emptyset$	Final line is false, such as $-15 = -20$. (See Example 5(c).)
Identity	Infinite; solution set {all real numbers}	Final line is true, such as $0 = 0$. (See Example 5(b).)

EXAMPLE 5 **Recognizing Conditional Equations, Identities, and Contradictions**

Solve each equation. Decide whether it is a *conditional equation*, an *identity*, or a *contradiction*.

(a)

$$5x - 9 = 4(x - 3)$$
$$5x - 9 = 4x - 12 \qquad \text{Distributive property}$$
$$5x - 9 - 4x = 4x - 12 - 4x \qquad \text{Subtract } 4x.$$
$$x - 9 = -12 \qquad \text{Combine like terms.}$$
$$x - 9 + 9 = -12 + 9 \qquad \text{Add 9.}$$
$$x = -3$$

The solution set, $\{-3\}$, has only one element, so $5x - 9 = 4(x - 3)$ is a conditional equation.

(b) $5x - 15 = 5(x - 3)$

Use the distributive property to clear parentheses on the right side.

$$5x - 15 = 5(x - 3)$$
$$5x - 15 = 5x - 15 \qquad \text{Distributive property}$$
$$5x - 15 - 5x + 15 = 5x - 15 - 5x + 15 \qquad \text{Subtract } 5x; \text{ add 15.}$$
$$\mathbf{0 = 0} \qquad \text{True}$$

The final line, the *true* statement $0 = 0$, indicates that the solution set is {all real numbers}, and the equation $5x - 15 = 5(x - 3)$ is an identity. (Notice that the first step yielded $5x - 15 = 5x - 15$, which is true for all values of x. We could have identified the equation as an identity at that point.)

(c)

$$5x - 15 = 5(x - 4)$$
$$5x - 15 = 5x - 20 \qquad \text{Distributive property}$$
$$5x - 15 - 5x = 5x - 20 - 5x \qquad \text{Subtract } 5x.$$
$$-15 = -20 \qquad \text{False}$$

Since the result, $-15 = -20$, is *false*, the equation has no solution. The solution set is $\emptyset$, so the equation $5x - 15 = 5(x - 4)$ is a contradiction.

> **Work Problem 6 at the Side.** ▶▶▶

6 Solve each equation. Decide whether it is a *conditional equation*, an *identity*, or a *contradiction*. Give the solution set.

(a) $5(x + 2) - 2(x + 1)$
$= 3x + 1$

(b) $\dfrac{x + 1}{3} + \dfrac{2x}{3} = x + \dfrac{1}{3}$

(c) $5(3x + 1) = x + 5$

ANSWERS
6. (a) contradiction; $\emptyset$
   (b) identity; {all real numbers}
   (c) conditional; {0}

Real-Data Applications

International Time Zones

Companies that operate globally, such as Coca-Cola and BP Amoco, must contend with time-zone differences between offices located in different countries. To ensure that everyone experiences daylight during morning and afternoon hours, the world is divided into 24 time zones. (Why 24?) Longitudinal lines are drawn between the North Pole and South Pole and are measured in degrees of longitude between 0° and 360°. The 0° longitudinal meridian passes through Greenwich, England (a suburb of London), and is called the **prime meridian.** The time at the prime meridian is called **Greenwich Mean Time (GMT).** The longitudinal meridians increase going west, and the 180° longitudinal meridian is called the **International Date Line.** When the International Date Line is crossed going west, the date is advanced one day; when it is crossed going east, the date becomes one day earlier. A map of the International Time Zones is shown here.

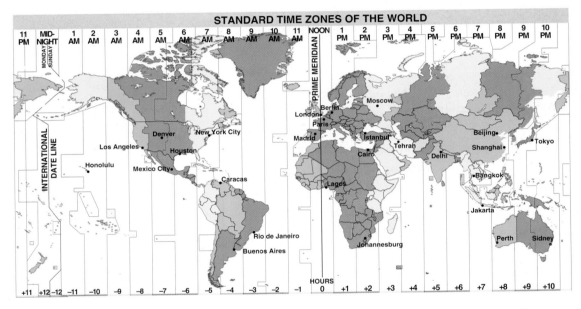

For Group Discussion

A London-based company employs a courier to deliver important documents between international offices. The courier adjusts her watch to local times as she travels between cities. For example, traveling from London to New York and then to Paris, she would have to adjust her watch $-5 + 6$ hr. In Paris, the time would be one hour ahead of GMT ($+1$). Write a similar integer expression to describe the changes in clock settings at each local time zone for the courier's trip. By how many hours does the time at the final location differ from GMT?

1. London to Tokyo to Cairo to Houston

2. London to Los Angeles to Caracas to Johannesburg to Bangkok to Paris

3. London to Houston to Honolulu to Tokyo (Remember the International Date Line.)

4. Determine one possible route, starting in London, that matches the time changes given by the integer expression $-4 - 4 + 5 + 5 + 6$.

2.1 Exercises

FOR EXTRA HELP

Addison-Wesley Math Tutor Center

MathXL

Digital Video Tutor CD 1 Videotape 1

Student's Solutions Manual

MyMathLab

InterAct Math Tutorial Software

Study Skills Workbook
Activity 3: Homework

1. Which equations are linear equations in x?

   A. $3x + x - 1 = 0$ B. $8 = x^2$

   C. $6x + 2 = 9$ D. $\frac{1}{2}x - \frac{1}{x} = 0$

2. Which of the equations in Exercise 1 are nonlinear equations in x? Explain why.

3. Decide whether 6 is a solution of $3(x + 4) = 5x$ by substituting 6 for x. If it is not a solution, explain why.

4. Use substitution to decide whether -2 is a solution of $5(x + 4) - 3(x + 6) = 9(x + 1)$. If it is not a solution, explain why.

5. The equation $4[x + (2 - 3x)] = 2(4 - 4x)$ is an identity. Let x represent the number of letters in your last name. Is this number a solution of this equation? Check your answer.

6. In Example 1, a student looked at the check and thought that $\{-11\}$ should be given as the solution set. Explain why this is not correct.

7. Identify each as an *expression* or an *equation*.

   (a) $3x = 6$

   (b) $3x + 6$

   (c) $5x + 6(x - 3) = 12x + 6$

   (d) $5x + 6(x - 3) - (12x + 6)$

8. Explain why $6x + 9 = 6x + 8$ cannot have a solution. (No work is necessary.)

9. The following "solution" contains a common student error. Identify it, and find the correct solution.

$$8x - 2(2x - 3) = 3x + 7$$

$8x - 4x - 6 = 3x + 7$	Distributive property
$4x - 6 = 3x + 7$	Combine like terms.
$x = 13$	Subtract $3x$; add 6.

10. When clearing parentheses in the expression

$$-5m - (2m - 4) + 5$$

on the right side of the equation in Exercise 35, the $-$ sign before the parenthesis acts like a factor representing what number? Clear parentheses and simplify this expression.

Solve and check each equation. See Examples 1 and 2.

11. $7x + 8 = 1$

12. $5x - 4 = 21$

13. $5x + 2 = 3x - 6$

14. $9p + 1 = 7p - 9$

15. $7x - 5x + 15 = x + 8$

16. $2x + 4 - x = 4x - 5$

17. $12w + 15w - 9 + 5 = -3w + 5 - 9$

18. $-4t + 5t - 8 + 4 = 6t - 4$

19. $3(2t - 4) = 20 - 2t$

20. $2(3 - 2x) = x - 4$

21. $-5(x + 1) + 3x + 2 = 6x + 4$

22. $5(x + 3) + 4x - 5 = 4 - 2x$

23. $2(x + 3) = -4(x + 1)$

24. $4(t - 9) = 8(t + 3)$

25. $3(2w + 1) - 2(w - 2) = 5$

26. $4(x - 2) + 2(x + 3) = 6$

27. $2x + 3(x - 4) = 2(x - 3)$

28. $6x - 3(5x + 2) = 4(1 - x)$

29. $6p - 4(3 - 2p) = 5(p - 4) - 10$

30. $-2k - 3(4 - 2k) = 2(k - 3) + 2$

31. $2[w - (2w + 4) + 3] = 2(w + 1)$

32. $4[2t - (3 - t) + 5] = -(2 + 7t)$

33. $-[2z - (5z + 2)] = 2 + (2z + 7)$

34. $-[6x - (4x + 8)] = 9 + (6x + 3)$

35. $-3m + 6 - 5(m - 1) = -5m - (2m - 4) + 5$

36. $4(k + 2) - 8k - 5 = -3k + 9 - 2(k + 6)$

37. $-3(x + 2) + 4(3x - 8) = 2(4x + 7) + 2(3x - 6)$

38. $-7(2x + 1) + 5(3x + 2) = 6(2x - 4) - (12x + 3)$

39. In order to solve the linear equation

$$\frac{8x}{3} - \frac{2x}{4} = -13,$$

we are allowed to multiply each side by the least common denominator of all the fractions in the equation. What is this least common denominator?

40. Suppose that in solving the equation

$$\frac{1}{3}x + \frac{1}{2}x = \frac{1}{6}x,$$

you begin by multiplying each side by 12, rather than the *least* common denominator, 6. Would you get the correct solution anyway? Explain.

41. To solve a linear equation with decimals, we multiply by a power of 10 so that all coefficients are integers. What is the smallest power of 10 that will accomplish this goal in each equation?

(a) $.05x + .12(x + 5000) = 940$ (Exercise 55)

(b) $.006(x + 2) = .007x + .009$ (Exercise 61)

42. The expression $.06(10 - x)(100)$ is equivalent to which of the following?

A. $.06 - .06x$ **B.** $60 - 6x$

C. $6 - 6x$ **D.** $6 - .06x$

Solve and check each equation. See Examples 3 and 4.

43. $\dfrac{m}{2} + \dfrac{m}{3} = 5$

44. $\dfrac{x}{5} - \dfrac{x}{4} = 1$

45. $\dfrac{3x}{4} + \dfrac{5x}{2} = 13$

46. $\dfrac{8x}{3} - \dfrac{2x}{4} = -13$

47. $\dfrac{1}{5}x - 2 = \dfrac{2}{3}x - \dfrac{2}{5}x$

48. $\dfrac{3}{4}x - \dfrac{1}{3}x = \dfrac{5}{6}x - 5$

49. $\dfrac{x - 8}{5} + \dfrac{8}{5} = -\dfrac{x}{3}$

50. $\dfrac{2r - 3}{7} + \dfrac{3}{7} = -\dfrac{r}{3}$

51. $\dfrac{3x - 1}{4} + \dfrac{x + 3}{6} = 3$

52. $\dfrac{3x + 2}{7} - \dfrac{x + 4}{5} = 2$

53. $\dfrac{4t + 1}{3} = \dfrac{t + 5}{6} + \dfrac{t - 3}{6}$

54. $\dfrac{2x + 5}{5} = \dfrac{3x + 1}{2} + \dfrac{-x + 7}{2}$

55. $.05x + .12(x + 5000) = 940$

56. $.09k + .13(k + 300) = 61$

57. $.02(50) + .08r = .04(50 + r)$

58. $.20(14,000) + .14t = .18(14,000 + t)$

59. $.05x + .10(200 - x) = .45x$

60. $.08x + .12(260 - x) = .48x$

61. $.006(x + 2) = .007x + .009$

62. $.004x + .006(50 - x) = .004(68)$

63. Explain the distinction between a conditional equation, an identity, and a contradiction.

64. A student tried to solve the equation $8x = 7x$ by dividing each side by x, obtaining $8 = 7$. He gave the solution set as $\emptyset$. Why is this incorrect?

65. Suppose you solve a linear equation and obtain, as your final result, an equation in Column I. Match each result with the solution set in Column II for the original equation.

I	II
(a) $5 = 5$	**A.** $\{0\}$
(b) $x = 0$	**B.** {all real numbers}
(c) $5 = 0$	**C.** $\emptyset$

66. Which one of the following linear equations does *not* have {all real numbers} as its solution set?

A. $4x = 5x - x$ **B.** $3(x + 4) = 3x + 12$

C. $4x = 3x$ **D.** $\dfrac{3}{4}x = .75x$

Decide whether each equation is a conditional equation, *an* identity, *or a* contradiction. *Give the solution set. See Example 5.*

67. $-2x + 5x - 9 = 3(x - 4) - 5$

68. $-6x + 2x - 11 = -2(2x - 3) + 4$

69. $-11x + 4(x - 3) + 6x = 4x - 12$

70. $3x - 5(x + 4) + 9 = -11 + 15x$

71. $-2(t + 3) - t - 4 = -3(t + 4) + 2$

72. $4(2d + 7) = 2d + 25 + 3(2d + 1)$

73. $7[2 - (3 + 4x)] - 2x = -9 + 2(1 - 15x)$

74. $4[6 - (1 + 2x)] + 10x = 2(10 - 3x) + 8x$

2.2 Formulas

A **mathematical model** is an equation or inequality that describes a real situation. Models for many applied problems already exist; they are called *formulas*. A **formula** is a mathematical equation in which variables are used to describe a relationship. Some formulas that we will be using are

$$d = rt, \quad I = prt, \quad \text{and} \quad P = 2L + 2W. \qquad \text{Formulas}$$

A list of some common formulas used in algebra is given inside the covers of this book.

OBJECTIVE 1 Solve a formula for a specified variable. In some applications, the appropriate formula may be solved for a different variable than the one to be found. For example, the formula $I = prt$ says that interest on a loan or investment equals principal (amount borrowed or invested) times rate (percent) times time at interest (in years). To determine how long it will take for an investment at a stated interest rate to earn a predetermined amount of interest, it would help to first solve the formula for t. This process is called **solving for a specified variable** or **solving a literal equation.**

The steps used in the following examples are very similar to those used in solving linear equations from **Section 2.1.** *When you are solving for a specified variable, the key is to treat that variable as if it were the only one; treat all other variables like numbers (constants).*

EXAMPLE 1 Solving for a Specified Variable

Solve the formula $I = prt$ for t.

We solve this formula for t by treating I, p, and r as constants (having fixed values) and treating t as the only variable. We first write the formula so that the variable for which we are solving, t, is on the left side. Then we use the properties of the previous section as follows.

$$prt = I$$
$$(pr)t = I \qquad \text{Associative property}$$
$$\frac{(pr)t}{pr} = \frac{I}{pr} \qquad \text{Divide by } pr.$$
$$t = \frac{I}{pr}$$

The result is a formula for t, time in years.

Work Problem 1 at the Side.

OBJECTIVES

1 Solve a formula for a specified variable.

2 Solve applied problems using formulas.

3 Solve percent problems.

1 Solve $I = prt$ for each given variable.

(a) p

(b) r

ANSWERS

1. (a) $p = \dfrac{I}{rt}$ **(b)** $r = \dfrac{I}{pt}$

② **(a)** Solve the formula

$$P = a + b + c$$

for a.

To solve for a specified variable, follow these steps.

Solving for a Specified Variable

Step 1 Transform so that all terms containing the specified variable are on one side of the equation and all terms without that variable are on the other side.

Step 2 If necessary, use the distributive property to combine the terms with the specified variable.* The result should be the product of a sum or difference and the variable.

Step 3 Divide each side by the factor that is the coefficient of the specified variable.

EXAMPLE 2 **Solving for a Specified Variable**

Solve the formula $P = 2L + 2W$ for W.

This formula gives the relationship between perimeter of a rectangle, P, length of the rectangle, L, and width of the rectangle, W. See Figure 1.

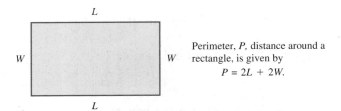

Perimeter, P, distance around a rectangle, is given by
$$P = 2L + 2W.$$

Figure 1

(b) Solve the formula

$$m = 2k + 3b$$

for k.

Solve the formula for W by isolating W on one side of the equals sign. To begin, subtract $2L$ from each side.

$$P = 2L + 2W$$

Step 1 $\quad P - 2L = 2L + 2W - 2L \qquad$ Subtract $2L$.

$$P - 2L = 2W$$

Step 2 is not needed here.

Step 3 $\qquad \dfrac{P - 2L}{2} = \dfrac{2W}{2} \qquad$ Divide by 2.

$$\dfrac{P - 2L}{2} = W \quad \text{or} \quad W = \dfrac{P - 2L}{2}$$

◀◀◀ **Work Problem 2 at the Side.**

CAUTION

In Step 3 of Example 2, you cannot simplify the fraction by dividing 2 into the term $2L$. The subtraction in the numerator must be done before the division.

$$\dfrac{P - 2L}{2} \neq P - L$$

ANSWERS

2. **(a)** $a = P - b - c$ **(b)** $k = \dfrac{m - 3b}{2}$

*Using the distributive property to write $ab + ac$ as $a(b + c)$ is called *factoring*. See **Chapter 7.**

EXAMPLE 3 **Solving a Formula with Parentheses**

The formula for the perimeter of a rectangle is sometimes written in the equivalent form $P = 2(L + W)$. Solve this form for W.

One way to begin is to use the distributive property on the right side of the equation to get $P = 2L + 2W$, which we would then solve as in Example 2. Another way to begin is to divide by the coefficient 2.

$$P = 2(L + W)$$

$$\frac{P}{2} = L + W \qquad \text{Divide by 2.}$$

$$\frac{P}{2} - L = W \quad \text{or} \quad W = \frac{P}{2} - L \qquad \text{Subtract } L.$$

We can show that this result is equivalent to our result in Example 2 by multiplying L by $\frac{2}{2}$.

$$\frac{P}{2} - \frac{2}{2}(L) = W \qquad \tfrac{2}{2} = 1, \text{ so } L = \tfrac{2}{2}(L).$$

$$\frac{P}{2} - \frac{2L}{2} = W$$

$$\frac{P - 2L}{2} = W \qquad \text{Subtract fractions.}$$

The final line agrees with the result in Example 2.

Work Problem 3 at the Side. ▶▶▶

A rectangular solid has the shape of a box, but is solid. See Figure 2. The labels H, W, and L represent the height, width, and length of the figure, respectively. The surface area of any solid three-dimensional figure is the total area of its surface. For a rectangular solid, the surface area A is

$$A = 2HW + 2LW + 2LH.$$

EXAMPLE 4 **Using the Distributive Property to Solve for a Specified Variable**

Given the surface area, height, and width of a rectangular solid, write a formula for the length.

To solve for the length L, treat L as the only variable and treat all other variables as constants.

$$A = 2HW + 2LW + 2LH$$

$$A - 2HW = 2LW + 2LH \qquad \text{Subtract } 2HW.$$

$$A - 2HW = L(2W + 2H) \qquad \text{Distributive property}$$

$$\frac{A - 2HW}{2W + 2H} = L \quad \text{or} \quad L = \frac{A - 2HW}{2W + 2H} \qquad \text{Divide by } 2W + 2H.$$

CAUTION

Be careful when working a problem like Example 4 to use the distributive property correctly. We must write the expression so that the specified variable is a *factor;* then we can divide by its coefficient in the final step.

Work Problem 4 at the Side. ▶▶▶

---

3 Solve the formula

$$y = \frac{1}{2}(x + 3)$$

for x.

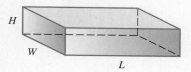

Figure 2

4 Solve the formula

$$A = 2HW + 2LW + 2LH$$

for W.

ANSWERS

3. $x = 2y - 3$

4. $W = \dfrac{A - 2LH}{2H + 2L}$

5 Solve each problem.

(a) A triangle has an area of 36 in.^{2} (square inches) and a base of 12 in. Find its height.

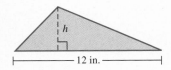

(b) The distance is 500 mi and the time is 20 hr. Find the rate.

(c) In 2003, Gil de Ferran won the Indianapolis 500 (mile) race with a speed of 156.291 mph. (*Source: World Almanac and Book of Facts*, 2004.) Find his time to the nearest thousandth.

OBJECTIVE 2 **Solve applied problems using formulas.**   The distance formula, $d = rt$, relates d, the distance traveled, r, the rate or speed, and t, the travel time.

EXAMPLE 5 **Finding Average Speed**

Janet Branson found that on average it took her $\frac{3}{4}$ hr each day to drive a distance of 15 mi to work. What was her average speed?

Find the speed r by solving $d = rt$ for r.

$$d = rt$$

$$\frac{d}{t} = \frac{rt}{t} \qquad \text{Divide by } t.$$

$$\frac{d}{t} = r \quad \text{or} \quad r = \frac{d}{t}$$

Notice that only Step 3 was needed to solve for r in this example. Now find the speed by substituting the given values of d and t into this formula.

$$r = \frac{15}{\frac{3}{4}} \qquad \text{Let } d = 15, t = \tfrac{3}{4}.$$

$$r = 15 \cdot \frac{4}{3} \qquad \text{Multiply by the reciprocal of } \tfrac{3}{4}.$$

$$r = 20$$

Her average speed was 20 mph. (That is, at times she may have traveled a little faster or slower than 20 mph, but overall her speed was 20 mph.)

Work Problem 5 at the Side.

OBJECTIVE 3 **Solve percent problems.**   An important everyday use of mathematics involves the concept of percent. Percent is written with the symbol %. The word **percent** means "per one hundred." One percent means "one per one hundred" or "one one-hundredth."

$$1\% = .01 \quad \text{or} \quad 1\% = \frac{1}{100}$$

Solving a Percent Problem

Let a represent a partial amount of b, the base, or whole amount. Then the following formula can be used in solving a percent problem.

$$\frac{\textbf{amount}}{\textbf{base}} = \frac{a}{b} = \textbf{percent (represented as a decimal)}$$

For example, if a class consists of 50 students and 32 are males, then the percent of males in the class is

$$\frac{\text{amount}}{\text{base}} = \frac{a}{b} = \frac{32}{50} \qquad \text{Let } a = 32, b = 50.$$

$$= .64$$

$$= 64\%.$$

ANSWERS
5. **(a)** 6 in.   **(b)** 25 mph   **(c)** 3.199 hr

EXAMPLE 6 **Solving Percent Problems**

(a) A 50-L mixture of acid and water contains 10 L of acid. What is the percent of acid in the mixture?

The given amount of the mixture is 50 L, and the part that is acid is 10 L. Let x represent the percent of acid. Then, the percent of acid in the mixture is

$$x = \frac{10}{50} = .20 \quad \text{or} \quad 20\%.$$

(b) If a savings account balance of $3550 earns 8% interest in one year, how much interest is earned?

Let x represent the amount of interest earned (that is, the part of the whole amount invested). Since $8\% = .08$, the equation is

$$\frac{x}{3550} = .08 \qquad \frac{a}{b} = \text{percent}$$

$$x = .08(3550) \qquad \text{Multiply by 3550.}$$

$$x = 284.$$

The interest earned is $284.

Work Problem 6 at the Side. ▶▶▶

❻ Solve each problem.

(a) A mixture of gasoline and oil contains 20 oz, 1 oz of which is oil. What percent of the mixture is oil?

(b) An automobile salesman earns an 8% commission on every car he sells. How much does he earn on a car that sells for $12,000?

EXAMPLE 7 **Interpreting Percents from a Graph**

In 2003, people in the United States spent an estimated $29.7 billion on their pets. Use the graph in Figure 3 to determine how much of this amount was spent on pet food.

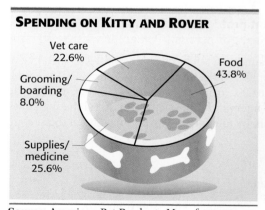

SPENDING ON KITTY AND ROVER

Vet care 22.6%
Grooming/boarding 8.0%
Supplies/medicine 25.6%
Food 43.8%

Source: American Pet Products Manufacturers Association Inc.

Figure 3

❼ Refer to Figure 3. How much was spent on pet supplies/medicine? Round your answer to the nearest tenth.

According to the graph, 43.8% was spent on food. Let x represent this amount in billions of dollars.

$$\frac{x}{29.7} = .438 \qquad 43.8\% = .438$$

$$x = .438(29.7) \qquad \text{Multiply by 29.7.}$$

$$x = 13.0 \qquad \text{Nearest tenth}$$

Therefore, about $13.0 billion was spent on pet food.

Work Problem 7 at the Side. ▶▶▶

ANSWERS
6. (a) 5% (b) $960
7. $7.6 billion

Real-Data Applications

Same (International) Time, Second Place

Employees of companies that operate globally must be able to converse by telephone with their colleagues around the world. It is sometimes challenging to find the common working hours at different global locations. Refer to the map of the International Time Zones. The time zone map is explored in the Focus-on application *International Time Zones* on p. 60.

For Group Discussion

Europeans use a 24-hr clock, or military time. For example, 03:45 is 3:45 A.M. and 15:45 is 3:45 P.M. At 15:45 in London, the time would be 8 hr earlier in Los Angeles or 7:45 A.M., since $15:45 - 8:00$ is 07:45. Assume that it is common for workdays to last from 8:00 A.M. to 6:00 P.M. worldwide (08:00–18:00). For each pair of cities, find the time interval during the workday and the workdays during the week that a person in the first city can talk by telephone with a colleague in the second city.

1. Moscow and Houston

   Moscow is _____ hr ahead of Houston.

   08:00–18:00 Moscow time corresponds to _____ Houston time.

   The overlap in time is _____ hr between _____ Moscow time and

   _____ Houston time.

   Monday–Friday in Moscow overlaps _____ in Houston.

2. Los Angeles and Tokyo

3. London and Los Angeles and London and Tokyo

2.2 Exercises

FOR EXTRA HELP

Addison-Wesley Math Tutor Center

Math XL MathXL

Digital Video Tutor CD 1 Videotape 1

Student's Solutions Manual

MyMathLab MyMathLab

Interactmath.com

RELATING CONCEPTS (EXERCISES 1–6) For Individual or Group Work

Consider the following equations:

First Equation

$$x = \frac{5x + 8}{3}$$

Second Equation

$$t = \frac{bt + k}{c} \quad (c \neq 0).$$

Solving the second equation for t requires the same logic as solving the first equation for x. When solving for t, we treat all other variables as though they were constants. **Work Exercises 1–6 in order,** *to see the "parallel logic" of solving for x and solving for t.*

1. **(a)** Clear the first equation of fractions by multiplying each side by 3.

   (b) Clear the second equation of fractions by multiplying each side by c.

2. **(a)** Transform so that the terms involving x are on the left side of the first equation by subtracting $5x$ from each side.

   (b) Transform so that the terms involving t are on the left side of the second equation by subtracting bt from each side.

3. **(a)** Combine like terms on the left side of the first equation. What property allows us to write $3x - 5x$ as $(3 - 5)x = -2x$?

   (b) Write the expression on the left side of the second equation so that t is a factor. What property allows us to do this?

4. **(a)** Divide each side of the first equation by the coefficient of x.

   (b) Divide each side of the second equation by the coefficient of t.

5. Look at your answer for the second equation. What restriction must be placed on the variables? Why is this necessary?

6. Write a short paragraph summarizing what you have learned in this group of exercises.

Solve each formula for the specified variable. See Examples 1–3.

7. $A = LW$ for W (area of a rectangle)

8. $d = rt$ for t (distance)

9. $P = 2L + 2W$ for L (perimeter of a rectangle)

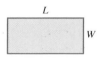

10. $A = bh$ for b (area of a parallelogram)

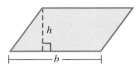

11. $V = LWH$ (volume of a rectangular solid)
   (a) for W **(b)** for H

12. $P = a + b + c$ (perimeter of a triangle)
   (a) for b **(b)** for c

13. $C = 2\pi r$ for r (circumference of a circle)

14. $A = \dfrac{1}{2} bh$ for h (area of a triangle)

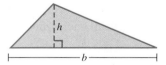

15. $A = \dfrac{1}{2}h(b + B)$ (area of a trapezoid)

   (a) for h **(b)** for B

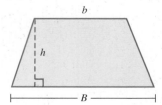

16. $S = 2\pi rh + 2\pi r^2$ for h
   (surface area of a right circular cylinder)

17. $F = \dfrac{9}{5}C + 32$ for C (Celsius to Fahrenheit)

18. $C = \dfrac{5}{9}(F - 32)$ for F (Fahrenheit to Celsius)

19. When a formula is solved for a particular variable, several different equivalent forms may be possible. If we solve $A = \frac{1}{2}bh$ for h, one possible correct answer is

$$h = \frac{2A}{b}.$$

Which one of the following is *not* equivalent to this?

A. $h = 2\left(\dfrac{A}{b}\right)$ **B.** $h = 2A\left(\dfrac{1}{b}\right)$

C. $h = \dfrac{A}{\frac{1}{2}b}$ **D.** $h = \dfrac{\frac{1}{2}A}{b}$

20. Suppose the formula

$$A = 2HW + 2LW + 2LH$$

is solved for L as follows.

$$A = 2HW + 2LW + 2LH$$
$$A - 2LW - 2HW = 2LH$$
$$\frac{A - 2LW - 2HW}{2H} = L$$

While there are no algebraic errors here, what is wrong with the final equation, if we are interested in solving for L?

Solve each equation for the specified variable. See Example 4.

21. $2k + ar = r - 3y$ for r

22. $4s + 7p = tp - 7$ for p

23. $w = \dfrac{3y - x}{y}$ for y

24. $c = \dfrac{-2t + 4}{t}$ for t

Solve each problem. See Example 5.

25. In 2003, Michael Waltrip won the Daytona 500 (mile) race with a speed of 133.870 mph. Find his time to the nearest thousandth. (*Source: World Almanac and Book of Facts*, 2004.)

26. In 2004, rain shortened the Indianapolis 500 race to 450 mi. It was won by Buddy Rice, who averaged 138.518 mph. What was his time to the nearest thousandth? (*Source:* indy500.com)

27. As of 2003, the highest temperature ever recorded in Chicago was 40°C. Find the corresponding Fahrenheit temperature. (*Source: World Almanac and Book of Facts*, 2004.)

28. The lowest temperature recorded in Burlington, Vermont, in 2002 was −4°F. Find the corresponding Celsius temperature. (*Source: World Almanac and Book of Facts*, 2004.)

29. The base of the Great Pyramid of Cheops is a square whose perimeter is 920 m. What is the length of each side of this square? (*Source: Atlas of Ancient Archaeology*, 1994.)

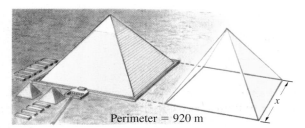

Perimeter = 920 m

30. The Peachtree Plaza Hotel in Atlanta is in the shape of a cylinder with radius 46 m and height 220 m. Find its volume to the nearest tenth. (*Hint:* Use the π key on your calculator.)

31. The circumference of a circle is 480π in. What is its radius? What is its diameter?

32. The radius of a circle is 2.5 in. What is its diameter? What is its circumference?

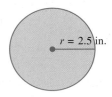

$r = 2.5$ in.

33. A cord of wood contains 128 ft³ (cubic feet) of wood. If a stack of wood is 4 ft wide and 4 ft high, how long must it be if it contains exactly 1 cord?

34. Give one set of possible dimensions for a stack of wood that contains 1.5 cords. (See Exercise 33.)

Solve each problem. See Example 6.

35. A mixture of alcohol and water contains a total of 36 oz of liquid. There are 9 oz of pure alcohol in the mixture. What percent of the mixture is water? What percent is alcohol?

36. A mixture of acid and water is 35% acid. If the mixture contains a total of 40 L, how many liters of pure acid are in the mixture? How many liters of pure water are in the mixture?

37. A real estate agent earned $6300 commission on a property sale of $210,000. What is her rate of commission?

38. A certificate of deposit for one year pays $221 simple interest on a principal of $3400. What is the interest rate being paid on this deposit?

When a consumer loan is paid off ahead of schedule, the finance charge is smaller than if the loan were paid off over its scheduled life. By one method, called the **rule of 78,** the amount of unearned interest (finance charge that need not be paid) is given by

where u is the amount of unearned interest (money saved) when a loan scheduled to run n payments is paid off k payments ahead of schedule. The total scheduled finance charge is f. Use this formula to solve Exercises 39–42.

39. Rhonda Alessi bought a new Ford and agreed to pay it off in 36 monthly payments. The total finance charge is $700. Find the unearned interest if she pays the loan off 4 payments ahead of schedule.

40. Charles Vosburg bought a car and agreed to pay it off in 36 monthly payments. The total finance charge on the loan was $600. With 12 payments remaining, Charles decided to pay the loan in full. Find the amount of unearned interest.

41. The finance charge on a loan taken out by Vic Denicola is $380.50. If there were 24 equal monthly installments needed to repay the loan, and the loan is paid in full with 8 months remaining, find the amount of unearned interest.

42. Adrian Ortega is scheduled to repay a loan in 24 equal monthly installments. The total finance charge on the loan is $450. With 9 payments remaining, he decides to repay the loan in full. Find the amount of unearned interest.

Exercises 43 and 44 deal with winning percentage in the team standings for major league baseball. Winning percentage (Pct.) is commonly expressed as a decimal rounded to the nearest thousandth. To find the winning percentage of a team, divide the number of wins (W) by the total number of games played (W + L).

43. At the start of play on May 11, 2004, the standings of the Central Division of the American League were as shown. Find the winning percentage of each team.

(a) Chicago (b) Minnesota

(c) Detroit (d) Cleveland

	W	L	Pct.
Chicago	17	13	
Minnesota	17	13	
Detroit	15	16	
Cleveland	13	18	
Kansas City	9	21	.300

44. Repeat Exercise 43 for the following standings for the Central Division of the National League.

(a) Houston (b) Chicago

(c) Cincinnati (d) St. Louis

	W	L	Pct.
Houston	20	11	
Chicago	18	13	
Cincinnati	16	15	
St. Louis	16	16	
Milwaukee	15	16	.484
Pittsburgh	12	17	.414

As mentioned in the chapter introduction, 106.7 million U.S. households owned at least one TV set in 2003. (Source: Nielsen Media Research.) Use this information to solve Exercises 45 and 46. Round answers to the nearest percent.

45. About 43.7 million U.S. households owned 3 or more TV sets in 2003. What percent of those owning at least one TV set was this?

46. About 51.2 million households that owned at least one TV set in 2003 received premium cable television. What percent of those owning at least one TV set received premium cable?

Television networks have been losing viewers to cable programming since 1982, as the two graphs show. Use these graphs to answer Exercises 47–50. See Example 7.

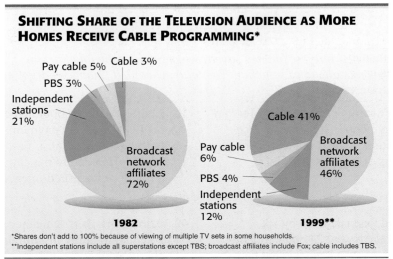

SHIFTING SHARE OF THE TELEVISION AUDIENCE AS MORE HOMES RECEIVE CABLE PROGRAMMING*

*Shares don't add to 100% because of viewing of multiple TV sets in some households.
**Independent stations include all superstations except TBS; broadcast affiliates include Fox; cable includes TBS.

Source: Nielsen Media Research, National Cable Television Association Report, Spring, 2000.

47. In a typical group of 50,000 television viewers, how many would have watched cable in 1982?

48. In 1982, how many of a typical group of 110,000 viewers watched independent stations?

49. How many of a typical group of 35,000 viewers watched cable in 1999?

50. In a typical group of 65,000 viewers, how many watched independent stations in 1999?

An average middle-income family will spend $161,430 to raise a child born in 2001 from birth to age 17. The graph shows the percents spent for various categories. Use the graph to answer Exercises 51–53. See Example 7.

51. To the nearest dollar, how much will be spent to provide housing for the child?

52. To the nearest dollar, how much will be spent for health care?

53. About $29,000 will be spent for food. To the nearest percent, what percent of the cost of raising a child from birth to age 17 is this? Does your answer agree with the percent shown in the graph?

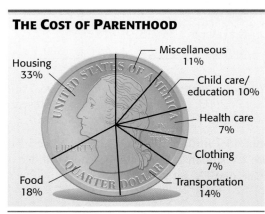

THE COST OF PARENTHOOD

Source: U.S. Department of Agriculture.

2.3 Applications of Linear Equations

OBJECTIVE 1 Translate from words to mathematical expressions.
Producing a mathematical model of a real situation often involves translating verbal statements into mathematical statements. Although the problems we will be working with are simple ones, the methods we use will also apply to more difficult problems later.

> **PROBLEM-SOLVING HINT**
> Usually there are key words and phrases in a verbal problem that translate into mathematical expressions involving addition, subtraction, multiplication, and division. Translations of some commonly used expressions follow.

TRANSLATING FROM WORDS TO MATHEMATICAL EXPRESSIONS

Verbal Expression	Mathematical Expression (where x and y are numbers)
Addition	
The **sum** of a number and 7	$x + 7$
6 **more than** a number	$x + 6$
3 **plus** a number	$3 + x$
24 **added to** a number	$x + 24$
A number **increased by** 5	$x + 5$
The **sum** of two numbers	$x + y$
Subtraction	
2 **less than** a number	$x - 2$
12 **minus** a number	$12 - x$
A number **decreased by** 12	$x - 12$
A number **subtracted from** 10	$10 - x$
The **difference between** two numbers	$x - y$
Multiplication	
16 **times** a number	$16x$
A number **multiplied by** 6	$6x$
$\frac{2}{3}$ **of** a number (used with fractions and percent)	$\frac{2}{3}x$
Twice (2 times) a number	$2x$
The **product** of two numbers	xy
Division	
The **quotient** of 8 and a number	$\frac{8}{x} \ (x \neq 0)$
A number **divided by** 13	$\frac{x}{13}$
The **ratio** of two numbers or the **quotient** of two numbers	$\frac{x}{y} \ (y \neq 0)$

> **Work Problem 1 at the Side.** ▶▶▶

OBJECTIVES

1 Translate from words to mathematical expressions.

2 Write equations from given information.

3 Distinguish between expressions and equations.

4 Use the six steps in solving an applied problem.

5 Solve percent problems.

6 Solve investment problems.

7 Solve mixture problems.

Study Skills Workbook
Activity 4: Note Taking

1 Translate each verbal expression as a mathematical expression. Use x as the variable.

(a) 9 added to a number

(b) The difference between 7 and a number

(c) Four times a number

(d) The quotient of 7 and a nonzero number

ANSWERS
1. **(a)** $9 + x$ or $x + 9$ **(b)** $7 - x$
   (c) $4x$ **(d)** $\frac{7}{x}$

2 Translate each verbal sentence into an equation. Use x as the variable.

(a) The sum of a number and 6 is 28.

CAUTION
Because subtraction and division are not commutative operations, be careful to correctly translate expressions involving them. For example, "2 less than a number" is translated as $x - 2$, *not* $2 - x$. "A number subtracted from 10" is expressed as $10 - x$, *not* $x - 10$.

For division, the number *by which* we are dividing is the denominator, and the number *into which* we are dividing is the numerator. For example, "a number divided by 13" and "13 divided into x" both translate as $\frac{x}{13}$. Similarly, "the quotient of x and y" is translated as $\frac{x}{y}$.

(b) If twice a number is decreased by 3, the result is 17.

OBJECTIVE 2 **Write equations from given information.** The symbol for equality, $=$, is often indicated by the word *is*. In fact, any words that indicate the idea of "sameness" translate to $=$.

EXAMPLE 1 **Translating Words into Equations**

Translate each verbal sentence into an equation.

(c) The product of a number and 7 is twice the number plus 12.

Verbal Sentence	Equation
Twice a number, **decreased by 3, is** 42.	$2x - 3 = 42$
If the **product of a number and 12** is decreased by 7, the result **is** 105.	$12x - 7 = 105$
The **quotient of a number and the number plus 4 is** 28.	$\dfrac{x}{x + 4} = 28$
The **quotient of a number and 4,** plus the number, **is** 10.	$\dfrac{x}{4} + x = 10$

(d) The quotient of a number and 6, added to twice the number, is 7.

◀◀ Work Problem 2 at the Side.

OBJECTIVE 3 **Distinguish between expressions and equations.** An expression translates as a phrase. An equation includes the $=$ symbol and translates as a sentence.

3 Decide whether each is an *expression* or an *equation*.

(a) $5x - 3(x + 2) = 7$

EXAMPLE 2 **Distinguishing between Expressions and Equations**

Decide whether each is an *expression* or an *equation*.

(a) $2(3 + x) - 4x + 7$
There is no equals sign, so this is an expression.

(b) $2(3 + x) - 4x + 7 = -1$
Because of the equals sign, this is an equation.

Note that the expression in part (a) simplifies to the expression $-2x + 13$, and the equation in part (b) has solution 7.

(b) $5x - 3(x + 2)$

◀◀ Work Problem 3 at the Side.

OBJECTIVE 4 **Use the six steps in solving an applied problem.** While there is no one method that will allow us to solve all types of applied problems, the following six steps are helpful.*

ANSWERS
2. (a) $x + 6 = 28$ (b) $2x - 3 = 17$
   (c) $7x = 2x + 12$ (d) $\dfrac{x}{6} + 2x = 7$
3. (a) equation   (b) expression

*Appendix A Strategies for Problem Solving introduces additional methods and tips for solving applied problems.

Solving an Applied Problem

Step 1 **Read** the problem, several times if necessary, until you *understand* what is given and what is to be found.

Step 2 **Assign a variable** to represent the unknown value, using diagrams or tables as needed. Write down what the variable represents. Express any other unknown values in terms of the variable.

Step 3 **Write an equation** using the variable expression(s).

Step 4 **Solve** the equation.

Step 5 **State the answer** to the problem. Does it seem reasonable?

Step 6 **Check** the answer in the words of the original problem.

4 Solve the problem.
The length of a rectangle is 5 cm more than its width. The perimeter is five times the width. What are the dimensions of the rectangle?

EXAMPLE 3 **Solving a Geometry Problem**

The length of a rectangle is 1 cm more than twice the width. The perimeter of the rectangle is 110 cm. Find the length and the width of the rectangle.

Step 1 **Read** the problem. We must find the length and width of the rectangle. The length is 1 cm more than twice the width, and the perimeter is 110 cm.

Step 2 **Assign a variable.** Let W = the width; then $1 + 2W$ = the length. Make a sketch, as in Figure 4.

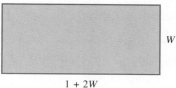

$$W$$

$$1 + 2W$$

Figure 4

Step 3 **Write an equation.** The perimeter of a rectangle is given by the formula $P = 2L + 2W$.

$$P = 2L + 2W$$
$$110 = 2(1 + 2W) + 2W \qquad \text{Let } L = 1 + 2W \text{ and } P = 110.$$

Step 4 **Solve** the equation obtained in Step 3.

$$110 = 2 + 4W + 2W \qquad \text{Distributive property}$$
$$110 = 2 + 6W \qquad \text{Combine like terms.}$$
$$110 - 2 = 2 + 6W - 2 \qquad \text{Subtract 2.}$$
$$108 = 6W$$
$$\frac{108}{6} = \frac{6W}{6} \qquad \text{Divide by 6.}$$
$$18 = W$$

Step 5 **State the answer.** The width of the rectangle is 18 cm and the length is $1 + 2(18) = 37$ cm.

Step 6 **Check** the answer by substituting these dimensions into the words of the original problem.

Work Problem 4 at the Side.

ANSWERS
4. width: 10 cm; length: 15 cm

5 Solve the problem.

At the end of the 2003 baseball season, Sammy Sosa and Barry Bonds had a lifetime total of 1197 home runs. Bonds had 119 more than Sosa. How many home runs did each player have? (*Source: World Almanac and Book of Facts*, 2004.)

EXAMPLE 4 Finding Unknown Numerical Quantities

Two outstanding major league pitchers in recent years are Randy Johnson and Pedro Martinez. In 2002, they combined for a total of 573 strikeouts. Johnson had 95 more strikeouts than Martinez. How many strikeouts did each pitcher have? (*Source: World Almanac and Book of Facts*, 2004.)

Step 1 **Read** the problem. We are asked to find the number of strikeouts each pitcher had.

Step 2 **Assign a variable** to represent the number of strikeouts for one of the men.

$$\text{Let } s = \text{the number of strikeouts for Pedro Martinez.}$$

We must also find the number of strikeouts for Randy Johnson. Since he had 95 more strikeouts than Martinez,

$$s + 95 = \text{the number of strikeouts for Johnson.}$$

Step 3 **Write an equation.** The sum of the numbers of strikeouts is 573, so

Martinez's strikeouts	+	Johnson's strikeouts	=	Total
↓		↓		↓
s	+	$(s + 95)$	=	573.

Step 4 **Solve** the equation.

$$s + (s + 95) = 573$$
$$2s + 95 = 573 \qquad \text{Combine like terms.}$$
$$2s + 95 - 95 = 573 - 95 \qquad \text{Subtract 95.}$$
$$2s = 478$$
$$\frac{2s}{2} = \frac{478}{2} \qquad \text{Divide by 2.}$$
$$s = 239$$

Step 5 **State the answer.** We let s represent the number of strikeouts for Martinez, so Martinez had 239. Also,

$$s + 95 = 239 + 95 = 334$$

is the number of strikeouts for Johnson.

Step 6 **Check.** 334 is 95 more than 239, and the sum of 239 and 334 is 573. The conditions of the problem are satisfied, and our answer checks.

CAUTION
A common error in solving applied problems is forgetting to answer all the questions asked in the problem. In Example 4, we were asked for the number of strikeouts *each* player had, so there was an extra step at the end in order to find the number Johnson had.

ANSWERS
5. Sosa: 539; Bonds: 658

◀◀◀ Work Problem 5 at the Side.

OBJECTIVE 5 Solve percent problems. Recall from **Section 2.2** that percent means "per one hundred," so 5% means .05, 14% means .14, and so on.

EXAMPLE 5 Solving a Percent Problem

In 2002 there were 301 long-distance area codes in the United States. This was an increase of 250% over the number when the area code plan originated in 1947. How many area codes were there in 1947? (*Source:* SBC Telephone Directory.)

Step 1 **Read** the problem. We are given that the number of area codes increased by 250% from 1947 to 2002, and there were 301 area codes in 2002. We must find the original number of area codes.

Step 2 **Assign a variable.** Let x represent the number of area codes in 1947.

$$250\% = 250(.01) = 2.5,$$

so $2.5x$ represents the number of codes added since then.

Step 3 **Write an equation** from the given information.

the number in 1947 + the increase = 301

$$x \quad + \quad 2.5x \quad = 301$$

Step 4 **Solve** the equation.

$$1x + 2.5x = 301 \qquad \text{Identity property}$$
$$3.5x = 301 \qquad \text{Combine like terms.}$$
$$x = 86 \qquad \text{Divide by 3.5.}$$

Step 5 **State the answer.** There were 86 area codes in 1947.

Step 6 **Check** that the increase, $301 - 86 = 215$, is 250% of 86.

CAUTION

Avoid two common errors that occur in solving problems like the one in Example 5.

1. Do not try to find 250% of 301 and subtract that amount from 301. The 250% should be applied to *the amount in 1947, not the amount in 2002.*

2. Do not write the equation as

$$x + 2.5 = 301. \qquad \text{Incorrect}$$

The percent must be multiplied by some amount; in this case, the amount is the number of area codes in 1947, giving $2.5x$.

Work Problem 6 at the Side. ▶▶▶

OBJECTIVE 6 Solve investment problems. We use linear equations to solve certain investment problems. The investment problems in this chapter deal with *simple interest.* In most real-world applications, *compound interest* (covered in a later chapter) is used.

6 Solve each problem.

(a) A number increased by 15% is 287.5. Find the number.

(b) Michelle Raymond was paid $162 for a week's work at her part-time job after 10% deductions for taxes. How much did she make before the deductions were made?

ANSWERS
6. **(a)** 250 **(b)** $180

7 Solve each problem.

(a) A woman invests $72,000 in two ways—some at 5% and some at 3%. Her total annual interest income is $3160. Find the amount she invests at each rate.

EXAMPLE 6 Solving an Investment Problem

After winning the state lottery, Mark LeBeau has $40,000 to invest. He will put part of the money in an account paying 4% interest and the remainder into stocks paying 6% interest. His accountant tells him that the total annual income from these investments should be $2040. How much should he invest at each rate?

Step 1 **Read** the problem again. We must find the two amounts.

Step 2 **Assign a variable.**

Let $x =$ the amount to invest at 4%;

$40,000 - x =$ the amount to invest at 6%.

The formula for interest is $I = prt$. Here the time, t, is 1 year. Make a table to organize the given information.

Rate (as a decimal)	Principal	Interest	
.04	x	$.04x$	
.06	$40,000 - x$	$.06(40,000 - x)$	
	40,000	2040	← Totals

Step 3 **Write an equation.** The last column of the table gives the equation.

interest at 4% + interest at 6% = total interest

$$.04x + .06(40,000 - x) = 2040$$

(b) A man has $34,000 to invest. He invests some at 5% and the balance at 4%. His total annual interest income is $1545. Find the amount he invests at each rate.

Step 4 **Solve** the equation. We do so without clearing decimals.

$$
\begin{aligned}
.04x + .06(40,000) - .06x &= 2040 && \text{Distributive property} \\
.04x + 2400 - .06x &= 2040 && \text{Multiply.} \\
-.02x + 2400 &= 2040 && \text{Combine like terms.} \\
-.02x &= -360 && \text{Subtract 2400.} \\
x &= 18,000 && \text{Divide by } -.02.
\end{aligned}
$$

Step 5 **State the answer.** Mark should invest $18,000 at 4%. At 6%, he should invest $40,000 - $18,000 = $22,000.

Step 6 **Check** by finding the annual interest at each rate; they should total $2040.

$.04(\$18,000) = \$720$ and $.06(\$22,000) = \$1320$

$\$720 + \$1320 = \$2040,$ as required.

Work Problem 7 at the Side.

PROBLEM-SOLVING HINT

In Example 6, we chose to let the variable represent the amount invested at 4%. Students often ask, "Can I let the variable represent the other unknown?" The answer is yes. The equation will be different, but in the end the two answers will be the same.

ANSWERS
7. (a) $50,000 at 5%; $22,000 at 3%
   (b) $18,500 at 5%; $15,500 at 4%

OBJECTIVE 7 Solve mixture problems. Mixture problems involving rates of concentration can be solved with linear equations.

EXAMPLE 7 Solving a Mixture Problem

A chemist must mix 8 L of a 40% acid solution with some 70% solution to get a 50% solution. How much of the 70% solution should be used?

Step 1 **Read** the problem. The problem asks for the amount of 70% solution to be used.

Step 2 **Assign a variable.** Let x = the number of liters of 70% solution to be used. The information in the problem is illustrated in Figure 5.

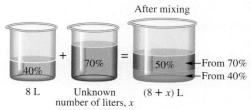

After mixing

8 L   Unknown $(8 + x)$ L
       number of liters, x

Figure 5

Use the given information to complete the following table.

Percent (as a decimal)	Number of Liters	Liters of Pure Acid
.40	8	$.40(8) = 3.2$
.70	x	$.70x$
.50	$8 + x$	$.50(8 + x)$

Sum must equal

The numbers in the right column were found by multiplying the strengths and the numbers of liters. The number of liters of pure acid in the 40% solution plus the number of liters of pure acid in the 70% solution must equal the number of liters of pure acid in the 50% solution.

Step 3 **Write an equation.**

$$3.2 + .70x = .50(8 + x)$$

Step 4 **Solve.**

$$3.2 + .70x = 4 + .50x \qquad \text{Distributive property}$$
$$.20x = .8 \qquad \text{Subtract 3.2 and .50}x.$$
$$x = 4 \qquad \text{Divide by .20.}$$

Step 5 **State the answer.** The chemist should use 4 L of the 70% solution.

Step 6 **Check.** 8 L of 40% solution plus 4 L of 70% solution is

$$8(.40) + 4(.70) = \mathbf{6\ L}$$

of acid. Similarly, $8 + 4$ or 12 L of 50% solution has

$$12(.50) = \mathbf{6\ L}$$

of acid in the mixture. The total amount of pure acid is 6 L both before and after mixing, so the answer checks.

Work Problem 8 at the Side.

8 Solve each problem.

(a) How many liters of a 10% solution should be mixed with 60 L of a 25% solution to get a 15% solution?

(b) How many pounds of candy worth $8 per lb should be mixed with 100 lb of candy worth $4 per lb to get a mixture that can be sold for $7 per lb?

ANSWERS
8. (a) 120 L **(b)** 300 lb

9 Solve each problem.

(a) How much pure acid should be added to 6 L of 30% acid to increase the concentration to 50% acid?

PROBLEM-SOLVING HINT

When pure water is added to a solution, remember that water is 0% of the chemical (acid, alcohol, etc.). Similarly, pure chemical is 100% chemical.

EXAMPLE 8 Solving a Mixture Problem When One Ingredient Is Pure

The octane rating of gasoline is a measure of its antiknock qualities. For a standard fuel, the octane rating is the percent of isooctane. How many liters of pure isooctane should be mixed with 200 L of 94% isooctane, referred to as 94 octane, to get a mixture that is 98% isooctane?

Step 1 **Read** the problem. The problem asks for the amount of pure isooctane.

Step 2 **Assign a variable.** Let $x =$ the number of liters of pure (100%) isooctane. Complete a table with the given information. Recall that $100\% = 100(.01) = 1$.

Percent (as a decimal)	Number of Liters	Liters of Pure Isooctane
1	x	x
.94	200	$.94(200)$
.98	$x + 200$	$.98(x + 200)$

Step 3 **Write an equation.** The equation comes from the last column of the table, as in Example 7.

$$x + .94(200) = .98(x + 200)$$

Step 4 **Solve.**

$x + .94(200) = .98x + .98(200)$	Distributive property
$x + 188 = .98x + 196$	Multiply.
$.02x = 8$	Subtract $.98x$ and 188.
$x = 400$	Divide by .02.

Step 5 **State the answer.** 400 L of isooctane are needed.

Step 6 **Check** by showing that $400 + .94(200) = .98(400 + 200)$.

(b) How much water must be added to 20 L of 50% antifreeze solution to reduce it to 40% antifreeze?

Work Problem 9 at the Side.

ANSWERS
9. (a) 2.4 L   (b) 5 L

2.3 Exercises

FOR EXTRA HELP

Addison-Wesley Math Tutor Center · MathXL  · Digital Video Tutor CD 1 Videotape 1 · Student's Solutions Manual · MyMathLab 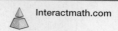 · Interactmath.com

*In each of the following, (**a**) translate as an expression and (**b**) translate as an equation or inequality. Use x to represent the number.*

1. (a) 12 more than a number _____
   (b) 12 is more than a number. _____

2. (a) 3 less than a number _____
   (b) 3 is less than a number. _____

3. (a) 4 smaller than a number _____
   (b) 4 is smaller than a number. _____

4. (a) 6 greater than a number _____
   (b) 6 is greater than a number. _____

5. Which one of the following is *not* a valid translation of "20% of a number"?

  A. $.20x$ **B.** $.2x$ **C.** $\dfrac{x}{5}$ **D.** $20x$

6. Explain why $13 - x$ is *not* a correct translation of "13 less than a number."

Translate each verbal phrase into a mathematical expression. Use x to represent the unknown number. See Example 1.

7. Twice a number, decreased by 13

8. The product of 6 and a number, decreased by 12

9. 12 increased by three times a number

10. 12 more than one-half of a number

11. The product of 8 and 12 less than a number

12. The product of 9 more than a number and 6 less than the number

13. The quotient of three times a number and 7

14. The quotient of 6 and five times a nonzero number

Use the variable x for the unknown, and write an equation representing the verbal sentence. Then solve the problem. See Example 1.

15. The sum of a number and 6 is -31. Find the number.

16. The sum of a number and -4 is 12. Find the number.

17. If the product of a number and -4 is subtracted from the number, the result is 9 more than the number. Find the number.

18. If the quotient of a number and 6 is added to twice the number, the result is 8 less than the number. Find the number.

19. When $\frac{2}{3}$ of a number is subtracted from 12, the result is 10. Find the number.

20. When 75% of a number is added to 6, the result is 3 more than the number. Find the number.

Decide whether each is an expression *or an* equation. *See Example 2.*

21. $5(x + 3) - 8(2x - 6)$

22. $-7(y + 4) + 13(y - 6)$

23. $5(x + 3) - 8(2x - 6) = 12$

24. $-7(y + 4) + 13(y - 6) = 18$

25. $\dfrac{r}{2} - \dfrac{r + 9}{6} - 8$

26. $\dfrac{r}{2} - \dfrac{r + 9}{6} = 8$

In Exercises 27 and 28, complete the six suggested problem-solving steps to solve each problem.

27. Two of the leading U.S. research universities are Massachusetts Institute of Technology (MIT) and Stanford University. In 2002, these two universities secured 230 patents on various inventions. Stanford secured 38 fewer patents than MIT. How many patents did each university secure? (*Source:* Association of University Technology Managers.)

Step 1 **Read** the problem carefully. What are you asked to find?

_____

_____

Step 2 **Assign a variable.** Let x = the number of patents MIT secured.

Then $x - 38 =$ _____

_____ .

Step 3 **Write an equation.**

_____ + _____ = 230

Step 4 **Solve** the equation.

$x =$ _____

Step 5 **State the answer.** MIT secured _____

patents, and Stanford secured _____

patents.

Step 6 **Check.** The number of Stanford

patents was _____ fewer than

the number of _____

and the total number of patents was

134 + _____ = _____ .

28. In a recent sample of book buyers, 70 more shopped at large chain bookstores than at small chain/independent bookstores. A total of 442 book buyers shopped at these two types of stores. How many buyers shopped at each type of bookstore? (*Source:* Book Industry Study Group.)

Step 1 **Read** the problem carefully. What are you asked to find?

_____

_____

Step 2 **Assign a variable.** Let x = the number of book buyers at large chain bookstores.

Then $x - 70 =$ _____

_____ .

Step 3 **Write an equation.**

_____ + _____ = 442

Step 4 **Solve** the equation.

$x =$ _____

Step 5 **State the answer.** There were _____

large chain bookstore shoppers and

_____ small chain/independent

shoppers.

Step 6 **Check.** The number of _____

_____ was

_____ more than the number of

_____ ,

and the total number of these shoppers was

256 + _____ = _____ .

Solve each problem. See Examples 3 and 4.

29. The John Hancock Center in Chicago has a rectangular base. The length of the base measures 65 ft less than twice the width. The perimeter of this base is 860 ft. What are the dimensions of the base?

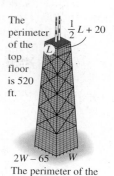

The perimeter of the top floor is 520 ft.

$\frac{1}{2}L + 20$

L

$2W - 65$ W

The perimeter of the base is 860 ft.

30. The Vietnam Veterans Memorial in Washington, D.C., is in the shape of two sides of an isosceles triangle. If the two walls of equal length were joined by a straight line of 438 ft, the perimeter of the resulting triangle would be 931.5 ft. Find the lengths of the two walls. (*Source:* Pamphlet obtained at Vietnam Veterans Memorial.)

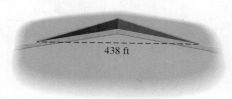

438 ft

31. The Bermuda Triangle supposedly causes trouble for aircraft pilots. It has a perimeter of 3075 mi. The shortest side measures 75 mi less than the middle side, and the longest side measures 375 mi more than the middle side. Find the lengths of the three sides.

32. The John Hancock Center (Exercise 29) tapers as it rises. The top floor is rectangular and has perimeter 520 ft. The width of the top floor measures 20 ft more than one-half its length. What are the dimensions of the top floor?

33. Galileo Galilei conducted experiments involving Italy's famous Leaning Tower of Pisa to investigate the relationship between an object's speed of fall and its weight. The Leaning Tower is 804 ft shorter than the Eiffel Tower in Paris, France. The two towers have a total height of 1164 ft. How tall is each tower? (*Source: Microsoft Encarta Encyclopedia 2002.*)

34. Mount Vernon, George Washington's home, had 327 thousand more visitors in 2003 than Monticello, Thomas Jefferson's home. Together the two presidents' homes had 1257 thousand visitors. How many visitors were there to each president's home? (*Source: USA Today*, May 14, 2004.)

35. In 2003, the New York Yankees and the New York Mets had the highest payrolls in major league baseball. The Mets' payroll was $32.8 million less than the Yankees' payroll, and the two payrolls totaled $266.6 million. What was the payroll for each team? (*Source:* Associated Press.)

36. Ted Williams and Rogers Hornsby were two great hitters. Together they got 5584 hits in their careers. Hornsby got 276 more hits than Williams. How many base hits did each get? (*Source:* Neft, D. S. and Cohen, R. M., *The Sports Encyclopedia: Baseball*, St. Martins Griffin; New York, 1997.)

Solve each percent problem. See Example 5.

37. Composite scores on the ACT exam rose from 20.6 in 1990 to 21.0 in 2001. To the nearest tenth, what percent increase was this? (*Source:* ACT, Inc.)

38. In 2001, the number of participants in the ACT exam was 1,070,000. In 1990, a total of 817,000 took the exam. To the nearest tenth, what percent increase was this? (*Source:* ACT, Inc.)

39. In 1990, the population of Miami, Florida, was 358,648. The 2002 population was 104.5% of the 1990 population. What was the 2002 population? (*Source:* U.S. Bureau of the Census.)

40. The consumer price index (CPI) in 2002 was 179.9. This represented an 18.1% increase from 1995. To the nearest tenth, what was the CPI in 1995? (*Source:* U.S. Bureau of Labor Statistics.)

41. At the end of a day, Jeff Hornsby found that the total cash register receipts at the motel where he works amounted to $2725. This included the 9% sales tax charged. Find the amount of the tax.

42. Fino Roverato sold his house for $159,000. He got this amount knowing that he would have to pay a 6% commission to his agent. What amount did he have after the agent was paid?

Solve each investment problem. See Example 6.

43. Carter Fenton earned $12,000 last year by giving tennis lessons. He invested part at 3% simple interest and the rest at 4%. He earned a total of $440 in interest. How much did he invest at each rate?

Rate (as a decimal)	Principal	Interest
.03	x	
.04		
	12,000	440

44. Linda Monroe won $60,000 on a slot machine in Las Vegas. She invested part at 2% simple interest and the rest at 3%. She earned a total of $1600 in interest. How much was invested at each rate?

Rate (as a decimal)	Principal	Interest
.02	x	
		1600

45. Julie Gasway invested some money at 4.5% simple interest and $1000 less than twice this amount at 3%. Her total annual income from the interest was $1020. How much was invested at each rate?

46. Terri Hoelker invested some money at 3.5% simple interest, and $5000 more than 3 times this amount at 4%. She earned $1440 in interest. How much did she invest at each rate?

CAUTION
It is a common error to write 300 as the distance for *each* car in Example 2. Three hundred miles is the *total* distance traveled.

As in Example 2, in general, the equation for a problem involving motion in opposite directions is of the form

partial distance + partial distance = total distance.

Work Problem 2 at the Side. ▶▶▶

EXAMPLE 3 **Solving a Motion Problem (Motion in the Same Direction)**

Jeff can bike to work in $\frac{3}{4}$ hr. By bus, the trip takes $\frac{1}{4}$ hr. If the bus travels 20 mph faster than Jeff rides his bike, how far is it to his workplace?

Step 1 **Read** the problem. We must find the distance between Jeff's home and his workplace.

Step 2 **Assign a variable.** Although the problem asks for a distance, it is easier here to let x be Jeff's speed when he rides his bike to work. Then the speed of the bus is $x + 20$. For the trip by bike,

$$d = rt = x \cdot \frac{3}{4} = \frac{3}{4}x,$$

and by bus,

$$d = rt = (x + 20) \cdot \frac{1}{4} = \frac{1}{4}(x + 20).$$

Summarize this information in a table.

	Rate	Time	Distance	
Bike	x	$\frac{3}{4}$	$\frac{3}{4}x$	⟵ Same
Bus	$x + 20$	$\frac{1}{4}$	$\frac{1}{4}(x + 20)$	⟵

Step 3 **Write an equation.** The key to setting up the correct equation is to realize that the distance in each case is the same. See Figure 7.

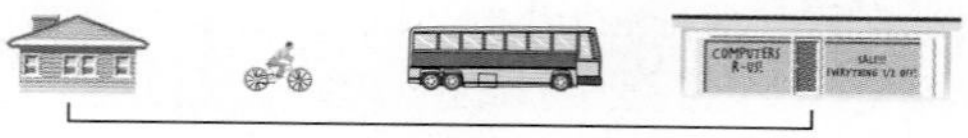

Home   Workplace

Figure 7

$$\frac{3}{4}x = \frac{1}{4}(x + 20) \qquad \text{The distance is the same.}$$

Step 4 **Solve.** $4\left(\frac{3}{4}x\right) = 4\left(\frac{1}{4}\right)(x + 20)$ Multiply by 4.

$$3x = x + 20 \qquad \text{Multiply; identity property}$$

$$2x = 20 \qquad \text{Subtract } x.$$

$$x = 10 \qquad \text{Divide by 2.}$$

Continued on Next Page

❷ Solve the problem.
Two cars leave the same location at the same time. One travels north at 60 mph and the other south at 45 mph. In how many hours will they be 420 mi apart?

ANSWERS
2. 4 hr

3 Solve the problem.

Elayn begins jogging at 5:00 A.M., averaging 3 mph. Clay leaves at 5:30 A.M., following her, averaging 5 mph. How long will it take him to catch up to her? (*Hint:* 30 min = $\frac{1}{2}$ hr.)

Step 5 **State the answer.** The required distance is given by

$$d = \frac{3}{4}x = \frac{3}{4}(10) = \frac{30}{4} = 7.5 \text{ mi.}$$

Step 6 **Check** by finding the distance using

$$d = \frac{1}{4}(x + 20) = \frac{1}{4}(10 + 20) = \frac{30}{4} = 7.5 \text{ mi,}$$

the same result.

As in Example 3, the equation for a problem involving motion in the same direction is often of the form

$$\text{one distance} = \text{other distance.}$$

> **PROBLEM-SOLVING HINT**
> In Example 3 it was easier to let the variable represent a quantity other than the one that we were asked to find. This is the case in some problems. It takes practice to learn when this approach is best, and practice means working lots of problems!

◀◀◀ Work Problem 3 at the Side.

4 Solve the problem.

One angle in a triangle is 15° larger than a second angle. The third angle is 25° larger than twice the second angle. Find the measure of each angle.

OBJECTIVE 3 **Solve problems involving the angles of a triangle.** An important result of Euclidean geometry (the geometry of the Greek mathematician Euclid) is that the sum of the angle measures of any triangle is 180°. This property is used in the next example.

EXAMPLE 4 **Finding Angle Measures**

Find the value of x, and determine the measure of each angle in Figure 8.

Step 1 **Read** the problem. We are asked to find the measure of each angle.

Step 2 **Assign a variable.** Let x represent the measure of one angle.

Step 3 **Write an equation.** The sum of the three measures shown in the figure must be 180°.

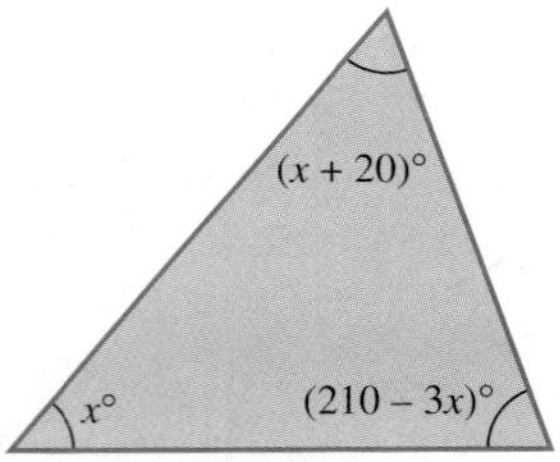

$$x + (x + 20) + (210 - 3x) = 180$$

Step 4 **Solve.**

$$-x + 230 = 180 \qquad \text{Combine like terms.}$$
$$-x = -50 \qquad \text{Subtract 230.}$$
$$x = 50 \qquad \text{Divide by } -1.$$

Figure 8 labels: $(x + 20)°$, $x°$, $(210 - 3x)°$

Figure 8

Step 5 **State the answer.** One angle measures 50°, another measures $x + 20 = 50 + 20 = 70°$, and the third measures $210 - 3x = 210 - 3(50) = 60°$.

Step 6 **Check.** Since $50° + 70° + 60° = 180°$, the answer is correct.

◀◀◀ Work Problem 4 at the Side.

ANSWERS

3. $\frac{3}{4}$ hr or 45 min

4. 35°, 50°, and 95°

2.4 Exercises

FOR EXTRA HELP

Addison-Wesley Math Tutor Center | MathXL | Digital Video Tutor CD 1 Videotape 1 | Student's Solutions Manual | MyMathLab | Interactmath.com

Solve each problem.

1. What amount of money is found in a coin hoard containing 38 nickels and 26 dimes?

2. The distance between Cape Town, South Africa, and Miami is 7700 mi. If a jet averages 480 mph between the two cities, what is its travel time in hours?

3. Tri Phong traveled from Louisville to Kansas City, a distance of 520 mi, in 10 hr. What was his rate in miles per hour?

4. A square has perimeter 40 in. What would be the perimeter of an equilateral triangle whose sides each measure the same length as the side of the square?

Write a short explanation in Exercises 5 and 6.

5. Read over Example 3 in this section. The solution of the equation is 10. Why is *10 mph* not the answer to the problem?

6. Suppose that you know that two angles of a triangle have equal measures, and the third angle measures 36°. Explain in a few words the strategy you would use to find the measures of the equal angles without actually writing an equation.

Solve each problem. See Example 1.

7. Otis Taylor has a box of coins that he uses when playing poker with his friends. The box currently contains 44 coins, consisting of pennies, dimes, and quarters. The number of pennies is equal to the number of dimes, and the total value is $4.37. How many of each denomination of coin does he have in the box?

Denomination	Number of Coins	Total Value
.01	x	$.01x$
	x	
.25		

8. Nana Nantambu found some coins while looking under her sofa pillows. There were equal numbers of nickels and quarters, and twice as many half-dollars as quarters. If she found $2.60 in all, how many of each denomination of coin did she find?

Denomination	Number of Coins	Total Value
.05	x	$.05x$
	x	
.50	$2x$	

9. Kim Falgout's daughter, Madeline, has a piggy bank with 47 coins. Some are quarters, and the rest are half-dollars. If the total value of the coins is $17.00, how many of each denomination does she have?

10. John Joslyn has a jar in his office that contains 39 coins. Some are pennies, and the rest are dimes. If the total value of the coins is $2.64, how many of each denomination does he have?

11. Dave Bowers collects U.S. gold coins. He has a collection of 41 coins. Some are $10 coins, and the rest are $20 coins. If the face value of the coins is $540, how many of each denomination does he have?

12. In the nineteenth century, the United States minted two-cent and three-cent pieces. Frances Steib has three times as many three-cent pieces as two-cent pieces, and the face value of these coins is $2.42. How many of each denomination does she have?

13. A total of 550 people attended a Kenny Loggins concert. Floor tickets cost $40 each, while balcony tickets cost $28 each. If a total of $20,800 was collected, how many of each type of ticket were sold?

14. The University of Nebraska (Omaha) production of *The Music Man* was a big success. For opening night, 410 tickets were sold. Students paid $3 each, while nonstudents paid $7 each. If a total of $1650 was collected, how many students and how many nonstudents attended?

In Exercises 15–18, find the rate based on the information provided. Round your answers to the nearest hundredth. All events were at the 2000 Summer Olympics in Sydney, Australia. (Source: http://espn.go.com/oly/summer00)

	Event	Participant	Distance	Time
15.	100-m hurdles, Women	Olga Shishigina, Kazakhstan	100 m	12.65 sec
16.	400-m hurdles, Women	Irina Privalova, Russia	400 m	53.02 sec
17.	400-m hurdles, Men	Angelo Taylor, USA	400 m	47.50 sec
18.	400-m dash, Men	Michael Johnson, USA	400 m	43.84 sec

Solve each problem. See Examples 2 and 3.

19. Two steamers leave a port on a river at the same time, traveling in opposite directions. Each is traveling 22 mph. How long will it take for them to be 110 mi apart?

	Rate	Time	Distance
First Steamer		t	
Second Steamer	22		
			110

20. A train leaves Kansas City, Kansas, and travels north at 85 km per hr. Another train leaves at the same time and travels south at 95 km per hr. How long will it take before they are 315 km apart?

	Rate	Time	Distance
First Train	85	t	
Second Train			
			315

21. Agents Mulder and Scully are driving to Georgia to investigate "Big Blue," a giant aquatic reptile reported to inhabit one of the local lakes. Mulder leaves Washington at 8:30 A.M. and averages 65 mph. His partner, Scully, leaves at 9:00 A.M., following the same path and averaging 68 mph. At what time will Scully catch up with Mulder?

	Rate	Time	Distance
Mulder			
Scully			

22. Lois and Clark are covering separate stories and have to travel in opposite directions. Lois leaves the *Daily Planet* at 8:00 A.M. and travels at 35 mph. Clark leaves at 8:15 A.M. and travels at 40 mph. At what time will they be 140 mi apart?

	Rate	Time	Distance
Lois			
Clark			

23. Latrella can get to school in 15 min if she rides her bike. It takes her 45 min if she walks. Her speed when walking is 10 mph slower than her speed when riding. What is her speed when she rides?

	Rate	Time	Distance
Riding			
Walking			

24. When Dewayne drives his car to work, the trip takes 30 min. When he rides the bus, it takes 45 min. The average speed of the bus is 12 mph less than his speed when driving. Find the distance he travels to work.

	Rate	Time	Distance
Car			
Bus			

25. Johnny leaves Memphis to visit his cousin, Anne Hoffman, in the town of Hornsby, TN, 80 mi away. He travels at an average speed of 50 mph. One-half hour later, Anne leaves to visit Johnny, traveling at an average speed of 60 mph. How long after Anne leaves will it be before they meet?

26. On an automobile trip, Aimee Cardella maintained a steady speed for the first two hours. Rush-hour traffic slowed her speed by 25 mph for the last part of the trip. The entire trip, a distance of 125 mi, took $2\frac{1}{2}$ hr. What was her speed during the first part of the trip?

Find the measure of each angle in the triangles shown. See Example 4.

27.

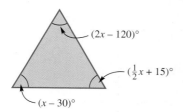

$(2x - 120)°$
$(\frac{1}{2}x + 15)°$
$(x - 30)°$

28.

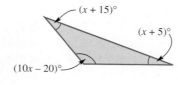

$(x + 15)°$
$(x + 5)°$
$(10x - 20)°$

29.

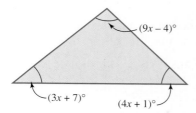

$(9x - 4)°$
$(3x + 7)°$
$(4x + 1)°$

30.

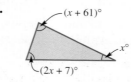

$(x + 61)°$
$x°$
$(2x + 7)°$

RELATING CONCEPTS (EXERCISES 31–34)   For Individual or Group Work

Consider the following two figures. **Work Exercises 31–34 in order.**

31. Solve for the measures of the unknown angles in Figure A.

32. Solve for the measure of the unknown angle marked $y°$ in Figure B.

33. Add the measures of the two angles you found in Exercise 31. How does the sum compare to the measure of the angle you found in Exercise 32?

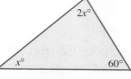
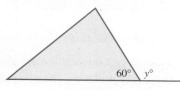

Figure A   Figure B

34. From Exercises 31–33, make a conjecture (an educated guess) about the relationship among the angles marked ①, ②, and ③ in the figure shown here.

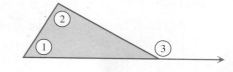

*In Exercises 35 and 36, the angles marked with variable expressions are called **vertical angles.** It is shown in geometry that vertical angles have equal measures. Find the measure of each angle.*

35.

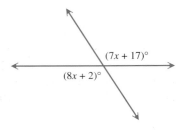

36.

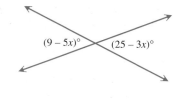

37. Two angles whose sum is equal to $90°$ are called **complementary angles.** Find the measures of the complementary angles shown in the figure.

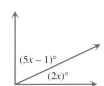

38. Two angles whose sum is equal to $180°$ are called **supplementary angles.** Find the measures of the supplementary angles shown in the figure.

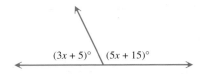

Another type of application often studied in introductory and intermediate algebra courses involves *consecutive integers.* **Consecutive integers** are integers that follow each other in counting order, such as 8, 9, and 10. Suppose we wish to solve the following problem:

Find three consecutive integers such that the sum of the first and third, increased by 3, is 50 more than the second.

Let x represent the first of the unknown integers. Then $x + 1$ will be the second, and $x + 2$ will be the third. The equation we need can be found by going back to the words of the original problem.

Sum of the first and third	increased by 3	is	50 more than the second.
↓	↓	↓	↓
$x + (x + 2)$	$+ 3$	$=$	$(x + 1) + 50$

The solution of this equation is 46, meaning that the first integer is $x = 46$, the second is $x + 1 = 47$, and the third is $x + 2 = 48$. The three integers are 46, 47, and 48. Check by substituting these numbers back into the words of the original problem.

Solve each problem involving consecutive integers.

39. Find three consecutive integers such that the sum of the first and twice the second is 17 more than twice the third.

40. Find four consecutive integers such that the sum of the first three is 54 more than the fourth.

41. If I add my current age to the age I will be next year on this date, the sum is 103 yr. How old will I be 10 years from today?

42. Two pages facing each other in this book have 189 as the sum of their page numbers. What are the two page numbers?

Summary Exercises on Solving Applied Problems

The applications that follow are of the various types introduced in this chapter. Use the strategies you have developed to solve each problem.

1. The length of a rectangle is 3 in. more than its width. If the length were decreased by 2 in. and the width were increased by 1 in., the perimeter of the resulting rectangle would be 24 in. Find the dimensions of the original rectangle.

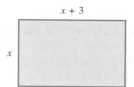

2. A farmer wishes to enclose a rectangular region with 210 m of fencing in such a way that the length is twice the width and the region is divided into two equal parts, as shown in the figure. What length and width should be used?

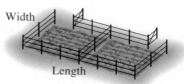

3. After a discount of 32%, the sale price for a *Harry Potter* Paperback Boxed Set (Books 1–4) by J. K. Rowling was $21.05. What was the regular price of the set of books to the nearest cent? (*Source:* amazon.com)

4. An electronics store offered a videodisc player for $255. This was the sale price, after the regular price had been discounted 40%. What was the regular price?

5. An amount of money is invested at 4% annual simple interest, and twice that amount is invested at 5%. The total annual interest is $112. How much is invested at each rate?

6. An amount of money is invested at 3% annual simple interest, and $2000 more than that amount is invested at 4%. The total annual interest is $920. How much is invested at each rate?

7. The popular television program *Frasier*, which concluded an 11-yr run on May 13, 2004, won 9 fewer than twice as many Emmy awards as *The Simpsons*. As of early 2004, the two series had won a total of 51 Emmys. How many Emmys had each series won? (*Source:* Academy of Television Arts and Sciences.)

8. As of May 2004, the two all-time top-grossing American movies were *Titanic* and *Star Wars: Episode IV—A New Hope*. *Titanic* grossed $139.8 million more than *Star Wars*. Together the two films brought in $1061.8 million. How much did each movie gross? (*Source: Variety*.)

9. Robert Cheruiyot from Kenya won the 2003 men's Boston marathon with a rate of 11.98 mph. The women's race was won by Svetlana Zakharova from Russia, who ran at 10.74 mph. Zakharova's time was .25 hr longer than Cheruiyot's. Find their winning times. (*Source: World Almanac and Book of Facts*, 2004.)

10. A newspaper recycling collection bin is in the shape of a box, 1.5 ft wide and 5 ft long. If the volume of the bin is 75 ft^3, find the height.

11. Joshua Rogers has a sheet of tin 12 cm by 16 cm. He plans to make a box by cutting equal squares out of each of the four corners and folding up the remaining edges. How large a square should he cut so that the finished box will have a length that is 5 cm less than twice the width?

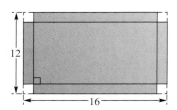

12. The perimeter of a triangle is 34 in. The middle side is twice as long as the shortest side. The longest side is 2 in. less than three times the shortest side. Find the lengths of the three sides.

13. The average cost of tuition and fees at a 4-yr public college or university during the 2001–2002 school year was $2367 more than at a 2-yr community college. Suppose a student plans to attend a local community college for 2 yr and then transfer to a public 4-yr university for 2 yr. The student expected to pay $10,250 tuition and fees for the 4 yr, assuming the 2001–2002 rates were locked in. What was the cost of tuition and fees during the 2001–2002 year at each type of school? (*Source:* National Center for Education Statistics.)

14. The sum of the smallest and largest of three consecutive integers is 32 more than the middle integer. What are the three integers?

15. Find the measure of each angle.

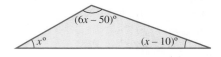

Chapter 2

SUMMARY

KEY TERMS

2.1 **linear (first-degree) equation in one variable**
A linear equation in one variable can be written in the form $Ax + B = C$, where A, B, and C are real numbers, with $A \neq 0$.

solution
A solution of an equation is a number that makes the equation true when substituted for the variable.

solution set
The solution set of an equation is the set of all its solutions.

equivalent equations
Equivalent equations are equations that have the same solution set.

conditional equation
An equation that has a finite (but nonzero) number of elements in its solution set is called a conditional equation.

contradiction
An equation that has no solution (that is, its solution set is $\emptyset$) is called a contradiction.

identity
An equation that is satisfied by every number is called an identity.

2.2 **mathematical model**
A mathematical model is an equation or inequality that describes a real situation.

formula
A formula is a mathematical equation in which variables are used to describe a relationship.

percent
One percent (1%) means "one per hundred."

2.4 **vertical angles**
Angles ① and ② shown in the figure are called vertical angles. They have equal measures.

complementary angles
Two angles whose sum is 90° are called complementary angles.

supplementary angles
Two angles whose sum is 180° are called supplementary angles.

consecutive integers
Two integers that differ by 1 are called consecutive integers.

TEST YOUR WORD POWER

See how well you have learned the vocabulary in this chapter. Answers, with examples, follow the Quick Review.

1. An **algebraic expression** is
   A. an expression that uses any of the four basic operations or the operations of raising to powers or taking roots on any collection of variables and numbers
   B. an expression that contains fractions
   C. an equation that uses any of the four basic operations or the operation of taking roots on any collection of variables and numbers
   D. an equation in algebra.

2. An **equation** is
   A. an algebraic expression
   B. an expression that contains fractions
   C. an expression that uses any of the four basic operations or the operations of raising to powers or taking roots on any collection of variables and numbers
   D. a statement that two algebraic expressions are equal.

3. A **solution set** is the set of numbers that
   A. make an expression undefined
   B. make an equation false
   C. make an equation true
   D. make an expression equal to 0.

QUICK REVIEW

Concepts	Examples

2.1 *Linear Equations in One Variable*

Addition and Multiplication Properties of Equality
The same number may be added to (or subtracted from) each side of an equation to obtain an equivalent equation. Similarly, the same nonzero number may be multiplied by or divided into each side of an equation to obtain an equivalent equation.

Solving a Linear Equation in One Variable

Step 1 Clear fractions.

Step 2 Simplify each side separately.

Step 3 Isolate the variable terms on one side.

Step 4 Isolate the variable.

Step 5 Check.

Examples

Solve the equation.

$$4(8 - 3t) = 32 - 8(t + 2)$$
$$32 - 12t = 32 - 8t - 16$$
$$32 - 12t = 16 - 8t$$
$$32 - 12t + \mathbf{12t} = 16 - 8t + \mathbf{12t}$$
$$32 = 16 + 4t$$
$$32 - \mathbf{16} = 16 + 4t - \mathbf{16}$$
$$16 = 4t$$
$$\frac{16}{4} = \frac{4t}{4}$$
$$4 = t$$

The solution set is {4}. This can be checked by substituting 4 for t in the original equation.

2.2 *Formulas*

Solving a Formula for a Specified Variable (Solving a Literal Equation)

Step 1 Transform so that all terms with the specified variable are on one side and all terms without that variable are on the other side.

Step 2 If necessary, use the distributive property to combine terms with the specified variable.

Step 3 Divide each side by the factor that is the coefficient of the specified variable.

Solve for h: $A = \dfrac{1}{2} bh$.

$$A = \frac{1}{2} bh$$
$$2A = 2\left(\frac{1}{2} bh\right) \qquad \text{Multiply by 2.}$$
$$2A = bh$$
$$\frac{2A}{b} = h \qquad \text{Divide by } b.$$

2.3 *Applications of Linear Equations*

Solving an Applied Problem

Step 1 Read the problem.

Step 2 Assign a variable.

How many liters of 30% alcohol solution and 80% alcohol solution must be mixed to obtain 100 L of 50% alcohol solution?

Let $\quad x =$ number of liters of 30% solution needed; then $100 - x =$ number of liters of 80% solution needed.

Summarize the information of the problem in a table.

Percent (as a decimal)	Liters of Solution	Liters of Pure Alcohol
.30	x	$.30x$
.80	$100 - x$	$.80(100 - x)$
.50	100	$.50(100)$

Concepts	Examples

2.3 *Applications of Linear Equations (continued)*
Step 3   Write an equation.

The equation is

$$.30x + .80(100 - x) = .50(100).$$

Step 4   Solve the equation.

The solution of the equation is 60.

Step 5   State the answer.

60 L of 30% solution and $100 - 60 = 40$ L of 80% solution are needed.

Step 6   Check.

$$.30(60) + .80(100 - 60) = 50 \text{ is true.}$$

2.4 *Further Applications of Linear Equations*
To solve a uniform motion problem, draw a sketch and make a table. Use the formula $d = rt$.

Two cars start from towns 400 mi apart and travel toward each other. They meet after 4 hr. Find the speed of each car if one travels 20 mph faster than the other.

Let $\quad x = $ speed of the slower car in miles per hour;
then $\quad x + 20 = $ speed of the faster car.

Use the information in the problem and $d = rt$ to complete the table.

	Rate	Time	Distance
Slower Car	x	4	$4x$
Faster Car	$x + 20$	4	$4(x + 20)$
		400	← Total

A sketch shows that the sum of the distances, $4x$ and $4(x + 20)$, must be 400.

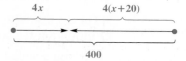

The equation is

$$4x + 4(x + 20) = 400.$$

Solving this equation gives $x = 40$. The slower car travels 40 mph, and the faster car travels $40 + 20 = 60$ mph.

Problems involving denominations of money and mixture problems are solved using methods similar to the one used for the mixture problem shown in the example above for **Section 2.3.**

ANSWERS TO TEST YOUR WORD POWER

1. A; *Examples:* $\dfrac{3y - 1}{2}, 6 + \sqrt{2x}, 4a^3b - c$

2. D; *Examples:* $2a + 3 = 7; 3y = -8, x^2 = 4$

3. C; *Example:* $\{8\}$ is the solution set of $2x + 5 = 21$.

Currency Exchange and Best Buys

When you travel to a foreign country, you can exchange dollars (in cash) for the local currency (in cash), or you can obtain cash using an ATM card, or you can charge purchases using a credit card. There are advantages and disadvantages to each method of currency exchange.

- *Currency Exchange Agency.* Travel agencies, banks, airport and rail stations, and some stores will house a currency exchange agency. You give cash in dollars in exchange for the local currency, but the agency charges different exchange rates for buying and selling dollars and will often impose commission fees.

- *ATM Cards.* You may need cash for routine purchases. You can use your ATM card to get local currency in cash from almost any bank. Typically, both your bank and the local bank will charge a fee for using the ATM. The exchange rate for that day is applied.

- *Credit Cards.* Typically, if you purchase goods in another currency, such as British pounds, then the exchange rate for the day that the transaction is recorded is applied. The credit card company will charge a fee for the currency exchange that can be very expensive.

For Group Discussion

Suppose you planned a two-week trip to England to see an English Premier League soccer match and to tour London. You budgeted $1000 for souvenirs, touring, and lunches for the entire two weeks.

The official exchange rate was $1 = £.6614.

You discovered in London that the currency exchange agency advertised the following rates.

"[The agency will] Buy pounds at 1.4509. Sell pounds at 1.6141. Fee: 2% to buy, 3% to sell, £2 minimum fee per transaction."

1. If you decided to exchange $1000 (U.S.) into British pounds at the *exchange agency*, how many British pounds would you receive (after the transaction fee)? How many British pounds did you anticipate that $1000 would have purchased, based on the exchange rate of $1 =£.6614?

2. If you decided to use your *ATM card* and withdrew £600, how much money was deducted from your U.S. account in dollars? Assume that the local bank charged a £1.50 fee and your bank charged a $1.25 fee.

3. You decide to buy an antique map for £600. You are trying to decide whether to use your *credit card* or get cash using your *ATM card*. The credit card charges a fee of 1.5% for currency exchange. Which is more economical?

4. The currency exchange agency makes a profit when changing money. Calculate the total amount (in dollars) that the agency would receive on the transaction to change $100 (U.S.) into British pounds and then to change the proceeds (after the transaction fee) back into dollars. What percent of the original $100 does the agency receive?

Chapter **2**

REVIEW EXERCISES

[2.1] *Solve each equation.*

1. $-(8 + 3x) + 5 = 2x + 6$

2. $-(r + 5) - (2 + 7r) + 8r = 3r - 8$

3. $\dfrac{m - 2}{4} + \dfrac{m + 2}{2} = 8$

4. $\dfrac{2q + 1}{3} - \dfrac{q - 1}{4} = 0$

5. $5(2x - 3) = 6(x - 1) + 4x$

6. $-3x + 2(4x + 5) = 10$

7. $-\dfrac{3}{4}x = -12$

8. $.05x + .03(1200 - x) = 42$

9. Which equation has $\{0\}$ as its solution set?

    A. $x - 5 = 5$ **B.** $4x = 5x$

    C. $x + 3 = -3$ **D.** $6x - 6 = 6$

10. Give the steps you would use to solve the equation $-2x + 5 = 7$.

Decide whether each equation is a conditional equation, *an* identity, *or a* contradiction. *Give the solution set.*

11. $7r - 3(2r - 5) + 5 + 3r = 4r + 20$

12. $8p - 4p - (p - 7) + 9p + 13 = 12p$

13. $-2r + 6(r - 1) + 3r - (4 - r) = -(r + 5) - 5$

[2.2] *Solve each formula for the indicated variable.*

14. $V = LWH$ for L

15. $A = \dfrac{1}{2}h(b + B)$ for b

16. $C = \pi d$ for d

Solve each problem.

17. A rectangular solid has a volume of 180 ft³. Its length is 6 ft and its width is 5 ft. Find its height.

18. The total number of AIDS cases reported in 1999 was 45,104. In 2000, this figure had decreased to 40,758. What percent decrease to the nearest tenth did this represent? (*Source:* U.S. Centers for Disease Control.)

19. Find the simple interest rate that Francesco Castellucio is earning, if a principal of $30,000 earns $7800 interest in 4 yr.

20. If the Fahrenheit temperature is 77°, what is the corresponding Celsius temperature?

21. The circle graph shows the projected racial composition of the U.S. workforce in the year 2006. The projected total number of people in the workforce for that year is 148,847,000. How many of these will be in the Hispanic category?

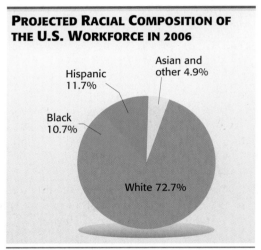

PROJECTED RACIAL COMPOSITION OF THE U.S. WORKFORCE IN 2006

Hispanic 11.7%
Asian and other 4.9%
Black 10.7%
White 72.7%

Source: U.S. Bureau of Labor Statistics.

22. The drum that Wade purchased has a circumference of 200π mm. Find the measure of its radius.

$C = 200\pi$ mm

[2.3] *Write each phrase as a mathematical expression, using x as the variable.*

23. One-third of a number, subtracted from 9

24. The product of 4 and a number, divided by 9 more than the number

Solve each problem.

25. The length of a rectangle is 3 m less than twice the width. The perimeter of the rectangle is 42 m. Find the length and width of the rectangle.

26. In a triangle with two sides of equal length, the third side measures 15 in. less than the sum of the two equal sides. The perimeter of the triangle is 53 in. Find the lengths of the three sides.

27. A candy clerk has three times as many kilograms of chocolate creams as peanut clusters. The clerk has 48 kg of the two candies altogether. How many kilograms of peanut clusters does the clerk have?

28. How many liters of a 20% solution of a chemical should be mixed with 15 L of a 50% solution to get a 30% mixture?

29. How much water should be added to 30 L of a 40% acid solution to reduce it to a 30% solution?

Percent (as a decimal)	Liters of Solution	Liters of Pure Acid
.40		
0	x	
.30		

30. Anna Mae Wood invested some money at 6% and $4000 less than this amount at 4%. Find the amount invested at each rate if her total annual interest income is $840.

Rate (as a decimal)	Principal	Interest
.06	x	
.04		

[2.4]

31. Which choice is the best *estimate* for the average speed of a trip of 405 mi that lasted 8.2 hr?

 A. 50 mph

 B. 30 mph

 C. 60 mph

 D. 40 mph

32. (a) A driver averaged 53 mph and took 10 hr to travel from Memphis to Chicago. What is the distance between Memphis and Chicago?

 (b) A small plane traveled from Warsaw to Rome, averaging 164 mph. The trip took 2 hr. What is the distance from Warsaw to Rome?

33. A passenger train and a freight train leave a town at the same time and go in opposite directions. They travel at 60 mph and 75 mph, respectively. How long will it take for them to be 297 mi apart?

	Rate	Time	Distance
Passenger Train	60	x	
Freight Train	75	x	

34. Two cars leave towns 230 km apart at the same time, traveling directly toward one another. One car travels 15 km per hr slower than the other. They pass one another 2 hr later. What are their speeds?

	Rate	Time	Distance
Faster Car	x	2	
Slower Car	$x - 15$	2	

35. An automobile averaged 45 mph for the first part of a trip and 50 mph for the second part. If the entire trip took 4 hr and covered 195 mi, for how long was the rate 45 mph?

36. An 85-mi trip to the beach took the Valenzuela family 2 hr. During the second hour, a rainstorm caused them to average 7 mph less than they traveled during the first hour. Find their average rate for the first hour.

MIXED REVIEW EXERCISES

Solve.

37. $(7 - 2k) + 3(5 - 3k) = k + 8$

38. $\dfrac{4x + 2}{4} + \dfrac{3x - 1}{8} = \dfrac{x + 6}{16}$

39. $-5(6p + 4) - 2p = -32p + 14$

40. The perimeter of a triangle is 68 in. The middle side is twice as long as the shortest side. The longest side is 4 in. less than three times the shortest side. Find the lengths of the three sides.

41. $ak + bt = 6t - sk$ for k

42. $5(2r - 3) + 7(2 - r) = 3(r + 2) - 7$

43. A square is such that if each side were increased by 4 in., the perimeter would be 8 in. less than twice the perimeter of the original square. Find the length of a side of the original square.

44. In the 2004 presidential election, John Kerry received 34 fewer electoral votes than George W. Bush. A total of 538 electoral votes were cast. How many electoral votes did each candidate receive? (*Source:* www.cnn.com)

45. Two cars start from the same point and travel in opposite directions. The car traveling west leaves 1 hr later than the car traveling east. The eastbound car travels 40 mph, and the westbound car travels 60 mph. When they are 240 mi apart, how long had each car traveled?

46. Some money is invested at 4% simple annual interest and $500 more than that amount is invested at 5%. After 1 year, a total of $142 interest was earned. How much was invested at each rate?

47. $.08x + .04(x + 200) = 188$

48. $Ax + By = C$ for x

Chapter 2

TEST

 Study Skills Workbook,
Activities 6 and 7:
Managing Time;
Test Preparation

Solve each equation.

1. $3(2x - 2) - 4(x + 6) = 4x + 8$

1. _____

2. $.08x + .06(x + 9) = 1.24$

2. _____

3. $\dfrac{x + 6}{10} + \dfrac{x - 4}{15} = 1$

3. _____

4. Decide whether each equation is a *conditional equation*, an *identity*, or a *contradiction*. Give its solution set.

(a) $3x - (2 - x) + 4x + 2 = 8x + 3$

(b) $\dfrac{x}{3} + 7 = \dfrac{5x}{6} - 2 - \dfrac{x}{2} + 9$

(c) $-4(2x - 6) = 5x + 24 - 7x$

4. (a) _____

(b) _____

(c) _____

5. Solve for v: $-16t^2 + vt - S = 0$.

5. _____

6. Solve for r: $ar + 2 = 3r - 6t$.

6. _____

Solve each problem.

7. The 2002 Daytona 500 (mile) race was won by Ward Burton, who averaged 142.971 mph. What was Burton's time, to the nearest thousandth?

7. _____

8. A certificate of deposit pays $2281.25 in simple interest for 1 year on a principal of $36,500. What is the rate of interest?

8. _____

9. Of the 38,123 offices, stations, and branches of the U.S. Postal Service in 2001, 27,876 were actually classified as post offices. What percent to the nearest tenth were classified as post offices? (*Source:* U.S. Postal Service.)

9. _____

10. _____

10. Tyler McGinnis invested some money at 3% simple interest and some at 5% simple interest. The total amount of his investments was $28,000, and the interest he earned during the first year was $1240. How much did he invest at each rate?

11. _____

11. Two cars leave from the same point at the same time, traveling in opposite directions. One travels 15 mph slower than the other. After 6 hr, they are 630 mi apart. Find the rate of each car.

12. _____

12. Find the measure of each angle.

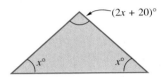

The formula

$$A = \frac{24f}{b(p + 1)}$$

gives the approximate annual interest rate for a consumer loan paid off with monthly payments. Here f is the finance charge on the loan, p is the number of payments, and b is the original amount of the loan. Use this formula to solve Problems 13 and 14.

13. _____

13. Find the approximate annual interest rate for an installment loan to be repaid in 24 monthly installments. The finance charge on the loan is $200, and the original loan balance is $1920.

14. _____

14. Find the approximate annual interest rate for an automobile loan to be repaid in 36 monthly installments. The finance charge on the loan is $740, and the amount financed is $3600. (Round to the nearest hundredth of a percent.)

15. _____

15. The circle graph shows the percents of various occupations in a representative sample of stockholders. Based on the figure, in a group of 5000 stockholders, how many would you expect to be white-collar workers?

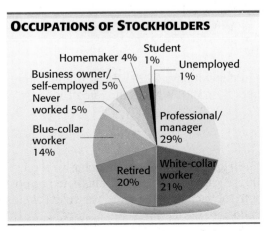

Source: Study by Peter D. Hart Research Associates for the Nasdaq Stock Market.

Cumulative Review Exercises

CHAPTERS 1–2

Let $A = \{-8, -\frac{2}{3}, -\sqrt{6}, 0, \frac{4}{5}, 9, \sqrt{36}\}$. Simplify the elements of A as necessary, and then list the elements that belong to each set listed in Exercises 1–6.

1. Natural numbers

2. Whole numbers

3. Integers

4. Rational numbers

5. Irrational numbers

6. Real numbers

Add or subtract, as indicated.

7. $-\dfrac{4}{3} - \left(-\dfrac{2}{7}\right)$

8. $|-4.2| + |5.6| - |-1.9|$

9. $(-2)^4 + (-2)^3$

10. $\sqrt{25} - \dfrac{\sqrt{100}}{2}$

Evaluate each expression.

11. $(-3)^5$

12. $\left(\dfrac{6}{7}\right)^3$

13. 4^6

14. -4^6

15. Which one of the following is not a real number: $-\sqrt{36}$ or $\sqrt{-36}$?

16. Which one of the following is undefined: $\dfrac{4-4}{4+4}$ or $\dfrac{4+4}{4-4}$?

Evaluate each expression if $a = 2$, $b = -3$, and $c = 4$.

17. $-3a + 2b - c$

18. $-2b^2 - c^2$

19. $-8(a^2 + b^3)$

20. $\dfrac{3a^3 - b}{4 + 3c}$

Simplify each expression.

21. $-7r + 5 - 13r + 12$

22. $-(3k + 8) - 2(4k - 7) + 3(8k + 12)$

Identify the property of real numbers illustrated in each equation.

23. $(a + b) + 4 = 4 + (a + b)$

24. $4x + 12x = (4 + 12)x$

25. $-9 + 9 = 0$

Solve each equation.

26. $-4x + 7(2x + 3) = 7x + 36$

27. $-\dfrac{3}{5}x + \dfrac{2}{3}x = 2$

28. $.06x + .03(100 + x) = 4.35$

29. $P = a + b + c$ for c

30. $4(2x - 6) + 3(x - 2) = 11x + 1$

31. $\dfrac{2}{3}x + \dfrac{5}{8}x = \dfrac{31}{24}x$

Solve each problem.

32. How much pure alcohol should be added to 7 L of 10% alcohol to increase the concentration to 30% alcohol?

33. A coin collection contains 29 coins. It consists of pennies, nickels, and quarters. The number of quarters is 4 fewer than the number of nickels, and the face value of the collection is $2.69. How many of each denomination are there in the collection?

34. Linda Casse invested some money at 5% simple annual interest and $2000 more than that amount at 6%. Her annual interest from the two investments totaled $670. How much did she invest at each rate?

35. Jack and Jill are running in the Fresh Water Fun Run. Jack runs at 7 mph and Jill runs at 5 mph. If they start at the same time, how long will it be before Jack is $\frac{1}{4}$ mi ahead of Jill?

36. Clark's rule, a formula used in reducing drug dosage according to weight from the recommended adult dosage to a child dosage, is

$$\dfrac{\text{weight of child in pounds}}{150} \times \text{adult dose} = \text{child's dose.}$$

Find a child's dosage if the child weighs 55 lb and the recommended adult dosage is 120 mg.

37. The body mass index, or BMI, of a person is given by the formula

$$\text{BMI} = \dfrac{704 \times (\text{weight in pounds})}{(\text{height in inches})^2}.$$

Ken Griffey, Jr., is listed as being 6 ft, 3 in. tall and weighing 205 lb. What is his BMI (to the nearest tenth)? (*Source: Reader's Digest*, October 1993.)

38. Since 1975, the number of daily newspapers has steadily declined. According to the table,

(a) by how much did the number of daily newspapers decrease between 1975 and 2000?

(b) by what *percent* to the nearest tenth did the number of daily newspapers decrease from 1975 to 2000?

Year	Number of Daily Newspapers
1975	1756
1980	1745
1985	1676
1990	1611
1995	1533
2000	1480

Source: Editor and Publisher Co.

39. Find the measure of each marked angle.

$(10x + 7)^\circ \quad (7x + 3)^\circ$

Linear Inequalities and Absolute Value

3.1 **Linear Inequalities in One Variable**

3.2 **Set Operations and Compound Inequalities**

3.3 **Absolute Value Equations and Inequalities**

Summary Exercises on Solving Linear and Absolute Value Equations and Inequalities

The cost of a college education has risen rapidly in the last decade. Average higher education tuition and fees for public institutions increased by 86% from the 1990–1991 school year to the 2001–2002 school year. (*Source:* National Center for Education Statistics, U.S. Department of Education.)

In Exercises 61–64 in Section 3.2, we apply the concepts of this chapter to college student expenses.

3.1 Linear Inequalities in One Variable

OBJECTIVES

1 Graph intervals on a number line.

2 Solve linear inequalities using the addition property.

3 Solve linear inequalities using the multiplication property.

4 Solve linear inequalities with three parts.

5 Solve applied problems using linear inequalities.

Study Skills Workbook
Activity 8: Study Cards

Solving inequalities is closely related to solving equations. In this section we introduce properties for solving inequalities.

Inequalities are algebraic expressions related by

$<$ "is less than,"

$\leq$ "is less than or equal to,"

$>$ "is greater than,"

$\geq$ "is greater than or equal to."

We solve an inequality by finding all real number solutions for it. For example, the solution set of $x \leq 2$ includes *all* real numbers that are less than or equal to 2, not just the integers less than or equal to 2. For example, -2.5, -1.7, -1, $\frac{1}{2}$, $\sqrt{2}$, $\frac{7}{4}$, and 2 are real numbers less than or equal to 2 and are therefore solutions of $x \leq 2$.

OBJECTIVE 1 Graph intervals on a number line. A good way to show the solution set of an inequality is by graphing. We graph all the real numbers satisfying $x \leq 2$ by placing a square bracket at 2 on a number line and drawing an arrow extending from the bracket to the left (to represent the fact that all numbers less than 2 are also part of the graph). The graph is shown in Figure 1.

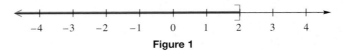

Figure 1

The set of numbers less than or equal to 2 is an example of an **interval** on the number line. To write intervals, we use **interval notation.** For example, using this notation, the interval of all numbers less than or equal to 2 is written $(-\infty, 2]$. The negative infinity symbol $-\infty$ does not indicate a number. It is used to show that the interval includes all real numbers less than 2. As on the number line, the square bracket indicates that 2 is included in the solution set. *A parenthesis is always used next to the infinity symbol.* The set of real numbers is written in interval notation as $(-\infty, \infty)$.

EXAMPLE 1 Graphing Intervals Written in Interval Notation on Number Lines

Write each inequality in interval notation and graph it.

(a) $x > -5$

The statement $x > -5$ says that x can represent any number greater than -5, but x cannot equal -5. This interval is written $(-5, \infty)$. We show this solution set on a number line by placing a parenthesis at -5 and drawing an arrow to the right, as in Figure 2. The parenthesis at -5 shows that -5 is *not* part of the graph.

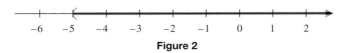

Figure 2

Continued on Next Page

(b) $-1 \leq x < 3$

This statement is read "-1 is less than or equal to x *and* x is less than 3." Thus, we want the set of numbers that are *between* -1 and 3, with -1 included and 3 excluded. In interval notation, we write the solution set as $[-1, 3)$, using a square bracket at -1 because it is part of the graph and a parenthesis at 3 because it is not part of the graph. The graph is shown in Figure 3.

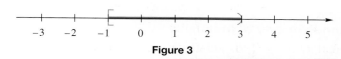

Figure 3

Work Problem 1 at the Side. ▶▶▶

We now summarize the various types of intervals.

Type of Interval	Set	Interval Notation	Graph
Open interval	$\{x \mid a < x\}$	(a, ∞)	
	$\{x \mid a < x < b\}$	(a, b)	
	$\{x \mid x < b\}$	$(-\infty, b)$	
	$\{x \mid x$ is a real number$\}$	$(-\infty, \infty)$	
Half-open interval	$\{x \mid a \leq x\}$	$[a, \infty)$	
	$\{x \mid a < x \leq b\}$	$(a, b]$	
	$\{x \mid a \leq x < b\}$	$[a, b)$	
	$\{x \mid x \leq b\}$	$(-\infty, b]$	
Closed interval	$\{x \mid a \leq x \leq b\}$	$[a, b]$	

An **inequality** says that two expressions are *not* equal. Solving inequalities is similar to solving equations.

Linear Inequality

A **linear inequality in one variable** can be written in the form

$$Ax + B < C,$$

where A, B, and C are real numbers, with $A \neq 0$.

(Throughout this section we give definitions and rules only for $<$, but they are also valid for $>$, $\leq$, and $\geq$.) Examples of linear inequalities include

$$x + 5 < 2, \quad t - 3 \geq 5, \quad \text{and} \quad 2k + 5 \leq 10. \quad \text{Linear inequalities}$$

❶ Write each inequality in interval notation and graph it.

(a) $x < -1$

(b) $x \geq -3$

(c) $-4 \leq x < 2$

ANSWERS
1. **(a)** $(-\infty, -1)$

$-4\,-3\,-2\,-1\ 0\ 1\ 2$

(b) $[-3, \infty)$

$-4\ -3\ -2\ -1\ 0$

(c) $[-4, 2)$

$-6\,-4\,-2\ 0\ 2\ 4$

2 Solve each inequality, check your solutions, and graph the solution set.

(a) $p + 6 < 8$

(b) $8x < 7x - 6$

OBJECTIVE 2 Solve linear inequalities using the addition property. We solve an inequality by finding all numbers that make the inequality true. Usually, an inequality has an infinite number of solutions. These solutions, like solutions of equations, are found by producing a series of simpler equivalent inequalities. **Equivalent inequalities** are inequalities with the same solution set. We use the **addition** and **multiplication properties of inequality** to produce equivalent inequalities.

Addition Property of Inequality

For all real numbers A, B, and C, the inequalities

$$A < B \quad \text{and} \quad A + C < B + C$$

are equivalent.

In words, adding the same number to each side of an inequality does not change the solution set.

EXAMPLE 2 Using the Addition Property of Inequality

Solve $x - 7 < -12$, and graph the solution set.

$$x - 7 < -12$$
$$x - 7 + 7 < -12 + 7 \qquad \text{Add 7.}$$
$$x < -5$$

Check: Substitute -5 for x in the *equation* $x - 7 = -12$. The result should be a true statement.

$$x - 7 = -12$$
$$-5 - 7 = -12 \qquad ? \quad \text{Let } x = -5.$$
$$-12 = -12 \qquad \text{True}$$

This shows that -5 is the boundary point. Now we test a number on each side of -5 to verify that numbers *less than* -5 make the *inequality* true. We choose -4 and -6.

$$x - 7 < -12$$

$-4 - 7 < -12$? Let $x = -4$.	$-6 - 7 < -12$? Let $x = -6$.
$-11 < -12$ False	$-13 < -12$ True
-4 is not in the solution set.	-6 is in the solution set.

The check confirms that $(-\infty, -5)$, graphed in Figure 4, is the solution set.

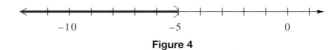

Figure 4

◄◄◄ Work Problem 2 at the Side.

As with equations, the addition property of inequality can be used to *subtract* the same number from each side of an inequality. For example, to solve the inequality $x + 4 > 10$, we subtract 4 from each side to get $x > 6$.

ANSWERS
2. (a) $(-\infty, 2)$

$-3\ -2\ -1\ \ 0\ \ 1\ \ 2\ \ 3$

(b) $(-\infty, -6)$

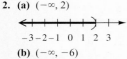

$-10\ -9\ -8\ -7\ -6\ -5\ -4$

EXAMPLE 3 **Using the Addition Property of Inequality**

Solve $14 + 2m \leq 3m$, and graph the solution set.

$$14 + 2m \leq 3m$$
$$14 + 2m - 2m \leq 3m - 2m \qquad \text{Subtract } 2m.$$
$$14 \leq m \qquad \text{Combine like terms.}$$

The inequality $14 \leq m$ (14 is less than or equal to m) can also be written $m \geq 14$ (m is greater than or equal to 14). Notice that in each case, the inequality symbol points to the lesser number, 14.

Check:
$$14 + 2m = 3m$$
$$14 + 2(14) = 3(14) \quad ? \qquad \text{Let } m = 14.$$
$$42 = 42 \qquad \text{True}$$

So 14 satisfies the equality part of $\leq$. Choose 10 and 15 as test points.

$$14 + 2m < 3m$$

$14 + 2(10) < 3(10)$? Let $m = 10$.	$14 + 2(15) < 3(15)$? Let $m = 15$.
$34 < 30$ False	$44 < 45$ True
10 is not in the solution set.	15 is in the solution set.

The check confirms that $[14, \infty)$ is the solution set. See Figure 5.

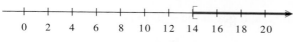

Figure 5

Work Problem 3 at the Side.

CAUTION
To avoid errors, rewrite an inequality such as $14 \leq m$ as $m \geq 14$ so that the variable is on the left, as in Example 3.

OBJECTIVE 3 **Solve linear inequalities using the multiplication property.** Solving an inequality such as $3x \leq 15$ requires dividing each side by 3 using the multiplication property of inequality. To see how this property works, start with the true statement

$$-2 < 5.$$

Multiply each side by, say, 8.

$$-2(8) < 5(8) \qquad \text{Multiply by 8.}$$
$$-16 < 40 \qquad \text{True}$$

The result is true. Start again with $-2 < 5$, and multiply each side by -8.

$$-2(-8) < 5(-8) \qquad \text{Multiply by } -8.$$
$$16 < -40 \qquad \text{False}$$

The result, $16 < -40$, is false. To make it true, we must change the direction of the inequality symbol to get

$$16 > -40. \qquad \text{True}$$

Work Problem 4 at the Side.

---

3 Solve $2k - 5 \geq 1 + k$, check, and graph the solution set.

4 Multiply both sides of each inequality by -5. Then insert the correct symbol, either $<$ or $>$, in the first blank, and fill in the other blanks in part (b).

(a) $7 < 8$

$$-35 \underline{\hspace{2cm}} -40$$

(b) $-1 > -4$

$$5 \underline{\hspace{2cm}} \underline{\hspace{2cm}}$$

ANSWERS
3. $[6, \infty)$

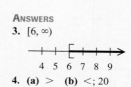

4. (a) $>$ **(b)** $<$; 20

5 Solve each inequality, check, and graph the solution set.

(a) $2x < -10$

(b) $-7k \geq 8$

(c) $-9m < -81$

As these examples suggest, multiplying each side of an inequality by a *negative* number reverses the direction of the inequality symbol. The same is true for dividing by a negative number since division is defined in terms of multiplication.

Multiplication Property of Inequality

For all real numbers A, B, and C, with $C \neq 0$,
(a) the inequalities

$$A < B \quad \text{and} \quad AC < BC$$

are equivalent **if $C > 0$;**
(b) the inequalities

$$A < B \quad \text{and} \quad AC > BC$$

are equivalent **if $C < 0$.**

In words, each side of an inequality may be multiplied (or divided) by a *positive* number without changing the direction of the inequality symbol. *Multiplying (or dividing) by a negative number requires that we reverse the inequality symbol.*

EXAMPLE 4 Using the Multiplication Property of Inequality

Solve each inequality, and graph the solution set.

(a) $5m \leq -30$

Use the multiplication property to divide each side by 5. *Since $5 > 0$, do not reverse the inequality symbol.*

$$5m \leq -30$$

$$\frac{5m}{5} \leq \frac{-30}{5} \qquad \text{Divide by 5.}$$

$$m \leq -6$$

Check that the solution set is the interval $(-\infty, -6]$, graphed in Figure 6.

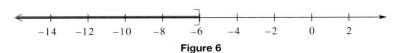

Figure 6

(b) $-4k \leq 32$

Divide each side by -4. *Since $-4 < 0$, reverse the inequality symbol.*

$$-4k \leq 32$$

$$\frac{-4k}{-4} \geq \frac{32}{-4} \qquad \text{Divide by } -4 \text{ and reverse the symbol.}$$

$$k \geq -8$$

Check the solution set. Figure 7 shows the graph of the solution set, $[-8, \infty)$.

Figure 7

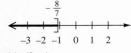
Work Problem 5 at the Side.

ANSWERS
5. **(a)** $(-\infty, -5)$

$-8\ -7\ -6\ -5\ -4$

(b) $\left(-\infty, -\frac{8}{7}\right]$

$-\frac{8}{7}$

$-3\ -2\ -1\ \ 0\ \ 1\ \ 2$

(c) $(9, \infty)$

$7\ \ 8\ \ 9\ \ 10\ \ 11\ \ 12$

The steps used in solving a linear inequality are given below.

Solving a Linear Inequality

Step 1 **Simplify each side separately.** Use the distributive property to clear parentheses and combine like terms as needed.

Step 2 **Isolate the variable terms on one side.** Use the addition property of inequality to get all terms with variables on one side of the inequality and all numbers on the other side.

Step 3 **Isolate the variable.** Use the multiplication property of inequality to change the inequality to the form $x < k$ or $x > k$.

CAUTION
Reverse the direction of the inequality symbol only when multiplying or dividing each side of an inequality by a negative number.

EXAMPLE 5 **Solving a Linear Inequality Using the Distributive Property**

Solve $-3(x + 4) + 2 \geq 7 - x$, and graph the solution set.

Step 1 $-3x - 12 + 2 \geq 7 - x$ Distributive property

$\qquad\quad -3x - 10 \geq 7 - x$

Step 2 $-3x - 10 + x \geq 7 - x + x$ Add x.

$\qquad\qquad -2x - 10 \geq 7$

$\qquad\quad -2x - 10 + \mathbf{10} \geq 7 + \mathbf{10}$ Add 10.

$\qquad\qquad\quad -2x \geq 17$

Step 3 $\dfrac{-2x}{-2} \leq \dfrac{17}{-2}$ Divide by -2; change $\geq$ to $\leq$.

$\qquad\qquad x \leq -\dfrac{17}{2}$

Figure 8 shows the graph of the solution set, $\left(-\infty, -\frac{17}{2}\right]$.

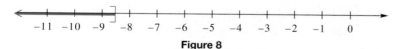

Figure 8

NOTE
In Example 5, if, after distributing, we add $3x$ to both sides of the inequality, we have

$$-3x - 10 + 3x \geq 7 - x + 3x \qquad \text{Add } 3x.$$

$$-10 \geq 2x + 7$$

$$-10 - 7 \geq 2x + 7 - 7 \qquad \text{Subtract 7.}$$

$$-17 \geq 2x$$

$$-\frac{17}{2} \geq x. \qquad \text{Divide by 2.}$$

The result "$-\frac{17}{2}$ is greater than or equal to x" means the same thing as "x is less than or equal to $-\frac{17}{2}$." Thus, the solution set is the same.

6 Solve, check, and graph the solution set of each inequality.

(a) $5 - 3(m - 1)$
$\leq 2(m + 3) + 1$

(b) $\frac{1}{4}(m + 3) + 2 \leq \frac{3}{4}(m + 8)$

EXAMPLE 6 Solving a Linear Inequality with Fractions

Solve $-\frac{2}{3}(r - 3) - \frac{1}{2} < \frac{1}{2}(5 - r)$, and graph the solution set.

To clear fractions, multiply each side by the least common denominator, 6.

$$-\frac{2}{3}(r - 3) - \frac{1}{2} < \frac{1}{2}(5 - r)$$

$$6\left[-\frac{2}{3}(r - 3) - \frac{1}{2}\right] < 6\left[\frac{1}{2}(5 - r)\right] \qquad \text{Multiply by 6.}$$

$$6\left[-\frac{2}{3}(r - 3)\right] - 6\left(\frac{1}{2}\right) < 6\left[\frac{1}{2}(5 - r)\right] \qquad \text{Distributive property}$$

$$-4(r - 3) - 3 < 3(5 - r)$$

Step 1 $\qquad -4r + 12 - 3 < 15 - 3r \qquad$ Distributive property

$$-4r + 9 < 15 - 3r$$

Step 2 $\qquad -4r + 9 + 3r < 15 - 3r + 3r \qquad$ Add $3r$.

$$-r + 9 < 15$$

$$-r + 9 - 9 < 15 - 9 \qquad \text{Subtract 9.}$$

$$-r < 6$$

Step 3 $\qquad -1(-r) > -1(6) \qquad$ Multiply by -1; change $<$ to $>$.

$$r > -6$$

Check that the solution set is $(-6, \infty)$. See Figure 9.

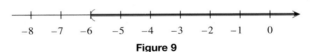

Figure 9

◀◀◀ Work Problem 6 at the Side.

OBJECTIVE 4 Solve linear inequalities with three parts.   For some applications, it is necessary to work with an inequality such as

$$3 < x + 2 < 8,$$

where $x + 2$ is *between* 3 and 8. To solve this inequality, we subtract 2 from each of the three parts of the inequality, giving

$$3 - 2 < x + 2 - 2 < 8 - 2$$

$$1 < x < 6.$$

Thus, x must be between 1 and 6 so that $x + 2$ will be between 3 and 8. The solution set, $(1, 6)$, is graphed in Figure 10.

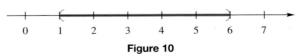

Figure 10

CAUTION
When inequalities have three parts, the order of the parts is important.
It would be *wrong* to write an inequality as $8 < x + 2 < 3$, since this would imply that $8 < 3$, a false statement. In general, three-part inequalities are written so that the symbols point in the same direction, and both point toward the lesser number.

ANSWERS

6. (a) $\left[\frac{1}{5}, \infty\right)$

(b) $\left[-\frac{13}{2}, \infty\right)$

EXAMPLE 7 **Solving a Three-Part Inequality**

Solve $-2 \le -3k - 1 \le 5$, and graph the solution set.

Begin by adding 1 to each of the three parts to isolate the variable term in the middle.

$$-2 + 1 \le -3k - 1 + 1 \le 5 + 1 \qquad \text{Add 1 to each part.}$$

$$-1 \le -3k \le 6$$

$$\frac{-1}{-3} \ge \frac{-3k}{-3} \ge \frac{6}{-3} \qquad \text{Divide each part by } -3; \text{ reverse the inequality symbols.}$$

$$\frac{1}{3} \ge k \ge -2$$

$$-2 \le k \le \frac{1}{3} \qquad \text{Rewrite in order based on the number line.}$$

Check that the solution set is $\left[-2, \frac{1}{3}\right]$, as shown in Figure 11.

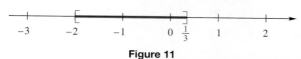

Figure 11

Work Problem 7 at the Side. ▶▶▶

Examples of the types of solution sets to be expected from solving linear equations and linear inequalities are shown below.

SOLUTIONS OF LINEAR EQUATIONS AND INEQUALITIES

Equation or Inequality	Typical Solution Set	Graph of Solution Set
Linear equation $5x + 4 = 14$	$\{2\}$	●―――→ at 2
Linear inequality $5x + 4 < 14$ or $5x + 4 > 14$	$(-\infty, 2)$ $(2, \infty)$	←―――) at 2 ; (―――→ at 2
Three-part inequality $-1 \le 5x + 4 \le 14$	$[-1, 2]$	[――] from −1 to 2

OBJECTIVE 5 **Solve applied problems using linear inequalities.** In addition to the familiar "is less than" and "is greater than," other expressions such as "is no more than" and "is at least" also indicate inequalities. The table below shows how to interpret these expressions.

Word Expression	Interpretation
a is at least b	$a \ge b$
a is no less than b	$a \ge b$
a is at most b	$a \le b$
a is no more than b	$a \le b$

7 Solve, check, and graph the solution set of each inequality.

(a) $-3 \le x - 1 \le 7$

_____

(b) $5 < 3x - 4 < 9$

_____

ANSWERS

7. **(a)** $[-2, 8]$

[――――] from −2 to 8

$-4\ -2\ \ 0\ \ 2\ \ 4\ \ 6\ \ 8\ \ 10$

(b) $\left(3, \frac{13}{3}\right)$

$\frac{13}{3}$

(――) from 3 to $\frac{13}{3}$

$2\ \ 3\ \ 4\ \ 5$

8 Solve the problem.

A rental company charges $5 to rent a leaf blower, plus $1.75 per hr. Dona Kenly can spend no more than $26 to blow leaves from her driveway and pool deck. What is the *maximum* amount of time she can use the rented leaf blower?

9 Solve the problem.

Wade has grades of 92, 90, and 84 on his first three tests. What grade must he make on his fourth test in order to keep an average of at least 90?

In Examples 8 and 9, we use the six problem-solving steps from **Section 2.3,** changing Step 3 to "Write an inequality" instead of "Write an equation."

EXAMPLE 8 **Using a Linear Inequality to Solve a Rental Problem**

A rental company charges $15 to rent a chain saw, plus $2 per hr. Jay Jenkins can spend no more than $35 to clear some logs from his yard. What is the *maximum* amount of time he can use the rented saw?

Step 1 **Read** the problem again.

Step 2 **Assign a variable.** Let h = the number of hours he can rent the saw.

Step 3 **Write an inequality.** He must pay $15, plus $2h$, to rent the saw for h hours, and this amount must be *no more than* $35.

$$\underbrace{15 + 2h}_{\text{Cost of renting}} \quad \underbrace{\leq}_{\text{is no more than}} \quad \underbrace{35}_{\text{35 dollars.}}$$

Step 4 **Solve.**
$$2h \leq 20 \qquad \text{Subtract 15.}$$
$$h \leq 10 \qquad \text{Divide by 2.}$$

Step 5 **State the answer.** He can use the saw for a maximum of 10 hr. (He may use it for less time, as indicated by the inequality $h \leq 10$.)

Step 6 **Check.** If Jay uses the saw for **10** hr, he will spend $15 + 2(\mathbf{10}) = 35$ dollars, the maximum amount.

◀◀ Work Problem 8 at the Side.

EXAMPLE 9 **Finding an Average Test Score**

Helen has scores of 88, 86, and 90 on her first three algebra tests. An average score of at least 90 will earn an A in the class. What possible scores on her fourth test will earn her an A average?

Let x represent the score on the fourth test. Her average score must be at least 90. To find the average of four numbers, add them and then divide by 4.

$$\underbrace{\frac{88 + 86 + 90 + x}{4}}_{\text{Average}} \quad \underbrace{\geq}_{\text{is at least}} \quad \underbrace{90}_{\text{90.}}$$

$$\frac{264 + x}{4} \geq 90 \qquad \text{Add the scores.}$$
$$264 + x \geq 360 \qquad \text{Multiply by 4.}$$
$$x \geq 96 \qquad \text{Subtract 264.}$$

She must score **96** or more on her fourth test.

Check: $\dfrac{88 + 86 + 90 + \mathbf{96}}{4} = \dfrac{360}{4} = 90$

A score of 96 or more will give an average of at least 90, as required.

◀◀ Work Problem 9 at the Side.

ANSWERS
8. 12 hr
9. at least 94

3.1 Exercises

FOR EXTRA HELP

Tutor Center — Addison-Wesley Math Tutor Center

Math XL — MathXL

Digital Video Tutor CD 2 Videotape 2

Student's Solutions Manual

MyMathLab — MathLab

Interactmath.com

Match each inequality with the correct graph or interval notation.

1. $x \leq 3$

A.

2. $x > 3$

B. (number line with open bracket at 3, marked 0 and 3)

3. $x < 3$

C. $(3, \infty)$

4. $x \geq 3$

D. $(-\infty, 3]$

5. $-3 \leq x \leq 3$

E. $(-3, 3)$

6. $-3 < x < 3$

F. $[-3, 3]$

7. Explain how you will determine whether to use parentheses or brackets when graphing the solution set of an inequality.

8. Describe the steps used to solve a linear inequality. Explain when it is necessary to reverse the inequality symbol.

Solve each inequality, giving its solution set in both interval and graph forms. Check your answers. See Examples 1–6.

9. $4x + 1 \geq 21$

10. $5t + 2 \geq 52$

11. $\dfrac{3k - 1}{4} > 5$

12. $\dfrac{5z - 6}{8} < 8$

13. $-4x < 16$

14. $-2m > 10$

15. $-\dfrac{3}{4}r \geq 30$

16. $-\dfrac{2}{3}x \leq 12$

17. $-1.3m \geq -5.2$

18. $-2.5x \leq -1.25$

19. $\dfrac{2k-5}{-4} > 5$

20. $\dfrac{3z-2}{-5} < 6$

21. $x + 4(2x - 1) \geq x$

22. $m - 2(m - 4) \leq 3m$

23. $-(4 + r) + 2 - 3r < -14$

24. $-(9 + k) - 5 + 4k \geq 4$

25. $-3(z - 6) > 2z - 2$

26. $-2(x + 4) \leq 6x + 16$

27. $\frac{2}{3}(3k - 1) \geq \frac{3}{2}(2k - 3)$

28. $\frac{7}{5}(10m - 1) < \frac{2}{3}(6m + 5)$

29. $-\frac{1}{4}(p + 6) + \frac{3}{2}(2p - 5) < 10$

30. $\frac{3}{5}(k - 2) - \frac{1}{4}(2k - 7) \leq 3$

RELATING CONCEPTS (EXERCISES 31–35)   For Individual or Group Work

Work Exercises 31–35 in order.

31. Solve the linear equation

$$5(x + 3) - 2(x - 4) = 2(x + 7),$$

and graph the solution set on a number line.

32. Solve the linear inequality

$$5(x + 3) - 2(x - 4) > 2(x + 7),$$

and graph the solution set on a number line.

33. Solve the linear inequality

$$5(x + 3) - 2(x - 4) < 2(x + 7),$$

and graph the solution set on a number line.

34. Graph all the solution sets of the equation and inequalities in Exercises 31–33 on the same number line. What set do you obtain?

35. Based on the results of Exercises 31–33, complete the following using a conjecture (educated guess): The solution set of

$$-3(x + 2) = 3x + 12$$

is $\{-3\}$, and the solution set of

$$-3(x + 2) < 3x + 12$$

is $(-3, \infty)$. Therefore the solution set of

$$-3(x + 2) > 3x + 12$$

is _____.

36. Which is the graph of $-2 < x$?

A. [number line with parenthesis at -2 opening right]

B. [number line with parenthesis at -2]

C. [number line with bracket at -2 opening right]

D. [number line with bracket at -2]

Solve each inequality, giving its solution set in both interval and graph forms. Check your answers. See Example 7.

37. $-4 < x - 5 < 6$

38. $-1 < x + 1 < 8$

39. $-9 \leq k + 5 \leq 15$

40. $-4 \leq m + 3 \leq 10$

41. $-6 \leq 2(z + 2) \leq 16$

42. $-15 < 3(p + 2) < -12$

43. $-16 < 3t + 2 < -10$

44. $-1 \leq \dfrac{2x - 5}{6} \leq 5$

45. $-3 \leq \dfrac{3m + 1}{4} \leq 3$

The weather forecast by time of day for one day of the U.S. Olympic Track and Field Trials, in Sacramento, California, is shown in the figure. Use this graph to work Exercises 46–49.

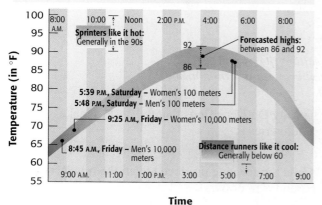

TRACKING THE HEAT
The forecast for the U.S. Olympic Track and Field Trials, by time of day. (Average temperature this time of year is a high of 93.5, low of 60.5.)

Source: Accuweather, Bee research.

46. Sprinters prefer Fahrenheit temperatures in the 90s. Using the upper boundary of the forecast, in what time period is the temperature expected to be at least 90°F?

47. Distance runners prefer cool temperatures. During what time period are temperatures predicted to be no more than 70°F? Use the lower forecast boundary.

48. What range of temperatures is predicted for the Women's 100-m event?

49. What range of temperatures is forecast for the Men's 10,000-m event?

Solve each problem. See Examples 8 and 9.

50. Amanda Pepper earned scores of 90 and 82 on her first two tests in English Literature. What score must she make on her third test to keep an average of 84 or greater?

51. Finley Westmoreland scored 92 and 96 on his first two tests in Methods in Teaching Mathematics. What score must he make on his third test to keep an average of 90 or greater?

52. A couple wishes to rent a car for one day while on vacation. Ford Automobile Rental wants $15.00 per day and 14¢ per mi, while Chevrolet-For-A-Day wants $14.00 per day and 16¢ per mi. After how many miles would the price to rent the Chevrolet exceed the price to rent a Ford?

53. Jane and Jim Saska went to Mobile, Alabama, for a week. They needed to rent a car, so they checked out two rental firms. Avis wanted $28 per day, with no mileage fee. Downtown Toyota wanted $108 per week and 14¢ per mi. How many miles would they have to drive before the Avis price is less than the Toyota price?

A product will produce a profit only when the revenue (R) from selling the product exceeds the cost (C) of producing it. Find the smallest whole number of units x that must be sold for each business to show a profit for the item described.

54. Peripheral Visions, Inc. finds that the cost to produce x studio-quality videotapes is

$$C = 20x + 100,$$

while the revenue produced from them is $R = 24x$ (C and R in dollars).

55. Speedy Delivery finds that the cost to make x deliveries is

$$C = 3x + 2300,$$

while the revenue produced from them is $R = 5.50x$ (C and R in dollars).

56. A BMI (body mass index) between 19 and 25 is considered healthy. Use the formula

$$BMI = \frac{704 \times (\text{weight in pounds})}{(\text{height in inches})^2}$$

to find the weight range w, to the nearest pound, that gives a healthy BMI for each height. (*Source: Washington Post.*)

(a) 72 in. **(b)** Your height in inches

57. To achieve the maximum benefit from exercising, the heart rate in beats per minute should be in the target heart rate zone (*THR*). For a person aged A, the formula is

$$.7(220 - A) \leq THR \leq .85(220 - A).$$

Find the *THR* to the nearest whole number for each age. (*Source:* Hockey, Robert V., *Physical Fitness: The Pathway to Healthful Living*, Times Mirror/Mosby College Publishing, 1989.)

(a) 35 **(b)** Your age

Find the unknown numbers in each description.

58. Six times a number is between -12 and 12.

59. Half a number is between -3 and 2.

60. When 1 is added to twice a number, the result is greater than or equal to 7.

61. If 8 is subtracted from a number, then the result is at least 5.

62. One third of a number is added to 6, giving a result of at least 3.

63. Three times a number, minus 5, is no more than 7.

3.2 Set Operations and Compound Inequalities

The table shows symptoms of an overactive thyroid and an underactive thyroid.

Underactive Thyroid	Overactive Thyroid
Sleepiness, s	Insomnia, i
Dry hands, d	Moist hands, m
Intolerance of cold, c	Intolerance of heat, h
Goiter, g	Goiter, g

Source: The Merck Manual of Diagnosis and Therapy,
16th Edition, Merck Research Laboratories, 1992.

Let N be the set of symptoms for an underactive thyroid, and let O be the set of symptoms for an overactive thyroid. Suppose we are interested in the set of symptoms that are found in *both* sets N and O. In this section we discuss the use of the words *and* and *or* as they relate to sets and inequalities.

OBJECTIVE 1 Find the intersection of two sets.   The intersection of two sets is defined using the word *and*.

Intersection of Sets

For any two sets A and B, the **intersection** of A and B, symbolized $A \cap B$, is defined as follows:

$$A \cap B = \{x \mid x \text{ is an element of } A \text{ and } x \text{ is an element of } B\}.$$

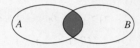

EXAMPLE 1 Finding the Intersection of Two Sets

Let $A = \{1, 2, 3, 4\}$ and $B = \{2, 4, 6\}$. Find $A \cap B$.
   The set $A \cap B$ contains those elements that belong to both A *and* B: the numbers 2 and 4. Therefore,

$$A \cap B = \{1, 2, 3, 4\} \cap \{2, 4, 6\} = \{2, 4\}.$$

Work Problem 1 at the Side. ▶▶▶

A **compound inequality** consists of two inequalities linked by a connective word such as *and* or *or*. Examples of compound inequalities are

$$x + 1 \leq 9 \quad \text{and} \quad x - 2 \geq 3$$

Compound inequalities

and $\qquad 2x > 4 \quad \text{or} \quad 3x - 6 < 5.$

OBJECTIVE 2 Solve compound inequalities with the word *and*.   Use the following steps.

Solving a Compound Inequality with *and*

Step 1   Solve each inequality in the compound inequality individually.

Step 2   Since the inequalities are joined with *and*, the solution set of the compound inequality will include all numbers that satisfy both inequalities in Step 1 (the intersection of the solution sets).

OBJECTIVES

1. Find the intersection of two sets.
2. Solve compound inequalities with the word *and*.
3. Find the union of two sets.
4. Solve compound inequalities with the word *or*.

1 List the elements in each set.

(a) $A \cap B$, if $A = \{3, 4, 5, 6\}$ and $B = \{5, 6, 7\}$

(b) $N \cap O$ (Refer to the thyroid table.)

ANSWERS
1. (a) $\{5, 6\}$ (b) $\{g\}$

② Solve each compound inequality, and graph the solution set.

(a) $x < 10$ and $x > 2$

(b) $x + 3 \leq 1$ and $x - 4 \geq -12$

EXAMPLE 2 Solving a Compound Inequality with *and*

Solve the compound inequality $x + 1 \leq 9$ and $x - 2 \geq 3$.

Step 1 Solve each inequality in the compound inequality individually.

$$x + 1 \leq 9 \quad \text{and} \quad x - 2 \geq 3$$
$$x + 1 - 1 \leq 9 - 1 \quad \text{and} \quad x - 2 + 2 \geq 3 + 2$$
$$x \leq 8 \quad \text{and} \quad x \geq 5$$

Step 2 Because the inequalities are joined with the word *and*, the solution set will include all numbers that satisfy both inequalities in Step 1 at the same time. Thus, the compound inequality is true whenever $x \leq 8$ and $x \geq 5$ are both true. The top graph in Figure 12 shows $x \leq 8$, and the bottom graph shows $x \geq 5$.

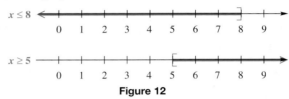

Figure 12

Find the intersection of the two graphs in Figure 12 to get the solution set of the compound inequality. The intersection of the two graphs in Figure 13 shows that the solution set in interval notation is $[5, 8]$.

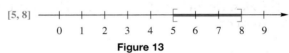

Figure 13

◀◀◀ Work Problem 2 at the Side.

---

③ Solve
$2x \geq x - 1$ and $3x \geq 3 + 2x$, and graph the solution set.

EXAMPLE 3 Solving a Compound Inequality with *and*

Solve the compound inequality $-3x - 2 > 5$ and $5x - 1 \leq -21$.

Step 1 Solve each inequality separately.

$$-3x - 2 > 5 \quad \text{and} \quad 5x - 1 \leq -21$$
$$-3x > 7 \quad \text{and} \quad 5x \leq -20$$
$$x < -\frac{7}{3} \quad \text{and} \quad x \leq -4$$

The graphs of $x < -\frac{7}{3}$ and $x \leq -4$ are shown in Figure 14.

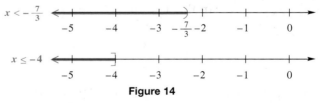

Figure 14

Step 2 Now find all values of x that satisfy both conditions; that is, the real numbers that are less than $-\frac{7}{3}$ and also less than or equal to -4. As shown by the graph in Figure 15, the solution set is $(-\infty, -4]$.

Figure 15

◀◀◀ Work Problem 3 at the Side.

ANSWERS
2. **(a)** $(2, 10)$

+‐(‐+‐+‐+‐+‐)‐+→
0 2 4 6 8 10 12

(b) $[-8, -2]$

+‐[‐+‐+‐]‐+→
−10 −8 −6 −4 −2 0

3. $[3, \infty)$

+‐+‐+‐+‐+‐[‐+‐+→
−1 0 1 2 3 4 5

EXAMPLE 4 **Solving a Compound Inequality with *and***

Solve $x + 2 < 5$ and $x - 10 > 2$.
First solve each inequality separately.

$$x + 2 < 5 \quad \text{and} \quad x - 10 > 2$$
$$x < 3 \quad \text{and} \quad x > 12$$

The graphs of $x < 3$ and $x > 12$ are shown in Figure 16.

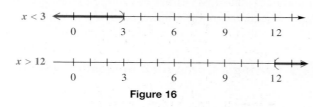

Figure 16

There is no number that is both less than 3 *and* greater than 12, so the given compound inequality has no solution. The solution set is $\emptyset$. See Figure 17.

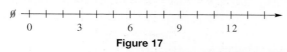

Figure 17

Work Problem 4 at the Side. ▶▶▶

OBJECTIVE **3** **Find the union of two sets.** The union of two sets is defined using the word *or*.

Union of Sets

For any two sets A and B, the **union** of A and B, symbolized $A \cup B$, is defined as follows:

$$A \cup B = \{x \mid x \text{ is an element of } A \textbf{ or } x \text{ is an element of } B\}.$$

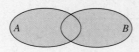

EXAMPLE 5 **Finding the Union of Two Sets**

Let $A = \{1, 2, 3, 4\}$ and $B = \{2, 4, 6\}$. Find $A \cup B$.
Begin by listing all the elements of set A: 1, 2, 3, 4. Then list any additional elements from set B. In this case the elements 2 and 4 are already listed, so the only additional element is 6. Therefore,

$$A \cup B = \{1, 2, 3, 4\} \cup \{2, 4, 6\}$$
$$= \{1, 2, 3, 4, 6\}.$$

The union consists of all elements in either A *or* B (or both).

NOTE

Although the elements 2 and 4 appeared in both sets A and B, they are written only once in $A \cup B$.

Work Problem 5 at the Side. ▶▶▶

4 Solve.

(a) $x < 5$ and $x > 5$

(b) $x + 2 > 3$ and
$2x + 1 < -3$

5 List the elements in each set.

(a) $A \cup B$, if $A = \{3, 4, 5, 6\}$
and $B = \{5, 6, 7\}$

(b) $N \cup O$ from the thyroid
table at the beginning of
this section

ANSWERS
4. **(a)** $\emptyset$ **(b)** $\emptyset$
5. **(a)** $\{3, 4, 5, 6, 7\}$ **(b)** $\{s, d, c, g, i, m, h\}$

6 Give each solution set in both interval and graph forms.

(a) $x + 2 > 3$ or
    $2x + 1 < -3$

⟶

(b) $x - 1 > 2$ or
    $3x + 5 < 2x + 6$

⟶

OBJECTIVE 4 Solve compound inequalities with the word _or_.   Use the following steps.

Solving a Compound Inequality with _or_

Step 1   Solve each inequality in the compound inequality individually.

Step 2   Since the inequalities are joined with _or,_ the solution set includes all numbers that satisfy either one of the two inequalities in Step 1 (the union of the solution sets).

EXAMPLE 6 Solving a Compound Inequality with _or_

Solve $6x - 4 < 2x$ or $-3x \leq -9$.

Step 1   Solve each inequality separately.

$$6x - 4 < 2x \quad \text{or} \quad -3x \leq -9$$
$$4x < 4$$
$$x < 1 \quad \text{or} \quad x \geq 3$$

The graphs of these two inequalities are shown in Figure 18.

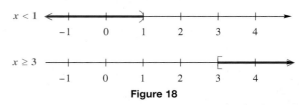

Figure 18

Step 2   Since the inequalities are joined with _or,_ find the union of the two solution sets. The union is shown in Figure 19 and is written

$$(-\infty, 1) \cup [3, \infty).$$

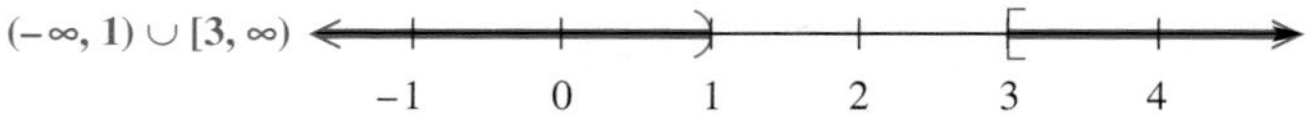

Figure 19

CAUTION

When inequalities are used to write the solution set in Example 6, it _must_ be written as

$$x < 1 \quad \text{or} \quad x \geq 3,$$

which keeps the numbers 1 and 3 in their order on the number line. Writing $3 \leq x < 1$ would imply that $3 \leq 1$, which is **_FALSE_**. There is no other way to write the solution set of such a union.

◄◄◄ **Work Problem 6 at the Side.**

ANSWERS
6. (a) $(-\infty, -2) \cup (1, \infty)$

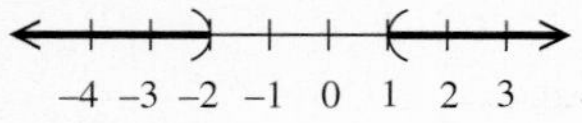

    −4 −3 −2 −1 0 1 2 3

(b) $(-\infty, 1) \cup (3, \infty)$

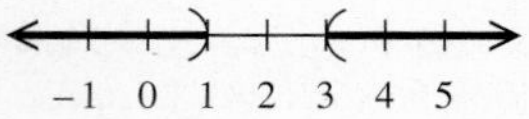

    −1 0 1 2 3 4 5

EXAMPLE 7 **Solving a Compound Inequality with *or***

Solve $-4x + 1 \geq 9$ or $5x + 3 \geq -12$.

First, solve each inequality separately.

$$-4x + 1 \geq 9 \quad \text{or} \quad 5x + 3 \geq -12$$
$$-4x \geq 8 \quad \text{or} \quad 5x \geq -15$$
$$x \leq -2 \quad \text{or} \quad x \geq -3$$

The graphs of these two inequalities are shown in Figure 20.

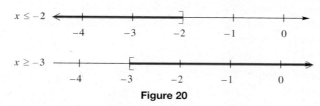

Figure 20

By taking the union, we obtain every real number as a solution, since every real number satisfies at least one of the two inequalities. The set of all real numbers is written in interval notation as $(-\infty, \infty)$ and graphed as in Figure 21.

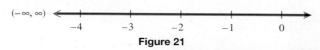

Figure 21

Work Problem 7 at the Side.

EXAMPLE 8 **Applying Intersection and Union**

The five highest domestic grossing films (adjusted for inflation) are listed in the table.

FIVE ALL-TIME HIGHEST GROSSING FILMS

Film	Admissions	Gross Income
Gone with the Wind	200,605,313	$972,900,000
Star Wars	178,119,595	$863,900,000
The Sound of Music	142,415,376	$690,700,000
E.T.	135,987,938	$659,500,000
The Ten Commandments	131,000,000	$635,400,000

Source: New York Times Almanac, 2001.

List the elements of the following sets.

(a) The set of top-five films with admissions greater than 180,000,000 *and* gross income greater than $800,000,000

The only film that satisfies both conditions is *Gone with the Wind*, so the set is

$$\{Gone \; with \; the \; Wind\}.$$

(b) The set of top-five films with admissions less than 170,000,000 *or* gross income greater than $700,000,000

Here, a film that satisfies at least one of the conditions is in the set. This set includes all five films:

$$\{Gone \; with \; the \; Wind, \; Star \; Wars, \; The \; Sound \; of \; Music, \; E.T.,$$
$$The \; Ten \; Commandments\}.$$

Work Problem 8 at the Side.

7 Solve.

(a) $2x + 1 \leq 9$ or $2x + 3 \leq 5$

(b) $3x - 2 \leq 13$ or $x + 5 \geq 7$

8 From Example 8, list the elements that satisfy each set.

(a) The set of films with admissions greater than 130,000,000 and gross income less than $500,000,000

(b) The set of films with admissions greater than 130,000,000 or gross income less than $500,000,000

ANSWERS
7. **(a)** $(-\infty, 4]$ **(b)** $(-\infty, \infty)$
8. **(a)** $\emptyset$ **(b)** {*Gone with the Wind, Star Wars, The Sound of Music, E.T., The Ten Commandments*}

Comparing Long-Distance Costs

Cellular phones are popular tools for both local and long-distance phone calls. Information about the rate plans offered by different cellular phone companies is readily available on the Internet. Using the Internet, you have found the following pricing schemes for both regular and cellular phones.

- The long-distance plan for the regular *in-home* phone costs $6.95 per month plus $.05 per min for long-distance calls both within your state or between states, with no limit to the number of minutes of call time.

- One option for the *cellular* phone is a flat monthly fee of $59.99 that includes 450 min of "anytime" local or long-distance calls.

Note: Basic phone rates are *not* included in the in-home plan, but since you intend to have an in-home phone anyway, you can disregard those costs. Also, calls in excess of the limits for the cellular plan are expensive: $.35 per minute over the maximum. You do *not* expect to exceed the number of minutes included in the basic cellular rate plan, so do not worry about those extra charges.

For Group Discussion

The question is "Which plan is more economical?" Let x represent the number of minutes of long-distance calls in a month.

1. Write an expression that represents the monthly costs for the in-home rate plan.

2. Write the expression that represents the monthly cost for the cellular rate plan.

3. The question is: "How many minutes of long-distance calls would you have to make in one month with the in-home phone to exceed the cost of the cellular phone plan?" Write a linear inequality that states that the in-home rate plan costs more than the cellular rate plan.

4. Solve the linear inequality and answer the question posed in Problem 3. What does your answer mean in terms of comparing phone costs?

5. Suppose you use the cellular phone plan for 450 min (the maximum number of minutes without incurring excess charges). How much more money would you pay compared to the in-home plan?

3.2 Exercises

FOR EXTRA HELP

Tutor Center Addison-Wesley Math Tutor Center

Math XL mathXL

Digital Video Tutor CD 2 Videotape 2

Student's Solutions Manual

MyMathLab MyMathLab

Interactmath.com

Decide whether each statement is true *or* false. *If it is* false, *explain why.*

1. The union of the solution sets of $2x + 1 = 3$, $2x + 1 > 3$, and $2x + 1 < 3$ is $(-\infty, \infty)$.

2. The intersection of the sets $\{x \mid x \geq 5\}$ and $\{x \mid x \leq 5\}$ is $\emptyset$.

3. The union of the sets $(-\infty, 6)$ and $(6, \infty)$ is $\{6\}$.

4. The intersection of the sets $[6, \infty)$ and $(-\infty, 6]$ is $\{6\}$.

Let $A = \{1, 2, 3, 4, 5, 6\}$, $B = \{1, 3, 5\}$, $C = \{1, 6\}$, *and* $D = \{4\}$. *Specify each set. See Examples 1 and 5.*

5. $A \cap D$

6. $B \cap C$

7. $B \cap \emptyset$

8. $A \cap \emptyset$

9. $A \cup B$

10. $B \cup D$

11. $B \cup C$

12. $C \cup B$

Two sets are specified by graphs. Graph the intersection of the two sets.

13.
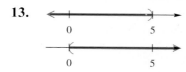

14.
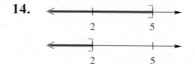

15.
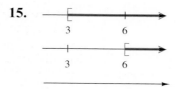

Two sets are specified by graphs. Graph the union of the two sets.

16.
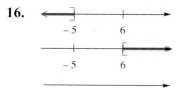

17.
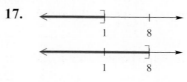

18.
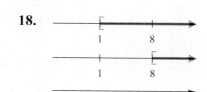

19. Give an example of intersection applied to a real-life situation.

20. A compound inequality uses one of the words *and* or *or*. Explain how you will determine whether to use *intersection* or *union* when graphing the solution set.

For each compound inequality, give the solution set in both interval and graph forms. See Examples 2–4.

21. $x < 2$ and $x > -3$

22. $x < 5$ and $x > 0$

23. $x \leq 2$ and $x \leq 5$

24. $x \geq 3$ and $x \geq 6$

25. $x \leq 3$ and $x \geq 6$

26. $x \leq -1$ and $x \geq 3$

27. $x - 3 \leq 6$ and $x + 2 \geq 7$

28. $x + 5 \leq 11$ and $x - 3 \geq -1$

29. $3x - 4 \leq 8$ and $4x - 1 \leq 15$

30. $7x + 6 \leq 48$ and $-4x \geq -24$

For each compound inequality, give the solution set in both interval and graph forms. See Examples 6 and 7.

31. $x \leq 1$ or $x \leq 8$

32. $x \geq 1$ or $x \geq 8$

33. $x \geq -2$ or $x \geq 5$

34. $x \leq -2$ or $x \leq 6$

35. $x + 3 \geq 1$ or $x - 8 \leq -4$

36. $x + 6 \geq 11$ or $x - 4 \leq 3$

37. $x + 2 > 7$ or $x - 1 < -6$

38. $x + 1 > 3$ or $x + 4 < 2$

39. $4x - 8 > 0$ or $4x - 1 < 7$

40. $3x < x + 12$ or $3x - 8 > 10$

Express each set in the simplest interval form.

41. $(-\infty, -1] \cap [-4, \infty)$

42. $[-1, \infty) \cap (-\infty, 9]$

43. $(-\infty, -6] \cap [-9, \infty)$

44. $(5, 11] \cap [6, \infty)$

45. $(-\infty, 3) \cup (-\infty, -2)$

46. $[-9, 1] \cup (-\infty, -3)$

47. $[3, 6] \cup (4, 9)$

48. $[-1, 2] \cup (0, 5)$

For each compound inequality, state whether intersection or union should be used. Then give the solution set in both interval and graph forms. See Examples 2, 3, 4, 6, and 7.

49. $x < -1$ and $x > -5$

50. $x > -1$ and $x < 7$

51. $x < 4$ or $x < -2$

52. $x < 5$ or $x < -3$

53. $x + 1 \geq 5$ and $x - 2 \leq 10$

54. $2x - 6 \leq -18$ and $2x \geq -18$

55. $-3x \leq -6$ or $-3x \geq 0$

56. $-8x \leq -24$ or $-5x \geq 15$

RELATING CONCEPTS (EXERCISES 57–60)   For Individual or Group Work

The figures represent the backyards of neighbors Luigi, Mario, Than, and Joe. Find the area and the perimeter of each yard. Suppose that each resident has 150 ft of fencing and enough sod to cover 1400 ft^2 of lawn.

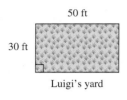

Luigi's yard

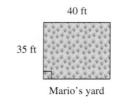

Mario's yard

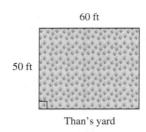

Than's yard

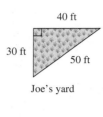

Joe's yard

Give the name or names of the residents whose yards satisfy each description.

57. The yard can be fenced *and* the yard can be sodded.

58. The yard can be fenced *and* the yard cannot be sodded.

59. The yard cannot be fenced *and* the yard can be sodded.

60. The yard cannot be fenced *and* the yard cannot be sodded.

Average expenses for full-time college students during a recent academic year are shown in the table.

COLLEGE EXPENSES (IN DOLLARS)

Type of Expense	Public Schools	Private Schools
Tuition and fees	2365	13,013
Board rates	2180	2742
Dormitory charges	2243	2990

Source: U.S. National Center for Education Statistics, *Digest of Education Statistics,* annual.

Use the table to list the elements of each set.

61. The set of expenses that are less than $2500 for public schools *and* are greater than $3000 for private schools

62. The set of expenses that are less than $2500 for public schools *and* are less than $3000 for private schools

63. The set of expenses that are less than $2200 for public schools *or* are greater than $3000 for private schools

64. The set of expenses that are greater than $10,000 *or* are less than $2000

3.3 Absolute Value Equations and Inequalities

In a production line, quality is controlled by randomly choosing items from the line and checking to see how selected measurements vary from the optimum measure. These differences are sometimes positive and sometimes negative, so they are expressed with absolute value. For example, a machine that fills quart milk cartons might be set to release 1 qt plus or minus 2 oz per carton. Then the number of ounces in each carton should satisfy the *absolute value inequality* $|x - 32| \le 2$, where x is the number of ounces.

OBJECTIVE **1** **Use the distance definition of absolute value.** In **Section 1.1** we saw that the absolute value of a number x, written $|x|$, represents the distance from x to 0 on the number line. For example, the solutions of $|x| = 4$ are 4 and -4, as shown in Figure 22.

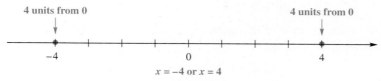

4 units from 0   4 units from 0

$x = -4$ or $x = 4$

Figure 22

Because absolute value represents distance from 0, it is reasonable to interpret the solutions of $|x| > 4$ to be all numbers that are *more* than 4 units from 0. The set $(-\infty, -4) \cup (4, \infty)$ fits this description. Figure 23 shows the graph of the solution set of $|x| > 4$. Because the graph consists of two separate intervals, the solution set is described using *or* as $x < -4$ or $x > 4$.

More than 4 units from 0   More than 4 units from 0

$x < -4$ or $x > 4$

Figure 23

The solution set of $|x| < 4$ consists of all numbers that are *less* than 4 units from 0 on the number line. Another way of thinking of this is to think of all numbers *between* -4 and 4. This set of numbers is given by $(-4, 4)$, as shown in Figure 24. Here, the graph shows that $-4 < x < 4$, which means $x > -4$ *and* $x < 4$.

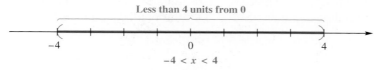

Less than 4 units from 0

$-4 < x < 4$

Figure 24

Work Problem 1 at the Side.))▶

The equation and inequalities just described are examples of **absolute value equations and inequalities.** They involve the absolute value of a variable expression and generally take the form

$$|ax + b| = k, \quad |ax + b| > k, \quad \text{or} \quad |ax + b| < k,$$

where k is a positive number. From Figures 22–24, we see that

$|x| = 4$ has the same solution set as $x = -4$ or $x = 4$,

$|x| > 4$ has the same solution set as $x < -4$ or $x > 4$,

$|x| < 4$ has the same solution set as $x > -4$ **and** $x < 4$.

OBJECTIVES

1 Use the distance definition of absolute value.

2 Solve equations of the form $|ax + b| = k$, for $k > 0$.

3 Solve inequalities of the form $|ax + b| < k$ and of the form $|ax + b| > k$, for $k > 0$.

4 Solve absolute value equations that involve rewriting.

5 Solve equations of the form $|ax + b| = |cx + d|$.

6 Solve special cases of absolute value equations and inequalities.

1 Graph the solution set of each equation or inequality.

(a) $|x| = 3$

(b) $|x| > 3$

(c) $|x| < 3$

ANSWERS

1. **(a)** number line marked $-3\ -2\ -1\ 0\ 1\ 2\ 3$

(b) number line marked $-3\ -2\ -1\ 0\ 1\ 2\ 3$

(c) number line marked $-3\ -2\ -1\ 0\ 1\ 2\ 3$

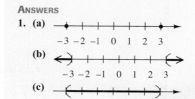

❷ Solve each equation, check, and graph the solution set.

(a) $|x + 2| = 3$

Thus, we can solve an absolute value equation or inequality by solving the appropriate compound equation or inequality.

Solving Absolute Value Equations and Inequalities

Let k be a positive real number, and p and q be real numbers.

1. To solve $|ax + b| = k$, solve the compound equation

$$ax + b = k \quad \text{or} \quad ax + b = -k.$$

The solution set is usually of the form $\{p, q\}$, which includes two numbers.

2. To solve $|ax + b| > k$, solve the compound inequality

$$ax + b > k \quad \text{or} \quad ax + b < -k.$$

The solution set is of the form $(-\infty, p) \cup (q, \infty)$, which consists of two separate intervals.

3. To solve $|ax + b| < k$, solve the three-part inequality

$$-k < ax + b < k.$$

The solution set is of the form (p, q), a single interval.

(b) $|3x - 4| = 11$

OBJECTIVE ❷ Solve equations of the form $|ax + b| = k$, for $k > 0$. Remember that because absolute value refers to distance from the origin, an absolute value equation will have two parts.

EXAMPLE 1 Solving an Absolute Value Equation

Solve $|2x + 1| = 7$.

For $|2x + 1|$ to equal 7, $2x + 1$ must be 7 units from 0 on the number line. This can happen only when $2x + 1 = 7$ or $2x + 1 = -7$. This is the first case in the preceding summary. Solve this compound equation as follows.

$$2x + 1 = 7 \quad \text{or} \quad 2x + 1 = -7$$
$$2x = 6 \quad \text{or} \quad 2x = -8$$
$$x = 3 \quad \text{or} \quad x = -4$$

Check by substituting 3 and then -4 in the original absolute value equation to verify that the solution set is $\{-4, 3\}$. The graph is shown in Figure 25.

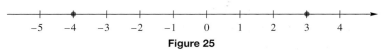

Figure 25

◀◀◀ **Work Problem 2 at the Side.**

ANSWERS
2. (a) $\{-5, 1\}$

$-5 \; -4 \; -3 \; -2 \; -1 \; 0 \; 1$

(b) $\left\{-\dfrac{7}{3}, 5\right\}$

$-\dfrac{7}{3} \quad 0 \quad 2 \quad 4 \; 5$

NOTE
Some people prefer to write the compound statements in parts 1 and 2 of the summary on the previous page as the equivalent forms

$$ax + b = k \quad \text{or} \quad -(ax + b) = k$$

and

$$ax + b > k \quad \text{or} \quad -(ax + b) > k.$$

These forms produce the same results.

OBJECTIVE 3 Solve inequalities of the form $|ax + b| < k$ and of the form $|ax + b| > k$, for $k > 0$.

EXAMPLE 2 Solving an Absolute Value Inequality with $>$

Solve $|2x + 1| > 7$.

By part 2 of the summary, this absolute value inequality is rewritten as

$$2x + 1 > 7 \quad \text{or} \quad 2x + 1 < -7,$$

because $2x + 1$ must represent a number that is *more* than 7 units from 0 on either side of the number line. Now, solve the compound inequality.

$$2x + 1 > 7 \quad \text{or} \quad 2x + 1 < -7$$
$$2x > 6 \quad \text{or} \quad 2x < -8$$
$$x > 3 \quad \text{or} \quad x < -4$$

Check these solutions. The solution set is $(-\infty, -4) \cup (3, \infty)$. See Figure 26. Notice that the graph consists of two intervals.

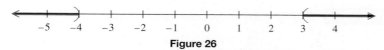
Figure 26

Work Problem 3 at the Side.

EXAMPLE 3 Solving an Absolute Value Inequality with $<$

Solve $|2x + 1| < 7$.

The expression $2x + 1$ must represent a number that is less than 7 units from 0 on either side of the number line. Another way of thinking of this is to realize that $2x + 1$ must be between -7 and 7. As part 3 of the summary shows, this is written as the three-part inequality

$$-7 < 2x + 1 < 7.$$
$$-8 < 2x < 6 \qquad \text{Subtract 1 from each part.}$$
$$-4 < x < 3 \qquad \text{Divide each part by 2.}$$

Check that the solution set is $(-4, 3)$, so the graph consists of the single interval shown in Figure 27.

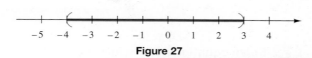
Figure 27

Work Problem 4 at the Side.

3 Solve each inequality, check, and graph the solution set.

(a) $|x + 2| > 3$

(b) $|3x - 4| \geq 11$

4 Solve each inequality, check, and graph the solution set.

(a) $|x + 2| < 3$

(b) $|3x - 4| \leq 11$

ANSWERS
3. (a) $(-\infty, -5) \cup (1, \infty)$

(b) $\left(-\infty, -\frac{7}{3}\right] \cup [5, \infty)$

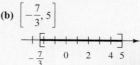

4. (a) $(-5, 1)$

(b) $\left[-\frac{7}{3}, 5\right]$

5 (a) Solve $|5x + 2| - 9 = -7$.

Look back at Figures 25, 26, and 27, with the graphs of $|2x + 1| = 7$, $|2x + 1| > 7$, and $|2x + 1| < 7$. If we find the union of the three sets, we get the set of all real numbers. This is because for any value of x, $|2x + 1|$ will satisfy one and only one of the following: it is equal to 7, greater than 7, or less than 7.

> **CAUTION**
> When solving absolute value equations and inequalities of the types in Examples 1, 2, and 3, remember the following.
>
> 1. The methods described apply when the constant is alone on one side of the equation or inequality and is *positive*.
>
> 2. Absolute value equations and absolute value inequalities of the form $|ax + b| > k$ translate into "or" compound statements.
>
> 3. Absolute value inequalities of the form $|ax + b| < k$ translate into "and" compound statements, which may be written as three-part inequalities.
>
> 4. An "or" statement *cannot* be written in three parts. It would be incorrect to use $-7 > 2x + 1 > 7$ in Example 2, because this would imply that $-7 > 7$, which is *false*.

(b) Solve $|x + 2| - 3 > 2$, and graph the solution set.

OBJECTIVE 4 Solve absolute value equations that involve rewriting. Sometimes an absolute value equation or inequality requires some rewriting before it can be set up as a compound statement, as shown in the next example.

(c) Solve, and graph the solution set.

$$|3x + 2| + 4 \leq 15$$

EXAMPLE 4 Solving an Absolute Value Equation That Requires Rewriting

Solve the equation $|x + 3| + 5 = 12$.

First, rewrite so that the absolute value expression is alone on one side of the equals sign by subtracting 5 from each side.

$$|x + 3| + 5 - 5 = 12 - 5 \qquad \text{Subtract 5.}$$
$$|x + 3| = 7$$

Now use the method shown in Example 1.

$$x + 3 = 7 \quad \text{or} \quad x + 3 = -7$$
$$x = 4 \quad \text{or} \qquad x = -10$$

Check that the solution set is $\{4, -10\}$ by substituting 4 and then -10 into the original equation.

We use a similar method to solve an absolute value *inequality* that requires rewriting.

Work Problem 5 at the Side.

OBJECTIVE 5 Solve equations of the form $|ax + b| = |cx + d|$. By definition, for two expressions to have the same absolute value, they must either be equal or be negatives of each other.

ANSWERS

5. (a) $\left\{ -\dfrac{4}{5}, 0 \right\}$

(b) $(-\infty, -7) \cup (3, \infty)$

$-7 \quad -4 \; -2 \quad 0 \qquad 3$

(c) $\left[-\dfrac{13}{3}, 3 \right]$

$-\dfrac{13}{3} \quad -2 \quad 0 \quad 2 \; 3$

Solving $|ax + b| = |cx + d|$

To solve an absolute value equation of the form

$$|ax + b| = |cx + d|,$$

solve the compound equation

$$ax + b = cx + d \quad \text{or} \quad ax + b = -(cx + d).$$

EXAMPLE 5 **Solving an Equation with Two Absolute Values**

Solve the equation $|z + 6| = |2z - 3|$.

This equation is satisfied either if $z + 6$ and $2z - 3$ are equal to each other, or if $z + 6$ and $2z - 3$ are negatives of each other. Thus,

$$z + 6 = 2z - 3 \quad \text{or} \quad z + 6 = -(2z - 3).$$

Solve each equation.

$$z + 6 = 2z - 3 \quad \text{or} \quad z + 6 = -2z + 3$$
$$9 = z \qquad\qquad\qquad 3z = -3$$
$$z = -1$$

Check that the solution set is $\{9, -1\}$.

Work Problem 6 at the Side. ▶▶▶

OBJECTIVE 6 Solve special cases of absolute value equations and inequalities. When a typical absolute value equation or inequality involves a *negative* constant or *0* alone on one side, as in Examples 6 and 7, we use the properties of absolute value to solve. Keep the following in mind.

Special Cases for Absolute Value

1. The absolute value of an expression can never be negative: $|a| \geq 0$ for all real numbers a.

2. The absolute value of an expression equals 0 only when the expression is equal to 0.

EXAMPLE 6 **Solving Special Cases of Absolute Value Equations**

Solve each equation.

(a) $|5r - 3| = -4$

See Case 1 in the preceding box. Since the absolute value of an expression can never be negative, there are no solutions for this equation. The solution set is $\emptyset$.

(b) $|7x - 3| = 0$

See Case 2 in the preceding box. The expression $7x - 3$ will equal 0 *only* if

$$7x - 3 = 0.$$

The solution of this equation is $\frac{3}{7}$. Thus, the solution set of the original equation is $\left\{\frac{3}{7}\right\}$, with just one element. Check by substitution.

Work Problem 7 at the Side. ▶▶▶

6 Solve each equation.

(a) $|k - 1| = |5k + 7|$

(b) $|4r - 1| = |3r + 5|$

7 Solve each equation.

(a) $|6x + 7| = -5$

(b) $\left|\dfrac{1}{4}x - 3\right| = 0$

ANSWERS

6. (a) $\{-1, -2\}$ **(b)** $\left\{-\dfrac{4}{7}, 6\right\}$

7. (a) $\emptyset$ **(b)** $\{12\}$

8 Solve.

(a) $|x| > -1$

(b) $|x| < -5$

(c) $|x + 2| \le 0$

EXAMPLE 7 Solving Special Cases of Absolute Value Inequalities

Solve each inequality.

(a) $|x| \ge -4$

The absolute value of a number is always greater than or equal to 0. Thus, $|x| \ge -4$ is true for *all* real numbers. The solution set is $(-\infty, \infty)$.

(b) $|x + 6| - 3 < -5$

Add 3 to each side to get the absolute value expression alone on one side.

$$|x + 6| < -2$$

There is no number whose absolute value is less than -2, so this inequality has no solution. The solution set is $\emptyset$.

(c) $|x - 7| + 4 \le 4$

Subtracting 4 from each side gives

$$|x - 7| \le 0.$$

The value of $|x - 7|$ will never be less than 0. However, $|x - 7|$ will equal 0 when $x = 7$. Therefore, the solution set is $\{7\}$.

Work Problem 8 at the Side.

ANSWERS
8. (a) $(-\infty, \infty)$ (b) $\emptyset$ (c) $\{-2\}$

3.3 Exercises

FOR EXTRA HELP

Tutor Center Addison-Wesley Math Tutor Center

Math XP MathXL

Digital Video Tutor CD 2 Videotape 2

Student's Solutions Manual

MyMathLab MyMathLab

Interactmath.com

Match each absolute value equation or inequality in Column I with the graph of its solution set in Column II.

I   **II**   **I**   **II**

1. $|x| = 5$ **A.** (graph, endpoints at -5 and 5)

$|x| < 5$ **B.** (graph, endpoints at -5 and 5)

$|x| > 5$ **C.** (graph, endpoints at -5 and 5)

$|x| \le 5$ **D.** (graph, endpoints at -5 and 5)

$|x| \ge 5$ **E.** (graph, points at -5 and 5)

2. $|x| = 9$ **A.** (graph, endpoints at -9 and 9)

$|x| > 9$ **B.** (graph, endpoints at -9 and 9)

$|x| \ge 9$ **C.** (graph, endpoints at -9 and 9)

$|x| < 9$ **D.** (graph, endpoints at -9 and 9)

$|x| \le 9$ **E.** (graph, points at -9 and 9)

3. Explain when to use *and* and when to use *or* if you are solving an absolute value equation or inequality of the form $|ax + b| = k$, $|ax + b| < k$, or $|ax + b| > k$, where k is a positive number.

4. How many solutions will $|ax + b| = k$ have if **(a)** $k = 0$;   **(b)** $k > 0$;   **(c)** $k < 0$?

Solve each equation. See Example 1.

5. $|x| = 12$ **6.** $|x| = 14$ **7.** $|4x| = 20$ **8.** $|5x| = 30$

9. $|x - 3| = 9$ **10.** $|p - 5| = 13$ **11.** $|2x + 1| = 9$ **12.** $|2x + 3| = 19$

13. $|4r - 5| = 17$ **14.** $|5t - 1| = 21$ **15.** $|2x + 5| = 14$ **16.** $|2x - 9| = 18$

17. $\left|\dfrac{1}{2}x + 3\right| = 2$ **18.** $\left|\dfrac{2}{3}q - 1\right| = 5$ **19.** $\left|1 - \dfrac{3}{4}k\right| = 7$ **20.** $\left|2 - \dfrac{5}{2}m\right| = 14$

Solve each inequality, and graph the solution set. See Example 2.

21. $|x| > 3$

22. $|x| > 5$

23. $|k| \geq 4$

24. $|r| \geq 6$

25. $|t + 2| > 10$

26. $|r + 5| > 20$

27. $|3x - 1| \geq 8$

28. $|4x + 1| \geq 21$

29. $|3 - x| > 5$

30. $|5 - x| > 3$

31. The graph of the solution set of $|2x + 1| = 9$ is given here.

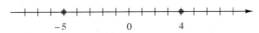

Without actually doing the algebraic work, graph the solution set of each inequality, referring to the graph above.

(a) $|2x + 1| < 9$

(b) $|2x + 1| > 9$

32. The graph of the solution set of $|3x - 4| < 5$ is given here.

Without actually doing the algebraic work, graph the solution set of the equation and the inequality, referring to the graph above.

(a) $|3x - 4| = 5$

(b) $|3x - 4| > 5$

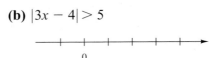

Solve each inequality, and graph the solution set. See Example 3. (Hint: Compare your answers to those in Exercises 21–30.)

33. $|x| \leq 3$

34. $|x| \leq 5$

35. $|k| < 4$

36. $|r| < 6$

37. $|t + 2| \leq 10$

38. $|r + 5| \leq 20$

39. $|3x - 1| < 8$

40. $|4x + 1| < 21$

41. $|3 - x| \leq 5$

42. $|5 - x| \leq 3$

Exercises 43–50 represent a sampling of the various types of absolute value equations and inequalities covered in Exercises 1–42. Decide which method of solution applies, find the solution set, and graph. See Examples 1–3.

43. $|-4 + k| > 9$

44. $|-3 + t| > 8$

45. $|7 + 2z| = 5$

46. $|9 - 3p| = 3$

47. $|3r - 1| \leq 11$

48. $|2s - 6| \leq 6$

49. $|-6x - 6| \leq 1$

50. $|-2x - 6| \leq 5$

Solve each equation or inequality. Give the solution set in set notation for equations and in interval notation for inequalities. See Example 4.

51. $|x| - 1 = 4$

52. $|x| + 3 = 10$

53. $|x + 4| + 1 = 2$

54. $|x + 5| - 2 = 12$

55. $|2x + 1| + 3 > 8$

56. $|6x - 1| - 2 > 6$

57. $|x + 5| - 6 \leq -1$

58. $|r - 2| - 3 \leq 4$

Solve each equation. See Example 5.

59. $|3x + 1| = |2x + 4|$

60. $|7x + 12| = |x - 8|$

61. $\left| m - \dfrac{1}{2} \right| = \left| \dfrac{1}{2} m - 2 \right|$

62. $\left| \dfrac{2}{3} r - 2 \right| = \left| \dfrac{1}{3} r + 3 \right|$

63. $|6x| = |9x + 1|$

64. $|13x| = |2x + 1|$

65. $|2p - 6| = |2p + 11|$

66. $|3x - 1| = |3x + 9|$

Solve each equation or inequality. See Examples 6 and 7.

67. $|12t - 3| = -8$

68. $|13w + 1| = -3$

69. $|4x + 1| = 0$

70. $|6r - 2| = 0$

71. $|2q - 1| < -6$

72. $|8n + 4| < -4$

73. $|x + 5| > -9$

74. $|x + 9| > -3$

75. $|7x + 3| \leq 0$

76. $|4x - 1| \leq 0$

77. $|5x - 2| \geq 0$

78. $|4 + 7x| \geq 0$

79. $|10z + 7| > 0$

80. $|4x + 1| > 0$

81. The 2001 recommended daily intake (RDI) of calcium for females aged 19–50 is 1000 mg/day. (*Source:* Food and Nutrition Board, National Academy of Sciences Institute of Medicine, 2001.) Actual vitamin needs vary from person to person. Write an absolute value inequality to express the RDI plus or minus 100 mg and solve it.

82. The average clotting time of blood is 7.45 sec with a variation of plus or minus 3.6 sec. Write this statement as an absolute value inequality and solve it.

RELATING CONCEPTS (EXERCISES 83–86) For Individual or Group Work

The ten tallest buildings in Kansas City, Missouri, are listed along with their heights.

Building	Height (in feet)
KCTV Tower	1042
One Kansas City Place	632
Transamerica Tower	591
Hyatt Regency	504
Power and Light Building	476
City Hall	443
Federal Office Building	433
1201 Walnut Street	427
Commerce Tower	407
City Center Square	404

Source: Marshall Gerometta and Rick Bronson, www.skyscrapers.com; Council on Tall Buildings and Urban Habitat, Lehigh University.

*Use this information to **work Exercises 83–86 in order.***

83. To find the average of a group of numbers, we add the numbers and then divide by the number of items added. Use a calculator to find the average of the heights.

84. Let k represent the average height of these buildings. If a height x satisfies the inequality

$$|x - k| < t,$$

then the height is said to be within t feet of the average. Using your result from Exercise 83, list the buildings that are within 50 ft of the average.

85. Repeat Exercise 84, but find the buildings that are within 75 ft of the average.

86. (a) Write an absolute value inequality that describes the height of a building that is *not* within 75 ft of the average.

(b) Solve the inequality you wrote in part (a).

(c) Use the result of part (b) to find the buildings that are not within 75 ft of the average.

(d) Confirm that your answer to part (c) makes sense by comparing it with your answer to Exercise 85.

Summary Exercises on Solving Linear and Absolute Value Equations and Inequalities

*Students often have difficulty distinguishing between the various types of equations and inequalities introduced in **Chapters 2 and 3**. This section of miscellaneous equations and inequalities provides practice in solving all such types. You might wish to refer to the boxes in these chapters that summarize the various methods of solution. Solve each equation or inequality.*

1. $4z + 1 = 49$

2. $|m - 1| = 6$

3. $6q - 9 = 12 + 3q$

4. $3p + 7 = 9 + 8p$

5. $|a + 3| = -4$

6. $2m + 1 \leq m$

7. $8r + 2 \geq 5r$

8. $4(a - 11) + 3a = 20a - 31$

9. $2q - 1 = -7$

10. $|3q - 7| - 4 = 0$

11. $6z - 5 \leq 3z + 10$

12. $|5z - 8| + 9 \geq 7$

13. $9x - 3(x + 1) = 8x - 7$

14. $|x| \geq 8$

15. $9x - 5 \geq 9x + 3$

16. $13p - 5 > 13p - 8$

17. $|q| < 5.5$

18. $4z - 1 = 12 + z$

19. $\dfrac{2}{3}x + 8 = \dfrac{1}{4}x$

20. $-\dfrac{5}{8}x \geq -20$

21. $\dfrac{1}{4}p < -6$

22. $7z - 3 + 2z = 9z - 8z$

23. $\dfrac{3}{5}q - \dfrac{1}{10} = 2$

24. $|r - 1| < 7$

25. $r + 9 + 7r = 4(3 + 2r) - 3$

26. $6 - 3(2 - p) < 2(1 + p) + 3$

27. $|2p - 3| > 11$

28. $\dfrac{x}{4} - \dfrac{2x}{3} = -10$

29. $|5a + 1| \le 0$

30. $5z - (3 + z) \ge 2(3z + 1)$

31. $-2 \le 3x - 1 \le 8$

32. $-1 \le 6 - x \le 5$

33. $|7z - 1| = |5z + 3|$

34. $|p + 2| = |p + 4|$

35. $|1 - 3x| \ge 4$

36. $\dfrac{1}{2} \le \dfrac{2}{3} r \le \dfrac{5}{4}$

37. $-(m + 4) + 2 = 3m + 8$

38. $\dfrac{p}{6} - \dfrac{3p}{5} = p - 86$

39. $-6 \le \dfrac{3}{2} - x \le 6$

40. $|5 - x| < 4$

41. $|x - 1| \ge -6$

42. $|2r - 5| = |r + 4|$

43. $8q - (1 - q) = 3(1 + 3q) - 4$

44. $8x - (x + 3) = -(2x + 1) - 12$

45. $|r - 5| = |r + 9|$

46. $|r + 2| < -3$

47. $2x + 1 > 5$ or $3x + 4 < 1$

48. $1 - 2x \ge 5$ and $7 + 3x \ge -2$

Chapter 3
SUMMARY

KEY TERMS

3.1   **interval**   An interval is a portion of a number line.

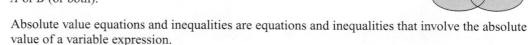
The interval $[-1, 3)$

    interval notation   The notation used to indicate an interval on the number line is called interval notation.

    inequality   An inequality is a mathematical statement that two expressions are not equal.

    linear inequality in one variable   A linear inequality in the variable x can be written in the form $Ax + B < C$, where A, B, and C are real numbers, with $A \neq 0$. (Other inequality symbols may be used.)

    equivalent inequalities   Equivalent inequalities are inequalities with the same solution set.

3.2   **intersection**   The intersection of two sets A and B is the set of elements that belong to both A and B.

    compound inequality   A compound inequality is formed by joining two inequalities with a connective word such as *and* or *or*.

    union   The union of two sets A and B is the set of elements that belong to either A or B (or both).

3.3   **absolute value equation; absolute value inequality**   Absolute value equations and inequalities are equations and inequalities that involve the absolute value of a variable expression.

NEW SYMBOLS

∞	infinity		$\cap$	set intersection
$-\infty$	negative infinity		$\cup$	set union
$(-\infty, \infty)$	the set of real numbers			

TEST YOUR WORD POWER

See how well you have learned the vocabulary in this chapter. Answers, with examples, follow the Quick Review.

1. An **inequality** is
   A. a statement that two algebraic expressions are equal
   B. a point on a number line
   C. an equation with no solutions
   D. a statement with algebraic expressions related by $<$, $\leq$, $>$, or $\geq$.

2. **Interval notation** is
   A. a portion of a number line
   B. a special notation for describing a point on a number line
   C. a way to use symbols to describe an interval on a number line
   D. a notation to describe unequal quantities.

3. The **intersection** of two sets A and B is the set of elements that belong
   A. to both A and B
   B. to either A or B, or both
   C. to either A or B, but not both
   D. to just A.

4. The **union** of two sets A and B is the set of elements that belong
   A. to both A and B
   B. to either A or B, or both
   C. to either A or B, but not both
   D. to just B.

QUICK REVIEW

Concepts	Examples

3.1   *Linear Inequalities in One Variable*
Solving Linear Inequalities in One Variable

Step 1   Simplify each side of the inequality by clearing parentheses and combining like terms.

Step 2   Use the addition property of inequality to get all terms with variables on one side and all terms without variables on the other side.

Step 3   Use the multiplication property of inequality to write the inequality in the form $x < k$ or $x > k$.

If an inequality is multiplied or divided by a negative number, the inequality symbol must be reversed.

To solve a three-part inequality, work with all three parts at the same time.

Solve $3(x + 2) - 5x \le 12$.

$$3x + 6 - 5x \le 12 \qquad \text{Distributive property}$$
$$-2x + 6 \le 12$$
$$-2x + 6 - 6 \le 12 - 6 \qquad \text{Subtract 6.}$$
$$-2x \le 6$$
$$\frac{-2x}{-2} \ge \frac{6}{-2} \qquad \text{Divide by } -2; \text{ change} \le \text{ to } \ge.$$
$$x \ge -3$$

The solution set $[-3, \infty)$ is graphed below.

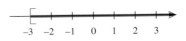

Solve $-4 < 2x + 3 \le 7$.

$$-4 - 3 < 2x + 3 - 3 \le 7 - 3 \qquad \text{Subtract 3.}$$
$$-7 < 2x \le 4$$
$$\frac{-7}{2} < \frac{2x}{2} \le \frac{4}{2} \qquad \text{Divide by 2.}$$
$$-\frac{7}{2} < x \le 2$$

The solution set $\left(-\frac{7}{2}, 2\right]$ is graphed below.

3.2   *Set Operations and Compound Inequalities*
Solving a Compound Inequality

Step 1   Solve each inequality in the compound inequality individually.

Step 2   If the inequalities are joined with *and*, the solution set is the intersection of the two individual solution sets.

If the inequalities are joined with *or*, the solution set is the union of the two individual solution sets.

Solve $x + 1 > 2$ and $2x < 6$.

$$x + 1 > 2 \quad \text{and} \quad 2x < 6$$
$$x > 1 \quad \text{and} \quad x < 3$$

The solution set is $(1, 3)$.

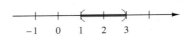

Solve $x \ge 4$ or $x \le 0$.
The solution set is $(-\infty, 0] \cup [4, \infty)$.

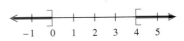

Concepts

3.3 *Absolute Value Equations and Inequalities*
Let k be a positive number.
To solve $|ax + b| = k,$ solve the compound equation

$$ax + b = k \quad \text{or} \quad ax + b = -k.$$

To solve $|ax + b| > k,$ solve the compound inequality

$$ax + b > k \quad \text{or} \quad ax + b < -k.$$

To solve $|ax + b| < k,$ solve the compound inequality

$$-k < ax + b < k.$$

To solve an absolute value equation of the form

$$|ax + b| = |cx + d|,$$

solve the compound equation

$$ax + b = cx + d \quad \text{or} \quad ax + b = -(cx + d).$$

Examples

Solve $|x - 7| = 3.$

$$x - 7 = 3 \quad \text{or} \quad x - 7 = -3$$
$$x = 10 \quad \text{or} \qquad x = 4$$

The solution set is $\{4, 10\}.$

Solve $|x - 7| > 3.$

$$x - 7 > 3 \quad \text{or} \quad x - 7 < -3$$
$$x > 10 \quad \text{or} \qquad x < 4$$

The solution set is $(-\infty, 4) \cup (10, \infty).$

Solve $|x - 7| < 3.$

$$-3 < x - 7 < 3$$
$$4 < x < 10 \qquad \text{Add 7 to each part.}$$

The solution set is $(4, 10).$

Solve $|x + 2| = |2x - 6|.$

$$x + 2 = 2x - 6 \quad \text{or} \quad x + 2 = -(2x - 6)$$
$$x = 8 \qquad\qquad x + 2 = -2x + 6$$
$$3x = 4$$
$$x = \frac{4}{3}$$

The solution set is $\{\frac{4}{3}, 8\}.$

ANSWERS TO TEST YOUR WORD POWER

1. D; *Examples:* $x < 5, 7 + 2k \geq 11, -5 < 2z - 1 \leq 3$
2. C; *Examples:* $(-\infty, 5], (1, \infty), [-3, 3)$
3. A; *Example:* If $A = \{2, 4, 6, 8\}$ and $B = \{1, 2, 3\}, A \cap B = \{2\}.$
4. B; *Example:* Using the preceding sets A and $B, A \cup B = \{1, 2, 3, 4, 6, 8\}.$

Real-Data Applications

What Do I Have to Average on My Tests to Get the Grade I Want?

On the first day of class, you are typically given a syllabus that describes the course requirements. If the syllabus includes a grading scale for homework, tests, projects, and final exam, then you should be able to predict the points you need on the final exam to earn a specific grade.

One intermediate algebra teacher bases final grades on points earned for activities as given in the Graded Classwork table on the left. To determine final grades, the teacher strictly adheres to the point ranges given in the Grade Distribution table on the right.

GRADED CLASSWORK

Activity	Points Available
Homework and vocabulary	45
Daily activities (scaled)	55
Lab participation and completion	100
Major exams (3 at 100 points)	300
Final exam	150
Total points	650

GRADE DISTRIBUTION

Grade	Points Required
A	585–650
B	520–584
C	455–519
IP*	< 455 and active
F	< 455 and inactive

* In Progress

Notice that exams account for 450 of the possible 650 points. The remaining 200 points should be fairly easy to earn by keeping up with the day-to-day course requirements.

Assumption: You earn a "baseline" number of points based on the following criteria.

1. You earn *all* of the homework and vocabulary points.

2. You earn a minimum of 50 points based on daily activities.

3. You earn a minimum of 90 lab participation and completion points.

For Group Discussion

1. Assume that you earn the baseline number of points. Let x represent the test points to be earned. Write and solve linear inequalities to find the minimum number of points that you need in test scores to earn grades no lower than A, B, and C. What "test average" is each minimum score? Round *up* to the nearest whole percent.

2. To keep your scholarship, you must earn a B in the course. Write a compound inequality to find the range of points that you need in test scores to earn a B average. Solve the inequality. What range of "test averages" are those minimum scores? Round *up* to the nearest whole percent.

3. Mark does not like to do the homework or participate in labs. Assume that Mark earns only 15 points in homework and vocabulary, 40 points in daily activities, and 50 points in lab participation. Write and solve linear inequalities to find the minimum number of points that Mark needs in test scores to earn grades no lower than A, B, and C. What "test average" is each minimum score? Round *up* to the nearest whole percent.

Chapter **3**

REVIEW EXERCISES

[3.1] *Solve each inequality. Give the solution set in both interval and graph forms.*

1. $-\dfrac{2}{3}x < 6$

2. $-5x - 4 \geq 11$

3. $\dfrac{6a + 3}{-4} < -3$

4. $\dfrac{9x + 5}{-3} > 3$

5. $5 - (6 - 4t) \geq 2t - 7$

6. $-6 \leq 2k \leq 24$

7. $8 \leq 3x - 1 < 14$

8. $-4 < 3 - 2z < 9$

Solve each problem.

9. The perimeter of a rectangular playground must be no greater than 120 m. The width of the playground must be 22 m. Find the possible lengths of the playground.

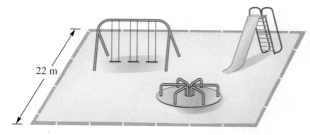

22 m

10. Etain O'Dea can rent a car from Ames for $48 per day plus 10¢ per mile, or from Hughes at $40 per day plus 15¢ per mile. She plans to use the car for 3 days. What number of miles would make Hughes cost at most as much as Ames?

11. To pass algebra, a student must have an average of at least 70% on five tests. On the first four tests, a student has grades of 75%, 79%, 64%, and 71%. What possible grades on the fifth test would guarantee a passing grade in the class?

12. While solving the inequality

$$10x + 2(x - 4) < 12x - 13,$$

a student did all the work correctly and obtained the statement $-8 < -13$. The student did not know what to do at this point, because the variable "disappeared." How would you explain to the student the interpretation of this result?

[3.2] *Let $A = \{a, b, c, d\}$, $B = \{a, c, e, f\}$, and $C = \{a, e, f, g\}$. Find each set.*

13. $A \cap B$ **14.** $A \cap C$ **15.** $B \cup C$ **16.** $A \cup C$

Solve each compound inequality. Give the solution set in both interval and graph forms.

17. $x > 6$ and $x < 9$

18. $x + 4 > 12$ and $x - 2 < 12$

19. $x > 5$ or $x \le -3$

20. $x \ge -2$ or $x < 2$

21. $x - 4 > 6$ and $x + 3 \le 10$

22. $-5x + 1 \ge 11$ or $3x + 5 \ge 26$

Express each union or intersection in simplest interval form.

23. $(-3, \infty) \cap (-\infty, 4)$

24. $(-\infty, 6) \cap (-\infty, 2)$

25. $(4, \infty) \cup (9, \infty)$

26. $(1, 2) \cup (1, \infty)$

27. The table shows the median weekly earnings of full-time workers by occupation for men and women.

Occupation	Men	Women
Managerial and professional	$1048	$753
Sales and office	$ 451	$365
Natural resources, construction, and maintenance	$ 616	$454
Installation, maintenance, and repair	$ 669	$656
Production, transportation, and material moving	$ 562	$399

Source: U.S. Bureau of Labor Statistics.

Give the occupation that satisfies each description.

(a) The median earnings for men are less than $600 *and* for women are greater than $370.

(b) The median earnings for men are greater than $600 *or* for women are less than $400.

[3.3] *Solve each absolute value equation.*

28. $|x| = 7$

29. $|x + 2| = 9$

30. $|3k - 7| = 8$

31. $|z - 4| = -12$

32. $|2k - 7| + 4 = 11$

33. $|4a + 2| - 7 = -3$

34. $|3p + 1| = |p + 2|$

35. $|2m - 1| = |2m + 3|$

Solve each absolute value inequality. Give the solution set in both interval and graph forms.

36. $|x| < 14$

37. $|-x + 6| \le 7$

38. $|2p + 5| \le 1$

39. $|x + 1| \ge -3$

40. $|5r - 1| > 9$

41. $|3x + 6| \geq 0$

MIXED REVIEW EXERCISES

Solve.

42. $(7 - 2x) + 3(5 - 3x) \geq x + 8$

43. $x < 5$ and $x \geq -4$

44. $\dfrac{3}{4}(a - 2) - \dfrac{1}{3}(5 - 2a) < -2$

45. To qualify for a company pension plan, an employee must average at least $1000 per month in earnings. During the first four months of the year, an employee made $900, $1200, $1040, and $760. What possible amounts earned during the fifth month will qualify the employee?

46. $-5r \geq -10$

47. $|7x - 2| > 9$

48. $|2x - 10| = 20$

49. $|m + 3| \leq 13$

50. $x \geq -2$ or $x < 4$

51. $|m - 1| = |2m + 3|$

In Exercises 52 and 53, sketch the graph of each solution set.

52. $x > 6$ and $x < 8$

53. $-5x + 1 \geq 11$ or $3x + 5 \geq 26$

54. If $k < 0$, what is the solution set of

(a) $|5x + 3| < k$,   (b) $|5x + 3| > k$,   (c) $|5x + 3| = k$?

Chapter 3

TEST

Study Skills Workbook
Activity 9: Taking a Test

1. What is the special rule that must be remembered when multiplying or dividing each side of an inequality by a negative number?

1. _____

Solve each inequality. Give the solution set in both interval and graph forms.

2. $4 - 6(x + 3) \leq -2 - 3(x + 6) + 3x$

2. ─────────────────→

3. $-\dfrac{4}{7}x > -16$

3. ─────────────────→

4. $-6 \leq \dfrac{4}{3}x - 2 \leq 2$

4. ─────────────────→

5. Which one of the following inequalities is equivalent to $x < -3$?

A. $-3x < 9$ **B.** $-3x > -9$ **C.** $-3x > 9$ **D.** $-3x < -9$

5. _____

6. The graph shows the number (in millions) of U.S. travelers to Europe. During which years were the numbers of travelers to Europe

(a) at least 9 million,   **(b)** less than 9 million,

(c) between 9 million and 12 million?

6. **(a)** _____

(b) _____

(c) _____

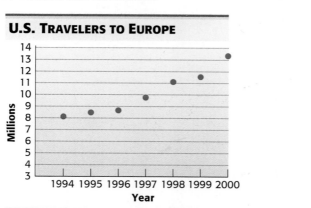

U.S. TRAVELERS TO EUROPE

Source: U.S. Office of Tourism Industries.

Solve each problem.

7. A student must have an average of at least 80% on the four tests in a course to get a B. The student had 83%, 76%, and 79% on the first three tests. What minimum percent on the fourth test would guarantee a B in the course?

7. _____

8. _____

8. A product will break even or produce a profit only if the revenue R (in dollars) from selling the product is at least the cost C (in dollars) of producing it. Suppose that the cost to produce x units of carpet is $C = 50x + 5000$, while the revenue is $R = 60x$. For what values of x is R at least equal to C?

9. (a) _____

    (b) _____

9. Let $A = \{1, 2, 5, 7\}$ and $B = \{1, 5, 9, 12\}$. Find

    (a) $A \cap B$,   **(b)** $A \cup B$.

10. _____

10. Solve $x \leq 2$ and $x \geq 2$.

Solve each compound or absolute value inequality. For Exercises 11–14, give the solution set in both interval and graph forms.

11. _____

11. $3k \geq 6$ and $k - 4 < 5$

12. _____

12. $-4x \leq -24$ or $4x - 2 < 10$

13. _____

13. $|4x + 3| \leq 7$

14. _____

14. $|5 - 6x| > 12$

15. _____

15. $|7 - x| \leq -1$

Solve each absolute value equation.

16. _____

16. $|3k - 2| + 1 = 8$

17. _____

17. $|3 - 5x| = |2x + 8|$

18. _____

18. $|4x + 3| + 5 = 4$

Cumulative Review Exercises

CHAPTERS 1–3

1. Write $\frac{108}{144}$ in lowest terms.

2. Is the statement *true* or *false*?

$$\frac{8(7) - 5(6 + 2)}{3 \cdot 5 + 1} \geq 1$$

Perform the indicated operations.

3. $\dfrac{5}{6} + \dfrac{1}{4} - \dfrac{7}{15}$

4. $\dfrac{9}{8} \cdot \dfrac{16}{3} \div \dfrac{5}{8}$

5. $9 - (-4) + (-2)$

6. $\dfrac{-4(9)(-2)}{-3^2}$

7. $|-7 - 1|(-4) + (-4)$

Evaluate each exponential expression.

8. $(-5)^3$

9. $\left(\dfrac{3}{2}\right)^4$

Evaluate each expression if $x = 2$, $y = -3$, and $z = 4$.

10. $-2y + 4(x - 3z)$

11. $\dfrac{3x^2 - y^2}{4z}$

Name each property illustrated.

12. $7(k + m) = 7k + 7m$

13. $3 + (5 + 2) = 3 + (2 + 5)$

14. Simplify $-4(k + 2) + 3(2k - 1)$ by combining terms.

Solve each equation, and check the solution.

15. $4 - 5(a + 2) = 3(a + 1) - 1$

16. $\dfrac{2}{3}x + \dfrac{3}{4}x = -17$

17. $\dfrac{2x + 3}{5} = \dfrac{x - 4}{2}$

18. $|3m - 5| = |m + 2|$

19. $3x + 4y = 24$ for y

20. $A = P(1 + ni)$ for n

Solve each inequality. Give the solution set in both interval and graph forms.

21. $3 - 2(x + 7) \leq -x + 3$

22. $-4 < 5 - 3x \leq 0$

23. $2x + 1 > 5$ or $2 - x > 2$

24. $|-7k + 3| \geq 4$

Solve each problem.

25. Sue Costa invested some money at 7% interest and the same amount at 10%. Her total interest for the year was $150 less than one-tenth of the total amount she invested. How much did she invest at each rate?

26. A dietician must use three foods, A, B, and C, in a diet. He must include twice as many grams of food A as food C, and 5 g of food B. The three foods must total at most 24 g. What is the largest amount of food C that the dietician can use?

27. Terry Harris got scores of 88% and 78% on her first two tests. What score must she make on her third test to keep an average of 80% or greater?

28. Two cars are 400 mi apart. Both start at the same time and travel toward one another. They meet 4 hr later. If the speed of one car is 20 mph faster than the other, what is the speed of each car?

29. Since 1990, the number of daily newspapers has steadily declined.

Year	Number of Daily Newspapers
1990	1611
1995	1533
1996	1520
1997	1509
1998	1489
1999	1483
2000	1480
2001	1468

Source: Statistical Abstract of the United States, 2002.

According to the table,

(a) by how much did the number of daily newspapers decrease between 1990 and 2001?

(b) by what *percent* did the number of daily newspapers decrease from 1990 to 2001?

30. For a woven hanging, Brian Altobello needs three pieces of yarn, which he will cut from a 40 cm piece. The longest piece is to be 3 times as long as the middle-sized piece, and the shortest piece is to be 5 cm shorter than the middle-sized piece. What lengths should he cut?

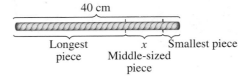

Graphs, Linear Equations, and Functions

4

4.1 **The Rectangular Coordinate System**

4.2 **Slope**

4.3 **Linear Equations in Two Variables**

4.4 **Linear Inequalities in Two Variables**

4.5 **Introduction to Functions**

4.6 **Variation**

In 1994, unleaded premium gasoline in the United States retailed for an average price of $1.31 per gallon. Ten years later in May 2004, this same gallon of gasoline cost an average of $2.19, an increase of more than 67%. Although this all-time high caused price shock at the gas pump for Americans, their European counterparts have paid much higher prices over the years. In May 2004, the same gallon of gasoline cost $5.46 per gallon in the United Kingdom and $5.05 in France. (*Source:* U.S. Energy Information Administration.)

In Example 6 of Section 4.3, we use the concepts of this chapter to write a *linear equation in two variables* that models the cost *y* to buy *x* gallons of gasoline.

4.1 The Rectangular Coordinate System

1 Plot ordered pairs.

2 Find ordered pairs that satisfy a given equation.

3 Graph lines.

4 Find x- and y-intercepts.

5 Recognize equations of horizontal and vertical lines.

There are many ways to present information graphically. The circle graph (or pie chart) in Figure 1(a) shows the cost breakdown for a gallon of regular unleaded gasoline in California. What contributes most to the cost?

Figure 1(b) shows a bar graph in which the heights of the bars represent the Btu (British thermal units) required to cool different-sized rooms. How many Btu are needed to cool a 1400 ft^2 room?

The line graph in Figure 1(c) shows personal spending (in billions of dollars) on medical care in the United States from 1997 through 2002. About how much was spent on medical care in 2002?

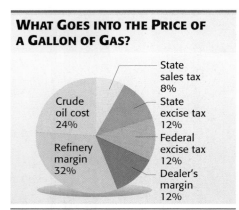

WHAT GOES INTO THE PRICE OF A GALLON OF GAS?

Source: California Energy Commission.

(a)

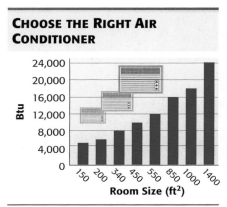

CHOOSE THE RIGHT AIR CONDITIONER

Source: Carey, Morris and James, *Home Improvement for Dummies*, IDG Books.

(b)

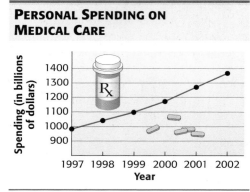

PERSONAL SPENDING ON MEDICAL CARE

Source: Bureau of Economic Analysis.

(c)

Figure 1

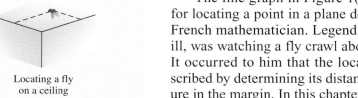

Locating a fly on a ceiling

The line graph in Figure 1(c) presents information based on a method for locating a point in a plane developed by René Descartes, a 17th-century French mathematician. Legend has it that Descartes, who was lying in bed ill, was watching a fly crawl about on the ceiling near a corner of the room. It occurred to him that the location of the fly on the ceiling could be described by determining its distances from the two adjacent walls. See the figure in the margin. In this chapter we use this insight to plot points and graph linear equations in two variables whose graphs are straight lines.

OBJECTIVE 1 Plot ordered pairs. Each of the pairs of numbers $(3, 1)$, $(-5, 6)$, and $(4, -1)$ is an example of an **ordered pair;** that is, a pair of numbers written within parentheses in which the order of the numbers is important. We graph an ordered pair using two perpendicular number lines that intersect at their 0 points, as shown in Figure 2. The common 0 point is called the **origin.** The position of any point in this plane is determined by referring to the horizontal number line, the **x-axis,** and the vertical number line, the **y-axis.** The first number in the ordered pair indicates the position relative to the x-axis, and the second number indicates the position relative to the y-axis. The x-axis and the y-axis make up a **rectangular** (or **Cartesian,** for Descartes) **coordinate system.**

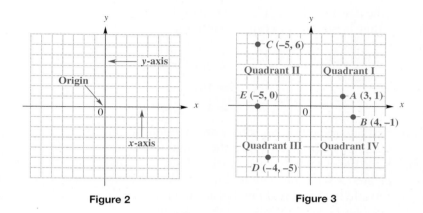

Figure 2   **Figure 3**

To locate, or **plot,** the point on the graph that corresponds to the ordered pair $(3, 1)$, we move three units from 0 to the right along the x-axis, and then one unit up parallel to the y-axis. The point corresponding to the ordered pair $(3, 1)$ is labeled A in Figure 3. Additional points are labeled B–E. The phrase "the point corresponding to the ordered pair $(3, 1)$" is often abbreviated as "the point $(3, 1)$." The numbers in the ordered pairs are called **components** and are the **coordinates** of the corresponding point.

We can relate this method of locating ordered pairs to the line graph in Figure 1(c). We move along the horizontal axis to a year, then up parallel to the vertical axis to find medical spending for that year. Thus, we can write the ordered pair $(2002, 1370)$ to indicate that in 2002, personal spending on medical care was about \$1370 billion.

CAUTION

The parentheses used to represent an ordered pair are also used to represent an open interval (introduced in **Section 3.1**). The context of the discussion tells whether ordered pairs or open intervals are being represented.

The four regions of the graph, shown in Figure 3, are called **quadrants I, II, III,** and **IV,** reading counterclockwise from the upper-right quadrant. The points on the x-axis and y-axis do not belong to any quadrant. For example, point E in Figure 3 belongs to no quadrant.

Work Problem 1 at the Side. ⟩⟩⟩

1 Plot each point. Name the quadrant (if any) in which each point is located.

(a) $(-4, 2)$

(b) $(3, -2)$

(c) $(-5, -6)$

(d) $(4, 6)$

(e) $(-3, 0)$

(f) $(0, -5)$

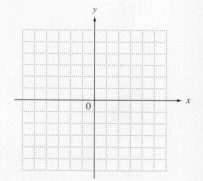

ANSWERS
1.

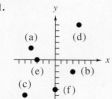

(a) II   **(b)** IV   **(c)** III   **(d)** I
(e) no quadrant   **(f)** no quadrant

2 (a) Complete each ordered pair for $3x - 4y = 12$.

$(0, \ \ )$

$(\ \ , 0)$

$(\ \ , -2)$

$(-4, \ \ )$

(b) Find one other ordered pair that satisfies the equation.

OBJECTIVE 2 Find ordered pairs that satisfy a given equation.
Each solution to an equation with two variables, such as $2x + 3y = 6$, will include two numbers, one for each variable. To keep track of which number goes with which variable, we write the solutions as ordered pairs. (If x and y are used as the variables, the x-value is given first.) For example, we can show that $(6, -2)$ is a solution of $2x + 3y = 6$ by substitution.

$$2x + 3y = 6$$
$$2(6) + 3(-2) = 6 \quad ? \quad \text{Let } x = 6, y = -2.$$
$$12 - 6 = 6 \quad ?$$
$$6 = 6 \quad \text{True}$$

Because the pair of numbers $(6, -2)$ makes the equation true, it is a solution. On the other hand, $(5, 1)$ is not a solution of the equation $2x + 3y = 6$ because

$$2x + 3y = 2(5) + 3(1)$$
$$= 10 + 3$$
$$= 13, \quad \textbf{\textit{not}} \quad 6.$$

To find ordered pairs that satisfy an equation, select any number for one of the variables, substitute it into the equation for that variable, and then solve for the other variable. Two other ordered pairs satisfying $2x + 3y = 6$ are $(0, 2)$ and $(3, 0)$. Since any real number could be selected for one variable and would lead to a real number for the other variable, linear equations in two variables have an infinite number of solutions.

EXAMPLE 1 Completing Ordered Pairs

Complete each ordered pair for $2x + 3y = 6$.

(a) $(-3, \ \ )$
We are given $x = -3$. We substitute into the equation to find y.

$$2x + 3y = 6$$
$$2(-3) + 3y = 6 \quad \text{Let } x = -3.$$
$$-6 + 3y = 6$$
$$3y = 12$$
$$y = 4$$

The ordered pair is $(-3, 4)$.

(b) $(\ \ , -4)$
Replace y with -4 in the equation to find x.

$$2x + 3y = 6$$
$$2x + 3(-4) = 6 \quad \text{Let } y = -4.$$
$$2x - 12 = 6$$
$$2x = 18$$
$$x = 9$$

The ordered pair is $(9, -4)$.

◀◀ **Work Problem 2 at the Side.**

ANSWERS

2. (a) $(0, -3), (4, 0), \left(\dfrac{4}{3}, -2\right), (-4, -6)$

(b) Many answers are possible; for example, $\left(-6, -\dfrac{15}{2}\right)$.

OBJECTIVE 3 **Graph lines.** The **graph of an equation** is the set of points corresponding to all ordered pairs that satisfy the equation. It gives a "picture" of the equation. Most equations in two variables are satisfied by an infinite number of ordered pairs, so their graphs include an infinite number of points.

To graph an equation, we plot a number of ordered pairs that satisfy the equation until we have enough points to suggest the shape of the graph. For example, to graph $2x + 3y = 6$, we plot all the ordered pairs found in Objective 2 and Example 1 on the previous page. These points, shown in a table of values and plotted in Figure 4(a), appear to lie on a straight line. If all the ordered pairs that satisfy the equation $2x + 3y = 6$ were graphed, they would form the straight line shown in Figure 4(b).

3 Graph $3x - 4y = 12$. Use the points from Problem 2 in the margin on the previous page.

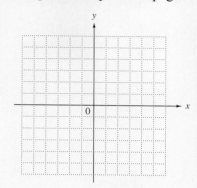

x	y
-3	4
0	2
3	0
6	-2
9	-4

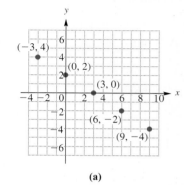

(a)

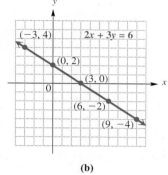

(b)

Figure 4

Work Problem 3 at the Side.

The equation $2x + 3y = 6$ is called a **first-degree equation** because it has no term with a variable to a power greater than one.

> *The graph of any first-degree equation in two variables is a straight line.*

Since first-degree equations with two variables have straight-line graphs, they are called *linear equations in two variables.*

Linear Equation in Two Variables

A **linear equation in two variables** can be written in the form

$$Ax + By = C,$$

where A, B, and C are real numbers (A and B not both 0). This form is called **standard form.**

OBJECTIVE 4 **Find x- and y-intercepts.** A straight line is determined if any two different points on the line are known, so finding two different points is enough to graph the line. Two useful points for graphing are the x- and y-intercepts. The **x-intercept** is the point (if any) where the line intersects the x-axis; likewise, the **y-intercept** is the point (if any) where the line intersects the y-axis.* See Figure 5.

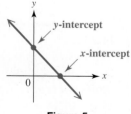

Figure 5

Answers
3.

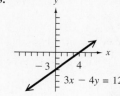

---
* Some texts define an intercept as a number, not a point.

4 Find the intercepts, and graph $2x - y = 4$.

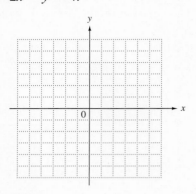

The y-value of the point where a line intersects the x-axis is 0. Similarly, the x-value of the point where a line intersects the y-axis is 0. This suggests a method for finding the x- and y-intercepts.

Finding Intercepts

When graphing the equation of a line,

let $y = 0$ to find the x-intercept;

let $x = 0$ to find the y-intercept.

EXAMPLE 2 Finding Intercepts

Find the x- and y-intercepts of $4x - y = -3$, and graph the equation.

We find the x-intercept by letting $y = 0$.

$$4x - y = -3$$
$$4x - 0 = -3 \quad \text{Let } y = 0.$$
$$4x = -3$$
$$x = -\frac{3}{4} \quad x\text{-intercept is } \left(-\tfrac{3}{4}, 0\right).$$

For the y-intercept, let $x = 0$.

$$4x - y = -3$$
$$4(0) - y = -3 \quad \text{Let } x = 0.$$
$$-y = -3$$
$$y = 3 \quad y\text{-intercept is } (0, 3).$$

The intercepts are the two points $\left(-\frac{3}{4}, 0\right)$ and $(0, 3)$. We show these ordered pairs in the table next to Figure 6 and use these points to draw the graph.

x	y
$-\frac{3}{4}$	0
0	3

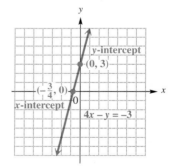

Figure 6

NOTE

While two points, such as the two intercepts in Figure 6, are sufficient to graph a straight line, *it is a good idea to use a third point to guard against errors.* Verify by substitution that $(-1, -1)$ also lies on the graph of $4x - y = -3$.

ANSWERS
4. x-intercept is $(2,0)$; y-intercept is $(0, -4)$.

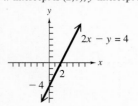

Work Problem 4 at the Side.

OBJECTIVE 5 Recognize equations of horizontal and vertical lines.
A graph can fail to have an x-intercept or a y-intercept, which is why the phrase "if any" was added when discussing intercepts.

EXAMPLE 3 Graphing a Horizontal Line

Graph $y = 2$.

Since y is always 2, there is no value of x corresponding to $y = 0$, so the graph has no x-intercept. The y-intercept is $(0, 2)$. The graph in Figure 7, shown with a table of ordered pairs, is a horizontal line.

x	y
−1	2
0	2
3	2

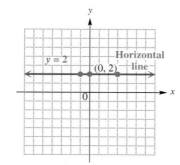

Figure 7

Work Problem 5 at the Side.

EXAMPLE 4 Graphing a Vertical Line

Graph $x + 1 = 0$.

The x-intercept is $(-1, 0)$. The standard form $1x + 0y = -1$ shows that every value of y leads to $x = -1$, so no value of y makes $x = 0$. The only way a straight line can have no y-intercept is if it is vertical, as in Figure 8.

x	y
−1	−4
−1	0
−1	5

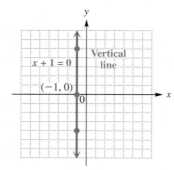

Figure 8

Work Problem 6 at the Side.

CAUTION

To avoid confusing equations of horizontal and vertical lines remember that

1. An equation with only the variable x will always intersect the x-axis and thus will be *vertical*.

2. An equation with only the variable y will always intersect the y-axis and thus will be *horizontal*.

5 Find the intercepts, and graph $y + 4 = 0$.

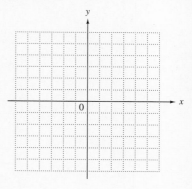

6 Find the intercepts, and graph the line $x = 2$.

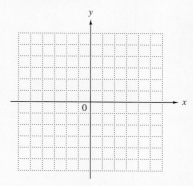

ANSWERS

5. no x-intercept; y-intercept is $(0, -4)$.

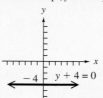

6. no y-intercept; x-intercept is $(2, 0)$.

7 Find the intercepts, and graph the line $3x - y = 0$.

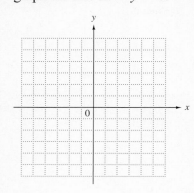

Some lines have both the x- and y-intercepts at the origin.

EXAMPLE 5 **Graphing a Line That Passes through the Origin**

Graph $x + 2y = 0$.

Find the x-intercept by letting $y = 0$.

$$x + 2y = 0$$
$$x + 2(0) = 0 \quad \text{Let } y = 0.$$
$$x + 0 = 0$$
$$x = 0 \quad x\text{-intercept is } (0, 0).$$

To find the y-intercept, let $x = 0$.

$$x + 2y = 0$$
$$0 + 2y = 0 \quad \text{Let } x = 0.$$
$$y = 0 \quad y\text{-intercept is } (0, 0).$$

Both intercepts are the same ordered pair, $(0, 0)$. (This means that the graph goes through the origin.) To find another point to graph the line, choose any nonzero number for x, say $x = 4$, and solve for y.

$$x + 2y = 0$$
$$4 + 2y = 0 \quad \text{Let } x = 4.$$
$$2y = -4$$
$$y = -2$$

This gives the ordered pair $(4, -2)$. These two points lead to the graph shown in Figure 9. As a check, verify that $(-2, 1)$ also lies on the line.

x	y
−2	1
0	0
4	−2

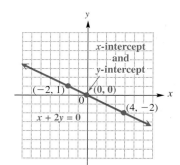

Figure 9

To find the additional point to graph, we could have chosen any number (except 0) for y instead of x.

Work Problem 7 at the Side.

ANSWERS
7. Both intercepts are $(0, 0)$.

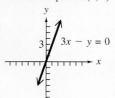

4.1 Exercises

FOR EXTRA HELP

Addison-Wesley Math Tutor Center

MathXL

Digital Video Tutor CD 2 Videotape 2

Student's Solutions Manual

MyMathLab

Interactmath.com

In Exercises 1 and 2, answer each question by locating ordered pairs on the graphs.

1. The graph shows the percent of women in math or computer science professions since 1970.

   (a) If (x, y) represents a point on the graph, what does x represent? What does y represent?

   (b) In what decade (10-year period) did the percent of women in math or computer science professions decrease?

   (c) When did the percent of women in math or computer science professions reach a maximum?

   (d) In what year was the percent of women in math or computer science professions about 27%?

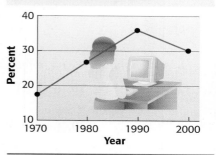

WOMEN IN MATH OR COMPUTER SCIENCE PROFESSIONS

Source: U.S. Bureau of the Census and Bureau of Labor Statistics.

2. The graph indicates personal spending in billions of dollars on medical care in the United States.

   (a) If (x, y) represents a point on the graph, what does x represent? What does y represent?

   (b) What was spending in 1999?

   (c) In what year was spending about $1270 billion?

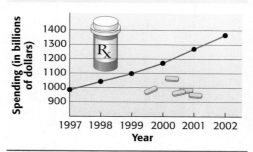

PERSONAL SPENDING ON MEDICAL CARE

Source: Bureau of Economic Analysis.

3. Observe the graphs in Exercises 1 and 2. If you were to use one of the other types of graphs mentioned in the opening paragraphs of this section to depict the information given there, which one would you choose?

4. What is another name for the rectangular coordinate system? After whom is it named?

Fill in each blank with the correct response.

5. The point with coordinates (0, 0) is called the _____ of a rectangular coordinate system.

6. For any value of x, the point $(x, 0)$ lies on the _____ -axis.

7. To find the x-intercept of a line, we let _____ equal 0 and solve for _____.

8. The equation _____ = 4 has a horizontal line as (x or y) its graph.

9. To graph a straight line, we must find a minimum of _____ points.

10. The point (_____, 4) is on the graph of $2x - 3y = 0$.

Name the quadrant, if any, in which each point is located.

11. (a) $(1, 6)$ **(b)** $(-4, -2)$ **12. (a)** $(-2, -10)$ **(b)** $(4, 8)$
(c) $(-3, 6)$ **(d)** $(7, -5)$ **(c)** $(-9, 12)$ **(d)** $(3, -9)$
(e) $(-3, 0)$ **(e)** $(0, -8)$

13. Use the given information to determine the possible quadrants in which the point (x, y) must lie. (*Hint:* Consider the signs of the coordinates in each quadrant, and the signs of their product and quotient.)

(a) $xy > 0$ **(b)** $xy < 0$

(c) $\dfrac{x}{y} < 0$ **(d)** $\dfrac{x}{y} > 0$

14. What must be true about the coordinates of any point that lies on an axis?

Locate each point on the rectangular coordinate system.

15. $(2, 3)$ **16.** $(-1, 2)$ **17.** $(-3, -2)$ **18.** $(1, -4)$

19. $(0, 5)$ **20.** $(-2, -4)$ **21.** $(-2, 4)$ **22.** $(3, 0)$

23. $(-2, 0)$ **24.** $(3, -3)$

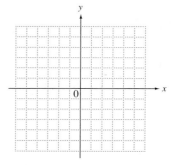

In each exercise, complete the given ordered pairs for the equation, and then graph the equation. See Example 1.

25. $x - y = 3$
$(0, \quad), (\quad, 0), (5, \quad), (2, \quad)$

26. $x - y = 5$
$(0, \quad), (\quad, 0), (1, \quad), (3, \quad)$

27. $x + 2y = 5$
$(0, \quad), (\quad, 0), (2, \quad), (\quad, 2)$

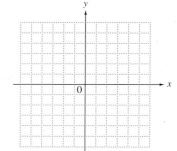 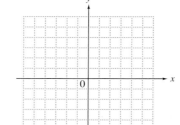 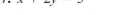

28. $x + 3y = -5$
$(0, \quad), (\quad, 0), (1, \quad), (\quad, -1)$

29. $4x - 5y = 20$
$(0, \quad), (\quad, 0), (2, \quad), (\quad, -3)$

30. $6x - 5y = 30$
$(0, \quad), (\quad, 0), (3, \quad), (\quad, -2)$

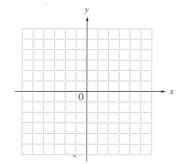 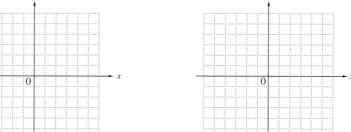

Find the x- and y-intercepts. Then graph each equation. See Examples 2–5.

31. $2x + 3y = 12$

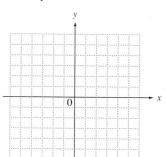

32. $5x + 2y = 10$

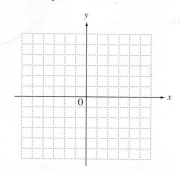

33. $x - 3y = 6$

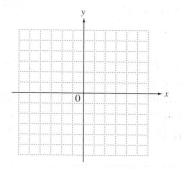

34. $x - 2y = -4$

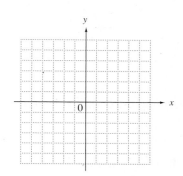

35. $3x - 7y = 9$

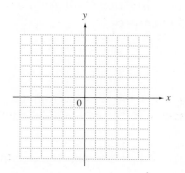

36. $5x + 6y = -10$

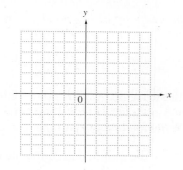

37. $y = 5$

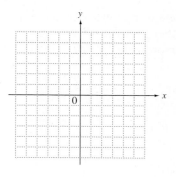

38. $y = -3$

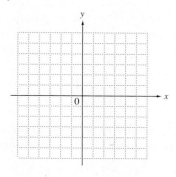

39. $x = 5$

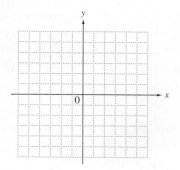

40. $x = -3$

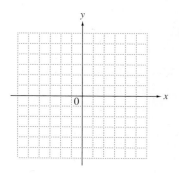

41. $x + 5y = 0$

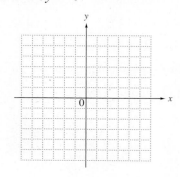

42. $x - 3y = 0$

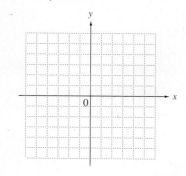

43. $2x = 3y$

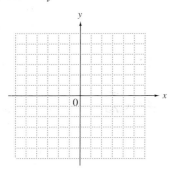

44. $3x = -4y$

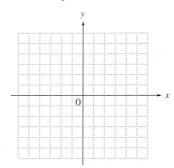

RELATING CONCEPTS (EXERCISES 45–56) For Individual or Group Work

*If the endpoints of a line segment are known, the coordinates of the midpoint of the segment can be found. The figure shows the coordinates of the points P and Q. Let $\overline{PQ}$ represent the line segment with endpoints at P and Q. To derive a formula for the midpoint of $\overline{PQ}$, **work Exercises 45–50 in order.***

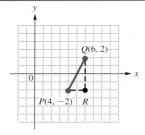

45. In the figure, R is the point with the same x-coordinate as Q and the same y-coordinate as P. Write the ordered pair that corresponds to R.

46. From the graph, determine the coordinates of the midpoint of $\overline{PR}$.

47. From the graph, determine the coordinates of the midpoint of $\overline{QR}$.

48. The x-coordinate of the midpoint M of $\overline{PQ}$ is the x-coordinate of the midpoint of $\overline{PR}$ and the y-coordinate is the y-coordinate of the midpoint of $\overline{QR}$. Write the ordered pair that corresponds to M.

49. The average of two numbers is found by dividing their sum by 2. Find the average of the x-coordinates of points P and Q. Find the average of the y-coordinates of points P and Q.

50. Compare your answers to Exercises 48 and 49. What connection is there between the coordinates of P and Q and the coordinates of M?

*The result of the preceding exercises leads to the **midpoint formula.** If the endpoints of a line segment $\overline{PQ}$ are (x_1, y_1) and (x_2, y_2), then its midpoint M has coordinates*

$$\left(\frac{x_1 + x_2}{2}, \frac{y_1 + y_2}{2} \right).$$

For example, the midpoint of the segment with endpoints $(4, -3)$ and $(6, -1)$ is

$$\left(\frac{4 + 6}{2}, \frac{-3 + (-1)}{2} \right) = \left(\frac{10}{2}, \frac{-4}{2} \right) = (5, -2).$$

Use the midpoint formula to find the midpoint of each segment with the given endpoints.

51. $(-8, 4)$ and $(-2, -6)$

52. $(5, 2)$ and $(-1, 8)$

53. $(3, -6)$ and $(6, 3)$

54. $(-10, 4)$ and $(7, 1)$

55. $(-9, 3)$ and $(9, 8)$

56. $(4, -3)$ and $(-1, 3)$

4.2 Slope

Slope (steepness) is used in many practical ways. The slope of a highway (sometimes called the *grade*) is often given as a percent. For example, a 10% (or $\frac{10}{100} = \frac{1}{10}$) slope means the highway rises 1 unit for every 10 horizontal units. Stairs and roofs have slopes too, as shown in Figure 10.

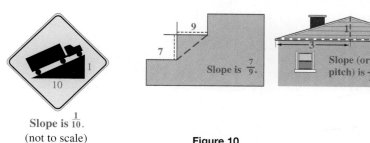

Slope is $\frac{1}{10}$.
(not to scale)

Slope is $\frac{7}{9}$.

Slope (or pitch) is $\frac{1}{3}$.

Figure 10

OBJECTIVES

1 Find the slope of a line given two points on the line.

2 Find the slope of a line given an equation of the line.

3 Graph a line given its slope and a point on the line.

4 Use slopes to determine whether two lines are parallel, perpendicular, or neither.

5 Solve problems involving average rate of change.

In each example mentioned, slope is the ratio of vertical change, or **rise**, to horizontal change, or **run**. A simple way to remember this is to think "slope is rise over run."

OBJECTIVE 1 **Find the slope of a line given two points on the line.** To obtain a formal definition of the slope of a line, we designate two different points on the line. To differentiate between the points, we write them as (x_1, y_1) and (x_2, y_2). See Figure 11. (The small numbers 1 and 2 in these ordered pairs are called *subscripts*. Read (x_1, y_1) as "*x*-sub-one, *y*-sub-one.")

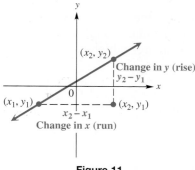

Figure 11

As we move along the line in Figure 11 from (x_1, y_1) to (x_2, y_2), the *y*-value changes (vertically) from y_1 to y_2, an amount equal to $y_2 - y_1$. As *y* changes from y_1 to y_2, the value of *x* changes (horizontally) from x_1 to x_2 by the amount $x_2 - x_1$. The ratio of the change in *y* to the change in *x* (the rise over the run) is called the *slope* of the line, with the letter *m* traditionally used for slope.

Slope Formula

The **slope** of the line through the distinct points (x_1, y_1) and (x_2, y_2) is

$$m = \frac{\text{rise}}{\text{run}} = \frac{\text{change in } y}{\text{change in } x} = \frac{y_2 - y_1}{x_2 - x_1} \quad (x_1 \neq x_2).$$

1 Use the information given for the walkway in the figure to find the following.

2 ft

10 ft

(a) The rise

(b) The run

(c) The slope

Work Problem 1 at the Side. ▶▶▶

ANSWERS

1. (a) 2 ft **(b)** 10 ft **(c)** $\frac{2}{10}$ or $\frac{1}{5}$

2 Find the slope of the line through each pair of points.

(a) $(-2, 7), (4, -3)$

EXAMPLE 1 Finding the Slope of a Line

Find the slope of the line through the points $(2, -1)$ and $(-5, 3)$.

If $(2, -1) = (x_1, y_1)$ and $(-5, 3) = (x_2, y_2)$, then

$$m = \frac{y_2 - y_1}{x_2 - x_1} = \frac{3 - (-1)}{-5 - 2} = \frac{4}{-7} = -\frac{4}{7}.$$

See Figure 12. On the other hand, if the pairs are reversed so that $(2, -1) = (x_2, y_2)$ and $(-5, 3) = (x_1, y_1)$, the slope is

$$m = \frac{-1 - 3}{2 - (-5)} = \frac{-4}{7} = -\frac{4}{7},$$

the same answer.

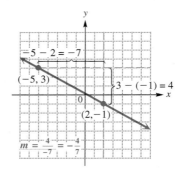

Figure 12

(b) $(1, 2), (8, 5)$

Example 1 suggests that the slope is the same no matter which point we consider first. Also, using similar triangles from geometry, we can show that the slope is the same no matter which two different points on the line we choose.

CAUTION

When calculating slope, *be careful to subtract the y-values and the x-values in the same order.*

Correct		**Incorrect**

$$\frac{y_2 - y_1}{x_2 - x_1} \quad \text{or} \quad \frac{y_1 - y_2}{x_1 - x_2} \qquad\qquad \frac{y_2 - y_1}{x_1 - x_2} \text{ or } \frac{y_1 - y_2}{x_2 - x_1}$$

(c) $(8, -4), (3, -2)$

Also, remember that *the change in y is the numerator and the change in x is the denominator.*

Work Problem 2 at the Side.

OBJECTIVE 2 Find the slope of a line given an equation of the line. When an equation of a line is given, one way to find the slope is to use the definition of slope by first finding two different points on the line.

ANSWERS

2. (a) $-\frac{5}{3}$ (b) $\frac{3}{7}$ (c) $-\frac{2}{5}$

EXAMPLE 2 Finding the Slope of a Line

Find the slope of the line $4x - y = -8$.

   The intercepts can be used as the two different points needed to find the slope. Let $y = 0$ to find that the x-intercept is $(-2, 0)$. Then let $x = 0$ to find that the y-intercept is $(0, 8)$. Use these two points in the slope formula. The slope is

$$m = \frac{\text{rise}}{\text{run}} = \frac{8 - 0}{0 - (-2)} = \frac{8}{2} = 4.$$

Work Problem 3 at the Side. ▶▶▶

EXAMPLE 3 Finding the Slopes of Horizontal and Vertical Lines

Find the slope of each line.

(a) $y = 2$

   Figure 7 in **Section 4.1** shows that the graph of $y = 2$ is a horizontal line. To find the slope, select two different points on the line, such as $(3, 2)$ and $(-1, 2)$, and use the slope formula.

$$m = \frac{\text{rise}}{\text{run}} = \frac{2 - 2}{3 - (-1)} = \frac{0}{4} = 0$$

In this case, the *rise* is 0, so the slope is 0.

(b) $x = -1$

   As shown in Figure 8 (**Section 4.1**), the graph of $x = -1$ (or $x + 1 = 0$) is a vertical line. Two points that satisfy the equation $x = -1$ are $(-1, 5)$ and $(-1, -4)$. Use these two points to find the slope.

$$m = \frac{\text{rise}}{\text{run}} = \frac{-4 - 5}{-1 - (-1)} = \frac{-9}{0}$$

Since division by 0 is undefined, the slope is undefined. This is why the definition of slope includes the restriction $x_1 \neq x_2$.

   Generalizing from Example 3, we can make the following statements about horizontal and vertical lines.

Slopes of Horizontal and Vertical Lines

The slope of a horizontal line is 0.

The slope of a vertical line is undefined.

Work Problem 4 at the Side. ▶▶▶

3 Find the slope of each line.

(a) $2x + y = 6$

(b) $3x - 4y = 12$

4 Find the slope of each line.

(a) $x = -6$

(b) $y + 5 = 0$

ANSWERS

3. (a) -2 **(b)** $\dfrac{3}{4}$ **4. (a)** undefined **(b)** 0

⑤ Find the slope of the graph of $2x - 5y = 8$.

The slope of a line can also be found directly from its equation. Look again at the equation $4x - y = -8$ from Example 2. Solve this equation for y.

$$4x - y = -8 \qquad \text{Equation from Example 2}$$
$$-y = -4x - 8 \quad \text{Subtract } 4x.$$
$$y = 4x + 8 \qquad \text{Multiply by } -1.$$

Notice that the slope, **4**, found using the slope formula in Example 2 is the same number as the coefficient of x in the equation $y = 4x + 8$. We will see in the next section that this always happens, *as long as the equation is solved for y.*

EXAMPLE 4 **Finding the Slope from an Equation**

Find the slope of the graph of $3x - 5y = 8$.
    Solve the equation for y.

$$3x - 5y = 8$$
$$-5y = -3x + 8 \quad \text{Subtract } 3x.$$
$$y = \frac{3}{5}x - \frac{8}{5} \quad \text{Divide by } -5.$$

The slope is given by the coefficient of x, so the slope is $\frac{3}{5}$.

◀◀◀ Work Problem 5 at the Side. ▶

OBJECTIVE ③ Graph a line given its slope and a point on the line.
Example 5 shows how to graph a straight line by using the slope and one point on the line.

EXAMPLE 5 **Using the Slope and a Point to Graph Lines**

Graph each line.

(a) With slope $\frac{2}{3}$ through the point $(-1, 4)$
    First locate the point $P(-1, 4)$ on a graph as shown in Figure 13. Then use the slope to find a second point. From the slope formula,

$$m = \frac{\text{change in } y}{\text{change in } x} = \frac{2}{3},$$

so move *up* 2 units and then 3 units to the *right* to locate another point on the graph (labeled R). The line through $P(-1, 4)$ and R is the required graph.

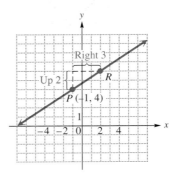

Figure 13

Continued on Next Page

ANSWERS

5. $\dfrac{2}{5}$

(b) Through (3, 1) with slope -4

Start by locating the point $P(3, 1)$ on a graph. Find a second point R on the line by writing -4 as $\frac{-4}{1}$ and using the slope formula.

$$m = \frac{\text{change in } y}{\text{change in } x} = \frac{-4}{1}$$

Move *down* 4 units from (3, 1), and then move 1 unit to the *right*. Draw a line through this second point R and $P(3, 1)$, as shown in Figure 14.

The slope also could be written as

$$m = \frac{\text{change in } y}{\text{change in } x} = \frac{4}{-1}.$$

In this case the second point R is located *up* 4 units and 1 unit to the *left*. Verify that this approach produces the same line.

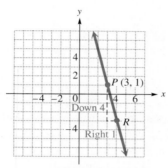

Figure 14

Work Problem 6 at the Side. ▶▶▶

In Example 5(a), the slope of the line is the *positive* number $\frac{2}{3}$. The graph of the line in Figure 13 goes up (rises) from left to right. The line in Example 5(b) has a *negative* slope, -4. As Figure 14 shows, its graph goes down (falls) from left to right. These facts suggest the following generalization.

A positive slope indicates that the line goes *up* (rises) from left to right.
A negative slope indicates that the line goes *down* (falls) from left to right.

Figure 15 shows lines of positive, 0, negative, and undefined slopes.

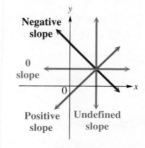

Figure 15

❻ Graph each line.

(a) Through $(1, -3)$;
$$m = -\frac{3}{4}$$

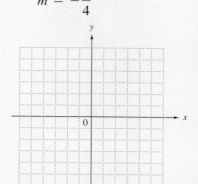

(b) Through $(-1, -4)$;
$$m = 2$$

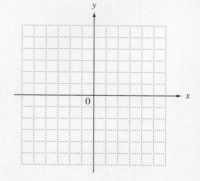

ANSWERS
6. (a)

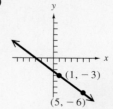

(b)

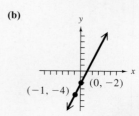

OBJECTIVE **4** **Use slopes to determine whether two lines are parallel, perpendicular, or neither.** The slopes of a pair of parallel or perpendicular lines are related in a special way. The slope of a line measures the steepness of the line. Since parallel lines have equal steepness, their slopes must be equal; also, lines with the same slope are parallel.

Slopes of Parallel Lines

Two nonvertical lines with the same slope are parallel.
Two nonvertical parallel lines have the same slope.

EXAMPLE 6 Determining whether Two Lines Are Parallel

Are the lines L_1, through $(-2, 1)$ and $(4, 5)$, and L_2, through $(3, 0)$ and $(0, -2)$, parallel?

$$\text{The slope of } L_1 \text{ is } \quad m_1 = \frac{5-1}{4-(-2)} = \frac{4}{6} = \frac{2}{3}.$$

$$\text{The slope of } L_2 \text{ is } \quad m_2 = \frac{-2-0}{0-3} = \frac{-2}{-3} = \frac{2}{3}.$$

Because the slopes are equal, the two lines are parallel.

To see how the slopes of perpendicular lines are related, consider a nonvertical line with slope $\frac{a}{b}$. If this line is rotated $90°$, the vertical change and the horizontal change are reversed and the slope is $-\frac{b}{a}$, since the horizontal change is now negative. See Figure 16. Thus, the slopes of perpendicular lines have product -1 and are negative reciprocals of each other. For example, if the slopes of two lines are $\frac{3}{4}$ and $-\frac{4}{3}$, then the lines are perpendicular because $\frac{3}{4}\left(-\frac{4}{3}\right) = -1$.

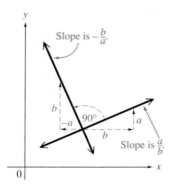

Figure 16

Slopes of Perpendicular Lines

If neither is vertical, perpendicular lines have slopes that are negative reciprocals; that is, their product is -1. Also, lines with slopes that are negative reciprocals are perpendicular.

EXAMPLE 7 Determining whether Two Lines Are Perpendicular

Are the lines with equations $2y = 3x - 6$ and $2x + 3y = -6$ perpendicular? Find the slope of each line by first solving each equation for y.

$$2y = 3x - 6$$
$$y = \frac{3}{2}x - 3$$
$$\uparrow$$
Slope

$$2x + 3y = -6$$
$$3y = -2x - 6$$
$$y = -\frac{2}{3}x - 2$$
$$\uparrow$$
Slope

Since the product of the slopes of the two lines is $\frac{3}{2}\left(-\frac{2}{3}\right) = -1$, the lines are perpendicular.

Work Problem 7 at the Side. ▶▶▶

OBJECTIVE 5 Solve problems involving average rate of change. We know that the slope of a line is the ratio of the change in y (vertical) to the change in x (horizontal). This idea can be applied to real-life situations. The slope gives the *average rate of change* in y per unit of change in x, where the value of y depends on the value of x.

EXAMPLE 8 Interpreting Slope as Average Rate of Change

The graph in Figure 17 approximates the percent of U.S. households owning multiple personal computers in the years 1997–2001. Find the average rate of change in percent per year.

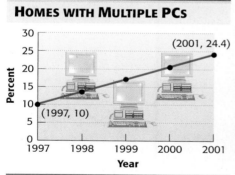

HOMES WITH MULTIPLE PCs

(2001, 24.4)

(1997, 10)

Year

Source: The Yankee Group.

Figure 17

To determine the average rate of change, we need two pairs of data. From the graph, if $x = 1997$, then $y = 10$ and if $x = 2001$, then $y = 24.4$, so we have the ordered pairs (1997, 10) and (2001, 24.4). By the slope formula,

$$\text{average rate of change} = \frac{y_2 - y_1}{x_2 - x_1} = \frac{24.4 - 10}{2001 - 1997} = \frac{14.4}{4} = 3.6.$$

This means that the number of U.S. households owning multiple computers *increased* by 3.6% each year from 1997 to 2001.

Work Problem 8 at the Side. ▶▶▶

7 Write *parallel, perpendicular,* or *neither* for each pair of two distinct lines.

(a) The line through $(-1, 2)$ and $(3, 5)$ and the line through $(4, 7)$ and $(8, 10)$

(b) The line through $(5, -9)$ and $(3, 7)$ and the line through $(0, 2)$ and $(8, 3)$

(c) $2x - y = 4$ and $2x + y = 6$

(d) $3x + 5y = 6$ and $5x - 3y = 2$

8 Use the ordered pairs (1997, 10) and (2000, 20.8), which are plotted in Figure 17, to find the average rate of change. How does it compare to the average rate of change found in Example 8?

ANSWERS
7. **(a)** parallel   **(b)** perpendicular
   (c) neither   **(d)** perpendicular
8. 3.6; It is the same.

9 In 1997, 36.4 percent of high school students smoked. In 2001, 28.5 percent of high school students smoked. Find the average rate of change in percent per year. (*Source: U.S. Centers for Disease Control and Prevention.*)

EXAMPLE 9 Interpreting Slope as Average Rate of Change

In 1997, sales of VCRs numbered 16.7 million. In 2002, sales of VCRs were 13.3 million. Find the average rate of change, in millions, per year. (*Source: The Gazette*, June 22, 2002.)

To use the slope formula, we need two ordered pairs. Here, if $x = 1997$, then $y = 16.7$ and if $x = 2002$, then $y = 13.3$, which gives the ordered pairs (1997, 16.7) and (2002, 13.3). (Note that y is in millions.)

$$\text{average rate of change} = \frac{13.3 - 16.7}{2002 - 1997} = \frac{-3.4}{5} = -.68$$

The graph in Figure 18 confirms that the line through the ordered pairs falls from left to right and therefore has negative slope. Thus, sales of VCRs *decreased* by .68 million each year from 1997 to 2002.

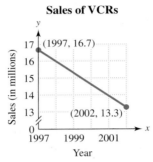

Sales of VCRs

Figure 18

Work Problem 9 at the Side.

ANSWERS
9. -1.975% per yr

4.2 Exercises

FOR EXTRA HELP Tutor Center Addison-Wesley Math Tutor Center | MathXL | Digital Video Tutor CD 2 Videotape 2 | Student's Solutions Manual | MyMathLab | Interactmath.com

1. A ski slope drops 30 ft for every horizontal 100 ft.

Which of the following express its slope? (There are several correct choices.)

A. $-.3$ **B.** $-\dfrac{3}{10}$ **C.** $-3\dfrac{1}{3}$ **D.** $-\dfrac{30}{100}$ **E.** $-\dfrac{10}{3}$

2. A hill has a slope of $-.05$. How many feet in the vertical direction correspond to a run of 50 ft?

Use the given figure to determine the slope of the line segment described, by counting the number of units of "rise," the number of units of "run," and then finding the quotient.

3. *AB*   **4.** *BC*   **5.** *CD*   **6.** *DE*

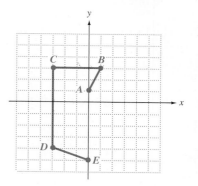

Find each slope using the slope formula.

7. $m = \dfrac{6 - 2}{5 - 3}$

8. $m = \dfrac{5 - 7}{-4 - 2}$

9. $m = \dfrac{4 - (-1)}{-3 - (-5)}$

10. $m = \dfrac{-6 - 0}{0 - (-3)}$

11. $m = \dfrac{-5 - (-5)}{3 - 2}$

12. $m = \dfrac{7 - (-2)}{-3 - (-3)}$

Find the slope of the line through each pair of points using the slope formula. See Example 1.

13. $(-2, -3)$ and $(-1, 5)$

14. $(-4, 3)$ and $(-3, -4)$

15. $(-4, 1)$ and $(2, 6)$

16. $(-3, -3)$ and $(5, 6)$

17. $(2, 4)$ and $(-4, 4)$

18. $(-6, 3)$ and $(2, 3)$

Find the slope of each line.

19.
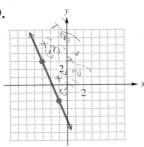

20.
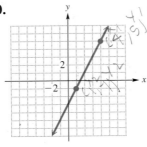

21.
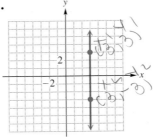

22.
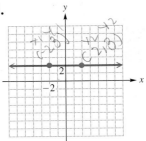

Based on the figure shown here, determine which line satisfies the given description.

23. The line has positive slope.

24. The line has negative slope.

25. The line has slope 0.

26. The line has undefined slope.

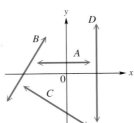

Find the slope of each line, and sketch the graph. See Examples 2–4.

27. $x + 2y = 4$

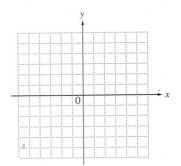

28. $x + 3y = -6$

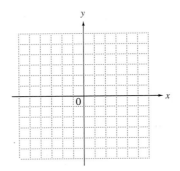

29. $-x + y = 4$

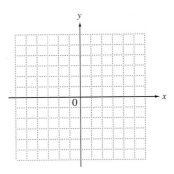

30. $-x + y = 6$

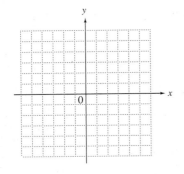

31. $6x + 5y = 30$

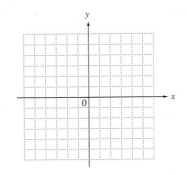

32. $3x + 4y = 12$

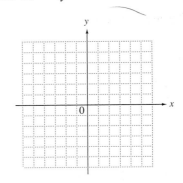

33. $x + 2 = 0$

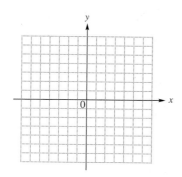

34. $x - 4 = 0$

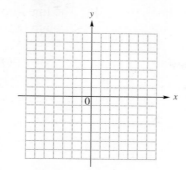

35. $y = 4x$

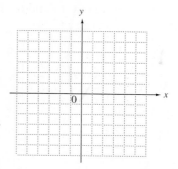

36. $y = -3x$

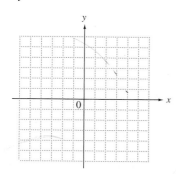

37. $y - 3 = 0$

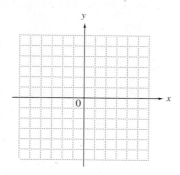

38. $y + 5 = 0$

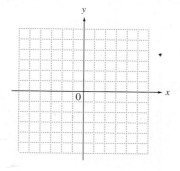

Use the method shown in Example 5 to graph each line.

39. Through $(-4, 2)$; $m = \dfrac{1}{2}$

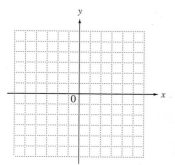

40. Through $(-2, -3)$; $m = \dfrac{5}{4}$

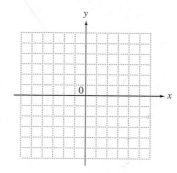

41. Through $(0, -2)$; $m = -\dfrac{2}{3}$

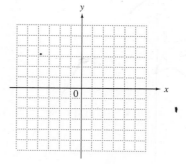

42. Through $(0, -4)$; $m = -\dfrac{3}{2}$

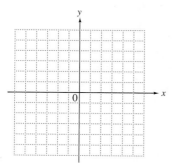

43. Through $(-1, -2)$; $m = 3$

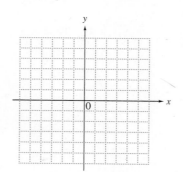

44. Through $(-2, -4)$; $m = 4$

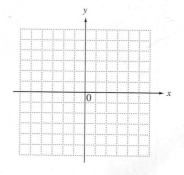

Decide whether each pair of lines is parallel, perpendicular, or neither. See Examples 6 and 7.

45. The line through $(4, 6)$ and $(-8, 7)$ and the line through $(-5, 5)$ and $(7, 4)$

46. The line through $(15, 9)$ and $(12, -7)$ and the line through $(8, -4)$ and $(5, -20)$

47. $2x + 5y = -7$ and $5x - 2y = 1$

48. $x + 4y = 7$ and $4x - y = 3$

49. $2x + y = 6$ and $x - y = 4$

50. $4x - 3y = 6$ and $3x - 4y = 2$

51. $3x = y$ and $2y - 6x = 5$

52. $x = 6$ and $6 - x = 8$

53. $2x + 5y = -8$ and $6 + 2x = 5y$

54. $4x + y = 0$ and $5x - 8 = 2y$

55. $4x - 3y = 8$ and $4y + 3x = 12$

56. $2x = y + 3$ and $2y + x = 3$

Use the concept of slope to solve each problem.

57. The upper deck at U.S. Cellular Field (formerly Comiskey Park) in Chicago has produced, among other complaints, displeasure with its steepness. It is 160 ft from home plate to the front of the upper deck and 250 ft from home plate to the back. The top of the upper deck is 63 ft above the bottom. What is its slope? (Consider the slope as a positive number.)

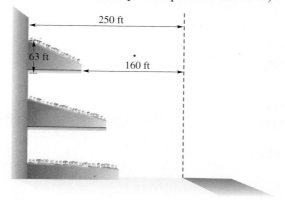

58. When designing the FleetCenter arena in Boston, architects designed the ramps leading up to the entrances so that circus elephants would be able to march up the ramps. The maximum grade (or slope) that an elephant will walk on is 13%. Suppose that such a ramp was constructed with a horizontal run of 150 ft. What would be the maximum vertical rise the architects could use?

Find and interpret the average rate of change illustrated in each graph.

59.

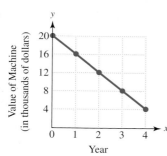

60.

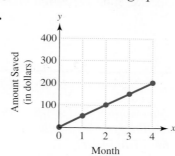

61.

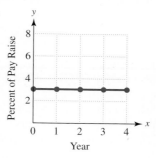

Use the idea of average rate of change to solve each problem. See Examples 8 and 9.

62. The graph provides a good approximation of the number of food stamp recipients (in millions) during 1996–2001.

(a) Use the given ordered pairs to find the average rate of change in food stamp recipients per year during this period.

(b) Interpret what a negative slope means in this situation.

FOOD STAMP RECIPIENTS

Source: U.S. Department of Agriculture.

63. The percent of tax returns filed electronically for the years 1997–2002 is shown in the graph.

   (a) Use the given ordered pairs to determine the average rate of change in the percent of tax returns filed electronically per year.

   (b) Explain how a positive rate of change is interpreted in this situation.

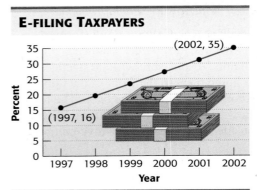

E-FILING TAXPAYERS

Source: Internal Revenue Service.

64. In 1997 when DVD players entered the market, .349 million (that is, 349,000) were sold. In 2002, sales of DVD players reached 15.5 million. Find and interpret the average rate of change in sales, in millions, per year. Round your answer to the nearest hundredth. (*Source: The Gazette,* June 22, 2002.)

65. When introduced in 1997, a DVD player sold for about $500. In 2002, the average price was $155. Find and interpret the average rate of change in price per year. (*Source: The Gazette,* June 22, 2002.)

RELATING CONCEPTS (EXERCISES 66–71)   For Individual or Group Work

*In these exercises we investigate a method of determining whether three points lie on the same straight line. (Such points are said to be **collinear**.) The points we consider are $A(3, 1)$, $B(6, 2)$, and $C(9, 3)$. **Work Exercises 66–71 in order.***

66. Find the slope of segment AB.

67. Find the slope of segment BC.

68. Find the slope of segment AC.

69. If slope of AB = slope of BC = slope of AC, then A, B, and C are collinear. Use the results of Exercises 66–68 to show that this statement is satisfied.

70. Use the slope formula to determine whether the points $(1, -2)$, $(3, -1)$, and $(5, 0)$ are collinear.

71. Repeat Exercise 70 for the points $(0, 6)$, $(4, -5)$, and $(-2, 12)$.

4.3 Linear Equations in Two Variables

OBJECTIVE 1 Write an equation of a line given its slope and *y*-intercept. In **Section 4.2** we found the slope of a line from the equation of the line by solving the equation for *y*. For example, we found that the slope of the line with equation

$$y = 4x + 8$$

is 4, the coefficient of x. What does the number **8** represent?

To find out, suppose a line has slope m and y-intercept $(0, b)$. We can find an equation of this line by choosing another point (x, y) on the line, as shown in Figure 19. Using the slope formula,

$$m = \frac{y - b}{x - 0}$$

$$m = \frac{y - b}{x}$$

$$mx = y - b \qquad \text{Multiply by } x.$$

$$mx + b = y \qquad \text{Add } b.$$

$$y = mx + b. \qquad \text{Rewrite.}$$

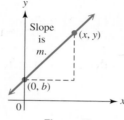

Figure 19

This last equation is called the *slope-intercept form* of the equation of a line, because we can identify the slope m and y-intercept $(0, b)$ at a glance. Thus, in the line with equation

$$y = 4x + 8,$$

the number 8 indicates that the y-intercept is $(0, 8)$.

Slope-Intercept Form

The **slope-intercept form** of the equation of a line with slope m and y-intercept $(0, b)$ is

$$y = mx + b.$$

Slope y-intercept is $(0, b)$.

EXAMPLE 1 Using the Slope-Intercept Form to Find an Equation of a Line

Find an equation of the line with slope $-\frac{4}{5}$ and y-intercept $(0, -2)$.

Here $m = -\frac{4}{5}$ and $b = -2$. Substitute these values into the slope-intercept form.

$$y = mx + b \qquad \text{Slope-intercept form}$$

$$y = -\frac{4}{5}x - 2 \qquad m = -\tfrac{4}{5}; b = -2$$

Work Problem 1 at the Side.

OBJECTIVES

1 Write an equation of a line given its slope and y-intercept.

2 Graph a line using its slope and y-intercept.

3 Write an equation of a line given its slope and a point on the line.

4 Write an equation of a line given two points on the line.

5 Write an equation of a line parallel or perpendicular to a given line.

6 Write an equation of a line that models real data.

1 Write an equation in slope-intercept form for each line with the given slope and y-intercept.

(a) Slope 2; y-intercept $(0, -3)$

(b) Slope $-\frac{2}{3}$; y-intercept $(0, 0)$

ANSWERS

1. (a) $y = 2x - 3$ **(b)** $y = -\dfrac{2}{3}x$

2 Graph each line using its slope and y-intercept.

(a) $y = 2x + 3$

(b) $3x + 4y = 8$

OBJECTIVE 2 Graph a line using its slope and y-intercept. If the equation of a line is written in slope-intercept form, we can use the slope and y-intercept to obtain its graph.

EXAMPLE 2 Graphing Lines Using Slope and y-Intercept

Graph each line using its slope and y-intercept.

(a) $y = 3x - 6$

Here $m = 3$ and $b = -6$. Plot the y-intercept $(0, -6)$. The slope 3 can be interpreted as

$$m = \frac{\text{rise}}{\text{run}} = \frac{\text{change in } y}{\text{change in } x} = \frac{3}{1}.$$

From $(0, -6)$, move *up* 3 units and to the *right* 1 unit, and plot a second point at $(1, -3)$. Join the two points with a straight line to obtain the graph in Figure 20.

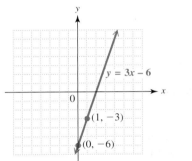

Figure 20

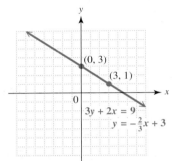

Figure 21

(b) $3y + 2x = 9$

Write the equation in slope-intercept form by solving for y.

$$3y + 2x = 9$$
$$3y = -2x + 9 \qquad \text{Subtract } 2x.$$
$$y = -\frac{2}{3}x + 3 \qquad \text{Slope-intercept form}$$

Slope $\longrightarrow$ $\longleftarrow$ y-intercept is $(0, 3)$.

To graph this equation, plot the y-intercept $(0, 3)$. The slope can be interpreted as either $\frac{-2}{3}$ or $\frac{2}{-3}$. Using $\frac{-2}{3}$, move from $(0, 3)$ *down* 2 units and to the *right* 3 units to locate the point $(3, 1)$. The line through these two points is the required graph. See Figure 21. (Verify that the point obtained using $\frac{2}{-3}$ as the slope is also on this line.)

Work Problem 2 at the Side.

NOTE
The slope-intercept form of a linear equation is the most useful for several reasons. Every linear equation (of a nonvertical line) has a *unique* (one and only one) slope-intercept form. In **Section 4.5** we study *linear functions,* which are defined using slope-intercept form. Also, this is the form we use when graphing a line with a graphing calculator.

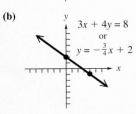

ANSWERS
2. **(a)** $y = 2x + 3$

(b) $3x + 4y = 8$ or $y = -\frac{3}{4}x + 2$

OBJECTIVE 3 **Write an equation of a line given its slope and a point on the line.** Let m represent the slope of a line and (x_1, y_1) represent a given point on the line. Let (x, y) represent any other point on the line. See Figure 22. Then by the slope formula,

$$m = \frac{y - y_1}{x - x_1}$$

$m(x - x_1) = y - y_1$ Multiply by $x - x_1$.

$y - y_1 = m(x - x_1)$.   Rewrite.

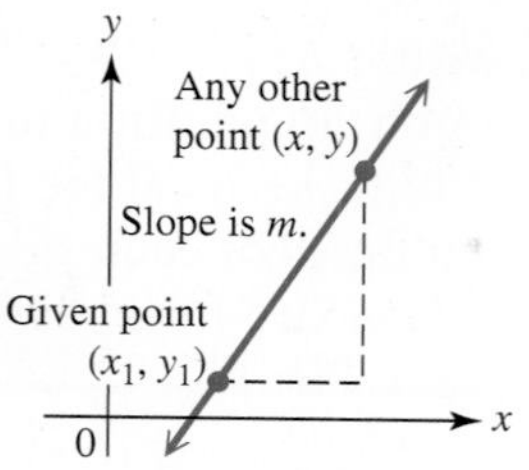

Figure 22

This last equation is the *point-slope form* of the equation of a line.

Point-Slope Form

The **point-slope form** of the equation of a line with slope m passing through the point (x_1, y_1) is

$$\overset{\text{Slope}}{\underset{\text{Given point}}{y - y_1 = m(x - x_1).}}$$

To use this form to write the equation of a line, we need to know the coordinates of a point (x_1, y_1) and the slope m of the line.

EXAMPLE 3 **Using the Point-Slope Form**

Find an equation of the line with slope $\frac{1}{3}$ passing through the point $(-2, 5)$.

Use the point-slope form of the equation of a line, with $(x_1, y_1) = (-2, 5)$ and $m = \frac{1}{3}$.

$y - y_1 = m(x - x_1)$ Point-slope form

$y - 5 = \frac{1}{3}[x - (-2)]$ $y_1 = 5, m = \frac{1}{3}, x_1 = -2$

$y - 5 = \frac{1}{3}(x + 2)$

$3y - 15 = x + 2$ Multiply by 3.

$-x + 3y = 17$ Subtract x; add 15.

In **Section 4.1,** we defined *standard form* for a linear equation as

$$Ax + By = C,$$

where A, B, and C are real numbers. Most often, however, A, B, and C are integers. In this case, let us agree that integers A, B, and C have no common factor (except 1) and $A \geq 0$. For example, the final equation in Example 3, $-x + 3y = 17$, is written in standard form as $x - 3y = -17$.

3 Write an equation of each line in standard form.

(a) Through $(-2, 7)$; $m = 3$

NOTE
The definition of "standard form" is not standard from one text to another. Any linear equation can be written in many different (all equally correct) forms. For example, the equation $2x + 3y = 8$ can be written as $2x = 8 - 3y$, $3y = 8 - 2x$, $x + \frac{3}{2}y = 4$, $4x + 6y = 16$, and so on. In addition to writing it in standard form $Ax + By = C$ with $A \geq 0$, let us agree that the form $2x + 3y = 8$ is preferred over any multiples of each side, such as $4x + 6y = 16$. (To write $4x + 6y = 16$ in standard form, divide each side by 2.)

◀◀◀ Work Problem 3 at the Side.

(b) Through $(1, 3)$; $m = -\dfrac{5}{4}$

OBJECTIVE 4 **Write an equation of a line given two points on the line.** To find an equation of a line when two points on the line are known, first use the slope formula to find the slope of the line. Then use the slope with either of the given points and the point-slope form of the equation of a line.

EXAMPLE 4 **Finding an Equation of a Line Given Two Points**

Find an equation of the line through the points $(-4, 3)$ and $(5, -7)$. Write the equation in standard form.

First find the slope by using the slope formula.

$$m = \frac{-7 - 3}{5 - (-4)} = -\frac{10}{9}$$

4 Write an equation in standard form for each line.

(a) Through $(-1, 2)$ and $(5, 7)$

Use either $(-4, 3)$ or $(5, -7)$ as (x_1, y_1) in the point-slope form of the equation of a line. If we choose $(-4, 3)$, then $-4 = x_1$ and $3 = y_1$.

$$y - y_1 = m(x - x_1) \qquad \text{Point-slope form}$$

$$y - 3 = -\frac{10}{9}[x - (-4)] \qquad y_1 = 3,\, m = -\tfrac{10}{9},\, x_1 = -4$$

$$y - 3 = -\frac{10}{9}(x + 4)$$

$$9y - 27 = -10x - 40 \qquad \text{Multiply by 9; distributive property}$$

$$10x + 9y = -13 \qquad \text{Standard form}$$

Verify that if $(5, -7)$ were used, the same equation would result.

◀◀◀ Work Problem 4 at the Side.

(b) Through $(-2, 6)$ and $(1, 4)$

A horizontal line has slope 0. From the point-slope form, the equation of a horizontal line through the point (a, b) is

$$y - y_1 = m(x - x_1) \qquad \text{Point-slope form}$$

$$y - b = 0(x - a) \qquad y_1 = b,\, m = 0,\, x_1 = a$$

$$y - b = 0 \qquad \text{Multiplication property of 0}$$

$$y = b. \qquad \text{Add } b.$$

Notice that the point-slope form does not apply to a vertical line, since the slope of a vertical line is undefined. A vertical line through the point (a, b) has equation $x = a$.

ANSWERS
3. (a) $3x - y = -13$ (b) $5x + 4y = 17$
4. (a) $5x - 6y = -17$ (b) $2x + 3y = 14$

In summary, horizontal and vertical lines have the following special equations.

> ### Equations of Horizontal and Vertical Lines
> The horizontal line through the point (a, b) has equation $y = b$.
> The vertical line through the point (a, b) has equation $x = a$.

Work Problem 5 at the Side. ▶▶▶

OBJECTIVE 5   Write an equation of a line parallel or perpendicular to a given line. As mentioned in the previous section, parallel lines have the same slope and perpendicular lines have slopes with product -1.

EXAMPLE 5   Finding Equations of Lines Parallel or Perpendicular to a Given Line

Find the equation in slope-intercept form of the line passing through the point $(-4, 5)$ and **(a)** parallel to the line $2x + 3y = 6$; **(b)** perpendicular to the line $2x + 3y = 6$.

(a) The slope of the line $2x + 3y = 6$ can be found by solving for y.

$$2x + 3y = 6$$
$$3y = -2x + 6 \qquad \text{Subtract } 2x.$$
$$y = -\frac{2}{3}x + 2 \qquad \text{Divide by 3.}$$
$$\uparrow \text{ Slope}$$

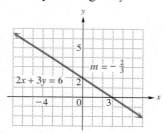

The slope is given by the coefficient of x, so $m = -\frac{2}{3}$. See the figure. Since parallel lines have the same slope, the required equation of the line through $(-4, 5)$ and parallel to $2x + 3y = 6$ must also have slope $-\frac{2}{3}$. To find this equation, use the point-slope form, with $(x_1, y_1) = (-4, 5)$ and $m = -\frac{2}{3}$.

$$y - 5 = -\frac{2}{3}[x - (-4)] \qquad y_1 = 5, m = -\tfrac{2}{3}, x_1 = -4$$

$$y - 5 = -\frac{2}{3}(x + 4)$$

$$y - 5 = -\frac{2}{3}x - \frac{8}{3} \qquad \text{Distributive property}$$

$$y = -\frac{2}{3}x - \frac{8}{3} + \frac{15}{3} \qquad \text{Add } 5 = \tfrac{15}{3}.$$

$$y = -\frac{2}{3}x + \frac{7}{3} \qquad \text{Combine like terms.}$$

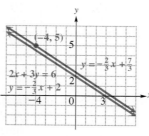

We did not clear fractions after the substitution step here because we want the equation in slope-intercept form—that is, solved for y. Both lines are shown in the figure.

——— Continued on Next Page

5 Write an equation for each line.

(a) Through $(8, -2)$; $m = 0$

(b) The vertical line through $(3, 5)$

ANSWERS
5. (a) $y = -2$ **(b)** $x = 3$

6 Write an equation in slope-intercept form of the line passing through the point $(-8, 3)$ and

(a) parallel to the line $2x - 3y = 10$.

(b) In part (a), the given line $2x + 3y = 6$ was written as

$$y = -\frac{2}{3}x + 2,$$

so the line has slope $-\frac{2}{3}$. To be perpendicular to the line $2x + 3y = 6$, a line must have a slope that is the negative reciprocal of $-\frac{2}{3}$, which is $\frac{3}{2}$. Use the point $(-4, 5)$ and slope $\frac{3}{2}$ in the point-slope form to get the equation of the perpendicular line shown in the figure.

$$y - 5 = \frac{3}{2}[x - (-4)] \qquad y_1 = 5, m = \frac{3}{2}, x_1 = -4$$

$$y - 5 = \frac{3}{2}(x + 4)$$

$$y - 5 = \frac{3}{2}x + 6 \qquad \text{Distributive property}$$

$$y = \frac{3}{2}x + 11 \qquad \text{Add 5.}$$

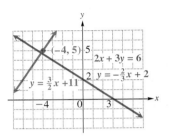

▶◀ **Work Problem 6 at the Side.**

A summary of the various forms of linear equations follows.

FORMS OF LINEAR EQUATIONS

Equation	Description	When to Use
$y = mx + b$	**Slope-Intercept Form** Slope is m. y-intercept is $(0, b)$.	The slope and y-intercept can be easily identified and used to quickly graph the equation.
$y - y_1 = m(x - x_1)$	**Point-Slope Form** Slope is m. Line passes through (x_1, y_1).	This form is ideal for finding the equation of a line if the slope and a point on the line or two points on the line are known.
$Ax + By = C$	**Standard Form** (A, B, and C integers, $A \geq 0$) Slope is $-\frac{A}{B}$ ($B \neq 0$). x-intercept is $(\frac{C}{A}, 0)$ ($A \neq 0$). y-intercept is $(0, \frac{C}{B})$ ($B \neq 0$).	The x- and y-intercepts can be found quickly and used to graph the equation. Slope must be calculated.
$y = b$	**Horizontal Line** Slope is 0. y-intercept is $(0, b)$.	If the graph intersects only the y-axis, then y is the only variable in the equation.
$x = a$	**Vertical Line** Slope is undefined. x-intercept is $(a, 0)$.	If the graph intersects only the x-axis, then x is the only variable in the equation.

(b) perpendicular to the line $2x - 3y = 10$.

OBJECTIVE 6 **Write an equation of a line that models real data.** We can use the information presented in this section to write equations of lines that mathematically describe, or *model*, real data if the given set of data changes at a fairly constant rate. In this case, the data fit a linear pattern, and the rate of change is the slope of the line.

ANSWERS

6. **(a)** $y = \frac{2}{3}x + \frac{25}{3}$ **(b)** $y = -\frac{3}{2}x - 9$

EXAMPLE 6 **Determining a Linear Equation to Describe Real Data**

Suppose it is time to fill your car with gasoline. At your local station, 89-octane gas is selling for $1.80 per gal.

(a) Write an equation that describes the cost y to buy x gallons of gas.

Experience has taught you that the total price you pay is determined by the number of gallons you buy multiplied by the price per gallon (in this case, $1.80). As you pump the gas, two sets of numbers spin by: the number of gallons pumped and the price for that number of gallons.

The table uses ordered pairs to illustrate this situation.

Number of Gallons Pumped	Price of This Number of Gallons
0	0($1.80) = $0.00
1	1($1.80) = $1.80
2	2($1.80) = $3.60
3	3($1.80) = $5.40
4	4($1.80) = $7.20

If we let x denote the number of gallons pumped, then the total price y in dollars can be found by the linear equation

Total price ⟶ ⟵ Number of gallons

$$y = 1.80x.$$

Theoretically, there are infinitely many ordered pairs (x, y) that satisfy this equation, but here we are limited to nonnegative values for x, since we cannot have a negative number of gallons. There is also a practical maximum value for x in this situation, which varies from one car to another. What determines this maximum value?

(b) You can also get a car wash at the gas station if you pay an additional $3.00. Write an equation that defines the price for gas and a car wash.

Since an additional $3.00 will be charged, you pay $1.80x + 3.00$ dollars for x gallons of gas and a car wash, described by

$$y = 1.8x + 3. \quad \text{Delete unnecessary 0s.}$$

(c) Interpret the ordered pairs $(5, 12)$ and $(10, 21)$ in relation to the equation from part (b).

The ordered pair $(5, 12)$ indicates that the price of 5 gal of gas and a car wash is $12.00. Similarly, $(10, 21)$ indicates that the price of 10 gal of gas and a car wash is $21.00.

Work Problem 7 at the Side.

NOTE

In Example 6(a), the ordered pair $(0, 0)$ satisfied the equation, so the linear equation has the form $y = mx$, where $b = 0$. If a situation involves an initial charge b plus a charge per unit m as in Example 6(b), the equation has the form $y = mx + b$, where $b \neq 0$.

7 **(a)** Suppose it costs $.10 per minute to make a long-distance call. Write an equation to describe the cost y to make an x-minute call.

(b) Suppose there is a flat rate of $.20 plus a charge of $.10 per minute to make a call. Write an equation that gives the cost y for a call of x minutes.

(c) Interpret the ordered pair $(15, 1.7)$ in relation to the equation from part (b).

ANSWERS

7. **(a)** $y = .1x$ (Note: $.10x = .1x$)
(b) $y = .1x + .2$
(c) The ordered pair $(15, 1.7)$ indicates that the price of a 15-minute call is $1.70.

8 The percent of mothers of children under 1 yr old who participated in the U.S. labor force is shown in the table for selected years.

Year	Percent
1980	38
1984	47
1988	51
1992	54
1998	59

Source: U.S. Bureau of the Census.

(a) Let $x = 0$ represent 1980, $x = 4$ represent 1984, and so on. Use the data for 1980 and 1998 to find an equation that models the data.

(b) Use the equation from part (a) to approximate the percentage of mothers of children under 1 yr old who participated in the U.S. labor force in 2000.

EXAMPLE 7 **Finding an Equation of a Line That Models Data**

Average annual tuition and fees for in-state students at public 4-year colleges are shown in the table for selected years and graphed as ordered pairs of points in the *scatter diagram* in Figure 23, where $x = 0$ represents 1990, $x = 4$ represents 1994, and so on, and y represents the cost in dollars.

Year	Cost (in dollars)
1990	2035
1994	2820
1996	3151
1998	3486
2000	3774

Source: U.S. National Center for Education Statistics.

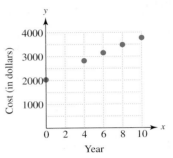

Figure 23

(a) Find an equation that models the data.

Since the points in Figure 23 lie approximately on a straight line, we can write a linear equation that models the relationship between year x and cost y. We choose two data points, $(0, 2035)$ and $(10, 3774)$, to find the slope of the line.

$$m = \frac{3774 - 2035}{10 - 0} = \frac{1739}{10} = 173.9$$

The slope 173.9 indicates that the cost of tuition and fees for in-state students at public 4-year colleges increased by about $174 per year from 1990 to 2000. We use this slope, the y-intercept $(0, \mathbf{2035})$, and the slope-intercept form to write an equation of the line. Thus,

$$y = \mathbf{173.9}x + \mathbf{2035}.$$

(b) Use the equation from part (a) to approximate the cost of tuition and fees at public 4-year colleges in 2002.

The value $x = 12$ corresponds to the year 2002, so we substitute 12 for x in the equation.

$$y = 173.9x + 2035$$
$$y = 173.9(\mathbf{12}) + 2035$$
$$y = 4121.8$$

According to the model, average tuition and fees for in-state students at public 4-year colleges in 2002 were about $4122.

NOTE
In Example 7, if we had chosen different data points, we would have found a slightly different equation. However, all such equations should yield similar results, since the data points are approximately linear.

◄◄◄ **Work Problem 8 at the Side.**

ANSWERS
8. (a) $y = 1.17x + 38$ **(b)** 61%

31. Through $(-5, 4)$; $m = \dfrac{1}{2}$

32. Through $(7, -2)$; $m = \dfrac{1}{4}$

33. Through $(-4, 12)$; horizontal

34. Through $(1, 5)$; horizontal

Write an equation that satisfies the given conditions.

35. Through $(9, 10)$; undefined slope

36. Through $(-2, 8)$; 0 slope

37. Through $(.5, .2)$; horizontal

38. Through $\left(\dfrac{5}{8}, \dfrac{2}{9}\right)$; vertical

Write the equation in standard form of the line through the given points. See Example 4.

39. $(3, 4)$ and $(5, 8)$

40. $(5, -2)$ and $(-3, 14)$

41. $(6, 1)$ and $(-2, 5)$

42. $(-2, 5)$ and $(-8, 1)$

43. $\left(-\dfrac{2}{5}, \dfrac{2}{5}\right)$ and $\left(\dfrac{4}{3}, \dfrac{2}{3}\right)$

44. $\left(\dfrac{3}{4}, \dfrac{8}{3}\right)$ and $\left(\dfrac{2}{5}, \dfrac{2}{3}\right)$

45. $(2, 5)$ and $(1, 5)$

46. $(-2, 2)$ and $(4, 2)$

47. $(7, 6)$ and $(7, -8)$

48. $(13, 5)$ and $(13, -1)$

Write the equation in slope-intercept form of the line satisfying the given conditions. See Example 5.

49. Through $(7, 2)$; parallel to $3x - y = 8$

50. Through $(4, 1)$; parallel to $2x + 5y = 10$

51. Through $(-2, -2)$; parallel to $-x + 2y = 10$

52. Through $(-1, 3)$; parallel to $-x + 3y = 12$

53. Through $(8, 5)$; perpendicular to $2x - y = 7$

54. Through $(2, -7)$; perpendicular to $5x + 2y = 18$

55. Through $(-2, 7)$; perpendicular to $x = 9$

56. Through $(8, 4)$; perpendicular to $x = -3$

Write an equation in the form $y = mx$ for each situation. Then give the three ordered pairs associated with the equation for x-values of 0, 5, and 10. See Example 6(a).

57. x represents the number of hours traveling at 45 mph, and y represents the distance traveled (in miles).

58. x represents the number of compact discs sold at $16 each, and y represents the total cost of the discs (in dollars).

59. x represents the number of gallons of gas sold at $2.00 per gal, and y represents the total cost of the gasoline (in dollars).

60. x represents the number of days a DVD movie is rented at $2.50 per day, and y represents the total charge for the rental (in dollars).

*For each situation, **(a)** write an equation in the form $y = mx + b$, **(b)** find and interpret the ordered pair associated with the equation for $x = 5$, and **(c)** answer the question. See Examples 6(b) and 6(c).*

61. A membership to the Midwest Athletic Club costs $99 plus $39 per month. (*Source:* Midwest Athletic Club.) Let x represent the number of months and y represent the cost. How much does the first year's membership cost?

62. For a family membership, the athletic club in Exercise 61 charges a membership fee of $159 plus $60 for each additional family member after the first. Let x represent the number of additional family members and y represent the cost. What is the membership fee for a four-person family?

63. A cell phone plan includes 900 anytime minutes for $50 per month, plus a one-time activation fee of $25. A Nokia 5165 cell phone is included at no additional charge. (*Source:* U.S. Cellular.) Let x represent the number of months of service and y represent the cost. If you sign a 2-yr contract, how much will this cell phone plan cost? (Assume that you never use more than the allotted number of minutes.)

64. Another cell phone plan includes 450 anytime minutes for $35 per month, plus $19.95 for a Nokia 5165 cell phone and $25 for a one-time activation fee. (*Source:* U.S. Cellular.) Let x represent the number of months of service and y represent the cost. If you sign a 1-yr contract, how much will this cell phone plan cost? (Assume that you never use more than the allotted number of minutes.)

65. A rental car costs $50 plus $.20 per mile. Let x represent the number of miles driven and y represent the total charge to the renter. How many miles was the car driven if the renter paid $84.60?

66. There is a $30 fee to rent a chain saw, plus $6 per day. Let x represent the number of days the saw is rented and y represent the charge to the user in dollars. If the total charge is $138, for how many days is the saw rented?

Solve each problem. In part (a), give equations in slope-intercept form. See Example 7.
(*Source for Exercises 67 and 68:* Jupiter Media Metrix.)

67. The percent of households that access the Internet by high-speed broadband is shown in the graph, where the year 2000 corresponds to $x = 0$.

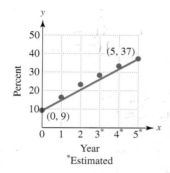

Year
*Estimated

(a) Use the ordered pairs from the graph to write an equation that models the data. What does the slope tell us in the context of this problem?

(b) Use the equation from part (a) to predict the percent of U.S. households that will access the Internet by broadband in 2006. Round your answer to the nearest percent.

68. The percent of U.S. households that access the Internet by dial-up is shown in the graph, where the year 2000 corresponds to $x = 0$.

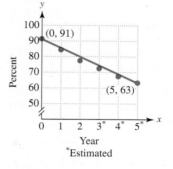

Year
*Estimated

(a) Use the ordered pairs from the graph to write an equation that models the data. What does the slope tell us in the context of this problem?

(b) Use the equation from part (a) to predict the percent of U.S. households that will access the Internet by dial-up in 2006. Round your answer to the nearest percent.

69. The number of post offices in the United States is shown in the bar graph.

(a) Use the information given for the years 1995 and 2000, letting $x = 5$ represent 1995, $x = 10$ represent 2000, and y represent the number of post offices, to write an equation that models the data.

(b) Use the equation to approximate the number of post offices in 1998. How does this result compare to the actual value, 27,952?

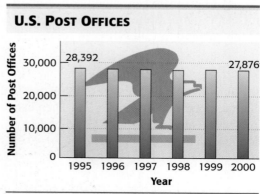

U.S. POST OFFICES

Source: U.S. Postal Service, *Annual Report of the Postmaster General.*

70. Median household income of African-Americans is shown in the bar graph.

(a) Use the information given for the years 1995 and 2000, letting $x = 5$ represent 1995, $x = 10$ represent 2000, and y represent the median income, to write an equation that models median household income.

(b) Use the equation to approximate the median income for 1997. How does your result compare to the actual value, $25,050?

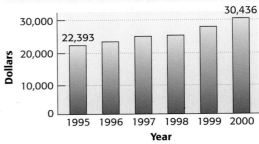

MEDIAN HOUSEHOLD INCOME FOR AFRICAN–AMERICANS

Source: U.S. Bureau of the Census.

RELATING CONCEPTS (EXERCISES 71–80)   For Individual or Group Work

*In **Section 2.2** we learned how formulas can be applied to problem solving. In Exercises 71–80, we will see how the formula that relates Celsius and Fahrenheit temperatures is derived.*
Work Exercises 71–80 in order.

71. There is a linear relationship between Celsius and Fahrenheit temperatures. When $C = 0°$,

$F = $ _____ $°$, and when $C = 100°$,

$F = $ _____ $°$.

72. Think of ordered pairs of temperatures (C, F), where C and F represent corresponding Celsius and Fahrenheit temperatures. The equation that relates the two scales has a straight-line graph that contains the two points determined in Exercise 71. What are these two points?

73. Find the slope of the line described in Exercise 72.

74. Now think of the point-slope form of the equation in terms of C and F, where C replaces x and F replaces y. Use the slope you found in Exercise 73 and one of the two points determined earlier, and find the equation that gives F in terms of C.

75. To obtain another form of the formula, use the equation you found in Exercise 74 and solve for C in terms of F.

76. For what temperature does $F = C$?

77. A quick way to estimate Fahrenheit temperature for a given Celsius temperature is to double C and add 30. Use this method to find F if $C = 15$.

78. Use the equation found in Exercise 74 to find F if $C = 15$. How does the answer compare with your answer to Exercise 77?

79. Use the method given in Exercise 77 to estimate the Fahrenheit temperature given $C = 30$. Then use the equation from Exercise 74 to find F when $C = 30$. How do the temperatures compare?

80. Explain why the method given in Exercise 77 to estimate Fahrenheit temperature gives a good approximation of $F = \frac{9}{5}C + 32$.

4.4 Linear Inequalities in Two Variables

OBJECTIVE 1 Graph linear inequalities in two variables. In **Section 3.1** we graphed linear inequalities in one variable on the number line. We now graph linear inequalities in two variables on a rectangular coordinate system.

OBJECTIVES

1. Graph linear inequalities in two variables.
2. Graph the intersection of two linear inequalities.
3. Graph the union of two linear inequalities.

Linear Inequality in Two Variables

An inequality that can be written as

$$Ax + By < C \quad \text{or} \quad Ax + By > C,$$

where A, B, and C are real numbers and A and B are not both 0, is a **linear inequality in two variables.**

The symbols $\leq$ and $\geq$ may replace $<$ and $>$ in the definition.

Consider the graph in Figure 24. The graph of the line $x + y = 5$ divides the points in the rectangular coordinate system into three sets:

1. Those points that lie on the line itself and satisfy the equation $x + y = 5$ [like $(0, 5)$, $(2, 3)$, and $(5, 0)$];

2. Those that lie in the half-plane above the line and satisfy the inequality $x + y > 5$ [like $(5, 3)$ and $(2, 4)$];

3. Those that lie in the half-plane below the line and satisfy the inequality $x + y < 5$ [like $(0, 0)$ and $(-3, -1)$].

The graph of the line $x + y = 5$ is called the **boundary line** for the inequalities $x + y > 5$ and $x + y < 5$. Graphs of linear inequalities in two variables are *regions* in the real number plane that may or may not include boundary lines.

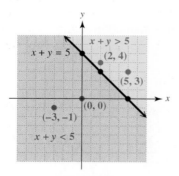

Figure 24

To graph a linear inequality in two variables, follow these steps.

Graphing a Linear Inequality

Step 1 **Draw the graph of the straight line that is the boundary.** Make the line solid if the inequality involves $\leq$ or $\geq$; make the line dashed if the inequality involves $<$ or $>$.

Step 2 **Choose a test point.** Choose any point not on the line, and substitute the coordinates of this point in the inequality.

Step 3 **Shade the appropriate region.** Shade the region that includes the test point if it satisfies the original inequality; otherwise, shade the region on the other side of the boundary line.

1 Graph each inequality.

(a) $x + y \le 4$

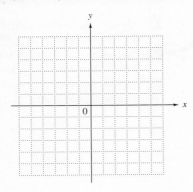

(b) $3x + y \ge 6$

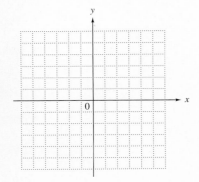

EXAMPLE 1 **Graphing a Linear Inequality**

Graph $3x + 2y \ge 6$.

Step 1 First graph the line $3x + 2y = 6$. The graph of this line, the boundary of the graph of the inequality, is shown in Figure 25.

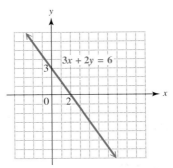

Figure 25

Step 2 The graph of the inequality $3x + 2y \ge 6$ includes the points of the boundary line $3x + 2y = 6$ and either the points *above* the line $3x + 2y = 6$ or the points *below* that line. To decide which, select any point not on the line $3x + 2y = 6$ as a test point. The origin, $(0, 0)$, is often a good choice. Substitute the values from the test point $(0, 0)$ for x and y in the inequality.

$$3x + 2y > 6$$
$$3(0) + 2(0) > 6 \quad ?$$
$$0 > 6 \qquad \text{False}$$

Step 3 Because the result is false, $(0, 0)$ does *not* satisfy the inequality, and so the solution set includes all points on the other side of the line. This region is shaded in Figure 26.

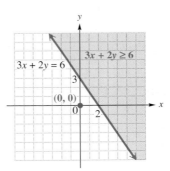

Figure 26

◀◀◀ Work Problem 1 at the Side.

If the inequality is written in the form $y > mx + b$ or $y < mx + b$, the inequality symbol indicates which half-plane to shade.

If $y > mx + b$, shade **above** the boundary line.

If $y < mx + b$, shade **below** the boundary line.

This method works only if the inequality is solved for y.

ANSWERS

1. (a)

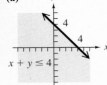

(b)

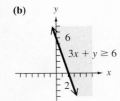

EXAMPLE 2 **Graphing a Linear Inequality**

Graph $x - 3y < 4$.

First graph the boundary line, shown in Figure 27. The points of the boundary line do not belong to the inequality $x - 3y < 4$ (because the inequality symbol is $<$, not $\leq$). For this reason, the line is dashed. Now solve the inequality for y.

$$x - 3y < 4$$
$$-3y < -x + 4 \qquad \text{Subtract } x.$$
$$y > \frac{x}{3} - \frac{4}{3} \qquad \text{Multiply by } -\tfrac{1}{3}; \text{ change } < \text{ to } >.$$

Because of the *is greater than* symbol, shade *above* the line. As a check, choose a test point not on the line, say $(1, 2)$, and substitute for x and y in the original inequality.

$$x - 3y < 4$$
$$1 - 3(2) < 4 \qquad ?$$
$$-5 < 4 \qquad \text{True}$$

This result agrees with the decision to shade above the line. The solution set, graphed in Figure 27, includes only those points in the shaded half-plane (not those on the line).

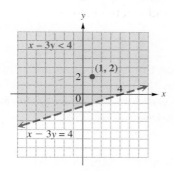

Figure 27

Work Problem 2 at the Side. ▶▶▶

OBJECTIVE 2 **Graph the intersection of two linear inequalities.** In **Section 3.2** we discussed how the words *and* and *or* are used with compound inequalities. In that section, the inequalities had one variable. Those ideas can be extended to include inequalities in two variables.

A pair of inequalities joined with the word *and* is interpreted as the intersection of the solution sets of the inequalities. ***The graph of the intersection of two or more inequalities is the region of the plane where all points satisfy all of the inequalities at the same time.***

2 Graph each inequality.

(a) $x - y > 2$

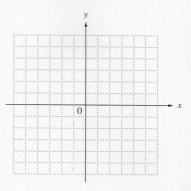

(b) $3x + 4y < 12$

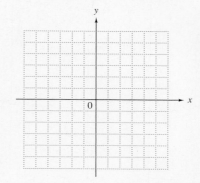

ANSWERS
2. **(a)**

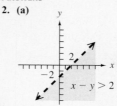

(b)

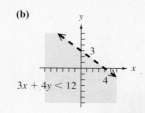

❸ Graph $x - y \leq 4$ and $x \geq -2$.

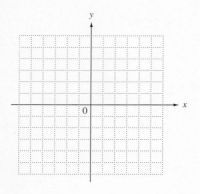

❹ Graph $7x - 3y < 21$ or $x > 2$.

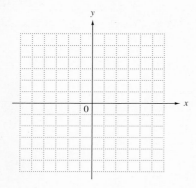

EXAMPLE 3 Graphing the Intersection of Two Inequalities

Graph $2x + 4y \geq 5$ and $x \geq 1$.

To begin, we graph each of the two inequalities $2x + 4y \geq 5$ and $x \geq 1$ separately. The graph of $2x + 4y \geq 5$ is shown in Figure 28(a), and the graph of $x \geq 1$ is shown in Figure 28(b).

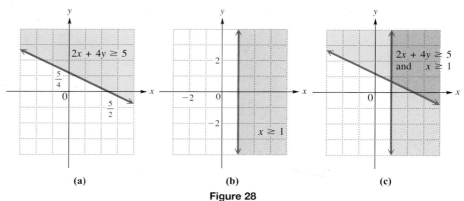

Figure 28

In practice, the two graphs in Figures 28(a) and 28(b) are graphed on the same axes. Then we use heavy shading to identify the intersection of the graphs, as shown in Figure 28(c). To check, we can use a test point from each of the four regions formed by the intersection of the boundary lines. Verify that only ordered pairs in the heavily shaded region satisfy both inequalities.

◀◀◀ **Work Problem 3 at the Side.**

OBJECTIVE ❸ Graph the union of two linear inequalities. When two inequalities are joined by the word *or,* we must find the union of the graphs of the inequalities. ***The graph of the union of two inequalities includes all of the points that satisfy either inequality.***

EXAMPLE 4 Graphing the Union of Two Inequalities

Graph $2x + 4y \geq 5$ or $x \geq 1$.

The graphs of the two inequalities are shown in Figures 28(a) and 28(b) in Example 3. The graph of the union is shown in Figure 29.

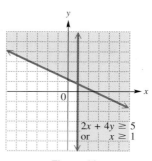

Figure 29

◀◀◀ **Work Problem 4 at the Side.**

ANSWERS

3.

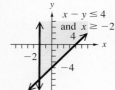

4.

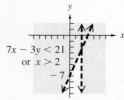

4.4 Exercises

FOR EXTRA HELP   Addison-Wesley Math Tutor Center   MathXL   Digital Video Tutor CD 2 Videotape 2   Student's Solutions Manual   MyMathLab   Interactmath.com

In each statement, fill in the first blank with one of the words solid *or* dashed. *Fill in the second blank with one of the words* above *or* below.

1. The boundary of the graph of $y \leq -x + 2$ will be a _____ line, and the shading will be _____ the line.

2. The boundary of the graph of $y < -x + 2$ will be a _____ line, and the shading will be _____ the line.

3. The boundary of the graph of $y > -x + 2$ will be a _____ line, and the shading will be _____ the line.

4. The boundary of the graph of $y \geq -x + 2$ will be a _____ line, and the shading will be _____ the line.

5. How is the boundary line $Ax + By = C$ used in graphing either $Ax + By < C$ or $Ax + By > C$?

6. Describe the two methods discussed in the text for deciding which region is the solution set of a linear inequality in two variables.

Graph each linear inequality. See Examples 1 and 2.

7. $x + y \leq 2$

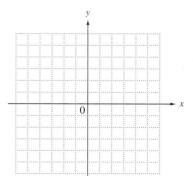

8. $x + y \leq -3$

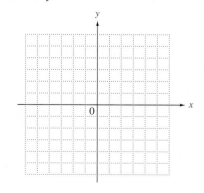

9. $4x - y < 4$

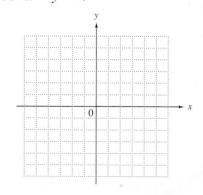

10. $3x - y < 3$

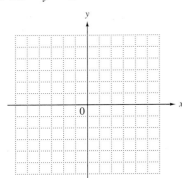

11. $x + 3y \geq -2$

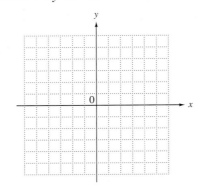

12. $x + 4y \geq -3$

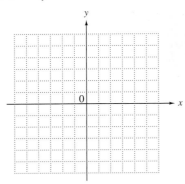

13. $x + y > 0$

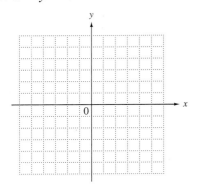

14. $x + 2y > 0$

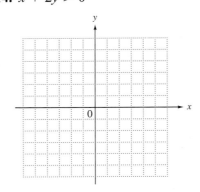

15. $x - 3y \leq 0$

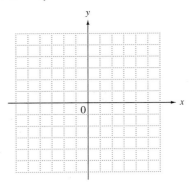

16. $x - 5y \leq 0$

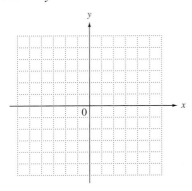

17. $y < x$

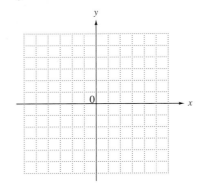

18. $y \leq 4x$

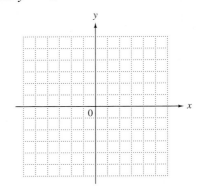

Graph the intersection of each pair of inequalities. See Example 3.

19. $x + y \leq 1$ and $x \geq 1$

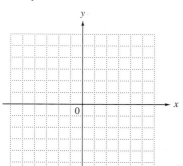

20. $x - y \geq 2$ and $x \geq 3$

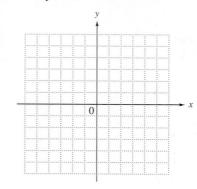

21. $2x - y \geq 2$ and $y < 4$

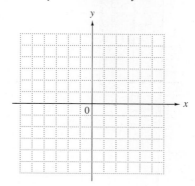

22. $3x - y \geq 3$ and $y < 3$

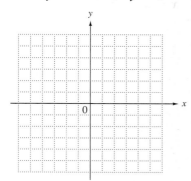

23. $x + y > -5$ and $y < -2$

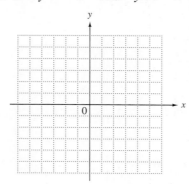

24. $6x - 4y < 10$ and $y > 2$

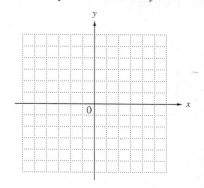

*Use the method described in **Section 3.3** to write each inequality as a compound inequality, and graph its solution set in the rectangular coordinate plane.*

25. $|x| \geq 3$

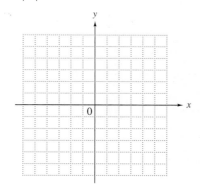

26. $|y| < 5$

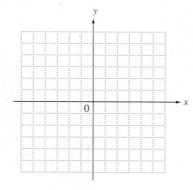

27. $|y + 1| < 2$

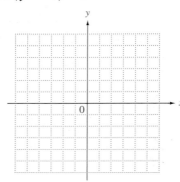

28. $|x - 2| \geq 1$

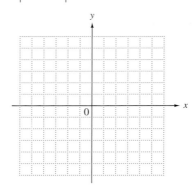

Graph the union of each pair of inequalities. See Example 4.

29. $x - y \geq 1$ or $y \geq 2$

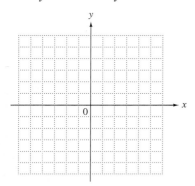

30. $x + y \leq 2$ or $y \geq 3$

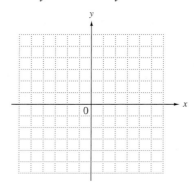

31. $x - 2 > y$ or $x < 1$

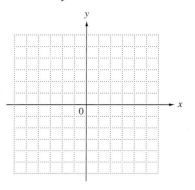

32. $x + 3 < y$ or $x > 3$

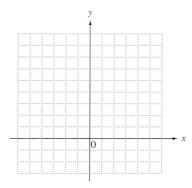

33. $3x + 2y < 6$ or $x - 2y > 2$

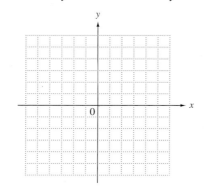

34. $x - y \geq 1$ or $x + y \leq 4$

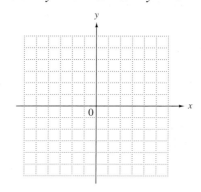

4.5 Introduction to Functions

We often describe one quantity in terms of another. Consider the following.

- The amount of your paycheck if you are paid hourly depends on the number of hours you worked.
- The cost at the gas station depends on the number of gallons of gas you pumped into your car.
- The distance traveled by a car moving at a constant speed depends on the time traveled.

We can use ordered pairs to represent these corresponding quantities. For example, we indicate the relationship between the amount of your paycheck and hours worked by writing ordered pairs in which the first number represents hours worked and the second number represents paycheck amount in dollars. Then the ordered pair (**5**, **40**) indicates that when you work **5** hr, your paycheck is $**40**. Similarly, the ordered pairs (10, 80) and (20, 160) show that working 10 hr results in an $80 paycheck and working 20 hr results in a $160 paycheck.

Work Problem 1 at the Side. ▶▶▶

Since the amount of your paycheck *depends* on the number of hours worked, your paycheck amount is called the *dependent variable,* and the number of hours worked is called the *independent variable.* Generalizing, if the value of the variable y depends on the value of the variable x, then y is the **dependent variable** and x is the **independent variable.**

Independent variable ⟶ ↓ ↓ ⟵ Dependent variable

$$(x, y)$$

OBJECTIVE 1 Define and identify relations and functions.   Since we can write related quantities using ordered pairs, a set of ordered pairs such as

$$\{(5, 40), (10, 80), (20, 160), (40, 320)\}$$

is called a *relation.*

Relation

A **relation** is any set of ordered pairs.

A special kind of relation, called a *function,* is very important in mathematics and its applications.

Function

A **function** is a relation in which, for each value of the first component of the ordered pairs, there is *exactly one value* of the second component.

OBJECTIVES

1. Define and identify relations and functions.
2. Find domain and range.
3. Identify functions defined by graphs and equations.
4. Use function notation.
5. Identify linear functions.

1 What would the ordered pair (40, 320) in the correspondence between number of hours worked and paycheck amount (in dollars) indicate?

ANSWERS

1. It indicates that when you work 40 hr, your paycheck is $320.

2 Determine whether each relation defines a function.

(a) $\{(0, 3), (-1, 2), (-1, 3)\}$

EXAMPLE 1 **Determining whether Relations Are Functions**

Tell whether each relation defines a function.

$$F = \{(1, 2), (-2, 4), (3, -1)\}$$
$$G = \{(-2, -1), (-1, 0), (0, 1), (1, 2), (2, 2)\}$$
$$H = \{(-4, 1), (-2, 1), (-2, 0)\}$$

Relations F and G are functions, because for each different x-value there is exactly one y-value. Notice that in G, the last two ordered pairs have the same y-value (1 is paired with 2, and 2 is paired with 2). This does not violate the definition of function, since the first components (x-values) are different and each is paired with only one second component (y-value).

In relation H, however, the last two ordered pairs have the *same x*-value paired with *two different y*-values (-2 is paired with both 1 and 0), so H is a relation but not a function. ***In a function, no two ordered pairs can have the same first component and different second components.***

Different y-values

$$H = \{(-4, 1), (\mathbf{-2, 1}), (\mathbf{-2, 0})\} \qquad \text{Not a function}$$

Same x-value

(b) $\{(2, -2), (4, -4), (6, -6)\}$

Work Problem 2 at the Side.

In a function, there is *exactly one* value of the dependent variable, the second component, for each value of the independent variable, the first component. This is what makes functions so important in applications.

Relations and functions can also be expressed as a correspondence or *mapping* from one set to another, as shown in Figure 30 for function F and relation H from Example 1. The arrow from 1 to 2 indicates that the ordered pair (1, 2) belongs to F—each first component is paired with exactly one second component. In the mapping for set H, which is not a function, the first component -2 is paired with two different second components, 1 and 0.

(c) $\{(-1, 5), (0, 5)\}$

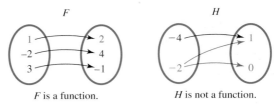

F is a function. H is not a function.

Figure 30

Since relations and functions are sets of ordered pairs, we can represent them using tables and graphs. A table and graph for function F is shown in Figure 31.

x	y
1	2
-2	4
3	-1

Graph of F

Figure 31

ANSWERS
2. (a) not a function (b) function
   (c) function

Finally, we can describe a relation or function using a rule that tells how to determine the dependent variable for a specific value of the independent variable. The rule may be given in words, such as "the dependent variable is twice the independent variable." Usually, however, the rule is given as an equation:

$$y = 2x.$$

Dependent variable   Independent variable

An equation is the most efficient way to define a relation or function.

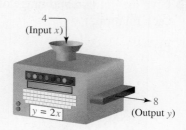

4 —
(Input x)

8
(Output y)

$y = 2x$

Function machine

> **NOTE**
> Another way to think of a function relationship is to think of the independent variable as an input and the dependent variable as an output. This is illustrated by the input-output (function) machine in the margin for the function defined by $y = 2x$.

OBJECTIVE 2 **Find domain and range.** For every relation, there are two important sets of elements called the *domain* and *range*.

Domain and Range

In a relation, the set of all values of the independent variable (x) is the **domain.** The set of all values of the dependent variable (y) is the **range.**

EXAMPLE 2 **Finding Domains and Ranges of Relations**

Give the domain and range of each relation. Tell whether the relation defines a function.

(a) $\{(3, -1), (4, 2), (4, 5), (6, 8)\}$

The domain, the set of x-values, is $\{3, 4, 6\}$; the range, the set of y-values, is $\{-1, 2, 5, 8\}$. This relation is not a function because the same x-value 4 is paired with two different y-values, 2 and 5.

(b)

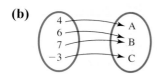

The domain of this relation is

$$\{4, 6, 7, -3\}.$$

The range is

$$\{A, B, C\}.$$

This mapping defines a function—each x-value corresponds to exactly one y-value.

(c)

x	y
-5	2
0	2
5	2

This is a table of ordered pairs, so the domain is the set of x-values, $\{-5, 0, 5\}$, and the range is the set of y-values, $\{2\}$. The table defines a function because each different x-value corresponds to exactly one y-value (even though it is the same y-value).

Work Problem 3 at the Side.

3 Give the domain and range of each relation. Does the relation define a function?

(a) $\{(4, 0), (4, 1), (4, 2)\}$

(b)

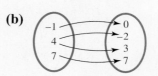

-1
4
7

0
-2
3
7

(c)

Year	Cell Phone Subscribers (in thousands)
1995	33,786
1996	44,043
1997	55,312
1998	69,209
1999	86,047

Source: Cellular Telecommunications Industry Association.

ANSWERS
3. **(a)** domain: $\{4\}$; range: $\{0, 1, 2\}$; No, the relation does not define a function.
   (b) domain: $\{-1, 4, 7\}$; range: $\{0, -2, 3, 7\}$; No, the relation does not define a function.
   (c) domain: $\{1995, 1996, 1997, 1998, 1999\}$; range: $\{33,786, 44,043, 55,312, 69,209, 86,047\}$; Yes, the relation defines a function.

4 Give the domain and range of each relation.

(a)

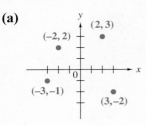

(b)

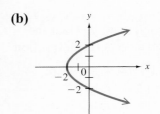

(c)

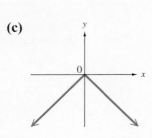

The graph of a relation gives a picture of the relation, which can be used to determine its domain and range.

EXAMPLE 3 Finding Domains and Ranges from Graphs

Give the domain and range of each relation.

(a)

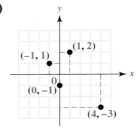

The domain is the set of x-values,

$$\{-1, 0, 1, 4\}.$$

The range is the set of y-values,

$$\{-3, -1, 1, 2\}.$$

(b)

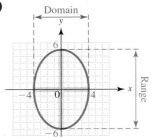

The x-values of the points on the graph include all numbers between -4 and 4, inclusive. The y-values include all numbers between -6 and 6, inclusive. Using interval notation,

the domain is $[-4, 4]$;

the range is $[-6, 6]$.

(c)

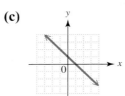

The arrowheads indicate that the line extends indefinitely left and right, as well as up and down. Therefore, both the domain and the range include all real numbers, written $(-\infty, \infty)$.

(d)

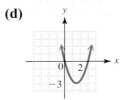

The arrowheads indicate that the graph extends indefinitely left and right, as well as upward. The domain is $(-\infty, \infty)$. Because there is a least y-value, -3, the range includes all numbers greater than or equal to -3, written $[-3, \infty)$.

Work Problem 4 at the Side.

Since relations are often defined by equations, such as $y = 2x + 3$ and $y^2 = x$, we must sometimes determine the domain of a relation from its equation. We assume the following agreement on the domain of a relation.

Agreement on Domain

The domain of a relation is assumed to be all real numbers that produce real numbers when substituted for the independent variable.

To illustrate this agreement, since any real number can be used as a replacement for x in $y = 2x + 3$, the domain of this function is the set of all real numbers. The function defined by $y = \frac{1}{x}$ has all real numbers except 0 as domain, since y is undefined if $x = 0$. In general, the domain of a function defined by an algebraic expression is all real numbers, except those numbers that lead to division by 0 or an even root of a negative number.

ANSWERS
4. (a) domain: $\{-3, -2, 2, 3\}$;
   range: $\{-2, -1, 2, 3\}$
   (b) domain: $[-2, \infty)$; range: $(-\infty, \infty)$
   (c) domain: $(-\infty, \infty)$; range: $(-\infty, 0]$

OBJECTIVE 3 **Identify functions defined by graphs and equations.**
Since each value of x in a function corresponds to only one value of y, any vertical line drawn through the graph of a function must intersect the graph in at most one point. This is the *vertical line test* for a function.

Vertical Line Test

If every vertical line intersects the graph of a relation in no more than one point, then the relation represents a function.

For example, the graph shown in Figure 32(a) is not the graph of a function since a vertical line intersects the graph in more than one point. The graph in Figure 32(b) does represent a function.

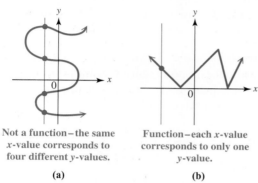

Not a function – the same x-value corresponds to four different y-values.

Function – each x-value corresponds to only one y-value.

(a)   **(b)**

Figure 32

EXAMPLE 4 **Using the Vertical Line Test**

Use the vertical line test to determine whether each relation graphed in Example 3 is a function.

(a)

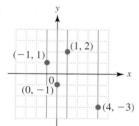

Function

(b)

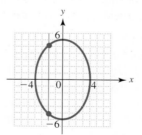

Not a function

(c)

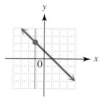

Function

(d)

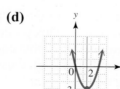

Function

The graphs in (a), (c), and (d) represent functions. The graph of the relation in (b) fails the vertical line test, since the same x-value corresponds to two different y-values; therefore, it is not the graph of a function.

Work Problem 5 at the Side.

5 Use the vertical line test to decide which graphs represent functions.

A.

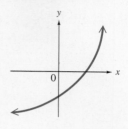

B.

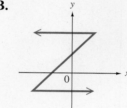

C.

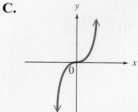

ANSWERS
5. A and C are graphs of functions.

6 Decide whether each relation defines a function, and give the domain.

(a) $y = 6x + 12$

(b) $y \le 4x$

(c) $y = -\sqrt{3x - 2}$

(d) $y^2 = 25x$

NOTE
Graphs that do not represent functions are still relations. *Remember that all equations and graphs represent relations and that all relations have a domain and range.*

It can be more difficult to decide whether a relation defined by an equation is a function. The next example gives some hints that may help.

EXAMPLE 5 Identifying Functions from Their Equations

Decide whether each relation defines a function and give the domain.

(a) $y = x + 4$

In the defining equation, $y = x + 4$, y is always found by adding 4 to x. Thus, each value of x corresponds to just one value of y and the relation defines a function; x can be any real number, so the domain is $(-\infty, \infty)$.

(b) $y = \sqrt{2x - 1}$

For any choice of x in the domain, there is exactly one corresponding value for y (the radical is a nonnegative number), so this equation defines a function. Since the equation involves a square root, the quantity under the radical sign cannot be negative. Thus,

$$2x - 1 \ge 0$$
$$2x \ge 1$$
$$x \ge \frac{1}{2},$$

and the domain of the function is $\left[\frac{1}{2}, \infty\right)$.

(c) $y^2 = x$

The ordered pairs $(16, 4)$ and $(16, -4)$ both satisfy this equation. Since one value of x, 16, corresponds to two values of y, 4 and -4, this equation does not define a function. Because x is equal to the square of y, the values of x must always be nonnegative. The domain of the relation is $[0, \infty)$.

(d) $y \le x - 1$

By definition, y is a function of x if every value of x leads to exactly one value of y. Here a particular value of x, say 1, corresponds to many values of y. The ordered pairs $(1, 0)$, $(1, -1)$, $(1, -2)$, $(1, -3)$, and so on, all satisfy the inequality. Thus, *an inequality never defines a function.* Any number can be used for x, so the domain is the set of real numbers, $(-\infty, \infty)$.

(e) $y = \dfrac{5}{x - 1}$

Given any value of x in the domain, we find y by subtracting 1, then dividing the result into 5. This process produces exactly one value of y for each value in the domain, so this equation defines a function. The domain includes all real numbers except those that make the denominator 0. We find these numbers by setting the denominator equal to 0 and solving for x.

$$x - 1 = 0$$
$$x = 1$$

The domain includes all real numbers *except* 1, written $(-\infty, 1) \cup (1, \infty)$.

Work Problem 6 at the Side.

ANSWERS
6. (a) yes; $(-\infty, \infty)$ **(b)** no; $(-\infty, \infty)$
(c) yes; $\left[\frac{2}{3}, \infty\right)$ **(d)** no; $[0, \infty)$

In summary, three variations of the definition of function are given here.

Variations of the Definition of Function

1. A **function** is a relation in which, for each value of the first component of the ordered pairs, there is exactly one value of the second component.

2. A **function** is a set of ordered pairs in which no first component is repeated.

3. A **function** is a rule or correspondence that assigns exactly one range value to each domain value.

OBJECTIVE 4 Use function notation. When a function f is defined with a rule or an equation using x and y for the independent and dependent variables, we say "y is a function of x" to emphasize that y *depends on* x. We use the notation

$$y = f(x),$$

called **function notation,** to express this and read $f(x)$ as "f of x." (In this special notation the parentheses do not indicate multiplication.) The letter f stands for *function*. For example, if $y = 9x - 5$, we can name this function f and write

$$f(x) = 9x - 5.$$

Note that $f(x)$ *is just another name for the dependent variable* y. For example, if $y = f(x) = 9x - 5$ and $x = 2$, then we find y, or $f(2)$, by replacing x with 2.

$$y = f(2) = 9 \cdot 2 - 5$$
$$= 18 - 5$$
$$= 13.$$

For function f, the statement "if $x = 2$, then $y = 13$" is represented by the ordered pair $(2, 13)$ and is abbreviated with function notation as

$$f(2) = 13.$$

Read $f(2)$ as "f of 2" or "f at 2." Also,

$$f(0) = 9 \cdot 0 - 5 = -5 \qquad \text{and} \qquad f(-3) = 9(-3) - 5 = -32.$$

These ideas can be illustrated as follows.

Name of the function

Defining expression

$$y \; = \; \boxed{f(x)} \; = \; \overbrace{9x - 5}$$

Value of the function   Name of the independent variable

CAUTION
The symbol $f(x)$ *does not* indicate "f times x," but represents the y-value for the indicated x-value. As just shown, $f(2)$ is the y-value that corresponds to the x-value 2.

7 Find $f(-3)$, $f(p)$, and $f(m + 1)$.

(a) $f(x) = 6x - 2$

(b) $f(x) = \dfrac{-3x + 5}{2}$

(c) $f(x) = \dfrac{1}{6}x - 1$

EXAMPLE 6 **Using Function Notation**

Let $f(x) = -x^2 + 5x - 3$. Find the following.

(a) $f(2)$

$$f(x) = -x^2 + 5x - 3$$
$$f(2) = -2^2 + 5 \cdot 2 - 3 \quad \text{Replace } x \text{ with 2.}$$
$$f(2) = -4 + 10 - 3$$
$$f(2) = 3$$

Since $f(2) = 3$, the ordered pair $(2, 3)$ belongs to f.

(b) $f(q)$

$$f(x) = -x^2 + 5x - 3$$
$$f(q) = -q^2 + 5q - 3 \quad \text{Replace } x \text{ with } q.$$

The replacement of one variable with another is important in later courses.

Sometimes letters other than f, such as g, h, or capital letters F, G, and H are used to name functions.

EXAMPLE 7 **Using Function Notation**

Let $g(x) = 2x + 3$. Find and simplify $g(a + 1)$.

$$g(x) = 2x + 3$$
$$g(a + 1) = 2(a + 1) + 3 \quad \text{Replace } x \text{ with } a + 1.$$
$$= 2a + 2 + 3$$
$$= 2a + 5$$

◀◀ Work Problem 7 at the Side.

Functions can be evaluated in a variety of ways, as shown in Example 8.

EXAMPLE 8 **Using Function Notation**

For each function, find $f(3)$.

(a) $f(x) = 3x - 7$
$$f(3) = 3(3) - 7$$
$$f(3) = 9 - 7$$
$$f(3) = 2$$

(b) $f = \{(-3, 5), (0, 3), (3, 1), (6, -1)\}$
We want $f(3)$, the y-value of the ordered pair where $x = 3$. As indicated by the ordered pair $(\mathbf{3}, \mathbf{1})$, when $x = 3$, $y = 1$, so $f(3) = 1$.

(c)

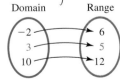

The domain element 3 is paired with 5 in the range, so $f(3) = 5$.

Continued on Next Page

ANSWERS
7. (a) -20; $6p - 2$; $6m + 4$

(b) 7; $\dfrac{-3p + 5}{2}$; $\dfrac{-3m + 2}{2}$

(c) $-\dfrac{3}{2}$; $\dfrac{1}{6}p - 1$; $\dfrac{1}{6}(m + 1) - 1$ or $\dfrac{1}{6}m - \dfrac{5}{6}$

(d)

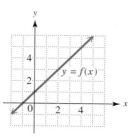

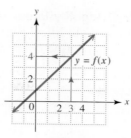

Figure 33

To evaluate $f(3)$, find 3 on the x-axis. See Figure 33. Then move up until the graph of f is reached. Moving horizontally to the y-axis gives 4 for the corresponding y-value. Thus, $f(3) = 4$.

Work Problem 8 at the Side. ▶▶▶

If a function f is defined by an equation with x and y, not with function notation, use the following steps to find $f(x)$.

Finding an Expression for $f(x)$

Step 1 Solve the equation for y.

Step 2 Replace y with $f(x)$.

EXAMPLE 9 Writing Equations Using Function Notation

Rewrite each equation using function notation. Then find $f(-2)$ and $f(a)$.

(a) $y = x^2 + 1$

This equation is already solved for y. Since $y = f(x)$,

$$f(x) = x^2 + 1.$$

To find $f(-2)$, let $x = -2$.

$$f(-2) = (-2)^2 + 1$$
$$= 4 + 1$$
$$= 5$$

Find $f(a)$ by letting $x = a$: $f(a) = a^2 + 1$.

(b) $x - 4y = 5$

First solve $x - 4y = 5$ for y. Then replace y with $f(x)$.

$$x - 4y = 5$$
$$x - 5 = 4y \qquad \text{Add } 4y; \text{ subtract 5.}$$
$$y = \frac{x - 5}{4} \quad \text{so} \quad f(x) = \frac{1}{4}x - \frac{5}{4}$$

Now find $f(-2)$ and $f(a)$.

$$f(-2) = \frac{1}{4}(-2) - \frac{5}{4} = -\frac{7}{4} \qquad \text{Let } x = -2.$$

$$f(a) = \frac{1}{4}a - \frac{5}{4} \qquad \text{Let } x = a.$$

Work Problem 9 at the Side. ▶▶▶

8 For each function, find $f(-2)$.

(a) $f(x) = -4x - 8$

(b) $f = \{(0, 5), (-1, 3), (-2, 1)\}$

(c)

x	$f(x)$
-4	16
-2	4
0	0
2	4
4	16

9 Rewrite each equation using function notation. Then find $f(-1)$.

(a) $y = \sqrt{x + 2}$

(b) $x^2 - 4y = 3$

ANSWERS

8. (a) 0   (b) 1   (c) 4

9. (a) $f(x) = \sqrt{x + 2}$; 1

(b) $f(x) = \dfrac{x^2 - 3}{4}$ or $f(x) = \dfrac{1}{4}x^2 - \dfrac{3}{4}$; $-\dfrac{1}{2}$

10 Graph each linear function. Give the domain and range.

(a) $f(x) = \dfrac{3}{4}x - 2$

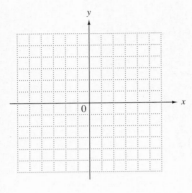

(b) $g(x) = 3$

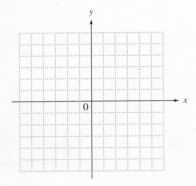

OBJECTIVE 5 Identify linear functions. Our first two-dimensional graphing was of straight lines. Linear equations (except for vertical lines with equations $x = a$) define *linear functions*.

> **Linear Function**
>
> A function that can be defined by
>
> $$f(x) = mx + b$$
>
> for real numbers m and b is a **linear function.**

Recall from **Section 4.3** that m is the slope of the line and $(0, b)$ is the y-intercept. In Example 9(b), we wrote the equation $x - 4y = 5$ as the linear function defined by

$$f(x) = \frac{1}{4}x - \frac{5}{4}.$$

Slope ⟶   ⟵ y-intercept is $\left(0, -\frac{5}{4}\right)$.

To graph this function, plot the y-intercept and use the definition of slope as $\frac{\text{rise}}{\text{run}}$ to find a second point on the line. Draw the straight line through the points to obtain the graph shown in Figure 34.

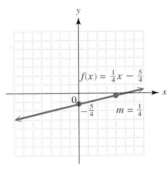

Figure 34

A linear function defined by $f(x) = b$ (whose graph is a horizontal line) is sometimes called a **constant function.** The domain of any linear function is $(-\infty, \infty)$. The range of a nonconstant linear function is $(-\infty, \infty)$, while the range of the constant function defined by $f(x) = b$ is $\{b\}$.

◀◀◀ **Work Problem 10 at the Side.**

ANSWERS

10. (a)

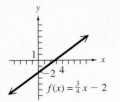

$f(x) = \frac{3}{4}x - 2$

domain: $(-\infty, \infty)$; range: $(-\infty, \infty)$

(b)

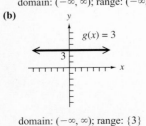

$g(x) = 3$

domain: $(-\infty, \infty)$; range: $\{3\}$

4.5 Exercises

FOR EXTRA HELP   Addison-Wesley Math Tutor Center   MathXL   Digital Video Tutor CD 3 Videotape 3   Student's Solutions Manual   MyMathLab   Interactmath.com

1. In an ordered pair of a relation, is the first element the independent or the dependent variable?

2. Give an example of a relation that is not a function, having domain $\{-3, 2, 6\}$ and range $\{4, 6\}$. (There are many possible correct answers.)

3. Explain what is meant by each term.
   (a) Relation
   (b) Domain of a relation
   (c) Range of a relation
   (d) Function

4. Describe the use of the vertical line test.

Decide whether each relation is a function, and give the domain and the range. Use the vertical line test in Exercises 17–22. See Examples 1–4.

5. $\{(5, 1), (3, 2), (4, 9), (7, 3)\}$

6. $\{(8, 0), (5, 4), (9, 3), (3, 9)\}$

7. $\{(2, 4), (0, 2), (2, 6)\}$

8. $\{(9, -2), (-3, 5), (9, 1)\}$

9. $\{(-3, 1), (4, 1), (-2, 7)\}$

10. $\{(-12, 5), (-10, 3), (8, 3)\}$

11. $\{(1, 1), (1, -1), (0, 0), (2, 4), (2, -4)\}$

12. $\{(2, 5), (3, 7), (4, 9), (5, 11)\}$

13.

14.
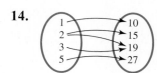

15.

x	y
1	5
1	2
1	−1
1	−4

16.

x	y
4	−3
2	−3
0	−3
−2	−3

17.

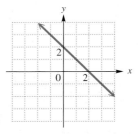

18.

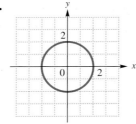

19.

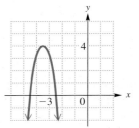

20.

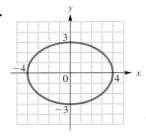

21.

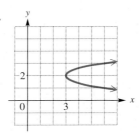

22.

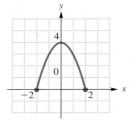

Decide whether each relation defines y as a function of x. Give the domain. See Example 5.

23. $y = x^2$ **24.** $y = x^3$ **25.** $x = y^6$ **26.** $x = y^4$

27. $y = 2x - 6$ **28.** $y = -6x + 8$ **29.** $x + y < 4$ **30.** $x - y < 3$

31. $y = \sqrt{x}$

32. $y = -\sqrt{x}$

33. $xy = 1$

34. $xy = -3$

35. $y = \sqrt{4x + 2}$

36. $y = \sqrt{9 - 2x}$

37. $y = \dfrac{2}{x - 9}$

38. $y = \dfrac{-7}{x - 16}$

39. Refer to the graph to answer the questions.

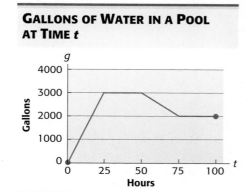

GALLONS OF WATER IN A POOL AT TIME t

(a) What numbers are possible values of the dependent variable?

(b) For how long is the water level increasing? decreasing?

(c) How many gallons are in the pool after 90 hr?

(d) Call this function g. What is $g(0)$? What does it mean in this example?

40. The graph shows the daily megawatts of electricity used on a record-breaking summer day in Sacramento, California.

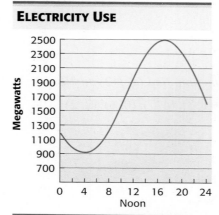

ELECTRICITY USE

Source: Sacramento Municipal Utility District.

(a) Is this the graph of a function?

(b) What is the domain?

(c) Estimate the number of megawatts used at 8 A.M.

(d) At what time was the most electricity used? the least electricity?

41. Give an example of a function from everyday life. (*Hint:* Fill in the blanks: _____ depends on _____, so _____ is a function of _____.)

42. Choose the correct response: The notation $f(3)$ means

A. the variable f times 3 or $3f$

B. the value of the dependent variable when the independent variable is 3

C. the value of the independent variable when the dependent variable is 3

D. f equals 3.

Let $f(x) = -3x + 4$ and $g(x) = -x^2 + 4x + 1$. Find the following. See Examples 6 and 7.

43. $f(0)$ **44.** $f(-3)$ **45.** $g(-2)$ **46.** $g(10)$

47. $f(p)$ **48.** $g(k)$ **49.** $f(-x)$ **50.** $g(-x)$

51. $f(x + 2)$ **52.** $g\left(-\dfrac{1}{x}\right)$ **53.** $g\left(\dfrac{p}{3}\right)$ **54.** $f(3t - 2)$

For each function, find (a) $f(2)$ and (b) $f(-1)$. See Example 8.

55. $f = \{(-1, 3), (4, 7), (0, 6), (2, 2)\}$ **56.** $f = \{(2, 5), (3, 9), (-1, 11), (5, 3)\}$

57.

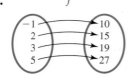

58.

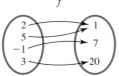

59.

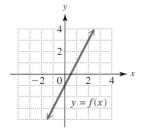

60.

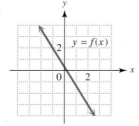

61. Fill in each blank with the correct response.

The equation $2x + y = 4$ has a straight _____ as its graph. One point that lies on the graph is $(3, \underline{\hspace{1cm}})$. If we solve the equation for y and use function notation, we have a _____ function defined by $f(x) = \underline{\hspace{1cm}}$. For this function, $f(3) = \underline{\hspace{1cm}}$, meaning that the point $(\underline{\hspace{1cm}}, \underline{\hspace{1cm}})$ lies on the graph of the function.

62. Which of the following defines a linear function?

A. $y = \dfrac{x - 5}{4}$ **B.** $y = \dfrac{1}{x}$

C. $y = x^2$ **D.** $y = \sqrt{x}$

An equation that defines y as a function of x is given. (a) Solve for y in terms of x, and replace y with the function notation f(x). (b) Find f(3). See Example 9.

63. $x + 3y = 12$

64. $x - 4y = 8$

65. $y + 2x^2 = 3$

66. $y - 3x^2 = 2$

67. $4x - 3y = 8$

68. $-2x + 5y = 9$

Graph each linear function. Give the domain and range. See Objective 5.

69. $f(x) = -2x + 5$

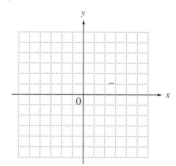

70. $g(x) = 4x - 1$

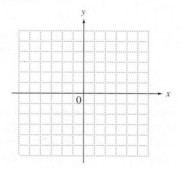

71. $h(x) = \dfrac{1}{2}x + 2$

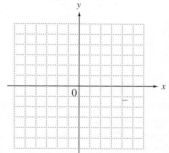

72. $F(x) = -\dfrac{1}{4}x + 1$

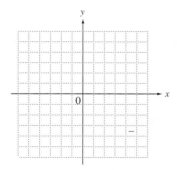

73. $g(x) = -4$

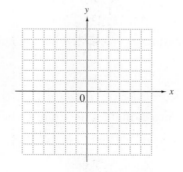

74. $f(x) = 5$

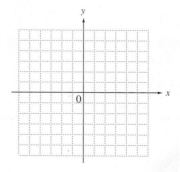

Solve each problem.

75. Suppose that a Yellow Cab driver charges $1.50 per mi.

(a) Fill in the table with the correct response for the price $f(x)$ he charges for a trip of x miles.

x	f(x)
0	
1	
2	
3	

(b) The linear function that gives a rule for the amount charged is $f(x) = $ _____.

(c) Graph this function for the domain {0, 1, 2, 3}.

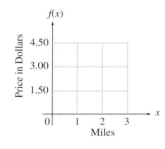

76. Suppose that a package weighing x pounds costs $f(x)$ dollars to mail to a given location, where $f(x) = 2.75x$.

(a) What is the value of $f(3)$?

(b) In your own words, describe what 3 and the value $f(3)$ mean in part (a), using the terms *independent variable* and *dependent variable*.

(c) How much would it cost to mail a 5-lb package? Write the answer using function notation.

Forensic scientists use the lengths of certain bones to calculate the height of a person. Two bones often used are the tibia (t), the bone from the ankle to the knee, and the femur (r), the bone from the knee to the hip socket. A person's height (h) is determined from the lengths of these bones using functions defined by the following formulas. All measurements are in centimeters.

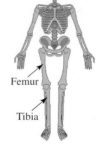

Functions for men: $h(r) = 69.09 + 2.24r$ or $h(t) = 81.69 + 2.39t$

Functions for women: $h(r) = 61.41 + 2.32r$ or $h(t) = 72.57 + 2.53t$

77. Find the height of a man with a femur measuring 56 cm.

78. Find the height of a man with a tibia measuring 40 cm.

79. Find the height of a woman with a femur measuring 50 cm.

80. Find the height of a woman with a tibia measuring 36 cm.

4.6 Variation

Certain types of functions are very common, especially in business and the physical sciences. These are functions where y depends on a multiple of x, or y depends on a number divided by x. In such situations, y is said to *vary directly as x* (in the first case) or *vary inversely as x* (in the second case). For example, by the distance formula, the distance traveled varies directly as the rate (or speed) and the time. The simple interest formula and the formulas for area and volume are other familiar examples of *direct variation*.

On the other hand, the force required to keep a car from skidding on a curve varies inversely as the radius of the curve. Another example of *inverse variation* is how travel time is inversely proportional to rate or speed.

OBJECTIVE 1 **Write an equation expressing direct variation.** The circumference of a circle is given by the formula $C = 2\pi r$, where r is the radius of the circle. See the figure. Circumference is always a constant multiple of the radius. (C is always found by multiplying r by the constant 2π.) Thus,

$C = 2\pi r$

As the *radius increases,* the *circumference increases.*

The reverse is also true.

As the *radius decreases,* the *circumference decreases.*

Because of this, the circumference is said to *vary directly* as the radius.

Direct Variation

y varies directly as x if there exists some constant k such that

$$y = kx.$$

Also, y is said to be **proportional to** x. The number k is called the **constant of variation.** In direct variation, for $k > 0$, as the value of x increases, the value of y also increases. Similarly, as x decreases, y decreases.

OBJECTIVE 2 **Find the constant of variation, and solve direct variation problems.** The direct variation equation $y = kx$ defines a linear function, where the constant of variation k is the slope of the line. For example, we wrote the equation

$$y = 1.80x$$

to describe the cost y to buy x gallons of gas in Example 6 of **Section 4.3.** The cost varies directly as, or is proportional to, the number of gallons of gas purchased. That is, as the number of gallons of gas increases, cost increases; also, as the number of gallons of gas decreases, cost decreases. The constant of variation k is 1.80, the cost of 1 gallon of gas.

1 Find the constant of variation, and write a direct variation equation.

(a) Suzanne Alley is paid a daily wage. One month she worked 17 days and earned $1334.50.

EXAMPLE 1 **Finding the Constant of Variation and the Variation Equation**

Gina Linko is paid an hourly wage. One week she worked 43 hr and was paid $795.50. How much does she earn per hour?

Let h represent the number of hours she works and P represent her corresponding pay. Then, P **varies directly as** h, so

$$P = kh.$$

Here k represents Gina's hourly wage. Since $P = 795.50$ when $h = 43$,

$$795.50 = 43k$$
$$k = \mathbf{18.50}. \qquad \text{Use a calculator.}$$

Her hourly wage is $18.50, and P and h are related by

$$P = \mathbf{18.50}h.$$

◀◀◀ Work Problem 1 at the Side.

(b) Distance varies directly as time (at a constant speed). A car travels 100 mi at a constant speed in 2 hr.

EXAMPLE 2 **Solving a Direct Variation Problem**

Hooke's law for an elastic spring states that the distance a spring stretches is proportional to the force applied. If a force of 150 newtons* stretches a certain spring 8 cm, how much will a force of 400 newtons stretch the spring?

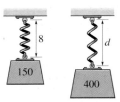

Figure 35

See Figure 35. If d is the distance the spring stretches and f is the force applied, then $d = kf$ for some constant k. Since a force of 150 newtons stretches the spring 8 cm, we can use these values to find k.

$$d = kf \qquad \text{Variation equation}$$
$$8 = k \cdot 150 \qquad \text{Let } d = 8 \text{ and } f = 150.$$
$$k = \frac{8}{150} \qquad \text{Find } k.$$
$$k = \frac{4}{75}$$

Substitute $\frac{4}{75}$ for k in the variation equation $d = kf$ to get

$$d = \frac{4}{75}f.$$

2 The charge (in dollars) to customers for electricity (in kilowatt-hours) varies directly as the number of kilowatt-hours used. It costs $52 to use 800 kilowatt-hours. Find the cost to use 1000 kilowatt-hours.

For a force of 400 newtons,

$$d = \frac{4}{75}(\mathbf{400}) = \frac{64}{3}. \qquad \text{Let } f = 400.$$

The spring will stretch $\frac{64}{3}$ cm if a force of 400 newtons is applied.

◀◀◀ Work Problem 2 at the Side.

ANSWERS
1. (a) $k = 78.50$; Let E represent her earnings for d days. Then $E = 78.50d$.
   (b) $k = 50$; Let d represent the distance traveled in h hours. Then $d = 50h$.
2. $65

*A newton is a unit of measure of force used in physics.

In summary, use the following steps to solve a variation problem.

Solving a Variation Problem

Step 1   Write the variation equation.

Step 2   Substitute the initial values and solve for k.

Step 3   Rewrite the variation equation with the value of k from Step 2.

Step 4   Substitute the remaining values, solve for the unknown, and find the required answer.

The direct variation equation $y = kx$ is a linear equation. However, other kinds of variation involve other types of equations. For example, one variable can be proportional to a power of another variable.

Direct Variation as a Power

y **varies directly as the *n*th power of *x*** if there exists a real number k such that

$$y = kx^n.$$

An example of direct variation as a power is the formula for the area of a circle, $A = \pi r^2$. Here, π is the constant of variation, and the area varies directly as the square of the radius.

EXAMPLE 3 **Solving a Direct Variation Problem**

The distance a body falls from rest varies directly as the square of the time it falls (disregarding air resistance). If a skydiver falls 64 ft in 2 sec, how far will she fall in 8 sec?

Step 1   If d represents the distance the skydiver falls and t the time it takes to fall, then d is a function of t, and

$$d = kt^2$$

for some constant k.

Step 2   To find the value of k, use the fact that the skydiver falls 64 ft in 2 sec.

$$d = kt^2 \qquad \text{Variation equation}$$
$$64 = k(2)^2 \qquad \text{Let } d = 64 \text{ and } t = 2.$$
$$k = 16 \qquad \text{Find } k.$$

Step 3   Using 16 for k, the variation equation becomes

$$d = 16t^2.$$

Step 4   Let $t = 8$ to find the number of feet the skydiver will fall in 8 sec.

$$d = 16(8)^2 = 1024 \qquad \text{Let } t = 8.$$

The skydiver will fall 1024 ft in 8 sec.

Work Problem 3 at the Side. ▶▶▶

---

3 The area of a circle varies directly as the square of its radius. A circle with radius 3 in. has area 28.278 in.^{2}.

(a) Write a variation equation and give the value of k.

(b) What is the area of a circle with radius 4.1 in.?

ANSWERS
3. **(a)** $A = kr^2$; 3.142
   (b) 52.817 in.^{2} (to the nearest thousandth)

OBJECTIVE **3** **Solve inverse variation problems.** In direct variation, where $k > 0$, as x increases, y increases. Similarly, as x decreases, y decreases. Another type of variation is *inverse variation*. With inverse variation, where $k > 0$, as one variable increases, the other variable decreases. For example, in a closed space, volume decreases as pressure increases, as illustrated by a trash compactor. See Figure 36. As the compactor presses down, the pressure on the trash increases; in turn, the trash occupies a smaller space.

As pressure on trash increases, volume of trash decreases.

Figure 36

Inverse Variation

y varies inversely as x if there exists a real number k such that

$$y = \frac{k}{x}.$$

Also, **y varies inversely as the nth power of x** if there exists a real number k such that

$$y = \frac{k}{x^n}.$$

The inverse variation equation also defines a function. Since x is in the denominator, these functions are *rational functions*. (See **Chapter 8.**) Another example of inverse variation comes from the distance formula. In its usual form, the formula is

$$d = rt.$$

Dividing each side by r gives

$$t = \frac{d}{r}.$$

Here, t (time) varies inversely as r (rate or speed), with d (distance) serving as the constant of variation. For example, if the distance between Chicago and Des Moines is 300 mi, then

$$t = \frac{300}{r}$$

and the values of r and t might be any of the following.

$$
\left.\begin{array}{l}
r = 50, t = 6 \\
r = 60, t = 5 \\
r = 75, t = 4
\end{array}\right\} \text{As } r \text{ increases, } t \text{ decreases.}
\qquad
\left.\begin{array}{l}
r = 30, t = 10 \\
r = 25, t = 12 \\
r = 20, t = 15
\end{array}\right\} \text{As } r \text{ decreases, } t \text{ increases.}
$$

If we *increase* the rate (speed) we drive, time *decreases*. If we *decrease* the rate (speed) we drive, time *increases*.

EXAMPLE 4 Solving an Inverse Variation Problem

The weight of an object above Earth varies inversely as the square of its distance from the center of Earth. A space shuttle in an elliptical orbit has a maximum distance from the center of Earth (*apogee*) of 6700 mi. Its minimum distance from the center of Earth (*perigee*) is 4090 mi. See Figure 37. If an astronaut in the shuttle weighs 57 lb at its apogee, what does the astronaut weigh at its perigee?

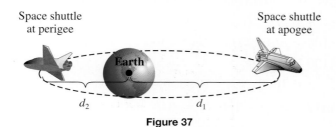

Space shuttle at perigee   Space shuttle at apogee

Earth

d_2 d_1

Figure 37

If w is the weight and d is the distance from the center of Earth, then

$$w = \frac{k}{d^2}$$

for some constant k. At the apogee the astronaut weighs 57 lb, and the distance from the center of Earth is 6700 mi. Use these values to find k.

$$57 = \frac{k}{(6700)^2} \qquad \text{Let } w = 57 \text{ and } d = 6700.$$

$$k = 57(6700)^2$$

Then the weight at the perigee with $d = 4090$ mi is

$$w = \frac{k}{d^2} = \frac{57(6700)^2}{(4090)^2} \approx 153 \text{ lb.} \qquad \text{Use a calculator.}$$

Work Problem 4 at the Side. ▶▶▶

OBJECTIVE 4 Solve joint variation problems. It is possible for one variable to depend on several others. If one variable varies directly as the *product* of several other variables (perhaps raised to powers), the first variable is said to *vary jointly* as the others.

Joint Variation

y **varies jointly as** *x* **and** *z* if there exists a real number k such that

$$y = kxz.$$

CAUTION

Note that *and* in the expression "*y* varies jointly as *x* **and** *z*" translates as the product

$$y = kxz.$$

The word *and* does not indicate addition here.

4 If the temperature is constant, the volume of a gas varies inversely as the pressure. For a certain gas, the volume is 10 cm^3 when the pressure is 6 kg per cm^2.

(a) Find the variation equation.

(b) Find the volume when the pressure is 12 kg per cm^2.

ANSWERS

4. (a) $V = \dfrac{60}{P}$ **(b)** 5 cm^3

5 The volume of a rectangular box of a given height is proportional to its width and length. A box with width 2 ft and length 4 ft has volume 12 ft³. Find the volume of a box with the same height that is 3 ft wide and 5 ft long.

EXAMPLE 5 Solving a Joint Variation Problem

The interest on a loan or an investment is given by the formula $I = prt$. Here, for a given principal p, the interest earned I varies jointly as the interest rate r and the time t that the principal is left at interest. If an investment earns $100 interest at 5% for 2 yr, how much interest will the same principal earn at 4.5% for 3 yr?

We use the formula $I = prt$, where p is the constant of variation because it is the same for both investments. For the first investment,

$$I = prt$$
$$100 = p(.05)(2) \qquad \text{Let } I = 100, r = .05, \text{ and } t = 2.$$
$$100 = .1p$$
$$p = 1000. \qquad \text{Divide by .1.}$$

Now we find I when $p = 1000$, $r = .045$, and $t = 3$.

$$I = 1000(.045)(3) = 135 \qquad \text{Let } p = 1000, r = .045, \text{ and } t = 3.$$

The interest will be $135.

Work Problem 5 at the Side.

6 The maximum load that a cylindrical column with a circular cross section can hold varies directly as the fourth power of the diameter of the cross section and inversely as the square of the height. A 9-m column 1 m in diameter will support 8 metric tons. How many metric tons can be supported by a column 12 m high and $\frac{2}{3}$ m in diameter?

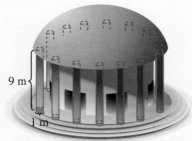

9 m

1 m

Load = 8 metric tons

OBJECTIVE 5 Solve combined variation problems. There are many combinations of direct and inverse variation, called **combined variation.**

EXAMPLE 6 Solving a Combined Variation Problem

Body mass index, or BMI, is used by physicians to assess a person's level of fatness. A BMI from 19 through 25 is considered desirable. BMI varies directly as an individual's weight in pounds and inversely as the square of the individual's height in inches. A person who weighs 118 lb and is 64 in. tall has a BMI of 20. (The BMI is rounded to the nearest whole number.) Find the BMI of a person who weighs 165 lb with a height of 70 in.

Let B represent the BMI, w the weight, and h the height. Then

$$B = \frac{kw}{h^2}. \qquad \begin{array}{l} \longleftarrow \text{ BMI varies directly as the weight.} \\ \longleftarrow \text{ BMI varies inversely as the square of the height.} \end{array}$$

To find k, let $B = 20$, $w = 118$, and $h = 64$.

$$20 = \frac{k(118)}{64^2}$$

$$k = \frac{20(64^2)}{118} \qquad \begin{array}{l}\text{Multiply by } 64^2; \\ \text{divide by 118.}\end{array}$$

$$k \approx 694 \qquad \text{Use a calculator.}$$

Now find B when $k = 694$, $w = 165$, and $h = 70$.

$$B = \frac{694(165)}{70^2} \approx 23 \qquad \begin{array}{l}\text{Nearest whole} \\ \text{number}\end{array}$$

The person's BMI is 23.

Work Problem 6 at the Side.

ANSWERS
5. 22.5 ft³
6. $\frac{8}{9}$ metric ton

4.6 Exercises

FOR EXTRA HELP   Addison-Wesley Math Tutor Center   MathXL   Digital Video Tutor CD 3 Videotape 3   Student's Solutions Manual   MyMathLab   Interactmath.com

Determine whether each equation represents direct, inverse, joint, *or* combined *variation.*

1. $y = \dfrac{3}{x}$

2. $y = \dfrac{8}{x}$

3. $y = 10x^2$

4. $y = 2x^3$

5. $y = 3xz^4$

6. $y = 6x^3z^2$

7. $y = \dfrac{4x}{wz}$

8. $y = \dfrac{6x}{st}$

Solve each problem. See Examples 2–5.

9. If x varies directly as y, and $x = 9$ when $y = 3$, find x when $y = 12$.

10. If x varies directly as y, and $x = 10$ when $y = 7$, find y when $x = 50$.

11. If z varies inversely as w, and $z = 10$ when $w = .5$, find z when $w = 8$.

12. If t varies inversely as s, and $t = 3$ when $s = 5$, find s when $t = 5$.

13. p varies jointly as q and r^2, and $p = 200$ when $q = 2$ and $r = 3$. Find p when $q = 5$ and $r = 2$.

14. f varies jointly as g^2 and h, and $f = 50$ when $g = 4$ and $h = 2$. Find f when $g = 3$ and $h = 6$.

15. For $k > 0$, if y varies directly as x, when x increases, y _____, and when x decreases, y _____.

16. For $k > 0$, if y varies inversely as x, when x increases, y _____, and when x decreases, y _____.

17. Explain the difference between inverse variation and direct variation.

18. What is meant by the constant of variation in a direct variation problem? If you were to graph the linear equation $y = kx$ for some constant k, what role would the value of k play in the graph?

Solve each problem involving variation. See Examples 1–6.

19. Todd bought 8 gal of gasoline and paid $13.59. To the nearest tenth of a cent, what is the price of gasoline per gallon?

20. Melissa gives horseback rides at Shadow Mountain Ranch. A 2.5-hr ride costs $50.00. What is the price per hour?

21. The volume of a can of tomatoes is proportional to the height of the can. If the volume of the can is 300 cm^3 when its height is 10.62 cm, find the volume of a can with height 15.92 cm.

22. The weight of an object on Earth is directly proportional to the weight of that same object on the moon. A 200-lb astronaut would weigh 32 lb on the moon. How much would a 50-lb dog weigh on the moon?

23. For a body falling freely from rest (disregarding air resistance), the distance the body falls varies directly as the square of the time. If an object is dropped from the top of a tower 576 ft high and hits the ground in 6 sec, how far did it fall in the first 4 sec?

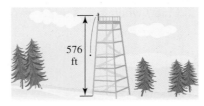

24. The amount of water emptied by a pipe varies directly as the square of the diameter of the pipe. For a certain constant water flow, a pipe emptying into a canal will allow 200 gal of water to escape in an hour. The diameter of the pipe is 6 in. How much water would a 12-in. pipe empty into the canal in an hour, assuming the same water flow?

25. The current in a simple electrical circuit is inversely proportional to the resistance. If the current is 20 amperes (an *ampere* is a unit for measuring current) when the resistance is 5 ohms, find the current when the resistance is 7.5 ohms.

26. The frequency (number of vibrations per second) of a vibrating guitar string varies inversely as its length. That is, a longer string vibrates fewer times in a second than a shorter string. Suppose a guitar string .65 m long vibrates 4.3 times per sec. What frequency would a string .5 m long have?

27. The amount of light (measured in foot-candles) produced by a light source varies inversely as the square of the distance from the source. If the illumination produced 1 m from a light source is 768 foot-candles, find the illumination produced 6 m from the same source.

28. The force with which Earth attracts an object above Earth's surface varies inversely with the square of the distance of the object from the center of Earth. If an object 4000 mi from the center of Earth is attracted with a force of 160 lb, find the force of attraction if the object were 6000 mi from the center of Earth.

29. For a given interest rate, simple interest varies jointly as principal and time. If $2000 left in an account for 4 yr earned interest of $280, how much interest would be earned in 6 yr?

30. The collision impact of an automobile varies jointly as its mass and the square of its speed. Suppose a 2000-lb car traveling at 55 mph has a collision impact of 6.1. What is the collision impact of the same car at 65 mph?

31. The force needed to keep a car from skidding on a curve varies inversely as the radius of the curve and jointly as the weight of the car and the square of the speed. If 242 lb of force keep a 2000-lb car from skidding on a curve of radius 500 ft at 30 mph, what force would keep the same car from skidding on a curve of radius 750 ft at 50 mph?

32. Natural gas provides 35.8% of U.S. energy. (*Source:* U.S. Energy Department.) The volume of gas varies inversely as the pressure and directly as the temperature. (Temperature must be measured in *Kelvin* (K), a unit of measurement used in physics.) If a certain gas occupies a volume of 1.3 L at 300 K and a pressure of 18 newtons per cm^2, find the volume at 340 K and a pressure of 24 newtons per cm^2.

33. The number of long-distance phone calls between two cities in a certain time period varies jointly as the populations of the cities, p_1 and p_2, and inversely as the distance between them. If 80,000 calls are made between two cities 400 mi apart, with populations of 70,000 and 100,000, how many calls are made between cities with populations of 50,000 and 75,000 that are 250 mi apart?

34. A body mass index from 27 through 29 carries a slight risk of weight-related health problems, while one of 30 or more indicates a great increase in risk. Use your own height and weight and the information in Example 6 to determine whether you are at risk.

Exercises 35 and 36 describe weight-estimation formulas that fishermen have used over the years. Girth *is the distance around the body of the fish.* (*Source: Sacramento Bee, November 9, 2000.*)

35. The weight of a bass varies jointly as its girth and the square of its length. A prize-winning bass weighed in at 22.7 lb and measured 36 in. long with 21 in. girth. How much would a bass 28 in. long with 18 in. girth weigh?

36. The weight of a trout varies jointly as its length and the square of its girth. One angler caught a trout that weighed 10.5 lb and measured 26 in. long with 18 in. girth. Find the weight of a trout that is 22 in. long with 15 in. girth.

RELATING CONCEPTS (EXERCISES 37–42)   For Individual or Group Work

A routine activity such as pumping gasoline can be related to many of the concepts studied in this chapter. Suppose that premium unleaded costs $1.75 per gal. **Work Exercises 37–42 in order.**

37. 0 gal of gasoline cost $0.00, while 1 gal costs $1.75. Represent these two pieces of information as ordered pairs of the form (gallons, price).

38. Use the information from Exercise 37 to find the slope of the line on which the two points lie.

39. Write the slope-intercept form of the equation of the line on which the two points lie.

40. Using function notation, if $f(x) = ax + b$ represents the line from Exercise 39, what are the values of a and b?

41. How does the value of a from Exercise 40 relate to gasoline in this situation? With relationship to the line, what do we call this number?

42. Why does the equation from Exercise 40 satisfy the conditions for direct variation? In the context of variation, what do we call the value of a?

Chapter 4

KEY TERMS

4.1	ordered pair	An ordered pair is a pair of numbers written in parentheses in which the order of the numbers is important.	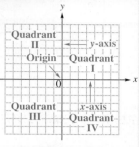
	origin	When two number lines intersect at a right angle, the origin is the common 0 point.	
	x-axis	The horizontal number line in a rectangular coordinate system is called the *x*-axis.	
	y-axis	The vertical number line in a rectangular coordinate system is called the *y*-axis.	Rectangular coordinate system
	rectangular (Cartesian) coordinate system	Two number lines that intersect at a right angle at their 0 points form a rectangular coordinate system, also called the Cartesian coordinate system.	
	plot	To plot an ordered pair is to locate it on a rectangular coordinate system.	
	components	The two numbers in an ordered pair are the components of the ordered pair.	
	coordinate	Each number in an ordered pair represents a coordinate of the corresponding point.	
	quadrant	A quadrant is one of the four regions in the plane determined by a rectangular coordinate system.	
	graph of an equation	The graph of an equation is the set of points corresponding to all ordered pairs that satisfy the equation.	
	first-degree equation	A first-degree equation has no term with a variable to a power greater than one.	
	linear equation in two variables	A first-degree equation with two variables is a linear equation in two variables.	
	x-intercept	The point where a line intersects the *x*-axis is the *x*-intercept.	
	y-intercept	The point where a line intersects the *y*-axis is the *y*-intercept.	
4.2	rise	The rise of a line is the vertical change between two points on the line.	
	run	The run of a line is the horizontal change between two points on the line.	
	slope	The ratio of the change in *y* compared to the change in *x* (rise/run) along a line is the slope of the line.	
4.4	linear inequality in two variables	A linear inequality in two variables is a first-degree inequality with two variables.	
	boundary line	In the graph of a linear inequality, the boundary line separates the region that satisfies the inequality from the region that does not satisfy the inequality.	
4.5	dependent variable	If the quantity *y* depends on *x*, then *y* is called the dependent variable in a relation between *x* and *y*.	
	independent variable	If *y* depends on *x*, then *x* is the independent variable in a relation between *x* and *y*.	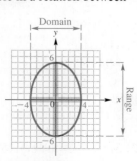
	relation	A relation is a set of ordered pairs of real numbers.	
	function	A function is a set of ordered pairs in which each value of the first component, *x*, corresponds to exactly one value of the second component, *y*.	
	domain	The domain of a relation is the set of first components (*x*-values) of the ordered pairs of the relation.	
	range	The range of a relation is the set of second components (*y*-values) of the ordered pairs of the relation.	Graph of a relation
	function notation	The function notation $f(x)$ is another way to represent the dependent variable *y* for the function *f*.	

KEY TERMS

linear function   A function that is defined by $f(x) = mx + b$ is a linear function.

constant function   A constant function is a linear function of the form $f(x) = b$, for a real number b.

NEW SYMBOLS

(a, b)   ordered pair

x_1 a specific value of the variable x (read "x sub one")

m   slope

f(x)   function of x (read "f of x")

TEST YOUR WORD POWER

See how well you have learned the vocabulary in this chapter. Answers, with examples, follow the Quick Review.

1. An **ordered pair** is a pair of numbers written
   A. in numerical order between brackets
   B. between parentheses or brackets
   C. between parentheses in which order is important
   D. between parentheses in which order does not matter.

2. The **coordinates** of a point are
   A. the numbers in the corresponding ordered pair
   B. the solution of an equation
   C. the values of the x- and y-intercepts
   D. the graph of the point.

3. A **linear equation in two variables** is an equation that can be written in the form
   A. $Ax + By < C$
   B. $ax = b$
   C. $y = x^2$
   D. $Ax + By = C$.

4. An **intercept** is
   A. the point where the x-axis and y-axis intersect
   B. a pair of numbers written between parentheses in which order matters
   C. one of the four regions determined by a rectangular coordinate system
   D. the point where a graph intersects the x-axis or the y-axis.

5. The **slope** of a line is
   A. the measure of the run over the rise of the line
   B. the distance between two points on the line
   C. the ratio of the change in y to the change in x along the line
   D. the horizontal change compared to the vertical change of two points on the line.

6. In a relationship between two variables x and y, the **independent variable** is
   A. x, if x depends on y
   B. x, if y depends on x
   C. either x or y
   D. the larger of x and y.

7. In a relationship between two variables x and y, the **dependent variable** is
   A. y, if y depends on x
   B. y, if x depends on y
   C. either x or y
   D. the smaller of x and y.

8. A **relation** is
   A. a set of ordered pairs
   B. the ratio of the change in y to the change in x along a line
   C. the set of all possible values of the independent variable
   D. all the second components of a set of ordered pairs.

9. A **function** is
   A. the numbers in an ordered pair
   B. a set of ordered pairs in which each x-value corresponds to exactly one y-value
   C. a pair of numbers written between parentheses in which order matters
   D. the set of all ordered pairs that satisfy an equation.

10. The **domain** of a function is
    A. the set of all possible values of the dependent variable y
    B. a set of ordered pairs
    C. the difference between the x-values
    D. the set of all possible values of the independent variable x.

11. The **range** of a function is
    A. the set of all possible values of the dependent variable y
    B. a set of ordered pairs
    C. the difference between the y-values
    D. the set of all possible values of the independent variable x.

QUICK REVIEW

Concepts	Examples
4.1 *The Rectangular Coordinate System*	
Finding Intercepts To find the x-intercept, let $y = 0$. To find the y-intercept, let $x = 0$.	The graph of $2x + 3y = 12$ has $\qquad\qquad x$-intercept $(6, 0)$ and $\qquad\qquad y$-intercept $(0, 4)$.
4.2 *Slope*	
If $x_1 \neq x_2$, then $$m = \frac{\text{rise}}{\text{run}} = \frac{\text{change in } y}{\text{change in } x} = \frac{y_2 - y_1}{x_2 - x_1}.$$ A horizontal line has 0 slope. A vertical line has undefined slope. Parallel lines have equal slopes.	For $2x + 3y = 12$, $$m = \frac{4 - 0}{0 - 6} = -\frac{2}{3}.$$ The graph of $y = -5$ has $m = 0$. The graph of $x = 3$ has undefined slope. $\begin{array}{c\|c} y = 2x + 5 & 4x - 2y = 6 \\ m = 2 & -2y = -4x + 6 \\ & y = 2x - 3 \\ & m = 2 \end{array}$ These lines are **parallel**.
The slopes of perpendicular lines are negative reciprocals (with a product of -1).	$\begin{array}{c\|c} y = 3x - 1 & x + 3y = 4 \\ m = 3 & 3y = -x + 4 \\ & y = -\dfrac{1}{3}x + \dfrac{4}{3} \\ & m = -\dfrac{1}{3} \end{array}$ These lines are **perpendicular**.
4.3 *Linear Equations in Two Variables*	
Slope-Intercept Form $y = mx + b$	$y = 2x + 3 \qquad m = 2$, y-intercept is $(0, 3)$.
Point-Slope Form $y - y_1 = m(x - x_1)$	$y - 3 = 4(x - 5) \qquad (5, 3)$ is on the line, $m = 4$.
Standard Form $Ax + By = C$	$2x - 5y = 8$
Horizontal Line $y = b$	$y = 4$
Vertical Line $x = a$	$x = -1$

Concepts	Examples
4.4 *Linear Inequalities in Two Variables*	
Graphing a Linear Inequality	Graph $2x - 3y \leq 6$.
Step 1 Draw the graph of the line that is the boundary. Make the line solid if the inequality involves $\leq$ or $\geq$; make the line dashed if the inequality involves $<$ or $>$.	Draw the graph of $2x - 3y = 6$. Use a solid line because the symbol $\leq$ is used. 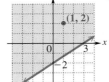
Step 2 Choose any point not on the line as a test point. Substitute the coordinates in the inequality.	Choose $(1, 2)$. $2(1) - 3(2) = 2 - 6 \leq 6$ True
Step 3 Shade the region that includes the test point if the test point satisfies the original inequality; otherwise, shade the region on the other side of the boundary line.	Shade the side of the line that includes $(1, 2)$.

4.5 *Introduction to Functions*	
To evaluate a function using function notation (that is, $f(x)$ notation) for a given value of x, substitute the value wherever x appears.	If $f(x) = x^2 - 7x + 12$, then $$f(1) = 1^2 - 7(1) + 12 = 6.$$
To write the equation that defines a function in function notation,	Write $2x + 3y = 12$ in function notation.
Step 1 Solve the equation for y.	$3y = -2x + 12$ Subtract $2x$. $$y = -\frac{2}{3}x + 4 \quad \text{Divide by 3.}$$
Step 2 Replace y with $f(x)$.	$$f(x) = -\frac{2}{3}x + 4$$

4.6 *Variation*	
If there is some constant k such that:	
$y = kx^n$, then y varies directly as, or is proportional to, x^n.	The area of a circle **varies directly as** the square of the radius: $A = kr^2$.
$y = \dfrac{k}{x^n}$, then y varies inversely as x^n.	Pressure **varies inversely as** volume: $P = \dfrac{k}{V}$.
$y = kxz$, then y varies jointly as x and z.	For a given principal, interest **varies jointly** as rate and time: $I = krt$.

ANSWERS TO TEST YOUR WORD POWER

1. C; *Examples:* $(0, 3), (3, 8), (4, 0)$
2. A; *Example:* The point associated with the ordered pair $(1, 2)$ has x-coordinate 1 and y-coordinate 2.
3. D; *Examples:* $3x + 2y = 6, x = y - 7, 4x = y$
4. D; *Example:* In Figure 4(b) of **Section 4.1,** the x-intercept is $(3, 0)$ and the y-intercept is $(0, 2)$.
5. C; *Example:* The line through $(3, 6)$ and $(5, 4)$ has slope $\dfrac{4 - 6}{5 - 3} = \dfrac{-2}{2} = -1.$
6. B; *Example:* See Answer 7, which follows.
7. A; *Example:* When borrowing money, the amount you borrow (independent variable) determines the size of your payments (dependent variable).
8. A; *Example:* The set $\{(2, 0), (4, 3), (6, 6), (8, 9)\}$ defines a relation.
9. B; *Example:* The relation given in Answer 8 is a function since each x-value corresponds to exactly one y-value.
10. D; *Example:* In the function in Answer 8, the domain is the set of x-values, $\{2, 4, 6, 8\}$.
11. A; *Example:* In the function in Answer 8, the range is the set of y-values, $\{0, 3, 6, 9\}$.

Chapter 4
REVIEW EXERCISES

[4.1] *Complete the given ordered pairs for each equation, and then graph the equation.*

1. $3x + 2y = 6$

$(0, \ ), (\ , 0), (\ , -2)$

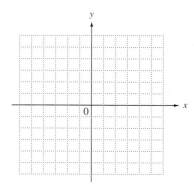

2. $x - y = 6$

$(2, \ ), (\ , -3), (1, \ ), (\ , -2)$

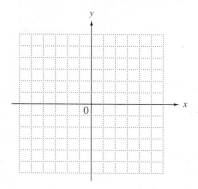

Find the x- and y-intercepts, and then graph each equation.

3. $4x + 3y = 12$

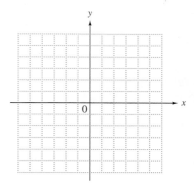

4. $5x + 7y = 15$

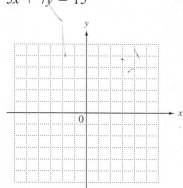

[4.2] *Find the slope of each line.*

5. Through $(-1, 2)$ and $(4, -6)$

6. $y = 2x + 3$

7. $-3x + 4y = 5$

8. $y = 4$

9. A line parallel to $3y = -2x + 5$

10. A line perpendicular to $3x - y = 6$

Tell whether the line has positive, negative, 0, *or* undefined *slope.*

11.

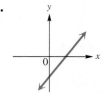

12.

13.

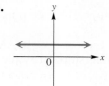

14.

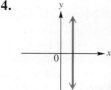

15. If the pitch of a roof is $\frac{1}{4}$, how many feet in the horizontal direction correspond to a rise of 3 ft?

16. Family income in the United States has steadily increased for many years (primarily due to inflation). In 1970 the median family income was about $10,000 per yr. In 2000 it was about $51,000 per yr. Find the average rate of change of median family income to the nearest dollar over that period. (*Source:* U.S. Bureau of the Census.)

[4.3] *Write an equation in slope-intercept form (if possible) for each line.*

17. Slope $\frac{3}{5}$; y-intercept $(0, -8)$

18. Slope $-\frac{1}{3}$; y-intercept $(0, 5)$

19. Slope 0; y-intercept $(0, 12)$

20. Undefined slope; through $(2, 7)$

21. Horizontal; through $(-1, 4)$

22. Vertical; through $(.3, .6)$

23. Through $(2, -5)$ and $(1, 4)$

24. Through $(-3, -1)$ and $(2, 6)$

25. Parallel to $4x - y = 3$ and through $(6, -2)$

26. Perpendicular to $2x - 5y = 7$ and through $(0, 1)$

27. The Midwest Athletic Club (**Section 4.3,** Exercises 61 and 62) offers two special membership plans. (*Source:* Midwest Athletic Club.) For each plan, write a linear equation in slope-intercept form and give the cost y in dollars of a 1-yr membership. Let x represent the number of months.

(a) Executive VIP/Gold membership: $159 fee plus $57 per month

(b) Executive Regular/Silver membership: $159 fee plus $47 per month

[4.4] *Graph each inequality.*

28. $3x - 2y \leq 12$

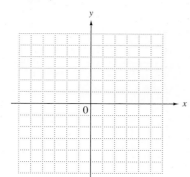

29. $5x - y > 6$

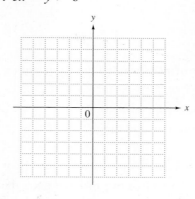

30. $x \geq 2$ or $y \geq 2$

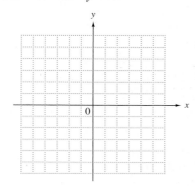

31. $2x + y \leq 1$ and $x \geq 2y$

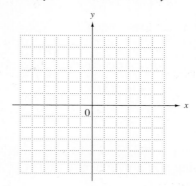

[4.5] *Give the domain and range of each relation. Identify any functions.*

32. $\{(-4, 2), (-4, -2), (1, 5), (1, -5)\}$

33.

34.

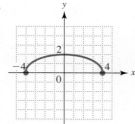

Determine whether each relation defines y as a function of x. Identify any linear functions. Give the domain in each case.

35. $y = 3x - 3$

36. $y < x + 2$

37. $y = |x - 4|$

38. $y = \sqrt{4x + 7}$ **39.** $x = y^2$ **40.** $y = \dfrac{7}{x - 36}$

41. The table shows profits for cable television station CNBC from 1994 through 1999.

(a) Does the table define a function?

(b) What are the domain and range?

(c) Call this function f. Give two ordered pairs that belong to f.

Year	Profit (in millions of dollars)
1994	40
1995	60
1996	80
1997	130
1998	180
1999	200

Source: Fortune, May 24, 1999, p. 142.

(d) Find $f(1994)$. What does it mean?

(e) If $f(x) = 180$, what does x equal?

Given $f(x) = -2x^2 + 3x - 6$, find each of the following.

42. $f(0)$ **43.** $f(3)$ **44.** $f(p)$ **45.** $f(-k)$

46. The equation $2x^2 - y = 0$ defines y as a function of x. Rewrite it using $f(x)$ notation, and find $f(3)$.

47. Suppose that $2x - 5y = 7$ defines a function. If $y = f(x)$, which one of the following defines the same function?

     A. $f(x) = \dfrac{7 - 2x}{5}$ **B.** $f(x) = \dfrac{-7 - 2x}{5}$

     C. $f(x) = \dfrac{-7 + 2x}{5}$ **D.** $f(x) = \dfrac{7 + 2x}{5}$

[4.6] Solve each variation problem.

48. In which one of the following does y vary inversely as x?

     A. $y = 2x$ **B.** $y = \dfrac{x}{3}$ **C.** $y = \dfrac{3}{x}$ **D.** $y = x^2$

49. If m varies inversely as p^2, and $m = 20$ when $p = 2$, find m when $p = 5$.

50. For the subject in a photograph to appear in the same perspective in the photograph as in real life, the viewing distance must be properly related to the amount of enlargement. For a particular camera, the viewing distance varies directly as the amount of enlargement. A picture taken with this camera that is enlarged 5 times should be viewed from a distance of 250 mm. Suppose a print 8.6 times the size of the negative is made. From what distance should it be viewed?

51. The volume of a rectangular box of a given height is proportional to its width and length. A box with width 4 ft and length 8 ft has volume 64 ft^3. Find the volume of a box with the same height that is 3 ft wide and 6 ft long.

Chapter 4

T E S T

Study Skills Workbook
Activity 10: Using Test Results

1. Find the slope of the line through $(6, 4)$ and $(-4, -1)$.

1. _____

For each line, find the slope and the x- and y-intercepts.

2. $3x - 2y = 13$

2. _____

3. $y = 5$

3. _____

4. Describe the graph of a line with undefined slope in a rectangular coordinate system.

4. _____

Find the x- and y-intercepts, and graph each equation.

5. _____

5. $4x - 3y = -12$

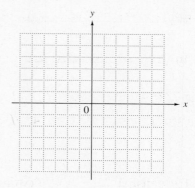

6. $y - 2 = 0$

6. _____

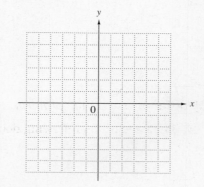

7. $y = -2x$

7. _____

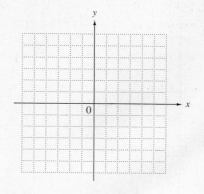

8.

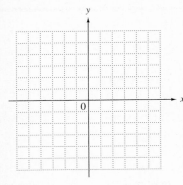

8. Graph $3x - 2y > 6$.

9. _____

10. _____

Write the equation of each line in slope-intercept form.

9. Through $(-3, 14)$; horizontal   **10.** Through $(4, -1)$; $m = -5$

11. (a) _____

(b) _____

11. Through $(-7, 2)$;

  (a) parallel to $3x + 5y = 6$

  (b) perpendicular to $y = 2x$

12. _____

_____

12. Which of the following is the graph of a function? Give its domain and range.

A.

B.

C.

D.

13. _____

_____

13. Which of the following does not define a function? Give its domain and range.

  A. $\{(0, 1), (-2, 3), (4, 8)\}$ **B.** $y = 2x - 6$

  C. $y = \sqrt{x + 2}$ **D.**

x	y
0	1
3	2
0	2
6	3

14. _____

14. If $f(x) = -x^2 + 2x - 1$, find $f(1)$ and $f(a)$.

Solve each problem.

15. _____

15. The current in a simple electrical circuit is inversely proportional to the resistance. If the current is 80 amps when the resistance is 30 ohms, find the current when the resistance is 12 ohms.

16. _____

16. The force of the wind blowing on a vertical surface varies jointly as the area of the surface and the square of the velocity. If a wind blowing at 40 mph exerts a force of 50 lb on a surface of 500 ft^2, how much force will a wind of 80 mph place on a surface of 2 ft^2?

Cumulative Review Exercises

CHAPTERS 1–4

Decide whether each statement is always true, sometimes true, *or* never true. *If the statement is* sometimes true, *give examples where it is true and where it is false.*

1. The absolute value of a negative number equals the additive inverse of the number.

2. The quotient of two integers with nonzero denominator is a rational number.

3. The sum of two negative numbers is positive.

4. The sum of a positive number and a negative number is 0.

Perform each operation.

5. $-|-2| - 4 + |-3| + 7$

6. $(-.8)^2$

7. $\sqrt{-64}$

8. $-\dfrac{2}{3}\left(-\dfrac{12}{5}\right)$

Simplify.

9. $-(-4m + 3)$

10. $3x^2 - 4x + 4 + 9x - x^2$

11. $\dfrac{3\sqrt{16} - (-1)7}{4 + (-6)}$

12. Write $-3 < x \le 5$ in interval notation.

13. Is $\sqrt{\dfrac{-2 + 4}{-5}}$ a real number?

Evaluate if $p = -4$, $q = -2$, and $r = 5$.

14. $-3(2q - 3p)$

15. $|p|^3 - |q^3|$

16. $\dfrac{\sqrt{r}}{-p + 2q}$

Solve.

17. $2z - 5 + 3z = 4 - (z + 2)$

18. $\dfrac{3a - 1}{5} + \dfrac{a + 2}{2} = -\dfrac{3}{10}$

19. $V = \dfrac{1}{3}\pi r^2 h$ for h

20. Two planes leave the Dallas-Fort Worth airport at the same time. One travels east at 550 mph, and the other travels west at 500 mph. Assuming no wind, how long will it take for the planes to be 2100 mi apart?

West ← ✈ Airport ✈ → East

21. Ms. Bell must take at least 30 units of a certain medication each day. She can get the medication from white pills or yellow pills, each of which contains 3 units of the drug. To provide other benefits, she needs to take twice as many of the yellow pills as white pills. Find the least number of white pills that will satisfy these requirements.

22. If each side of a square were increased by 4 in., the perimeter would be 8 in. less than twice the perimeter of the original square. Find the length of a side of the original square.

23. How are the solution sets of a linear equation and the two associated inequalities related?

Solve.

24. $3 - 2(m + 3) < 4m$

25. $2k + 4 < 10$ and $3k - 1 > 5$

26. $2k + 4 > 10$ or $3k - 1 < 5$

27. $|5x + 3| = 13$

28. $|x + 2| < 9$

29. $|2x - 5| \geq 9$

30. Complete the ordered pairs $(0, \quad)$, $(\quad, 0)$, and $(2, \quad)$ for the equation $3x - 4y = 12$.

31. Graph $-4x + 2y = 8$, and give the intercepts.

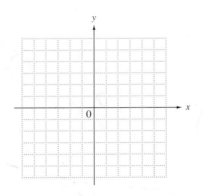

Find the slope of each line.

32. Through $(-5, 8)$ and $(-1, 2)$

33. Parallel to $y = -\dfrac{1}{2}x + 5$

34. Perpendicular to $4x - 3y = 12$

Write an equation in slope-intercept form for each line.

35. Slope $-\dfrac{3}{4}$; y-intercept $(0, -1)$

36. Horizontal; through $(2, -2)$

37. Through $(4, -3)$ and $(1, 1)$

38. For the function defined by $f(x) = -4x + 10$,
   (a) what is the domain?
   (b) what is $f(-3)$?

Use the graph to answer Exercises 39 and 40.

39. What is the slope of the line segment joining the points for 1992 and 2000?

40. Which one of the two line segments shown has a greater slope?

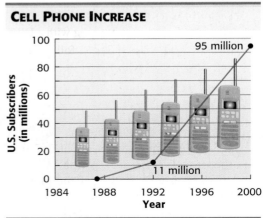

CELL PHONE INCREASE

Source: Cellular Telecommunications Industry Association, Intel Corp.

Systems of Linear Equations

5

5.1 **Systems of Linear Equations in Two Variables**

5.2 **Systems of Linear Equations in Three Variables**

5.3 **Applications of Systems of Linear Equations**

5.4 **Solving Systems of Linear Equations by Matrix Methods**

During the last decade of the twentieth century, the number of people living with AIDS in various racial groups in the United States followed linear patterns, as shown in the accompanying graph. At some year during that decade, the number of people in two of these groups was the same. The graph can be used to determine that year and number by finding the coordinates of the point of intersection of the two lines. See Exercises 1 and 2 of the Chapter 5 Test.

The process of determining the point of intersection of two lines is the idea behind solving a *system of linear equations in two variables*. This chapter illustrates methods of finding such points.

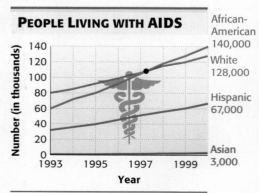

PEOPLE LIVING WITH AIDS

Source: U.S. Centers for Disease Control and Prevention.

5.1 Systems of Linear Equations in Two Variables

OBJECTIVES

1 Solve linear systems by graphing.

2 Decide whether an ordered pair is a solution of a linear system.

3 Solve linear systems (with two equations and two variables) by substitution.

4 Solve linear systems (with two equations and two variables) by elimination.

5 Solve special systems.

As technology continues to improve, the sale of digital cameras increases, while that of conventional cameras decreases. This can be seen in Figure 1, which illustrates this growth and decline using a graph. The two straight-line graphs intersect where the two types of cameras had the same sales.

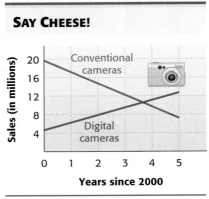

SAY CHEESE!

Source: Consumer Electronics Association.

Figure 1

We could use a linear equation to model the graph of digital camera sales and another linear equation to model the graph of conventional camera sales. Such a set of equations is called a **system of equations,** in this case a **linear system of equations.** The point where the graphs in Figure 1 intersect is a solution of each of the individual equations. It is also the solution of the linear system of equations.

OBJECTIVE 1 Solve linear systems by graphing. The **solution set of a system of equations** contains all ordered pairs that satisfy all the equations of the system *at the same time*. An example of a linear system is

$$x + y = 5$$
$$2x - y = 4.$$

Linear system of equations

One way to find the solution set of a linear system of equations is to graph each equation and find the point where the graphs intersect.

EXAMPLE 1 Solving a System by Graphing

Solve the system of equations by graphing.

$$x + y = 5 \quad (1)$$
$$2x - y = 4 \quad (2)$$

When we graph these linear equations as shown in Figure 2, the graph suggests that the point of intersection is the ordered pair (3, 2).

Continued on Next Page

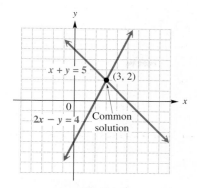

Figure 2

To be sure that (3, 2) is a solution of *both* equations, we check by substituting 3 for x and 2 for y in each equation.

$x + y = 5$	(1)	$2x - y = 4$	(2)
$3 + 2 = 5$	?	$2(3) - 2 = 4$	?
$5 = 5$	True	$6 - 2 = 4$	?
		$4 = 4$	True

Since (3, 2) makes both equations true, {(3, 2)} is the solution set of the system.

Work Problem 1 at the Side. ▶▶▶

⌨ **Calculator Tip** A graphing calculator can be used to solve a system. Each equation must be solved for y before being entered in the calculator. The point of intersection of the graphs, which is the solution of the system, can then be displayed. Consult your owner's manual for details.

OBJECTIVE 2 Decide whether an ordered pair is a solution of a linear system. To decide if an ordered pair is a solution of a system, we substitute the ordered pair in both equations of the system, just as we did when we checked the solution in Example 1.

EXAMPLE 2 Deciding whether an Ordered Pair Is a Solution

Decide whether the given ordered pair is a solution of the given system.

(a) $x + y = 6$
$4x - y = 14$; (4, 2)

Replace x with 4 and y with 2 in each equation of the system.

$x + y = 6$		$4x - y = 14$	
$4 + 2 = 6$	?	$4(4) - 2 = 14$	?
$6 = 6$	True	$16 - 2 = 14$	?
		$14 = 14$	True

Since (4, 2) makes both equations true, (4, 2) is a solution of the system.

Continued on Next Page

---

1 Solve each system by graphing.

(a) $x - y = 3$ (1)
$2x - y = 4$ (2)

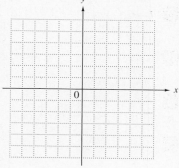

(b) $2x + \ y = -5$ (1)
$-x + 3y = 6$ (2)

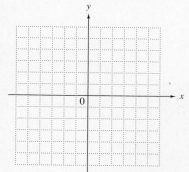

ANSWERS
1. (a) $\{(1, -2)\}$

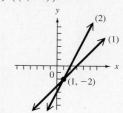

(b) $\{(-3, 1)\}$

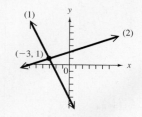

2 Are the given ordered pairs solutions of the given systems?

(a) $2x + y = -6$
   $x + 3y = 2$; $(-4, 2)$

(b) $3x + 2y = 11$
   $x + 5y = 36$; $(-1, 7)$

$3x + 2y = 11$		$x + 5y = 36$	
$3(-1) + 2(7) = 11$	?	$-1 + 5(7) = 36$	?
$-3 + 14 = 11$	?	$-1 + 35 = 36$	?
$11 = 11$	True	$34 = 36$	False

The ordered pair $(-1, 7)$ is not a solution of the system, since it does not make *both* equations true.

◀◀◀ **Work Problem 2 at the Side.**

Since the graph of a linear equation is a straight line, there are three possibilities for the solution set of a linear system in two variables.

Graphs of Linear Systems in Two Variables

1. The two graphs intersect in a single point. The coordinates of this point give the only solution of the system. In this case the system is **consistent,** and the equations are **independent.** This is the most common case. See Figure 3(a).

2. The graphs are parallel lines. In this case the system is **inconsistent;** that is, there is no solution common to both equations of the system, and the solution set is ∅. See Figure 3(b).

3. The graphs are the same line. In this case the equations are **dependent,** since any solution of one equation of the system is also a solution of the other. The solution set is an infinite set of ordered pairs representing the points on the line. See Figure 3(c).

(b) $9x - y = -4$
   $4x + 3y = 11$; $(-1, 5)$

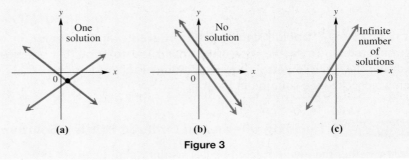

Figure 3

OBJECTIVE 3 Solve linear systems (with two equations and two variables) by substitution. Since it can be difficult to read exact coordinates, especially if they are not integers, from a graph, we usually use algebraic methods to solve systems. One such method, the **substitution method,** is most useful for solving linear systems in which one equation is solved or can be easily solved for one variable in terms of the other.

EXAMPLE 3 Solving a System by Substitution

Solve the system.

$$2x - y = 6 \quad (1)$$
$$x = y + 2 \quad (2)$$

Since equation (2) is solved for x, substitute $y + 2$ for x in equation (1).

Continued on Next Page

ANSWERS
2. **(a)** yes **(b)** no

$$2x - y = 6 \quad (1)$$
$$2(y + 2) - y = 6 \quad \text{Let } x = y + 2.$$
$$2y + 4 - y = 6 \quad \text{Distributive property}$$
$$y + 4 = 6 \quad \text{Combine terms.}$$
$$y = 2 \quad \text{Subtract 4.}$$

We found y. Now find x by substituting 2 for y in equation (2).

$$x = y + 2 = 2 + 2 = 4$$

Thus, $x = 4$ and $y = 2$, giving the ordered pair $(4, 2)$. Check this solution in both equations of the original system. The solution set is $\{(4, 2)\}$.

CAUTION
Be careful! Even though we found y first in Example 3, *the x-coordinate is always written first in the ordered pair solution of a system.*

Work Problem 3 at the Side. ▶▶▶

The substitution method is summarized as follows.

Solving a Linear System by Substitution

Step 1 **Solve one of the equations for either variable.** If one of the variable terms has coefficient 1 or -1, choose it, since the substitution method is usually easier this way.

Step 2 **Substitute** for that variable in the other equation. The result should be an equation with just one variable.

Step 3 **Solve** the equation from Step 2.

Step 4 **Find the other value.** Substitute the result from Step 3 into the equation from Step 1 to find the value of the other variable.

Step 5 **Check** the solution in both of the original equations. Then write the solution set.

EXAMPLE 4 **Solving a System by Substitution**

Solve the system.

$$3x + 2y = 13 \quad (1)$$
$$4x - y = -1 \quad (2)$$

Step 1 First solve one of the equations for x or y. Since the coefficient of y in equation (2) is -1, it is easiest to solve for y in equation (2).

$$4x - y = -1 \quad (2)$$
$$-y = -1 - 4x \quad \text{Subtract } 4x.$$
$$y = 1 + 4x \quad \text{Multiply by } -1.$$

Step 2 Substitute $1 + 4x$ for y in equation (1).

$$3x + 2y = 13 \quad (1)$$
$$3x + 2(1 + 4x) = 13 \quad \text{Let } y = 1 + 4x.$$

Continued on Next Page

3 Solve by substitution.

(a) $7x - 2y = -2$
$\quad y = 3x$

(b) $5x - 3y = -6$
$\quad x = 2 - y$

ANSWERS
3. (a) $\{(-2, -6)\}$ (b) $\{(0, 2)\}$

4 Solve by substitution.

(a) $3x - y = 10$
$2x + 5y = 1$

(b) $4x - 5y = -11$
$x + 2y = 7$

Step 3 Solve for x.

$$3x + 2(1 + 4x) = 13 \quad \text{From Step 2}$$
$$3x + 2 + 8x = 13 \quad \text{Distributive property}$$
$$11x = 11 \quad \text{Combine terms; subtract 2.}$$
$$x = 1 \quad \text{Divide by 11.}$$

Step 4 Now solve for y. From Step 1, $y = 1 + 4x$, so

$$y = 1 + 4(1) = 5. \quad \text{Let } x = 1.$$

Step 5 Check the solution $(1, 5)$ in both equations (1) and (2).

$3x + 2y = 13$	(1)	$4x - y = -1$	(2)
$3(1) + 2(5) = 13$	?	$4(1) - 5 = -1$	?
$3 + 10 = 13$	?	$4 - 5 = -1$	?
$13 = 13$	True	$-1 = -1$	True

The solution set is $\{(1, 5)\}$.

> ◀◀ **Work Problem 4 at the Side.**

OBJECTIVE 4 Solve linear systems (with two equations and two variables) by elimination. Another algebraic method, the **elimination method**, involves combining the two equations in a system so that one variable is eliminated. This is done using the following logic:

$$\text{If} \quad a = b \text{ and } c = d, \quad \text{then} \quad a + c = b + d.$$

EXAMPLE 5 Solving a System by Elimination

Solve the system.

$$2x + 3y = -6 \quad (1)$$
$$4x - 3y = 6 \quad (2)$$

Notice that adding the equations together will eliminate the variable y.

$$
\begin{array}{ll}
2x + 3y = -6 & (1) \\
\underline{4x - 3y = 6} & (2) \\
6x = 0 & \text{Add.} \\
x = 0 & \text{Solve for } x.
\end{array}
$$

To find y, substitute 0 for x in either equation (1) or equation (2).

$$
\begin{array}{ll}
2x + 3y = -6 & (1) \\
2(0) + 3y = -6 & \text{Let } x = 0. \\
0 + 3y = -6 & \\
3y = -6 & \\
y = -2 &
\end{array}
$$

The solution of the system is $(0, -2)$. Check by substituting 0 for x and -2 for y in both equations of the original system. The solution set is $\{(0, -2)\}$.

> ◀◀ **Work Problem 5 at the Side.**

5 Solve by elimination.

(a) $3x - y = -7$
$2x + y = -3$

(b) $-2x + 3y = -10$
$2x + 2y = 5$

By adding the equations in Example 5, we eliminated the variable y because the coefficients of the y-terms were opposites. In many cases the coefficients will *not* be opposites, and we must transform one or both equations so that the coefficients of one pair of variable terms are opposites.

ANSWERS
4. (a) $\{(3, -1)\}$ (b) $\{(1, 3)\}$

5. (a) $\{(-2, 1)\}$ (b) $\left\{\left(\frac{7}{2}, -1\right)\right\}$

Solving a Linear System by Elimination

Step 1 **Write both equations in standard form** $Ax + By = C$.

Step 2 **Make the coefficients of one pair of variable terms opposites.** Multiply one or both equations by appropriate numbers so that the sum of the coefficients of either the x- or y-terms is 0.

Step 3 **Add** the new equations to eliminate a variable. The sum should be an equation with just one variable.

Step 4 **Solve** the equation from Step 3 for the remaining variable.

Step 5 **Find the other value.** Substitute the result of Step 4 into either of the original equations and solve for the other variable.

Step 6 **Check** the solution in both of the original equations. Then write the solution set.

6 Solve by elimination.

(a) $x + 3y = 8$
     $2x - 5y = -17$

EXAMPLE 6 **Solving a System by Elimination**

Solve the system.

$$5x - 2y = 4 \quad (1)$$
$$2x + 3y = 13 \quad (2)$$

Step 1 Both equations are in standard form.

Step 2 Suppose that you wish to eliminate the variable x. One way to do this is to multiply equation (1) by 2 and equation (2) by -5.

$$10x - 4y = 8 \qquad \text{2 times each side of equation (1)}$$
$$-10x - 15y = -65 \qquad \text{-5 times each side of equation (2)}$$

Step 3 Now add.

$$
\begin{array}{r}
10x - 4y = 8 \\
\underline{-10x - 15y = -65} \\
-19y = -57 \qquad \text{Add.}
\end{array}
$$

(b) $6x - 2y = -21$
     $-3x + 4y = 36$

Step 4 Solve for y. $y = 3$ Divide by -19.

Step 5 To find x, substitute 3 for y in either equation (1) or (2). Substituting in equation (2) gives

$$2x + 3y = 13 \quad (2)$$
$$2x + 3(3) = 13 \quad \text{Let } y = 3.$$
$$2x + 9 = 13$$
$$2x = 4 \quad \text{Subtract 9.}$$
$$x = 2. \quad \text{Divide by 2.}$$

(c) $2x + 3y = 19$
     $3x - 7y = -6$

Step 6 The solution is $(2, 3)$. To check, substitute 2 for x and 3 for y in both equations (1) and (2).

$$
\begin{array}{ll}
5x - 2y = 4 \quad (1) & 2x + 3y = 13 \quad (2) \\
5(2) - 2(3) = 4 \quad ? & 2(2) + 3(3) = 13 \quad ? \\
10 - 6 = 4 \quad ? & 4 + 9 = 13 \quad ? \\
4 = 4 \quad \text{True} & 13 = 13 \quad \text{True}
\end{array}
$$

The solution set is $\{(2, 3)\}$.

Work Problem 6 at the Side.

ANSWERS

6. (a) $\{(-1, 3)\}$ (b) $\left\{\left(-\dfrac{2}{3}, \dfrac{17}{2}\right)\right\}$
   (c) $\{(5, 3)\}$

7 Solve each system.

(a) $\dfrac{1}{3}x - \dfrac{1}{2}y = \dfrac{1}{6}$

$\quad 3x - 2y = 9$

EXAMPLE 7 **Solving a System with Fractional Coefficients**

Solve the system.

$$5x - 2y = 4 \qquad (1)$$
$$\frac{1}{2}x + \frac{3}{4}y = \frac{13}{4} \qquad (2)$$

If an equation in a system has fractional coefficients, as in equation (2), first multiply by the least common denominator to clear the fractions.

$$4\left(\frac{1}{2}x + \frac{3}{4}y\right) = 4 \cdot \frac{13}{4} \qquad \text{Multiply equation (2) by the LCD, 4.}$$

$$4 \cdot \frac{1}{2}x + 4 \cdot \frac{3}{4}y = 4 \cdot \frac{13}{4} \qquad \text{Distributive property}$$

$$2x + 3y = 13 \qquad \text{Equivalent to equation (2)}$$

The system of equations becomes

$$5x - 2y = 4 \qquad (1)$$
$$2x + 3y = 13, \qquad \text{Equation (2) with fractions cleared}$$

which is identical to the system we solved in Example 6. The solution set is $\{(2, 3)\}$. To confirm this, check the solution in both equations (1) and (2).

◀◀◀ Work Problem 7 at the Side.

(b) $\dfrac{x}{5} + \dfrac{2y}{3} = -\dfrac{8}{5}$

$\quad 3x - \ y = \ \ 9$

NOTE

If an equation in a system contains decimal coefficients, it is best to first clear the decimals by multiplying by 10, 100, or 1000, depending on the number of decimal places. Then solve the system. For example, we multiply *each side* of the equation

$$.5x + .75y = 3.25$$

by 100 to get the equivalent equation

$$50x + 75y = 325.$$

OBJECTIVE 5 **Solve special systems.**   As we saw in Figures 3(b) and (c), some systems of linear equations have no solution or an infinite number of solutions.

EXAMPLE 8 **Solving a System of Dependent Equations**

Solve the system.

$$2x - \ y = 3 \qquad (1)$$
$$6x - 3y = 9 \qquad (2)$$

We multiply equation (1) by -3, and then add the result to equation (2).

$$-6x + 3y = -9 \qquad \text{-3 times each side of equation (1)}$$
$$\underline{6x - 3y = \ \ 9} \qquad (2)$$
$$0 = \ 0 \qquad \text{True}$$

Continued on Next Page

ANSWERS

7. (a) $\{(5, 3)\}$ (b) $\{(2, -3)\}$

Adding these equations gives the true statement $0 = 0$. In the original system, we could get equation (2) from equation (1) by multiplying equation (1) by 3. Because of this, equations (1) and (2) are equivalent and have the same graph, as shown in Figure 4. The equations are dependent. The solution set is the set of all points on the line with equation $2x - y = 3$, written

$$\{(x, y) \mid 2x - y = 3\}$$

and read "the set of all ordered pairs (x, y), such that $2x - y = 3$."

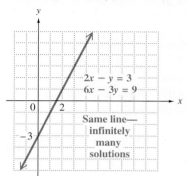

Figure 4

NOTE
When a system has an infinite number of solutions, as in Example 8, either equation of the system could be used to write the solution set. We prefer to use the equation (in standard form) with coefficients that are integers having no common factor (except 1).

Work Problem 8 at the Side. ▶▶▶

EXAMPLE 9 **Solving an Inconsistent System**

Solve the system.

$$x + 3y = 4 \quad (1)$$
$$-2x - 6y = 3 \quad (2)$$

Multiply equation (1) by 2, and then add the result to equation (2).

$$2x + 6y = 8 \qquad \text{Equation (1) multiplied by 2}$$
$$\underline{-2x - 6y = 3} \qquad (2)$$
$$0 = 11 \qquad \text{False}$$

The result of the addition step is a false statement, which indicates that the system is inconsistent. As shown in Figure 5, the graphs of the equations of the system are parallel lines. There are no ordered pairs that satisfy both equations, so there is no solution for the system. The solution set is $\emptyset$.

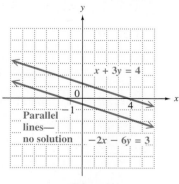

Figure 5

Work Problem 9 at the Side. ▶▶▶

8 Solve the system. Then graph both equations.

$$2x + y = 6 \qquad (1)$$
$$-8x - 4y = -24 \qquad (2)$$

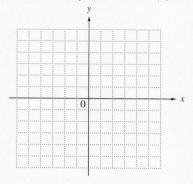

9 Solve the system. Then graph both equations.

$$4x - 3y = 8 \qquad (1)$$
$$8x - 6y = 14 \qquad (2)$$

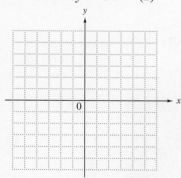

ANSWERS
8. $\{(x, y) \mid 2x + y = 6\}$

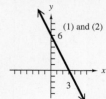

9. $\emptyset$

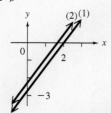

10 Write the equations of Example 8 in slope-intercept form. Use function notation.

The results of Examples 8 and 9 are generalized as follows.

Special Cases of Linear Systems

If both variables are eliminated when a system of linear equations is solved, then

1. there is no solution if the resulting statement is *false;*

2. there are infinitely many solutions if the resulting statement is *true*.

Slopes and y-intercepts can be used to decide if the graphs of a system of equations are parallel lines or if they coincide. In Example 8, writing each equation in slope-intercept form shows that both lines have slope 2 and y-intercept $(0, -3)$, so the graphs are the same line and the system has an infinite number of solutions.

Work Problem 10 at the Side.

In Example 9, both equations have slope $-\frac{1}{3}$, but the y-intercepts are $(0, \frac{4}{3})$ and $(0, -\frac{1}{2})$, showing that the graphs are two distinct parallel lines. Thus, the system has $\emptyset$ as its solution set.

Work Problem 11 at the Side.

11 Write the equations of Example 9 in slope-intercept form. Use function notation.

ANSWERS

10. Both equations are $f(x) = 2x - 3$.

11. $f(x) = -\frac{1}{3}x + \frac{4}{3}; f(x) = -\frac{1}{3}x - \frac{1}{2}$

5.1 Exercises

FOR EXTRA HELP

Addison-Wesley Math Tutor Center · MathXL · Digital Video Tutor CD 3 / Videotape 3 · Student's Solutions Manual · MyMathLab · Interactmath.com

Fill in the blanks with the correct responses.

1. If $(3, -6)$ is a solution of a linear system in two variables, then substituting _____ for x and _____ for y leads to true statements in *both* equations.

2. A solution of a system of independent linear equations in two variables is a(n) _____.

3. If the solution process leads to a false statement such as $0 = 5$ when solving a system, the solution set is _____.

4. If the solution process leads to a true statement such as $0 = 0$ when solving a system, the system has _____ equations.

5. If the two lines forming a system have the same slope and different y-intercepts, the system has _____ solution(s). (how many?)

6. If the two lines forming a system have different slopes, the system has _____ solution(s). (how many?)

7. Which ordered pair could possibly be a solution of the graphed system of equations? Why?
   A. $(3, 3)$
   B. $(-3, 3)$
   C. $(-3, -3)$
   D. $(3, -3)$

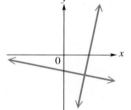

8. Which ordered pair could possibly be a solution of the graphed system of equations? Why?
   A. $(3, 0)$
   B. $(-3, 0)$
   C. $(0, 3)$
   D. $(0, -3)$

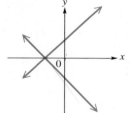

9. Match each system with the correct graph.

   (a) $x + y = 6$
   $x - y = 0$

   (b) $x + y = -6$
   $x - y = 0$

   (c) $x + y = 0$
   $x - y = -6$

   (d) $x + y = 0$
   $x - y = 6$

   A.

   B.

   C.

   D.

Solve each system by graphing. See Example 1.

10. $x + y = 4$
$2x - y = 2$

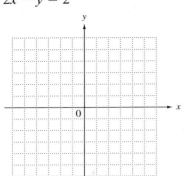

11. $x + y = -5$
$-2x + y = 1$

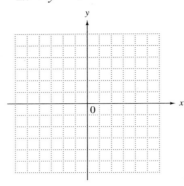

12. $x - 4y = -4$
$3x + y = 1$

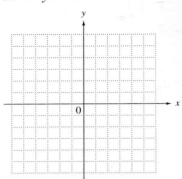

Decide whether the given ordered pair is a solution of the given system. See Example 2.

13. $x + y = 6$
$x - y = 4$; $(5, 1)$

14. $x - y = 17$
$x + y = -1$; $(8, -9)$

15. $2x - y = 8$
$3x + 2y = 20$; $(5, 2)$

16. $3x - 5y = -12$
$x - y = 1$; $(-1, 2)$

Solve each system by substitution. See Examples 3, 4, and 7.

17. $4x + y = 6$
$y = 2x$

18. $2x - y = 6$
$y = 5x$

19. $3x - 4y = -22$
$-3x + y = 0$

20. $-3x + y = -5$
$x + 2y = 0$

21. $-x - 4y = -14$
$2x = y + 1$

22. $-3x - 5y = -17$
$4x = y - 8$

23. $5x - 4y = 9$
$3 - 2y = -x$

24. $6x - y = -9$
$4 + 7x = -y$

25. $x = 3y + 5$
$x = \dfrac{3}{2}y$

26. $x = 6y - 2$
$x = \dfrac{3}{4}y$

27. $\dfrac{1}{2}x + \dfrac{1}{3}y = 3$
$y = 3x$

28. $\dfrac{1}{4}x - \dfrac{1}{5}y = 9$
$y = 5x$

Solve each system by elimination. If the system is inconsistent or has dependent equations, say so. See Examples 5–9.

29. $2x - 5y = 11$
   $3x + y = 8$

30. $-2x + 3y = 1$
   $-4x + y = -3$

31. $3x + 4y = -6$
   $5x + 3y = 1$

32. $4x + 3y = 1$
   $3x + 2y = 2$

33. $3x + 3y = 0$
   $4x + 2y = 3$

34. $8x + 4y = 0$
   $4x - 2y = 2$

35. $7x + 2y = 6$
   $-14x - 4y = -12$

36. $x - 4y = 2$
   $4x - 16y = 8$

37. $\dfrac{x}{2} + \dfrac{y}{3} = -\dfrac{1}{3}$
   $\dfrac{x}{2} + 2y = -7$

38. $\dfrac{x}{5} + y = \dfrac{6}{5}$
   $\dfrac{x}{10} + \dfrac{y}{3} = \dfrac{5}{6}$

39. $5x - 5y = 3$
   $x - y = 12$

40. $2x - 3y = 7$
   $-4x + 6y = 14$

Write each equation in slope-intercept form, and then tell how many solutions the system has. Do not actually solve.

41. $3x + 7y = 4$
   $6x + 14y = 3$

42. $-x + 2y = 8$
   $4x - 8y = 1$

43. $2x = -3y + 1$
   $6x = -9y + 3$

44. $5x = -2y + 1$
   $10x = -4y + 2$

45. Assuming you want to minimize the amount of work required, tell whether you would use the substitution or elimination method to solve each system. Explain your answers. *Do not actually solve.*

   (a) $6x - y = 5$
      $y = 11x$

   (b) $3x + y = -7$
      $x - y = -5$

   (c) $3x - 2y = 0$
      $9x + 8y = 7$

Solve each system by the method of your choice. (For Exercises 46–48, see your answers for Exercise 45.)

46. $6x - y = 5$
   $y = 11x$

47. $3x + y = -7$
   $x - y = -5$

48. $3x - 2y = 0$
   $9x + 8y = 7$

49. $2x + 3y = 10$
$-3x + y = 18$

50. $3x - 5y = 7$
$2x + 3y = 30$

51. $\frac{1}{2}x - \frac{1}{8}y = -\frac{1}{4}$
$-4x + y = 2$

52. $\frac{1}{6}x + \frac{1}{3}y = 8$
$\frac{1}{4}x + \frac{1}{2}y = 12$

53. $.3x + .2y = .4$
$.5x + .4y = .7$

54. $.2x + .5y = 6$
$.4x + y = 9$

Answer the questions in Exercises 55–58 by observing the graphs provided.

55. The figure shows graphs that represent supply and demand for a certain brand of low-fat frozen yogurt at various prices per half-gallon (in dollars).

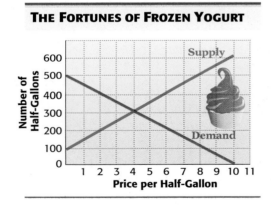

THE FORTUNES OF FROZEN YOGURT

(a) At what price does supply equal demand?

(b) For how many half-gallons does supply equal demand?

(c) What are the supply and demand at a price of $2 per half-gallon?

56. La Bronda Jones compared the monthly payments she would incur for two types of mortgages: fixed-rate and variable-rate. Her observations led to the following graphs.

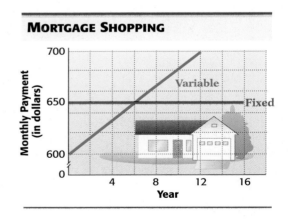

MORTGAGE SHOPPING

(a) For which years would the monthly payment be more for the fixed-rate mortgage than for the variable-rate mortgage?

(b) In what year would the payments be the same, and what would those payments be?

57. If the rates of growth between 1990 and 2000 continue, the populations of Houston, Phoenix, Dallas, and Philadelphia will follow the trends indicated in the graph.

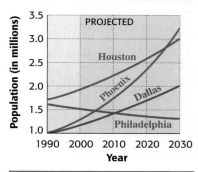

THE GROWTH GAME

Size of cities if the rate of population growth from 1990 to 2000 continues:

Source: U.S. Census Bureau, *Chronicle* research.

(a) Which cities will experience population growth?

(b) Which city will experience population decline?

(c) Rank the city populations from least to greatest for the year 2000.

(d) In which year will the population of Dallas equal that of Philadephia? About what will this population be?

(e) Write as an ordered pair (year, population in millions) the point at which Houston and Phoenix will have the same population.

58. The graph shows how the production of vinyl LPs, audiocassettes, and compact discs (CDs) changed over the final years of the twentieth century.

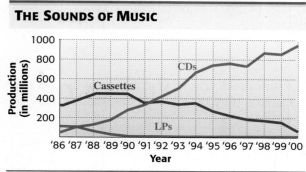

THE SOUNDS OF MUSIC

Source: Recording Industry Association of America.

(a) In what year did cassette production and CD production reach equal levels? What was that level?

(b) Express the point of intersection of the graphs of LP production and CD production as an ordered pair of the form (year, production level).

(c) Between what years did cassette production first stabilize and remain fairly constant?

(d) Describe the trend in CD production from 1986 through 2000. If a straight line were used to approximate its graph, would the line have positive, negative, or 0 slope?

(e) If a straight line were used to approximate the graph of cassette production from 1990 through 2000, would the line have positive, negative, or 0 slope? Explain.

(f) During the years 2001 and 2002, CD production decreased. What might explain this?

Use the graph given in Figure 1 at the beginning of this section (repeated here) to work Exercises 59–62.

59. For which years during the period 2000–2004 were sales of digital cameras less than sales of conventional cameras?

60. Estimate the year in which sales for the two types of cameras were the same. About what was this sales figure?

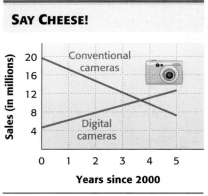

SAY CHEESE!

Source: Consumer Electronics Association.

61. If $x = 0$ represents 2000 and $x = 4$ represents 2004, sales (y) in millions of dollars can be modeled by the linear equations in the following system.

$$2.5x + y = 19.4 \quad \text{Conventional cameras}$$
$$-1.7x + y = 4.4 \quad \text{Digital cameras}$$

Solve this system using the substitution method. Express values as fractions in lowest terms. Write the solution as an ordered pair of the form (year, sales).

62. Find approximations for the x- and y-values, and interpret your answer for Exercise 61. How does it compare to your estimate from Exercise 60?

RELATING CONCEPTS (EXERCISES 63–66) For Individual or Group Work

Work Exercises 63–66 in order to see the connections between systems of linear equations and the graphs of linear functions.

63. Use elimination or substitution to solve the system.

$$3x + y = 6 \quad (1)$$
$$-2x + 3y = 7 \quad (2)$$

64. For equation (1) in the system of Exercise 63, solve for y and rename it $f(x)$. What special kind of function is f?

65. For equation (2) in the system of Exercise 63, solve for y and rename it $g(x)$. What special kind of function is g?

66. Use the result of Exercise 63 to fill in the blanks with the appropriate responses:

Because the graphs of f and g are straight lines that are neither parallel nor coincide,

they intersect in exactly _____ point. The coordinates of the point are

(_____ , _____). Using function notation, this is given by $f($_____$) =$ _____

and $g($_____$) =$ _____ .

5.2 Systems of Linear Equations in Three Variables

A solution of an equation in three variables, such as

$$2x + 3y - z = 4,$$

is called an **ordered triple** and is written (x, y, z). For example, the ordered triple $(0, 1, -1)$ is a solution of the equation, because

$$2(0) + 3(1) - (-1) = 0 + 3 + 1 = 4.$$

Verify that another solution of this equation is $(10, -3, 7)$.

In the rest of this chapter, the term *linear equation* is extended to equations of the form

$$Ax + By + Cz + \ldots + Dw = K,$$

where not all the coefficients $A, B, C, \ldots, D$ equal 0. For example,

$$2x + 3y - 5z = 7 \quad \text{and} \quad x - 2y - z + 3u - 2w = 8$$

are linear equations, the first with three variables and the second with five variables.

OBJECTIVE 1 Understand the geometry of systems of three equations in three variables.   In this section, we discuss the solution of a system of linear equations in three variables, such as

$$
\begin{aligned}
4x + 8y + \ z &= 2 \\
x + 7y - 3z &= -14 \\
2x - 3y + 2z &= 3.
\end{aligned}
$$

Theoretically, a system of this type can be solved by graphing. However, the graph of a linear equation with three variables is a *plane,* not a line. Since the graph of each equation of the system is a plane, which requires three-dimensional graphing, this method is not practical. However, it does illustrate the number of solutions possible for such systems, as shown in Figure 6.

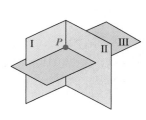

A single solution

(a)

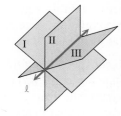

Points of a line in common

(b)

All points in common

(c)

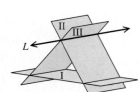

No points in common

(d)

No points in common

(e)

No points in common

(f)

Figure 6

OBJECTIVES

1 Understand the geometry of systems of three equations in three variables.

2 Solve linear systems (with three equations and three variables) by elimination.

3 Solve linear systems where some of the equations have missing terms.

4 Solve special systems.

Figure 6 on the preceding page illustrates the following cases.

Graphs of Linear Systems in Three Variables

1. The three planes may meet at a single, common point that is the solution of the system. See Figure 6(a).

2. The three planes may have the points of a line in common so that the infinite set of points that satisfy the equation of the line is the solution of the system. See Figure 6(b).

3. The three planes may coincide so that the solution of the system is the set of all points on a plane. See Figure 6(c).

4. The planes may have no points common to all three so that there is no solution of the system. See Figures 6(d), (e), and (f).

OBJECTIVE 2 **Solve linear systems (with three equations and three variables) by elimination.** Is it possible to solve a system of three equations in three variables such as the one that follows?

$$4x + 8y + z = 2$$
$$x + 7y - 3z = -14$$
$$2x - 3y + 2z = 3$$

Graphing to find the solution set of such a system is impractical, so these systems are solved with an extension of the elimination method from **Section 5.1,** summarized as follows.

Solving a Linear System in Three Variables

Step 1 **Eliminate a variable.** Use the elimination method to eliminate any variable from any two of the original equations. The result is an equation in two variables.

Step 2 **Eliminate the same variable again.** Eliminate the *same* variable from any *other* two equations. The result is an equation in the same two variables as in Step 1.

Step 3 **Eliminate a different variable and solve.** Use the elimination method to eliminate a second variable from the two equations in two variables that result from Steps 1 and 2. The result is an equation in one variable that gives the value of that variable.

Step 4 **Find a second value.** Substitute the value of the variable found in Step 3 into either of the equations in two variables to find the value of the second variable.

Step 5 **Find a third value.** Use the values of the two variables from Steps 3 and 4 to find the value of the third variable by substituting into any of the original equations.

Step 6 **Check** the solution in all of the original equations. Then write the solution set.

EXAMPLE 1 Solving a System in Three Variables

Solve the system.

$$4x + 8y + z = 2 \qquad (1)$$
$$x + 7y - 3z = -14 \qquad (2)$$
$$2x - 3y + 2z = 3 \qquad (3)$$

Step 1 As before, the elimination method involves eliminating a variable from the sum of two equations. The choice of which variable to eliminate is arbitrary. Suppose we decide to begin by eliminating z. To do this, we multiply equation (1) by 3 and then add the result to equation (2).

$$12x + 24y + 3z = 6 \qquad \text{Multiply each side of (1) by 3.}$$
$$\underline{x + 7y - 3z = -14} \qquad (2)$$
$$13x + 31y = -8 \qquad \text{Add.} \quad (4)$$

Step 2 Equation (4) has only two variables. To get another equation without z, we multiply equation (1) by -2 and add the result to equation (3). It is essential at this point to *eliminate the same variable, z.*

$$-8x - 16y - 2z = -4 \qquad \text{Multiply each side of (1) by } -2.$$
$$\underline{2x - 3y + 2z = 3} \qquad (3)$$
$$-6x - 19y = -1 \qquad \text{Add.} \quad (5)$$

Step 3 Now we solve the resulting system of equations (4) and (5) for x and y.

$$13x + 31y = -8 \qquad (4)$$
$$-6x - 19y = -1 \qquad (5)$$

This step is possible only if the *same* variable is eliminated in Steps 1 and 2.

$$78x + 186y = -48 \qquad \text{Multiply each side of (4) by 6.}$$
$$\underline{-78x - 247y = -13} \qquad \text{Multiply each side of (5) by 13.}$$
$$-61y = -61 \qquad \text{Add.}$$
$$y = 1$$

Step 4 We substitute 1 for y in either equation (4) or (5). Choosing (5) gives

$$-6x - 19y = -1 \qquad (5)$$
$$-6x - 19(1) = -1 \qquad \text{Let } y = 1.$$
$$-6x - 19 = -1$$
$$-6x = 18$$
$$x = -3.$$

Step 5 We substitute -3 for x and 1 for y in any one of the three original equations to find z. Choosing (1) gives

$$4x + 8y + z = 2 \qquad (1)$$
$$4(-3) + 8(1) + z = 2 \qquad \text{Let } x = -3 \text{ and } y = 1.$$
$$-4 + z = 2$$
$$z = 6.$$

Continued on Next Page

1 Check that the solution $(-3, 1, 6)$ satisfies equations (2) and (3) of Example 1.

(a) $x + 7y - 3z = -14$ (2)

Does the solution satisfy equation (2)?

(b) $2x - 3y + 2z = 3$ (3)

Does the solution satisfy equation (3)?

2 Solve each system.

(a) $\begin{aligned} x + y + z &= 2 \\ x - y + 2z &= 2 \\ -x + 2y - z &= 1 \end{aligned}$

(b) $\begin{aligned} 2x + y + z &= 9 \\ -x - y + z &= 1 \\ 3x - y + z &= 9 \end{aligned}$

Step 6 It appears that the ordered triple $(-3, 1, 6)$ is the only solution of the system. We must check that the solution satisfies all three original equations of the system. For equation (1),

$$4x + 8y + z = 2 \qquad (1)$$
$$4(-3) + 8(1) + 6 = 2 \qquad ?$$
$$-12 + 8 + 6 = 2 \qquad ?$$
$$2 = 2. \qquad \text{True}$$

◀◀◀ Work Problem 1 at the Side.

Because $(-3, 1, 6)$ also satisfies equations (2) and (3), the solution set is $\{(-3, 1, 6)\}$.

◀◀◀ Work Problem 2 at the Side.

OBJECTIVE 3 Solve linear systems where some of the equations have missing terms. If a linear system has an equation missing a term or terms, one elimination step can be omitted.

EXAMPLE 2 Solving a System of Equations with Missing Terms

Solve the system.

$$\begin{aligned} 6x - 12y &= -5 \qquad (1) \\ 8y + z &= 0 \qquad (2) \\ 9x - z &= 12 \qquad (3) \end{aligned}$$

Since equation (3) is missing the variable y, a good way to begin the solution is to eliminate y again using equations (1) and (2).

$$\begin{aligned} 12x - 24y &= -10 \qquad \text{Multiply each side of (1) by 2.} \\ 24y + 3z &= 0 \qquad \text{Multiply each side of (2) by 3.} \\ \hline 12x + 3z &= -10 \qquad \text{Add. (4)} \end{aligned}$$

Use this result, together with equation (3), $9x - z = 12$, to eliminate z. Multiply equation (3) by 3. This gives

$$\begin{aligned} 27x - 3z &= 36 \qquad \text{Multiply each side of (3) by 3.} \\ 12x + 3z &= -10 \qquad (4) \\ \hline 39x &= 26 \qquad \text{Add.} \end{aligned}$$

$$x = \frac{26}{39} = \frac{2}{3}.$$

Substituting into equation (3) gives

$$9x - z = 12 \qquad (3)$$
$$9\left(\frac{2}{3}\right) - z = 12 \qquad \text{Let } x = \tfrac{2}{3}.$$
$$6 - z = 12$$
$$z = -6.$$

———— Continued on Next Page

ANSWERS
1. (a) yes **(b)** yes
2. (a) $\{(-1, 1, 2)\}$ **(b)** $\{(2, 1, 4)\}$

Substituting -6 for z in equation (2) gives

$$8y + z = 0 \qquad (2)$$
$$8y - 6 = 0 \qquad \text{Let } z = -6.$$
$$8y = 6$$
$$y = \frac{3}{4}.$$

Thus, $x = \frac{2}{3}$, $y = \frac{3}{4}$, and $z = -6$. Check these values in each of the original equations of the system to verify that the solution set of the system is $\{(\frac{2}{3}, \frac{3}{4}, -6)\}$.

Work Problem 3 at the Side. ▶▶▶

OBJECTIVE 4 Solve special systems. Linear systems with three variables may be inconsistent or may include dependent equations. The next examples illustrate these cases.

EXAMPLE 3 Solving an Inconsistent System with Three Variables

Solve the system.

$$2x - 4y + 6z = 5 \qquad (1)$$
$$-x + 3y - 2z = -1 \qquad (2)$$
$$x - 2y + 3z = 1 \qquad (3)$$

Eliminate x by adding equations (2) and (3) to get the equation

$$y + z = 0.$$

Now, *eliminate x again,* using equations (1) and (3).

$$-2x + 4y - 6z = -2 \qquad \text{Multiply each side of (3) by } -2.$$
$$\underline{2x - 4y + 6z = 5} \qquad (1)$$
$$0 = 3 \qquad \text{False}$$

The resulting false statement indicates that equations (1) and (3) have no common solution. Thus, the system is inconsistent and the solution set is $\emptyset$. The graph of this system would show these two planes parallel to one another.

NOTE
If a false statement results when adding as in Example 3, it is not necessary to go any further with the solution. Since two of the three planes are parallel, it is not possible for the three planes to have any common points.

Work Problem 4 at the Side. ▶▶▶

3 Solve each system.

(a) $\quad x - y = 6$
$\qquad 2y + 5z = 1$
$\qquad 3x - 4z = 8$

(b) $5x - y = 26$
$\qquad 4y + 3z = -4$
$\qquad x + z = 5$

4 Solve each system.

(a) $\quad 3x - 5y + 2z = 1$
$\qquad 5x + 8y - z = 4$
$\qquad -6x + 10y - 4z = 5$

(b) $7x - 9y + 2z = 0$
$\qquad y + z = 0$
$\qquad 8x - z = 0$

ANSWERS
3. (a) $\{(4, -2, 1)\}$ **(b)** $\{(5, -1, 0)\}$
4. (a) $\emptyset$ **(b)** $\{(0, 0, 0)\}$

5 Solve the system.

$$x - y + z = 4$$
$$-3x + 3y - 3z = -12$$
$$2x - 2y + 2z = 8$$

EXAMPLE 4 **Solving a System of Dependent Equations with Three Variables**

Solve the system.

$$2x - 3y + 4z = 8 \qquad (1)$$

$$-x + \frac{3}{2}y - 2z = -4 \qquad (2)$$

$$6x - 9y + 12z = 24 \qquad (3)$$

Multiplying each side of equation (1) by 3 gives equation (3). Multiplying each side of equation (2) by -6 also gives equation (3). Because of this, the equations are dependent. All three equations have the same graph, as illustrated in Figure 6(c). The solution set is written

$$\{(x, y, z) \mid 2x - 3y + 4z = 8\}.$$

Although any one of the three equations could be used to write the solution set, we use the equation with coefficients that are integers with no common factor (except 1), as we did in **Section 5.1.**

Work Problem 5 at the Side.

ANSWERS
5. $\{(x, y, z) \mid x - y + z = 4\}$

5.2 Exercises

FOR EXTRA HELP

Tutor Center — Addison-Wesley Math Tutor Center

MathXL MathXL

Digital Video Tutor CD 3 Videotape 3

Student's Solutions Manual

MyMathLab MyMathLab

Interactmath.com

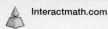

1. Explain what the following statement means: The solution set of the system

$$2x + y + z = 3$$
$$3x - y + z = -2$$
$$4x - y + 2z = 0$$

   is $\{(-1, 2, 3)\}$.

2. The two equations

$$x + y + z = 6$$
$$2x - y + z = 3$$

   have a common solution of $(1, 2, 3)$. Which equation would complete a system of three linear equations in three variables having solution set $\{(1, 2, 3)\}$?

   A. $3x + 2y - z = 1$ **B.** $3x + 2y - z = 4$
   C. $3x + 2y - z = 5$ **D.** $3x + 2y - z = 6$

Solve each system of equations. See Example 1.

3. $2x - 5y + 3z = -1$
   $x + 4y - 2z = 9$
   $x - 2y - 4z = -5$

4. $x + 3y - 6z = 7$
   $2x - y + z = 1$
   $x + 2y + 2z = -1$

5. $3x + 2y + z = 8$
   $2x - 3y + 2z = -16$
   $x + 4y - z = 20$

6. $-3x + y - z = -10$
   $-4x + 2y + 3z = -1$
   $2x + 3y - 2z = -5$

7. $-x + 2y + 6z = 2$
   $3x + 2y + 6z = 6$
   $x + 4y - 3z = 1$

8. $2x + y + 2z = 1$
   $x + 2y + z = 2$
   $x - y - z = 0$

9. $2x + 5y + 2z = 0$
$4x - 7y - 3z = 1$
$3x - 8y - 2z = -6$

10. $5x - 2y + 3z = -9$
$4x + 3y + 5z = 4$
$2x + 4y - 2z = 14$

11. $x + y - z = -2$
$2x - y + z = -5$
$-x + 2y - 3z = -4$

12. $x + 2y + 3z = 1$
$-x - y + 3z = 2$
$-6x + y + z = -2$

Solve each system of equations. See Example 2.

13. $2x - 3y + 2z = -1$
$x + 2y + z = 17$
$2y - z = 7$

14. $2x - y + 3z = 6$
$x + 2y - z = 8$
$2y + z = 1$

15. $4x + 2y - 3z = 6$
$x - 4y + z = -4$
$-x + 2z = 2$

16. $2x + 3y - 4z = 4$
$x - 6y + z = -16$
$-x + 3z = 8$

17. $2x + y = 6$
$3y - 2z = -4$
$3x - 5z = -7$

18. $4x - 8y = -7$
$4y + z = 7$
$-8x + z = -4$

19. Using your immediate surroundings, give an example of three planes that

(a) intersect in a single point;

(b) do not intersect;

(c) intersect in infinitely many points.

20. Suppose that a system has infinitely many ordered triple solutions of the form (x, y, z) such that

$$x + y + 2z = 1.$$

Give three specific ordered triples that are solutions of the system.

Solve each system of equations. See Examples 1, 3, and 4.

21. $\begin{aligned} 2x + 2y - 6z &= 5 \\ -3x + y - z &= -2 \\ -x - y + 3z &= 4 \end{aligned}$

22. $\begin{aligned} -2x + 5y + z &= -3 \\ 5x + 14y - z &= -11 \\ 7x + 9y - 2z &= -5 \end{aligned}$

23. $\begin{aligned} -5x + 5y - 20z &= -40 \\ x - y + 4z &= 8 \\ 3x - 3y + 12z &= 24 \end{aligned}$

24. $\begin{aligned} x + 4y - z &= 3 \\ -2x - 8y + 2z &= -6 \\ 3x + 12y - 3z &= 9 \end{aligned}$

25. $\begin{aligned} 2x + y - z &= 6 \\ 4x + 2y - 2z &= 12 \\ -x - \frac{1}{2}y + \frac{1}{2}z &= -3 \end{aligned}$

26. $\begin{aligned} 2x - 8y + 2z &= -10 \\ -x + 4y - z &= 5 \\ \frac{1}{8}x - \frac{1}{2}y + \frac{1}{8}z &= -\frac{5}{8} \end{aligned}$

27. $\begin{aligned} x + y - 2z &= 0 \\ 3x - y + z &= 0 \\ 4x + 2y - z &= 0 \end{aligned}$

28. $\begin{aligned} 2x + 3y - z &= 0 \\ x - 4y + 2z &= 0 \\ 3x - 5y - z &= 0 \end{aligned}$

29.
$$x + 5y - 2z = -1$$
$$-2x + 8y + z = -4$$
$$3x - y + 5z = 19$$

30.
$$x + 3y + z = 2$$
$$4x + y + 2z = -4$$
$$5x + 2y + 3z = -2$$

RELATING CONCEPTS (EXERCISES 31–38) For Individual or Group Work

Suppose that on a distant planet a function of the form
$$f(x) = ax^2 + bx + c \quad (a \neq 0)$$
describes the height in feet of a projectile x seconds after it has been projected upward.
Work Exercises 31–38 in order *to see how this can be related to a system of three equations in three variables a, b, and c.*

31. After 1 sec, the height of a certain projectile is 128 ft. Thus, $f(1) = 128$. Use this information to find one equation in the variables a, b, and c. (*Hint:* Substitute 1 for x and 128 for $f(x)$.)

32. After 1.5 sec, the height is 140 ft. Find a second equation in a, b, and c.

33. After 3 sec, the height is 80 ft. Find a third equation in a, b, and c.

34. Write a system of three equations in a, b, and c, based on your answers in Exercises 31–33. Solve the system.

35. What is the function f for this particular projectile?

36. In the function f written in Exercise 35, the _____ of the projectile is a function of the _____ elapsed since it was projected.

37. What was the initial height of the projectile? (*Hint:* Find $f(0)$.)

38. The projectile reaches its maximum height in 1.625 sec. Find its maximum height.

5.3 Applications of Systems of Linear Equations

Many applied problems involve more than one unknown quantity. Although some problems with two unknowns can be solved using just one variable, it is often easier to use two variables. To solve a problem with two unknowns, we must write two equations that relate the unknown quantities. The system formed by the pair of equations can then be solved using the methods of this chapter.

Problems that can be solved by writing a system of equations have been of interest historically. The following problem, which is given in the exercises for this section, first appeared in a Hindu work that dates back to about A.D. 850.

> The mixed price of 9 citrons [a lemonlike fruit shown in the photo] and 7 fragrant wood apples is 107; again, the mixed price of 7 citrons and 9 fragrant wood apples is 101. O you arithmetician, tell me quickly the price of a citron and the price of a wood apple here, having distinctly separated those prices well.

The following steps, based on the six-step problem-solving method first introduced in **Section 2.3,** give a strategy for solving applied problems using more than one variable.

OBJECTIVES

1. Solve problems using two variables.
2. Solve money problems using two variables.
3. Solve mixture problems using two variables.
4. Solve distance-rate-time problems using two variables.
5. Solve problems with three variables using a system of three equations.

Solving an Applied Problem by Writing a System of Equations

Step 1 **Read** the problem, several times if necessary, until you understand what is given and what is to be found.

Step 2 **Assign variables** to represent the unknown values, using diagrams or tables as needed. *Write down* what each variable represents.

Step 3 **Write a system of equations** that relates the unknowns.

Step 4 **Solve** the system of equations.

Step 5 **State the answer** to the problem. Does it seem reasonable?

Step 6 **Check** the answer in the words of the original problem.

OBJECTIVE 1 Solve problems using two variables.   Problems about the perimeter of a geometric figure often involve two unknowns and can be solved using systems of equations.

❶ Solve the problem.
   The length of the foundation of a rectangular house is to be 6 m more than its width. Find the length and width of the house if the perimeter must be 48 m.

EXAMPLE 1 **Finding the Dimensions of a Soccer Field**

Unlike football, where the dimensions of a playing field cannot vary, a rectangular soccer field may have a width between 50 and 100 yd and a length between 100 and 130 yd. Suppose that one particular field has a perimeter of 320 yd. Its length measures 40 yd more than its width. What are the dimensions of this field? (*Source: Microsoft Encarta Encyclopedia 2002.*)

Step 1 **Read** the problem again. We must find the dimensions of the field.

Step 2 **Assign variables.** Let L = the length and W = the width. Figure 7 shows a soccer field with these variables as labels.

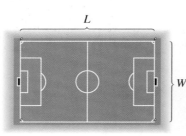

Figure 7

Step 3 **Write a system of equations.** Because the perimeter is 320 yd, we find one equation by using the perimeter formula:

$$2L + 2W = 320.$$

Because the length is 40 yd more than the width, we have

$$L = W + 40.$$

The system is, therefore,

$$2L + 2W = 320 \quad (1)$$
$$L = W + 40. \quad (2)$$

Step 4 **Solve** the system of equations. Since equation (2) is solved for L, we can use the substitution method. We substitute $W + 40$ for L in equation (1), and solve for W.

$$
\begin{aligned}
2L + 2W &= 320 && (1)\\
2(W + 40) + 2W &= 320 && \text{Let } L = W + 40.\\
2W + 80 + 2W &= 320 && \text{Distributive property}\\
4W + 80 &= 320 && \text{Combine terms.}\\
4W &= 240 && \text{Subtract 80.}\\
W &= 60 && \text{Divide by 4.}
\end{aligned}
$$

Let $W = 60$ in the equation $L = W + 40$ to find L.

$$L = 60 + 40 = 100$$

Step 5 **State the answer.** The length is **100** yd, and the width is **60** yd. Both dimensions are within the ranges given in the problem.

Step 6 **Check.** The perimeter is $2(100) + 2(60) = 320$ yd, and the length, 100 yd, is indeed 40 yd more than the width, since $100 - 40 = 60$. The answer is correct.

ANSWERS
1. length: 15 m; width: 9 m

◀◀◀ Work Problem 1 at the Side.

OBJECTIVE 2 Solve money problems using two variables. Professional sport ticket prices increase annually. Average per-ticket prices in three of the four major sports (football, basketball, and hockey) now exceed $40.00.

EXAMPLE 2 Solving a Problem about Ticket Prices

It was reported in March 2004 that during the National Hockey League and National Basketball Association seasons, two hockey tickets and one basketball ticket purchased at their average prices would have cost $126.77. One hockey ticket and two basketball tickets would have cost $128.86. What were the average ticket prices for the two sports? (*Source:* Team Marketing Report, Chicago.)

Step 1   **Read** the problem again. There are two unknowns.

Step 2   **Assign variables.** Let h represent the average price for a hockey ticket and b represent the average price for a basketball ticket.

Step 3   **Write a system of equations.** Because two hockey tickets and one basketball ticket cost a total of $126.77, one equation for the system is

$$2h + b = 126.77.$$

By similar reasoning, the second equation is

$$h + 2b = 128.86.$$

Therefore, the system is

$$2h + b = 126.77 \quad (1)$$
$$h + 2b = 128.86. \quad (2)$$

Step 4   **Solve** the system of equations. To eliminate h, multiply equation (2) by -2 and add.

$$
\begin{array}{ll}
2h + b = 126.77 & (1) \\
\underline{-2h - 4b = -257.72} & \text{Multiply each side of (2) by } -2. \\
-3b = -130.95 & \text{Add.} \\
b = 43.65 & \text{Divide by } -3.
\end{array}
$$

To find the value of h, let $b = 43.65$ in equation (2).

$$
\begin{array}{ll}
h + 2b = 128.86 & (2) \\
h + 2(\mathbf{43.65}) = 128.86 & \text{Let } b = 43.65. \\
h + 87.30 = 128.86 & \text{Multiply.} \\
h = 41.56 & \text{Subtract } 87.30.
\end{array}
$$

Step 5   **State the answer.** The average price for one basketball ticket was $43.65. For one hockey ticket, the average price was $41.56.

Step 6   **Check** that these values satisfy the conditions stated in the problem.

Work Problem 2 at the Side. ⟩⟩⟩

OBJECTIVE 3 Solve mixture problems using two variables. We solved mixture problems in **Section 2.3** using one variable. For many mixture problems we can use more than one variable and a system of equations.

---

2 Solve the problem.
   For recent Major League Baseball and National Football League seasons, based on average ticket prices, three baseball tickets and two football tickets would have cost $159.50, while two baseball tickets and one football ticket would have cost $89.66. What were the average ticket prices for the two sports? (*Source:* Team Marketing Report, Chicago.)

ANSWERS
2. baseball: $19.82; football: $50.02

3 Solve each problem.

(a) A grocer has some $4 per lb coffee and some $8 per lb coffee, which he will mix to make 50 lb of $5.60 per lb coffee. How many pounds of each should be used?

(b) Some 40% ethyl alcohol solution is to be mixed with some 80% solution to get 200 L of a 50% solution. How many liters of each should be used?

EXAMPLE 3 Solving a Mixture Problem

How many ounces each of 5% hydrochloric acid and 20% hydrochloric acid must be combined to get 10 oz of solution that is 12.5% hydrochloric acid?

Step 1 **Read** the problem. Two solutions of different strengths are being mixed together to get a specific amount of a solution with an "in-between" strength.

Step 2 **Assign variables.** Let x represent the number of ounces of 5% solution and y represent the number of ounces of 20% solution. Use a table to summarize the information from the problem.

Percent (as a Decimal)	Ounces of Solution	Ounces of Pure Acid
5% = .05	x	.05x
20% = .20	y	.20y
12.5% = .125	10	(.125)10

Figure 8 also illustrates what is happening in the problem.

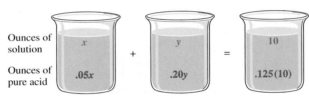

Figure 8

Step 3 **Write a system of equations.** When the x ounces of 5% solution and the y ounces of 20% solution are combined, the total number of ounces is 10, so

$$x + y = 10. \quad (1)$$

The ounces of acid in the 5% solution (.05x) plus the ounces of acid in the 20% solution (.20y) should equal the total ounces of acid in the mixture, which is (.125)10, or 1.25. That is,

$$.05x + .20y = 1.25. \quad (2)$$

Notice that these equations can be quickly determined by reading down in the table or using the labels in Figure 8.

Step 4 **Solve** the system of equations (1) and (2). Eliminate x by first multiplying equation (2) by 100 to clear it of decimals and then multiplying equation (1) by -5.

$$
\begin{array}{ll}
5x + 20y = 125 & \text{Multiply each side of (2) by 100.} \\
\underline{-5x - 5y = -50} & \text{Multiply each side of (1) by } -5. \\
15y = 75 & \text{Add.} \\
y = 5 &
\end{array}
$$

Because $y = 5$ and $x + y = 10$, x is also 5.

Step 5 **State the answer.** The desired mixture will require 5 oz of the 5% solution and 5 oz of the 20% solution.

Step 6 **Check** that these values satisfy both equations of the system.

◀◀◀ Work Problem 3 at the Side.

ANSWERS
3. (a) 30 lb of $4; 20 lb of $8
(b) 150 L of 40%; 50 L of 80%

OBJECTIVE 4 **Solve distance-rate-time problems using two variables.**
Motion problems require the distance formula, $d = rt$, where d is distance, r is rate (or speed), and t is time. These applications often lead to systems of equations.

EXAMPLE 4 **Solving a Motion Problem**

A car travels 250 km in the same time that a truck travels 225 km. If the speed of the car is 8 km per hr faster than the speed of the truck, find both speeds.

Step 1 **Read** the problem again. Given the distances traveled, we need to find the speed of each vehicle.

Step 2 **Assign variables.**

$$\text{Let } x = \text{ the speed of the car}$$
$$\text{and } y = \text{ the speed of the truck.}$$

As in Example 3, a table helps organize the information. Fill in the given information for each vehicle (in this case, distance) and use the assigned variables for the unknown speeds (rates).

	d	r	t
Car	250	x	
Truck	225	y	

To get an expression for time, solve the distance formula, $d = rt$, for t. Since $\frac{d}{r} = t$, the two times can be written as $\frac{250}{x}$ and $\frac{225}{y}$.

Step 3 **Write a system of equations.** The problem states that the car travels 8 km per hr faster than the truck. Since the two speeds are x and y,

$$x = y + 8.$$

Both vehicles travel for the same time, so from the table

$$\frac{250}{x} = \frac{225}{y}.$$

This is not a linear equation. Multiplying each side by xy gives

$$250y = 225x,$$

which is linear. The system is

$$x = y + 8 \qquad (1)$$
$$250y = 225x. \qquad (2)$$

Step 4 **Solve** the system of equations by substitution. Replace x with $y + 8$ in equation (2).

$$250y = 225x \qquad\qquad (2)$$
$$250y = 225(y + 8) \qquad \text{Let } x = y + 8.$$
$$250y = 225y + 1800 \qquad \text{Distributive property}$$
$$25y = 1800 \qquad\qquad \text{Subtract 225y.}$$
$$y = 72 \qquad\qquad \text{Divide by 25.}$$

Because $x = y + 8$, the value of x is $72 + 8 = 80$.

Continued on Next Page

❹ Solve the problem.

A train travels 600 mi in the same time that a truck travels 520 mi. Find the speed of each vehicle if the train's average speed is 8 mph faster than the truck's.

❺ Solve the system of equations from Example 5.

$$x - 3y = 0 \quad (1)$$
$$x - z = 5 \quad (2)$$
$$259x + 299y + 329z = 6607 \quad (3)$$

Step 5 **State the answer.** The car's speed is 80 km per hr, and the truck's speed is 72 km per hr.

Step 6 **Check.** This is especially important since one of the equations had variable denominators.

$$\text{Car: } t = \frac{d}{r} = \frac{250}{80} = 3.125$$
$$\text{Truck: } t = \frac{d}{r} = \frac{225}{72} = 3.125$$
Times are equal.

Since $80 - 72 = 8$, the conditions of the problem are satisfied.

▶◀ **Work Problem 4 at the Side.**

OBJECTIVE ❺ Solve problems with three variables using a system of three equations. To solve such problems, we extend the method used for two unknowns to three variables and three equations.

EXAMPLE 5 **Solving a Problem Involving Prices**

At Panera Bread, a loaf of honey wheat bread costs $2.59, a loaf of sunflower bread costs $2.99, and a loaf of French bread costs $3.29. On a recent day, three times as many loaves of honey wheat were sold as sunflower. The number of loaves of French bread sold was 5 less than the number of loaves of honey wheat sold. Total receipts for these breads were $66.07. How many loaves of each type of bread were sold? (*Source:* Panera Bread menu.)

Step 1 **Read** the problem again. There are three unknowns in this problem.

Step 2 **Assign variables** to represent the three unknowns.

Let x = the number of loaves of honey wheat,
y = the number of loaves of sunflower,
and z = the number of loaves of French bread.

Step 3 **Write a system of three equations.** Since three times as many loaves of honey wheat were sold as sunflower,

$$x = 3y, \quad \text{or} \quad x - 3y = 0. \quad (1)$$

Also,

Number of loaves of French bread	equals	5 less than the number of loaves of honey wheat.
↓	↓	↓
z	$=$	$x - 5,$

so $x - z = 5. \quad (2)$

Multiplying the cost of a loaf of each kind of bread by the number of loaves of that kind sold and adding gives the total receipts.

$$2.59x + 2.99y + 3.29z = 66.07$$

Multiply each side of this equation by 100 to clear it of decimals.

$$259x + 299y + 329z = 6607 \quad (3)$$

Step 4 **Solve** the system of three equations using the method shown in **Section 5.2.**

▶◀ **Work Problem 5 at the Side.**

ANSWERS
4. train: 60 mph; truck: 52 mph
5. $\{(12, 4, 7)\}$

Continued on Next Page

Step 5 **State the answer.** The solution set is $\{(12, 4, 7)\}$, meaning that 12 loaves of honey wheat, 4 loaves of sunflower, and 7 loaves of French bread were sold.

Step 6 **Check.** Since $12 = 3 \cdot 4$, the number of loaves of honey wheat is three times the number of loaves of sunflower. Also, $12 - 7 = 5$, so the number of loaves of French bread is 5 less than the number of loaves of honey wheat. Multiply the appropriate cost per loaf by the number of loaves sold and add the results to check that total receipts were $66.07.

Work Problem 6 at the Side. ▶▶▶

EXAMPLE 6 Solving a Business Production Problem

A company produces three color television sets, models X, Y, and Z. Each model X set requires 2 hr of electronics work, 2 hr of assembly time, and 1 hr of finishing time. Each model Y requires 1, 3, and 1 hr of electronics, assembly, and finishing time, respectively. Each model Z requires 3, 2, and 2 hr of the same work, respectively. There are 100 hr available for electronics, 100 hr available for assembly, and 65 hr available for finishing per week. How many of each model should be produced each week if all available time must be used?

Step 1 **Read** the problem again. There are three unknowns.

Step 2 **Assign variables.**

Let $\quad x =$ the number of model X produced per week,

$\quad y =$ the number of model Y produced per week,

and $\quad z =$ the number of model Z produced per week.

Organize the information in a table.

	Each Model X	Each Model Y	Each Model Z	Totals
Hours of Electronics Work	2	1	3	100
Hours of Assembly Time	2	3	2	100
Hours of Finishing Time	1	1	2	65

Step 3 **Write a system of three equations.** The x model X sets require $2x$ hr of electronics, the y model Y sets require $1y$ (or y) hr of electronics, and the z model Z sets require $3z$ hr of electronics. Since 100 hr are available for electronics,

$$2x + y + 3z = 100. \quad (1)$$

Similarly, from the fact that 100 hr are available for assembly,

$$2x + 3y + 2z = 100, \quad (2)$$

and the fact that 65 hr are available for finishing leads to the equation

$$x + y + 2z = 65. \quad (3)$$

Notice that by reading across the table, we can easily determine the coefficients and constants in the equations of the system.

Continued on Next Page

6 Solve the problem.

A department store has three kinds of perfume: cheap, better, and best. It has 10 more bottles of cheap than better, and 3 fewer bottles of best than better. Each bottle of cheap costs $8, better costs $15, and best costs $32. The total value of all the perfume is $589. How many bottles of each are there?

ANSWERS
6. 21 bottles of cheap; 11 of better; 8 of best

7 Solve the problem.

A paper mill makes newsprint, bond, and copy machine paper. Each ton of newsprint requires 3 tons of recycled paper and 1 ton of wood pulp. Each ton of bond requires 2 tons of recycled paper, 4 tons of wood pulp, and 3 tons of rags. A ton of copy machine paper requires 2 tons of recycled paper, 3 tons of wood pulp, and 2 tons of rags. The mill has 4200 tons of recycled paper, 5800 tons of wood pulp, and 3900 tons of rags. How much of each kind of paper can be made from these supplies?

Step 4 **Solve** the system

$$2x + y + 3z = 100$$
$$2x + 3y + 2z = 100$$
$$x + y + 2z = 65$$

to find $x = 15$, $y = 10$, and $z = 20$.

Step 5 **State the answer.** The company should produce 15 model X, 10 model Y, and 20 model Z sets per week.

Step 6 **Check** that these values satisfy the conditions of the problem.

◀◀◀ **Work Problem 7 at the Side.**

ANSWERS
7. 400 tons of newsprint; 900 tons of bond; 600 tons of copy machine paper

5.3 Exercises

FOR EXTRA HELP

Addison-Wesley Math Tutor Center

Math XP MathXL

Digital Video Tutor CD 3 Videotape 3

Student's Solutions Manual

MyMathLab MyMathLab

Interactmath.com

Solve each problem. See Example 1.

1. During the 2003 Major League Baseball regular season, the Minnesota Twins played 162 games. They won 18 more games than they lost. What was their win-loss record that year?

2. Refer to Exercise 1. During the same 162-game season, the Detroit Tigers lost 76 more games than they won. What was the team's win-loss record?

2003 MLB FINAL STANDINGS AMERICAN LEAGUE CENTRAL

Team	W	L
Minnesota	—	—
Chicago	86	76
Kansas City	83	79
Cleveland	68	94
Detroit	—	—

Source: CBS.SportsLine.com.

3. Venus and Serena measured a tennis court and found that it was 42 ft longer than it was wide and had a perimeter of 228 ft. What were the length and the width of the tennis court?

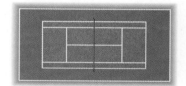

4. Carmelo and LeBron found that the width of their basketball court was 44 ft less than the length. If the perimeter was 288 ft, what were the length and the width of their court?

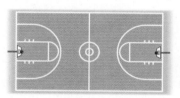

5. The two biggest Fortune 500 companies in 2003 were Wal-Mart and ExxonMobil. ExxonMobil's revenue was $46 billion less than that of Wal-Mart. Total revenue for the two companies was $472 billion. What was the revenue for each company? (*Source:* Fortune 500.)

6. In 2002, U.S. exports to Canada were $63,453 million more than exports to Mexico. Together, exports to these two countries totaled $258,393 million. How much were exports to each country? (*Source:* Office of Trade and Economic Analysis, U.S. Department of Commerce.)

In Exercises 7 and 8, find the measures of the angles marked x and y. Remember that (1) the sum of the measures of the angles of a triangle is 180°, (2) supplementary angles have a sum of 180°, and (3) vertical angles have equal measures.

7.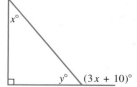

8.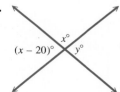

The Fan Cost Index (FCI) represents the cost of four average-price tickets, four small soft drinks, two small beers, four hot dogs, parking for one car, two game programs, and two souvenir caps to a sporting event. (Source: www.teammarketing.com) Use the concept of FCI in Exercises 9 and 10. See Example 2.

9. In 2004, the FCI prices for the National Hockey League and the National Basketball Association totaled $514.91. The hockey FCI was $7.61 less than that of basketball. What were the FCIs for these sports?

10. In 2004, the FCI prices for Major League Baseball and the National Football League totaled $457.27. The football FCI was $146.23 more than that of baseball. What were the FCIs for these sports?

Solve each problem. See Example 2.

11. Andrew McGinnis works at Arby's. During one particular day he sold 15 Junior Roast Beef sandwiches and 10 Big Montana sandwiches, totaling $64.25. Another day he sold 30 Junior Roast Beef sandwiches and 5 Big Montana sandwiches, totaling $65.65. How much did each type of sandwich cost? (*Source:* Arby's menu.)

12. Tokyo and New York are among the most expensive cities worldwide for business travelers. Using average costs per day for each city (which includes room, meals, laundry, and two taxi fares), 2 days in Tokyo and 3 days in New York cost $2015. Four days in Tokyo and 2 days in New York cost $2490. What is the average cost per day for each city? (*Source:* ECA International.)

The formulas $p = br$ (percentage = base $\times$ rate) and $I = prt$ (simple interest = principal $\times$ rate $\times$ time) are used in the applications in Exercises 17–24. To prepare to use these formulas, answer the questions in Exercises 13 and 14.

13. If a container of liquid contains 60 oz of solution, what is the number of ounces of pure acid if the given solution contains the following acid concentrations?

(a) 10%   **(b)** 25%   **(c)** 40%   **(d)** 50%

14. If $5000 is invested in an account paying simple annual interest, how much interest will be earned during the first year at the following rates?

(a) 2%   **(b)** 3%   **(c)** 4%   **(d)** 3.5%

15. If a pound of turkey costs $.99, how much will x pounds cost?

16. If a ticket to the movie *Shrek 2* costs $8 and y tickets are sold, how much is collected from the sale?

Solve each problem. See Example 3.

17. How many gallons each of 25% alcohol and 35% alcohol should be mixed to get 20 gal of 32% alcohol?

Percent (as a Decimal)	Gallons of Solution	Gallons of Pure Alcohol
25% = .25	x	
35% = .35	y	
32% =	20	

18. How many liters each of 15% acid and 33% acid should be mixed to get 120 L of 21% acid?

Percent (as a Decimal)	Liters of Solution	Liters of Pure Acid
15% = .15	x	
33% =	y	
21% =	120	

19. Pure acid is to be added to a 10% acid solution to obtain 54 L of a 20% acid solution. What amounts of each should be used? (*Hint:* Pure acid is 100% acid.)

20. A truck radiator holds 36 L of fluid. How much pure antifreeze must be added to a mixture that is 4% antifreeze to fill the radiator with a mixture that is 20% antifreeze?

21. A party mix is made by adding nuts that sell for $2.50 per kg to a cereal mixture that sells for $1 per kg. How much of each should be added to get 30 kg of a mix that will sell for $1.70 per kg?

	Price per Kilogram	Number of Kilograms	Value
Nuts	2.50	x	
Cereal	1.00	y	
Mixture	1.70		

22. A popular fruit drink is made by mixing fruit juices. Such a drink with 50% juice is to be mixed with another drink that is 30% juice to get 200 L of a drink that is 45% juice. How much of each should be used?

	Percent (as a Decimal)	Liters of Drink	Liters of Pure Juice
50% Juice	.50	x	
30% Juice	.30	y	
Mixture	.45		

23. A total of $3000 is invested, part at 2% simple interest and part at 4%. If the total annual return from the two investments is $100, how much is invested at each rate?

Rate (as a Decimal)	Principal	Interest
.02	x	$.02x$
.04	y	$.04y$
	3000	100

24. An investor must invest a total of $15,000 in two accounts, one paying 4% annual simple interest, and the other 3%. If he wants to earn $550 annual interest, how much should he invest at each rate?

Rate (as a Decimal)	Principal	Interest
.04	x	
.03	y	
	15,000	

The formula $d = rt$ (distance = rate $\times$ time) is used in the applications in Exercises 27–30. To prepare to use this formula, answer the questions in Exercises 25 and 26.

25. If the speed of a boat in still water is 10 mph, and the speed of the current of a river is x mph, what is the speed of the boat

(a) going upstream (that is, against the current);

(b) going downstream (that is, with the current)?

26. If the speed of a killer whale is 25 mph and the whale swims for y hours, how many miles does the whale travel?

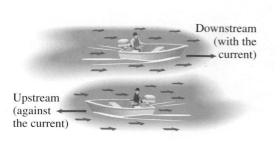

Downstream (with the current)

Upstream (against the current)

Solve each problem. See Example 4.

27. A freight train and an express train leave towns 390 km apart, traveling toward one another. The freight train travels 30 km per hr slower than the express train. They pass one another 3 hr later. What are their speeds?

	r	t	d
Freight Train	x	3	
Express Train	y	3	

28. A train travels 150 km in the same time that a plane covers 400 km. If the speed of the plane is 20 km per hr less than 3 times the speed of the train, find both speeds.

29. In his motorboat, Bill Ruhberg travels upstream at top speed to his favorite fishing spot, a distance of 36 mi, in 2 hr. Returning, he finds that the trip downstream, still at top speed, takes only 1.5 hr. Find the speed of Bill's boat and the speed of the current.

	r	t	d
Upstream	$x - y$	2	
Downstream	$x + y$		

30. Traveling for 3 hr into a steady headwind, a plane flies 1650 mi. The pilot determines that flying *with* the same wind for 2 hr, he could make a trip of 1300 mi. Find the speed of the plane and the speed of the wind.

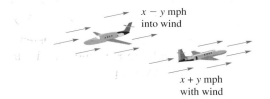

$x - y$ mph into wind

$x + y$ mph with wind

Solve each problem. See Examples 1–6.

31. In 2002, the top-grossing North American concert tour was that of Paul McCartney. McCartney and the second-place Rolling Stones together took in $191.2 million from ticket sales. If the Rolling Stones took in $15.4 million less than McCartney, how much did each act generate?

32. Pam Snow plans to mix pecan clusters that sell for $3.60 per lb with chocolate truffles that sell for $7.20 per lb to get a mixture that she can sell in Valentine boxes for $4.95 per lb. How much of the $3.60 clusters and the $7.20 truffles should she use to create 80 lb of the mix?

	Price per Pound	Number of Pounds	Value
Pecan Clusters		x	
Chocolate Truffles		y	
Valentine Mixture		80	

33. Tickets to a production of *Romeo and Juliet* at Miami Dade Community College cost $5 for general admission or $4 with a student ID. If 184 people paid to see a performance and $812 was collected, how many of each type of ticket were sold?

34. At a business meeting at Panera Bread, the bill (without tax) for two cappuccinos and three house lattes was $14.55. At another table, the bill for one cappuccino and two house lattes was $8.77. How much did each type of beverage cost? (*Source:* Panera Bread menu.)

35. The mixed price of 9 citrons and 7 fragrant wood apples is 107; again, the mixed price of 7 citrons and 9 fragrant wood apples is 101. O you arithmetician, tell me quickly the price of a citron and the price of a wood apple here, having distinctly separated those prices well. (*Source:* Hindu work, A.D. 850.) (*Hint:* "Mixed price" refers to the price of a mixture of the two fruits.)

36. Braving blizzard conditions on the planet Hoth, Luke Skywalker sets out at top speed in his snow speeder for a rebel base 4800 mi away. He travels into a steady headwind and makes the trip in 3 hr. Returning, he finds that the trip back, still at top speed but now with a tailwind, takes only 2 hr. Find the top speed of Luke's snow speeder and the speed of the wind.

	r	t	d
Into Headwind			
With Tailwind			

Solve each problem involving three unknowns. See Examples 5 and 6. (In Exercises 37–40, remember that the sum of the measures of the angles of a triangle is 180°.)

37. In the figure, $z = x + 10$ and $x + y = 100$. Determine a third equation involving x, y, and z, and then find the measures of the three angles.

38. In the figure, x is 10 less than y and x is 20 less than z. Write a system of equations and find the measures of the three angles.

39. In a certain triangle, the measure of the second angle is 10° more than three times the first. The third angle measure is equal to the sum of the measures of the other two. Find the measures of the three angles.

40. The measure of the largest angle of a triangle is 12° less than the sum of the measures of the other two. The smallest angle measures 58° less than the largest. Find the measures of the angles.

41. The perimeter of a triangle is 70 cm. The longest side is 4 cm less than the sum of the other two sides. Twice the shortest side is 9 cm less than the longest side. Find the length of each side of the triangle.

42. The perimeter of a triangle is 56 in. The longest side measures 4 in. less than the sum of the other two sides. Three times the shortest side is 4 in. more than the longest side. Find the lengths of the three sides.

43. In football, a team is credited with a win, a loss, or a tie. The 2002 Pittsburgh Steelers of the National Football League played a total of 16 regular-season games. They won 5 more than they lost, and they tied 4 fewer than half the number of games they won. What were their numbers of wins, losses, and ties? (*Source:* nfl.com)

44. In the 2002 Winter Olympics in Salt Late City, Utah, the United States earned 3 fewer gold medals than silver. The number of bronze medals earned was 15 fewer than twice the number of silver medals. The United States earned a total of 34 medals. How many of each kind of medal did the United States earn? (*Source:* news.bbc.co.uk/winterolympics 2002)

45. The Harlem Globetrotter Web site shop offers jerseys, bobblehead sets, and basketballs. One jersey, two bobblehead sets, and one basketball cost $302. Two jerseys, one bobblehead set, and three basketballs cost $284. Buying one of each item costs $182. What is the price for each of these items? (*Source:* www.secure.harlemglobetrotters.com)

46. Three kinds of tickets are available for a Third Day concert: "up close," "in the middle," and "far out." "Up close" tickets cost $10 more than "in the middle" tickets, while "in the middle" tickets cost $10 more than "far out" tickets. Twice the cost of an "up close" ticket is $20 more than 3 times the cost of a "far out" ticket. Find the price of each kind of ticket.

47. A hardware supplier manufactures three kinds of clamps, types A, B, and C. Production restrictions require it to make 10 units more type C clamps than the total of the other types and twice as many type B clamps as type A. The shop must produce a total of 490 units of clamps per day. How many units of each type can be made per day?

48. A Mardi Gras trinket manufacturer supplies three wholesalers, A, B, and C. The output from a day's production is 320 cases of trinkets. She must send wholesaler A three times as many cases as she sends B, and she must send wholesaler C 160 cases fewer than she provides A and B together. How many cases should she send to each wholesaler to distribute the entire day's production to them?

5.4 Solving Systems of Linear Equations by Matrix Methods

OBJECTIVE 1 Define a matrix. An ordered array of numbers such as

$$\text{Rows} \begin{array}{c} \text{Columns} \\ \begin{bmatrix} 2 & 3 & 5 \\ 7 & 1 & 2 \end{bmatrix} \end{array} \quad \text{Matrix}$$

OBJECTIVES

1 Define a matrix.

2 Write the augmented matrix for a system.

3 Use row operations to solve a system with two equations.

4 Use row operations to solve a system with three equations.

5 Use row operations to solve special systems.

is called a **matrix.** The numbers are called **elements** of the matrix. Matrices (the plural of *matrix*) are named according to the number of **rows** and **columns** they contain. The rows are read horizontally, and the columns are read vertically. For example, the first row in the preceding matrix is 2 3 5 and the first column is $\begin{smallmatrix} 2 \\ 7 \end{smallmatrix}$. This matrix is a 2×3 (read "two by three") matrix because it has 2 rows and 3 columns. The number of rows is given first, and then the number of columns. Two other examples follow.

$$\begin{bmatrix} -1 & 0 \\ 1 & -2 \end{bmatrix} \quad \begin{array}{c} 2 \times 2 \\ \text{matrix} \end{array} \qquad \begin{bmatrix} 8 & -1 & -3 \\ 2 & 1 & 6 \\ 0 & 5 & -3 \\ 5 & 9 & 7 \end{bmatrix} \quad \begin{array}{c} 4 \times 3 \\ \text{matrix} \end{array}$$

A **square matrix** is one that has the same number of rows as columns. The 2×2 matrix is a square matrix.

Calculator Tip Figure 9 shows how a graphing calculator displays the preceding two matrices. Work with matrices is made much easier by using technology when available. Consult your owner's manual for details.

```
[A]
    [[-1 0 ]
     [1  -2]]
```

```
[B]
    [[8 -1 -3]
     [2  1  6 ]
     [0  5 -3]
     [5  9  7 ]]
```

Figure 9

In this section, we discuss a method of solving linear systems that uses matrices. The advantage of this new method is that it can be done by a graphing calculator or a computer, allowing large systems of equations to be solved easily.

OBJECTIVE 2 Write the augmented matrix for a system. To begin, we write an *augmented matrix* for the system. An **augmented matrix** has a vertical bar that separates the columns of the matrix into two groups. For example, to solve the system

$$x - 3y = 1$$
$$2x + y = -5,$$

start with the augmented matrix

$$\left[\begin{array}{cc|c} 1 & -3 & 1 \\ 2 & 1 & -5 \end{array}\right]. \quad \text{Augmented matrix}$$

System of equations:

$$x - 3y = 1$$
$$2x + y = -5$$

Augmented matrix:

$$\begin{bmatrix} 1 & -3 & 1 \\ 2 & 1 & -5 \end{bmatrix}$$

$\uparrow$ $\uparrow$

Coefficients   Constants
of the
variables

Place the coefficients of the variables to the left of the bar, and the constants to the right. The bar separates the coefficients from the constants. The matrix is just a shorthand way of writing the system of equations, so the rows of the augmented matrix can be treated the same as the equations of a system of equations.

We know that exchanging the position of two equations in a system does not change the system. Also, multiplying any equation in a system by a nonzero number does not change the system. Comparable changes to the augmented matrix of a system of equations produce new matrices that correspond to systems with the same solutions as the original system.

The following **row operations** produce new matrices that lead to systems having the same solutions as the original system.

Matrix Row Operations

1. Any two rows of the matrix may be interchanged.

2. The numbers in any row may be multiplied by any nonzero real number.

3. Any row may be transformed by adding to the numbers of the row the product of a real number and the corresponding numbers of another row.

Examples of these row operations follow.

Row operation 1:

$$\begin{bmatrix} 2 & 3 & 9 \\ 4 & 8 & -3 \\ 1 & 0 & 7 \end{bmatrix} \text{ becomes } \begin{bmatrix} 1 & 0 & 7 \\ 4 & 8 & -3 \\ 2 & 3 & 9 \end{bmatrix}.$$

Interchange row 1 and row 3.

Row operation 2:

$$\begin{bmatrix} 2 & 3 & 9 \\ 4 & 8 & -3 \\ 1 & 0 & 7 \end{bmatrix} \text{ becomes } \begin{bmatrix} 6 & 9 & 27 \\ 4 & 8 & -3 \\ 1 & 0 & 7 \end{bmatrix}.$$

Multiply the numbers in row 1 by 3.

Row operation 3:

$$\begin{bmatrix} 2 & 3 & 9 \\ 4 & 8 & -3 \\ 1 & 0 & 7 \end{bmatrix} \text{ becomes } \begin{bmatrix} 0 & 3 & -5 \\ 4 & 8 & -3 \\ 1 & 0 & 7 \end{bmatrix}.$$

Multiply the numbers in row 3 by -2; add them to the corresponding numbers in row 1.

The third row operation corresponds to the way we eliminated a variable from a pair of equations in the previous sections.

OBJECTIVE 3 Use row operations to solve a system with two equations. Row operations can be used to rewrite a matrix. The goal is a matrix in the form

$$\begin{bmatrix} 1 & a & b \\ 0 & 1 & c \end{bmatrix} \quad \text{or} \quad \begin{bmatrix} 1 & a & b & c \\ 0 & 1 & d & e \\ 0 & 0 & 1 & f \end{bmatrix}$$

for systems with two or three equations, respectively. Notice that there are 1s down the diagonal from upper left to lower right and 0s below the 1s. A matrix written this way is said to be in **row echelon form.** When these matrices are rewritten as systems of equations, the value of one variable is known, and the rest can be found by substitution. The following examples illustrate this method.

EXAMPLE 1 **Using Row Operations to Solve a System with Two Variables**

Use row operations to solve the system.

$$x - 3y = 1$$
$$2x + y = -5$$

We start with the augmented matrix of the system.

$$\begin{bmatrix} 1 & -3 & | & 1 \\ 2 & 1 & | & -5 \end{bmatrix}$$

Now we use the various row operations to change this matrix into one that leads to a system that is easier to solve.

It is best to work by columns. We start with the first column and make sure that there is a 1 in the first row, first column position. There is already a 1 in this position. Next, we get 0 in every position below the first. To get a 0 in row two, column one, we use the third row operation and add to the numbers in row two the result of multiplying each number in row one by -2. (We abbreviate this as $-2R_1 + R_2$.) Row one remains unchanged.

$$\begin{bmatrix} 1 & -3 & | & 1 \\ 2 + 1(-2) & 1 + -3(-2) & | & -5 + 1(-2) \end{bmatrix}$$

Original number -2 times number
from row two   from row one

$$\begin{bmatrix} 1 & -3 & | & 1 \\ 0 & 7 & | & -7 \end{bmatrix} \quad -2R_1 + R_2$$

The matrix now has a 1 in the first position of column one, with 0 in every position below the first.

Now we go to column two. A 1 is needed in row two, column two. We get this 1 by using the second row operation, multiplying each number of row two by $\frac{1}{7}$.

$$\begin{bmatrix} 1 & -3 & | & 1 \\ 0 & 1 & | & -1 \end{bmatrix} \quad \frac{1}{7}R_2$$

This augmented matrix leads to the system of equations

$$1x - 3y = 1 \qquad \qquad x - 3y = 1$$
$$0x + 1y = -1 \quad \text{or} \qquad y = -1.$$

From the second equation, $y = -1$. We substitute -1 for y in the first equation to get

$$x - 3y = 1$$
$$x - 3(-1) = 1$$
$$x + 3 = 1$$
$$x = -2.$$

The solution set of the system is $\{(-2, -1)\}$. Check this solution by substitution in both equations of the system.

Work Problem 1 at the Side. ▶▶▶

1 Use row operations to solve the system.

$$x - 2y = 9$$
$$3x + y = 13$$

ANSWERS
1. $\{(5, -2)\}$

▦ **Calculator Tip** If the augmented matrix of the system in Example 1 is entered as matrix A in a graphing calculator (Figure 10(a)) and the row echelon form of the matrix is found (Figure 10(b)), the system becomes

$$x + \frac{1}{2}y = -\frac{5}{2}$$
$$y = -1.$$

While this system looks different from the one we obtained in Example 1, it is equivalent, since its solution set is also $\{(-2, -1)\}$.

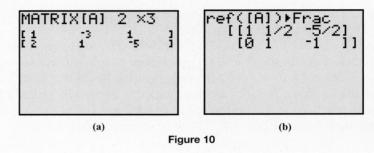

(a)   (b)

Figure 10

OBJECTIVE 4 Use row operations to solve a system with three equations. As before, we use row operations to get 1s down the diagonal from left to right and all 0s below each 1.

EXAMPLE 2 Using Row Operations to Solve a System with Three Variables

Use row operations to solve the system.

$$x - y + 5z = -6$$
$$3x + 3y - z = 10$$
$$x + 3y + 2z = 5$$

Start by writing the augmented matrix of the system.

$$\left[\begin{array}{ccc|c} 1 & -1 & 5 & -6 \\ 3 & 3 & -1 & 10 \\ 1 & 3 & 2 & 5 \end{array}\right]$$

This matrix already has 1 in row one, column one. Next get 0s in the rest of column one. First, add to row two the results of multiplying each number of row one by -3. This gives the matrix

$$\left[\begin{array}{ccc|c} 1 & -1 & 5 & -6 \\ 0 & 6 & -16 & 28 \\ 1 & 3 & 2 & 5 \end{array}\right]. \qquad -3R_1 + R_2$$

Now add to the numbers in row three the results of multiplying each number of row one by -1.

$$\left[\begin{array}{ccc|c} 1 & -1 & 5 & -6 \\ 0 & 6 & -16 & 28 \\ 0 & 4 & -3 & 11 \end{array}\right] \qquad -1R_1 + R_3$$

Continued on Next Page

Get 1 in row two, column two by multiplying each number in row two by $\frac{1}{6}$.

$$\begin{bmatrix} 1 & -1 & 5 & | & -6 \\ 0 & 1 & -\frac{8}{3} & | & \frac{14}{3} \\ 0 & 4 & -3 & | & 11 \end{bmatrix} \quad \frac{1}{6}R_2$$

Introduce 0 in row three, column two by adding to row three the results of multiplying each number in row two by -4.

$$\begin{bmatrix} 1 & -1 & 5 & | & -6 \\ 0 & 1 & -\frac{8}{3} & | & \frac{14}{3} \\ 0 & 0 & \frac{23}{3} & | & -\frac{23}{3} \end{bmatrix} \quad -4R_2 + R_3$$

Finally, obtain 1 in row three, column three by multiplying each number in row three by $\frac{3}{23}$.

$$\begin{bmatrix} 1 & -1 & 5 & | & -6 \\ 0 & 1 & -\frac{8}{3} & | & \frac{14}{3} \\ 0 & 0 & 1 & | & -1 \end{bmatrix} \quad \frac{3}{23}R_3$$

This final matrix gives the system of equations

$$x - y + 5z = -6$$
$$y - \frac{8}{3}z = \frac{14}{3}$$
$$z = -1.$$

Substitute -1 for z in the second equation, $y - \frac{8}{3}z = \frac{14}{3}$, to get $y = 2$. Finally, substitute 2 for y and -1 for z in the first equation, $x - y + 5z = -6$, to get $x = 1$. The solution set of the original system is $\{(1, 2, -1)\}$. Check by substitution in the original system.

Work Problem 2 at the Side. ▶▶▶

OBJECTIVE 5 **Use row operations to solve special systems.**

EXAMPLE 3 **Recognizing Inconsistent Systems or Dependent Equations**

Use row operations to solve each system.

(a) $\quad 2x - 3y = 8$
$\quad\;\; -6x + 9y = 4$

$$\begin{bmatrix} 2 & -3 & | & 8 \\ -6 & 9 & | & 4 \end{bmatrix} \quad \text{Write the augmented matrix.}$$

$$\begin{bmatrix} 1 & -\frac{3}{2} & | & 4 \\ -6 & 9 & | & 4 \end{bmatrix} \quad \frac{1}{2}R_1$$

$$\begin{bmatrix} 1 & -\frac{3}{2} & | & 4 \\ 0 & 0 & | & 28 \end{bmatrix} \quad 6R_1 + R_2$$

The corresponding system of equations is

$$x - \frac{3}{2}y = 4$$
$$0 = 28, \quad \text{False}$$

which has no solution and is inconsistent. The solution set is $\emptyset$.

—— **Continued on Next Page**

② Use row operations to solve the system.
$$2x - y + z = 7$$
$$x - 3y - z = 7$$
$$-x + y - 5z = -9$$

ANSWERS
2. $\{(2, -2, 1)\}$

3 Use row operations to solve each system.

(a)
$$x - y = 2$$
$$-2x + 2y = 2$$

(b)
$$-10x + 12y = 30$$
$$5x - 6y = -15$$

$$\begin{bmatrix} -10 & 12 & | & 30 \\ 5 & -6 & | & -15 \end{bmatrix} \quad \text{Write the augmented matrix.}$$

$$\begin{bmatrix} 1 & -\frac{6}{5} & | & -3 \\ 5 & -6 & | & -15 \end{bmatrix} \quad -\frac{1}{10}R_1$$

$$\begin{bmatrix} 1 & -\frac{6}{5} & | & -3 \\ 0 & 0 & | & 0 \end{bmatrix} \quad -5R_1 + R_2$$

The corresponding system is

$$x - \frac{6}{5}y = -3$$
$$0 = 0, \quad \text{True}$$

which has dependent equations. Using the second equation of the original system, we write the solution set as

$$\{(x, y) \mid 5x - 6y = -15\}.$$

Work Problem 3 at the Side.

(b)
$$x - y = 2$$
$$-2x + 2y = -4$$

ANSWERS
3. (a) $\emptyset$ (b) $\{(x, y) \mid x - y = 2\}$

5.4 Exercises

FOR EXTRA HELP | Tutor Center Addison-Wesley Math Tutor Center | MathXL | Digital Video Tutor CD 3 Videotape 3 | Student's Solutions Manual | MyMathLab MyMathLab | Interactmath.com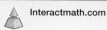

1. Consider the matrix $\begin{bmatrix} -2 & 3 & 1 \\ 0 & 5 & -3 \\ 1 & 4 & 8 \end{bmatrix}$, and answer the following.

(a) What are the elements of the second row?

(b) What are the elements of the third column?

(c) Is this a square matrix? Explain.

(d) Give the matrix obtained by interchanging the first and third rows.

(e) Give the matrix obtained by multiplying the first row by $-\frac{1}{2}$.

(f) Give the matrix obtained by multiplying the third row by 3 and adding it to the first row.

2. Give the dimensions of each matrix.

(a) $\begin{bmatrix} 3 & -7 \\ 4 & 5 \\ -1 & 0 \end{bmatrix}$
(b) $\begin{bmatrix} 4 & 9 & 0 \\ -1 & 2 & -4 \end{bmatrix}$
(c) $\begin{bmatrix} 6 & 3 \\ -2 & 5 \\ 4 & 10 \\ 1 & -11 \end{bmatrix}$

Complete the steps in the matrix solution of each system by filling in the blanks. Give the final system and the solution set. See Example 1.

3. $4x + 8y = 44$
$2x - y = -3$

$\begin{bmatrix} 4 & 8 & | & 44 \\ 2 & -1 & | & -3 \end{bmatrix}$

$\begin{bmatrix} 1 & \underline{\quad} & | & \underline{\quad} \\ 2 & -1 & | & -3 \end{bmatrix}$ $\frac{1}{4}R_1$

$\begin{bmatrix} 1 & 2 & | & 11 \\ 0 & \underline{\quad} & | & \underline{\quad} \end{bmatrix}$ $-2R_1 + R_2$

$\begin{bmatrix} 1 & 2 & | & 11 \\ 0 & 1 & | & \underline{\quad} \end{bmatrix}$ $-\frac{1}{5}R_2$

4. $2x - 5y = -1$
$3x + y = 7$

$\begin{bmatrix} 2 & -5 & | & -1 \\ 3 & 1 & | & 7 \end{bmatrix}$

$\begin{bmatrix} 1 & -\frac{5}{2} & | & \underline{\quad} \\ 3 & 1 & | & 7 \end{bmatrix}$ $\frac{1}{2}R_1$

$\begin{bmatrix} 1 & -\frac{5}{2} & | & -\frac{1}{2} \\ 0 & \underline{\quad} & | & \underline{\quad} \end{bmatrix}$ $-3R_1 + R_2$

$\begin{bmatrix} 1 & -\frac{5}{2} & | & -\frac{1}{2} \\ 0 & 1 & | & \underline{\quad} \end{bmatrix}$ $\frac{2}{17}R_2$

Use row operations to solve each system. See Examples 1 and 3.

5. $x + y = 5$
$x - y = 3$

6. $x + 2y = 7$
$x - y = -2$

7. $2x + 4y = 6$
$3x - y = 2$

8. $4x + 5y = -7$
$x - y = 5$

9. $3x + 4y = 13$
$2x - 3y = -14$

10. $5x + 2y = 8$
$3x - y = 7$

11. $-4x + 12y = 36$
$x - 3y = 9$

12. $2x - 4y = 8$
$-3x + 6y = 5$

13. $2x + y = 4$
$4x + 2y = 8$

14. $-3x - 4y = 1$
$6x + 8y = -2$

15. $\dfrac{1}{2}x + \dfrac{1}{3}y = 0$
$\dfrac{2}{3}x + \dfrac{3}{4}y = 0$

16. $1.2x + .3y = 0$
$2.9x - .6y = 0$

Complete the steps in the matrix solution of each system by filling in the blanks. Give the final system and the solution set. See Example 2.

17. $x + y - z = -3$
$2x + y + z = 4$
$5x - y + 2z = 23$

$$\begin{bmatrix} 1 & 1 & -1 & | & -3 \\ 2 & 1 & 1 & | & 4 \\ 5 & -1 & 2 & | & 23 \end{bmatrix}$$

$$\begin{bmatrix} 1 & 1 & -1 & | & -3 \\ 0 & \_\_ & \_\_ & | & \_\_ \\ 0 & \_\_ & \_\_ & | & \_\_ \end{bmatrix} \quad \begin{array}{l} -2R_1 + R_2 \\ -5R_1 + R_3 \end{array}$$

$$\begin{bmatrix} 1 & 1 & -1 & | & -3 \\ 0 & 1 & \_\_ & | & \_\_ \\ 0 & -6 & 7 & | & 38 \end{bmatrix} \quad -1R_2$$

$$\begin{bmatrix} 1 & 1 & -1 & | & -3 \\ 0 & 1 & -3 & | & -10 \\ 0 & 0 & \_\_ & | & \_\_ \end{bmatrix} \quad 6R_2 + R_3$$

$$\begin{bmatrix} 1 & 1 & -1 & | & -3 \\ 0 & 1 & -3 & | & -10 \\ 0 & 0 & 1 & | & \_\_ \end{bmatrix} \quad -\tfrac{1}{11}R_3$$

18. $2x + y + 2z = 11$
$2x - y - z = -3$
$3x + 2y + z = 9$

$$\begin{bmatrix} 2 & 1 & 2 & | & 11 \\ 2 & -1 & -1 & | & -3 \\ 3 & 2 & 1 & | & 9 \end{bmatrix}$$

$$\begin{bmatrix} 1 & \_\_ & \_\_ & | & \_\_ \\ 2 & -1 & -1 & | & -3 \\ 3 & 2 & 1 & | & 9 \end{bmatrix} \quad \tfrac{1}{2}R_1$$

$$\begin{bmatrix} 1 & \tfrac{1}{2} & 1 & | & \tfrac{11}{2} \\ 0 & \_\_ & \_\_ & | & \_\_ \\ 0 & \_\_ & \_\_ & | & \_\_ \end{bmatrix} \quad \begin{array}{l} -2R_1 + R_2 \\ -3R_1 + R_3 \end{array}$$

$$\begin{bmatrix} 1 & \tfrac{1}{2} & 1 & | & \tfrac{11}{2} \\ 0 & 1 & \_\_ & | & \_\_ \\ 0 & \tfrac{1}{2} & -2 & | & -\tfrac{15}{2} \end{bmatrix} \quad -\tfrac{1}{2}R_2$$

$$\begin{bmatrix} 1 & \tfrac{1}{2} & 1 & | & \tfrac{11}{2} \\ 0 & 1 & \tfrac{3}{2} & | & 7 \\ 0 & 0 & \_\_ & | & \_\_ \end{bmatrix} \quad -\tfrac{1}{2}R_2 + R_3$$

$$\begin{bmatrix} 1 & \tfrac{1}{2} & 1 & | & \tfrac{11}{2} \\ 0 & 1 & \tfrac{3}{2} & | & 7 \\ 0 & 0 & 1 & | & \_\_ \end{bmatrix} \quad -\tfrac{4}{11}R_3$$

Use row operations to solve each system. See Examples 2 and 3.

19. $x + y - 3z = 1$
$2x - y + z = 9$
$3x + y - 4z = 8$

20. $2x + 4y - 3z = -18$
$3x + y - z = -5$
$x - 2y + 4z = 14$

21. $x + y - z = 6$
$2x - y + z = -9$
$x - 2y + 3z = 1$

22. $x + 3y - 6z = 7$
$2x - y + 2z = 0$
$x + y + 2z = -1$

23. $x - y = 1$
$y - z = 6$
$x + z = -1$

24. $x + y = 1$
$2x - z = 0$
$y + 2z = -2$

25. $4x + 8y + 4z = 9$
$x + 3y + 4z = 10$
$5x + 10y + 5z = 12$

26. $x + 2y + 3z = -2$
$2x + 4y + 6z = -5$
$x - y + 2z = 6$

27. $x - 2y + z = 4$
$3x - 6y + 3z = 12$
$-2x + 4y - 2z = -8$

28. $x + 3y + z = 1$
$2x + 6y + 2z = 2$
$3x + 9y + 3z = 3$

29. $5x + 3y - z = 0$
$2x - 3y + z = 0$
$x + 4y - 2z = 0$

30. $4x + 5y - z = 0$
$7x - 5y + z = 0$
$x + 3y - 2z = 0$

Chapter 5

SUMMARY

5.1	**system of equations**	Two or more equations that are to be solved at the same time form a system of equations.
	linear system	A linear system is a system of equations that contains only linear equations.
	solution set of a system	All ordered pairs that satisfy all the equations of a system at the same time make up the solution set of the system.
	consistent system	A system is consistent if it has a solution.
	independent equations	Independent equations are equations whose graphs are different lines.
	inconsistent system	A system is inconsistent if it has no solution.
	dependent equations	Dependent equations are equations whose graphs are the same line.
5.4	**matrix**	A matrix is a rectangular array of numbers, consisting of horizontal **rows** and vertical **columns.**
	elements of a matrix	The numbers in a matrix are its elements.
	square matrix	A square matrix is a matrix that has the same number of rows as columns.
	augmented matrix	An augmented matrix is a matrix that has a vertical bar that separates the columns of the matrix into two groups.
	row echelon form	If a matrix is written with 1s down the diagonal from upper left to lower right and 0s below the 1s, it is said to be in row echelon form.

(x, y, z) ordered triple $\begin{bmatrix} a & b & c \\ d & e & f \end{bmatrix}$ **matrix with 2 rows, 3 columns (2 × 3)**

See how well you have learned the vocabulary in this chapter. Answers, with examples, follow the Quick Review.

1. A **system of equations** consists of
   A. at least two equations with different variables
   B. two or more equations that have an infinite number of solutions
   C. two or more equations that are to be solved at the same time
   D. two or more inequalities that are to be solved.

2. The **solution set of a system of equations in two variables** is
   A. all ordered pairs that satisfy one equation of the system

   B. all ordered pairs that satisfy all the equations of the system at the same time
   C. any ordered pair that satisfies one or more equations of the system
   D. the set of values that make all the equations of the system false.

3. An **inconsistent system** is a system of equations
   A. with one solution
   B. with no solution
   C. with an infinite number of solutions
   D. that have the same graph.

4. **Dependent equations**
   A. have different graphs
   B. have no solution
   C. have one solution
   D. are different forms of the same equation.

5. A **matrix** is
   A. an ordered pair of numbers
   B. an array of numbers with the same number of rows and columns
   C. a pair of numbers written between brackets
   D. a rectangular array of numbers.

299

QUICK REVIEW

Concepts	Examples

5.1 *Systems of Linear Equations in Two Variables*
Solving a Linear System by Substitution

Solve by substitution.

$$4x - y = 7 \quad (1)$$
$$3x + 2y = 30 \quad (2)$$

Step 1 Solve one of the equations for either variable.

Solve for y in equation (1).

$$y = 4x - 7$$

Step 2 Substitute for that variable in the other equation. The result should be an equation with just one variable.

Substitute $4x - 7$ for y in equation (2), and solve for x.

$$3x + 2y = 30 \quad (2)$$
$$3x + 2(4x - 7) = 30$$

Step 3 Solve the equation from Step 2.

$$3x + 8x - 14 = 30$$
$$11x - 14 = 30$$
$$11x = 44$$
$$x = 4$$

Step 4 Find the value of the other variable by substituting the result from Step 3 into the equation from Step 1.

Substitute 4 for x in the equation $y = 4x - 7$ to find that $y = \mathbf{9}$.

Step 5 Check the solution in both of the original equations. Then write the solution set.

Check to see that $\{(4, 9)\}$ is the solution set.

Solving a Linear System by Elimination

Solve by elimination.

Step 1 Write both equations in standard form.

$$5x + y = 2 \quad (1)$$
$$2x - 3y = 11 \quad (2)$$

Step 2 Make the coefficients of one pair of variable terms opposites.

To eliminate y, multiply equation (1) by 3, and add the result to equation (2).

Step 3 Add the new equations. The sum should be an equation with just one variable.

$$15x + 3y = 6$$
$$\underline{2x - 3y = 11} \quad (2)$$
$$17x \qquad = 17$$

Step 4 Solve the equation from Step 3.

$$x = 1$$

Step 5 Find the value of the other variable by substituting the result of Step 4 into either of the original equations.

Let $x = 1$ in equation (1), and solve for y.

$$5(1) + y = 2$$
$$y = -3$$

Step 6 Check the solution in both of the original equations. Then write the solution set.

Check to verify that $\{(1, -3)\}$ is the solution set.

Concepts	Examples

5.2 *Systems of Linear Equations in Three Variables*
Solving a Linear System in Three Variables

Step 1   Use the elimination method to eliminate any variable from any two of the original equations.

Step 2   Eliminate the *same* variable from any *other* two equations.

Step 3   Eliminate a second variable from the two equations in two variables that result from Steps 1 and 2. The result is an equation in one variable that gives the value of that variable.

Step 4   Substitute the value of the variable found in Step 3 into either of the equations in two variables to find the value of the second variable.

Step 5   Use the values of the two variables from Steps 3 and 4 to find the value of the third variable by substituting into any of the original equations.

Step 6   Check the solution in all of the original equations. Then write the solution set.

5.3 *Applications of Systems of Linear Equations*
Use the six-step problem-solving method.

Step 1   Read the problem carefully.

Step 2   Assign variables.

Step 3   Write a system of equations that relates the unknowns.

Step 4   Solve the system.

Step 5   State the answer.

Step 6   Check.

Examples

Solve the system.

$$x + 2y - z = 6 \quad (1)$$
$$x + y + z = 6 \quad (2)$$
$$2x + y - z = 7 \quad (3)$$

Add equations (1) and (2); z is eliminated and the result is $2x + 3y = 12$.

Eliminate z again by adding equations (2) and (3) to get $3x + 2y = 13$. Now solve the system

$$2x + 3y = 12 \quad (4)$$
$$3x + 2y = 13. \quad (5)$$

To eliminate x, multiply equation (4) by -3 and equation (5) by 2.

$$-6x - 9y = -36$$
$$\underline{6x + 4y = 26}$$
$$-5y = -10$$
$$y = 2$$

Let $y = 2$ in equation (4).

$$2x + 3(2) = 12$$
$$2x + 6 = 12$$
$$2x = 6$$
$$x = 3$$

Let $y = 2$ and $x = 3$ in any of the original equations to find $z = 1$.

Check. The solution set is $\{(3, 2, 1)\}$.

The perimeter of a rectangle is 18 ft. The length is 3 ft more than twice the width. What are the dimensions of the rectangle?

Let x represent the length and y represent the width. From the perimeter formula, one equation is $2x + 2y = 18$. From the problem, another equation is $x = 3 + 2y$. Solve the system

$$2x + 2y = 18$$
$$x = 3 + 2y$$

to get $x = 7$ and $y = 2$. The length is 7 ft, and the width is 2 ft. Since the perimeter is

$$2(7) + 2(2) = 18, \quad \text{and} \quad 3 + 2(2) = 7,$$

the solution checks.

Concepts	*Examples*

5.4 *Solving Systems of Linear Equations by Matrix Methods*

Matrix Row Operations

1. Any two rows of the matrix may be interchanged.

$$\begin{bmatrix} 1 & 5 & 7 \\ 3 & 9 & -2 \\ 0 & 6 & 4 \end{bmatrix} \text{ becomes } \begin{bmatrix} 3 & 9 & -2 \\ 1 & 5 & 7 \\ 0 & 6 & 4 \end{bmatrix} \quad \text{Interchange } R_1 \text{ and } R_2.$$

2. The numbers in any row may be multiplied by any nonzero real number.

$$\begin{bmatrix} 1 & 5 & 7 \\ 3 & 9 & -2 \\ 0 & 6 & 4 \end{bmatrix} \text{ becomes } \begin{bmatrix} 1 & 5 & 7 \\ 1 & 3 & -\frac{2}{3} \\ 0 & 6 & 4 \end{bmatrix} \quad \tfrac{1}{3}R_2$$

3. Any row may be transformed by adding to the numbers of the row the product of a real number and the numbers of another row.

$$\begin{bmatrix} 1 & 5 & 7 \\ 3 & 9 & -2 \\ 0 & 6 & 4 \end{bmatrix} \text{ becomes } \begin{bmatrix} 1 & 5 & 7 \\ 0 & -6 & -23 \\ 0 & 6 & 4 \end{bmatrix} \quad -3R_1 + R_2$$

A system can be solved by matrix methods. Write the augmented matrix, and use row operations to obtain a matrix in row echelon form.

Solve using row operations.

$$x + 3y = 7$$
$$2x + y = 4$$

$$\begin{bmatrix} 1 & 3 & | & 7 \\ 2 & 1 & | & 4 \end{bmatrix} \quad \text{Augmented matrix}$$

$$\begin{bmatrix} 1 & 3 & | & 7 \\ 0 & -5 & | & -10 \end{bmatrix} \quad -2R_1 + R_2$$

$$\begin{bmatrix} 1 & 3 & | & 7 \\ 0 & 1 & | & 2 \end{bmatrix} \quad -\tfrac{1}{5}R_2$$

$$x + 3y = 7$$
$$y = 2$$

When $y = 2$, $x + 3(2) = 7$, so $x = 1$. The solution set is $\{(1, 2)\}$.

ANSWERS TO TEST YOUR WORD POWER

1. C; *Example:* $\begin{array}{l} 3x - y = 3 \\ 2x + y = 7 \end{array}$

2. B; *Example:* The ordered pair (2, 3) satisfies both equations of the system in Answer 1, so $\{(2, 3)\}$ is the solution set of the system.

3. B; *Example:* The equations of two parallel lines form an inconsistent system. Their graphs never intersect, so the system has no solution.

4. D; *Example:* The equations $4x - y = 8$ and $8x - 2y = 16$ are dependent because their graphs are the same line.

5. D; *Examples:* $\begin{bmatrix} 3 & -1 & 0 \\ 4 & 2 & 1 \end{bmatrix}, \begin{bmatrix} 1 & 2 \\ 4 & 3 \end{bmatrix}$

Chapter **5**

REVIEW EXERCISES

[5.1] **1.** Solve the system by graphing.

$$x + 3y = 8$$
$$2x - y = 2$$

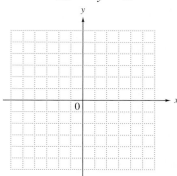

2. The graph shows the trends during the years 1974–2000 relating to bachelor's degrees awarded in the United States.

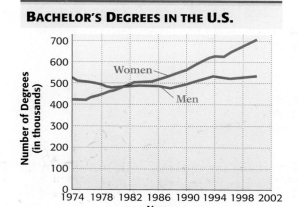

BACHELOR'S DEGREES IN THE U.S.

Source: U.S. National Center for Education Statistics, *Digest of Education Statistics*, annual.

(a) Between what years shown on the horizontal axis did the number of degrees for men and women reach equal numbers?

(b) When the number of degrees for men and women reached equal numbers, what was that number (approximately)?

Solve each system using the substitution method.

3. $3x + y = -4$
$x = \dfrac{2}{3}y$

4. $9x - y = -4$
$y = x + 4$

5. $-5x + 2y = -2$
$x + 6y = 26$

Solve each system using the elimination method.

6. $6x + 5y = 4$
$-4x + 2y = 8$

7. $\dfrac{x}{6} + \dfrac{y}{6} = -\dfrac{1}{2}$
$x - y = -9$

8. $4x + 5y = 9$
$3x + 7y = -1$

9. $-3x + y = 6$
$2y = 12 + 6x$

10. $5x - 4y = 2$
$-10x + 8y = 7$

Suppose that two linear equations are graphed on the same set of coordinate axes.
Sketch what the graph might look like if the system has the given description.

11. The system has a single solution.

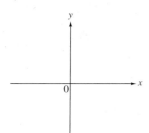

12. The system has no solution.

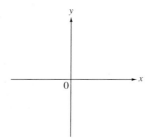

13. The system has infinitely many solutions.

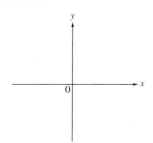

14. Without doing any algebraic work, explain why the system

$$y = 3x + 2$$
$$y = 3x - 4$$

has $\emptyset$ as its solution set. Answer based only on your knowledge of the graphs of the two lines.

[5.2] *Solve each system of equations.*

15.
$$2x + 3y - z = -16$$
$$x + 2y + 2z = -3$$
$$-3x + y + z = -5$$

16.
$$3x - y - z = -8$$
$$4x + 2y + 3z = 15$$
$$-6x + 2y + 2z = 10$$

17.
$$4x - y = 2$$
$$3y + z = 9$$
$$x + 2z = 7$$

[5.3] *Solve each problem using a system of equations.*

18. A regulation National Hockey League ice rink has perimeter 570 ft. The length is 30 ft longer than twice the width. What are the dimensions of an NHL ice rink? (*Source: Microsoft Encarta Encyclopedia 2002.*)

19. On a 6-day business trip, Gina Rivera rented a car for $53 per day at weekday rates and $35 per day at weekend rates. If her total rental bill was $264, how many days did she rent at each rate? (*Source: Enterprise.*)

20. A plane flies 560 mi in 1.75 hr traveling with the wind. The return trip later against the same wind takes the plane 2 hr. Find the speed of the plane and the speed of the wind.

	r	t	d
With Wind	$x + y$	1.75	
Against Wind		2	

21. Sweet's Candy Store is offering a special mix for Valentine's Day. Ms. Sweet will mix some $2 per lb nuts with some $1 per lb chocolate candy to get 100 lb of mix, which she will sell at $1.30 per lb. How many pounds of each should she use?

	Price per Pound	Number of Pounds	Value
Nuts		x	
Chocolate		y	
Mixture		100	

22. A biologist wants to grow two types of algae, green and brown. She has 15 kg of nutrient X and 26 kg of nutrient Y. A vat of green algae needs 2 kg of nutrient X and 3 kg of nutrient Y, while a vat of brown algae needs 1 kg of nutrient X and 2 kg of nutrient Y. How many vats of each type of algae should she grow in order to use all the nutrients?

23. The sum of the measures of the angles of a triangle is 180°. The largest angle measures 10° less than the sum of the other two. The measure of the middle-sized angle is the average of the other two. Find the measures of the three angles.

24. How many liters each of 8%, 10%, and 20% hydrogen peroxide should be mixed together to get 8 L of 12.5% solution, if the amount of 8% solution used must be 2 L more than the amount of 20% solution used?

25. In the great baseball year of 1961, Yankee teammates Mickey Mantle, Roger Maris, and John Blanchard combined for 136 home runs. Mantle hit 7 fewer than Maris. Maris hit 40 more than Blanchard. What were the home run totals for each player? (*Source: Neft, David S. and Richard M. Cohen, The Sports Encyclopedia: Baseball 2003.*)

[5.4] *Solve each system using row operations.*

26. $2x + 5y = -4$
$4x - y = 14$

27. $6x + 3y = 9$
$-7x + 2y = 17$

28. $x + 2y - z = 1$
$3x + 4y + 2z = -2$
$-2x - y + z = -1$

MIXED REVIEW EXERCISES

Solve by any method.

29. $\dfrac{2}{3}x + \dfrac{1}{6}y = \dfrac{19}{2}$
$\dfrac{1}{3}x - \dfrac{2}{9}y = 2$

30. $2x - 5y = 8$
$3x + 4y = 10$

31. $x = 7y + 10$
$2x + 3y = 3$

32.
$$x + 4y = 17$$
$$-3x + 2y = -9$$

33.
$$-7x + 3y = 12$$
$$5x + 2y = 8$$

34.
$$2x + 5y - z = 12$$
$$-x + y - 4z = -10$$
$$-8x - 20y + 4z = 31$$

35. To make a 10% acid solution for chemistry class, Xavier wants to mix some 5% solution with 10 L of 20% solution. How many liters of 5% solution should he use?

Percent (as a Decimal)	Liters of Solution	Liters of Pure Acid

36. In the 2004 Summer Olympics in Athens, Greece, the top three medal-winning countries were the United States, Russia, and China, with a combined total of 258 medals. The United States won 11 more medals than Russia, while China won 29 fewer medals than Russia. How many medals did each country win? (*Source: The Gazette*, August 30, 2004.)

RELATING CONCEPTS (EXERCISES 37–41) For Individual or Group Work

Thus far in this text we have studied only linear *equations. In later chapters we will study the graphs of other kinds of equations. One such graph is a* circle, *which has an equation of the form*

$$x^2 + y^2 + ax + by + c = 0.$$

It is a fact from geometry that given three noncollinear *points (that is, points that do not all lie on the same straight line), there will be a circle that contains them. For example, the points* $(4, 2)$, $(-5, -2)$, *and* $(0, 3)$ *lie on the circle whose equation is shown in the figure.* **Work Exercises 37–41 in order** *to find an equation of the circle passing through the points* $(2, 1)$, $(-1, 0)$, *and* $(3, 3)$.

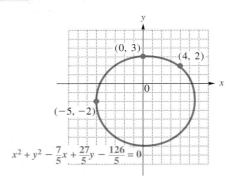

$$x^2 + y^2 - \frac{7}{5}x + \frac{27}{5}y - \frac{126}{5} = 0$$

37. Let $x = 2$ and $y = 1$ in the equation $x^2 + y^2 + ax + by + c = 0$ to find an equation in a, b, and c.

38. Let $x = -1$ and $y = 0$ to find a second equation in a, b, and c.

39. Let $x = 3$ and $y = 3$ to find a third equation in a, b, and c.

40. Solve the system of equations formed by your answers in Exercises 37–39 to find the values of a, b, and c. What is the equation of the circle?

41. Explain why the relation whose graph is a circle is not a function.

Chapter 5

T E S T

The graph shows the numbers of people in the United States of two differ-
ent races living with AIDS during the period 1993–2000. Use the graph
to answer the questions in Exercises 1 and 2.

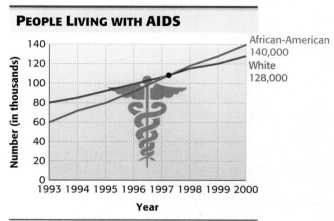

PEOPLE LIVING WITH AIDS

African-American 140,000
White 128,000

Source: U.S. Centers for Disease Control.

1. In what year were the numbers the same?

2. How many people of each race were living with AIDS that year?

3. Use a graph to solve the system.

$$x + y = 7$$
$$x - y = 5$$

Solve each system using substitution.

4. $2x - 3y = 24$

$y = -\dfrac{2}{3}x$

5. $12x - 5y = 8$

$3x = \dfrac{5}{4}y + 2$

6. $3x - y = -8$
$2x + 6y = 3$

1. _____

2. _____

3. _____

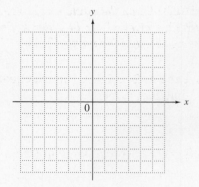

4. _____

5. _____

6. _____

7. _____

8. _____

9. _____

10. _____

11. _____

Solve each system using elimination.

7. $3x + y = 12$
   $2x - y = 3$

8. $-5x + 2y = -4$
   $6x + 3y = -6$

9. $3x + 4y = 8$
   $8y = 7 - 6x$

10. $3x + 5y + 3z = 2$
    $6x + 5y + z = 0$
    $3x + 10y - 2z = 6$

11. $4x + y + z = 11$
    $x - y - z = 4$
    $y + 2z = 0$

Solve each problem using a system of equations.

12. _____

12. Julia Roberts is one of the biggest box-office stars in Hollywood. As of June 2004, her two top-grossing domestic films, *Ocean's Eleven* and *Runaway Bride,* together earned $335.5 million. If *Runaway Bride* grossed $31.3 million less than *Ocean's Eleven,* how much did each film gross? (*Source:* www.the-numbers.com)

13. _____

13. Two cars start from points 420 mi apart and travel toward each other. They meet after 3.5 hr. Find the average speed of each car if one travels 30 mph slower than the other.

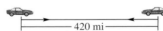
420 mi

14. _____

14. A chemist needs 12 L of a 40% alcohol solution. She must mix a 20% solution and a 50% solution. How many liters of each will be required to obtain what she needs?

15. _____

15. A local electronics store will sell 7 AC adaptors and 2 rechargeable flashlights for $86, or 3 AC adaptors and 4 rechargeable flashlights for $84. What is the price of a single AC adaptor and a single rechargeable flashlight?

16. _____

16. The owner of a tea shop wants to mix three kinds of tea to make 100 oz of a mixture that will sell for $.83 per oz. He uses Orange Pekoe, which sells for $.80 per oz, Irish Breakfast, for $.85 per oz, and Earl Grey, for $.95 per oz. If he wants to use twice as much Orange Pekoe as Irish Breakfast, how much of each kind of tea should he use?

Solve each system using row operations.

17. _____

17. $3x + 2y = 4$
    $5x + 5y = 9$

18. $x + 3y + 2z = 11$
    $3x + 7y + 4z = 23$
    $5x + 3y - 5z = -14$

18. _____

Cumulative Review Exercises
CHAPTERS 1–5

Evaluate.

1. $(-3)^4$

2. -3^4

3. $-(-3)^4$

4. $\sqrt{.49}$

5. $-\sqrt{.49}$

6. $\sqrt{-.49}$

Evaluate if $x = -4$, $y = 3$, and $z = 6$.

7. $|2x| + y^2 - z^3$

8. $-5(x^3 - y^3)$

9. $\dfrac{2x^2 - x + z}{y^2 - z}$

Solve each equation.

10. $7(2x + 3) - 4(2x + 1) = 2(x + 1)$

11. $.04x + .06(x - 1) = 1.04$

12. $ax + by = cx + d$ for x

13. $|6x - 8| = 4$

Solve each inequality.

14. $\dfrac{2}{3}x + \dfrac{5}{12}x \le 20$

15. $|3x + 2| \le 4$

16. $|12t + 7| \ge 0$

17. A recent survey measured public recognition of the most popular contemporary advertising slogans. Complete the results shown in the table if 2500 people were surveyed.

Slogan (product or company)	Percent Recognition (nearest tenth of a percent)	Actual Number Who Recognized Slogan (nearest whole number)
Please Don't Squeeze the . . . (Charmin)	80.4%	
The Breakfast of Champions (Wheaties)	72.5%	
The King of Beers (Budweiser)		1570
Like a Good Neighbor (State Farm)		1430

(Other slogans included "You're in Good Hands" (Allstate), "Snap, Crackle, Pop" (Rice Krispies), and "The Un-Cola" (7-Up).)
Source: Department of Integrated Marketing Communications, Northwestern University.

Solve each problem.

18. A jar contains only pennies, nickels, and dimes. The number of dimes is 1 more than the number of nickels, and the number of pennies is 6 more than the number of nickels. How many of each denomination can be found in the jar, if the total value is $4.80?

19. Two angles of a triangle have the same measure. The measure of the third angle is 4° less than twice the measure of each of the equal angles. Find the measures of the three angles.

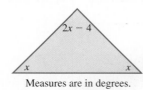

Measures are in degrees.

In Exercises 20–25, point A has coordinates $(-2, 6)$ and point B has coordinates $(4, -2)$.

20. What is the equation of the horizontal line through A?

21. What is the equation of the vertical line through B?

22. What is the slope of AB?

23. What is the slope of a line perpendicular to line AB?

24. What is the standard form of the equation of line AB?

25. Write the equation of the line in the form of a linear function.

26. Graph the linear function whose graph has slope $\frac{2}{3}$ and passes through the point $(-1, -3)$.

27. Graph the inequality $-3x - 2y \leq 6$.

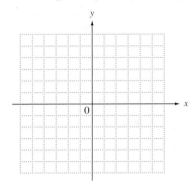

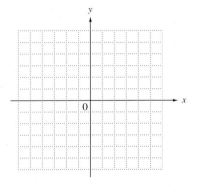

Solve by any method.

28. $-2x + 3y = -15$
$\quad\;\; 4x - \;\; y = 15$

29. $\quad x + y + z = 10$
$\quad\;\; x - y - z = 0$
$\quad -x + y - z = -4$

Solve each problem using a system of equations.

30. A grocer plans to mix candy that sells for $1.20 per lb with candy that sells for $2.40 per lb to get a mixture that he plans to sell for $1.65 per lb. How much of the $1.20 and $2.40 candy should he use if he wants 80 lb of the mix?

31. A small company took out three loans totaling $25,000. The company was able to borrow some of the money at 8%. It borrowed $2000 more than $\frac{1}{2}$ the amount of the 8% loan at 10%, and the rest at 9%. The total annual interest was $2220. How much did the company borrow at each rate?

The graph shows a company's costs to produce computer parts and the revenue from the sale of those parts.

32. At what production level does the cost equal the revenue? What is the revenue at that point?

33. Profit is revenue less cost. Estimate the profit on the sale of 1100 parts.

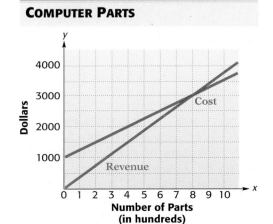

COMPUTER PARTS

Exponents, Polynomials, and Polynomial Functions

6

6.1 Integer Exponents and Scientific Notation

6.2 Adding and Subtracting Polynomials

6.3 Polynomial Functions

6.4 Multiplying Polynomials

6.5 Dividing Polynomials

In 1980 MasterCard International Incorporated first began offering debit cards in an effort to challenge Visa, the leader in credit card transactions at that time. Now immensely popular, debit cards draw money from consumers' bank accounts rather than from established lines of credit. By 2005, it is estimated that 269 million debit cards will be in use. (*Source: Microsoft Encarta Encyclopedia 2002*; HSN Consultants Inc.)

We introduced the concept of function in Section 4.5 and extend our work to include *polynomial functions* in this chapter. In Exercise 11 of Section 6.3, we use a polynomial function to model the number of bank debit cards issued.

6.1 Integer Exponents and Scientific Notation

OBJECTIVES

1 Use the product rule for exponents.

2 Define 0 and negative exponents.

3 Use the quotient rule for exponents.

4 Use the power rules for exponents.

5 Simplify exponential expressions.

6 Use the rules for exponents with scientific notation.

Recall from **Section 1.3** that we use exponents to write products of repeated factors. For example,

$$2^5 \text{ is defined as } 2 \cdot 2 \cdot 2 \cdot 2 \cdot 2 = 32.$$

The number 5, the *exponent,* shows that the *base* 2 appears as a factor 5 times. The quantity 2^5 is called an *exponential* or a *power.* We read 2^5 as "2 to the fifth power" or "2 to the fifth."

OBJECTIVE 1 **Use the product rule for exponents.** There are several useful rules that simplify work with exponents. For example, the product $2^5 \cdot 2^3$ can be simplified as follows.

$$2^5 \cdot 2^3 = \overbrace{(2 \cdot 2 \cdot 2 \cdot 2 \cdot 2)(2 \cdot 2 \cdot 2)}^{5 + 3 = 8} = 2^8$$

This result, that products of exponential expressions with the *same base* are found by adding exponents, is generalized as the **product rule for exponents.**

Product Rule for Exponents

If m and n are natural numbers and a is any real number, then

$$a^m \cdot a^n = a^{m+n}.$$

In words, when multiplying powers of like bases, keep the same base and add the exponents.

---

1 Apply the product rule for exponents, if possible, in each case.

(a) $m^8 \cdot m^6$

(b) $r^7 \cdot r$

(c) $k^4 k^3 k^6$

(d) $m^5 \cdot p^4$

(e) $(-4a^3)(6a^2)$

(f) $(-5p^4)(-9p^5)$

To see that the product rule is true, use the definition of an exponent.

$$a^m = \underbrace{a \cdot a \cdot a \cdots a}_{a \text{ appears as a factor } m \text{ times.}} \qquad a^n = \underbrace{a \cdot a \cdot a \cdots a}_{a \text{ appears as a factor } n \text{ times.}}$$

From this, $a^m \cdot a^n = \underbrace{a \cdot a \cdot a \cdots a}_{m \text{ factors}} \cdot \underbrace{a \cdot a \cdot a \cdots a}_{n \text{ factors}}$

$$= \underbrace{a \cdot a \cdot a \cdots a}_{(m+n) \text{ factors}}$$

$$a^m \cdot a^n = a^{m+n}.$$

EXAMPLE 1 **Using the Product Rule for Exponents**

Apply the product rule for exponents, if possible, in each case.

(a) $3^4 \cdot 3^7 = 3^{4+7} = 3^{11}$ (b) $5^3 \cdot 5 = 5^3 \cdot 5^1 = 5^{3+1} = 5^4$

(c) $y^3 \cdot y^8 \cdot y^2 = y^{3+8+2} = y^{13}$

(d) $(5y^2)(-3y^4) = 5(-3)y^2 y^4$ Associative and commutative properties

$\qquad\qquad\quad = -15y^{2+4}$ Multiply; product rule

$\qquad\qquad\quad = -15y^6$

(e) $(7p^3q)(2p^5q^2) = 7(2)p^3 p^5 q q^2 = 14p^8 q^3$

(f) $x^2 \cdot y^4$

Because the bases are not the same, the product rule does not apply.

ANSWERS

1. (a) m^{14} (b) r^8 (c) k^{13}
(d) The product rule does not apply.
(e) $-24a^5$ (f) $45p^9$

◀◀◀ Work Problem 1 at the Side.

CAUTION
Be careful in problems like Example 1(a) not to multiply the bases. Notice that $3^4 \cdot 3^7 = 3^{11}$, *not* 9^{11}. *Keep the same base and add the exponents.*

OBJECTIVE 2 Define 0 and negative exponents. So far we have discussed only positive exponents. Now we define 0 as an exponent. Suppose we multiply 4^2 by 4^0. By the product rule, extended to whole numbers,

$$4^2 \cdot 4^0 = 4^{2+0} = 4^2.$$

For the product rule to hold true, 4^0 must equal 1, and so we define a^0 this way for any nonzero real number a.

Zero Exponent
If a is any nonzero real number, then
$$a^0 = 1.$$

The expression 0^0 is undefined.*

EXAMPLE 2 Using 0 as an Exponent

Evaluate each expression.

(a) $6^0 = 1$ **(b)** $(-6)^0 = 1$ Base is -6.

(c) $-6^0 = -(6^0) = -1$ Base is 6.   **(d)** $-(-6)^0 = -1$

(e) $5^0 + 12^0 = 1 + 1 = 2$ **(f)** $(8k)^0 = 1, \quad k \neq 0$

Work Problem 2 at the Side. ▶▶▶

We should define negative exponents so that the rules for exponents are valid. With this in mind, using the product rule should give

$$8^2 \cdot 8^{-2} = 8^{2+(-2)} = 8^0 = 1.$$

This indicates that 8^{-2} is the reciprocal of 8^2. But $\dfrac{1}{8^2}$ is the reciprocal of 8^2, and a number can have only one reciprocal. Therefore, it is reasonable to conclude that $8^{-2} = \dfrac{1}{8^2}$. We can generalize and make the following definition.

Negative Exponent
For any natural number n and any nonzero real number a,
$$a^{-n} = \frac{1}{a^n}.$$

With this definition, the expression a^n is meaningful for any integer exponent n and any nonzero real number a.

---

*In advanced treatments, 0^0 is called an *indeterminate form*.

2 Evaluate each expression.

(a) 29^0

(b) $(-29)^0$

(c) $-(-29)^0$

(d) -29^0

(e) $8^0 - 15^0$

(f) $(-15p^5)^0, \quad p \neq 0$

ANSWERS
2. (a) 1   **(b)** 1   **(c)** -1 **(d)** -1
    (e) 0   **(f)** 1

3 In parts (a)–(f), write the expressions with only positive exponents. In parts (g) and (h), simplify each expression.

(a) 6^{-3}

CAUTION

A negative exponent does not indicate that an expression represents a negative number. Negative exponents lead to reciprocals.

Expression	Example	
a^{-m}	$3^{-2} = \dfrac{1}{3^2} = \dfrac{1}{9}$	Not negative
$-a^{-m}$	$-3^{-2} = -\dfrac{1}{3^2} = -\dfrac{1}{9}$	Negative

(b) 8^{-1}

EXAMPLE 3 Using Negative Exponents

In parts (a)–(f), write the expressions with only positive exponents. In parts (g) and (h), simplify each expression.

(a) $2^{-3} = \dfrac{1}{2^3}$ **(b)** $6^{-1} = \dfrac{1}{6^1} = \dfrac{1}{6}$

(c) $(2x)^{-4}, \quad x \neq 0$

(c) $(5z)^{-3} = \dfrac{1}{(5z)^3}, \quad z \neq 0$ **(d)** $5z^{-3} = 5\left(\dfrac{1}{z^3}\right) = \dfrac{5}{z^3}, \quad z \neq 0$

(e) $-m^{-2} = -\dfrac{1}{m^2}, \quad m \neq 0$ **(f)** $(-m)^{-2} = \dfrac{1}{(-m)^2}, \quad m \neq 0$

(d) $7r^{-6}, \quad r \neq 0$

(g) $3^{-1} + 4^{-1} = \dfrac{1}{3} + \dfrac{1}{4} = \dfrac{4}{12} + \dfrac{3}{12} = \dfrac{7}{12}$ $\quad \frac{1}{3} \cdot \frac{4}{4} = \frac{4}{12}; \frac{1}{4} \cdot \frac{3}{3} = \frac{3}{12}$

(h) $5^{-1} - 2^{-1} = \dfrac{1}{5} - \dfrac{1}{2} = \dfrac{2}{10} - \dfrac{5}{10} = -\dfrac{3}{10}$

(e) $-q^{-4}, \quad q \neq 0$

CAUTION

In Example 3(g), note that $3^{-1} + 4^{-1} \neq (3 + 4)^{-1}$. The expression on the left is equal to $\frac{7}{12}$, as shown in the solution, while the expression on

(f) $(-q)^{-4}, \quad q \neq 0$

the right is $7^{-1} = \frac{1}{7}$. Similar reasoning can be applied to part (h).

◀◀◀ **Work Problem 3 at the Side.**

(g) $3^{-1} + 5^{-1}$

EXAMPLE 4 Using Negative Exponents

Evaluate each expression.

(a) $\dfrac{1}{2^{-3}} = \dfrac{1}{\dfrac{1}{2^3}} = 1 \div \dfrac{1}{2^3} = 1 \cdot \dfrac{2^3}{1} = 2^3 = 8$

(h) $4^{-1} - 2^{-1}$

(b) $\dfrac{2^{-3}}{3^{-2}} = \dfrac{\dfrac{1}{2^3}}{\dfrac{1}{3^2}} = \dfrac{1}{2^3} \div \dfrac{1}{3^2} = \dfrac{1}{2^3} \cdot \dfrac{3^2}{1} = \dfrac{3^2}{2^3} = \dfrac{9}{8}$

ANSWERS

3. **(a)** $\dfrac{1}{6^3}$ **(b)** $\dfrac{1}{8}$ **(c)** $\dfrac{1}{(2x)^4}$ **(d)** $\dfrac{7}{r^6}$

(e) $-\dfrac{1}{q^4}$ **(f)** $\dfrac{1}{(-q)^4}$ **(g)** $\dfrac{8}{15}$ **(h)** $-\dfrac{1}{4}$

Example 4 suggests the following generalizations.

Special Rules for Negative Exponents

If $a \neq 0$ and $b \neq 0$, then $\quad \dfrac{1}{a^{-n}} = a^n \quad$ and $\quad \dfrac{a^{-n}}{b^{-m}} = \dfrac{b^m}{a^n}.$

Work Problem 4 at the Side. ▶▶▶

OBJECTIVE 3 Use the quotient rule for exponents. A quotient, such as $\dfrac{a^8}{a^3}$, can be simplified in much the same way as a product. (In all quotients of this type, assume that the denominator is not 0.) Using the definition of an exponent,

$$\frac{a^8}{a^3} = \frac{a \cdot a \cdot a \cdot a \cdot a \cdot a \cdot a \cdot a}{a \cdot a \cdot a} = a \cdot a \cdot a \cdot a \cdot a = a^5.$$

Notice that $8 - 3 = 5$. In the same way,

$$\frac{a^3}{a^8} = \frac{a \cdot a \cdot a}{a \cdot a \cdot a \cdot a \cdot a \cdot a \cdot a \cdot a} = \frac{1}{a^5} = a^{-5}.$$

Here, $3 - 8 = -5$. These examples suggest the **quotient rule for exponents.**

Quotient Rule for Exponents

If a is any nonzero real number and m and n are integers, then

$$\frac{a^m}{a^n} = a^{m-n}.$$

In words, when dividing powers of like bases, keep the same base and subtract the exponent of the denominator from the exponent of the numerator.

EXAMPLE 5 Using the Quotient Rule for Exponents

Apply the quotient rule for exponents, if possible, and write each result using only positive exponents.

Numerator exponent ↓
Denominator exponent ↓

(a) $\dfrac{3^7}{3^2} = 3^{7-2} = 3^5$
  ↑ Minus sign

(b) $\dfrac{p^6}{p^2} = p^{6-2} = p^4, \quad p \neq 0$

(c) $\dfrac{k^7}{k^{12}} = k^{7-12} = k^{-5} = \dfrac{1}{k^5}, \quad k \neq 0$

(d) $\dfrac{2^7}{2^{-3}} = 2^{7-(-3)} = 2^{7+3} = 2^{10}$

(e) $\dfrac{8^{-2}}{8^5} = 8^{-2-5} = 8^{-7} = \dfrac{1}{8^7}$

(f) $\dfrac{6}{6^{-1}} = \dfrac{6^1}{6^{-1}} = 6^{1-(-1)} = 6^2$

(g) $\dfrac{z^{-5}}{z^{-8}} = z^{-5-(-8)} = z^3, \quad z \neq 0$

(h) $\dfrac{a^3}{b^4}, \quad b \neq 0$
The quotient rule does not apply because the bases are different.

Work Problem 5 at the Side. ▶▶▶

---

4 Evaluate each expression.

(a) $\dfrac{1}{4^{-3}}$

(b) $\dfrac{3^{-3}}{9^{-1}}$

5 Apply the quotient rule for exponents, if possible, and write each result using only positive exponents.

(a) $\dfrac{4^8}{4^6}$

(b) $\dfrac{x^{12}}{x^3}, \quad x \neq 0$

(c) $\dfrac{r^5}{r^8}, \quad r \neq 0$

(d) $\dfrac{2^8}{2^{-4}}$

(e) $\dfrac{6^{-3}}{6^4}$

(f) $\dfrac{8}{8^{-1}}$

(g) $\dfrac{t^{-4}}{t^{-6}}, \quad t \neq 0$

(h) $\dfrac{x^3}{y^5}, \quad y \neq 0$

ANSWERS

4. (a) 64 **(b)** $\dfrac{1}{3}$

5. (a) 4^2 **(b)** x^9 **(c)** $\dfrac{1}{r^3}$ **(d)** 2^{12}

(e) $\dfrac{1}{6^7}$ **(f)** 8^2 **(g)** t^2

(h) The quotient rule does not apply.

6 Use one or more power rules to simplify each expression.

(a) $(r^5)^4$

(b) $\left(\dfrac{3}{4}\right)^3$

(c) $(9x)^3$

(d) $(5r^6)^3$

(e) $\left(\dfrac{-3n^4}{m}\right)^3$, $m \neq 0$

OBJECTIVE 4 Use the power rules for exponents. The expression $(3^4)^2$ can be simplified as

$$(3^4)^2 = 3^4 \cdot 3^4 = 3^{4+4} = 3^8,$$

where $4 \cdot 2 = 8$. This example suggests the first **power rule for exponents.** The other two power rules can be demonstrated with similar examples.

Power Rules for Exponents

If a and b are real numbers and m and n are integers, then

(a) $(a^m)^n = a^{mn}$, **(b)** $(ab)^m = a^m b^m$, and **(c)** $\left(\dfrac{a}{b}\right)^m = \dfrac{a^m}{b^m}$ $(b \neq 0)$.

In words,

(a) to raise a power to a power, multiply exponents;

(b) to raise a product to a power, raise each factor to that power; and

(c) to raise a quotient to a power, raise the numerator and the denominator to that power.

EXAMPLE 6 Using the Power Rules for Exponents

Use one or more power rules to simplify each expression.

(a) $(p^8)^3 = p^{8 \cdot 3} = p^{24}$

(b) $\left(\dfrac{2}{3}\right)^4 = \dfrac{2^4}{3^4} = \dfrac{16}{81}$

(c) $(3y)^4 = 3^4 y^4 = 81y^4$

(d) $(6p^7)^2 = 6^2 p^{7 \cdot 2} = 6^2 p^{14} = 36p^{14}$

(e) $\left(\dfrac{-2m^5}{z}\right)^3 = \dfrac{(-2)^3 \, m^{5 \cdot 3}}{z^3} = \dfrac{(-2)^3 m^{15}}{z^3} = \dfrac{-8m^{15}}{z^3}$, $z \neq 0$

◀◀◀ Work Problem 6 at the Side.

The reciprocal of a^n is $\dfrac{1}{a^n} = \left(\dfrac{1}{a}\right)^n$. Also, by definition, a^n and a^{-n} are reciprocals since

$$a^n \cdot a^{-n} = a^n \cdot \dfrac{1}{a^n} = 1.$$

Thus, since both are reciprocals of a^n,

$$a^{-n} = \left(\dfrac{1}{a}\right)^n.$$

Some examples of this result are

$$6^{-3} = \left(\dfrac{1}{6}\right)^3 \quad \text{and} \quad \left(\dfrac{1}{3}\right)^{-2} = 3^2.$$

ANSWERS

6. **(a)** r^{20} **(b)** $\dfrac{27}{64}$ **(c)** $729x^3$
   (d) $125r^{18}$ **(e)** $\dfrac{-27n^{12}}{m^3}$

This discussion can be generalized as follows.

More Special Rules for Negative Exponents

If $a \neq 0$ and $b \neq 0$ and n is an integer, then

$$a^{-n} = \left(\frac{1}{a}\right)^n \quad \text{and} \quad \left(\frac{a}{b}\right)^{-n} = \left(\frac{b}{a}\right)^n.$$

In words, any nonzero number raised to the negative nth power is equal to the reciprocal of that number raised to the nth power.

EXAMPLE 7 **Using Negative Exponents with Fractions**

Write each expression with only positive exponents and then evaluate.

(a) $\left(\frac{3}{7}\right)^{-2} = \left(\frac{7}{3}\right)^2 = \frac{49}{9}$

(b) $\left(\frac{4}{5}\right)^{-3} = \left(\frac{5}{4}\right)^3 = \frac{125}{64}$

Work Problem 7 at the Side. ▶▶▶

The definitions and rules of this section are summarized here.

Definitions and Rules for Exponents

For all integers m and n and all real numbers a and b, the following rules apply.

Product Rule $a^m \cdot a^n = a^{m+n}$

Quotient Rule $\dfrac{a^m}{a^n} = a^{m-n} \quad (a \neq 0)$

Zero Exponent $a^0 = 1 \quad (a \neq 0)$

Negative Exponent $a^{-n} = \dfrac{1}{a^n} \quad (a \neq 0)$

Power Rules $(a^m)^n = a^{mn}$

$(ab)^m = a^m b^m$

$\left(\dfrac{a}{b}\right)^m = \dfrac{a^m}{b^m} \quad (b \neq 0)$

Special Rules $\dfrac{1}{a^{-n}} = a^n \quad (a \neq 0)$ $\dfrac{a^{-n}}{b^{-m}} = \dfrac{b^m}{a^n} \quad (a, b \neq 0)$

$a^{-n} = \left(\dfrac{1}{a}\right)^n \quad (a \neq 0)$ $\left(\dfrac{a}{b}\right)^{-n} = \left(\dfrac{b}{a}\right)^n \quad (a, b \neq 0)$

OBJECTIVE 5 **Simplify exponential expressions.** With the rules of exponents developed so far in this section, we can simplify expressions that involve one or more rules, as shown in Example 8 on the next page.

---

7 Write each expression with only positive exponents and then evaluate.

(a) $\left(\frac{3}{4}\right)^{-3}$

(b) $\left(\frac{5}{6}\right)^{-2}$

ANSWERS

7. (a) $\left(\dfrac{4}{3}\right)^3; \dfrac{64}{27}$ (b) $\left(\dfrac{6}{5}\right)^2; \dfrac{36}{25}$

8 Simplify each expression so that no negative exponents appear in the final result. Assume all variables represent nonzero real numbers.

(a) $5^4 \cdot 5^{-6}$

(b) $x^{-4} \cdot x^{-6} \cdot x^8$

(c) $(5^{-3})^{-2}$

(d) $(y^{-2})^7$

(e) $\dfrac{a^{-3}b^5}{a^4 b^{-2}}$

(f) $(3^2 k^{-4})^{-1}$

(g) $\left(\dfrac{2y}{x^3}\right)^2 \left(\dfrac{4y}{x}\right)^{-1}$

EXAMPLE 8 Using the Definitions and Rules for Exponents

Simplify each expression so that no negative exponents appear in the final result. Assume all variables represent nonzero real numbers.

(a) $3^2 \cdot 3^{-5} = 3^{2+(-5)} = 3^{-3} = \dfrac{1}{3^3}$ or $\dfrac{1}{27}$

(b) $x^{-3} \cdot x^{-4} \cdot x^2 = x^{-3+(-4)+2} = x^{-5} = \dfrac{1}{x^5}$

(c) $(4^{-2})^{-5} = 4^{(-2)(-5)} = 4^{10}$

(d) $(x^{-4})^6 = x^{(-4)6} = x^{-24} = \dfrac{1}{x^{24}}$

(e) $\dfrac{x^{-4}y^2}{x^2 y^{-5}} = \dfrac{x^{-4}}{x^2} \cdot \dfrac{y^2}{y^{-5}}$

$= x^{-4-2} \cdot y^{2-(-5)}$

$= x^{-6}y^7$

$= \dfrac{y^7}{x^6}$

(f) $(2^3 x^{-2})^{-2} = (2^3)^{-2} \cdot (x^{-2})^{-2}$

$= 2^{-6}x^4$

$= \dfrac{x^4}{2^6}$ or $\dfrac{x^4}{64}$

(g) $\left(\dfrac{3x^2}{y}\right)^2 \left(\dfrac{4x^3}{y^{-2}}\right)^{-1} = \dfrac{3^2 (x^2)^2}{y^2} \cdot \dfrac{y^{-2}}{4x^3}$ Combination of rules

$= \dfrac{9x^4}{y^2} \cdot \dfrac{y^{-2}}{4x^3}$ Power rule

$= \dfrac{9}{4}x^{4-3}y^{-2-2} = \dfrac{9x}{4y^4}$ Quotient rule; $a^{-n} = \left(\dfrac{1}{a}\right)^n$

NOTE
There is often more than one way to simplify expressions like those in Example 8. For instance, we could simplify Example 8(e) as follows.

$\dfrac{x^{-4}y^2}{x^2 y^{-5}} = \dfrac{y^5 y^2}{x^4 x^2} = \dfrac{y^7}{x^6}$ Use $\dfrac{a^{-n}}{b^{-m}} = \dfrac{b^m}{a^n}$; product rule

◄◄ **Work Problem 8 at the Side.**

OBJECTIVE 6 Use the rules for exponents with scientific notation. The number of one-celled organisms that will sustain a whale for a few hours is 400,000,000,000,000, and the shortest wavelength of visible light is approximately .0000004 m. It is often simpler to write these numbers using *scientific notation.*

In scientific notation, a number is written with the decimal point after the first nonzero digit and multiplied by a power of 10.

Scientific Notation

A number is written in **scientific notation** when it is expressed in the form

$$a \times 10^n$$

where $1 \le |a| < 10$, and n is an integer.

ANSWERS

8. (a) $\dfrac{1}{5^2}$ or $\dfrac{1}{25}$ (b) $\dfrac{1}{x^2}$ (c) 5^6

(d) $\dfrac{1}{y^{14}}$ (e) $\dfrac{b^7}{a^7}$ (f) $\dfrac{k^4}{3^2}$ or $\dfrac{k^4}{9}$ (g) $\dfrac{y}{x^5}$

For example, in scientific notation,

$$8000 = 8 \times 1000 = 8 \times 10^3.$$

The following numbers are not in scientific notation.

$.230 \times 10^4$ 46.5×10^{-3}

.230 is less than 1.   46.5 is greater than 10.

To write a number in scientific notation, use the following steps. (If the number is negative, ignore the negative sign, go through these steps, and then attach a negative sign to the result.)

Converting to Scientific Notation

Step 1   **Position the decimal point.** Place a caret, ^, to the right of the first nonzero digit, where the decimal point will be placed.

Step 2   **Determine the numeral for the exponent.** Count the number of digits from the decimal point to the caret. This number gives the absolute value of the exponent on 10.

Step 3   **Determine the sign for the exponent.** Decide whether multiplying by 10^n should make the result of Step 1 larger or smaller. The exponent should be positive to make the result larger; it should be negative to make the result smaller.

It is helpful to remember that for $n \geq 1$, $10^{-n} < 1$ and $10^n \geq 10$.

EXAMPLE 9   **Writing Numbers in Scientific Notation**

Write each number in scientific notation.

(a) 820,000

Place a caret to the right of the 8 (the first nonzero digit) to mark the new location of the decimal point.

$$8_\wedge 20,000$$

Count from the decimal point, which is understood to be after the last 0, to the caret.

$$8.20,000. \leftarrow \text{Decimal point}$$

Count 5 places.

Since the number 8.2 is to be made larger, the exponent on 10 is positive.

$$820,000 = 8.2 \times 10^5$$

(b) .0000072

Count from left to right.

$$.000007.2$$

6 places

Since the number 7.2 is to be made smaller, the exponent on 10 is negative.

$$.0000072 = 7.2 \times 10^{-6}$$

Work Problem 9 at the Side. ▶▶▶

9 Write each number in scientific notation.

(a) 400,000

(b) 29,800,000

(c) −6083

(d) .00172

(e) .0000000503

ANSWERS

9. **(a)** 4×10^5 **(b)** 2.98×10^7
    (c) -6.083×10^3 **(d)** 1.72×10^{-3}
    (e) 5.03×10^{-8}

⑩ Write each number in standard notation.

(a) 4.98×10^5

(b) 6.8×10^{-7}

(c) -5.372×10^0

Converting from Scientific Notation

Multiplying a number by a positive power of 10 makes the number larger, so move the decimal point to the right n places if n is positive in 10^n.

Multiplying by a negative power of 10 makes a number smaller, so move the decimal point to the left $|n|$ places if n is negative.

If n is 0, leave the decimal point where it is.

EXAMPLE 10 Converting from Scientific Notation to Standard Notation

Write each number in standard notation.

(a) 6.93×10^7

$$6.9300000. \qquad \text{Attach 0s as necessary.}$$
7 places

We moved the decimal point 7 places to the right. (We had to attach five 0s.)

$$6.93 \times 10^7 = 69,300,000$$

(b) 4.7×10^{-6}

$$.000004.7$$
6 places

We moved the decimal point 6 places to the left.

$$4.7 \times 10^{-6} = .0000047$$

(c) $-1.083 \times 10^0 = -1.083 \times 1 = -1.083$

◀◀◀ Work Problem 10 at the Side.

When problems require operations with numbers that are very large and/or very small, and a calculator is not available, we can write the numbers in scientific notation and perform the calculations using the rules for exponents.

EXAMPLE 11 Using Scientific Notation in Computation

Evaluate $\dfrac{1,920,000 \times .0015}{.000032 \times 45,000}$.

$$\frac{1,920,000 \times .0015}{.000032 \times 45,000} = \frac{1.92 \times 10^6 \times 1.5 \times 10^{-3}}{3.2 \times 10^{-5} \times 4.5 \times 10^4} \qquad \text{Express all numbers in scientific notation.}$$

$$= \frac{1.92 \times 1.5 \times 10^6 \times 10^{-3}}{3.2 \times 4.5 \times 10^{-5} \times 10^4} \qquad \text{Commutative property}$$

$$= \frac{1.92 \times 1.5 \times 10^3}{3.2 \times 4.5 \times 10^{-1}} \qquad \text{Product rule}$$

$$= \frac{1.92 \times 1.5}{3.2 \times 4.5} \times 10^4 \qquad \text{Quotient rule}$$

$$= .2 \times 10^4 \qquad \text{Simplify.}$$

$$= (2 \times 10^{-1}) \times 10^4$$

$$= 2 \times 10^3 \quad \text{or} \quad 2000$$

◀◀◀ Work Problem 11 at the Side.

⑪ Evaluate

$$\frac{200,000 \times .0003}{.06 \times 4,000,000}.$$

Answers
10. **(a)** 498,000 **(b)** .00000068 **(c)** -5.372
11. 2.5×10^{-4} or .00025

Calculator Tip To enter numbers in scientific notation, you can use the (EE) or (EXP) key on a scientific calculator. For instance, to work Example 11 using a popular model calculator with an (EE) key, enter the following symbols.

1.92 (EE) 6 × 1.5 (EE) 3 (+/-) ÷ (() 3.2 (EE) 5 (+/-) × 4.5 (EE) 4 ()) =

The (EXP) key is used in exactly the same way. Notice that the negative exponent −3 is entered by pressing 3, then (+/-). (*Keystrokes vary among different models of calculators,* so you should refer to your owner's manual if this sequence does not apply to your particular model.)

12 The distance to the sun is 9.3×10^7 mi. How long would it take a rocket, traveling at 3.2×10^3 mph, to reach the sun? (*Hint:* $t = \frac{d}{r}$.)

EXAMPLE 12 **Using Scientific Notation to Solve Problems**

In 1990, the national health care expenditure was $695.6 billion. By 2000, this figure had risen by a factor of 1.9; that is, it almost doubled in only 10 years. (*Source:* U.S. Centers for Medicare & Medicaid Services.)

(a) Write the 1990 health care expenditure using scientific notation.

$$695.6 \text{ billion} = 695.6 \times 10^9$$
$$= (6.956 \times 10^2) \times 10^9$$
$$= 6.956 \times 10^{11} \qquad \text{Product rule}$$

In 1990, the expenditure was $\$6.956 \times 10^{11}$.

(b) What was the expenditure in 2000?
Multiply the result in part (a) by 1.9.

$$(6.956 \times 10^{11}) \times 1.9 = (1.9 \times 6.956) \times 10^{11} \qquad \text{Commutative and associative properties}$$
$$= 13.216 \times 10^{11} \qquad \text{Round to three decimal places.}$$
$$= 1.3216 \times 10^{12} \qquad \text{Scientific notation}$$

The 2000 expenditure was about $1,321,600,000,000 (over $1 trillion).

Work Problem 12 at the Side. ▶▶▶

ANSWERS
12. approximately 2.9×10^4 hr

Real-Data Applications

Earthquake Intensities Measured by the Richter Scale

Charles F. Richter devised a scale in 1935 to compare the intensities, or relative power, of earthquakes. The **intensity** of an earthquake is measured relative to the intensity of a standard **zero-level** earthquake of intensity I_0. The relationship is equivalent to $I = I_0 \times 10^R$, where R is the **Richter scale** measure. For example, if an earthquake has magnitude 5.0 on the Richter scale, then its intensity is calculated as $I = I_0 \times 10^{5.0} = I_0 \times 100,000$, which is 100,000 times as intense as a zero-level earthquake. The following diagram illustrates the intensities of earthquakes and their Richter scale magnitudes.

Intensity	I_0	$I_0 \times 10^1$	$I_0 \times 10^2$	$I_0 \times 10^3$	$I_0 \times 10^4$	$I_0 \times 10^5$	$I_0 \times 10^6$	$I_0 \times 10^7$	$I_0 \times 10^8$
Richter Scale	0	1	2	3	4	5	6	7	8

To compare two earthquakes to each other, a ratio of the intensities is calculated. For example, to compare an earthquake that measures 8.0 on the Richter scale to one that measures 5.0, simply find the ratio of the intensities:

$$\frac{\text{intensity } 8.0}{\text{intensity } 5.0} = \frac{I_0 \times 10^{8.0}}{I_0 \times 10^{5.0}} = \frac{10^8}{10^5} = 10^{8-5} = 10^3 = 1000.$$

Therefore an earthquake that measures 8.0 on the Richter Scale is 1000 times as intense as one that measures 5.0.

For Group Discussion

The table gives Richter scale measurements for several earthquakes.

	Earthquake	Richter Scale Measurement
1960	Concepción, Chile	9.5
1906	San Francisco, California	8.3
1939	Erzincan, Turkey	8.0
1998	Sumatra, Indonesia	7.0
1998	Adana, Turkey	6.3

Source: World Almanac and Book of Facts, 2004.

1. Compare the intensity of the 1939 Erzincan earthquake to the 1998 Sumatra earthquake.

2. Compare the intensity of the 1998 Adana earthquake to the 1906 San Francisco earthquake.

3. Compare the intensity of the 1939 Erzincan earthquake to the 1998 Adana earthquake.

4. Suppose an earthquake measures 7.2 on the Richter scale. How would the intensity of a second earthquake compare if its Richter scale measure differed by $+3.0$? By -1.0?

6.1 Exercises

FOR EXTRA HELP

Tutor Center Addison-Wesley Math Tutor Center   MathXL   Digital Video Tutor CD 3 Videotape 3   Student's Solutions Manual   MyMathLab MyMathLab   Interactmath.com

Decide whether each expression has been simplified correctly. If not, correct it.

1. $(ab)^2 = ab^2$

2. $(5x)^3 = 5^3 x^3$

3. $\left(\dfrac{4}{a}\right)^3 = \dfrac{4^3}{a}$ $(a \neq 0)$

4. $y^2 \cdot y^6 = y^{12}$

5. $x^3 \cdot x^4 = x^7$

6. $xy^0 = 0$ $(y \neq 0)$

Apply the product rule for exponents, if possible, in each case. See Example 1.

7. $13^4 \cdot 13^8$

8. $9^6 \cdot 9^4$

9. $x^3 \cdot x^5 \cdot x^9$

10. $y^4 \cdot y^5 \cdot y^6$

11. $(-3w^5)(9w^3)$

12. $(-5x^2)(3x^4)$

13. $(2x^2y^5)(9xy^3)$

14. $(8s^4t)(3s^3t^5)$

15. $r^2 \cdot s^4$

16. $p^3 \cdot q^2$

In Exercises 17 and 18, match the expression in Column I with its equivalent expression in Column II. Choices may be used once, more than once, or not at all.* See Example 2.

	I	II		I	II
17.	(a) 9^0	**A.** 0	**18.**	(a) $2x^0$	**A.** 0
	(b) -9^0	**B.** 1		(b) $-2x^0$	**B.** 1
	(c) $(-9)^0$	**C.** -1		(c) $(2x)^0$	**C.** -1
	(d) $-(-9)^0$	**D.** 9		(d) $(-2x)^0$	**D.** 2
		E. -9			**E.** -2

Evaluate. Assume all variables represent nonzero numbers. See Example 2.

19. 25^0

20. 14^0

21. -7^0

22. -10^0

23. $(-15)^0$

24. $(-20)^0$

25. $-4^0 - m^0$

26. $-8^0 - k^0$

*The authors thank Mitchel Levy of Broward Community College for his suggestions for Exercises 17, 18, 27, 28, 53, and 54.

In Exercises 27 and 28, match the expression in Column I with its equivalent expression in Column II. Choices may be used once, more than once, or not at all. See Example 3.

	I		II			I		II

27. (a) 4^{-2} **A.** 16 **28. (a)** 5^{-3} **A.** 125

   (b) -4^{-2} **B.** $\dfrac{1}{16}$ **(b)** -5^{-3} **B.** -125

   (c) $(-4)^{-2}$ **C.** -16 **(c)** $(-5)^{-3}$ **C.** $\dfrac{1}{125}$

   (d) $-(-4)^{-2}$ **D.** $-\dfrac{1}{16}$ **(d)** $-(-5)^{-3}$ **D.** $-\dfrac{1}{125}$

Write each expression with only positive exponents. Assume all variables represent nonzero numbers. In Exercises 41–44, simplify each expression. See Example 3.

29. 5^{-4} **30.** 7^{-2} **31.** 8^{-1} **32.** 12^{-1}

33. $(4x)^{-2}$ **34.** $(5t)^{-3}$ **35.** $4x^{-2}$ **36.** $5t^{-3}$

37. $-a^{-3}$ **38.** $-b^{-4}$ **39.** $(-a)^{-4}$ **40.** $(-b)^{-6}$

41. $5^{-1} + 6^{-1}$ **42.** $2^{-1} + 8^{-1}$ **43.** $8^{-1} - 3^{-1}$ **44.** $6^{-1} - 4^{-1}$

Evaluate each expression. See Examples 4 and 7.

45. $\dfrac{1}{4^{-2}}$ **46.** $\dfrac{1}{3^{-3}}$ **47.** $\dfrac{2^{-2}}{3^{-3}}$ **48.** $\dfrac{3^{-3}}{2^{-2}}$

49. $\left(\dfrac{2}{3}\right)^{-3}$ **50.** $\left(\dfrac{3}{2}\right)^{-3}$ **51.** $\left(\dfrac{4}{5}\right)^{-2}$ **52.** $\left(\dfrac{5}{4}\right)^{-2}$

In Exercises 53 and 54, match the expression in Column I with its equivalent expression in Column II. Choices may be used once, more than once, or not at all.

	I		II			I		II

53. (a) $\left(\dfrac{1}{3}\right)^{-1}$ **A.** $\dfrac{1}{3}$ **54. (a)** $\left(\dfrac{2}{5}\right)^{-2}$ **A.** $\dfrac{25}{4}$

   (b) $\left(-\dfrac{1}{3}\right)^{-1}$ **B.** 3 **(b)** $\left(-\dfrac{2}{5}\right)^{-2}$ **B.** $-\dfrac{25}{4}$

   (c) $-\left(\dfrac{1}{3}\right)^{-1}$ **C.** $-\dfrac{1}{3}$ **(c)** $-\left(\dfrac{2}{5}\right)^{-2}$ **C.** $\dfrac{4}{25}$

   (d) $-\left(-\dfrac{1}{3}\right)^{-1}$ **D.** -3 **(d)** $-\left(-\dfrac{2}{5}\right)^{-2}$ **D.** $-\dfrac{4}{25}$

Apply the quotient rule for exponents, if applicable, and write each result using only positive exponents. Assume all variables represent nonzero numbers. See Example 5.

55. $\dfrac{4^8}{4^6}$

56. $\dfrac{5^9}{5^7}$

57. $\dfrac{x^{12}}{x^8}$

58. $\dfrac{y^{14}}{y^{10}}$

59. $\dfrac{r^7}{r^{10}}$

60. $\dfrac{y^8}{y^{12}}$

61. $\dfrac{6^4}{6^{-2}}$

62. $\dfrac{7^5}{7^{-3}}$

63. $\dfrac{6^{-3}}{6^7}$

64. $\dfrac{5^{-4}}{5^2}$

65. $\dfrac{7}{7^{-1}}$

66. $\dfrac{8}{8^{-1}}$

67. $\dfrac{r^{-3}}{r^{-6}}$

68. $\dfrac{s^{-4}}{s^{-8}}$

69. $\dfrac{x^3}{y^2}$

70. $\dfrac{y^5}{t^3}$

Use one or more power rules to simplify each expression. Assume all variables represent nonzero numbers. See Example 6.

71. $(x^3)^6$

72. $(y^5)^4$

73. $\left(\dfrac{3}{5}\right)^3$

74. $\left(\dfrac{4}{3}\right)^2$

75. $(4t)^3$

76. $(5t)^4$

77. $(-6x^2)^3$

78. $(-2x^5)^5$

79. $\left(\dfrac{-4m^2}{t}\right)^3$

80. $\left(\dfrac{-5n^4}{r^2}\right)^3$

Simplify each expression so that no negative exponents appear in the final result. Assume all variables represent nonzero numbers. See Example 8.

81. $3^5 \cdot 3^{-6}$

82. $4^4 \cdot 4^{-6}$

83. $a^{-3}a^2a^{-4}$

84. $k^{-5}k^{-3}k^4$

85. $(k^2)^{-3}k^4$

86. $(x^3)^{-4}x^5$

87. $-4r^{-2}(r^4)^2$

88. $-2m^{-1}(m^3)^2$

89. $(5a^{-1})^4(a^2)^{-3}$

90. $(3p^{-4})^2(p^3)^{-1}$

91. $(z^{-4}x^3)^{-1}$

92. $(y^{-2}z^4)^{-3}$

93. $\dfrac{(p^{-2})^3}{5p^4}$

94. $\dfrac{(m^4)^{-1}}{9m^3}$

95. $\dfrac{4a^5(a^{-1})^3}{(a^{-2})^{-2}}$

96. $\dfrac{12k^{-2}(k^{-3})^{-4}}{6k^5}$

97. $\dfrac{(-y^{-4})^2}{6(y^{-5})^{-1}}$

98. $\dfrac{2(-m^{-1})^{-4}}{9(m^{-3})^2}$

99. $\dfrac{(2k)^2m^{-5}}{(km)^{-3}}$

100. $\dfrac{(3rs)^{-2}}{3^2r^2s^{-4}}$

101. $\left(\dfrac{3k^{-2}}{k^4}\right)^{-1}\cdot\dfrac{2}{k}$

102. $\left(\dfrac{7m^{-2}}{m^{-3}}\right)^{-2}\cdot\dfrac{m^3}{4}$

103. $\left(\dfrac{2p}{q^2}\right)^3\left(\dfrac{3p^4}{q^{-4}}\right)^{-1}$

104. $\left(\dfrac{5z^3}{2a^2}\right)^{-3}\left(\dfrac{8a^{-1}}{15z^{-2}}\right)^{-3}$

Write each number in scientific notation. See Example 9.

105. 530

106. 1600

107. .830

108. .0072

109. .00000692

110. .875

111. $-38,500$

112. $-976,000,000$

Write each number in standard notation. See Example 10.

113. 7.2×10^4

114. 8.91×10^2

115. 2.54×10^{-3}

116. 5.42×10^{-4}

117. -6×10^4

118. -9×10^3

119. 1.2×10^{-5}

120. 2.7×10^{-6}

Use the rules for exponents to find each value. See Example 11.

121. $\dfrac{3 \times 10^{-2}}{12 \times 10^3}$

122. $\dfrac{5 \times 10^{-3}}{25 \times 10^2}$

123. $\dfrac{.05 \times 1600}{.0004}$

124. $\dfrac{.003 \times 40,000}{.00012}$

Solve each problem. See Example 12.

125. The U.S. budget first passed **$1,000,000,000** in 1917. Seventy years later in 1987 it exceeded **$1,000,000,000,000** for the first time. President George W. Bush's budget request for fiscal 2003 was **$2,128,000,000,000.** If stacked in dollar bills, this amount would stretch **144,419** mi, almost two-thirds of the distance to the moon. Write the four boldfaced numbers in scientific notation. (*Source: The Gazette*, February 5, 2002.)

126. In 1970, Wal-Mart had **1500** employees. In 1997, Wal-Mart became the largest private employer in the United States, with **680,000** employees. In 1999, Wal-Mart became the largest private employer in the world, with **1,100,000** employees. By 2007, the company is expected to have **2,200,000** employees. Write these four numbers in scientific notation. (*Source:* Wal-Mart.)

127. In 2000, the population of the United States was 281.4 million. (*Source:* U.S. Bureau of the Census.)

  (a) Write the 2000 population using scientific notation.

  (b) Write $1 trillion, that is, $1,000,000,000,000, using scientific notation.

  (c) Using your answers from parts (a) and (b), calculate how much each person in the United States in the year 2000 would have had to contribute in order to make someone a trillionaire. Write this amount in standard notation to the nearest dollar.

128. On October 28, 1998, IBM announced a computer capable of 3.9×10^8 operations per second. This was 15,000 times faster than the normal desktop computer at that time. What was the number of operations that the normal desktop could do? (*Source:* IBM.)

129. In the early years of the Powerball Lottery, a player would choose five numbers from 1 through 49 and one number from 1 through 42. It can be shown that there are about 8.009×10^7 different ways to do this. Suppose that a group of 2000 persons decided to purchase tickets for all these numbers and each ticket cost $1.00. How much should each person have expected to pay? (*Source:* www.powerball.com)

130. The speed of light is approximately 3×10^{10} cm per sec. How long does it take light to travel 9×10^{12} cm?

131. The average distance from Earth to the sun is 9.3×10^7 mi. How long would it take a rocket, traveling at 2.9×10^3 mph, to reach the sun?

132. A *light-year* is the distance that light travels in one year. Find the number of miles in a light-year if light travels 1.86×10^5 mi per sec.

133. (a) The planet Mercury has an average distance from the sun of 3.6×10^7 mi, while the average distance of Venus from the sun is 6.7×10^7 mi. How long would it take a spacecraft traveling at 1.55×10^3 mph to travel from Venus to Mercury? (Give your answer in hours, in standard notation.)

(b) Use the information from part (a) to find the number of days it would take the spacecraft to travel from Venus to Mercury. Round your answer to the nearest whole number of days.

134. When the distance between the centers of the moon and Earth is 4.60×10^8 m, an object on the line joining the centers of the moon and Earth exerts the same gravitational force on each when it is 4.14×10^8 m from the center of Earth. How far is the object from the center of the moon at that point?

135. In some cases, $-a^n$ and $(-a)^n$ do give the same result for $a > 0$. Using $a = 2$ and $n = 2, 3, 4,$ and 5, draw a conclusion as to when they are equal and when they are opposites.

136. Your friend evaluated $4^5 \cdot 4^2$ as 16^7. Explain to him why his answer is incorrect.

137. In your own words, describe how to rewrite a fraction raised to a negative power as a fraction raised to a positive power.

138. Explain in your own words how to raise a power to a power.

6.2 Adding and Subtracting Polynomials

OBJECTIVE 1 Know the basic definitions for polynomials. Just as whole numbers are the basis of arithmetic, *polynomials* are fundamental in algebra. To understand polynomials, we must review several words from **Section 1.4.** A **term** is a number, a variable, or the product or quotient of a number and one or more variables raised to powers. Examples of terms include

$$4x, \quad \frac{1}{2}m^5 \left(\text{or } \frac{m^5}{2} \right), \quad -7z^9, \quad 6x^2z, \quad \frac{5}{3x^2}, \quad \text{and} \quad 9. \quad \text{Terms}$$

The number in the product is called the **numerical coefficient,** or just the **coefficient.*** In the term $8x^3$, the coefficient is **8.** In the term $-4p^5$, it is -4. The coefficient of the term k is understood to be 1. The coefficient of $-r$ is -1. In the term $\frac{x}{3}$, the coefficient is $\frac{1}{3}$ since $\frac{x}{3} = \frac{1x}{3} = \frac{1}{3}x$.

> **Work Problem 1 at the Side.**

Recall that any combination of variables or constants (numerical values) joined by the basic operations of addition, subtraction, multiplication, and division (except by 0), or raising to powers or taking roots is called an **algebraic expression.** The simplest kind of algebraic expression is a *polynomial*.

Polynomial

A **polynomial** is a term or a finite sum of terms in which all variables have whole number exponents and no variables appear in denominators or under radicals.

Examples of polynomials include

$$3x - 5, \quad 4m^3 - 5m^2p + 8, \quad \text{and} \quad -5t^2s^3. \quad \text{Polynomials}$$

Even though the expression $3x - 5$ involves subtraction, it is a sum of terms since it could be written as $3x + (-5)$.

Some examples of expressions that are not polynomials are

$$x^{-1} + 3x^{-2}, \quad \sqrt{9 - x}, \quad \text{and} \quad \frac{1}{x}. \quad \text{Not polynomials}$$

The first of these is not a polynomial because it has negative integer exponents, the second because it involves a variable under a radical, and the third because it contains a variable in the denominator.

Most of the polynomials used in this book contain only one variable. A polynomial containing only the variable x is called a **polynomial in x.** A polynomial in one variable is written in **descending powers** of the variable if the exponents on the variable decrease from left to right. For example,

$$x^5 - 6x^2 + 12x - 5$$

is a polynomial in descending powers of x. The term -5 in this polynomial can be thought of as $-5x^0$, since $-5x^0 = -5(1) = -5$.

> **Work Problem 2 at the Side.**

---

* More generally, any factor in a term is the coefficient of the product of the remaining factors. For example, $3x^2$ is the coefficient of y in the term $3x^2y$, and $3y$ is the coefficient of x^2 in $3x^2y$.

OBJECTIVES

1 Know the basic definitions for polynomials.

2 Find the degree of a polynomial.

3 Add and subtract polynomials.

1 Identify each coefficient.

(a) $-9m^5$

(b) $12y^2x$

(c) x

(d) $-y$

(e) $\dfrac{z}{4}$

2 Write each polynomial in descending powers.

(a) $-4 + 9y + y^3$

(b) $-3z^4 + 2z^3 + z^5 - 6z$

(c) $-12m^{10} + 8m^9 + 10m^{12}$

ANSWERS

1. (a) -9 (b) 12 (c) 1 (d) -1 (e) $\dfrac{1}{4}$
2. (a) $y^3 + 9y - 4$ (b) $z^5 - 3z^4 + 2z^3 - 6z$
   (c) $10m^{12} - 12m^{10} + 8m^9$

❸ Identify each polynomial as a *trinomial, binomial, monomial,* or *none of these.*

(a) $12m^4 - 6m^2$

(b) $-6y^3 + 2y^2 - 8y$

(c) $3a^5$

(d) $-2k^{10} + 2k^9 - 8k^5 + 2k$

❹ Give the degree of each polynomial.

(a) $9y^4 + 8y^3 - 6$

(b) $-12m^7 + 11m^3 + m^9$

(c) $-2k$

(d) 10

(e) $3mn^2 + 2m^3n$

Some polynomials with a specific number of terms are so common that they are given special names. A polynomial with exactly three terms is a **trinomial,** and a polynomial with exactly two terms is a **binomial.** A single-term polynomial is a **monomial.** The table that follows gives examples.

Type of Polynomial	Examples
Monomial	$5x, \quad 7m^9, \quad -8, \quad x^2y^2$
Binomial	$3x^2 - 6, \quad 11y + 8, \quad 5a^2b + 3a$
Trinomial	$y^2 + 11y + 6, \quad 8p^3 - 7p + 2m, \quad -3 + 2k^5 + 9z^4$
None of these	$p^3 - 5p^2 + 2p - 5, \quad -9z^3 + 5c^3 + 2m^5 + 11r^2 - 7r$

◀◀◀ **Work Problem 3 at the Side.**

OBJECTIVE 2 Find the degree of a polynomial. The **degree of a term** with one variable is the exponent on the variable. For example, the degree of $2x^3$ is **3**, the degree of $-x^4$ is **4**, and the degree of $17x$ (that is, $17x^1$) is **1**. The degree of a term in more than one variable is defined to be the sum of the exponents on the variables. For example, the degree of $5x^3y^7$ is **10**, because $3 + 7 = 10$.

The greatest degree of any term in a polynomial is called the **degree of the polynomial.** In most cases, we will be interested in finding the degree of a polynomial in one variable. For example, $4x^3 - 2x^2 - 3x + 7$ has degree **3**, because the greatest degree of any term is 3 (the degree of $4x^3$).

The table shows several polynomials and their degrees.

Polynomial	Degree
$9x^2 - 5x + 8$	2
$17m^9 + 18m^{14} - 9m^3$	14
$5x$	1, because $5x = 5x^1$
-2	0, because $-2 = -2x^0$ (Any nonzero constant has degree 0.)
$5a^2b^5$	7, because $2 + 5 = 7$
$13xy^4 + x^3y^9 + 7xy$	12, because the degrees of the terms are 5, 12, and 2; 12 is the greatest.

NOTE
The number 0 has no degree, since 0 times a variable to any power is 0.

◀◀◀ **Work Problem 4 at the Side.**

OBJECTIVE 3 Add and subtract polynomials. We use the distributive property to simplify polynomials by combining terms. For example,

$$x^3 + 4x^2 + 5x^2 - 1 = x^3 + (4 + 5)x^2 - 1 \quad \text{Distributive property}$$
$$= x^3 + 9x^2 - 1.$$

On the other hand, the terms in the polynomial $4x + 5x^2$ cannot be combined. As these examples suggest, only terms containing exactly the same variables to the same powers may be combined. As mentioned in **Section 1.4,** such terms are called **like terms.**

ANSWERS
3. (a) binomial (b) trinomial
   (c) monomial (d) none of these
4. (a) 4 (b) 9 (c) 1 (d) 0 (e) 4

CAUTION
Remember that only like terms can be combined.

EXAMPLE 1 Combining Like Terms

Combine terms.

(a) $-5y^3 + 8y^3 - y^3 = (-5 + 8 - 1)y^3 = 2y^3$

(b) $6x + 5y - 9x + 2y = 6x - 9x + 5y + 2y$ Associative and commutative properties

$$= -3x + 7y$$ Combine like terms.

Since $-3x$ and $7y$ are unlike terms, no further simplification is possible.

(c) $5x^2y - 6xy^2 + 9x^2y + 13xy^2 = 5x^2y + 9x^2y - 6xy^2 + 13xy^2$

$$= 14x^2y + 7xy^2$$

Work Problem 5 at the Side. ▶▶▶

We use the following rule to add two polynomials.

Adding Polynomials
To add two polynomials, combine like terms.

Polynomials can be added horizontally or vertically.

EXAMPLE 2 Adding Polynomials

Add: $(3a^5 - 9a^3 + 4a^2) + (-8a^5 + 8a^3 + 2)$.

Use the commutative and associative properties to rearrange the polynomials so that like terms are together. Then use the distributive property to combine like terms.

$$(3a^5 - 9a^3 + 4a^2) + (-8a^5 + 8a^3 + 2)$$
$$= 3a^5 - 8a^5 - 9a^3 + 8a^3 + 4a^2 + 2$$
$$= -5a^5 - a^3 + 4a^2 + 2$$ Combine like terms.

Add these same two polynomials vertically by placing like terms in columns.

$$\begin{array}{r} 3a^5 - 9a^3 + 4a^2 \\ -8a^5 + 8a^3 + 2 \\ \hline -5a^5 - a^3 + 4a^2 + 2 \end{array}$$

Work Problem 6 at the Side. ▶▶▶

In **Section 1.2,** we defined subtraction of real numbers as

$$a - b = a + (-b).$$

That is, we add the first number (minuend) and the negative (or opposite) of the second (subtrahend). We can give a similar definition for subtraction of polynomials by defining the **negative of a polynomial** as that polynomial with the sign of every coefficient changed.

---

5 Combine terms.

(a) $11x + 12x - 7x - 3x$

(b) $11p^5 + 4p^5 - 6p^3 + 8p^3$

(c) $2y^2z^4 + 3y^4 + 5y^4 - 9y^4z^2$

6 Add, using both the horizontal and vertical methods.

(a) $(12y^2 - 7y + 9)$
    $+ (-4y^2 - 11y + 5)$

(b) $\begin{array}{r} -6r^5 + 2r^3 - r^2 \\ 8r^5 - 2r^3 + 5r^2 \\ \hline \end{array}$

ANSWERS
5. (a) $13x$ (b) $15p^5 + 2p^3$
   (c) $2y^2z^4 + 8y^4 - 9y^4z^2$
6. (a) $8y^2 - 18y + 14$ (b) $2r^5 + 4r^2$

7 Subtract, using both the horizontal and vertical methods.

(a) $(6y^3 - 9y^2 + 8)$
$\quad - (2y^3 + y^2 + 5)$

Subtracting Polynomials

To subtract two polynomials, add the first polynomial and the negative of the *second* polynomial.

EXAMPLE 3 **Subtracting Polynomials**

Subtract: $(-6m^2 - 8m + 5) - (-5m^2 + 7m - 8)$.

Change every sign in the second polynomial and add.

$(-6m^2 - 8m + 5) - (-5m^2 + 7m - 8)$

$\quad = -6m^2 - 8m + 5 + 5m^2 - 7m + 8 \qquad$ Definition of subtraction

$\quad = -6m^2 + 5m^2 - 8m - 7m + 5 + 8 \qquad$ Rearrange terms.

$\quad = -m^2 - 15m + 13 \qquad$ Combine like terms.

Check by adding the sum, $-m^2 - 15m + 13$, to the second polynomial. The result should be the first polynomial.

To subtract these two polynomials vertically, write the first polynomial above the second, lining up like terms in columns.

$$-6m^2 - 8m + 5$$
$$-5m^2 + 7m - 8$$

Change all the signs in the second polynomial, and add.

$$\begin{array}{r} -6m^2 - \ 8m + \ 5 \\ + 5m^2 - \ 7m + \ 8 \\ \hline -m^2 - 15m + 13 \end{array}$$

Change all signs.

Add in columns.

◀◀◀ Work Problem 7 at the Side.

(b) $\quad 6y^3 - 2y^2 + \ 5y$
$\quad -2y^3 + 8y^2 - 11y$

ANSWERS
7. (a) $4y^3 - 10y^2 + 3$
$\quad$ **(b)** $8y^3 - 10y^2 + 16y$

6.2 Exercises

FOR EXTRA HELP   Addison-Wesley Math Tutor Center   MathXL   Digital Video Tutor CD 3 Videotape 3   Student's Solutions Manual   MyMathLab   Interactmath.com

*We defined a polynomial written in descending powers in the text. Sometimes we write a polynomial in **ascending powers,** with the degree of the terms increasing from left to right. Decide whether each polynomial is written in* descending *powers,* ascending *powers, or* neither.

1. $2x^3 + x - 3x^2$

2. $3x^5 + x^4 - 2x^3 + x$

3. $4p^3 - 8p^5 + p^7$

4. $q^2 + 3q^4 - 2q + 1$

5. $-m^3 + 5m^2 + 3m + 10$

6. $4 - x + 3x^2$

Give the coefficient and the degree of each term.

7. $7z$

8. $3r$

9. $-15p^2$

10. $-27k^3$

11. x^4

12. y^6

13. $-mn^5$

14. $-a^5b$

Identify each polynomial as a monomial, binomial, trinomial, *or* none of these. *Give the degree of each.*

Polynomial	Type	Degree	Polynomial	Type	Degree
15. 24			**16.** 5		
17. $7m - 21$			**18.** $-x^2 + 3x^5$		
19. $2r^3 + 3r^2 + 5r$			**20.** $5z^2 - 5z + 7$		
21. $-6p^4q - 3p^3q^2 + 2pq^3 - q^4$			**22.** $8s^3t - 3s^2t^2 + 2st^3 + 9$		

Combine terms. See Example 1.

23. $5z^4 + 3z^4$

24. $8r^5 - 2r^5$

25. $-m^3 + 2m^3 + 6m^3$

26. $3p^4 + 5p^4 - 2p^4$

27. $x + x + x + x + x$

28. $z - z - z + z$

29. $m^4 - 3m^2 + m$

30. $5a^5 + 2a^4 - 9a^3$

31. $y^2 + 7y - 4y^2$

32. $2c^2 - 4 + 8 - c^2$

33. $2k + 3k^2 + 5k^2 - 7$

34. $4x^2 + 2x - 6x^2 - 6$

35. $n^4 - 2n^3 + n^2 - 3n^4 + n^3$

36. $2q^3 + 3q^2 - 4q - q^3 + 5q^2$

Add or subtract as indicated. See Examples 2 and 3.

37. Add.
$$-12p^2 + 4p - 1$$
$$\underline{3p^2 + 7p - 8}$$

38. Add.
$$-6y^3 + 8y + 5$$
$$\underline{9y^3 + 4y - 6}$$

39. Subtract.
$$12a + 15$$
$$\underline{7a - 3}$$

40. Subtract.
$$-3b + 6$$
$$\underline{-2b + 8}$$

41. Subtract.
$$6m^2 - 11m + 5$$
$$\underline{-8m^2 + 2m - 1}$$

42. Subtract.
$$-4z^2 + 2z - 1$$
$$\underline{3z^2 - 5z + 2}$$

43. Add.
$$12z^2 - 11z + 8$$
$$5z^2 + 16z - 2$$
$$\underline{-4z^2 + 5z - 9}$$

44. Add.
$$-6m^3 + 2m^2 + 5m$$
$$8m^3 + 4m^2 - 6m$$
$$\underline{-3m^3 + 2m^2 - 7m}$$

45. Add.
$$6y^3 - 9y^2 \quad\;\; + 8$$
$$\underline{4y^3 + 2y^2 + 5y}$$

46. Add.
$$-7r^8 + 2r^6 - r^5$$
$$\underline{\qquad 3r^6 \qquad + 5}$$

47. Subtract.
$$-5a^4 \quad + 8a^2 - 9$$
$$\underline{6a^3 + a^2 + 2}$$

48. Subtract.
$$- 2m^3 + 8m^2$$
$$\underline{m^4 - m^3 \qquad + 2m}$$

49. $(3r + 8) - (2r - 5)$

50. $(2d + 7) - (3d - 1)$

51. $(5x^2 + 7x - 4) + (3x^2 - 6x + 2)$

52. $(4k^3 + k^2 + k) + (2k^3 - 4k^2 - 3k)$

53. $(2a^2 + 3a - 1) - (4a^2 + 5a + 6)$

54. $(q^4 - 2q^2 + 10) - (3q^4 + 5q^2 - 5)$

55. $(z^5 + 3z^2 + 2z) - (4z^5 + 2z^2 - 5z)$

56. $(5t^3 - 3t^2 + 2t) - (4t^3 + 2t^2 + 3t)$

6.3 Polynomial Functions

OBJECTIVE 1 Recognize and evaluate polynomial functions. In **Chapter 4** we studied linear (first-degree polynomial) functions, defined as $f(x) = mx + b$. Now we consider more general polynomial functions.

OBJECTIVES

1 Recognize and evaluate polynomial functions.

2 Use a polynomial function to model data.

3 Add and subtract polynomial functions.

4 Graph basic polynomial functions.

Polynomial Function

A **polynomial function of degree n** is defined by

$$f(x) = a_n x^n + a_{n-1} x^{n-1} + \cdots + a_1 x + a_0,$$

for real numbers $a_n, a_{n-1}, \ldots, a_1$, and a_0, where $a_n \neq 0$ and n is a whole number.

Another way of describing a polynomial function is to say that it is a function defined by a polynomial in one variable, consisting of one or more terms. It is usually written in descending powers of the variable, and its degree is the degree of the polynomial that defines it.

Suppose that the polynomial $3x^2 - 5x + 7$ defines function f. Then

$$f(x) = 3x^2 - 5x + 7.$$

If $x = -2$, then $f(x) = 3x^2 - 5x + 7$ takes on the value

$$\begin{aligned}
f(-2) &= 3(-2)^2 - 5(-2) + 7 \qquad \text{Let } x = -2. \\
&= 3 \cdot 4 + 10 + 7 \\
&= \mathbf{29}.
\end{aligned}$$

Thus, $f(-2) = \mathbf{29}$ and the ordered pair $(-2, 29)$ belongs to f.

1 Let $f(x) = -x^2 + 5x - 11$. Find each value.

(a) $f(1)$

(b) $f(-4)$

EXAMPLE 1 Evaluating Polynomial Functions

Let $f(x) = 4x^3 - x^2 + 5$. Find each value.

(a) $f(3)$

$$\begin{aligned}
f(x) &= 4x^3 - x^2 + 5 \\
f(3) &= 4 \cdot 3^3 - 3^2 + 5 \qquad \text{Substitute 3 for } x. \\
&= 4 \cdot 27 - 9 + 5 \qquad \text{Order of operations} \\
&= 108 - 9 + 5 \\
&= 104
\end{aligned}$$

(b) $\begin{aligned}[t] f(-4) &= 4 \cdot (-4)^3 - (-4)^2 + 5 \qquad \text{Let } x = -4; \text{ use parentheses.} \\
&= 4 \cdot (-64) - 16 + 5 \qquad \text{Be careful with signs.} \\
&= -267 \end{aligned}$

(c) $f(0)$

While f is the most common letter used to represent functions, recall that other letters such as g and h are also used. The capital letter P is often used for polynomial functions. Note that the function defined as $P(x) = 4x^3 - x^2 + 5$ yields the same ordered pairs as the function f in Example 1.

Work Problem 1 at the Side. ▶▶▶

ANSWERS
1. **(a)** -7 **(b)** -47 **(c)** -11

2 Use the function in Example 2 to approximate the number of households expected to pay at least one bill on-line each month in 2006.

OBJECTIVE 2 Use a polynomial function to model data. Polynomial functions can be used to approximate data. They are usually valid for small intervals, and they allow us to predict (with caution) what might happen for values just outside the intervals. These intervals are often periods of years, as shown in Example 2.

EXAMPLE 2 Using a Polynomial Model to Approximate Data

The number of U.S. households estimated to see and pay at least one bill on-line each month during the years 2000 through 2006 can be modeled by the polynomial function defined by

$$P(x) = .808x^2 + 2.625x + .502,$$

where $x = 0$ corresponds to the year 2000, $x = 1$ corresponds to 2001, and so on, and $P(x)$ is in millions. Use this function to approximate the number of households expected to pay at least one bill on-line each month in 2005.

Since $x = 5$ corresponds to 2005, we must find $P(5)$.

$$P(x) = .808x^2 + 2.625x + .502$$
$$P(5) = .808(5)^2 + 2.625(5) + .502 \qquad \text{Let } x = 5.$$
$$= 33.827 \qquad\qquad\qquad\qquad \text{Evaluate.}$$

Thus, in 2005 about 33.83 million households are expected to pay at least one bill on-line each month.

Work Problem 2 at the Side.

OBJECTIVE 3 Add and subtract polynomial functions. The operations of addition, subtraction, multiplication, and division are also defined for functions. For example, businesses use the equation "profit equals revenue minus cost," written using function notation as

$$P(x) = R(x) - C(x),$$

Profit function   Revenue function   Cost function

where x is the number of items produced and sold. Thus, the profit function is found by subtracting the cost function from the revenue function.

We define the following **operations on functions.**

Adding and Subtracting Functions

If $f(x)$ and $g(x)$ define functions, then

$$(f + g)(x) = f(x) + g(x) \qquad \text{Sum function}$$

and
$$(f - g)(x) = f(x) - g(x). \qquad \text{Difference function}$$

In each case, the domain of the new function is the intersection of the domains of $f(x)$ and $g(x)$.

ANSWERS
2. 45.34 million

EXAMPLE 3 **Adding and Subtracting Functions**

For the polynomial functions defined by

$$f(x) = x^2 - 3x + 7 \quad \text{and} \quad g(x) = -3x^2 - 7x + 7,$$

find **(a)** the sum and **(b)** the difference.

(a) $(f + g)(x) = f(x) + g(x)$ Use the definition.

$\qquad = (x^2 - 3x + 7) + (-3x^2 - 7x + 7)$ Substitute.

$\qquad = -2x^2 - 10x + 14$ Add the polynomials.

(b) $(f - g)(x) = f(x) - g(x)$ Use the definition.

$\qquad = (x^2 - 3x + 7) - (-3x^2 - 7x + 7)$ Substitute.

$\qquad = (x^2 - 3x + 7) + (3x^2 + 7x - 7)$ Change subtraction to addition.

$\qquad = 4x^2 + 4x$ Add.

▶ **Work Problem 3 at the Side.** ▷▷▷

EXAMPLE 4 **Adding and Subtracting Functions**

For the polynomial functions defined by

$$f(x) = 10x^2 - 2x \quad \text{and} \quad g(x) = 2x,$$

find each of the following.

(a) $(f + g)(2)$

$\qquad (f + g)(2) = f(2) + g(2)$ Use the definition.

$\qquad\qquad = [10(2)^2 - 2(2)] + 2(2)$ Substitute.

$\qquad\qquad = 40$

Alternatively, we could first find $(f + g)(x)$.

$\qquad (f + g)(x) = f(x) + g(x)$ Use the definition.

$\qquad\qquad = (10x^2 - 2x) + 2x$ Substitute.

$\qquad\qquad = 10x^2$ Combine like terms.

Then,

$\qquad (f + g)(2) = 10(2)^2$

$\qquad\qquad = 40.$ The result is the same.

(b) $(f - g)(x)$ and $(f - g)(1)$

$\qquad (f - g)(x) = f(x) - g(x)$ Use the definition.

$\qquad\qquad = (10x^2 - 2x) - 2x$ Substitute.

$\qquad\qquad = 10x^2 - 4x$ Combine like terms.

Then,

$\qquad (f - g)(1) = 10(1)^2 - 4(1)$ Substitute.

$\qquad\qquad = 6.$

Confirm that $f(1) - g(1)$ gives the same result.

▶ **Work Problem 4 at the Side.** ▷▷▷

---

3 Let

$$f(x) = 3x^2 + 8x - 6$$

and $\quad g(x) = -4x^2 + 4x - 8.$

Find each function.

(a) $(f + g)(x)$

(b) $(f - g)(x)$

4 For

$$f(x) = 18x^2 - 24x$$

and $\quad g(x) = 3x,$

find each of the following.

(a) $(f + g)(x)$ and $(f + g)(-1)$

(b) $(f - g)(x)$ and $(f - g)(1)$

ANSWERS
3. **(a)** $-x^2 + 12x - 14$
    (b) $7x^2 + 4x + 2$
4. **(a)** $18x^2 - 21x; \ 39$
    (b) $18x^2 - 27x; \ -9$

OBJECTIVE 4 Graph basic polynomial functions. Functions were introduced in **Section 4.5.** Recall that each input (or x-value) of a function results in one output (or y-value). The simplest polynomial function is the **identity function,** defined by $f(x) = x$. The domain (set of x-values) of this function is all real numbers, $(-\infty, \infty)$, and it pairs each real number with itself. Therefore, the range (set of y-values) is also $(-\infty, \infty)$. Its graph is a straight line, as first seen in **Chapter 4.** (Notice that a *linear function* is a specific kind of polynomial function.) Figure 1 shows its graph and a table of selected ordered pairs.

x	$f(x) = x$
-2	-2
-1	-1
0	0
1	1
2	2

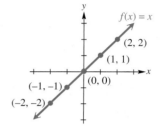

Figure 1

Another polynomial function, defined by $f(x) = x^2$, is the **squaring function.** For this function, every real number is paired with its square. The input can be any real number, so the domain is $(-\infty, \infty)$. Since the square of any real number is nonnegative, the range is $[0, \infty)$. Its graph is a *parabola*. Figure 2 shows the graph and a table of selected ordered pairs.

x	$f(x) = x^2$
-2	4
-1	1
0	0
1	1
2	4

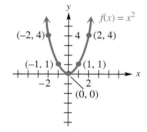

Figure 2

The **cubing function** is defined by $f(x) = x^3$. Every real number is paired with its cube. The domain and the range are both $(-\infty, \infty)$. Its graph is neither a line nor a parabola. See Figure 3 and the table of ordered pairs. (Polynomial functions of degree 3 and greater are studied in detail in more advanced courses.)

x	$f(x) = x^3$
-2	-8
-1	-1
0	0
1	1
2	8

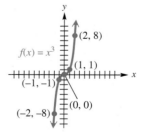

Figure 3

EXAMPLE 5 Graphing Variations of the Identity, Squaring, and Cubing Functions

Graph each function by creating a table of ordered pairs. Give the domain and the range of each function by observing the graphs.

(a) $f(x) = 2x$

To find each range value, multiply the domain value by 2. Plot the points and join them with a straight line. See Figure 4. Both the domain and the range are $(-\infty, \infty)$.

x	f(x) = 2x
−2	−4
−1	−2
0	0
1	2
2	4

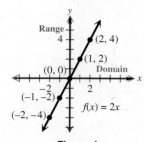

Figure 4

(b) $f(x) = -x^2$

For each input x, square it and then take its opposite. Plotting and joining the points gives a parabola that opens down. See the table and Figure 5. The domain is $(-\infty, \infty)$, and the range is $(-\infty, 0]$.

x	f(x) = −x²
−2	−4
−1	−1
0	0
1	−1
2	−4

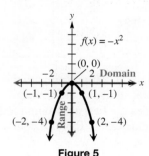

Figure 5

(c) $f(x) = x^3 - 2$

For this function, cube the input and then subtract 2 from the result. The graph is that of the cubing function *shifted* 2 units down. See the table and Figure 6. The domain and the range are both $(-\infty, \infty)$.

x	f(x) = x³ − 2
−2	−10
−1	−3
0	−2
1	−1
2	6

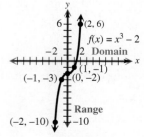

Figure 6

Work Problem 5 at the Side.

5 Graph $f(x) = -2x^2$. Give the domain and the range.

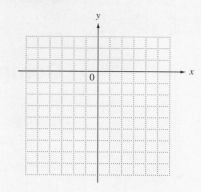

ANSWERS

5.

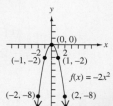

domain: $(-\infty, \infty)$; range: $(-\infty, 0]$

Real-Data Applications

Comparing Mathematical Models

The number of individuals (in thousands) covered by private or governmental health insurance in the United States from 1994 through 1998 can be modeled by the polynomial function defined by

$$P(x) = 1206x + 222{,}448, \quad \text{Linear model}$$

where x represents years since 1994. The number of individuals (in thousands) can also be modeled by

$$Q(x) = 12x^2 + 1159x + 222{,}471 \qquad \text{Quadratic model}$$

and $\quad R(x) = 104x^3 - 613x^2 + 2054x + 222{,}347, \quad$ Cubic model

again where x represents years since 1994.

For Group Discussion

1. Use each model to determine the number of individuals (in thousands) covered by private or governmental health insurance in the United States for each year from 1994 through 1998. Complete the table.

	Number of Individuals Insured (in thousands)		
Year	Using $P(x)$	Using $Q(x)$	Using $R(x)$
1994			
1995			
1996			
1997			
1998			

2. The actual data are given in the table. Compare the data you entered in the table above to the actual data. Which model provides the best approximation of (that is, *best fits*) the actual data? Explain your answer.

Year	Number of Individuals Insured (in thousands)
1994	222,387
1995	223,733
1996	225,077
1997	225,646
1998	227,462

Source: U.S. Bureau of the Census, Health Insurance Historical Table 1.

3. Use the models to make predictions.
   (a) Predict the number of individuals (in thousands) who were covered by private or governmental health insurance in 1999 using P, Q, and R.

   (b) The actual number of individuals (in thousands) insured in 1999 was 231,533. How do your predictions compare to the actual number? Which model gives the best approximation?

   (c) Use each model to predict the number of individuals (in thousands) insured in 2002. What do you notice about your predictions?

   (d) Discuss the validity of using these models to predict the number of individuals (in thousands) covered by private or governmental insurance in years beyond 1998.

6.3 **Exercises**

FOR EXTRA HELP

Tutor Center Addison-Wesley Math Tutor Center

Math XL MathXL

Digital Video Tutor CD 3 Videotape 3

Student's Solutions Manual

MyMathLab MyMathLab

Interactmath.com

*For each polynomial function, find **(a)** f(−1) and **(b)** f(2). See Example 1.*

1. $f(x) = 6x - 4$

2. $f(x) = -2x + 5$

3. $f(x) = x^2 - 3x + 4$

4. $f(x) = 3x^2 + x - 5$

5. $f(x) = 5x^4 - 3x^2 + 6$

6. $f(x) = -4x^4 + 2x^2 - 1$

7. $f(x) = -x^2 + 2x^3 - 8$

8. $f(x) = -x^2 - x^3 + 11x$

Solve each problem. See Example 2.

9. The number of airports in the United States during the period from 1980 through 1999 can be approximated by the polynomial function defined by
$$f(x) = -2.1852x^2 + 236.28x + 15{,}191,$$
where $x = 0$ represents 1980, $x = 1$ represents 1981, and so on. Use this function to approximate the number of airports in each given year. (*Source:* U.S. Federal Aviation Administration.)

(a) 1980   **(b)** 1990   **(c)** 1999

10. The percent of births to unmarried women during the period from 1990 through 2000 can be approximated by the polynomial function defined by
$$f(x) = -.066x^2 + 1.12x + 28.3,$$
where $x = 0$ represents 1990, $x = 1$ represents 1991, and so on. Use this function to approximate the percent (to the nearest tenth) of births to unmarried women in each given year. (*Source:* National Center for Health Statistics.)

(a) 1990   **(b)** 1995   **(c)** 2000

11. The number of bank debit cards issued during the period from 1990 through 2000 can be modeled by the polynomial function defined by
$$P(x) = -.31x^3 + 5.8x^2 - 15x + 9,$$
where $x = 0$ corresponds to the year 1990, $x = 1$ corresponds to 1991, and so on, and $P(x)$ is in millions. Use this function to approximate the number of bank debit cards issued in each given year. Round answers to the nearest million. (*Source: Statistical Abstract of the United States*, 2000.)

(a) 1990

(b) 1996

(c) 1999

12. The total number of cases on the docket of the U.S. Supreme Court during the period from 1980 to 2000 can be modeled by the polynomial function defined by
$$P(x) = -.0325x^3 + 4.524x^2 + 102.5x + 5128,$$
where $x = 0$ represents 1980, $x = 1$ represents 1981, and so on. Use this function to approximate the number of cases on the docket of the U.S. Supreme Court in each given year. (*Source:* Office of the Clerk, Supreme Court of the United States.)

(a) 1980

(b) 1990

(c) 2000

For each pair of functions, find (a) $(f + g)(x)$ and (b) $(f - g)(x)$. See Example 3.

13. $f(x) = 5x - 10, g(x) = 3x + 7$

14. $f(x) = -4x + 1, g(x) = 6x + 2$

15. $f(x) = 4x^2 + 8x - 3, g(x) = -5x^2 + 4x - 9$

16. $f(x) = 3x^2 - 9x + 10, g(x) = -4x^2 + 2x + 12$

Let $f(x) = x^2 - 9$, $g(x) = 2x$, and $h(x) = x - 3$. Find each of the following. See Example 4.

17. $(f + g)(x)$

18. $(f - g)(x)$

19. $(f + g)(3)$

20. $(f - g)(-3)$

21. $(f - h)(x)$

22. $(f + h)(x)$

23. $(f - h)(-3)$

24. $(f + h)(-2)$

25. $(g + h)(-10)$

26. $(g - h)(10)$

27. $(g - h)(-3)$

28. $(g + h)\left(\dfrac{1}{3}\right)$

Graph each function by creating a table of ordered pairs. Give the domain and the range. See Example 5.

29. $f(x) = -2x + 1$

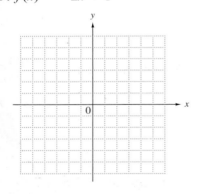

30. $f(x) = 3x + 2$

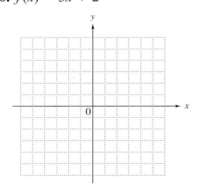

31. $f(x) = -3x^2$

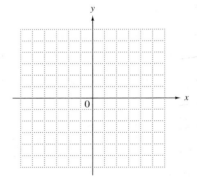

32. $f(x) = \dfrac{1}{2}x^2$

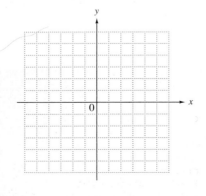

33. $f(x) = x^3 + 1$

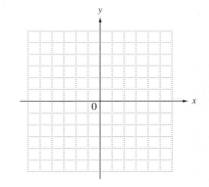

34. $f(x) = -x^3 + 2$

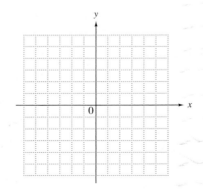

6.4 Multiplying Polynomials

OBJECTIVE 1 Multiply terms. Recall that the product of the two terms $3x^4$ and $5x^3$ is found by using the commutative and associative properties, along with the rules for exponents.

$$(3x^4)(5x^3) = 3 \cdot 5 \cdot x^4 \cdot x^3$$
$$= 15x^{4+3}$$
$$= 15x^7$$

EXAMPLE 1 Multiplying Monomials

Find each product.

(a) $-4a^3(3a^5) = -4(3)a^3 \cdot a^5 = -12a^8$

(b) $2m^2z^4(8m^3z^2) = 2(8)m^2 \cdot m^3 \cdot z^4 \cdot z^2 = 16m^5z^6$

Work Problem 1 at the Side.

OBJECTIVE 2 Multiply any two polynomials. We use the distributive property to extend this process to find the product of any two polynomials.

EXAMPLE 2 Multiplying Polynomials

Find each product.

(a) $-2(8x^3 - 9x^2)$

$$-2(8x^3 - 9x^2) = -2(8x^3) - 2(-9x^2) \quad \text{Distributive property}$$
$$= -16x^3 + 18x^2$$

(b) $5x^2(-4x^2 + 3x - 2) = 5x^2(-4x^2) + 5x^2(3x) + 5x^2(-2)$
$$= -20x^4 + 15x^3 - 10x^2$$

(c) $(3x - 4)(2x^2 + x)$

Use the distributive property to multiply each term of $2x^2 + x$ by $3x - 4$.

$$(3x - 4)(2x^2 + x) = (3x - 4)(2x^2) + (3x - 4)(x)$$

Here $3x - 4$ has been treated as a single expression so that the distributive property could be used. Now use the distributive property two more times.

$$= 3x(2x^2) + (-4)(2x^2) + (3x)(x) + (-4)(x)$$
$$= 6x^3 - 8x^2 + 3x^2 - 4x$$
$$= 6x^3 - 5x^2 - 4x$$

(d) $2x^2(x + 1)(x - 3) = 2x^2[(x + 1)(x) + (x + 1)(-3)]$
$$= 2x^2[x^2 + x - 3x - 3]$$
$$= 2x^2(x^2 - 2x - 3)$$
$$= 2x^4 - 4x^3 - 6x^2$$

Work Problem 2 at the Side.

OBJECTIVES

1. Multiply terms.
2. Multiply any two polynomials.
3. Multiply binomials.
4. Find the product of the sum and difference of two terms.
5. Find the square of a binomial.
6. Multiply polynomial functions.

1 Find each product.

(a) $-6m^5(2m^4)$

(b) $8k^3y(9ky^3)$

2 Find each product.

(a) $-2r(9r - 5)$

(b) $3p^2(5p^3 + 2p^2 - 7)$

(c) $(4a - 5)(3a + 6)$

(d) $3x^3(x + 4)(x - 6)$

ANSWERS
1. (a) $-12m^9$ (b) $72k^4y^4$
2. (a) $-18r^2 + 10r$ (b) $15p^5 + 6p^4 - 21p^2$
   (c) $12a^2 + 9a - 30$ (d) $3x^5 - 6x^4 - 72x^3$

3 Find each product.

(a) $2m - 5$
$\underline{3m + 4}$

It is often easier to multiply polynomials by writing them vertically.

EXAMPLE 3 **Multiplying Polynomials Vertically**

Find each product.

(a) $(5a - 2b)(3a + b)$

$$
\begin{array}{r}
5a \phantom{{}-2b} - 2b \\
\underline{3a \phantom{{}+} + \phantom{{}b}b} \\
5ab - 2b^2 \quad \longleftarrow \quad b(5a - 2b) \\
\underline{15a^2 - 6ab \phantom{{}-2b^2} \quad \longleftarrow \quad 3a(5a - 2b)} \\
15a^2 - ab - 2b^2 \quad \text{Combine like terms.}
\end{array}
$$

(b) $(3m^3 - 2m^2 + 4)(3m - 5)$

$$
\begin{array}{r}
3m^3 - 2m^2 + 4 \\
\underline{3m - 5} \\
-15m^3 + 10m^2 - 20 \quad -5(3m^3 - 2m^2 + 4) \\
\underline{9m^4 - 6m^3 + 12m \phantom{{}- 20}} \quad 3m(3m^3 - 2m^2 + 4) \\
9m^4 - 21m^3 + 10m^2 + 12m - 20 \quad \text{Combine like terms.}
\end{array}
$$

Work Problem 3 at the Side.

OBJECTIVE 3 **Multiply binomials.** When working with polynomials, the product of two binomials occurs repeatedly. There is a shortcut method for finding these products. Recall that a binomial has just two terms, such as $3x - 4$ or $2x + 3$. We can find the product of these binomials using the distributive property as follows.

$$
\begin{aligned}
(3x - 4)(2x + 3) &= 3x(2x + 3) - 4(2x + 3) \\
&= 3x(2x) + 3x(3) - 4(2x) - 4(3) \\
&= 6x^2 + 9x - 8x - 12
\end{aligned}
$$

(b) $5a^3 - 6a^2 + 2a - 3$
$\underline{2a - 5}$

Before combining like terms to find the simplest form of the answer, let us check the origin of each of the four terms in the sum. First, $6x^2$ is the product of the two *first* terms.

$$(3x - 4)(2x + 3) \qquad 3x(2x) = 6x^2 \qquad \textbf{First terms}$$

To get $9x$, the *outer* terms are multiplied.

$$(3x - 4)(2x + 3) \qquad 3x(3) = 9x \qquad \textbf{Outer terms}$$

The term $-8x$ comes from the *inner* terms.

$$(3x - 4)(2x + 3) \qquad -4(2x) = -8x \qquad \textbf{Inner terms}$$

Finally, -12 comes from the *last* terms.

$$(3x - 4)(2x + 3) \qquad -4(3) = -12 \qquad \textbf{Last terms}$$

The product is found by combining these four results.

$$
\begin{aligned}
(3x - 4)(2x + 3) &= 6x^2 + 9x - 8x - 12 \\
&= 6x^2 + x - 12
\end{aligned}
$$

To keep track of the order of multiplying these terms, we use the initials FOIL (**F**irst, **O**uter, **I**nner, **L**ast). All the steps of the FOIL method can be done as follows. Try to do as many of these steps as possible mentally.

ANSWERS
3. (a) $6m^2 - 7m - 20$
   (b) $10a^4 - 37a^3 + 34a^2 - 16a + 15$

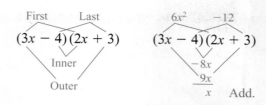

4 Use the FOIL method to find each product.

(a) $(3z + 2)(z + 1)$

EXAMPLE 4 Using the FOIL Method

Use the FOIL method to find each product.

(a) $(4m - 5)(3m + 1)$

First terms $(4m - 5)(3m + 1)$ $4m(3m) = 12m^2$

Outer terms $(4m - 5)(3m + 1)$ $4m(1) = 4m$

(b) $(5r - 3)(2r - 5)$

Inner terms $(4m - 5)(3m + 1)$ $-5(3m) = -15m$

Last terms $(4m - 5)(3m + 1)$ $-5(1) = -5$

Simplify by combining the four terms.

$$\quad\quad\quad\quad\quad F \quad\quad O \quad\quad I \quad\quad L$$
$$(4m - 5)(3m + 1) = 12m^2 + 4m - 15m - 5$$
$$= 12m^2 - 11m - 5$$

The procedure can be written in compact form as follows.

(c) $(4p + 5q)(3p - 2q)$

$$(4m - 5)(3m + 1)$$

$$12m^2$$

$$-5$$

$$-15m$$

$$4m$$

$$-11m \quad \text{Add.}$$

Combine these four results to get $12m^2 - 11m - 5$.

(d) $(4y - z)(2y + 3z)$

$$\text{First} \quad \text{Outer} \quad \text{Inner} \quad \text{Last}$$
$$\downarrow \quad\quad \downarrow \quad\quad \downarrow \quad\quad \downarrow$$

(b) $(6a - 5b)(3a + 4b) = 18a^2 + 24ab - 15ab - 20b^2$
$$= 18a^2 + 9ab - 20b^2$$

(c) $(2k + 3z)(5k - 3z) = 10k^2 + 9kz - 9z^2$ FOIL

Work Problem 4 at the Side. ▶▶▶

(e) $(8r + 1)(8r - 1)$

OBJECTIVE 4 **Find the product of the sum and difference of two terms.** Some types of binomial products occur frequently. For example, the product of the sum and difference of the same two terms, x and y, is

$$(x + y)(x - y) = x^2 - xy + xy - y^2 \quad \text{FOIL}$$
$$= x^2 - y^2.$$

Product of the Sum and Difference of Two Terms

The **product of the sum and difference of the two terms** x and y is the difference of the squares of the terms.

$$(x + y)(x - y) = x^2 - y^2$$

ANSWERS
4. (a) $3z^2 + 5z + 2$ (b) $10r^2 - 31r + 15$
   (c) $12p^2 + 7pq - 10q^2$
   (d) $8y^2 + 10yz - 3z^2$ (e) $64r^2 - 1$

5 Find each product.

(a) $(m + 5)(m - 5)$

(b) $(x - 4y)(x + 4y)$

(c) $(7m - 2n)(7m + 2n)$

(d) $4y^2(y + 7)(y - 7)$

EXAMPLE 5 **Multiplying the Sum and Difference of Two Terms**

Find each product.

(a) $(p + 7)(p - 7) = p^2 - 7^2 \qquad (x + y)(x - y) = x^2 - y^2$
$$= p^2 - 49$$

(b) $(2r + 5)(2r - 5) = (2r)^2 - 5^2$
$$= 2^2 r^2 - 25$$
$$= 4r^2 - 25$$

(c) $(6m + 5n)(6m - 5n) = (6m)^2 - (5n)^2$
$$= 36m^2 - 25n^2$$

(d) $2x^3(x + 3)(x - 3) = 2x^3(x^2 - 9)$
$$= 2x^5 - 18x^3$$

◀◀◀ **Work Problem 5 at the Side.**

OBJECTIVE 5 **Find the square of a binomial.** Another special binomial product is the *square of a binomial*. To find the square of $x + y$, or $(x + y)^2$, multiply $x + y$ by itself.

$$(x + y)(x + y) = x^2 + xy + xy + y^2 \qquad \text{FOIL}$$
$$= x^2 + 2xy + y^2$$

A similar result is true for the square of a difference.

Square of a Binomial

The **square of a binomial** is the sum of the square of the first term, twice the product of the two terms, and the square of the last term.

$$(x + y)^2 = x^2 + 2xy + y^2$$
$$(x - y)^2 = x^2 - 2xy + y^2$$

EXAMPLE 6 **Squaring Binomials**

Find each product.

(a) $(m + 7)^2 = m^2 + 2 \cdot m \cdot 7 + 7^2 \qquad (x + y)^2 = x^2 + 2xy + y^2$
$$= m^2 + 14m + 49$$

(b) $(p - 5)^2 = p^2 - 2 \cdot p \cdot 5 + 5^2 \qquad (x - y)^2 = x^2 - 2xy + y^2$
$$= p^2 - 10p + 25$$

(c) $(2p + 3v)^2 = (2p)^2 + 2(2p)(3v) + (3v)^2$
$$= 4p^2 + 12pv + 9v^2$$

(d) $(3r - 5s)^2 = (3r)^2 - 2(3r)(5s) + (5s)^2$
$$= 9r^2 - 30rs + 25s^2$$

ANSWERS
5. (a) $m^2 - 25$ (b) $x^2 - 16y^2$
(c) $49m^2 - 4n^2$ (d) $4y^4 - 196y^2$

CAUTION
As the products in the formula for the square of a binomial show,

$$(x + y)^2 \neq x^2 + y^2.$$

More generally,

$$(x + y)^n \neq x^n + y^n \quad (n \neq 1).$$

> **Work Problem 6 at the Side.** ▶▶▶

We can use the patterns for the special products with more complicated products, as the following example shows.

EXAMPLE 7 **Multiplying More Complicated Binomials**

Use special products to find each product.

(a) $[(3p - 2) + 5q][(3p - 2) - 5q]$
$= (3p - 2)^2 - (5q)^2$ Product of sum and difference of terms
$= 9p^2 - 12p + 4 - 25q^2$ Square both quantities.

(b) $[(2z + r) + 1]^2 = (2z + r)^2 + 2(2z + r)(1) + 1^2$ Square of a binomial
$= 4z^2 + 4zr + r^2 + 4z + 2r + 1$ Square again; use the distributive property.

(c) $(x + y)^3 = (x + y)^2(x + y)$
$= (x^2 + 2xy + y^2)(x + y)$ Square $x + y$.
$= x^3 + 2x^2y + xy^2 + x^2y + 2xy^2 + y^3$
$= x^3 + 3x^2y + 3xy^2 + y^3$

(d) $(2a + b)^4 = (2a + b)^2(2a + b)^2$
$= (4a^2 + 4ab + b^2)(4a^2 + 4ab + b^2)$ Square $2a + b$.
$= 16a^4 + 16a^3b + 4a^2b^2 + 16a^3b + 16a^2b^2$
$\quad + 4ab^3 + 4a^2b^2 + 4ab^3 + b^4$
$= 16a^4 + 32a^3b + 24a^2b^2 + 8ab^3 + b^4$

> **Work Problem 7 at the Side.** ▶▶▶

OBJECTIVE 6 **Multiply polynomial functions.** In **Section 6.3** we saw how functions can be added and subtracted. Functions can also be multiplied.

Multiplying Functions
If $f(x)$ and $g(x)$ define functions, then

$$(fg)(x) = f(x) \cdot g(x). \quad \text{Product function}$$

The domain of the product function is the intersection of the domains of $f(x)$ and $g(x)$.

6 Find each product.

(a) $(a + 2)^2$

(b) $(2m - 5)^2$

(c) $(y + 6z)^2$

(d) $(3k - 2n)^2$

7 Find each product.

(a) $[(m - 2n) - 3]$
$\cdot [(m - 2n) + 3]$

(b) $[(k - 5h) + 2]^2$

(c) $(p + 2q)^3$

(d) $(x + 2)^4$

ANSWERS
6. **(a)** $a^2 + 4a + 4$
    (b) $4m^2 - 20m + 25$
    (c) $y^2 + 12yz + 36z^2$
    (d) $9k^2 - 12kn + 4n^2$
7. **(a)** $m^2 - 4mn + 4n^2 - 9$
    (b) $k^2 - 10kh + 25h^2 + 4k - 20h + 4$
    (c) $p^3 + 6p^2q + 12pq^2 + 8q^3$
    (d) $x^4 + 8x^3 + 24x^2 + 32x + 16$

8 For
$$f(x) = 2x + 7$$
and $g(x) = x^2 - 4$,

find $(fg)(x)$ and $(fg)(2)$.

EXAMPLE 8 **Multiplying Polynomial Functions**

For $f(x) = 3x + 4$ and $g(x) = 2x^2 + x$, find $(fg)(x)$ and $(fg)(-1)$.

$$
\begin{aligned}
(fg)(x) &= f(x) \cdot g(x) && \text{Use the definition.}\\
&= (3x + 4)(2x^2 + x)\\
&= 6x^3 + 3x^2 + 8x^2 + 4x && \text{FOIL}\\
&= 6x^3 + 11x^2 + 4x && \text{Combine like terms.}
\end{aligned}
$$

Then

$$
\begin{aligned}
(fg)(-1) &= 6(-1)^3 + 11(-1)^2 + 4(-1) && \text{Let } x = -1.\\
&= -6 + 11 - 4\\
&= 1.
\end{aligned}
$$

(Another way to find $(fg)(-1)$ is to find $f(-1)$ and $g(-1)$ and then multiply the results. Verify this by showing that $f(-1) \cdot g(-1)$ equals 1. This follows from the definition.)

Work Problem 8 at the Side.

CAUTION
Write the product $f(x) \cdot g(x)$ as $(fg)(x)$, *not* $f(g(x))$, which has a different mathematical meaning as discussed in **Section 12.1.**

ANSWERS
8. $2x^3 + 7x^2 - 8x - 28; 0$

6.4 Exercises

FOR EXTRA HELP

Tutor Center Addison-Wesley Math Tutor Center

Math XL MathXL

Digital Video Tutor CD 3 Videotape 3

Student's Solutions Manual

MyMathLab MyMathLab

Interactmath.com

Find each product. See Examples 1–3.

1. $-8m^3(3m^2)$

2. $4p^2(-5p^4)$

3. $3x(-2x + 5)$

4. $5y(-6y - 1)$

5. $-q^3(2 + 3q)$

6. $-3a^4(4 - a)$

7. $6k^2(3k^2 + 2k + 1)$

8. $5r^3(2r^2 - 3r - 4)$

9. $(2m + 3)(3m^2 - 4m - 1)$

10. $(4z - 2)(z^2 + 3z + 5)$

11. $4x^3(x - 3)(x + 2)$

12. $2y^5(y - 8)(y + 2)$

13. $(2y + 3)(3y - 4)$

14. $(5m - 3)(2m + 6)$

15. $5m - 3n$
$5m + 3n$

16. $2k + 6q$
$2k - 6q$

17. $-b^2 + 3b + 3$
$2b + 4$

18. $-r^2 - 4r + 8$
$3r - 2$

19. $2z^3 - 5z^2 + 8z - 1$
$4z + 3$

20. $3z^4 - 2z^3 + z - 5$
$2z - 5$

21. $2p^2 + 3p + 6$
$3p^2 - 4p - 1$

22. $5y^2 - 2y + 4$
$2y^2 + y + 3$

Use the FOIL method to find each product. See Example 4.

23. $(m + 5)(m - 8)$

24. $(p - 6)(p + 4)$

25. $(4k + 3)(3k - 2)$

26. $(5w + 2)(2w + 5)$

27. $(z - w)(3z + 4w)$

28. $(s + t)(2s - 5t)$

29. $(6c - d)(2c + 3d)$

30. $(2m - n)(3m + 5n)$

31. $(.2x + 1.3)(.5x - .1)$

32. $(.5y - .4)(.1y + 2.1)$

33. $\left(3r + \dfrac{1}{4}y\right)(r - 2y)$

34. $\left(5w - \dfrac{2}{3}z\right)(w + 5z)$

35. Describe the FOIL method in your own words.

36. Explain why the product of the sum and difference of two terms is not a trinomial.

Find each product. See Example 5.

37. $(2p - 3)(2p + 3)$

38. $(3x - 8)(3x + 8)$

39. $(5m - 1)(5m + 1)$

40. $(6y + 3)(6y - 3)$

41. $(3a + 2c)(3a - 2c)$

42. $(5r - 4s)(5r + 4s)$

43. $\left(4x - \dfrac{2}{3}\right)\left(4x + \dfrac{2}{3}\right)$

44. $\left(3t + \dfrac{5}{4}\right)\left(3t - \dfrac{5}{4}\right)$

45. $(4m + 7n^2)(4m - 7n^2)$

46. $(2k^2 + 6h)(2k^2 - 6h)$

47. $5y^3(y + 2)(y - 2)$

48. $3x^3(x - 4)(x + 4)$

Find each square. See Example 6.

49. $(y - 5)^2$

50. $(a - 3)^2$

51. $(2p + 7)^2$

52. $(3z + 8)^2$

53. $(4n - 3m)^2$

54. $(5r - 7s)^2$

55. $\left(k - \dfrac{5}{7}p\right)^2$

56. $\left(q - \dfrac{3}{4}r\right)^2$

57. Explain how the expressions $(x + y)^2$ and $x^2 + y^2$ differ.

58. Explain how you can find the product $101 \cdot 99$ using the special product

$$(a + b)(a - b) = a^2 - b^2.$$

Find each product. See Example 7.

59. $[(5x + 1) + 6y]^2$

60. $[(3m - 2) + p]^2$

61. $[(2a + b) - 3][(2a + b) + 3]$

62. $[(m + p) + 5][(m + p) - 5]$

63. $[(2h - k) + j][(2h - k) - j]$

64. $[(3m - y) + z][(3m - y) - z]$

65. $(x + 2)^3$

66. $(z - 3)^3$

67. $(5r - s)^3$

68. $(x + 3y)^3$

69. $(q - 2)^4$

70. $(m - p)^4$

RELATING CONCEPTS (EXERCISES 71–78) For Individual or Group Work

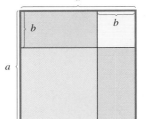

Consider the figure. **Work Exercises 71–78 in order.**

71. What is the length of each side of the blue square in terms of a and b?

72. What is the formula for the area of a square? Use the formula to write an expression, in the form of a product, for the area of the blue square.

73. Each green rectangle has an area of _____. Therefore, the total area in green is represented by the polynomial _____.

74. The yellow square has an area of _____.

75. The area of the entire colored region is represented by _____, because each side of the entire colored region has length _____.

76. The area of the blue square is equal to the area of the entire colored region minus the total area of the green squares minus the area of the yellow square. Write this as a simplified polynomial in a and b.

77. What must be true about the expressions for the area of the blue square you found in Exercises 72 and 76?

78. Write a statement of equality based on your answer in Exercise 77. How does this reinforce one of the main ideas of this section?

For each pair of functions, find the product $(fg)(x)$. See Example 8.

79. $f(x) = 2x, g(x) = 5x - 1$

80. $f(x) = 3x, g(x) = 6x - 8$

81. $f(x) = x + 1, g(x) = 2x - 3$

82. $f(x) = x - 7, g(x) = 4x + 5$

83. $f(x) = 2x - 3, g(x) = 4x^2 + 6x + 9$

84. $f(x) = 3x + 4, g(x) = 9x^2 - 12x + 16$

Let $f(x) = x^2 - 9$, $g(x) = 2x$, and $h(x) = x - 3$. Find each of the following. See Example 8.

85. $(fg)(2)$

86. $(fh)(1)$

87. $(fh)(-1)$

88. $(gh)(-3)$

89. $(fg)(-2)$

Show that each statement is false by replacing x with 2 and y with 3. Then, rewrite each statement with the correct product.

90. $(x + y)^2 = x^2 + y^2$

91. $(x + y)^3 = x^3 + y^3$

92. $(x + y)^4 = x^4 + y^4$

6.5 Dividing Polynomials

OBJECTIVE 1 Divide a polynomial by a monomial. We now discuss polynomial division, beginning with division by a monomial. (Recall that a monomial is a single term, such as $8x$, $-9m^4$, or $11y^2$.)

Dividing by a Monomial

To divide a polynomial by a monomial, divide each term in the polynomial by the monomial, and then write each quotient in lowest terms.

EXAMPLE 1 Dividing a Polynomial by a Monomial

Divide.

(a) $\dfrac{15x^2 - 12x + 6}{3} = \dfrac{15x^2}{3} - \dfrac{12x}{3} + \dfrac{6}{3}$ Divide each term by 3.

$= 5x^2 - 4x + 2$ Write in lowest terms.

Check this answer by multiplying it by the divisor, 3. You should get $15x^2 - 12x + 6$ as the result.

$$3\underbrace{(5x^2 - 4x + 2)}_{} = \underbrace{15x^2 - 12x + 6}_{}$$

Divisor Quotient   Original polynomial

(b) $\dfrac{5m^3 - 9m^2 + 10m}{5m^2} = \dfrac{5m^3}{5m^2} - \dfrac{9m^2}{5m^2} + \dfrac{10m}{5m^2}$ Divide each term by $5m^2$.

$= m - \dfrac{9}{5} + \dfrac{2}{m}$ Write in lowest terms.

This result is not a polynomial. (Why?) The quotient of two polynomials need not be a polynomial.

(c) $\dfrac{8xy^2 - 9x^2y + 6x^2y^2}{x^2y^2} = \dfrac{8xy^2}{x^2y^2} - \dfrac{9x^2y}{x^2y^2} + \dfrac{6x^2y^2}{x^2y^2}$

$= \dfrac{8}{x} - \dfrac{9}{y} + 6$

> **Work Problem 1 at the Side.** ▶▶▶

OBJECTIVE 2 Divide a polynomial by a polynomial of two or more terms. This process is similar to that for dividing whole numbers.

EXAMPLE 2 Dividing a Polynomial by a Polynomial

Divide $\dfrac{2m^2 + m - 10}{m - 2}$.

Write the problem, making sure that both polynomials are written in descending powers of the variables.

$$m - 2\overline{)2m^2 + m - 10}$$

Continued on Next Page

OBJECTIVES

1 Divide a polynomial by a monomial.

2 Divide a polynomial by a polynomial of two or more terms.

3 Divide polynomial functions.

❶ Divide.

(a) $\dfrac{12p + 30}{6}$

(b) $\dfrac{9y^3 - 4y^2 + 8y}{2y^2}$

(c) $\dfrac{8a^2b^2 - 20ab^3}{4a^3b}$

ANSWERS

1. (a) $2p + 5$ **(b)** $\dfrac{9y}{2} - 2 + \dfrac{4}{y}$

(c) $\dfrac{2b}{a} - \dfrac{5b^2}{a^2}$

2 Divide.

(a) $\dfrac{2r^2 + r - 21}{r - 3}$

Divide the first term of $2m^2 + m - 10$ by the first term of $m - 2$.

Since $\dfrac{2m^2}{m} = 2m$, place this result above the division line.

$$\begin{array}{r} \mathbf{2m} \quad\quad\quad\quad \\ m - 2\overline{)2m^2 + m - 10} \end{array}$$ $\longleftarrow$ Result of $\frac{2m^2}{m}$

Multiply $m - 2$ and $2m$, and write the result below $2m^2 + m - 10$.

$$\begin{array}{r} 2m \quad\quad\quad\quad \\ m - 2\overline{)2m^2 + m - 10} \\ \underline{2m^2 - 4m} \quad\quad\quad \end{array}$$ $\longleftarrow$ $2m(m - 2) = 2m^2 - 4m$

Now subtract by mentally changing the signs on $2m^2 - 4m$ and *adding*.

$$\begin{array}{r} 2m \quad\quad\quad\quad \\ m - 2\overline{)2m^2 + m - 10} \\ \underline{2m^2 - 4m} \quad\quad\quad \\ 5m \quad\quad\quad \end{array}$$ $\longleftarrow$ Subtract. The difference is $5m$.

Bring down -10 and continue by dividing $5m$ by m.

$$\begin{array}{r} 2m + 5 \\ m - 2\overline{)2m^2 + m - 10} \\ \underline{2m^2 - 4m} \quad\quad\quad \\ 5m - 10 \\ \underline{5m - 10} \\ 0 \end{array}$$

$\longleftarrow$ $\frac{5m}{m} = 5$

$\longleftarrow$ Bring down -10.

$\longleftarrow$ $5(m - 2) = 5m - 10$

$\longleftarrow$ Subtract. The difference is 0.

Finally, $(2m^2 + m - 10) \div (m - 2) = 2m + 5$. Check by multiplying $m - 2$ and $2m + 5$. The result should be $2m^2 + m - 10$.

Work Problem 2 at the Side.

(b) $\dfrac{2k^2 + 17k + 30}{2k + 5}$

EXAMPLE 3 Dividing a Polynomial with a Missing Term

Divide $3x^3 - 2x + 5$ by $x - 3$.

Make sure that $3x^3 - 2x + 5$ is in descending powers of the variable. Add a term with 0 coefficient as a placeholder for the missing x^2-term.

$$\begin{array}{r} \text{Missing term} \\ \downarrow \quad\quad\quad\quad\quad \\ x - 3\overline{)3x^3 + 0x^2 - 2x + 5} \end{array}$$

Start with $\dfrac{3x^3}{x} = 3x^2$.

$$\begin{array}{r} 3x^2 \quad\quad\quad\quad\quad\quad \\ x - 3\overline{)3x^3 + 0x^2 - 2x + 5} \\ \underline{3x^3 - 9x^2} \quad\quad\quad\quad\quad \end{array}$$

$\longleftarrow$ $\frac{3x^3}{x} = 3x^2$

$\longleftarrow$ $3x^2(x - 3)$

Subtract by mentally changing the signs on $3x^3 - 9x^2$ and adding.

$$\begin{array}{r} 3x^2 \quad\quad\quad\quad\quad\quad \\ x - 3\overline{)3x^3 + 0x^2 - 2x + 5} \\ \underline{3x^3 - 9x^2} \quad\quad\quad\quad\quad \\ 9x^2 \quad\quad\quad\quad\quad \end{array}$$

$\longleftarrow$ Subtract.

Bring down the next term.

$$\begin{array}{r} 3x^2 \quad\quad\quad\quad\quad\quad \\ x - 3\overline{)3x^3 + 0x^2 - 2x + 5} \\ \underline{3x^3 - 9x^2} \quad\quad\quad\quad\quad \\ 9x^2 - 2x \quad\quad\quad \end{array}$$

$\longleftarrow$ Bring down $-2x$.

ANSWERS

2. (a) $2r + 7$ (b) $k + 6$

Continued on Next Page

In the next step, $\dfrac{9x^2}{x} = 9x$.

$$
\begin{array}{r}
3x^2 + 9x \qquad\qquad \longleftarrow \ \frac{9x^2}{x} = 9x \\
x - 3\overline{)3x^3 + 0x^2 - 2x + 5} \\
\underline{3x^3 - 9x^2 \qquad\qquad} \\
9x^2 - 2x \qquad \\
\underline{9x^2 - 27x \qquad} \longleftarrow \ 9x(x-3) \\
25x + 5 \quad \longleftarrow \ \text{Subtract; bring down 5.}
\end{array}
$$

Finally, $\dfrac{25x}{x} = 25$.

$$
\begin{array}{r}
3x^2 + 9x + 25 \qquad \longleftarrow \ \frac{25x}{x} = 25 \\
x - 3\overline{)3x^3 + 0x^2 - 2x + 5} \\
\underline{3x^3 - 9x^2 \qquad\qquad\quad} \\
9x^2 - 2x \qquad\quad \\
\underline{9x^2 - 27x \qquad\quad} \\
25x + 5 \quad \\
\underline{25x - 75 \quad} \longleftarrow \ 25(x-3) \\
\mathbf{80} \quad \longleftarrow \ \text{Remainder}
\end{array}
$$

Write the remainder, 80, as the numerator of the fraction $\frac{80}{x-3}$. In summary,

$$
\frac{3x^3 - 2x + 5}{x - 3} = 3x^2 + 9x + 25 + \frac{80}{x - 3}.
$$

Check by multiplying $x - 3$ and $3x^2 + 9x + 25$ and adding 80. The result should be $3x^3 - 2x + 5$.

CAUTION

Remember to write $\dfrac{\text{remainder}}{\text{divisor}}$ as part of the quotient.

Work Problem 3 at the Side. ▶▶▶

EXAMPLE 4 **Performing a Division with a Fractional Coefficient in the Quotient**

Divide $2p^3 + 5p^2 + p - 2$ by $2p + 2$.

$$
\begin{array}{r}
\overset{\frac{3p^2}{2p} = \frac{3}{2}p}{} \\
p^2 + \dfrac{3}{2}p - 1 \qquad\quad \\
2p + 2\overline{)2p^3 + 5p^2 + p - 2} \\
\underline{2p^3 + 2p^2 \qquad\qquad} \\
3p^2 + p \qquad\quad \\
\underline{3p^2 + 3p \qquad\quad} \\
-2p - 2 \\
\underline{-2p - 2} \\
0
\end{array}
$$

Since the remainder is 0, the quotient is $p^2 + \frac{3}{2}p - 1$.

Work Problem 4 at the Side. ▶▶▶

❸ Divide.

$$
\frac{3k^3 + 9k - 14}{k - 2}
$$

❹ Divide $2p^3 + 7p^2 + 9p + 2$ by $2p + 2$.

ANSWERS

3. $3k^2 + 6k + 21 + \dfrac{28}{k - 2}$

4. $p^2 + \dfrac{5}{2}p + 2 + \dfrac{-2}{2p + 2}$

5 Divide.

(a) $\dfrac{3r^5 - 15r^4 - 2r^3 + 19r^2 - 7}{3r^2 - 2}$

(b) $\dfrac{4x^4 - 7x^2 + x + 5}{2x^2 - x}$

EXAMPLE 5 Dividing by a Polynomial with a Missing Term

Divide $6r^4 + 9r^3 + 2r^2 - 8r + 7$ by $3r^2 - 2$.

Write $3r^2 - 2$ as $3r^2 + 0r - 2$ and divide as usual.

$$
\begin{array}{r}
2r^2 + 3r + 2 \\
3r^2 + 0r - 2 \overline{)6r^4 + 9r^3 + 2r^2 - 8r + 7} \\
\underline{6r^4 + 0r^3 - 4r^2} \\
9r^3 + 6r^2 - 8r \\
\underline{9r^3 + 0r^2 - 6r} \\
6r^2 - 2r + 7 \\
\underline{6r^2 + 0r - 4} \\
-2r + 11
\end{array}
$$

Missing term ↑

Since the degree of the remainder, $-2r + 11$, is less than the degree of the divisor, $3r^2 - 2$, the process is now finished. The result is written

$$2r^2 + 3r + 2 + \frac{-2r + 11}{3r^2 - 2}.$$

Work Problem 5 at the Side.

CAUTION
When dividing a polynomial by a polynomial of two or more terms:

1. Be sure the terms in both polynomials are in descending powers.
2. Write any missing terms with 0 placeholders.

OBJECTIVE 3 Divide polynomial functions.

Dividing Functions

If $f(x)$ and $g(x)$ define functions, then

$$\left(\frac{f}{g}\right)(x) = \frac{f(x)}{g(x)}. \qquad \text{Quotient function}$$

The domain of the quotient function is the intersection of the domains of $f(x)$ and $g(x)$, excluding any values of x for which $g(x) = 0$.

6 For

$$f(x) = 2x^2 + 17x + 30$$

and $g(x) = 2x + 5$,

find $\left(\frac{f}{g}\right)(x)$ and $\left(\frac{f}{g}\right)(-1)$.

EXAMPLE 6 Dividing Polynomial Functions

For $f(x) = 2x^2 + x - 10$ and $g(x) = x - 2$, find $\left(\frac{f}{g}\right)(x)$ and $\left(\frac{f}{g}\right)(-3)$.

$$\left(\frac{f}{g}\right)(x) = \frac{f(x)}{g(x)} = \frac{2x^2 + x - 10}{x - 2}$$

This quotient, found in Example 2, with x replacing m, is $2x + 5$, so

$$\left(\frac{f}{g}\right)(x) = 2x + 5, \quad x \neq 2.$$

Then $\left(\frac{f}{g}\right)(-3) = 2(-3) + 5 = -1.$ Let $x = -3$.

(Which is easier to find here: $\left(\frac{f}{g}\right)(-3)$ or $\frac{f(-3)}{g(-3)}$?)

Work Problem 6 at the Side.

ANSWERS

5. (a) $r^3 - 5r^2 + 3 + \dfrac{-1}{3r^2 - 2}$

   (b) $2x^2 + x - 3 + \dfrac{-2x + 5}{2x^2 - x}$

6. $x + 6, \quad x \neq -\dfrac{5}{2}; \; 5$

6.5 Exercises

FOR EXTRA HELP   Addison-Wesley Math Tutor Center   MathXL   Digital Video Tutor CD 3 Videotape 3   Student's Solutions Manual   MyMathLab   Interactmath.com

Divide. See Example 1.

1. $\dfrac{15x^3 - 10x^2 + 5}{5}$

2. $\dfrac{27m^4 - 18m^3 + 9m}{9}$

3. $\dfrac{9y^2 + 12y - 15}{3y}$

4. $\dfrac{80r^2 - 40r + 10}{10r}$

5. $\dfrac{15m^3 + 25m^2 + 30m}{5m^2}$

6. $\dfrac{64x^3 - 72x^2 + 12x}{8x^3}$

7. $\dfrac{14m^2n^2 - 21mn^3 + 28m^2n}{14m^2n}$

8. $\dfrac{24h^2k + 56hk^2 - 28hk}{16h^2k^2}$

Divide. See Examples 2–5.

9. $\dfrac{y^2 + 3y - 18}{y + 6}$

10. $\dfrac{q^2 + 4q - 32}{q - 4}$

11. $\dfrac{3t^2 + 17t + 10}{3t + 2}$

12. $\dfrac{2k^2 - 3k - 20}{2k + 5}$

13. $\dfrac{p^2 + 2p + 20}{p + 6}$

14. $\dfrac{x^2 + 11x + 16}{x + 8}$

15. $(2z^3 - 5z^2 + 6z - 15) \div (2z - 5)$

16. $(3p^3 + p^2 + 18p + 6) \div (3p + 1)$

17. $(4x^3 + 9x^2 - 10x + 3) \div (4x + 1)$

18. $(10z^3 - 26z^2 + 17z - 13) \div (5z - 3)$

19. $\dfrac{14x + 6x^3 - 15 - 19x^2}{3x^2 - 2x + 4}$

20. $\dfrac{37m - 18m^2 - 13 + 8m^3}{2m^2 - 3m + 6}$

21. $(3x^3 - x + 4) \div (x - 2)$

22. $(4x^3 - 3x - 2) \div (x + 1)$

23. $(2x^3 - 11x^2 + 28) \div (x - 5)$

24. $(3x^3 - 4x + 2) \div (x - 1)$

25. $\dfrac{4k^4 + 6k^3 + 3k - 1}{2k^2 + 1}$

26. $\dfrac{6y^4 + 9y^3 + 10y^2 + 6y + 4}{3y^2 + 2}$

27. $(9z^4 - 13z^3 + 23z^2 - 10z + 8) \div (z^2 - z + 2)$

28. $(2q^4 + 5q^3 - 11q^2 + 11q - 20) \div (2q^2 - q + 2)$

29. $\left(2x^2 - \dfrac{7}{3}x - 1 \right) \div (3x + 1)$

30. $\left(m^2 + \dfrac{7}{2}m + 3 \right) \div (2m + 3)$

31. $\left(3a^2 - \dfrac{23}{4}a - 5 \right) \div (4a + 3)$

32. $\left(3q^2 + \dfrac{19}{5}q - 3 \right) \div (5q - 2)$

For each pair of functions, find the quotient $\left(\frac{f}{g}\right)(x)$ and give any x-values that are not in the domain of the quotient function. See Example 6.

33. $f(x) = 10x^2 - 2x, g(x) = 2x$

34. $f(x) = 18x^2 - 24x, g(x) = 3x$

35. $f(x) = 2x^2 - x - 3, g(x) = x + 1$

36. $f(x) = 4x^2 - 23x - 35, g(x) = x - 7$

37. $f(x) = 8x^3 - 27, g(x) = 2x - 3$

38. $f(x) = 27x^3 + 64, g(x) = 3x + 4$

Let $f(x) = x^2 - 9$, $g(x) = 2x$, and $h(x) = x - 3$. Find each of the following. See Example 6.

39. $\left(\dfrac{f}{g}\right)(x)$

40. $\left(\dfrac{f}{h}\right)(x)$

41. $\left(\dfrac{f}{g}\right)(2)$

42. $\left(\dfrac{f}{h}\right)(1)$

43. $\left(\dfrac{h}{g}\right)(x)$

44. $\left(\dfrac{f}{h}\right)(-3)$

45. $\left(\dfrac{h}{g}\right)(3)$

46. $\left(\dfrac{f}{g}\right)(-1)$

Solve each problem.

47. The volume of a box is $2p^3 + 15p^2 + 28p$ ft. The height is p ft and the length is $p + 4$ ft. Find an expression for the width.

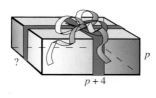

48. Suppose a car goes $2m^3 + 15m^2 + 13m - 63$ km in $2m + 9$ hr. Find an expression for its rate.

Chapter 6

SUMMARY

KEY TERMS

6.2

term — A term is a number, a variable, or the product or quotient of a number and one or more variables raised to powers.

coefficient (numerical coefficient) — A coefficient is a factor in a term (usually used for the numerical factor).

algebraic expression — An algebraic expression is any combination of variables or constants (numerical values) joined by the basic operations of addition, subtraction, multiplication, and division (except by 0), or raising to powers or taking roots.

polynomial — A polynomial is a term or a finite sum of terms in which all variables have whole number exponents and no variables appear in denominators.

polynomial in x — A polynomial in x is a polynomial containing only the variable x.

descending powers — A polynomial in one variable is written in descending powers if the exponents on the variable in the terms decrease from left to right.

trinomial — A trinomial is a polynomial with exactly three terms.

binomial — A binomial is a polynomial with exactly two terms.

monomial — A monomial is a polynomial with exactly one term.

degree of a term — The degree of a term with one variable is the exponent on that variable.

degree of a polynomial — The degree of a polynomial is the greatest degree of any of the terms in the polynomial.

negative of a polynomial — The negative of a polynomial is obtained by changing the sign of every coefficient in the polynomial.

6.3

polynomial function of degree n — A function defined by $f(x) = a_n x^n + a_{n-1} x^{n-1} + \cdots + a_1 x + a_0$, where $a_n \neq 0$ and n is a whole number, is a polynomial function of degree n.

identity function — The simplest polynomial function is the identity function, defined by $f(x) = x$.

squaring function — The polynomial function defined by $f(x) = x^2$ is called the squaring function.

cubing function — The polynomial function defined by $f(x) = x^3$ is called the cubing function.

TEST YOUR WORD POWER

See how well you have learned the vocabulary in this chapter. Answers, with examples, follow the Quick Review.

1. A **polynomial** is an algebraic expression made up of
   A. a term or a finite product of terms with positive coefficients and exponents
   B. the sum of two or more terms with whole number coefficients and exponents
   C. the product of two or more terms with positive exponents
   D. a term or a finite sum of terms with real coefficients and whole number exponents.

2. A **monomial** is a polynomial with
   A. only one term
   B. exactly two terms
   C. exactly three terms
   D. more than three terms.

3. A **binomial** is a polynomial with
   A. only one term
   B. exactly two terms
   C. exactly three terms
   D. more than three terms.

4. A **trinomial** is a polynomial with
   A. only one term
   B. exactly two terms
   C. exactly three terms
   D. more than three terms.

5. **FOIL** is a method for
   A. adding two binomials
   B. adding two trinomials
   C. multiplying two binomials
   D. multiplying two trinomials.

QUICK REVIEW

Concepts	*Examples*

6.1 *Integer Exponents and Scientific Notation*

Definitions and Rules for Exponents

Product Rule: $a^m \cdot a^n = a^{m+n}$

Quotient Rule: $\dfrac{a^m}{a^n} = a^{m-n}$ $(a \neq 0)$

Zero Exponent: $a^0 = 1$ $(a \neq 0)$

Negative Exponent: $a^{-n} = \dfrac{1}{a^n}$ $(a \neq 0)$

Power Rules: $(a^m)^n = a^{mn}, \quad (ab)^m = a^m b^m$

$$\left(\frac{a}{b}\right)^n = \frac{a^n}{b^n} \quad (b \neq 0)$$

Special Rules: $\dfrac{1}{a^{-n}} = a^n$ $(a \neq 0)$

$$\frac{a^{-n}}{b^{-m}} = \frac{b^m}{a^n} \quad (a, b \neq 0)$$

$$a^{-n} = \left(\frac{1}{a}\right)^n \quad (a \neq 0)$$

$$\left(\frac{a}{b}\right)^{-n} = \left(\frac{b}{a}\right)^n \quad (a, b \neq 0)$$

Apply the rules of exponents.

$$3^4 \cdot 3^2 = 3^6$$

$$\frac{2^5}{2^3} = 2^2$$

$$27^0 = 1, \quad (-5)^0 = 1$$

$$5^{-2} = \frac{1}{5^2}$$

$$(6^3)^4 = 6^{12}, \quad (5p)^4 = 5^4 p^4$$

$$\left(\frac{2}{3}\right)^5 = \frac{2^5}{3^5}$$

$$\frac{1}{x^{-3}} = x^3$$

$$\frac{r^{-3}}{t^{-4}} = \frac{t^4}{r^3}$$

$$4^{-3} = \left(\frac{1}{4}\right)^3$$

$$\left(\frac{4}{7}\right)^{-2} = \left(\frac{7}{4}\right)^2$$

Scientific Notation

A number is in scientific notation when it is written as a product of a number between 1 and 10 (inclusive of 1) and an integer power of 10.

Write 23,500,000,000 in scientific notation.

$$23,500,000,000 = 2.35 \times 10^{10}$$

Write 4.3×10^{-6} in standard notation.

$$4.3 \times 10^{-6} = .0000043$$

6.2 *Adding and Subtracting Polynomials*

Add or subtract polynomials by combining like terms.

$$(x^2 - 2x + 3) + (2x^2 - 8) = 3x^2 - 2x - 5$$
$$(5x^4 + 3x^2) - (7x^4 + x^2 - x) = -2x^4 + 2x^2 + x$$

6.3 *Polynomial Functions*

The graph of $f(x) = x$ is a line, and the graph of $f(x) = x^2$ is a parabola. The graph of $f(x) = x^3$ is neither of these. They define the identity, squaring, and cubing functions, respectively.

Graph the identity, squaring, and cubing functions.

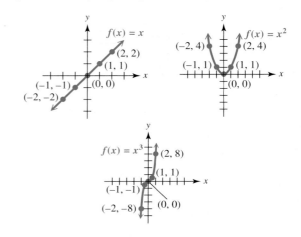

Concepts	Examples

6.4 Multiplying Polynomials

To multiply two polynomials, multiply each term of one by each term of the other.

$$(x^3 + 3x)(4x^2 - 5x + 2)$$
$$= 4x^5 + 12x^3 - 5x^4 - 15x^2 + 2x^3 + 6x$$
$$= 4x^5 - 5x^4 + 14x^3 - 15x^2 + 6x$$

To multiply two binomials, use the **FOIL** method. Multiply the **First** terms, the **Outer** terms, the **Inner** terms, and the **Last** terms. Then add these products.

$$(2x + 3)(x - 7) = 2x(x) + 2x(-7) + 3x + 3(-7)$$
$$= 2x^2 - 14x + 3x - 21$$
$$= 2x^2 - 11x - 21$$

Special Products
$$(x + y)(x - y) = x^2 - y^2$$
$$(x + y)^2 = x^2 + 2xy + y^2$$
$$(x - y)^2 = x^2 - 2xy + y^2$$

$$(3m + 8)(3m - 8) = 9m^2 - 64$$
$$(5a + 3b)^2 = 25a^2 + 30ab + 9b^2$$
$$(2k - 1)^2 = 4k^2 - 4k + 1$$

6.5 Dividing Polynomials

Dividing by a Monomial

To divide a polynomial by a monomial, divide each term in the polynomial by the monomial, and then write each fraction in lowest terms.

$$\frac{2x^3 - 4x^2 + 6x - 8}{2x} = \frac{2x^3}{2x} - \frac{4x^2}{2x} + \frac{6x}{2x} - \frac{8}{2x}$$
$$= x^2 - 2x + 3 - \frac{4}{x}$$

Dividing by a Polynomial

Use the "long division" process.

Divide $\dfrac{m^3 - m^2 + 2m + 5}{m + 1}$.

$$\begin{array}{r} m^2 - 2m + 4 \\ m + 1{\overline{\smash{\big)}\,m^3 - m^2 + 2m + 5}} \\ \underline{m^3 + m^2} \\ -2m^2 + 2m \\ \underline{-2m^2 - 2m} \\ 4m + 5 \\ \underline{4m + 4} \\ 1 \leftarrow \text{Remainder} \end{array}$$

The answer is $m^2 - 2m + 4 + \dfrac{1}{m + 1}$.

ANSWERS TO TEST YOUR WORD POWER

1. D; *Example:* $5x^3 + 2x^2 - 7$
2. A; *Examples:* $-4, 2t^3, 15a^2b$
3. B; *Example:* $3t^3 + 5t$
4. C; *Example:* $2a^2 - 3ab + b^2$ F   O   I   L
5. C; *Example:* $(m + 4)(m - 3) = m(m) + m(-3) + 4m + 4(-3)$
$$= m^2 + m - 12$$

Reply All: The Computer Network Manager's Nightmare

Most businesses, including colleges, rely on e-mail messaging for internal communications. The computer network manager is responsible for ensuring that e-mail is a reliable and efficient system for maintaining that communication. One of the biggest problems involves the Reply All option in most e-mail software. If a message was sent to 10 addresses, the Reply All option sends a copy of the answer to each address in the list. In comparison, the Reply option sends an answer to only one address. A computer virus creates an even worse problem.

For Group Discussion

Assume that the discussion about e-mail messages relates only to those described in the following problem.

1. A committee of 10 faculty members is formed to recommend faculty salary increases for the next year. One member is selected as the committee chair. The chair requests that e-mail discussions be sent to all 10 committee members (using the Reply All option).

   (a) The committee chair sends an e-mail message to every committee member, including himself, to which all other committee members reply. How many relevant e-mail messages are in each person's mailbox? How many e-mail messages has the computer network processed? (*Hint:* Think about the number of e-mails that would be sent if the committee had only 2 members, 3 members, 4 members, etc.)

   (b) The committee chair forwards an e-mail message from the chancellor to the committee members suggesting that summer salaries must be reduced for the faculty to receive a pay raise for the next year. A heated debate ensues. Counting the original e-mail message about the chancellor's comment, the chairman counts eight relevant e-mail messages in his mailbox. How many e-mail messages has the computer network processed?

   (c) Suppose the committee chair inadvertently sent the original message to the "All College Faculty" group, a list of 150 names set up by the computer manager primarily for administrators to notify the faculty about policy. The chairman counted eight relevant e-mail messages in his mailbox, each sent using Reply All. How many e-mail messages would the computer network have processed?

2. The "Love Bug" computer virus spreads by sending copies of itself to each e-mail address in the recipient's address book. Suppose that each person has 50 e-mail addresses in his or her address book and that it takes a virus 10 sec to process. Also assume that each recipient's computer automatically receives and processes the computer virus. (It is more reasonable to assume that a virus is processed only after the recipient "opens" the message.) How many messages has the computer network processed after 1 hr?

3. It is illegal to send a chain letter using the U.S. postal service. A chain letter requests that you send copies of it to, say, 10 people. Discuss why you think chain letters are outlawed.

Chapter **6**

REVIEW EXERCISES

[6.1] *Simplify. Write answers with only positive exponents. Assume all variables represent positive real numbers.*

1. 4^3

2. $\left(\dfrac{1}{3}\right)^4$

3. $(-5)^3$

4. $\dfrac{2}{(-3)^{-2}}$

5. $\left(\dfrac{2}{3}\right)^{-4}$

6. $\left(\dfrac{5}{4}\right)^{-2}$

7. $5^{-1} + 6^{-1}$

8. $-3^0 + 3^0$

9. $(-3x^4 y^3)(4x^{-2} y^5)$

10. $\dfrac{6m^{-4} n^3}{-3mn^2}$

11. $\dfrac{(5p^{-2} q)(4p^5 q^{-3})}{2p^{-5} q^5}$

12. $\dfrac{x^{-2} y^{-4}}{x^{-4} y^{-2}}$

13. $(3^{-4})^2$

14. $(x^{-4})^{-2}$

15. $(xy^{-3})^{-2}$

16. $(z^{-3})^3 z^{-6}$

17. $(5m^{-3})^2 (m^4)^{-3}$

18. $\dfrac{(3r)^2 r^4}{r^{-2} r^{-3}} (9r^{-3})^{-2}$

19. $\left(\dfrac{5z^{-3}}{z^{-1}}\right)\left(\dfrac{5}{z^2}\right)$

20. $\left(\dfrac{6m^{-4}}{m^{-9}}\right)^{-1}\left(\dfrac{m^{-2}}{16}\right)$

21. $\left(\dfrac{3r^5}{5r^{-3}}\right)^{-2}\left(\dfrac{9r^{-1}}{2r^{-5}}\right)^3$

Write in scientific notation.

22. 13,450

23. .0000000765

24. .138

25. In 2000, the total population of the United States was **281,400,000.** Of this amount, **50,454** Americans were centenarians, that is, age **100** or older. Write the three boldfaced numbers using scientific notation. (*Source:* U.S. Bureau of the Census.)

Write without scientific notation.

26. 1.21×10^6

27. 5.8×10^{-3}

Use scientific notation to compute. Give answers in both scientific notation and standard form.

28. $\dfrac{16 \times 10^4}{8 \times 10^8}$

29. $\dfrac{6 \times 10^{-2}}{4 \times 10^{-5}}$

30. $\dfrac{.0000000164}{.0004}$

31. $\dfrac{.0009 \times 12{,}000{,}000}{400{,}000}$

[6.2] *Give the numerical coefficient for each term.*

32. $14p^5$

33. $-z$

*For each polynomial, **(a)** write in descending powers, **(b)** identify as monomial, binomial, trinomial, or none of these, and **(c)** give the degree.*

34. $9k + 11k^3 - 3k^2$

35. $14m^6 + 9m^7$

36. $-7q^5r^3$

37. Give an example of a polynomial in the variable x such that it has degree 5, is lacking a third-degree term, and is in descending powers of the variable.

Add or subtract as indicated.

38. Add.

$$3x^2 - 5x + 6$$
$$-4x^2 + 2x - 5$$

39. Subtract.

$$-5y^3 \qquad\quad + 8y - 3$$
$$\qquad\quad 4y^2 + 2y + 9$$

40. $(4a^3 - 9a + 15) - (-2a^3 + 4a^2 + 7a)$

41. $(3y^2 + 2y - 1) + (5y^2 - 11y + 6)$

42. Find the perimeter of the triangle.

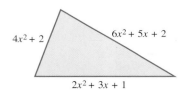

[6.3]

43. For the polynomial function defined by $f(x) = -2x^2 + 5x + 7$, find each value.

(a) $f(-2)$ (b) $f(3)$

44. For $f(x) = 2x + 3$ and $g(x) = 5x^2 - 3x + 2$, find each of the following.

(a) $(f + g)(x)$ (b) $(f - g)(x)$ (c) $(f + g)(-1)$ (d) $(f - g)(-1)$

45. The number of people, in millions, enrolled in Health Maintenance Organizations (HMOs) during the period from 1990 through 2000 can be modeled by the polynomial function defined by

$$f(x) = .241x^2 + 3.26x + 30.0,$$

where $x = 0$ corresponds to 1990, $x = 1$ corresponds to 1991, and so on. Use this model to approximate the number of people enrolled in each given year. (*Source:* Interstudy; U.S. National Center for Health Statistics.)

(a) 1990   (b) 1995   (c) 2000

Graph each polynomial function defined as follows.

46. $f(x) = -2x + 5$ **47.** $f(x) = x^2 - 6$ **48.** $f(x) = -x^3 + 1$

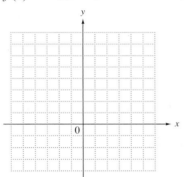   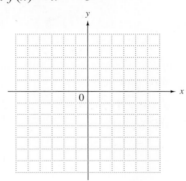   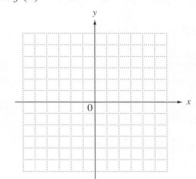

[6.4] *Find each product.*

49. $-6k(2k^2 + 7)$ **50.** $(7y - 8)(2y + 3)$ **51.** $(3w - 2t)(2w - 3t)$

52. $(2p^2 + 6p)(5p^2 - 4)$ **53.** $(3z^3 - 2z^2 + 4z - 1)(3z - 2)$ **54.** $(6r^2 - 1)(6r^2 + 1)$

55. $\left(z + \dfrac{3}{5}\right)\left(z - \dfrac{3}{5}\right)$ **56.** $(4m + 3)^2$ **57.** $(2x + 5)^3$

[6.5] *Divide.*

58. $\dfrac{4y^3 - 12y^2 + 5y}{4y}$ **59.** $\dfrac{2p^3 + 9p^2 + 27}{2p - 3}$ **60.** $\dfrac{5p^4 + 15p^3 - 33p^2 - 9p + 18}{5p^2 - 3}$

MIXED REVIEW EXERCISES

61. Match each expression (a)–(i) in Column I with its equivalent expression A–I in Column II. Choices may be used once, more than once, or not at all.

I

(a) 4^{-2} (f) -4^0

(b) -4^2 (g) $-4^0 + 4^0$

(c) 4^0 (h) $-4^0 - 4^0$

(d) $(-4)^0$ (i) $4^{-2} + 4^{-1}$

(e) $(-4)^{-2}$

II

A. $\dfrac{1}{16}$ F. $\dfrac{5}{16}$

B. 0 G. -16

C. 1 H. -2

D. $-\dfrac{1}{16}$ I. none of these

E. -1

62. In a recent year, the estimated population of Luxembourg was 3.92×10^5. The population density was 400 people per mi^2. Based on this information, what is the area of Luxembourg to the nearest square mile?

Perform the indicated operations, and then simplify. Write answers with only positive exponents. Assume all variables represent nonzero real numbers.

63. $(4x + 1)(2x - 3)$

64. $\dfrac{6^{-1} y^3 (y^2)^{-2}}{6y^{-4}(y^{-1})}$

65. $(y^6)^{-5}(2y^{-3})^{-4}$

66. $(2x - 9)^2$

67. $\dfrac{20y^3x^3 + 15y^4x + 25yx^4}{10yx^2}$

68. $7p^5(3p^4 + p^3 + 2p^2)$

69. $\dfrac{(-z^{-2})^3}{5(z^{-3})^{-1}}$

70. $\dfrac{x^3 + 7x^2 + 7x - 15}{x + 5}$

71. $(-5 + 11w) + (6 + 5w) + (-15 - 8w^2)$

72. $(2k - 1) - (3k^2 - 2k + 6)$

Chapter **6**

TEST

1. Match each expression in Column I with its equivalent expression from Column II. Choices may be used once, more than once, or not at all.

I	II
(a) 7^{-2}	**A.** 1
(b) 7^0	**B.** $\dfrac{1}{9}$
(c) -7^0	**C.** $\dfrac{1}{49}$
(d) $(-7)^0$	**D.** -1
(e) -7^2	**E.** -49
(f) $7^{-1} + 2^{-1}$	**F.** $\dfrac{9}{14}$
(g) $(7 + 2)^{-1}$	**G.** $\dfrac{2}{7}$
(h) $\dfrac{7^{-1}}{2^{-1}}$	**H.** 0
(i) $(-7)^{-2}$	**I.** none of these

Simplify. Write answers with only positive exponents. Assume all variables represent nonzero real numbers.

2. $(3x^{-2}y^3)^{-2}(4x^3y^{-4})$

3. $\dfrac{36r^{-4}(r^2)^{-3}}{6r^4}$

4. $\left(\dfrac{4p^2}{q^4}\right)^3\left(\dfrac{6p^8}{q^{-8}}\right)^{-2}$

5. $(-2x^4y^{-3})^0(-4x^{-3}y^{-8})^2$

6. Write 9.1×10^{-7} without using scientific notation.

7. Use scientific notation to simplify $\dfrac{2,500,000 \times .00003}{.05 \times 5,000,000}$. Write the answer in both scientific notation and standard form.

8. If $f(x) = -2x^2 + 5x - 6$ and $g(x) = 7x - 3$, find each of the following.

    (a) $f(4)$ **(b)** $(f + g)(x)$ **(c)** $(f - g)(x)$ **(d)** $(f - g)(-2)$

9. Graph the function defined by $f(x) = -2x^2 + 3$.

(a) _____ (b) _____ (c) _____

(d) _____ (e) _____ (f) _____

1. (g) _____ (h) _____ (i) _____

2. _____

3. _____

4. _____

5. _____

6. _____

7. _____

8. (a) _____

   (b) _____

   (c) _____

   (d) _____

9.

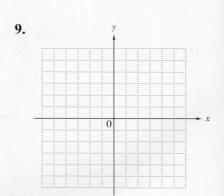

10. (a) _____

    (b) _____

    (c) _____

10. The number of medical doctors, in thousands, in the United States during the period from 1990 through 2000 can be modeled by the polynomial function defined by

$$f(x) = -.197x^2 + 21.7x + 615,$$

where $x = 0$ corresponds to 1990, $x = 1$ corresponds to 1991, and so on. Use this model to approximate the number of doctors to the nearest thousand in each given year. (*Source:* American Medical Association.)

    (a) 1990 **(b)** 1996 **(c)** 2000

Perform the indicated operations.

11. _____

11. $(4x^3 - 3x^2 + 2x - 5) - (3x^3 + 11x + 8) + (x^2 - x)$

12. _____

12. $(5x - 3)(2x + 1)$ **13.** $(2m - 5)(3m^2 + 4m - 5)$

13. _____

14. _____

14. $(6x + y)(6x - y)$ **15.** $(3k + q)^2$

15. _____

16. _____

16. $[2y + (3z - x)][2y - (3z - x)]$ **17.** $\dfrac{16p^3 - 32p^2 + 24p}{4p^2}$

17. _____

18. _____

18. $(x^3 + 3x^2 - 4) \div (x - 1)$

19. (a) _____

    (b) _____

19. If $f(x) = x^2 + 3x + 2$ and $g(x) = x + 1$, find **(a)** $(fg)(x)$ and **(b)** $(fg)(-2)$.

20. (a) _____

    (b) _____

20. Use $f(x)$ and $g(x)$ from Problem 19 to find **(a)** $\left(\dfrac{f}{g}\right)(x)$ and **(b)** $\left(\dfrac{f}{g}\right)(-2)$.

Cumulative Review Exercises

CHAPTERS 1–6

Match each number in Column I with the choice or choices of sets of numbers in Column II to which the number belongs.

I   **II**

1. 34   **2.** 0   **A.** Natural numbers   **B.** Whole numbers

3. 2.16   **4.** $-\sqrt{36}$ **C.** Integers   **D.** Rational numbers

5. $\sqrt{13}$ **6.** $-\dfrac{4}{5}$ **E.** Irrational numbers   **F.** Real numbers

Evaluate.

7. $9 \cdot 4 - 16 \div 4$ **8.** $-|8 - 13| - |-4| + |-9|$

Solve.

9. $-5(8 - 2z) + 4(7 - z) = 7(8 + z) - 3$ **10.** $3(x + 2) - 5(x + 2) = -2x - 4$

11. $2(m + 5) - 3m + 1 > 5$ **12.** $|3x - 1| = 2$ **13.** $|3z + 1| \geq 7$

14. A recent survey polled teens about the most important inventions of the twentieth century. Complete the results shown in the table if 1500 teens were surveyed.

Most Important Invention	Percent	Actual Number
Personal computer		480
Pacemaker	26%	
Wireless communication	18%	
Television		150

Source: Lemelson-MIT Program.

15. Find the measure of each angle of the triangle.

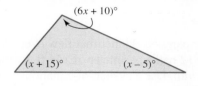

16. Find the slope of the line through $(-4, 5)$ and $(2, -3)$. Then write an equation of the line in standard form.

Graph each equation or inequality.

17. $-3x + 4y = 12$ **18.** $y \leq 2x - 6$ **19.** $3x + 2y < 0$

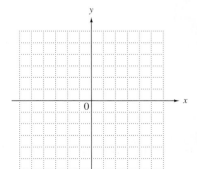   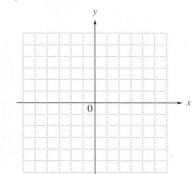   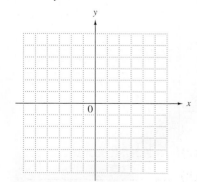

20. The graph shows the annual number of twin births in the United States for selected years.

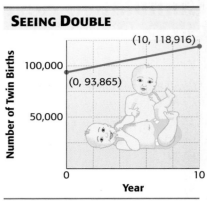

SEEING DOUBLE

Source: National Center for Health Statistics.

(a) Use the information given on the graph to find and interpret the average rate of change in the number of twin births per year.

(b) If $x = 0$ represents 1990, use your answer from part (a) to write an equation of the line in slope-intercept form that models the annual number of twin births for the years 1990 through 2000.

(c) Use the equation from part (b) to approximate the number of twin births in 2002.

21. Give the domain and range of the relation $\{(-4, -2), (-1, 0), (2, 0), (5, 2)\}$. Does this relation define a function?

Solve each system.

22. $3x - 4y = 1$
$2x + 3y = 12$

23. $3x - 2y = 4$
$-6x + 4y = 7$

24. $x + 3y - 6z = 7$
$2x - y + z = 1$
$x + 2y + 2z = -1$

25. The Star-Spangled Banner that flew over Fort McHenry during the War of 1812 had a perimeter of 144 ft. Its length measured 12 ft more than its width. Use a system of equations to find the dimensions of this flag, which is displayed in the Smithsonian Institution's Museum of American History in Washington, D.C. (*Source:* National Park Service brochure.)

Simplify. Write answers with only positive exponents. Assume all variables represent positive real numbers.

26. $\left(\dfrac{2m^3 n}{p^2}\right)^3$

27. $\dfrac{x^{-6} y^3 z^{-1}}{x^7 y^{-4} z}$

28. $(2m^{-2} n^3)^{-3}$

Perform the indicated operations.

29. $2(3x^2 - 8x + 1) - 4(x^2 - 3x - 9)$

30. $(3x + 2y)(5x - y)$

31. $(8m + 5n)(8m - 5n)$

32. $\dfrac{m^3 - 3m^2 + 5m - 3}{m - 1}$

Factoring

7.1 Greatest Common Factors; Factoring by Grouping

7.2 Factoring Trinomials

7.3 Special Factoring

Summary Exercises on Factoring

7.4 Solving Equations by Factoring

Factoring is used to solve quadratic equations, which have many useful applications. An important one is to express the distance a falling or propelled object travels in a specific time. Such equations are used in astronomy and the space program to describe the motion of objects in space.

In Section 7.4 we use the concepts of this chapter to explore how to find the heights of objects after they are propelled or dropped.

7.1 Greatest Common Factors; Factoring by Grouping

OBJECTIVES

1 Factor out the greatest common factor.

2 Factor by grouping.

 Study Skills Workbook
Activity 8: Study Cards

1 Factor out the greatest common factor.

(a) $7k + 28$

(b) $32m + 24$

(c) $8a - 9$

(d) $5z + 5$

Writing a polynomial as the product of two or more simpler polynomials is called **factoring** the polynomial. For example, the product of $3x$ and $5x - 2$ is $15x^2 - 6x$, and $15x^2 - 6x$ can be factored as the product $3x(5x - 2)$.

$$3x(5x - 2) = 15x^2 - 6x \qquad \text{Multiplying}$$
$$15x^2 - 6x = 3x(5x - 2) \qquad \text{Factoring}$$

Notice that both multiplying and factoring use the distributive property, but in opposite directions. Factoring "undoes" or reverses multiplying.

OBJECTIVE 1 Factor out the greatest common factor. The first step in factoring a polynomial is to find the *greatest common factor* for the terms of the polynomial. The **greatest common factor (GCF)** is the largest term that is a factor of all terms in the polynomial. For example, the greatest common factor for $8x + 12$ is 4, since 4 is the largest term that is a factor of (that is, divides into) both $8x$ and 12. Using the distributive property,

$$8x + 12 = 4(2x) + 4(3)$$
$$= 4(2x + 3).$$

As a check, multiply 4 and $2x + 3$. The result should be $8x + 12$. Using the distributive property this way is called *factoring out the greatest common factor*.

EXAMPLE 1 Factoring Out the Greatest Common Factor

Factor out the greatest common factor.

(a) $9z - 18$
Since 9 is the GCF, factor 9 from each term.
$$9z - 18 = 9 \cdot z - 9 \cdot 2$$
$$= 9(z - 2)$$
Check: $9(z - 2) = 9z - 18$ Original polynomial

(b) $56m + 35p = 7(8m + 5p)$

(c) $2y + 5$ There is no common factor other than 1.

(d) $12 + 24z = 12 \cdot 1 + 12 \cdot 2z$
$$= 12(1 + 2z) \qquad \text{12 is the GCF.}$$
Check: $12(1 + 2z) = 12(1) + 12(2z)$ Distributive property
$$= 12 + 24z \qquad \text{Original polynomial}$$

CAUTION
In Example 1 (d), remember to write the factor 1. *Always check answers by multiplying.*

▶◀◀ **Work Problem 1 at the Side.**

ANSWERS
1. (a) $7(k + 4)$ (b) $8(4m + 3)$
   (c) cannot be factored   (d) $5(z + 1)$

EXAMPLE 2 **Factoring Out the Greatest Common Factor**

Factor out the greatest common factor.

(a) $9x^2 + 12x^3$

The numerical part of the GCF is 3, the largest number that divides into both 9 and 12. For the variable parts, x^2 and x^3, use the least exponent that appears on x; here the least exponent is 2. The GCF is $3x^2$.

$$9x^2 + 12x^3 = 3x^2(3) + 3x^2(4x)$$
$$= 3x^2(3 + 4x)$$

(b) $32p^4 - 24p^3 + 40p^5 = 8p^3(4p) + 8p^3(-3) + 8p^3(5p^2) \quad \text{GCF} = 8p^3$
$$= 8p^3(4p - 3 + 5p^2)$$

(c) $3k^4 - 15k^7 + 24k^9 = 3k^4(1 - 5k^3 + 8k^5)$

Check by multiplying:

$$3k^4(1 - 5k^3 + 8k^5) = 3k^4(1) + 3k^4(-5k^3) + 3k^4(8k^5)$$
$$= 3k^4 - 15k^7 + 24k^9 \quad \text{Original polynomial}$$

(d) $24m^3n^2 - 18m^2n + 6m^4n^3$

The numerical part of the GCF is 6. Here 2 is the least exponent that appears on m, while 1 is the least exponent on n. The GCF is $6m^2n$.

$$24m^3n^2 - 18m^2n + 6m^4n^3 = 6m^2n(4mn) + 6m^2n(-3) + 6m^2n(m^2n^2)$$
$$= 6m^2n(4mn - 3 + m^2n^2)$$

(e) $25x^2y^3 + 30y^5 - 15x^4y^7 = 5y^3(5x^2 + 6y^2 - 3x^4y^4)$

> **Work Problem 2 at the Side.** ▶▶▶

A greatest common factor need not be a monomial. The next example shows a binomial greatest common factor.

EXAMPLE 3 **Factoring Out a Binomial Factor**

Factor out the greatest common factor.

(a) $(x - 5)(x + 6) + (x - 5)(2x + 5)$

The greatest common factor here is $x - 5$.

$$(x - 5)(x + 6) + (x - 5)(2x + 5) = (x - 5)[(x + 6) + (2x + 5)]$$
$$= (x - 5)(x + 6 + 2x + 5)$$
$$= (x - 5)(3x + 11)$$

(b) $z^2(m + n) + x^2(m + n) = (m + n)(z^2 + x^2)$

(c) $p(r + 2s) - q^2(r + 2s) = (r + 2s)(p - q^2)$

(d) $(p - 5)(p + 2) - (p - 5)(3p + 4)$

$$= (p - 5)[(p + 2) - (3p + 4)] \quad \text{Factor out } p - 5.$$
$$= (p - 5)[p + 2 - 3p - 4] \quad \text{Be careful with signs.}$$
$$= (p - 5)[-2p - 2] \quad \text{Combine terms.}$$
$$= (p - 5)[-2(p + 1)] \quad \text{Look for a common factor.}$$
$$= -2(p - 5)(p + 1)$$

> **Work Problem 3 at the Side.** ▶▶▶

2 Factor out the greatest common factor.

(a) $16y^4 + 8y^3$

(b) $14p^2 - 9p^3 + 6p^4$

(c) $15z^2 + 45z^5 - 60z^6$

(d) $4x^2z - 2xz + 8z^2$

(e) $12y^5x^2 + 8y^3x^3$

(f) $5m^4x^3 + 15m^5x^6 - 20m^4x^6$

3 Factor out the greatest common factor.

(a) $(a + 2)(a - 3)$
$\quad + (a + 2)(a + 6)$

(b) $(y - 1)(y + 3)$
$\quad - (y - 1)(y + 4)$

(c) $k^2(a + 5b) + m^2(a + 5b)$

(d) $r^2(y + 6) + r^2(y + 3)$

ANSWERS
2. **(a)** $8y^3(2y + 1)$ **(b)** $p^2(14 - 9p + 6p^2)$
   (c) $15z^2(1 + 3z^3 - 4z^4)$
   (d) $2z(2x^2 - x + 4z)$
   (e) $4y^3x^2(3y^2 + 2x)$
   (f) $5m^4x^3(1 + 3mx^3 - 4x^3)$
3. **(a)** $(a + 2)(2a + 3)$
   (b) $(y - 1)(-1)$, or $-y + 1$
   (c) $(a + 5b)(k^2 + m^2)$
   (d) $r^2(2y + 9)$

4 Factor each polynomial in two ways.

(a) $-k^2 + 3k$

When the coefficient of the term of greatest degree is negative, it is sometimes preferable to factor out the -1 that is understood along with the GCF.

EXAMPLE 4 Factoring Out a Negative Common Factor

Factor $-a^3 + 3a^2 - 5a$ in two ways.

First, a could be used as the common factor, giving

$$-a^3 + 3a^2 - 5a = a(-a^2) + a(3a) + a(-5) \qquad \text{Factor out } a.$$
$$= a(-a^2 + 3a - 5).$$

Because of the leading negative sign, $-a$ could be used as the common factor.

$$-a^3 + 3a^2 - 5a = -a(a^2) + (-a)(-3a) + (-a)(5) \qquad \text{Factor out } -a.$$
$$= -a(a^2 - 3a + 5)$$

Either answer is correct.

(b) $-6r^3 - 5r^2 + 14r$

> **NOTE**
> Example 4 showed two ways of factoring a polynomial. Sometimes there may be a reason to prefer one of these forms over the other, but both are correct. The answer section in this book will usually give the form where the common factor has a positive coefficient.

Work Problem 4 at the Side.

OBJECTIVE 2 **Factor by grouping.** Sometimes the terms of a polynomial have a greatest common factor of 1, but it still may be possible to factor the polynomial by using a process called *factoring by grouping*. We usually factor by grouping when a polynomial has more than three terms. For example, to factor the polynomial

$$ax - ay + bx - by,$$

5 Factor $6p - 6q + rp - rq$.

group the terms as follows.

Terms with common factors

$$(ax - ay) + (bx - by)$$

Then factor $ax - ay$ as $a(x - y)$ and factor $bx - by$ as $b(x - y)$.

$$ax - ay + bx - by = (ax - ay) + (bx - by)$$
$$= a(x - y) + b(x - y)$$

The common factor is $x - y$. The final factored form is

$$ax - ay + bx - by = (x - y)(a + b).$$

Work Problem 5 at the Side.

ANSWERS
4. (a) $k(-k + 3)$ or $-k(k - 3)$
   (b) $r(-6r^2 - 5r + 14)$ or $-r(6r^2 + 5r - 14)$
5. $(p - q)(6 + r)$

EXAMPLE 5 **Factoring by Grouping**

Factor $3x - 3y - ax + ay$.

Grouping terms gives

$$(3x - 3y) + (-ax + ay) = 3(x - y) + a(-x + y).$$

There is no simple common factor here. However, if we factor out $-a$ instead of a in the second group of terms, we get

$$(3x - 3y) + (-ax + ay) = 3(x - y) - a(x - y),$$
$$= (x - y)(3 - a).$$

Check by multiplying.

$$(x - y)(3 - a) = 3x - ax - 3y + ay \quad \text{FOIL}$$
$$= 3x - 3y - ax + ay \quad \text{Rearrange terms.}$$

This final product is the original polynomial.

Work Problem 6 at the Side. ▶▶▶

⑥ Factor $xy - 2y - 4x + 8$.

NOTE

In Example 5, different grouping would lead to the factored form

$$(a - 3)(y - x).$$

Verify by multiplying that this form is also correct.

The steps used in factoring by grouping are listed here.

⑦ Factor $2xy + 3y + 2x + 3$.

Factoring by Grouping

Step 1 **Group terms.** Collect the terms into groups so that each group has a common factor.

Step 2 **Factor within the groups.** Factor out the common factor in each group.

Step 3 **Factor the entire polynomial.** If each group now has a common factor, factor it out. If not, try a different grouping.

Always check the factored form by multiplying.

EXAMPLE 6 **Factoring by Grouping**

Factor $6ax + 12bx + a + 2b$ by grouping.

$$6ax + 12bx + a + 2b = (6ax + 12bx) + (a + 2b) \quad \text{Group terms.}$$

Now factor $6x$ from the first group, and use the identity property of multiplication to introduce the factor 1 in the second group.

$$= 6x(a + 2b) + 1(a + 2b)$$
$$= (a + 2b)(6x + 1) \quad \text{Factor out } a + 2b.$$

Again, as in Example 1(d), remember to write the 1. *Check* by multiplying.

Work Problem 7 at the Side. ▶▶▶

ANSWERS
6. $(x - 2)(y - 4)$
7. $(2x + 3)(y + 1)$

8 Factor.

(a) $mn + 6 + 2n + 3m$

EXAMPLE 7 Rearranging Terms before Factoring by Grouping

Factor $p^2q^2 - 10 - 2q^2 + 5p^2$.

Neither the first two terms nor the last two terms have a common factor except 1. Rearrange and group the terms as follows.

$$(p^2q^2 - 2q^2) + (5p^2 - 10) \qquad \text{Rearrange and group the terms.}$$
$$= q^2(p^2 - 2) + 5(p^2 - 2) \qquad \text{Factor out the common factors.}$$
$$= (p^2 - 2)(q^2 + 5) \qquad \text{Factor out } p^2 - 2.$$

Check: $(p^2 - 2)(q^2 + 5) = p^2q^2 + 5p^2 - 2q^2 - 10 \qquad$ FOIL
$$= p^2q^2 - 10 - 2q^2 + 5p^2 \qquad \text{Original polynomial}$$

CAUTION
In Example 7, do not stop at the step

$$q^2(p^2 - 2) + 5(p^2 - 2).$$

This expression is *not in factored form* because it is a *sum* of two terms, $q^2(p^2 - 2)$ and $5(p^2 - 2)$, not a product.

Work Problem 8 at the Side.

(b) $4y - zx + yx - 4z$

ANSWERS
8. (a) $(m + 2)(n + 3)$ (b) $(y - z)(x + 4)$

7.1 Exercises

FOR EXTRA HELP

Tutor Center Addison-Wesley Math Tutor Center

Math XL MathXL

Digital Video Tutor CD 3 Videotape 3

Student's Solutions Manual

MyMathLab

Interactmath.com

1. Explain in your own words what it means to factor a polynomial.

2. What is the first step in attempting to factor a polynomial?

Find the greatest common factor for each list of terms.

3. $9m^3, 3m^2, 15m$

4. $4a^2, 6ab, 2a^3$

5. $6m(r + t)^2, 3p(r + t)^4$

6. $7z^2(m + n)^4, 9z^3(m + n)^5$

7. Which one of the following has the greatest common factor of $6x^3y^4 - 12x^5y^2 + 24x^4y^8$ as one of the factors?
   A. $6x^3y^2(y^2 - 2x^2 + 4xy^6)$
   B. $6xy(x^2y^3 - 2x^4y + 4x^3y^7)$
   C. $2x^3y^2(3y^2 - 6x^2 + 12xy^6)$
   D. $6x^2y^2(xy^2 - 2x^3 + 4x^2y^6)$

8. When directed to factor the polynomial $4x^2y^5 - 8xy^3$ completely, a student responded with $2xy^3(2xy^2 - 4)$. When the teacher did not give him full credit, he complained because when his factors are multiplied, the product is the original polynomial. Was the teacher justified in her grading? Why or why not?

Factor out the greatest common factor. See Examples 1–4.

9. $10x - 30$

10. $15y - 60$

11. $8s + 16t$

12. $35p + 70q$

13. $6 + 12r$

14. $9 + 18m$

15. $8k^3 + 24k$

16. $9z^4 + 27z$

17. $3xy - 5xy^2$

18. $5h^2j + 7hj$

19. $-4p^3q^4 - 2p^2q^5$

20. $-3z^5w^2 - 18z^3w^4$

21. $21x^5 + 35x^4 - 14x^3$

22. $18k^3 - 36k^4 + 48k^5$

23. $15a^2c^3 - 25ac^2 + 5a^2c$

24. $15y^3z^3 + 27y^2z^4 - 36yz^5$

25. $-27m^3p^5 + 5r^4s^3 - 8x^5z^4$

26. $-50r^4t^2 + 81x^3y^3 - 49p^2q^4$

27. $(m - 4)(m + 2) + (m - 4)(m + 3)$

28. $(z - 5)(z + 7) + (z - 5)(z + 9)$

29. $(2z - 1)(z + 6) - (2z - 1)(z - 5)$

30. $(3x + 2)(x - 4) - (3x + 2)(x + 8)$

31. $5(2 - x)^3 - (2 - x)^4 + 4(2 - x)^2$

32. $3(5 - x)^4 + 2(5 - x)^3 - (5 - x)^2$

Factor each polynomial twice. First use a common factor with a positive coefficient, and then use a common factor with a negative coefficient. See Example 4.

33. $-r^3 + 3r^2 + 5r$

34. $-t^4 + 8t^3 - 12t$

35. $-12s^5 + 48s^4$

36. $-16y^4 + 64y^3$

37. $-2x^5 + 6x^3 + 4x^2$

38. $-5a^3 + 10a^4 - 15a^5$

Factor by grouping. See Examples 5–7.

39. $mx + 3qx + my + 3qy$

40. $2k + 2h + jk + jh$

41. $10m + 2n + 5mk + nk$

42. $3ma + 3mb + 2ab + 2b^2$

43. $4 - 2q - 6p + 3pq$

44. $20 + 5m + 12n + 3mn$

45. $p^2 - 4zq + pq - 4pz$

46. $r^2 - 9tw + 3rw - 3rt$

47. $6y^2 + 9y + 4xy + 6x$

48. $8xy + 12x^2 + 10y + 15x$

49. $m^3 + 4m^2 - 6m - 24$

50. $2a^3 + a^2 - 14a - 7$

51. $-3a^3 - 3ab^2 + 2a^2b + 2b^3$

52. $-16m^3 + 4m^2p^2 - 4mp + p^3$

53. $4 + xy - 2y - 2x$

54. $2ab^2 - 4 - 8b^2 + a$

55. $8 + 9y^4 - 6y^3 - 12y$

56. $x^3y^2 - 3 - 3y^2 + x^3$

Factor out the variable that is raised to the lesser exponent. (For example, in Exercise 57, factor out m^{-5}.)

57. $3m^{-5} + m^{-3}$

58. $k^{-2} + 2k^{-4}$

59. $3p^{-3} + 2p^{-2}$

7.2 Factoring Trinomials

OBJECTIVE 1 **Factor trinomials when the coefficient of the squared term is 1.** We begin by finding the product of $x + 3$ and $x - 5$.

$$(x + 3)(x - 5) = x^2 - 5x + 3x - 15$$
$$= x^2 - 2x - 15$$

By this result, the factored form of $x^2 - 2x - 15$ is $(x + 3)(x - 5)$.

Multiplying

Factored form $\longrightarrow (x + 3)(x - 5) = x^2 - 2x - 15 \longleftarrow$ Product

Factoring

Since multiplying and factoring are operations that "undo" each other, factoring trinomials involves using FOIL backwards. As shown here, the x^2-term comes from multiplying x and x, and -15 comes from multiplying 3 and -5.

Product of x and x is x^2.

$$(x + 3)(x - 5) = x^2 - 2x - 15$$

Product of 3 and -5 is -15.

We find the $-2x$ in $x^2 - 2x - 15$ by multiplying the outer terms, multiplying the inner terms, and adding.

Outer terms: $x(-5) = -5x$

$$(x + 3)(x - 5)$$

Add to get $-2x$.

Inner terms: $3 \cdot x = 3x$

Based on this example, follow these steps to factor a trinomial $x^2 + bx + c$, where 1 is the coefficient of the squared term.

Factoring $x^2 + bx + c$

Step 1 **Find pairs whose product is c.** Find all pairs of integers whose product is the third term of the trinomial, c.

Step 2 **Find pairs whose sum is b.** Choose the pair whose sum is the coefficient of the middle term, b.

If there are no such integers, the polynomial cannot be factored. A polynomial that cannot be factored with integer coefficients is a **prime polynomial.**

Some examples of prime polynomials are

$$x^2 + x + 2, \quad x^2 - x - 1, \quad \text{and} \quad 2x^2 + x + 7. \qquad \text{Prime polynomials}$$

OBJECTIVES

1 Factor trinomials when the coefficient of the squared term is 1.

2 Factor trinomials when the coefficient of the squared term is not 1.

3 Use an alternative method for factoring trinomials.

4 Factor by substitution.

❶ Factor each polynomial.

(a) $p^2 + 6p + 5$

(b) $a^2 + 9a + 20$

(c) $k^2 - k - 6$

(d) $b^2 - 7b + 10$

(e) $y^2 - 8y + 6$

❷ Factor each polynomial.

(a) $m^2 + 2mn - 8n^2$

(b) $z^2 - 7zx + 9x^2$

EXAMPLE 1 Factoring Trinomials in $x^2 + bx + c$ Form

Factor each polynomial.

(a) $y^2 + 2y - 35$

Step 1 Find pairs of numbers whose product is -35.

$$-35(1)$$
$$35(-1)$$
$$7(-5)$$
$$5(-7)$$

Step 2 Write sums of those numbers.

$$-35 + 1 = -34$$
$$35 + (-1) = 34$$
$$7 + (-5) = 2 \leftarrow \text{Coefficient of the middle term}$$
$$5 + (-7) = -2$$

The required numbers are 7 and -5, so

$$y^2 + 2y - 35 = (y + 7)(y - 5).$$

Check by finding the product of $y + 7$ and $y - 5$.

(b) $r^2 + 8r + 12$

Look for two numbers with a product of 12 and a sum of 8. Of all pairs of numbers having a product of 12, only the pair 6 and 2 has a sum of 8. Therefore,

$$r^2 + 8r + 12 = (r + 6)(r + 2).$$

Because of the commutative property, it would be equally correct to write $(r + 2)(r + 6)$. *Check* by multiplying.

EXAMPLE 2 Recognizing a Prime Polynomial

Factor $m^2 + 6m + 7$.

Look for two numbers whose product is 7 and whose sum is 6. Only two pairs of integers, 7 and 1 and -7 and -1, give a product of 7. Neither of these pairs has a sum of 6, so $m^2 + 6m + 7$ cannot be factored with integer coefficients and is prime.

◀◀ Work Problem 1 at the Side.

Factoring a trinomial that has more than one variable uses a similar process.

EXAMPLE 3 Factoring a Trinomial in Two Variables

Factor $p^2 + 6ap - 16a^2$.

Look for two expressions whose product is $-16a^2$ and whose sum is $6a$. The quantities $8a$ and $-2a$ have the necessary product and sum, so

$$p^2 + 6ap - 16a^2 = (p + 8a)(p - 2a).$$

Check: $(p + 8a)(p - 2a) = p^2 - 2ap + 8ap - 16a^2$ FOIL

$$= p^2 + 6ap - 16a^2 \quad \text{Original polynomial}$$

◀◀ Work Problem 2 at the Side.

Sometimes a trinomial will have a common factor that should be factored out first.

ANSWERS
1. (a) $(p + 1)(p + 5)$ (b) $(a + 5)(a + 4)$
   (c) $(k - 3)(k + 2)$ (d) $(b - 5)(b - 2)$
   (e) prime
2. (a) $(m - 2n)(m + 4n)$ (b) prime

EXAMPLE 4 **Factoring a Trinomial with a Common Factor**

Factor $16y^3 - 32y^2 - 48y$.

Start by factoring out the greatest common factor, $16y$.

$$16y^3 - 32y^2 - 48y = \mathbf{16y}(y^2 - 2y - 3)$$

To factor $y^2 - 2y - 3$, look for two integers whose product is -3 and whose sum is -2. The necessary integers are -3 and 1, so

$$16y^3 - 32y^2 - 48y = 16y(y - 3)(y + 1).$$

CAUTION

When factoring, always look for a common factor first. Remember to write the common factor as part of the answer.

Work Problem 3 at the Side. ▶▶▶

OBJECTIVE 2 **Factor trinomials when the coefficient of the squared term is not 1.** We can use a generalization of the method shown in Objective 1 to factor a trinomial of the form $ax^2 + bx + c$, where $a \neq 1$. To factor $3x^2 + 7x + 2$, for example, we first identify the values of a, b, and c.

$$\begin{array}{ccc} ax^2 & + \; bx & + \; c \\ \downarrow & \downarrow & \downarrow \\ 3x^2 & + \; 7x & + \; 2 \end{array}$$

$$a = 3, \quad b = 7, \quad c = 2$$

The product ac is $3 \cdot 2 = 6$, so we must find integers having a product of 6 and a sum of 7 (since the middle term has coefficient $b = 7$). The necessary integers are 1 and 6, so we write $7x$ as $1x + 6x$, or $x + 6x$. Thus,

$$3x^2 + 7x + 2 = 3x^2 + \underbrace{x + 6x}_{x + 6x = 7x} + 2$$

$$= (3x^2 + x) + (6x + 2) \qquad \text{Factor by grouping.}$$
$$= x(3x + 1) + 2(3x + 1)$$
$$= (3x + 1)(x + 2).$$

EXAMPLE 5 **Factoring a Trinomial in $ax^2 + bx + c$ Form**

Factor $12r^2 - 5r - 2$.

Since $a = 12$, $b = -5$, and $c = -2$, the product ac is -24. The two integers whose product is -24 and whose sum is b, -5, are 3 and -8.

$$12r^2 - 5r - 2 = 12r^2 + 3r - 8r - 2 \qquad \text{Write } -5r \text{ as } 3r - 8r.$$
$$= 3r(4r + 1) - 2(4r + 1) \qquad \text{Factor by grouping.}$$
$$= (4r + 1)(3r - 2) \qquad \text{Factor out the common factor.}$$

Work Problem 4 at the Side. ▶▶▶

OBJECTIVE 3 **Use an alternative method for factoring trinomials.** An alternative approach, the method of trying repeated combinations and using FOIL, is especially helpful when the product ac is large.

❸ Factor $5m^4 - 5m^3 - 100m^2$.

❹ Factor each trinomial.

(a) $3y^2 - 11y - 4$

(b) $6k^2 - 19k + 10$

ANSWERS
3. $5m^2(m - 5)(m + 4)$
4. (a) $(y - 4)(3y + 1)$ (b) $(2k - 5)(3k - 2)$

5 Use the method of Example 6 to factor each trinomial.

(a) $10x^2 + 17x + 3$

(b) $16y^2 - 34y - 15$

(c) $8t^2 - 13t + 5$

EXAMPLE 6 Factoring Trinomials in $ax^2 + bx + c$ Form

Factor each polynomial.

(a) $3x^2 + 7x + 2$

To factor this trinomial we use an alternative method. The goal is to find the correct numbers to fill in the blanks.

$$3x^2 + 7x + 2 = (\underline{\quad}x + \underline{\quad})(\underline{\quad}x + \underline{\quad})$$

Addition signs are used since all the signs in the trinomial indicate addition. The first two expressions have a product of $3x^2$, so they must be $3x$ and x.

$$3x^2 + 7x + 2 = (3x + \underline{\quad})(x + \underline{\quad})$$

The product of the two last terms must be 2, so the numbers must be 2 and 1. There is a choice. The 2 could be used with the $3x$ or with the x. Only one of these choices can give the correct middle term, $7x$. Use the FOIL method to check each one.

$$\overset{3x}{\overbrace{(3x + 2)(x + 1)}} \qquad \overset{6x}{\overbrace{(3x + 1)(x + 2)}}$$
$$\underset{2x}{\underbrace{\qquad}} \qquad\qquad \underset{x}{\underbrace{\qquad}}$$

$$3x + 2x = 5x \qquad\qquad 6x + x = 7x$$

Wrong middle term $\qquad$ Correct middle term

Therefore, $3x^2 + 7x + 2 = (3x + 1)(x + 2)$. (Compare to the method on the preceding page.)

(b) $12r^2 - 5r - 2$

To reduce the number of trials, we note that the trinomial has no common factor (except 1). This means that neither of its factors can have a common factor. We should keep this in mind as we choose factors. We try 4 and 3 for the two first terms.

$$12r^2 - 5r - 2 = (4r\underline{\quad})(3r\underline{\quad})$$

The factors of -2 are -2 and 1 or 2 and -1. Try both possibilities.

$$(4r - 2)(3r + 1) \qquad \overset{8r}{\overbrace{(4r - 1)(3r + 2)}}$$
$$\qquad\qquad\qquad\qquad \underset{-3r}{\underbrace{\qquad}}$$

Wrong: $4r - 2$ has a common factor of 2. This cannot be correct, since 2 is not a factor of $12r^2 - 5r - 2$.

$$8r - 3r = 5r$$

Wrong middle term

The middle term on the right is $5r$, instead of the $-5r$ that is needed. We get $-5r$ by interchanging the signs in the factors.

$$\overset{-8r}{\overbrace{(4r + 1)(3r - 2)}}$$
$$\underset{3r}{\underbrace{\qquad}}$$

$$-8r + 3r = -5r$$

Correct middle term

Thus, $12r^2 - 5r - 2 = (4r + 1)(3r - 2)$. (Compare to Example 5.)

◀◀◀ Work Problem 5 at the Side.

ANSWERS

5. (a) $(5x + 1)(2x + 3)$
   (b) $(8y + 3)(2y - 5)$
   (c) $(8t - 5)(t - 1)$

This alternative method of factoring a trinomial $ax^2 + bx + c$, $a \neq 1$, is summarized here.

Factoring $ax^2 + bx + c$

Step 1 **Find pairs whose product is a.** Write all pairs of integer factors of the coefficient of the squared term, a.

Step 2 **Find pairs whose product is c.** Write all pairs of integer factors of the last term, c.

Step 3 **Choose inner and outer terms.** Use FOIL and various combinations of the factors from Steps 1 and 2 until the necessary middle term is found.

If no such combinations exist, the trinomial is prime.

EXAMPLE 7 **Factoring a Trinomial in Two Variables**

Factor $18m^2 - 19mx - 12x^2$.

There is no common factor (except 1). Follow the steps to factor the trinomial. There are many possible factors of both 18 and -12. Try 6 and 3 for 18 and -3 and 4 for -12.

$$(6m - 3x)(3m + 4x) \qquad\bigg|\qquad (6m + 4x)(3m - 3x)$$

Wrong: common factor | Wrong: common factors

Since 6 and 3 do not work in this situation, try 9 and 2 instead, with -4 and 3 as factors of -12.

$$(9m + 3x)(2m - 4x) \qquad\bigg|\qquad (9m - 4x)(2m + 3x)$$

$27mx$

$-8mx$

Wrong: common factors | $27mx + (-8mx) = 19mx$

The result on the right differs from the correct middle term only in sign, so interchange the signs in the factors. *Check* by multiplying.

$$18m^2 - 19mx - 12x^2 = (9m + 4x)(2m - 3x)$$

Work Problem 6 at the Side. ▶▶▶

EXAMPLE 8 **Factoring $ax^2 + bx + c$, with $a < 0$**

Factor $-3x^2 + 16x + 12$.

While it is possible to factor this trinomial directly, it is helpful to first factor out -1. Then proceed as in the earlier examples.

$$-3x^2 + 16x + 12 = -1(3x^2 - 16x - 12)$$
$$= -1(3x + 2)(x - 6)$$
$$= -(3x + 2)(x - 6)$$

This factored form can be written in other ways. Two of them are

$$(-3x - 2)(x - 6) \quad \text{and} \quad (3x + 2)(-x + 6).$$

Verify that these both give the original trinomial when multiplied.

Work Problem 7 at the Side. ▶▶▶

6 Factor each trinomial.

(a) $7p^2 + 15pq + 2q^2$

(b) $6m^2 + 7mn - 5n^2$

(c) $12z^2 - 5zy - 2y^2$

(d) $8m^2 + 18mx - 5x^2$

7 Factor each trinomial.

(a) $-6r^2 + 13r + 5$

(b) $-8x^2 + 10x - 3$

ANSWERS
6. **(a)** $(7p + q)(p + 2q)$
   (b) $(3m + 5n)(2m - n)$
   (c) $(3z - 2y)(4z + y)$
   (d) $(4m - x)(2m + 5x)$
7. **(a)** $-(2r - 5)(3r + 1)$
   (b) $-(4x - 3)(2x - 1)$

8 Factor each trinomial.

(a) $2m^3 - 4m^2 - 6m$

(b) $12r^4 + 6r^3 - 90r^2$

(c) $30y^5 - 55y^4 - 50y^3$

9 Factor each polynomial.

(a) $6(a-1)^2 + (a-1) - 2$

(b) $8(z+5)^2 - 2(z+5) - 3$

(c) $15(m-4)^2 - 11(m-4) + 2$

10 Factor each trinomial.

(a) $y^4 + y^2 - 6$

(b) $2p^4 + 7p^2 - 15$

(c) $6r^4 - 13r^2 + 5$

ANSWERS
8. (a) $2m(m+1)(m-3)$
   (b) $6r^2(r+3)(2r-5)$
   (c) $5y^3(2y-5)(3y+2)$
9. (a) $(2a-3)(3a-1)$
   (b) $(4z+17)(2z+11)$
   (c) $(3m-13)(5m-22)$
10. (a) $(y^2-2)(y^2+3)$
    (b) $(2p^2-3)(p^2+5)$
    (c) $(3r^2-5)(2r^2-1)$

EXAMPLE 9 Factoring a Trinomial with a Common Factor

Factor $16y^3 + 24y^2 - 16y$.

$$16y^3 + 24y^2 - 16y = 8y(2y^2 + 3y - 2) \qquad \text{GCF} = 8y$$
$$= 8y(2y-1)(y+2) \qquad \text{Remember the common factor.}$$

Work Problem 8 at the Side.

OBJECTIVE 4 Factor by substitution. Sometimes we can factor a more complicated polynomial by substituting a variable for an expression.

EXAMPLE 10 Factoring a Polynomial Using Substitution

Factor $2(x+3)^2 + 5(x+3) - 12$.

Since the binomial $x + 3$ appears to powers 2 and 1, we let the substitution variable represent $x + 3$. We may choose any letter we wish except x. We choose y to equal $x + 3$.

$$2(x+3)^2 + 5(x+3) - 12 = 2y^2 + 5y - 12 \qquad \text{Let } y = x + 3.$$
$$= (2y-3)(y+4) \qquad \text{Factor.}$$

Now we replace y with $x + 3$ to get

$$2(x+3)^2 + 5(x+3) - 12 = [2(x+3) - 3][(x+3) + 4]$$
$$= (2x + 6 - 3)(x + 7)$$
$$= (2x + 3)(x + 7).$$

CAUTION
Remember to make the final substitution of $x + 3$ for y in Example 10.

Work Problem 9 at the Side.

EXAMPLE 11 Factoring a Trinomial in $ax^4 + bx^2 + c$ Form

Factor $6y^4 + 7y^2 - 20$.

The variable y appears to powers in which the larger exponent is twice the smaller exponent. We can let a substitution variable equal the smaller power. Here, we let $m = y^2$.

$$6y^4 + 7y^2 - 20 = 6(y^2)^2 + 7y^2 - 20$$
$$= 6m^2 + 7m - 20 \qquad \text{Subtitute.}$$
$$= (3m-4)(2m+5) \qquad \text{Factor.}$$
$$= (3y^2-4)(2y^2+5) \qquad m = y^2$$

NOTE
Some students feel comfortable factoring polynomials like the one in Example 11 directly, without using the substitution method.

Work Problem 10 at the Side.

7.2 Exercises

FOR EXTRA HELP

Addison-Wesley Math Tutor Center | Math XL MathXL | Digital Video Tutor CD 3 Videotape 3 | Student's Solutions Manual | MyMathLab | Interactmath.com

1. Which one of the following is *not* a valid way of starting the process of factoring $12x^2 + 29x + 10$?

   A. $(12x \quad)(x \quad)$ **B.** $(4x \quad)(3x \quad)$

   C. $(6x \quad)(2x \quad)$ **D.** $(8x \quad)(4x \quad)$

2. Which one of the following is the completely factored form of $2x^6 - 5x^5 - 3x^4$?

   A. $x^4(2x + 1)(x - 3)$ **B.** $x^4(2x - 1)(x + 3)$

   C. $(2x^5 + x^4)(x - 3)$ **D.** $x^3(2x^2 + x)(x - 3)$

3. Which one of the following is the completely factored form of $4x^2 - 4x - 24$?

   A. $4(x - 2)(x + 3)$ **B.** $4(x + 2)(x + 3)$

   C. $4(x + 2)(x - 3)$ **D.** $4(x - 2)(x - 3)$

4. Which one of the following is *not* a factored form of $-x^2 + 16x - 60$?

   A. $(x - 10)(-x + 6)$ **B.** $(-x - 10)(x + 6)$

   C. $(-x + 10)(x - 6)$ **D.** $-1(x - 10)(x - 6)$

Factor each trinomial. See Examples 1–9.

5. $y^2 + 7y - 30$

6. $z^2 + 2z - 24$

7. $p^2 - p - 56$

8. $k^2 - 11k + 30$

9. $-m^2 + 16m - 60$

10. $-p^2 + 6p + 27$

11. $a^2 - 2ab - 35b^2$

12. $z^2 + 8zw + 15w^2$

13. $y^2 - 3yq - 15q^2$

14. $k^2 - 11hk + 28h^2$

15. $x^2y^2 + 11xy + 18$

16. $p^2q^2 - 5pq - 18$

17. $-6m^2 - 13m + 15$

18. $-15y^2 + 17y + 18$

19. $10x^2 + 3x - 18$

20. $8k^2 + 34k + 35$

21. $20k^2 + 47k + 24$

22. $27z^2 + 42z - 5$

23. $15a^2 - 22ab + 8b^2$

24. $15p^2 + 24pq + 8q^2$

25. $36m^2 - 60m + 25$

26. $25r^2 - 90r + 81$

27. $40x^2 + xy + 6y^2$

28. $14c^2 - 17cd - 6d^2$

29. $6x^2z^2 + 5xz - 4$

30. $8m^2n^2 - 10mn + 3$

31. $24x^2 + 42x + 15$

32. $36x^2 + 18x - 4$

33. $-15a^2 - 70a + 120$

34. $-12a^2 - 10a + 42$

35. $-11x^3 + 110x^2 - 264x$

36. $-9k^3 - 36k^2 + 189k$

37. $2x^3y^3 - 48x^2y^4 + 288xy^5$

38. $6m^3n^2 - 24m^2n^3 - 30mn^4$

Factor each trinomial. See Example 10.

39. $10(k + 1)^2 - 7(k + 1) + 1$

40. $4(m - 5)^2 - 4(m - 5) - 15$

41. $3(m + p)^2 - 7(m + p) - 20$

42. $4(x - y)^2 - 23(x - y) - 6$

43. $a^2(a + b)^2 - ab(a + b)^2 - 6b^2(a + b)^2$

44. $m^2(m - p) + mp(m - p) - 2p^2(m - p)$

Factor each trinomial. See Example 11.

45. $2x^4 - 9x^2 - 18$

46. $6z^4 + z^2 - 1$

47. $16x^4 + 16x^2 + 3$

48. $9r^4 + 9r^2 + 2$

49. $12p^6 - 32p^3r + 5r^2$

50. $2y^6 + 7xy^3 + 6x^2$

RELATING CONCEPTS (EXERCISES 51–56)   For Individual or Group Work

If the terms of a polynomial have no common factor, then none of the terms of its factors can have a common factor, as seen in Examples 6 and 7. ***Work Exercises 51–56 in order.***

51. Is 2 a factor of the composite number 45?

52. List all positive integer factors of 45. Is 2 a factor of any of these factors?

53. Is 5 a factor of $10x^2 + 29x + 10$?

54. Factor $10x^2 + 29x + 10$. Is 5 a factor of either of its factors?

55. Suppose that k is an odd integer and you are asked to factor $2x^2 + kx + 8$. Why is $2x + 4$ not a possible choice for a factor in factoring this polynomial?

56. The polynomial $12y^2 - 11y - 15$ can be factored using the methods of this section. Explain why $3y + 15$ cannot be one of its factors.

7.3 Special Factoring

OBJECTIVE 1 Factor a difference of squares. The special products introduced in **Section 6.4** are used in reverse when factoring. Recall that the product of the sum and difference of two terms leads to a **difference of squares,** a pattern that occurs often when factoring.

Difference of Squares

$$x^2 - y^2 = (x + y)(x - y)$$

EXAMPLE 1 Factoring Differences of Squares

Factor each polynomial.

(a) $4a^2 - 64$

There is a common factor of 4.

$$4a^2 - 64 = 4(a^2 - 16) \qquad \text{Factor out the common factor.}$$
$$= 4(a + 4)(a - 4) \qquad \text{Factor the difference of squares.}$$

(b) $16m^2 - 49p^2 = (4m)^2 - (7p)^2 = (4m + 7p)(4m - 7p)$

with the pattern $A^2 - B^2 = (A + B)(A - B)$

(c) $81k^2 - (a + 2)^2 = (9k)^2 - (a + 2)^2 = (9k + \overline{a + 2})(9k - [a + 2])$
$$= (9k + a + 2)(9k - a - 2)$$

with the pattern $A^2 - B^2 = (A + B)(A - B)$

We could have used the method of substitution here.

(d) $x^4 - 81 = (x^2 + 9)(x^2 - 9) \qquad \text{Factor the difference of squares.}$
$$= (x^2 + 9)(x + 3)(x - 3) \qquad \text{Factor } x^2 - 9.$$

Work Problem 1 at the Side. ▶▶▶

> **CAUTION**
> *Assuming no greatest common factor except 1, it is not possible to factor (with real numbers) a sum of squares,* such as $x^2 + 9$ in Example 1(d). In particular, $x^2 + y^2 \neq (x + y)^2$, as shown next.

OBJECTIVE 2 Factor a perfect square trinomial. Two other special products from **Section 6.4** lead to the following rules for factoring.

Perfect Square Trinomial

$$x^2 + 2xy + y^2 = (x + y)^2$$
$$x^2 - 2xy + y^2 = (x - y)^2$$

OBJECTIVES

1 Factor a difference of squares.

2 Factor a perfect square trinomial.

3 Factor a difference of cubes.

4 Factor a sum of cubes.

1 Factor each polynomial.

(a) $2x^2 - 18$

(b) $9a^2 - 16b^2$

(c) $(m + 3)^2 - 49z^2$

(d) $y^4 - 16$

ANSWERS
1. **(a)** $2(x + 3)(x - 3)$
(b) $(3a - 4b)(3a + 4b)$
(c) $(m + 3 + 7z)(m + 3 - 7z)$
(d) $(y^2 + 4)(y + 2)(y - 2)$

2 Identify any perfect square trinomials.

(a) $z^2 + 12z + 36$

(b) $2x^2 - 4x + 4$

(c) $9a^2 + 12ab + 16b^2$

3 Factor each polynomial.

(a) $49z^2 - 14zk + k^2$

(b) $9a^2 + 48ab + 64b^2$

(c) $(k + m)^2 - 12(k + m) + 36$

(d) $x^2 - 2x + 1 - y^2$

Because the trinomial $x^2 + 2xy + y^2$ is the square of $x + y$, it is called a **perfect square trinomial.** In this pattern, both the first and the last terms of the trinomial must be perfect squares. In the factored form $(x + y)^2$, twice the product of the first and the last terms must give the middle term of the trinomial. It is important to understand these patterns in terms of words, since they occur with many different symbols (other than x and y).

$$4m^2 + 20m + 25$$
Perfect square trinomial
since $20m = 2(2m)(5)$

$$p^2 - 8p + 64$$
Not a perfect square trinomial;
middle term should be $16p$ or $-16p$.

▐▐◀ **Work Problem 2 at the Side.**

EXAMPLE 2 **Factoring Perfect Square Trinomials**

Factor each polynomial.

(a) $144p^2 - 120p + 25$

Here $144p^2 = (12p)^2$ and $25 = 5^2$. The sign on the middle term is $-$, so if $144p^2 - 120p + 25$ is a perfect square trinomial, the factored form will have to be

$$(12p - 5)^2.$$

Take twice the product of the two terms to see if this is correct.

$$2(12p)(-5) = -120p$$

This is the middle term of the given trinomial, so

$$144p^2 - 120p + 25 = (12p - 5)^2.$$

(b) $4m^2 + 20mn + 49n^2$

If this is a perfect square trinomial, it will equal $(2m + 7n)^2$. By the pattern described earlier, if multiplied out, this squared binomial has a middle term of $2(2m)(7n) = 28mn$, which *does not equal* $20mn$. Verify that this trinomial cannot be factored by the methods of the previous section either. It is prime.

(c) $(r + 5)^2 + 6(r + 5) + 9 = [(r + 5) + 3]^2$
$$= (r + 8)^2,$$

since $2(r + 5)(3) = 6(r + 5)$, the middle term.

(d) $m^2 - 8m + 16 - p^2$

Since there are four terms, we will use factoring by grouping. The first three terms here form a perfect square trinomial. Group them together, and factor as follows.

$$(m^2 - 8m + 16) - p^2 = (m - 4)^2 - p^2$$

The result is the difference of squares. Factor again to get

$$= (m - 4 + p)(m - 4 - p).$$

▐▐◀ **Work Problem 3 at the Side.**

Perfect square trinomials, of course, can be factored using the general methods shown earlier for other trinomials. The patterns given here provide "shortcuts."

ANSWERS
2. (a) perfect square trinomial
   (b) not a perfect square trinomial
   (c) not a perfect square trinomial
3. (a) $(7z - k)^2$ (b) $(3a + 8b)^2$
   (c) $[(k + m) - 6]^2$ or $(k + m - 6)^2$
   (d) $(x - 1 + y)(x - 1 - y)$

OBJECTIVE 3 **Factor a difference of cubes.** A difference of cubes, such as $x^3 - y^3$, can be factored as follows.

4 Factor each polynomial.

(a) $x^3 - 1000$

Difference of Cubes

$$x^3 - y^3 = (x - y)(x^2 + xy + y^2)$$

We could check this pattern by finding the product of $x - y$ and $x^2 + xy + y^2$.

EXAMPLE 3 **Factoring Differences of Cubes**

Factor each polynomial.

(a) $m^3 - 8 = m^3 - 2^3$

$\qquad\qquad = (m - 2)(m^2 + 2m + 2^2)$

$\qquad\qquad = (m - 2)(m^2 + 2m + 4)$

Check:
$$(m - 2)(m^2 + 2m + 4)$$
with m^3 and -8 from the outer product, and $-2m$ from the inner:

— Opposite of the product of the cube roots gives the middle term.

(b) $8k^3 - y^3$

(b) $27x^3 - 8y^3 = (3x)^3 - (2y)^3$

$\qquad\qquad = (3x - 2y)[(3x)^2 + (3x)(2y) + (2y)^2]$

$\qquad\qquad = (3x - 2y)(9x^2 + 6xy + 4y^2)$

(c) $1000k^3 - 27n^3 = (10k)^3 - (3n)^3$

$\qquad\qquad = (10k - 3n)[(10k)^2 + (10k)(3n) + (3n)^2]$

$\qquad\qquad = (10k - 3n)(100k^2 + 30kn + 9n^2)$

▶ **Work Problem 4 at the Side.** ▶▶▶

OBJECTIVE 4 **Factor a sum of cubes.** While the binomial $x^2 + y^2$ (a sum of *squares*) cannot be factored with real numbers, a **sum of cubes,** such as $x^3 + y^3$, is factored as follows.

(c) $27m^3 - 64$

Sum of Cubes

$$x^3 + y^3 = (x + y)(x^2 - xy + y^2)$$

To verify this result, find the product of $x + y$ and $x^2 - xy + y^2$. Compare this pattern with the pattern for a difference of cubes.

NOTE
The sign of the second term in the binomial factor of a sum or difference of cubes is *always the same* as the sign in the original polynomial. In the trinomial factor, the first and last terms are *always positive;* the sign of the middle term is *the opposite of* the sign of the second term in the binomial factor.

ANSWERS
4. (a) $(x - 10)(x^2 + 10x + 100)$
   (b) $(2k - y)(4k^2 + 2ky + y^2)$
   (c) $(3m - 4)(9m^2 + 12m + 16)$

5 Factor each polynomial.

(a) $8p^3 + 125$

EXAMPLE 4 **Factoring Sums of Cubes**

Factor each polynomial.

(a) $r^3 + 27 = r^3 + 3^3$
$$= (r + 3)(r^2 - 3r + 3^2)$$
$$= (r + 3)(r^2 - 3r + 9)$$

(b) $27z^3 + 125 = (3z)^3 + 5^3$
$$= (3z + 5)[(3z)^2 - (3z)(5) + 5^2]$$
$$= (3z + 5)(9z^2 - 15z + 25)$$

(c) $125t^3 + 216s^6 = (5t)^3 + (6s^2)^3$
$$= (5t + 6s^2)[(5t)^2 - (5t)(6s^2) + (6s^2)^2]$$
$$= (5t + 6s^2)(25t^2 - 30ts^2 + 36s^4)$$

(d) $3x^3 + 192 = 3(x^3 + 64)$
$$= 3(x + 4)(x^2 - 4x + 16)$$

(b) $27m^3 + 125n^3$

CAUTION

A common error is to think that the xy-term has a coefficient of 2 when factoring the sum or difference of cubes. Since there is no coefficient of 2, the trinomials $x^2 + xy + y^2$ and $x^2 - xy + y^2$ cannot be factored further.

◀◀◀ **Work Problem 5 at the Side.**

The special types of factoring in this section are summarized here. *These should be memorized.*

(c) $2x^3 + 2000$

Special Types of Factoring

Difference of Squares	$x^2 - y^2 = (x + y)(x - y)$
Perfect Square Trinomial	$x^2 + 2xy + y^2 = (x + y)^2$
	$x^2 - 2xy + y^2 = (x - y)^2$
Difference of Cubes	$x^3 - y^3 = (x - y)(x^2 + xy + y^2)$
Sum of Cubes	$x^3 + y^3 = (x + y)(x^2 - xy + y^2)$

ANSWERS
5. (a) $(2p + 5)(4p^2 - 10p + 25)$
   (b) $(3m + 5n)(9m^2 - 15mn + 25n^2)$
   (c) $2(x + 10)(x^2 - 10x + 100)$

7.3 Exercises

FOR EXTRA HELP

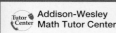 Addison-Wesley Math Tutor Center

 MathXL

 Digital Video Tutor CD 4 Videotape 4

Student's Solutions Manual

MyMathLab

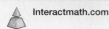 Interactmath.com

1. Which of the following binomials are differences of squares?

   A. $64 - m^2$ **B.** $2x^2 - 25$

   C. $k^2 + 9$ **D.** $4z^4 - 49$

2. Which of the following binomials are sums or differences of cubes?

   A. $64 + y^3$ **B.** $125 - p^6$

   C. $9x^3 + 125$ **D.** $(x + y)^3 - 1$

3. Which of the following trinomials are perfect squares?

   A. $x^2 - 8x - 16$ **B.** $4m^2 + 20m + 25$

   C. $9z^4 + 30z^2 + 25$ **D.** $25a^2 - 45a + 81$

4. Of the twelve polynomials listed in Exercises 1–3, which ones can be factored using the methods of this section?

5. The binomial $9x^2 + 81$ is an example of the sum of two squares that can be factored. Under what conditions can the sum of two squares be factored?

6. Insert the correct signs in the blanks.

   (a) $8 + t^3 = (2 \_\_ t)(4 \_\_ 2t \_\_ t^2)$

   (b) $z^3 - 1 = (z \_\_ 1)(z^2 \_\_ z \_\_ 1)$

Factor each polynomial. See Examples 1–4.

7. $p^2 - 16$ **8.** $k^2 - 9$ **9.** $25x^2 - 4$

10. $36m^2 - 25$ **11.** $18a^2 - 98b^2$ **12.** $32c^2 - 98d^2$

13. $64m^4 - 4y^4$ **14.** $243x^4 - 3t^4$ **15.** $(y + z)^2 - 81$

16. $(h + k)^2 - 9$ **17.** $16 - (x + 3y)^2$ **18.** $64 - (r + 2t)^2$

19. $(p + q)^2 - (p - q)^2$ **20.** $(a + b)^2 - (a - b)^2$ **21.** $k^2 - 6k + 9$

22. $x^2 + 10x + 25$ **23.** $4z^2 + 4zw + w^2$ **24.** $9y^2 + 6yz + z^2$

25. $16m^2 - 8m + 1 - n^2$ **26.** $25c^2 - 20c + 4 - d^2$ **27.** $4r^2 - 12r + 9 - s^2$

28. $9a^2 - 24a + 16 - b^2$ **29.** $x^2 - y^2 + 2y - 1$ **30.** $-k^2 - h^2 + 2kh + 4$

31. $98m^2 + 84mn + 18n^2$ **32.** $80z^2 - 40zw + 5w^2$ **33.** $(p + q)^2 + 2(p + q) + 1$

34. $(x + y)^2 + 6(x + y) + 9$

35. $(a - b)^2 + 8(a - b) + 16$

36. $(m - n)^2 + 4(m - n) + 4$

37. $8x^3 - y^3$

38. $z^3 + 125p^3$

39. $64g^3 + 27h^3$

40. $27a^3 - 8b^3$

41. $24n^3 + 81p^3$

42. $250x^3 - 16y^3$

43. $(y + z)^3 - 64$

44. $(p - q)^3 + 125$

45. $m^6 - 125$

46. $x^6 + 729$

47. $k^6 + (k + 3)^3$

48. $(a + b)^3 - (a - b)^3$

RELATING CONCEPTS (EXERCISES 49–54) For Individual or Group Work

The binomial $x^6 - y^6$ may be considered either as a difference of squares or a difference of cubes. **Work Exercises 49–54 in order.**

49. Factor $x^6 - y^6$ by first factoring as a difference of squares. Then factor further by considering one of the factors as a sum of cubes and the other factor as a difference of cubes.

50. Based on your answer in Exercise 49, fill in the blank with the correct factors so that $x^6 - y^6$ is factored completely:

$x^6 - y^6 = (x - y)(x + y)$ _____ .

51. Factor $x^6 - y^6$ by first factoring as a difference of cubes. Then factor further by considering one of the factors as a difference of squares.

52. Based on your answer in Exercise 51, fill in the blank with the correct factor so that $x^6 - y^6$ is factored:

$x^6 - y^6 = (x - y)(x + y)$ _____ .

53. Notice that the factor you wrote in the blank in Exercise 52 is a fourth-degree polynomial, while the two factors you wrote in the blank in Exercise 50 are both second-degree polynomials. What must be true about the product of the two factors you wrote in the blank in Exercise 50? Verify this.

54. If you have a choice of factoring as a difference of squares or a difference of cubes, how should you start to more easily obtain the factored form of the polynomial? Base the answer on your results in Exercises 49–53 and the methods of factoring explained in this section.

Summary Exercises on Factoring

A polynomial is completely factored when the polynomial is in the form described below.

1. The polynomial is written as a product of prime polynomials with integer coefficients.

2. None of the polynomial factors can be factored further, except that a monomial factor need not be factored completely.

Factoring a Polynomial

Step 1 **Factor out any common factor.**

Step 2 **If the polynomial is a binomial,** check to see if it is the difference of squares, the difference of cubes, or the sum of cubes.

If the polynomial is a trinomial, check to see if it is a perfect square trinomial. If it is not, factor as in **Section 7.2.**

If the polynomial has more than three terms, try to factor by grouping.

Step 3 ***Check the factored form by multiplying.***

Factor each polynomial.

1. $100a^2 - 9b^2$

2. $10r^2 + 13r - 3$

3. $18p^5 - 24p^3 + 12p^6$

4. $15x^2 - 20x$

5. $x^2 + 2x - 35$

6. $9 - a^2 + 2ab - b^2$

7. $225p^2 + 256$

8. $x^3 - 100$

9. $6b^2 - 17b - 3$

10. $k^2 - 6k + 16$

11. $18m^3n + 3m^2n^2 - 6mn^3$

12. $6t^2 + 19tu - 77u^2$

13. $2p^2 + 11pq + 15q^2$

14. $9m^2 - 45m + 18m^3$

15. $4k^2 + 28kr + 49r^2$

16. $54m^3 - 2000$

17. $mn - 2n + 5m - 10$

18. $9m^2 - 30mn + 25n^2 - p^2$

19. $x^3 + 3x^2 - 9x - 27$

20. $56k^3 - 875$

21. $9r^2 + 100$

22. $8p^3 - 125$

23. $6k^2 - k - 1$

24. $27m^2 + 144mn + 192n^2$

25. $x^4 - 625$

26. $125m^6 + 216$

27. $ab + 6b + ac + 6c$

28. $p^3 + 64$

29. $4y^2 - 8y$

30. $6a^4 - 11a^2 - 10$

31. $14z^2 - 3zk - 2k^2$

32. $12z^3 - 6z^2 + 18z$

33. $256b^2 - 400c^2$

34. $z^2 - zp + 20p^2$

35. $1000z^3 + 512$

36. $64m^2 - 25n^2$

37. $10r^2 + 23rs - 5s^2$

38. $12k^2 - 17kq - 5q^2$

39. $32x^2 + 16x^3 - 24x^5$

40. $48k^4 - 243$

41. $14x^2 - 25xq - 25q^2$

42. $5p^2 - 10p$

43. $y^2 + 3y - 10$

44. $b^2 - 7ba - 18a^2$

45. $2a^3 + 6a^2 - 4a$

46. $12m^2rx + 4mnrx + 40n^2rx$

47. $18p^2 + 53pr - 35r^2$

48. $21a^2 - 5ab - 4b^2$

49. $(x - 2y)^2 - 4$

50. $(3m - n)^2 - 25$

51. $(5r + 2s)^2 - 6(5r + 2s) + 9$

52. $(p + 8q)^2 - 10(p + 8q) + 25$

53. $z^4 - 9z^2 + 20$

54. $21m^4 - 32m^2 - 5$

7.4 Solving Equations by Factoring

The equations that we have solved so far in this book have been linear equations. Recall from **Section 2.1** that in a linear equation, the greatest power of the variable is 1. To solve equations of degree greater than 1, other methods must be developed. One of these methods involves factoring.

OBJECTIVES

1 Learn and use the zero-factor property.

2 Solve applied problems that require the zero-factor property.

OBJECTIVE 1 **Learn and use the zero-factor property.** Some equations can be solved by factoring. Solving equations by factoring depends on a special property of the number 0, called the **zero-factor property**.

Zero-Factor Property

If two numbers have a product of 0, then at least one of the numbers must be 0. That is, if $ab = 0$, then either $a = 0$ or $b = 0$.

To prove the zero-factor property, we first assume $a \neq 0$. (If a does equal 0, then the property is proved already.) If $a \neq 0$, then $\frac{1}{a}$ exists, and each side of $ab = 0$ can be multiplied by $\frac{1}{a}$ to get

$$\frac{1}{a} \cdot ab = \frac{1}{a} \cdot 0$$

$$b = 0.$$

Thus, if $a \neq 0$, then $b = 0$, and the property is proved.

CAUTION

If $ab = 0$, then $a = 0$ or $b = 0$. However, if $ab = 6$, for example, it is not necessarily true that $a = 6$ or $b = 6$; in fact, it is very likely that *neither* $a = 6$ *nor* $b = 6$. *The zero-factor property works only for a product equal to 0.*

EXAMPLE 1 **Using the Zero-Factor Property to Solve an Equation**

Solve $(x + 6)(2x - 3) = 0$.

Here the product of $x + 6$ and $2x - 3$ is 0. By the zero-factor property, this can be true only if

$$x + 6 = 0 \quad \text{or} \quad 2x - 3 = 0.$$

Solve these two equations.

$$x + 6 = 0 \quad \text{or} \quad 2x - 3 = 0$$
$$x = -6 \qquad\qquad 2x = 3$$
$$x = \frac{3}{2}$$

The solutions are $x = -6$ or $x = \frac{3}{2}$.

Continued on Next Page

1 Solve each equation.

(a) $(3x + 5)(x + 1) = 0$

Check the solutions 6 and $\frac{3}{2}$ by substitution in the original equation.

If $x = -6$, then

$$(x + 6)(2x - 3) = 0$$
$$(-6 + 6)[2(-6) - 3] = 0 \quad ?$$
$$0(-15) = 0. \quad \text{True}$$

If $x = \frac{3}{2}$, then

$$(x + 6)(2x - 3) = 0$$
$$\left(\frac{3}{2} + 6\right)\left(2 \cdot \frac{3}{2} - 3\right) = 0 \quad ?$$
$$\frac{15}{2}(0) = 0. \quad \text{True}$$

Both solutions check; the solution set is $\left\{-6, \frac{3}{2}\right\}$.

◀◀◀ Work Problem 1 at the Side.

Since the product $(x + 6)(2x - 3)$ equals $2x^2 + 9x - 18$, the equation of Example 1 has a squared term and is an example of a *quadratic equation*. A quadratic equation has degree 2.

> **Quadratic Equation**
>
> An equation that can be written in the form
>
> $$ax^2 + bx + c = 0,$$
>
> where $a \neq 0$, is a **quadratic equation.** This form is called **standard form.**

Quadratic equations are discussed in more detail in **Chapter 10.**

The steps involved in solving a quadratic equation by factoring are summarized below.

(b) $(3x + 11)(5x - 2) = 0$

> **Solving a Quadratic Equation by Factoring**
>
> *Step 1* **Write in standard form.** Rewrite the equation if necessary so that one side is 0.
>
> *Step 2* **Factor** the polynomial.
>
> *Step 3* **Use the zero-factor property.** Set each variable factor equal to 0.
>
> *Step 4* **Find the solution(s).** Solve each equation formed in Step 3.
>
> *Step 5* **Check** each solution in the *original* equation.

EXAMPLE 2 **Solving a Quadratic Equation by Factoring**

Solve each equation.

(a) $2x^2 + 3x = 2$

Step 1
$$2x^2 + 3x = 2$$
$$2x^2 + 3x - 2 = 0 \qquad \text{Standard form}$$

Step 2
$$(x + 2)(2x - 1) = 0 \qquad \text{Factor.}$$

Step 3
$$x + 2 = 0 \quad \text{or} \quad 2x - 1 = 0 \qquad \text{Zero-factor property}$$

Step 4
$$x = -2 \qquad\qquad 2x = 1 \qquad \text{Solve each equation.}$$
$$x = \frac{1}{2}$$

Continued on Next Page

ANSWERS

1. (a) $\left\{-\frac{5}{3}, -1\right\}$ **(b)** $\left\{-\frac{11}{3}, \frac{2}{5}\right\}$

Step 5 Check each solution in the original equation.

If $x = -2$, then

$$2x^2 + 3x = 2$$
$$2(-2)^2 + 3(-2) = 2 \quad ?$$
$$2(4) - 6 = 2 \quad ?$$
$$8 - 6 = 2 \quad ?$$
$$2 = 2. \qquad \text{True}$$

If $x = \frac{1}{2}$, then

$$2x^2 + 3x = 2$$
$$2\left(\frac{1}{2}\right)^2 + 3\left(\frac{1}{2}\right) = 2 \quad ?$$
$$2\left(\frac{1}{4}\right) + \frac{3}{2} = 2 \quad ?$$
$$\frac{1}{2} + \frac{3}{2} = 2 \quad ?$$
$$2 = 2. \qquad \text{True}$$

Because both solutions check, the solution set is $\{-2, \frac{1}{2}\}$.

(b)
$$4x^2 = 4x - 1$$
$$4x^2 - 4x + 1 = 0 \qquad \text{Standard form}$$
$$(2x - 1)^2 = 0 \qquad \text{Factor.}$$
$$2x - 1 = 0 \qquad \text{Zero-factor property}$$
$$2x = 1$$
$$x = \frac{1}{2}$$

There is only one solution because the trinomial is a perfect square. The solution set is $\{\frac{1}{2}\}$.

Work Problem 2 at the Side. ▶▶▶

EXAMPLE 3 **Solving a Quadratic Equation with a Missing Term**

Solve $5x^2 - 25x = 0$.

This quadratic equation has a missing term. Comparing it with the standard form $ax^2 + bx + c = 0$ shows that $c = 0$. The zero-factor property can still be used.

$$5x^2 - 25x = 0$$
$$5x(x - 5) = 0 \qquad \text{Factor.}$$
$$5x = 0 \quad \text{or} \quad x - 5 = 0 \qquad \text{Zero-factor property}$$
$$x = 0 \quad \text{or} \qquad x = 5$$

The solutions are 0 and 5, as can be verified by substituting in the original equation. The solution set is $\{0, 5\}$.

CAUTION
Remember to include 0 as a solution of the equation in Example 3.

Work Problem 3 at the Side. ▶▶▶

❷ Solve each equation.

(a) $3x^2 - x = 4$

(b) $25x^2 = -20x - 4$

❸ Solve each equation.

(a) $x^2 = -12x$

(b) $t^2 - 16 = 0$

ANSWERS

2. (a) $\left\{-1, \frac{4}{3}\right\}$ (b) $\left\{-\frac{2}{5}\right\}$

3. (a) $\{-12, 0\}$ (b) $\{-4, 4\}$

4 Solve.

$$(x + 6)(x - 2) = -8 + x$$

EXAMPLE 4 Solving an Equation That Requires Rewriting

Solve $(2q + 1)(q + 1) = 2(1 - q) + 6$.

$$(2q + 1)(q + 1) = 2(1 - q) + 6$$

$2q^2 + 3q + 1 = 2 - 2q + 6$	Multiply on each side.
$2q^2 + 3q + 1 = 8 - 2q$	Add on the right.
$2q^2 + 5q - 7 = 0$	Standard form
$(2q + 7)(q - 1) = 0$	Factor.
$2q + 7 = 0$ or $q - 1 = 0$	Zero-factor property
$\quad 2q = -7 \qquad q = 1$	
$\quad q = -\dfrac{7}{2}$	

Check that the solution set is $\{-\frac{7}{2}, 1\}$.

Work Problem 4 at the Side.

The zero-factor property can be extended to solve certain polynomial equations of degree 3 or higher, as shown in the next example.

EXAMPLE 5 Solving an Equation of Degree 3

Solve $-x^3 + x^2 = -6x$.

Start by adding $6x$ to each side to get 0 on the right side.

$-x^3 + x^2 + 6x = 0$	
$x^3 - x^2 - 6x = 0$	Multiply by -1.
$x(x^2 - x - 6) = 0$	Factor out x.
$x(x + 2)(x - 3) = 0$	Factor the trinomial.

Use the zero-factor property, extended to include the three variable factors.

$$x = 0 \quad \text{or} \quad x + 2 = 0 \quad \text{or} \quad x - 3 = 0$$
$$x = -2 \qquad x = 3$$

Check that the solution set is $\{-2, 0, 3\}$.

Work Problem 5 at the Side.

5 Solve.

$$3x^3 + x^2 = 4x$$

OBJECTIVE 2 Solve applied problems that require the zero-factor property. An application may lead to a quadratic equation. We continue to use the six-step problem-solving method introduced in **Section 2.3.**

EXAMPLE 6 Using a Quadratic Equation in an Application

A piece of sheet metal is in the shape of a parallelogram. The longer sides of the parallelogram are each 8 m longer than the distance between them. The area of the parallelogram is 48 m². Find the length of the longer sides and the distance between them.

Step 1 **Read** the problem again. There will be two answers.

Step 2 **Assign a variable.** Let x represent the distance between the longer sides. Then $x + 8$ is the length of each longer side. See Figure 1.

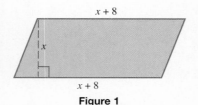

$x + 8$

x

$x + 8$

Figure 1

ANSWERS
4. $\{-4, 1\}$
5. $\left\{-\dfrac{4}{3}, 0, 1\right\}$

Continued on Next Page

Step 3 **Write an equation.** The area of a parallelogram is given by $A = bh$, where b is the length of the longer side and h is the distance between the longer sides. Here $b = x + 8$ and $h = x$.

$$A = bh$$
$$48 = (x + 8)x \qquad \text{Let } A = 48, b = x + 8, h = x.$$

Step 4 **Solve.**

$48 = x^2 + 8x$	Distributive property
$0 = x^2 + 8x - 48$	Standard form
$0 = (x + 12)(x - 4)$	Factor.
$x + 12 = 0 \quad$ or $\quad x - 4 = 0$	Zero-factor property
$x = -12 \quad$ or $\qquad x = 4$	

Step 5 **State the answer.** A distance cannot be negative, so reject -12 as a solution. The only possible solution is 4, so the distance between the longer sides is 4 m. The length of the longer sides is $4 + 8 = 12$ m.

Step 6 **Check.** The length of the longer sides is 8 m more than the distance between them, and the area is $4 \cdot 12 = 48$ m², so the answer checks.

CAUTION
A solution of the equation may not satisfy the physical requirements of the application, as in Example 6. Reject such solutions.

Work Problem 6 at the Side. ▶▶▶

A function defined by a quadratic polynomial is called a *quadratic function*. (See **Chapter 10.**) The next example uses such a function.

EXAMPLE 7 Using a Quadratic Function in an Application

Quadratic functions are used to describe the height a falling object or a propelled object reaches in a specific time. For example, if a small rocket is launched vertically upward from ground level with an initial velocity of 128 ft per sec, then its height in feet after t seconds is a function defined by

$$h(t) = -16t^2 + 128t,$$

if air resistance is neglected. After how many seconds will the rocket be 220 ft above the ground?

We must let $h(t) = 220$ and solve for t.

$220 = -16t^2 + 128t$	Let $h(t) = 220$.
$16t^2 - 128t + 220 = 0$	Standard form
$4t^2 - 32t + 55 = 0$	Divide by 4.
$(2t - 5)(2t - 11) = 0$	Factor.
$2t - 5 = 0 \quad$ or $\quad 2t - 11 = 0$	Zero-factor property
$t = 2.5 \quad$ or $\qquad t = 5.5$	

The rocket will reach a height of 220 ft twice: on its way up at 2.5 sec and again on its way down at 5.5 sec.

Work Problem 7 at the Side. ▶▶▶

6 Solve the problem.
Carl is planning to build a rectangular deck along the back of his house. He wants the area of the deck to be 60 m², and the width to be 1 m less than half the length. What length and width should he use?

7 Solve the problem.
How long will it take the rocket in Example 7 to reach a height of 256 ft?

ANSWERS
6. length: 12 m; width: 5 m
7. 4 sec

Idle Prime Time

A positive integer greater than 1 is a **prime number** if its only factors are 1 and itself. Every positive integer can be written as a product of prime numbers in a unique way, except for the order of the factors. Finding new primes has intrigued people from ancient Greece to modern times. The *Great Internet Mersenne Prime Search* is a consortium headed by George Woltman and Scott Kurowski that has discovered seven world-record primes. On May 15, 2004, Josh Findley discovered the prime number $2^{24,036,583} - 1$. His calculation took over two weeks on his 2.4 GHz Pentium 4 computer. This prime number is nearly a million digits larger than the last prime found and has 7,235,733 decimal digits when written out. (*Source:* http://www.mersenne.org/prime.htm)

Prime numbers are essential in the development of unbreakable codes that, in an era of Internet commerce, ensure security in transmitting and storing computer data.

The oldest known method for finding prime numbers is the Sieve of Eratosthenes, similar to the version shown below. Numbers that are not prime (composite numbers) are eliminated and only the prime numbers are left. Begin with 2. Two is prime but multiples of 2 are not, so delete the remaining numbers in Column 2 and all of Columns 4 and 6. Three is prime, but multiples of 3 are not, so delete the remaining numbers in Column 3. Examine the remaining numbers and eliminate any that are composite (such as 25 or 91). The prime numbers are highlighted.

For Group Discussion

1. **Twin primes** occur in pairs that differ by 2. List all the twin primes from the table.

2. Observe that all prime numbers larger than 3 are in Columns 1 and 5. Each number in Column 5 is 1 less than a multiple of 6, and therefore has the form $6n - 1$. Each number in Column 1 has a similar structure, $6n + 1$. Show that the larger of each of these twin primes, found in the year 2000, has the form $6n + 1$:

   $$1,693,965 \times 2^{66,443} \pm 1$$

   and

   $$4,648,619,711,505 \times 2^{60,000} \pm 1.$$

   (*Hint:* Show that the leading term is divisible by both 2 and 3.)

SIEVE OF ERATOSTHENES

Col 1	Col 2	Col 3	Col 4	Col 5	Col 6
1	2	3	4	5	6
7	8	9	10	11	12
13	14	15	16	17	18
19	20	21	22	23	24
25	26	27	28	29	30
31	32	33	34	35	36
37	38	39	40	41	42
43	44	45	46	47	48
49	50	51	52	53	54
55	56	57	58	59	60
61	62	63	64	65	66
67	68	69	70	71	72
73	74	75	76	77	78
79	80	81	82	83	84
85	86	87	88	89	90
91	92	93	94	95	96
97	98	99	100	101	102

3. **Mersenne primes,** named for the 17th-century French monk Marin Mersenne, have the form $2^p - 1$, where p is a prime number. Not all such numbers are prime. Show that $2^{11} - 1$ is composite and $2^5 - 1$ is prime.

4. A **Sophie Germain prime,** named for an 18th-century French mathematician, is an odd prime p for which $2p + 1$ is also prime. For example, 5 is a Sophie Germain prime since 11 (which is $2 \cdot 5 + 1$) is prime, but 13 is not since 27 (which is $2 \cdot 13 + 1$) is composite. List the Sophie Germain primes from the table.

400

7.4 Exercises

FOR EXTRA HELP

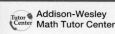 Addison-Wesley Math Tutor Center

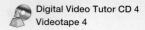 MathXL

 Digital Video Tutor CD 4 Videotape 4

Student's Solutions Manual

MyMathLab

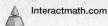 Interactmath.com

1. Explain in your own words how the zero-factor property is used in solving a quadratic equation.

2. One of the following equations is *not* in proper form for using the zero-factor property. Which one is it? Explain why it is not in proper form.
   A. $(x + 2)(x - 6) = 0$
   B. $x(3x - 7) = 0$
   C. $3t(t + 8)(t - 9) = 0$
   D. $y(y - 3) + 6(y - 3) = 0$

Solve each equation using the zero-factor property. See Example 1.

3. $(x + 10)(x - 5) = 0$

4. $(x + 7)(x + 3) = 0$

5. $(3k + 8)(2k - 5) = 0$

6. $(2q + 5)(3q - 4) = 0$

Solve each equation. See Examples 2–4.

7. $m^2 - 3m - 10 = 0$

8. $x^2 + x - 12 = 0$

9. $z^2 + 9z + 18 = 0$

10. $x^2 - 18x + 80 = 0$

11. $2x^2 = 7x + 4$

12. $2x^2 = 3 - x$

13. $15k^2 - 7k = 4$

14. $12x^2 + 4x = 5$

15. $16x^2 + 24x = -9$

16. $49x^2 + 14x = -1$

17. $(5x + 1)(x + 3) = -2(5x + 1)$

18. $(3x + 1)(x - 3) = 2 + 3(x + 5)$

19. $6m^2 - 36m = 0$

20. $-3m^2 + 27m = 0$

21. $-3m^2 + 27 = 0$

22. $-2x^2 + 8 = 0$

23. $4p^2 - 16 = 0$

24. $9x^2 - 81 = 0$

25. $(x - 3)(x + 5) = -7$

26. $(x + 8)(x - 2) = -21$

27. $(2x + 1)(x - 3) = 6x + 3$

28. $(3x + 2)(x - 3) = 7x - 1$

29. $(x + 3)(x - 6) = (2x + 2)(x - 6)$

30. $(2x + 1)(x + 5) = (x + 11)(x + 3)$

Solve each equation. See Example 5.

31. $2x^3 - 9x^2 - 5x = 0$

32. $6x^3 - 13x^2 - 5x = 0$

33. $9t^3 = 16t$

34. $25x^3 = 64x$

35. $2r^3 + 5r^2 - 2r - 5 = 0$

36. $2p^3 + p^2 - 98p - 49 = 0$

37. A student tried to solve the equation in Exercise 33 by first dividing each side by t, obtaining $9t^2 = 16$. She then solved the resulting equation by the zero-factor property to get the solution set $\{-\frac{4}{3}, \frac{4}{3}\}$. What was incorrect about her procedure?

38. Without actually solving each equation, determine which one of the following has 0 in its solution set.

A. $4x^2 - 25 = 0$ **B.** $x^2 + 2x - 3 = 0$

C. $6x^2 + 9x + 1 = 0$ **D.** $x^3 + 4x^2 = 3x$

Solve each problem. See Examples 6 and 7.

39. A garden has an area of 320 ft². Its length is 4 ft more than its width. What are the dimensions of the garden?

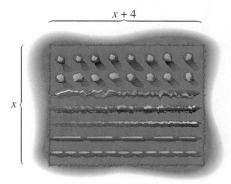

40. A square mirror has sides measuring 2 ft less than the sides of a square painting. If the difference between their areas is 32 ft², find the lengths of the sides of the mirror and the painting.

41. A sign has the shape of a triangle. The length of the base is 3 m less than the height. What are the measures of the base and the height, if the area is 44 m²?

42. The base of a parallelogram is 7 ft more than the height. If the area of the parallelogram is 60 ft², what are the measures of the base and the height?

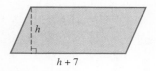

43. A farmer has 300 ft of fencing and wants to enclose a rectangular area of 5000 ft². What dimensions should she use?

44. A rectangular landfill has an area of 30,000 ft². Its length is 200 ft more than its width. What are the dimensions of the landfill?

45. A box with no top is to be constructed from a piece of cardboard whose length measures 6 in. more than its width. The box is to be formed by cutting squares that measure 2 in. on each side from the four corners and then folding up the sides. If the volume of the box will be 110 in.^{3}, what are the dimensions of the piece of cardboard?

46. The surface area of the box with open top shown in the figure is 161 in.^{2}. Find the dimensions of the base. (*Hint:* The surface area is a function defined by $S(x) = x^2 + 16x$.)

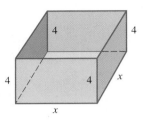

47. Refer to Example 7. After how many seconds will the rocket be 240 ft above the ground? 112 ft above the ground?

48. If an object is propelled upward with an initial velocity of 64 ft per sec from a height of 80 ft, then its height in feet t seconds after it is propelled is a function defined by

$$f(t) = -16t^2 + 64t + 80.$$

How long after it is propelled will it hit the ground? (*Hint:* When it hits the ground, its height is 0 ft.)

49. If a rock is dropped from a building 576 ft high, then its distance in feet from the ground t seconds later is a function defined by

$$f(t) = -16t^2 + 576.$$

How long after it is dropped will it hit the ground?

50. If a baseball is dropped from a helicopter 625 ft above the ground, then its distance in feet from the ground t seconds later is a function defined by

$$f(t) = -16t^2 + 625.$$

How long after it is dropped will it hit the ground?

Chapter 7

SUMMARY

KEY TERMS

7.1 greatest common factor The product of the largest common numerical factor and each variable factor of least degree common to every term in a polynomial is the greatest common factor of the terms of the polynomial.

7.2 prime polynomial A polynomial that cannot be factored with integer coefficients is a prime polynomial.

7.4 standard form of a quadratic equation An equation that can be written in the form $ax^2 + bx + c = 0$, where $a \neq 0$, is a quadratic equation. This form is called standard form.

TEST YOUR WORD POWER

See how well you have learned the vocabulary in this chapter. Answers, with examples, follow the Quick Review.

1. **Factoring** is
   A. a method of multiplying polynomials
   B. the process of writing a polynomial as a product
   C. the answer in a multiplication problem
   D. a way to add the terms of a polynomial.

2. A **difference of squares** is a binomial
   A. that can be factored as the difference of two cubes
   B. that cannot be factored

   C. that is squared
   D. that can be factored as the product of the sum and difference of two terms.

3. A **perfect square trinomial** is a trinomial
   A. that can be factored as the square of a binomial
   B. that cannot be factored
   C. that is multiplied by a binomial
   D. where all terms are perfect squares.

4. A **quadratic equation** is a polynomial equation of
   A. degree one
   B. degree two
   C. degree three
   D. degree four.

5. The **zero-factor property** is used to
   A. factor a perfect square trinomial
   B. factor by grouping
   C. solve a polynomial equation of degree 2 or more
   D. solve a linear equation.

QUICK REVIEW

Concepts	Examples

7.1 *Greatest Common Factors: Factoring by Grouping*

The Greatest Common Factor
The product of the largest common numerical factor and each common variable raised to the least exponent that appears on that variable in any term is the greatest common factor of the terms of the polynomial.

Factor $4x^2y - 50xy^2 = 2^2x^2y - 2 \cdot 5^2xy^2$.
The greatest common factor is $2xy$.

$$4x^2y - 50xy^2 = 2xy(2x - 25y)$$

Factoring by Grouping
Group the terms so that each group has a common factor. Factor out the common factor in each group. If the groups now have a common factor, factor it out. If not, try a different grouping.

Factor by grouping.

$$5a - 5b - ax + bx = (5a - 5b) + (-ax + bx)$$
$$= 5(a - b) - x(a - b)$$
$$= (a - b)(5 - x)$$

405

Concepts	Examples
7.2 *Factoring Trinomials* To factor a trinomial, choose factors of the first term and factors of the last term. Then, place them in a pair of parentheses of this form: $$(\qquad)(\qquad).$$ Try various combinations of the factors until the correct middle term of the trinomial is found.	Factor $15x^2 + 14x - 8$. The factors of 15 are 5 and 3, and 15 and 1. The factors of -8 are -4 and 2, 4 and -2, -1 and 8, and 1 and -8. Various combinations of these factors lead to the correct factorization. $$15x^2 + 14x - 8 = (5x - 2)(3x + 4).$$ Check by multiplying, using the FOIL method.
7.3 *Special Factoring* **Difference of Squares** $$x^2 - y^2 = (x + y)(x - y)$$ **Perfect Square Trinomials** $$x^2 + 2xy + y^2 = (x + y)^2$$ $$x^2 - 2xy + y^2 = (x - y)^2$$ **Difference of Cubes** $$x^3 - y^3 = (x - y)(x^2 + xy + y^2)$$ **Sum of Cubes** $$x^3 + y^3 = (x + y)(x^2 - xy + y^2)$$	$$4m^2 - 25n^2 = (2m)^2 - (5n)^2$$ $$= (2m + 5n)(2m - 5n)$$ $$9y^2 + 6y + 1 = (3y + 1)^2$$ $$16p^2 - 56p + 49 = (4p - 7)^2$$ $$8 - 27a^3 = (2 - 3a)(4 + 6a + 9a^2)$$ $$64z^3 + 1 = (4z + 1)(16z^2 - 4z + 1)$$
7.4 *Solving Equations by Factoring* *Step 1* Rewrite the equation if necessary so that one side is 0. *Step 2* Factor the polynomial. *Step 3* Set each factor equal to 0. *Step 4* Solve each equation from Step 3. *Step 5* Check each solution.	Solve. $\quad 2x^2 + 5x = 3$ $$2x^2 + 5x - 3 = 0 \quad \text{Standard form}$$ $$(x + 3)(2x - 1) = 0$$ $$x + 3 = 0 \quad \text{or} \quad 2x - 1 = 0$$ $$x = -3 \qquad\qquad 2x = 1$$ $$x = \frac{1}{2}$$ A check verifies that the solution set is $\{-3, \frac{1}{2}\}$.

ANSWERS TO TEST YOUR WORD POWER

1. B; *Example:* $x^2 - 5x - 14 = (x - 7)(x + 2)$
2. D; *Example:* $b^2 - 49$ is the difference of the squares b^2 and 7^2. It can be factored as $(b + 7)(b - 7)$.
3. A; *Example:* $a^2 + 2a + 1$ is a perfect square trinomial; its factored form is $(a + 1)^2$.
4. B; *Examples:* $x^2 - 3x + 2 = 0$, $x^2 - 9 = 0$, $2m^2 = 6m + 8$
5. C; *Example:* Use the zero-factor property to write $(x + 4)(x - 2) = 0$ as $x + 4 = 0$ or $x - 2 = 0$, then solve each linear equation to find the solution set $\{-4, 2\}$.

Chapter 7
R E V I E W   E X E R C I S E S

[7.1] *Factor out the greatest common factor.*

1. $21y^2 + 35y$

2. $12q^2b + 8qb^2 - 20q^3b^2$

3. $(x + 3)(4x - 1) - (x + 3)(3x + 2)$

4. $(z + 1)(z - 4) + (z + 1)(2z + 3)$

Factor by grouping.

5. $4m + nq + mn + 4q$

6. $x^2 + 5y + 5x + xy$

7. $2m + 6 - am - 3a$

8. $2am - 2bm - ap + bp$

[7.2] *Factor completely.*

9. $3p^2 - p - 4$

10. $12r^2 - 5r - 3$

11. $10m^2 + 37m + 30$

12. $10k^2 - 11kh + 3h^2$

13. $9x^2 + 4xy - 2y^2$

14. $24x - 2x^2 - 2x^3$

15. $2k^4 - 5k^2 - 3$

16. $p^2(p + 2)^2 + p(p + 2)^2 - 6(p + 2)^2$

[7.3] *Factor completely.*

17. $16x^2 - 25$

18. $9t^2 - 49$

19. $x^2 + 14x + 49$

20. $9k^2 - 12k + 4$

21. $r^3 + 27$

22. $125x^3 - 1$

23. $m^6 - 1$

24. $x^8 - 1$

25. $x^2 + 6x + 9 - 25y^2$

[7.4] *Solve each equation.*

26. $(x + 1)(5x + 2) = 0$

27. $p^2 - 5p + 6 = 0$

28. $6z^2 = 5z + 50$

29. $6r^2 + 7r = 3$

30. $-4m^2 + 36 = 0$

31. $6x^2 + 9x = 0$

32. $(2x + 1)(x - 2) = -3$

33. $x^2 - 8x + 16 = 0$

34. $2x^3 - x^2 - 28x = 0$

Solve each problem.

35. A triangular wall brace creates the shape of a right triangle. One of the perpendicular sides is 1 ft longer than twice the other. The area enclosed by the triangle is 10.5 ft^2. Find the shorter of the perpendicular sides.

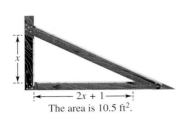

$2x + 1$
The area is 10.5 ft^2.

36. A rectangular parking lot has a length 20 ft more than its width. Its area is 2400 ft^2. What are the dimensions of the lot?

$W + 20$

W

The area is 2400 ft^2.

A rock is propelled directly upward from ground level. After t seconds, its height in feet is given by $f(t) = -16t^2 + 256t$ (if air resistance is neglected).

37. When will the rock return to the ground?

38. After how many seconds will it be 240 ft above the ground?

39. Why does the question in Exercise 38 have two answers?

MIXED REVIEW EXERCISES

Factor completely.

40. $30a + am - am^2$

41. $8 - a^3$

42. $9x^2 + 13xy - 3y^2$

43. $15y^3 + 20y^2$

Solve.

44. $5x^2 - 17x - 12 = 0$

45. $x^3 - x = 0$

46. When Europeans arrived in America, many native Americans of the Northeast lived in *longhouses* that sheltered several related families. The rectangular floor area of a typical Huron longhouse was about 2750 ft^2. The length was 85 ft greater than the width. What were the dimensions of the floor?

Chapter 7

TEST

Factor.

1. $11z^2 - 44z$

1. _____

2. $10x^2y^5 - 5x^2y^3 - 25x^5y^3$

2. _____

3. $3x + by + bx + 3y$

3. _____

4. $-2x^2 - x + 36$

4. _____

5. $6x^2 + 11x - 35$

5. _____

6. $4p^2 + 3pq - q^2$

6. _____

7. $16a^2 + 40ab + 25b^2$

7. _____

8. $x^2 + 2x + 1 - 4z^2$

8. _____

9. $a^3 + 2a^2 - ab^2 - 2b^2$

9. _____

10. $9k^2 - 121j^2$

10. _____

11. $y^3 - 216$

11. _____

12. $6k^4 - k^2 - 35$

12. _____

13. _____

13. $27x^6 + 1$

14. _____

14. $-x^2 + x + 30$

15. _____

15. $(t^2 + 3)^2 + 4(t^2 + 3) - 5$

16. _____

16. Explain why $(x^2 + 2y)p + 3(x^2 + 2y)$ is not in factored form. Then factor the polynomial.

17. _____

17. Which one of the following is *not* a factored form of $-x^2 - x + 12$?
   A. $(3 - x)(x + 4)$ **B.** $-(x - 3)(x + 4)$
   C. $(-x + 3)(x + 4)$ **D.** $(x - 3)(-x + 4)$

Solve each equation.

18. _____

18. $3x^2 + 8x = -4$

19. _____

19. $3x^2 - 5x = 0$

20. _____

20. $5m(m - 1) = 2(1 - m)$

Solve each problem.

21. _____

21. The area of the rectangle shown is 40 in.^{2}. Find the length and the width of the rectangle.

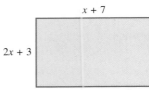

The area is 40 in.^{2}.

22. _____

22. A ball is propelled upward from ground level. After t seconds, its height in feet is a function defined by $f(t) = -16t^2 + 96t$. After how many seconds will it reach a height of 128 ft?

Cumulative Review Exercises

CHAPTERS 1–7

Simplify each expression.

1. $-2(m - 3)$

2. $-(-4m + 3)$

3. $3x^2 - 4x + 4 + 9x - x^2$

Evaluate if $p = -4$, $q = -2$, and $r = 5$.

4. $-3(2q - 3p)$

5. $8r^2 + q^2$

6. $\dfrac{\sqrt{r}}{-p + 2q}$

7. $\dfrac{rp + 6r^2}{p^2 + q - 1}$

Solve.

8. $2z - 5 + 3z = 4 - (z + 2)$

9. $\dfrac{3a - 1}{5} + \dfrac{a + 2}{2} = -\dfrac{3}{10}$

10. $-\dfrac{4}{3}d \geq -5$

11. $3 - 2(m + 3) < 4m$

12. $2k + 4 < 10$ and $3k - 1 > 5$

13. $2k + 4 > 10$ or $3k - 1 < 5$

14. $|5x + 3| - 10 = 3$

15. $|x + 2| < 9$

16. $|2y - 5| \geq 9$

17. $V = lwh$ for h

18. Two planes leave the Dallas-Fort Worth airport at the same time. One travels east at 550 mph, and the other travels west at 500 mph. Assuming no wind, how long will it take for the planes to be 2100 mi apart?

	r	t	d
Eastbound plane	550	x	
Westbound plane	500	x	

19. Graph $4x + 2y = -8$.

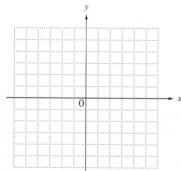

20. Find the slope of the line through the points $(-4, 8)$ and $(-2, 6)$.

21. What is the slope of the line shown here?

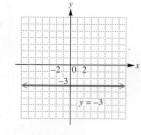

Use the function defined by $f(x) = 2x + 7$ to find the following.

22. $f(-4)$

23. The x-intercept of its graph

24. The y-intercept of its graph

Solve each system.

25. $3x - 2y = -7$
$2x + 3y = 17$

26. $2x + 3y - 6z = 5$
$8x - y + 3z = 7$
$3x + 4y - 3z = 7$

Perform the indicated operations. Assume variables represent nonzero real numbers.

27. $(3x^2y^{-1})^{-2}(2x^{-3}y)^{-1}$

28. $\dfrac{5m^{-2}y^3}{3m^{-3}y^{-1}}$

Perform the indicated operations.

29. $(3x^3 + 4x^2 - 7) - (2x^3 - 8x^2 + 3x)$

30. $(7x + 3y)^2$

31. $(2p + 3)(5p^2 - 4p - 8)$

Factor.

32. $16w^2 + 50wz - 21z^2$

33. $4x^2 - 4x + 1 - y^2$

34. $4y^2 - 36y + 81$

35. $100x^4 - 81$

36. $8p^3 + 27$

Solve.

37. $(p + 4)(2p + 3)(p - 1) = 0$

38. $9q^2 = 6q - 1$

39. A sign is to have the shape of a triangle with a height 3 ft greater than the length of the base. How long should the base be if the area is to be 14 ft²?

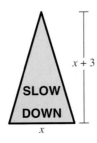

40. A game board has the shape of a rectangle. The longer sides are each 2 in. longer than the distance between them. The area of the board is 288 in.². Find the length of the longer sides and the distance between them.

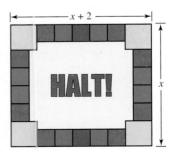

Rational Expressions and Functions

8

8.1 **Rational Expressions and Functions; Multiplying and Dividing**

8.2 **Adding and Subtracting Rational Expressions**

8.3 **Complex Fractions**

8.4 **Equations with Rational Expressions and Graphs**

Summary Exercises on Rational Expressions and Equations

8.5 **Applications of Rational Expressions**

Americans have been car crazy ever since the first automobiles hit the road early in the twentieth century. Today there are about 213.5 million vehicles in the United States driving on 3.4 million miles of paved roadways. There is even a museum devoted exclusively to our four-wheeled passion and its influence on our lives and culture. The Museum of Automobile History in Syracuse, N.Y., features some 200 years of automobile memorabilia, including rare advertising pieces, designer drawings, and Hollywood movie posters. (*Source: Home and Away*, May/June 2002.)

In Exercises 67 and 68 of Section 8.2, we use a *rational expression* to determine the cost of restoring a vintage automobile.

8.1 Rational Expressions and Functions; Multiplying and Dividing

OBJECTIVES

1. Define rational expressions.
2. Define rational functions and describe their domains.
3. Write rational expressions in lowest terms.
4. Multiply rational expressions.
5. Find reciprocals for rational expressions.
6. Divide rational expressions.

OBJECTIVE 1 Define rational expressions. In arithmetic, a rational number is the quotient of two integers, with the denominator not 0. In algebra, a **rational expression** or *algebraic fraction* is the quotient of two polynomials, again with the denominator not 0. For example,

$$\frac{x}{y}, \quad \frac{-a}{4}, \quad \frac{m+4}{m-2}, \quad \frac{8x^2 - 2x + 5}{4x^2 + 5x}, \quad \text{and} \quad x^5 \left(\text{or } \frac{x^5}{1} \right) \quad \text{Rational expressions}$$

are all rational expressions. In other words, rational expressions are the elements of the set

$$\left\{ \frac{P}{Q} \,\middle|\, P \text{ and } Q \text{ are polynomials, with } Q \neq 0 \right\}.$$

OBJECTIVE 2 Define rational functions and describe their domains. A function that is defined by a rational expression is called a **rational function** and has the form

$$f(x) = \frac{P(x)}{Q(x)}, \quad \text{where } Q(x) \neq 0.$$

The domain of a rational function includes all real numbers except those that make $Q(x)$, that is, the denominator, equal to 0. For example, the domain of

$$f(x) = \frac{2}{\underbrace{x-5}_{\text{Cannot equal 0}}}$$

includes all real numbers except 5, because 5 would make the denominator equal to 0.

Figure 1 shows a graph of the function defined by

$$f(x) = \frac{2}{x-5}.$$

Notice that the graph does not exist when $x = 5$. It does not intersect the dashed vertical line whose equation is $x = 5$. This line is an *asymptote*. We will discuss graphs of rational functions in more detail in **Section 8.4.**

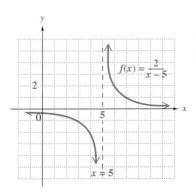

Figure 1

EXAMPLE 1 **Finding Numbers That Are Not in the Domains of Rational Functions**

Find all numbers that are not in the domain of each rational function. Then give the domain using set notation.

(a) $f(x) = \dfrac{3}{7x - 14}$

The only values that cannot be used are those that make the denominator 0. To find these values, set the denominator equal to 0 and solve the resulting equation.

$$7x - 14 = 0$$
$$7x = 14 \quad \text{Add 14.}$$
$$x = 2 \quad \text{Divide by 7.}$$

The number 2 cannot be used as a replacement for x. The domain of f includes all real numbers except 2, written using set notation as $\{x \mid x \neq 2\}$.

(b) $g(x) = \dfrac{3 + x}{x^2 - 4x + 3}$

Set the denominator equal to 0, and solve the equation.

$$x^2 - 4x + 3 = 0$$
$$(x - 1)(x - 3) = 0 \quad \text{Factor.}$$
$$x - 1 = 0 \quad \text{or} \quad x - 3 = 0 \quad \text{Zero-factor property}$$
$$x = 1 \quad \text{or} \quad x = 3$$

The domain of g includes all real numbers except 1 and 3, written $\{x \mid x \neq 1, 3\}$.

(c) $h(x) = \dfrac{8x + 2}{3}$

The denominator, 3, can never be 0, so the domain of h includes all real numbers, written $(-\infty, \infty)$.

(d) $f(x) = \dfrac{2}{x^2 + 4}$

Setting $x^2 + 4$ equal to 0 leads to $x^2 = -4$. There is no real number whose square is -4. Therefore, any real number can be used, and as in part (c), the domain of f includes all real numbers $(-\infty, \infty)$.

> **Work Problem 1 at the Side.** ▶▶▶

OBJECTIVE 3 Write rational expressions in lowest terms. In arithmetic, we write the fraction $\frac{15}{20}$ in lowest terms by dividing the numerator and denominator by 5 to get $\frac{3}{4}$. We write rational expressions in lowest terms in a similar way, using the **fundamental property of rational numbers.**

Fundamental Property of Rational Numbers

If $\frac{a}{b}$ is a rational number and if c is any nonzero real number, then

$$\frac{a}{b} = \frac{ac}{bc}.$$

In words, the numerator and denominator of a rational number may either be multiplied or divided by the same nonzero number without changing the value of the rational number.

---

1 Find all numbers that are not in the domain of each rational function. Then give the domain using set notation.

(a) $f(x) = \dfrac{x + 4}{x - 6}$

(b) $f(x) = \dfrac{x + 6}{x^2 - x - 6}$

(c) $f(x) = \dfrac{3 + 2x}{5}$

(d) $f(x) = \dfrac{2}{x^2 + 1}$

ANSWERS
1. **(a)** 6; $\{x \mid x \neq 6\}$ **(b)** $-2, 3$; $\{x \mid x \neq -2, 3\}$
   (c) none; The domain includes all real numbers $(-\infty, \infty)$.
   (d) none; The domain includes all real numbers $(-\infty, \infty)$.

In the fundamental property, $\frac{a}{b} = \frac{ac}{bc}$. Since $\frac{c}{c}$ is equivalent to 1, the fundamental property is based on the identity property of multiplication.

A rational expression is a quotient of two polynomials. Since the value of a polynomial is a real number for every value of the variable for which it is defined, any statement that applies to rational numbers will also apply to rational expressions. We use the following steps to write rational expressions in lowest terms.

Writing a Rational Expression in Lowest Terms

Step 1 **Factor** both numerator and denominator to find their greatest common factor (GCF).

Step 2 **Apply the fundamental property.**

EXAMPLE 2 **Writing Rational Expressions in Lowest Terms**

Write each rational expression in lowest terms.

(a) $\dfrac{8k}{16} = \dfrac{k \cdot 8}{2 \cdot 8} = \dfrac{k}{2} \cdot 1 = \dfrac{k}{2}$ Factor; apply the fundamental property.

(b) $\dfrac{8 + k}{16}$

The numerator cannot be factored, so this expression cannot be simplified further and is in lowest terms.

(c) $\dfrac{a^2 - a - 6}{a^2 + 5a + 6} = \dfrac{(a - 3)(a + 2)}{(a + 3)(a + 2)}$ Factor the numerator and the denominator.

$\qquad = \dfrac{a - 3}{a + 3} \cdot 1 \qquad \dfrac{a + 2}{a + 2} = 1$

$\qquad = \dfrac{a - 3}{a + 3}$ Lowest terms

(d) $\dfrac{y^2 - 4}{2y + 4} = \dfrac{(y + 2)(y - 2)}{2(y + 2)}$ Factor the difference of squares in the numerator; factor the denominator.

$\qquad = \dfrac{y - 2}{2}$ Lowest terms

(e) $\dfrac{x^3 - 27}{x - 3} = \dfrac{(x - 3)(x^2 + 3x + 9)}{x - 3}$ Factor the difference of cubes.

$\qquad = x^2 + 3x + 9$ Lowest terms

(f) $\dfrac{pr + qr + ps + qs}{pr + qr - ps - qs} = \dfrac{(pr + qr) + (ps + qs)}{(pr + qr) - (ps + qs)}$ Group terms.

$\qquad = \dfrac{r(p + q) + s(p + q)}{r(p + q) - s(p + q)}$ Factor within groups.

$\qquad = \dfrac{(p + q)(r + s)}{(p + q)(r - s)}$ Factor by grouping.

$\qquad = \dfrac{r + s}{r - s}$ Lowest terms

CAUTION
Be careful! When using the fundamental property of rational numbers, *only common factors may be divided.* For example,

$$\frac{y-2}{2} \neq y \quad \text{and} \quad \frac{y-2}{2} \neq y-1$$

because the 2 in $y-2$ is not a *factor* of the numerator. *Remember to factor before writing a fraction in lowest terms.*

Work Problem 2 at the Side. ▶▶▶

In the rational expression from Example 2(c),

$$\frac{a^2-a-6}{a^2+5a+6}, \quad \text{or} \quad \frac{(a-3)(a+2)}{(a+3)(a+2)},$$

a can take any value except -3 or -2 since these values make the denominator 0. In the simplified rational expression

$$\frac{a-3}{a+3},$$

a cannot equal -3. Because of this,

$$\frac{a^2-a-6}{a^2+5a+6} = \frac{a-3}{a+3}$$

for all values of a except -3 or -2. From now on such statements of equality will be made with the understanding that they apply only for those real numbers that make neither denominator equal 0. We will no longer state such restrictions.

EXAMPLE 3 **Writing Rational Expressions in Lowest Terms**

Write each rational expression in lowest terms.

(a) $\dfrac{m-3}{3-m}$

In this rational expression, the numerator and denominator are opposites. The given expression can be written in lowest terms by writing the denominator as $-1(m-3)$, giving

$$\frac{m-3}{3-m} = \frac{m-3}{-1(m-3)} = \frac{1}{-1} = -1.$$

The numerator could have been rewritten instead to get the same result.

(b) $\dfrac{r^2-16}{4-r} = \dfrac{(r+4)(r-4)}{4-r}$ Factor the difference of squares in the numerator.

$$= \frac{(r+4)(r-4)}{-1(r-4)} \quad \text{Write } 4-r \text{ as } -1(r-4).$$

$$= \frac{r+4}{-1} \quad \text{Fundamental property}$$

$$= -(r+4) \quad \text{or} \quad -r-4 \quad \text{Lowest terms}$$

---

2 Write each rational expression in lowest terms.

(a) $\dfrac{y^2+2y-3}{y^2-3y+2}$

(b) $\dfrac{3y+9}{y^2-9}$

(c) $\dfrac{y+2}{y^2+4}$

(d) $\dfrac{1+p^3}{1+p}$

(e) $\dfrac{3x+3y+rx+ry}{5x+5y-rx-ry}$

ANSWERS

2. **(a)** $\dfrac{y+3}{y-2}$ **(b)** $\dfrac{3}{y-3}$
   (c) already in lowest terms
   (d) $1-p+p^2$ **(e)** $\dfrac{3+r}{5-r}$

3 Write each rational expression in lowest terms.

(a) $\dfrac{y - 2}{2 - y}$

As shown in Examples 3(a) and (b), the quotient $\frac{a}{-a}$ $(a \neq 0)$ can be simplified as

$$\frac{a}{-a} = \frac{a}{-1(a)} = \frac{1}{-1} = -1.$$

The following statement summarizes this result.

> In general, if the numerator and the denominator of a rational expression are opposites, the expression equals -1.

Based on this result, the following are true:

$$\frac{q - 7}{7 - q} = -1 \quad \text{and} \quad \frac{-5a + 2b}{5a - 2b} = -1.$$

Numerator and denominator in each expression are opposites.

However, the expression

$$\frac{r - 2}{r + 2} \quad \text{Numerator and denominator are } not \text{ opposites.}$$

(b) $\dfrac{8 - b}{8 + b}$

cannot be simplifed further.

◀◀◀ Work Problem 3 at the Side.

OBJECTIVE 4 Multiply rational expressions. To multiply rational expressions, follow these steps.

> **Multiplying Rational Expressions**
>
> *Step 1* **Factor** all numerators and denominators as completely as possible.
>
> *Step 2* **Apply the fundamental property.**
>
> *Step 3* **Multiply** remaining factors in the numerator and remaining factors in the denominator. Leave the denominator in factored form.
>
> *Step 4* **Check** to be sure the product is in lowest terms.

(c) $\dfrac{p - 2}{4 - p^2}$

EXAMPLE 4 Multiplying Rational Expressions

Multiply.

(a) $\dfrac{5p - 5}{p} \cdot \dfrac{3p^2}{10p - 10} = \dfrac{5(p - 1)}{p} \cdot \dfrac{3p \cdot p}{2 \cdot 5(p - 1)}$ Factor.

$= \dfrac{1}{1} \cdot \dfrac{3p}{2}$ Fundamental property

$= \dfrac{3p}{2}$ Multiply.

Continued on Next Page

ANSWERS
3. (a) -1 (b) already in lowest terms
(c) $\dfrac{-1}{2 + p}$

(b) $\dfrac{k^2 + 2k - 15}{k^2 - 4k + 3} \cdot \dfrac{k^2 - k}{k^2 + k - 20} = \dfrac{(k + 5)(k - 3)}{(k - 3)(k - 1)} \cdot \dfrac{k(k - 1)}{(k + 5)(k - 4)}$

$$= \dfrac{k}{k - 4}$$

(c) $(p - 4) \cdot \dfrac{3}{5p - 20} = \dfrac{p - 4}{1} \cdot \dfrac{3}{5p - 20}$ Write $p - 4$ as $\dfrac{p - 4}{1}$.

$$= \dfrac{p - 4}{1} \cdot \dfrac{3}{5(p - 4)} \quad \text{Factor.}$$

$$= \dfrac{3}{5} \quad \begin{array}{l}\text{Fundamental property;} \\ \text{multiply.}\end{array}$$

(d) $\dfrac{x^2 + 2x}{x + 1} \cdot \dfrac{x^2 - 1}{x^3 + x^2} = \dfrac{x(x + 2)}{x + 1} \cdot \dfrac{(x + 1)(x - 1)}{x^2(x + 1)}$ Factor.

$$= \dfrac{(x + 2)(x - 1)}{x(x + 1)} \quad \begin{array}{l}\text{Multiply;} \\ \text{lowest terms.}\end{array}$$

(e) $\dfrac{x - 6}{x^2 - 12x + 36} \cdot \dfrac{x^2 - 3x - 18}{x^2 + 7x + 12} = \dfrac{x - 6}{(x - 6)^2} \cdot \dfrac{(x + 3)(x - 6)}{(x + 3)(x + 4)}$ Factor.

$$= \dfrac{1}{x + 4} \quad \begin{array}{l}\text{Lowest} \\ \text{terms}\end{array}$$

Remember to include 1 in the numerator when all other factors are eliminated using the fundamental property.

> **Work Problem 4 at the Side.** ▶▶▶

OBJECTIVE 5 Find reciprocals for rational expressions. The rational numbers $\frac{a}{b}$ and $\frac{c}{d}$ are reciprocals of each other if they have a product of 1. The **reciprocal** of a rational expression is defined in the same way: *Two rational expressions are reciprocals of each other if they have a product of 1.* Recall that 0 has no reciprocal. The table shows several rational expressions and their reciprocals. In the first two cases, check that the product of the rational expression and its reciprocal is 1.

Rational Expression	Reciprocal
$\dfrac{5}{k}$	$\dfrac{k}{5}$
$\dfrac{m^2 - 9m}{2}$	$\dfrac{2}{m^2 - 9m}$
$\dfrac{0}{4}$	undefined

The examples in the table suggest the following procedure.

Finding the Reciprocal

To find the reciprocal of a nonzero rational expression, interchange the numerator and denominator of the expression.

> **Work Problem 5 at the Side.** ▶▶▶

---

4 Multiply.

(a) $\dfrac{2r + 4}{5r} \cdot \dfrac{3r}{5r + 10}$

(b) $\dfrac{c^2 + 2c}{c^2 - 4} \cdot \dfrac{c^2 - 4c + 4}{c^2 - c}$

(c) $\dfrac{m^2 - 16}{m + 2} \cdot \dfrac{1}{m + 4}$

(d)

$\dfrac{x - 3}{x^2 + 2x - 15} \cdot \dfrac{x^2 - 25}{x^2 + 3x - 40}$

5 Find each reciprocal.

(a) $\dfrac{-3}{r}$

(b) $\dfrac{7}{y + 8}$

(c) $\dfrac{a^2 + 7a}{2a - 1}$

(d) $\dfrac{0}{-5}$

ANSWERS

4. (a) $\dfrac{6}{25}$ (b) $\dfrac{c - 2}{c - 1}$ (c) $\dfrac{m - 4}{m + 2}$ (d) $\dfrac{1}{x + 8}$

5. (a) $\dfrac{r}{-3}$ (b) $\dfrac{y + 8}{7}$ (c) $\dfrac{2a - 1}{a^2 + 7a}$

    (d) There is no reciprocal.

6 Divide.

(a) $\dfrac{16k^2}{5} \div \dfrac{3k}{10}$

OBJECTIVE **6** **Divide rational expressions.** Dividing rational expressions is like dividing rational numbers.

Dividing Rational Expressions

To divide two rational expressions, multiply the first (the dividend) by the reciprocal of the second (the divisor).

EXAMPLE 5 **Dividing Rational Expressions**

Divide.

(a) $\dfrac{2z}{9} \div \dfrac{5z^2}{18} = \dfrac{2z}{9} \cdot \dfrac{18}{5z^2}$ Multiply by the reciprocal of the divisor.

$= \dfrac{2z}{9} \cdot \dfrac{2 \cdot 9}{5z^2}$ Factor.

$= \dfrac{4}{5z}$ Multiply; lowest terms

(b) $\dfrac{5p + 2}{6} \div \dfrac{15p + 6}{5}$

(b) $\dfrac{8k - 16}{3k} \div \dfrac{3k - 6}{4k^2} = \dfrac{8k - 16}{3k} \cdot \dfrac{4k^2}{3k - 6}$ Multiply by the reciprocal.

$= \dfrac{8(k - 2)}{3k} \cdot \dfrac{4k^2}{3(k - 2)}$ Factor.

$= \dfrac{32k}{9}$ Multiply; lowest terms

(c) $\dfrac{5m^2 + 17m - 12}{3m^2 + 7m - 20} \div \dfrac{5m^2 + 2m - 3}{15m^2 - 34m + 15}$

$= \dfrac{5m^2 + 17m - 12}{3m^2 + 7m - 20} \cdot \dfrac{15m^2 - 34m + 15}{5m^2 + 2m - 3}$ Definition of division

$= \dfrac{(5m - 3)(m + 4)}{(m + 4)(3m - 5)} \cdot \dfrac{(3m - 5)(5m - 3)}{(5m - 3)(m + 1)}$ Factor.

$= \dfrac{5m - 3}{m + 1}$ Lowest terms

(c)

$\dfrac{y^2 - 2y - 3}{y^2 + 4y + 4} \div \dfrac{y^2 - 1}{y^2 + y - 2}$

Work Problem 6 at the Side.

ANSWERS

6. (a) $\dfrac{32k}{3}$ (b) $\dfrac{5}{18}$ (c) $\dfrac{y - 3}{y + 2}$

8.1 Exercises

FOR EXTRA HELP

Tutor Center — Addison-Wesley Math Tutor Center

MathXL — MathXL

Digital Video Tutor CD 4 — Videotape 4

Student's Solutions Manual

MyMathLab — MyMathLab

Interactmath.com

Rational expressions can often be written in lowest terms in seemingly different ways. For example,

$$\frac{y - 3}{-5} \quad and \quad \frac{-y + 3}{5}$$

look different, but we get the second expression by multiplying the first by -1 in both the numerator and denominator. To practice recognizing equivalent rational expressions, match the expressions in Exercises 1–6 with their equivalents in Choices A–F.

1. $\dfrac{x - 3}{x + 4}$ **2.** $\dfrac{x + 3}{x - 4}$ **3.** $\dfrac{x - 3}{x - 4}$ **4.** $\dfrac{x + 3}{x + 4}$ **5.** $\dfrac{3 - x}{x + 4}$ **6.** $\dfrac{x + 3}{4 - x}$

A. $\dfrac{-x - 3}{4 - x}$ **B.** $\dfrac{-x - 3}{-x - 4}$ **C.** $\dfrac{3 - x}{-x - 4}$ **D.** $\dfrac{-x + 3}{-x + 4}$ **E.** $\dfrac{x - 3}{-x - 4}$ **F.** $\dfrac{-x - 3}{x - 4}$

7. In Example 1(a), we showed that the domain of the rational function defined by $f(x) = \dfrac{3}{7x - 14}$ does not include 2. Explain in your own words why this is so. In general, how do we find the value or values excluded from the domain of a rational function?

8. The domain of the rational function defined by $g(x) = \dfrac{x + 1}{x^2 + 3}$ includes all real numbers. Explain.

Find all numbers that are not in the domain of each function. Then give the domain using set notation. See Example 1.

9. $f(x) = \dfrac{x}{x - 7}$ **10.** $f(x) = \dfrac{x}{x + 3}$ **11.** $f(x) = \dfrac{6x - 5}{7x + 1}$ **12.** $f(x) = \dfrac{8x - 3}{2x + 7}$

13. $f(x) = \dfrac{12x + 3}{x}$ **14.** $f(x) = \dfrac{9x + 8}{x}$ **15.** $f(x) = \dfrac{3x + 1}{2x^2 + x - 6}$ **16.** $f(x) = \dfrac{2x + 4}{3x^2 + 11x - 42}$

17. $f(x) = \dfrac{x + 2}{14}$ **18.** $f(x) = \dfrac{x - 9}{26}$ **19.** $f(x) = \dfrac{2x^2 - 3x + 4}{3x^2 + 8}$ **20.** $f(x) = \dfrac{9x^2 - 8x + 3}{4x^2 + 1}$

Study Skills Workbook
Activity 11: Mind Maps

21. **(a)** Identify the two *terms* in the numerator and the two *terms* in the denominator of the rational expression $\dfrac{x^2 + 4x}{x + 4}$.

(b) Describe the steps you would use to write this rational expression in lowest terms. (*Hint:* It simplifies to x.)

22. Only one of the following rational expressions can be simplified. Which one is it?

A. $\dfrac{x^2 + 2}{x^2}$ **B.** $\dfrac{x^2 + 2}{2}$

C. $\dfrac{x^2 + y^2}{y^2}$ **D.** $\dfrac{x^2 - 5x}{x}$

23. Only one of the following rational expressions is *not* equivalent to $\dfrac{x - 3}{4 - x}$. Which one is it?

A. $\dfrac{3 - x}{x - 4}$ **B.** $\dfrac{x + 3}{4 + x}$

C. $-\dfrac{3 - x}{4 - x}$ **D.** $-\dfrac{x - 3}{x - 4}$

24. Which two of the following rational expressions equal -1?

A. $\dfrac{2x + 3}{2x - 3}$ **B.** $\dfrac{2x - 3}{3 - 2x}$

C. $\dfrac{2x + 3}{3 + 2x}$ **D.** $\dfrac{2x + 3}{-2x - 3}$

Write each rational expression in lowest terms. See Example 2.

25. $\dfrac{x^2 (x + 1)}{x (x + 1)}$

26. $\dfrac{y^3 (y - 4)}{y^2 (y - 4)}$

27. $\dfrac{(x + 4)(x - 3)}{(x + 5)(x + 4)}$

28. $\dfrac{(2x + 7)(x - 1)}{(2x + 3)(2x + 7)}$

29. $\dfrac{4x(x + 3)}{8x^2(x - 3)}$

30. $\dfrac{5y^2 (y + 8)}{15y (y - 8)}$

31. $\dfrac{3x + 7}{3}$

32. $\dfrac{4x - 9}{4}$

33. $\dfrac{6m + 18}{7m + 21}$

34. $\dfrac{5r - 20}{3r - 12}$

35. $\dfrac{3z^2 + z}{18z + 6}$

36. $\dfrac{2x^2 - 5x}{16x - 40}$

37. $\dfrac{2t + 6}{t^2 - 9}$

38. $\dfrac{5s - 25}{s^2 - 25}$

39. $\dfrac{x^2 + 2x - 15}{x^2 + 6x + 5}$

40. $\dfrac{y^2 - 5y - 14}{y^2 + y - 2}$

41. $\dfrac{8x^2 - 10x - 3}{8x^2 - 6x - 9}$

42. $\dfrac{12x^2 - 4x - 5}{8x^2 - 6x - 5}$

43. $\dfrac{a^3 + b^3}{a + b}$

44. $\dfrac{r^3 - s^3}{r - s}$

45. $\dfrac{2c^2 + 2cd - 60d^2}{2c^2 - 12cd + 10d^2}$

46. $\dfrac{3s^2 - 9st - 54t^2}{3s^2 - 6st - 72t^2}$

47. $\dfrac{ac - ad + bc - bd}{ac - ad - bc + bd}$

48. $\dfrac{2xy + 2xw + y + w}{2xy + y - 2xw - w}$

Write each rational expression in lowest terms. See Example 3.

49. $\dfrac{7 - b}{b - 7}$

50. $\dfrac{r - 13}{13 - r}$

51. $\dfrac{x^2 - y^2}{y - x}$

52. $\dfrac{m^2 - n^2}{n - m}$

53. $\dfrac{(a - 3)(x + y)}{(3 - a)(x - y)}$

54. $\dfrac{(8 - p)(x + 2)}{(p - 8)(x - 2)}$

55. $\dfrac{5k - 10}{20 - 10k}$

56. $\dfrac{7x - 21}{63 - 21x}$

57. $\dfrac{a^2 - b^2}{a^2 + b^2}$

58. $\dfrac{p^2 + q^2}{p^2 - q^2}$

Multiply or divide as indicated. See Examples 4 and 5.

59. $\dfrac{(x + 2)(x + 1)}{(x + 3)(x - 2)} \cdot \dfrac{(x + 3)(x + 4)}{(x + 2)(x + 1)}$

60. $\dfrac{(x + 3)(x - 4)}{(x - 4)(x + 2)} \cdot \dfrac{(x + 5)(x - 6)}{(x + 3)(x - 6)}$

61. $\dfrac{(2x + 3)(x - 4)}{(x + 8)(x - 4)} \div \dfrac{(x - 4)(x + 2)}{(x - 4)(x + 8)}$

62. $\dfrac{(6x + 5)(x - 3)}{(x + 9)(x - 1)} \div \dfrac{(x - 3)(2x + 7)}{(x - 1)(x + 9)}$

63. $\dfrac{7t + 7}{-6} \div \dfrac{4t + 4}{15}$

64. $\dfrac{8z - 16}{-20} \div \dfrac{3z - 6}{40}$

65. $\dfrac{4x}{8x + 4} \cdot \dfrac{14x + 7}{6}$

66. $\dfrac{12x - 20}{5x} \cdot \dfrac{6}{9x - 15}$

67. $\dfrac{p^2 - 25}{4p} \cdot \dfrac{2}{5 - p}$

68. $\dfrac{a^2 - 1}{4a} \cdot \dfrac{2}{1 - a}$

69. $\dfrac{m^2 - 49}{m + 1} \div \dfrac{7 - m}{m}$

70. $\dfrac{k^2 - 4}{3k^2} \div \dfrac{2 - k}{11k}$

71. $\dfrac{12x - 10y}{3x + 2y} \cdot \dfrac{6x + 4y}{10y - 12x}$

72. $\dfrac{9s - 12t}{2s + 2t} \cdot \dfrac{3s + 3t}{4t - 3s}$

73. $\dfrac{x^2 - 25}{x^2 + x - 20} \cdot \dfrac{x^2 + 7x + 12}{x^2 - 2x - 15}$

74. $\dfrac{t^2 - 49}{t^2 + 4t - 21} \cdot \dfrac{t^2 + 8t + 15}{t^2 - 2t - 35}$

75. $\dfrac{6x^2 + 5xy - 6y^2}{12x^2 - 11xy + 2y^2} \div \dfrac{4x^2 - 12xy + 9y^2}{8x^2 - 14xy + 3y^2}$

76. $\dfrac{8a^2 - 6ab - 9b^2}{6a^2 - 5ab - 6b^2} \div \dfrac{4a^2 + 11ab + 6b^2}{9a^2 + 12ab + 4b^2}$

77. $\dfrac{3k^2 + 17kp + 10p^2}{6k^2 + 13kp - 5p^2} \div \dfrac{6k^2 + kp - 2p^2}{6k^2 - 5kp + p^2}$

78. $\dfrac{16c^2 + 24cd + 9d^2}{16c^2 - 16cd + 3d^2} \div \dfrac{16c^2 - 9d^2}{16c^2 - 24cd + 9d^2}$

79. $\left(\dfrac{6k^2 - 13k - 5}{k^2 + 7k} \div \dfrac{2k - 5}{k^3 + 6k^2 - 7k} \right) \cdot \dfrac{k^2 - 5k + 6}{3k^2 - 8k - 3}$

80. $\left(\dfrac{2x^3 + 3x^2 - 2x}{3x - 15} \div \dfrac{2x^3 - x^2}{x^2 - 3x - 10} \right) \cdot \dfrac{5x^2 - 10x}{3x^2 + 12x + 12}$

8.2 Adding and Subtracting Rational Expressions

OBJECTIVE 1 Add and subtract rational expressions with the same denominator. The following steps, used to add or subtract rational numbers, are also used to add or subtract rational expressions.

OBJECTIVES

1. Add and subtract rational expressions with the same denominator.
2. Find a least common denominator.
3. Add and subtract rational expressions with different denominators.

Adding or Subtracting Rational Expressions

Step 1 **If the denominators are the same,** add or subtract the numerators. Place the result over the common denominator.

If the denominators are different, first find the least common denominator. Write all rational expressions with this LCD, and then add or subtract the numerators. Place the result over the common denominator.

Step 2 **Simplify.** Write all answers in lowest terms.

① Add or subtract.

(a) $\dfrac{3m}{8} + \dfrac{5n}{8}$

EXAMPLE 1 **Adding and Subtracting Rational Expressions with the Same Denominator**

Add or subtract as indicated.

(a) $\dfrac{3y}{5} + \dfrac{x}{5} = \dfrac{3y + x}{5}$ ← Add the numerators.
                     ← Keep the common denominator.

The denominators of these rational expressions are the same, so just add the numerators, and place the sum over the common denominator.

(b) $\dfrac{7}{3a} + \dfrac{10}{3a}$

(b) $\dfrac{7}{2r^2} - \dfrac{11}{2r^2} = \dfrac{7 - 11}{2r^2}$ Subtract the numerators; keep the common denominator.

$= \dfrac{-4}{2r^2}$

$= -\dfrac{2}{r^2}$ Lowest terms

(c) $\dfrac{2}{y^2} - \dfrac{5}{y^2}$

(c) $\dfrac{m}{m^2 - p^2} + \dfrac{p}{m^2 - p^2} = \dfrac{m + p}{m^2 - p^2}$ Add the numerators; keep the common denominator.

$= \dfrac{m + p}{(m + p)(m - p)}$ Factor.

$= \dfrac{1}{m - p}$ Lowest terms

(d) $\dfrac{a}{a + b} + \dfrac{b}{a + b}$

(d) $\dfrac{4}{x^2 + 2x - 8} + \dfrac{x}{x^2 + 2x - 8} = \dfrac{4 + x}{x^2 + 2x - 8}$ Add.

$= \dfrac{4 + x}{(x - 2)(x + 4)}$ Factor.

$= \dfrac{1}{x - 2}$ Lowest terms

(e) $\dfrac{2y - 1}{y^2 + y - 2} - \dfrac{y}{y^2 + y - 2}$

Work Problem 1 at the Side. ▶▶▶

ANSWERS

1. (a) $\dfrac{3m + 5n}{8}$ (b) $\dfrac{17}{3a}$

    (c) $-\dfrac{3}{y^2}$ (d) 1   (e) $\dfrac{1}{y + 2}$

2 Find the LCD for each group of denominators.

(a) $5k^3s$, $10ks^4$

OBJECTIVE 2 Find a least common denominator. We add or subtract rational expressions with different denominators by first writing them with a common denominator, usually the **least common denominator (LCD).**

> ### Finding the Least Common Denominator
>
> *Step 1* **Factor** each denominator.
>
> *Step 2* **Find the least common denominator.** The LCD is the product of all different factors from each denominator, with each factor raised to the *greatest* power that occurs in any denominator.

(b) $3 - x$, $9 - x^2$

EXAMPLE 2 Finding Least Common Denominators

Assume that the given expressions are denominators of fractions. Find the LCD for each group.

(a) $5xy^2$, $2x^3y$

Each denominator is already factored.

$$5xy^2 = 5 \cdot x \cdot y^2$$
$$2x^3y = 2 \cdot x^3 \cdot y$$

(c) z, $z + 6$

$$\text{LCD} = 5 \cdot 2 \cdot x^3 \cdot y^2 \quad \leftarrow \begin{array}{l}\text{Greatest exponent on } x \text{ is 3.}\\ \text{Greatest exponent on } y \text{ is 2.}\end{array}$$
$$= 10x^3y^2$$

(b) $k - 3$, k

Each denominator is already factored. The LCD, an expression divisible by *both* $k - 3$ and k, is

$$k(k - 3).$$

It is usually best to leave a least common denominator in factored form.

(d) $2y^2 - 3y - 2$, $2y^2 + 3y + 1$

(c) $y^2 - 2y - 8$, $y^2 + 3y + 2$

Factor the denominators.

$$\left.\begin{array}{l} y^2 - 2y - 8 = (y - 4)(y + 2) \\ y^2 + 3y + 2 = (y + 2)(y + 1) \end{array}\right\} \text{Factor.}$$

The LCD, divisible by both polynomials, is $(y - 4)(y + 2)(y + 1)$.

(d) $8z - 24$, $5z^2 - 15z$

$$\left.\begin{array}{l} 8z - 24 = 8(z - 3) \\ 5z^2 - 15z = 5z(z - 3) \end{array}\right\} \text{Factor.}$$

(e) $x^2 - 2x + 1$, $x^2 - 4x + 3$, $4x - 4$

The LCD is $8 \cdot 5z \cdot (z - 3) = 40z(z - 3)$.

(e) $m^2 + 5m + 6$, $m^2 + 4m + 4$, $2m + 6$

$$\left.\begin{array}{l} m^2 + 5m + 6 = (m + 3)(m + 2) \\ m^2 + 4m + 4 = (m + 2)^2 \\ 2m + 6 = 2(m + 3) \end{array}\right\} \text{Factor.}$$

The LCD is $2(m + 3)(m + 2)^2$.

◀◀◀ Work Problem 2 at the Side.

ANSWERS

2. (a) $10k^3s^4$ **(b)** $(3 + x)(3 - x)$
(c) $z(z + 6)$ **(d)** $(y - 2)(2y + 1)(y + 1)$
(e) $4(x - 3)(x - 1)^2$

OBJECTIVE 3 Add and subtract rational expressions with different denominators. Before adding or subtracting two rational expressions, we write each expression with the least common denominator by multiplying its numerator and denominator by the factors needed to get the LCD. This procedure is valid because we are multiplying each rational expression by a form of 1, the identity element for multiplication.

Adding or subtracting rational expressions follows the same procedure as that used for rational numbers. Consider the sum $\frac{7}{15} + \frac{5}{12}$. The LCD for 15 and 12 is 60. Multiply $\frac{7}{15}$ by $\frac{4}{4}$ (a form of 1) and multiply $\frac{5}{12}$ by $\frac{5}{5}$ (another form of 1) so that each fraction has denominator 60. Then add the numerators.

$$\frac{7}{15} + \frac{5}{12} = \frac{7 \cdot 4}{15 \cdot 4} + \frac{5 \cdot 5}{12 \cdot 5} \qquad \text{Fundamental property}$$

$$= \frac{28}{60} + \frac{25}{60}$$

$$= \frac{28 + 25}{60} \qquad \text{Add the numerators.}$$

$$= \frac{53}{60}$$

EXAMPLE 3 **Adding and Subtracting Rational Expressions with Different Denominators**

Add or subtract as indicated.

(a) $\dfrac{5}{2p} + \dfrac{3}{8p}$

The LCD for $2p$ and $8p$ is $8p$. To write the first rational expression with a denominator of $8p$, multiply by $\frac{4}{4}$.

$$\frac{5}{2p} + \frac{3}{8p} = \frac{5 \cdot 4}{2p \cdot 4} + \frac{3}{8p} \qquad \text{Fundamental property}$$

$$= \frac{20}{8p} + \frac{3}{8p}$$

$$= \frac{20 + 3}{8p} \qquad \text{Add the numerators.}$$

$$= \frac{23}{8p}$$

(b) $\dfrac{6}{r} - \dfrac{5}{r - 3}$

Write each rational expression with the LCD, $r(r - 3)$.

$$\frac{6}{r} - \frac{5}{r - 3} = \frac{6(r - 3)}{r(r - 3)} - \frac{r \cdot 5}{r(r - 3)} \qquad \text{Fundamental property}$$

$$= \frac{6r - 18}{r(r - 3)} - \frac{5r}{r(r - 3)} \qquad \text{Distributive and commutative properties}$$

$$= \frac{6r - 18 - 5r}{r(r - 3)} \qquad \text{Subtract the numerators.}$$

$$= \frac{r - 18}{r(r - 3)} \qquad \text{Combine terms in the numerator.}$$

Work Problem 3 at the Side. ▶▶▶

---

3 Add or subtract.

(a) $\dfrac{6}{7} + \dfrac{1}{5}$

(b) $\dfrac{8}{3k} - \dfrac{2}{9k}$

(c) $\dfrac{2}{y} - \dfrac{1}{y + 4}$

ANSWERS

3. **(a)** $\dfrac{37}{35}$ **(b)** $\dfrac{22}{9k}$ **(c)** $\dfrac{y + 8}{y(y + 4)}$

4 Subtract.

(a) $\dfrac{5x + 7}{2x + 7} - \dfrac{-x - 14}{2x + 7}$

CAUTION

One of the most common sign errors in algebra occurs when a rational expression with two or more terms in the numerator is being subtracted. In this situation, *the subtraction sign must be distributed to every term in the numerator of the fraction that follows it.* Study Example 4 carefully to see how this is done.

EXAMPLE 4 Using the Distributive Property When Subtracting Rational Expressions

Subtract.

(a) $\dfrac{7x}{3x + 1} - \dfrac{x - 2}{3x + 1}$

The denominators are the same for both rational expressions. The subtraction sign must be applied to *both* terms in the numerator of the second rational expression. Notice the careful use of the distributive property here.

$$\dfrac{7x}{3x + 1} - \dfrac{x - 2}{3x + 1} = \dfrac{7x - (x - 2)}{3x + 1} \qquad \text{Subtract the numerators;}$$
$$\text{keep the common denominator.}$$

$$= \dfrac{7x - x + 2}{3x + 1} \qquad \text{Distributive property;}$$
$$\text{be careful with signs.}$$

$$= \dfrac{6x + 2}{3x + 1} \qquad \text{Combine terms in the numerator.}$$

$$= \dfrac{2(3x + 1)}{3x + 1} \qquad \text{Factor the numerator.}$$

$$= 2 \qquad \text{Lowest terms}$$

(b) $\dfrac{2}{r - 2} - \dfrac{r}{r - 1}$

(b) $\dfrac{1}{q - 1} - \dfrac{1}{q + 1}$

$$= \dfrac{1(q + 1)}{(q - 1)(q + 1)} - \dfrac{1(q - 1)}{(q + 1)(q - 1)} \qquad \begin{array}{l}\text{The LCD is } (q - 1)(q + 1);\\ \text{fundamental property}\end{array}$$

$$= \dfrac{(q + 1) - (q - 1)}{(q - 1)(q + 1)} \qquad \text{Subtract.}$$

$$= \dfrac{q + 1 - q + 1}{(q - 1)(q + 1)} \qquad \text{Distributive property}$$

$$= \dfrac{2}{(q - 1)(q + 1)} \qquad \begin{array}{l}\text{Combine terms in the}\\ \text{numerator.}\end{array}$$

Work Problem 4 at the Side.

In some problems, rational expressions to be added or subtracted have denominators that are opposites of each other, such as

$$\dfrac{y}{y - 2} + \dfrac{8}{2 - y}. \qquad \text{Denominators are opposites.}$$

The next example illustrates how to proceed in such a problem.

ANSWERS

4. (a) 3 (b) $\dfrac{-r^2 + 4r - 2}{(r - 2)(r - 1)}$

EXAMPLE 5 Adding Rational Expressions with Denominators That Are Opposites

Add.

$$\frac{y}{y-2} + \frac{8}{2-y}$$

To get a common denominator of $y-2$, multiply the second expression by -1 in both the numerator and the denominator.

$$\frac{y}{y-2} + \frac{8}{2-y} = \frac{y}{y-2} + \frac{8(-1)}{(2-y)(-1)}$$

$$= \frac{y}{y-2} + \frac{-8}{y-2}$$

$$= \frac{y-8}{y-2} \quad \text{Add the numerators.}$$

If we had used $2-y$ as the common denominator and rewritten the first expression, we would have obtained

$$\frac{8-y}{2-y},$$

an equivalent answer. Verify this.

Work Problem 5 at the Side.

5 Add or subtract as indicated.

(a) $\dfrac{8}{x-4} + \dfrac{2}{4-x}$

(b) $\dfrac{9}{2x-9} - \dfrac{4}{9-2x}$

EXAMPLE 6 Adding and Subtracting Three Rational Expressions

Add and subtract as indicated.

$$\frac{3}{x-2} + \frac{5}{x} - \frac{6}{x^2-2x}$$

The denominator of the third rational expression factors as $x(x-2)$, which is the LCD for the three rational expressions.

$$\frac{3}{x-2} + \frac{5}{x} - \frac{6}{x^2-2x}$$

$$= \frac{3x}{x(x-2)} + \frac{5(x-2)}{x(x-2)} - \frac{6}{x(x-2)} \quad \text{Fundamental property}$$

$$= \frac{3x + 5(x-2) - 6}{x(x-2)} \quad \text{Add and subtract the numerators.}$$

$$= \frac{3x + 5x - 10 - 6}{x(x-2)} \quad \text{Distributive property}$$

$$= \frac{8x - 16}{x(x-2)} \quad \text{Combine terms in the numerator.}$$

$$= \frac{8(x-2)}{x(x-2)} \quad \text{Factor the numerator.}$$

$$= \frac{8}{x} \quad \text{Lowest terms}$$

Work Problem 6 at the Side.

6 Add and subtract as indicated.

$$\frac{4}{x-5} + \frac{-2}{x} - \frac{10}{x^2-5x}$$

ANSWERS

5. (a) $\dfrac{6}{x-4}$ or $\dfrac{-6}{4-x}$ (b) $\dfrac{13}{2x-9}$ or $\dfrac{-13}{9-2x}$

6. $\dfrac{2}{x-5}$

7 Subtract.

$$\frac{-a}{a^2 + 3a - 4} - \frac{4a}{a^2 + 7a + 12}$$

EXAMPLE 7 Subtracting Rational Expressions

Subtract.

$$\frac{m + 4}{m^2 - 2m - 3} - \frac{2m - 3}{m^2 - 5m + 6}$$

$$= \frac{m + 4}{(m - 3)(m + 1)} - \frac{2m - 3}{(m - 3)(m - 2)} \qquad \text{Factor each denominator.}$$

The LCD is $(m - 3)(m + 1)(m - 2)$.

$$= \frac{(m + 4)(m - 2)}{(m - 3)(m + 1)(m - 2)} - \frac{(2m - 3)(m + 1)}{(m - 3)(m - 2)(m + 1)} \qquad \begin{array}{l}\text{Fundamental}\\ \text{property}\end{array}$$

$$= \frac{(m + 4)(m - 2) - (2m - 3)(m + 1)}{(m - 3)(m + 1)(m - 2)} \qquad \text{Subtract.}$$

$$= \frac{m^2 + 2m - 8 - (2m^2 - m - 3)}{(m - 3)(m + 1)(m - 2)} \qquad \text{Multiply in the numerator.}$$

$$= \frac{m^2 + 2m - 8 - 2m^2 + m + 3}{(m - 3)(m + 1)(m - 2)} \qquad \begin{array}{l}\text{Distributive property;}\\ \text{be careful with signs.}\end{array}$$

$$= \frac{-m^2 + 3m - 5}{(m - 3)(m + 1)(m - 2)} \qquad \begin{array}{l}\text{Combine terms in the}\\ \text{numerator.}\end{array}$$

If we try to factor the numerator, we find that this rational expression is in lowest terms.

Work Problem 7 at the Side.

8 Add.

$$\frac{4}{p^2 - 6p + 9} + \frac{1}{p^2 + 2p - 15}$$

EXAMPLE 8 Adding Rational Expressions

Add.

$$\frac{5}{x^2 + 10x + 25} + \frac{2}{x^2 + 7x + 10}$$

$$= \frac{5}{(x + 5)^2} + \frac{2}{(x + 5)(x + 2)} \qquad \text{Factor each denominator.}$$

The LCD is $(x + 5)^2(x + 2)$.

$$= \frac{5(x + 2)}{(x + 5)^2(x + 2)} + \frac{2(x + 5)}{(x + 5)^2(x + 2)} \qquad \text{Fundamental property}$$

$$= \frac{5(x + 2) + 2(x + 5)}{(x + 5)^2(x + 2)} \qquad \text{Add.}$$

$$= \frac{5x + 10 + 2x + 10}{(x + 5)^2(x + 2)} \qquad \text{Distributive property}$$

$$= \frac{7x + 20}{(x + 5)^2(x + 2)} \qquad \begin{array}{l}\text{Combine terms in the}\\ \text{numerator.}\end{array}$$

Work Problem 8 at the Side.

ANSWERS

7. $\dfrac{-5a^2 + a}{(a + 4)(a - 1)(a + 3)}$

8. $\dfrac{5p + 17}{(p - 3)^2(p + 5)}$

8.2 Exercises

FOR EXTRA HELP   Addison-Wesley Math Tutor Center   MathXL   Digital Video Tutor CD 4 Videotape 4   Student's Solutions Manual   MyMathLab   Interactmath.com

1. Write an explanation for adding or subtracting rational expressions that have a common denominator.

2. Write an explanation for adding or subtracting rational expressions that have different denominators.

Add or subtract as indicated. Write all answers in lowest terms. See Example 1.

3. $\dfrac{7}{t} + \dfrac{2}{t}$

4. $\dfrac{5}{r} + \dfrac{9}{r}$

5. $\dfrac{11}{5x} - \dfrac{1}{5x}$

6. $\dfrac{7}{4y} - \dfrac{3}{4y}$

7. $\dfrac{5x + 4}{6x + 5} + \dfrac{x + 1}{6x + 5}$

8. $\dfrac{6y + 12}{4y + 3} + \dfrac{2y - 6}{4y + 3}$

9. $\dfrac{x^2}{x + 5} - \dfrac{25}{x + 5}$

10. $\dfrac{y^2}{y + 6} - \dfrac{36}{y + 6}$

11. $\dfrac{4}{p^2 + 7p + 12} + \dfrac{p}{p^2 + 7p + 12}$

12. $\dfrac{5}{x^2 + x - 20} + \dfrac{x}{x^2 + x - 20}$

13. $\dfrac{a^3}{a^2 + ab + b^2} - \dfrac{b^3}{a^2 + ab + b^2}$

14. $\dfrac{p^3}{p^2 - pq + q^2} + \dfrac{q^3}{p^2 - pq + q^2}$

Assume that the expressions given are denominators of fractions. Find the least common denominator (LCD) for each group. See Example 2.

15. $18x^2y^3,\quad 24x^4y^5$

16. $24a^3b^4,\quad 18a^5b^2$

17. $z - 2,\quad z$

18. $k + 3,\quad k$

19. $2y + 8,\quad y + 4$

20. $3r - 21,\quad r - 7$

21. $x^2 - 81,\quad x^2 + 18x + 81$

22. $y^2 - 16,\quad y^2 - 8y + 16$

23. $m + n,\quad m - n,\quad m^2 - n^2$

24. $r + s,\quad r - s,\quad r^2 - s^2$

25. $x^2 - 3x - 4,\quad x + x^2$

26. $y^2 - 8y + 12,\quad y^2 - 6y$

27. $2t^2 + 7t - 15,\quad t^2 + 3t - 10$

28. $s^2 - 3s - 4,\quad 3s^2 + s - 2$

29. $2y + 6,\quad y^2 - 9,\quad y$

30. $9x + 18,\quad x^2 - 4,\quad x$

31. One student added two rational expressions and obtained the answer $\dfrac{3}{5 - y}$. Another student obtained the answer $\dfrac{-3}{y - 5}$ for the same problem. Is it possible that both answers are correct? Explain.

32. What is *wrong* with the following work?

$$\frac{x}{x + 2} - \frac{4x - 1}{x + 2} = \frac{x - 4x - 1}{x + 2} = \frac{-3x - 1}{x + 2}$$

Add or subtract as indicated. Write all answers in lowest terms. See Examples 3–8.

33. $\dfrac{8}{t} + \dfrac{7}{3t}$

34. $\dfrac{5}{x} + \dfrac{9}{4x}$

35. $\dfrac{5}{12x^2 y} - \dfrac{11}{6xy}$

36. $\dfrac{7}{18a^3 b^2} - \dfrac{2}{9ab}$

37. $\dfrac{1}{x - 1} - \dfrac{1}{x}$

38. $\dfrac{3}{x - 3} - \dfrac{1}{x}$

39. $\dfrac{3a}{a + 1} + \dfrac{2a}{a - 3}$

40. $\dfrac{2x}{x + 4} + \dfrac{3x}{x - 7}$

41. $\dfrac{17y + 3}{9y + 7} - \dfrac{-10y - 18}{9y + 7}$

42. $\dfrac{7x + 8}{3x + 2} - \dfrac{x + 4}{3x + 2}$

43. $\dfrac{2}{4 - x} + \dfrac{5}{x - 4}$

44. $\dfrac{3}{2 - t} + \dfrac{1}{t - 2}$

45. $\dfrac{w}{w - z} - \dfrac{z}{z - w}$

46. $\dfrac{a}{a - b} - \dfrac{b}{b - a}$

47. $\dfrac{5}{12 + 4x} - \dfrac{7}{9 + 3x}$

48. $\dfrac{3}{10x + 15} - \dfrac{8}{12x + 18}$

49. $\dfrac{4x}{x - 1} - \dfrac{2}{x + 1} - \dfrac{4}{x^2 - 1}$

50. $\dfrac{4}{x + 3} - \dfrac{x}{x - 3} - \dfrac{18}{x^2 - 9}$

51. $\dfrac{15}{y^2 + 3y} + \dfrac{2}{y} + \dfrac{5}{y + 3}$

52. $\dfrac{7}{t - 2} - \dfrac{6}{t^2 - 2t} - \dfrac{3}{t}$

53. $\dfrac{5}{x - 2} + \dfrac{1}{x} + \dfrac{2}{x^2 - 2x}$

54. $\dfrac{5x}{x - 3} + \dfrac{2}{x} + \dfrac{6}{x^2 - 3x}$

55. $\dfrac{3x}{x + 1} + \dfrac{4}{x - 1} - \dfrac{6}{x^2 - 1}$

56. $\dfrac{5x}{x + 3} + \dfrac{x + 2}{x} - \dfrac{6}{x^2 + 3x}$

57. $\dfrac{4}{x + 1} + \dfrac{1}{x^2 - x + 1} - \dfrac{12}{x^3 + 1}$

58. $\dfrac{5}{x + 2} + \dfrac{2}{x^2 - 2x + 4} - \dfrac{60}{x^3 + 8}$

59. $\dfrac{2x + 4}{x + 3} + \dfrac{3}{x} - \dfrac{6}{x^2 + 3x}$

60. $\dfrac{4x + 1}{x + 5} - \dfrac{2}{x} + \dfrac{10}{x^2 + 5x}$

61. $\dfrac{3}{x^2 - 5x + 6} - \dfrac{2}{x^2 - 4x + 4}$

62. $\dfrac{2}{m^2 - 4m + 4} + \dfrac{3}{m^2 + m - 6}$

63. $\dfrac{3}{x^2 + 4x + 4} + \dfrac{7}{x^2 + 5x + 6}$

64. $\dfrac{5}{x^2 + 6x + 9} - \dfrac{2}{x^2 + 4x + 3}$

65. $\dfrac{5x}{x^2 + xy - 2y^2} - \dfrac{3x}{x^2 + 5xy - 6y^2}$

66. $\dfrac{6x}{6x^2 + 5xy - 4y^2} - \dfrac{2y}{9x^2 - 16y^2}$

A concours d'elegance is a competition in which a maximum of 100 points is awarded to a car based on its general attractiveness. The function defined by the rational expression

$$c(x) = \frac{1010}{49(101 - x)} - \frac{10}{49}$$

approximates the cost, in thousands of dollars, of restoring a car so that it will win x points.

Use this information to work Exercises 67 and 68.

67. Simplify the expression for $c(x)$ by performing the indicated subtraction.

68. Use the simplified expression to determine how much it would cost to win 95 points.

RELATING CONCEPTS (EXERCISES 69–74)   For Individual or Group Work

In Example 6 we showed that

$$\frac{3}{x - 2} + \frac{5}{x} - \frac{6}{x^2 - 2x} \quad \text{simplifies to} \quad \frac{8}{x}.$$

Algebra is, in a sense, a generalized form of arithmetic. **Work Exercises 69–74 in order,** *to see how the algebra in this example is related to the arithmetic of common fractions.*

69. Perform the following operations, and express your answer in lowest terms.

$$\frac{3}{7} + \frac{5}{9} - \frac{6}{63}$$

70. Substitute 9 for x in the given problem from Example 6. Compare this problem to the one given in Exercise 69. What do you notice?

71. Now substitute 9 for x in the answer given in Example 6. Do your results agree with the result you obtained in Exercise 69?

72. Replace x in the problem from Example 6 with the number of letters in your last name, assuming that this number is not 2. If your last name has two letters, let $x = 3$. Now predict the answer to your problem. Verify that your prediction is correct.

73. Why will $x = 2$ not work for the problem from Example 6?

74. What other value of x is not allowed in the problem given from Example 6?

8.3 Complex Fractions

A **complex fraction** is an expression having a fraction in the numerator, denominator, or both. Examples of complex fractions include

$$\frac{1+\dfrac{1}{x}}{2}, \quad \frac{\dfrac{4}{y}}{6-\dfrac{3}{y}}, \quad \text{and} \quad \frac{\dfrac{m^2-9}{m+1}}{\dfrac{m+3}{m^2-1}}. \qquad \text{Complex fractions}$$

OBJECTIVES

1 Simplify complex fractions by simplifying the numerator and denominator. (Method 1)
2 Simplify complex fractions by multiplying by a common denominator. (Method 2)
3 Compare the two methods of simplifying complex fractions.
4 Simplify rational expressions with negative exponents.

OBJECTIVE 1 Simplify complex fractions by simplifying the numerator and denominator. (Method 1)   There are two different methods for simplifying complex fractions.

Simplifying a Complex Fraction: Method 1

Step 1 Simplify the numerator and denominator separately.

Step 2 Divide by multiplying the numerator by the reciprocal of the denominator.

Step 3 Simplify the resulting fraction, if possible.

In Step 2, we are treating the complex fraction as a quotient of two rational expressions and dividing. Before performing this step, be sure that both the numerator and denominator are single fractions.

EXAMPLE 1 Simplifying Complex Fractions by Method 1

Use Method 1 to simplify each complex fraction.

(a) $\dfrac{\dfrac{x+1}{x}}{\dfrac{x-1}{2x}}$

Both the numerator and the denominator are already simplified, so divide by multiplying the numerator by the reciprocal of the denominator.

$$\frac{\dfrac{x+1}{x}}{\dfrac{x-1}{2x}} = \frac{x+1}{x} \div \frac{x-1}{2x} \qquad \text{Write as a division problem.}$$

$$= \frac{x+1}{x} \cdot \frac{2x}{x-1} \qquad \text{Multiply by the reciprocal of } \tfrac{x-1}{2x}.$$

$$= \frac{2x(x+1)}{x(x-1)} \qquad \text{Multiply.}$$

$$= \frac{2(x+1)}{x-1} \qquad \text{Simplify.}$$

Continued on Next Page

1 Use Method 1 to simplify each complex fraction.

(a) $\dfrac{\dfrac{a+2}{5a}}{\dfrac{a-3}{7a}}$

(b) $\dfrac{2+\dfrac{1}{k}}{2-\dfrac{1}{k}}$

(c) $\dfrac{\dfrac{r^2-4}{4}}{1+\dfrac{2}{r}}$

(b) $\dfrac{2+\dfrac{1}{y}}{3-\dfrac{2}{y}} = \dfrac{\dfrac{2y}{y}+\dfrac{1}{y}}{\dfrac{3y}{y}-\dfrac{2}{y}} = \dfrac{\dfrac{2y+1}{y}}{\dfrac{3y-2}{y}}$ Simplify the numerator and denominator. (Step 1)

$= \dfrac{2y+1}{y} \div \dfrac{3y-2}{y}$ Write as a division problem.

$= \dfrac{2y+1}{y} \cdot \dfrac{y}{3y-2}$ Multiply by the reciprocal of $\frac{3y-2}{y}$. (Step 2)

$= \dfrac{2y+1}{3y-2}$ Multiply and simplify. (Step 3)

Work Problem 1 at the Side.

OBJECTIVE 2 Simplify complex fractions by multiplying by a common denominator. (Method 2) The second method for simplifying complex fractions uses the identity property of multiplication.

Simplifying a Complex Fraction: Method 2

Step 1 Multiply the numerator and denominator of the complex fraction by the least common denominator of the fractions in the numerator and the fractions in the denominator of the complex fraction.

Step 2 Simplify the resulting fraction, if possible.

EXAMPLE 2 Simplifying Complex Fractions by Method 2

Use Method 2 to simplify each complex fraction.

(a) $\dfrac{2+\dfrac{1}{y}}{3-\dfrac{2}{y}}$

Multiply the numerator and denominator by the LCD of all the fractions in the numerator and denominator of the complex fraction. (This is the same as multiplying by 1.) Here the LCD is y.

$\dfrac{2+\dfrac{1}{y}}{3-\dfrac{2}{y}} = \dfrac{2+\dfrac{1}{y}}{3-\dfrac{2}{y}} \cdot 1 = \dfrac{\left(2+\dfrac{1}{y}\right)\cdot y}{\left(3-\dfrac{2}{y}\right)\cdot y}$ Multiply the numerator and denominator by y, since $\frac{y}{y}=1$. (Step 1)

$= \dfrac{2\cdot y + \dfrac{1}{y}\cdot y}{3\cdot y - \dfrac{2}{y}\cdot y}$ Distributive property

$= \dfrac{2y+1}{3y-2}$ Simplify. (Step 2)

Compare this method with that used in Example 1(b).

Continued on Next Page

ANSWERS

1. (a) $\dfrac{7(a+2)}{5(a-3)}$ (b) $\dfrac{2k+1}{2k-1}$ (c) $\dfrac{r(r-2)}{4}$

(b) $\dfrac{2p + \dfrac{5}{p-1}}{3p - \dfrac{2}{p}} = \dfrac{\left(2p + \dfrac{5}{p-1}\right) \cdot p(p-1)}{\left(3p - \dfrac{2}{p}\right) \cdot p(p-1)}$

Multiply the numerator and denominator by the LCD, $p(p-1)$.

$= \dfrac{2p[p(p-1)] + \dfrac{5}{p-1} \cdot p(p-1)}{3p[p(p-1)] - \dfrac{2}{p} \cdot p(p-1)}$

Distributive property

$= \dfrac{2p[p(p-1)] + 5p}{3p[p(p-1)] - 2(p-1)}$

$= \dfrac{2p^3 - 2p^2 + 5p}{3p^3 - 3p^2 - 2p + 2}$

Multiply; lowest terms

Work Problem 2 at the Side. ▶▶▶

OBJECTIVE 3 Compare the two methods of simplifying complex fractions. Choosing whether to use Method 1 or Method 2 to simplify a complex fraction is usually a matter of preference. Some students prefer one method over the other, while other students feel comfortable with both methods and rely on practice with many examples to determine which method they will use on a particular problem.

In the next example, we illustrate how to simplify a complex fraction using both methods so that you can observe the processes and decide for yourself the pros and cons of each method.

EXAMPLE 3   Simplifying Complex Fractions Using Both Methods

Use both Method 1 and Method 2 to simplify each complex fraction.

Method 1	Method 2

Method 1

(a) $\dfrac{\dfrac{2}{x-3}}{\dfrac{5}{x^2-9}}$

$= \dfrac{\dfrac{2}{x-3}}{\dfrac{5}{(x-3)(x+3)}}$

$= \dfrac{2}{x-3} \div \dfrac{5}{(x-3)(x+3)}$

$= \dfrac{2}{x-3} \cdot \dfrac{(x-3)(x+3)}{5}$

$= \dfrac{2(x+3)}{5}$

Method 2

(a) $\dfrac{\dfrac{2}{x-3}}{\dfrac{5}{x^2-9}}$

$= \dfrac{\dfrac{2}{x-3} \cdot (x-3)(x+3)}{\dfrac{5}{(x-3)(x+3)} \cdot (x-3)(x+3)}$

$= \dfrac{2(x+3)}{5}$

Continued on Next Page

2 Use Method 2 to simplify each complex fraction.

(a) $\dfrac{\dfrac{5}{y} + 6}{\dfrac{8}{3y} - 1}$

(b) $\dfrac{\dfrac{1}{y} + \dfrac{1}{y-1}}{\dfrac{1}{y} - \dfrac{2}{y-1}}$

ANSWERS

2. **(a)** $\dfrac{15 + 18y}{8 - 3y}$ **(b)** $\dfrac{2y-1}{-y-1}$ or $\dfrac{1-2y}{y+1}$

3 Use both methods to simplify each complex fraction.

(a) $\dfrac{\dfrac{5}{y+2}}{\dfrac{-3}{y^2-4}}$

(b) $\dfrac{\dfrac{1}{a}-\dfrac{1}{b}}{\dfrac{1}{a^2}-\dfrac{1}{b^2}}$

Method 1	Method 2

Method 1

(b) $\dfrac{\dfrac{1}{x}+\dfrac{1}{y}}{\dfrac{1}{x^2}-\dfrac{1}{y^2}}$

$= \dfrac{\dfrac{y}{xy}+\dfrac{x}{xy}}{\dfrac{y^2}{x^2y^2}-\dfrac{x^2}{x^2y^2}}$

$= \dfrac{\dfrac{y+x}{xy}}{\dfrac{y^2-x^2}{x^2y^2}}$

$= \dfrac{y+x}{xy} \div \dfrac{y^2-x^2}{x^2y^2}$

$= \dfrac{y+x}{xy} \cdot \dfrac{x^2y^2}{(y-x)(y+x)}$

$= \dfrac{xy}{y-x}$

Method 2

(b) $\dfrac{\dfrac{1}{x}+\dfrac{1}{y}}{\dfrac{1}{x^2}-\dfrac{1}{y^2}}$

$= \dfrac{\left(\dfrac{1}{x}+\dfrac{1}{y}\right)\cdot x^2y^2}{\left(\dfrac{1}{x^2}-\dfrac{1}{y^2}\right)\cdot x^2y^2}$

$= \dfrac{\left(\dfrac{1}{x}\right)x^2y^2+\left(\dfrac{1}{y}\right)x^2y^2}{\left(\dfrac{1}{x^2}\right)x^2y^2-\left(\dfrac{1}{y^2}\right)x^2y^2}$

$= \dfrac{xy^2+x^2y}{y^2-x^2}$

$= \dfrac{xy(y+x)}{(y+x)(y-x)}$

$= \dfrac{xy}{y-x}$

◀◀◀ Work Problem 3 at the Side.

OBJECTIVE 4 Simplify rational expressions with negative exponents. Rational expressions and complex fractions sometimes involve negative exponents. To simplify such expressions, we begin by rewriting the expressions with only positive exponents.

4 Simplify each expression, using only positive exponents in the answer.

(a) $\dfrac{r^{-2}-s^{-1}}{4r^{-1}+s^{-2}}$

(b) $\dfrac{b^{-4}}{b^{-5}+2}$

EXAMPLE 4 Simplifying a Rational Expression with Negative Exponents

Simplify $\dfrac{m^{-1}+p^{-2}}{2m^{-2}-p^{-1}}$, using only positive exponents in the answer.

First write the expression with only positive exponents.

$$\dfrac{m^{-1}+p^{-2}}{2m^{-2}-p^{-1}} = \dfrac{\dfrac{1}{m}+\dfrac{1}{p^2}}{\dfrac{2}{m^2}-\dfrac{1}{p}} \qquad \text{Definition of negative exponent}$$

Note that the 2 in $2m^{-2}$ is *not* raised to the -2 power (since m is the base for the exponent -2), so $2m^{-2} = \dfrac{2}{m^2}$. Simplify the complex fraction using Method 2, multiplying numerator and denominator by the LCD, m^2p^2.

$$\dfrac{\dfrac{1}{m}+\dfrac{1}{p^2}}{\dfrac{2}{m^2}-\dfrac{1}{p}} = \dfrac{m^2p^2\left(\dfrac{1}{m}+\dfrac{1}{p^2}\right)}{m^2p^2\left(\dfrac{2}{m^2}-\dfrac{1}{p}\right)} = \dfrac{m^2p^2\cdot\dfrac{1}{m}+m^2p^2\cdot\dfrac{1}{p^2}}{m^2p^2\cdot\dfrac{2}{m^2}-m^2p^2\cdot\dfrac{1}{p}} = \dfrac{mp^2+m^2}{2p^2-m^2p}$$

◀◀◀ Work Problem 4 at the Side.

ANSWERS
3. (Both methods give the same answers.)
   (a) $\dfrac{5(y-2)}{-3}$ (b) $\dfrac{ab}{b+a}$
4. (a) $\dfrac{s^2-r^2s}{4rs^2+r^2}$ (b) $\dfrac{b}{1+2b^5}$

8.3 Exercises

FOR EXTRA HELP

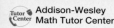 Addison-Wesley Math Tutor Center

 MathXL

 Digital Video Tutor CD 4 Videotape 4

Student's Solutions Manual

MyMathLab MyMathLab

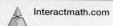 Interactmath.com

1. Explain in your own words Method 1 for simplifying complex fractions.

2. Method 2 for simplifying complex fractions says that we can multiply both the numerator and the denominator of the complex fraction by the same nonzero expression. What property of real numbers from **Section 1.4** justifies this method?

Use either method to simplify each complex fraction. See Examples 1–3.

3. $\dfrac{\dfrac{12}{x-1}}{\dfrac{6}{x}}$

4. $\dfrac{\dfrac{24}{t+4}}{\dfrac{6}{t}}$

5. $\dfrac{\dfrac{k+1}{2k}}{\dfrac{3k-1}{4k}}$

6. $\dfrac{\dfrac{1-r}{4r}}{\dfrac{-1-r}{8r}}$

7. $\dfrac{\dfrac{4z^2 x^4}{9}}{\dfrac{12x^2z^5}{15}}$

8. $\dfrac{\dfrac{3y^2x^3}{8}}{\dfrac{9y^3x^4}{16}}$

9. $\dfrac{\dfrac{1}{x}+1}{-\dfrac{1}{x}+1}$

10. $\dfrac{\dfrac{2}{k}-1}{\dfrac{2}{k}+1}$

11. $\dfrac{\dfrac{3}{x}+\dfrac{3}{y}}{\dfrac{3}{x}-\dfrac{3}{y}}$

12. $\dfrac{\dfrac{4}{t}-\dfrac{4}{s}}{\dfrac{4}{t}+\dfrac{4}{s}}$

13. $\dfrac{\dfrac{8x-24y}{10}}{\dfrac{x-3y}{5x}}$

14. $\dfrac{\dfrac{10x-5y}{12}}{\dfrac{2x-y}{6y}}$

15. $\dfrac{\dfrac{x^2-16y^2}{xy}}{\dfrac{1}{y}-\dfrac{4}{x}}$

16. $\dfrac{\dfrac{2}{s}-\dfrac{3}{t}}{\dfrac{4t^2-9s^2}{st}}$

17. $\dfrac{y-\dfrac{y-3}{3}}{\dfrac{4}{9}+\dfrac{2}{3y}}$

18. $\dfrac{p - \dfrac{p+2}{4}}{\dfrac{3}{4} - \dfrac{5}{2p}}$

19. $\dfrac{\dfrac{x+2}{x} + \dfrac{1}{x+2}}{\dfrac{5}{x} + \dfrac{x}{x+2}}$

20. $\dfrac{\dfrac{y+3}{y} - \dfrac{4}{y-1}}{\dfrac{y}{y-1} + \dfrac{1}{y}}$

RELATING CONCEPTS (EXERCISES 21–26)   For Individual or Group Work

Simplifying a complex fraction by Method 1 is a good way to review the methods of adding, subtracting, multiplying, and dividing rational expressions. Method 2 gives a good review of the fundamental property of rational expressions. Refer to the following complex fraction, and **work Exercises 21–26 in order.**

$$\dfrac{\dfrac{4}{m} + \dfrac{m+2}{m-1}}{\dfrac{m+2}{m} - \dfrac{2}{m-1}}$$

21. Add the fractions in the numerator.

22. Subtract as indicated in the denominator.

23. Divide your answer from Exercise 21 by your answer from Exercise 22.

24. Go back to the original complex fraction and find the least common denominator of all denominators.

25. Multiply the numerator and denominator of the complex fraction by your answer from Exercise 24.

26. Your answers for Exercises 23 and 25 should be the same. Write an explanation comparing the two methods. Which method do you prefer? Explain why.

Simplify each expression, using only positive exponents in the answer. See Example 4.

27. $\dfrac{1}{x^{-2} + y^{-2}}$

28. $\dfrac{1}{p^{-2} - q^{-2}}$

29. $\dfrac{x^{-2} + y^{-2}}{x^{-1} + y^{-1}}$

30. $\dfrac{x^{-1} - y^{-1}}{x^{-2} - y^{-2}}$

31. $\dfrac{x^{-1} + 2y^{-1}}{2y + 4x}$

32. $\dfrac{a^{-2} - 4b^{-2}}{3b - 6a}$

8.4 Equations with Rational Expressions and Graphs

In **Section 8.1**, we defined the domain of a rational expression as the set of all possible values of the variable. Any value that makes the denominator 0 is excluded.

OBJECTIVE 1 Determine the domain of a rational equation.   The **domain of a rational equation** is the intersection (overlap) of the domains of the rational expressions in the equation.

EXAMPLE 1 Determining the Domains of Rational Equations

Find the domain of each equation.

(a) $\dfrac{2}{x} - \dfrac{3}{2} = \dfrac{7}{2x}$

The domains of the three rational terms of the equation are, in order, $\{x \mid x \neq 0\}$, $(-\infty, \infty)$, and $\{x \mid x \neq 0\}$. The intersection of these three domains is all real numbers except 0, which may be written $\{x \mid x \neq 0\}$.

(b) $\dfrac{2}{x - 3} - \dfrac{3}{x + 3} = \dfrac{12}{x^2 - 9}$

The domains of these three terms are, respectively, $\{x \mid x \neq 3\}$, $\{x \mid x \neq -3\}$, and $\{x \mid x \neq \pm 3\}$. The domain of the equation is the intersection of the three domains, all real numbers except 3 and -3, written $\{x \mid x \neq \pm 3\}$.

> **Work Problem 1 at the Side.** ▶▶▶

1 Find the domain of each equation.

(a) $\dfrac{3}{x} + \dfrac{1}{2} = \dfrac{5}{6x}$

OBJECTIVE 2 Solve rational equations.   The easiest way to solve most equations involving rational expressions is to multiply all terms in the equation by the least common denominator. This step will clear the equation of all denominators. *We can do this only with equations, not expressions.*

CAUTION
When each side of an equation is multiplied by a *variable* expression, the resulting "solutions" may not satisfy the original equation. *You must either determine and observe the domain or check all potential solutions in the original equation. It is wise to do both.*

(b)

$$\dfrac{4}{x - 5} - \dfrac{2}{x + 5} = \dfrac{1}{x^2 - 25}$$

EXAMPLE 2 Solving an Equation with Rational Expressions

Solve $\dfrac{2}{x} - \dfrac{3}{2} = \dfrac{7}{2x}$.

The domain, which excludes 0, was found in Example 1(a).

$$2x\left(\dfrac{2}{x} - \dfrac{3}{2}\right) = 2x\left(\dfrac{7}{2x}\right) \qquad \text{Multiply by the LCD, } 2x.$$

$$2x\left(\dfrac{2}{x}\right) - 2x\left(\dfrac{3}{2}\right) = 2x\left(\dfrac{7}{2x}\right) \qquad \text{Distributive property}$$

$$4 - 3x = 7 \qquad \text{Multiply.}$$

$$-3x = 3 \qquad \text{Subtract 4.}$$

$$x = -1 \qquad \text{Divide by } -3.$$

Continued on Next Page

ANSWERS
1. **(a)** $\{x \mid x \neq 0\}$ **(b)** $\{x \mid x \neq \pm 5\}$

2 Solve $-\dfrac{3}{20} + \dfrac{2}{x} = \dfrac{5}{4x}$.

Check: Replace x with -1 in the original equation.

$$\dfrac{2}{x} - \dfrac{3}{2} = \dfrac{7}{2x} \qquad \text{Original equation}$$

$$\dfrac{2}{-1} - \dfrac{3}{2} = \dfrac{7}{2(-1)} \qquad ? \quad \text{Let } x = -1.$$

$$-2 - \dfrac{3}{2} = -\dfrac{7}{2} \qquad ?$$

$$-\dfrac{7}{2} = -\dfrac{7}{2} \qquad \text{True}$$

The solution set is $\{-1\}$.

Work Problem 2 at the Side.

EXAMPLE 3 **Solving an Equation with No Solution**

Solve $\dfrac{2}{x-3} - \dfrac{3}{x+3} = \dfrac{12}{x^2-9}$.

Using the result from Example 1(b), we know that the domain excludes 3 and -3, since these values make one or more of the denominators in the equation equal 0. Multiply each side by the LCD, $(x+3)(x-3)$.

$$(x+3)(x-3)\left(\dfrac{2}{x-3} - \dfrac{3}{x+3}\right) = (x+3)(x-3)\left(\dfrac{12}{x^2-9}\right)$$

$$(x+3)(x-3)\left(\dfrac{2}{x-3}\right) - (x+3)(x-3)\left(\dfrac{3}{x+3}\right)$$

$$= (x+3)(x-3)\left(\dfrac{12}{x^2-9}\right) \quad \begin{array}{l}\text{Distributive}\\\text{property}\end{array}$$

$$2(x+3) - 3(x-3) = 12 \qquad \text{Multiply.}$$

$$2x + 6 - 3x + 9 = 12 \qquad \text{Distributive property}$$

$$-x + 15 = 12 \qquad \text{Combine terms.}$$

$$-x = -3 \qquad \text{Subtract 15.}$$

$$x = 3 \qquad \text{Divide by } -1.$$

3 Solve each equation.

(a) $\dfrac{3}{x+1} = \dfrac{1}{x-1} - \dfrac{2}{x^2-1}$

Since 3 is not in the domain, it cannot be a solution of the equation. Substituting 3 in the original equation shows why.

Check:

$$\dfrac{2}{x-3} - \dfrac{3}{x+3} = \dfrac{12}{x^2-9} \qquad \text{Original equation}$$

$$\dfrac{2}{3-3} - \dfrac{3}{3+3} = \dfrac{12}{3^2-9} \qquad ? \quad \text{Let } x = 3.$$

$$\dfrac{2}{0} - \dfrac{3}{6} = \dfrac{12}{0} \qquad ?$$

(b) $\dfrac{1}{x-3} + \dfrac{1}{x+3} = \dfrac{6}{x^2-9}$

Since division by 0 is undefined, the given equation has no solution, and the solution set is $\emptyset$.

Work Problem 3 at the Side.

ANSWERS
2. $\{5\}$
3. (a) $\emptyset$ (b) $\emptyset$

EXAMPLE 4 Solving an Equation with Rational Expressions

Solve $\dfrac{3}{p^2 + p - 2} - \dfrac{1}{p^2 - 1} = \dfrac{7}{2(p^2 + 3p + 2)}$.

Factor each denominator to find the LCD, $2(p - 1)(p + 2)(p + 1)$. The domain excludes $1, -2,$ and -1. Multiply each side by the LCD.

$$2(p - 1)(p + 2)(p + 1)\left(\dfrac{3}{(p + 2)(p - 1)} - \dfrac{1}{(p + 1)(p - 1)}\right)$$

$$= 2(p - 1)(p + 2)(p + 1)\left(\dfrac{7}{2(p + 2)(p + 1)}\right)$$

$2 \cdot 3(p + 1) - 2(p + 2) = 7(p - 1)$	Distributive property
$6p + 6 - 2p - 4 = 7p - 7$	Distributive property
$4p + 2 = 7p - 7$	Combine terms
$9 = 3p$	Subtract $4p$; add 7.
$3 = p$	Divide by 3.

Note that 3 is in the domain; substitute 3 for p in the original equation to check that the solution set is $\{3\}$.

Work Problem 4 at the Side. ▶▶▶

EXAMPLE 5 Solving an Equation That Leads to a Quadratic Equation

Solve $\dfrac{2}{3x + 1} = \dfrac{1}{x} - \dfrac{6x}{3x + 1}$.

Since the denominator $3x + 1$ cannot equal 0, $-\frac{1}{3}$ is excluded from the domain, as is 0. Multiply each side by the LCD, $x(3x + 1)$.

$$x(3x + 1)\left(\dfrac{2}{3x + 1}\right) = x(3x + 1)\left[\dfrac{1}{x} - \dfrac{6x}{3x + 1}\right]$$

$$x(3x + 1)\left(\dfrac{2}{3x + 1}\right) = x(3x + 1)\left(\dfrac{1}{x}\right) - x(3x + 1)\left(\dfrac{6x}{3x + 1}\right)$$

Distributive property

$$2x = 3x + 1 - 6x^2$$

Write this quadratic equation in standard form with 0 on the right side.

$6x^2 - 3x + 2x - 1 = 0$	
$6x^2 - x - 1 = 0$	Standard form
$(3x + 1)(2x - 1) = 0$	Factor.
$3x + 1 = 0 \quad$ or $\quad 2x - 1 = 0$	Zero-factor property
$x = -\dfrac{1}{3} \quad$ or $\quad x = \dfrac{1}{2}$	

Because $-\frac{1}{3}$ is not in the domain of the equation, it is not a solution. Check that the solution set is $\{\frac{1}{2}\}$.

Work Problem 5 at the Side. ▶▶▶

4 Solve

$$\dfrac{4}{x^2 + x - 6} - \dfrac{1}{x^2 - 4} = \dfrac{2}{x^2 + 5x + 6}.$$

5 Solve

$$\dfrac{1}{x + 4} + \dfrac{x}{x - 4} = \dfrac{-8}{x^2 - 16}.$$

ANSWERS
4. $\{-9\}$
5. $\{-1\}$

6 Graph each rational function, and give the equations of the vertical and horizontal asymptotes.

(a) $f(x) = -\dfrac{1}{x}$

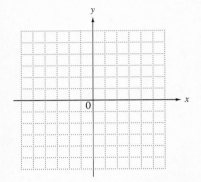

(b) $f(x) = \dfrac{2}{x+3}$

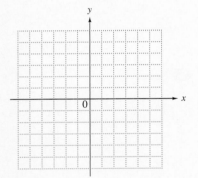

ANSWERS
6. **(a)** vertical asymptote: $x = 0$; horizontal
asymptote: $y = 0$

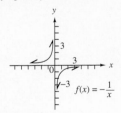

$f(x) = -\dfrac{1}{x}$

(b) vertical asymptote: $x = -3$; horizontal
asymptote: $y = 0$

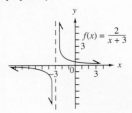

$f(x) = \dfrac{2}{x+3}$

OBJECTIVE 3 Recognize the graph of a rational function. As mentioned in **Section 8.1**, a function defined by a rational expression is a *rational function*. Because one or more values of x may be excluded from the domain of most rational functions, their graphs are often *discontinuous*. That is, there will be one or more breaks in the graph. For example, we use point plotting and observing the domain to graph the simple rational function defined by

$$f(x) = \frac{1}{x}.$$

The domain of this function includes all real numbers except 0. Thus, there will be no point on the graph with $x = 0$. The vertical line with equation $x = 0$ is called a **vertical asymptote** of the graph. The horizontal line with equation $y = 0$ is called a **horizontal asymptote.** We show some typical ordered pairs in the table for both negative and positive x-values.

x	-3	-2	-1	$-\frac{1}{2}$	$-\frac{1}{4}$	$-\frac{1}{10}$	$\frac{1}{10}$	$\frac{1}{4}$	$\frac{1}{2}$	1	2	3
y	$-\frac{1}{3}$	$-\frac{1}{2}$	-1	-2	-4	-10	10	4	2	1	$\frac{1}{2}$	$\frac{1}{3}$

Notice that the closer positive values of x are to 0, the larger y is. Similarly, the closer negative values of x are to 0, the smaller (more negative) y is. Using this observation, excluding 0 from the domain, and plotting the points in the table, we obtain the graph in Figure 2.

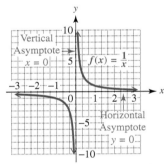

Figure 2

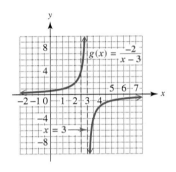

Figure 3

The graph of

$$g(x) = \frac{-2}{x-3}$$

is shown in Figure 3. Some ordered pairs are shown in the table.

x	-2	-1	0	1	2	2.5	2.75	3.25	3.5	4	5	6
y	$\frac{2}{5}$	$\frac{1}{2}$	$\frac{2}{3}$	1	2	4	8	-8	-4	-2	-1	$-\frac{2}{3}$

There is no point on the graph for $x = 3$ because 3 is excluded from the domain. The dashed line $x = 3$ represents the vertical asymptote and is not part of the graph. As suggested by the points from the table, the graph gets closer to the vertical asymptote as the x-values get closer to 3. Again, $y = 0$ is a horizontal asymptote.

Work Problem 6 at the Side.

8.4 Exercises

FOR EXTRA HELP 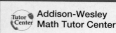 Addison-Wesley Math Tutor Center 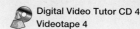 MathXL   Digital Video Tutor CD 4 Videotape 4 Student's Solutions Manual 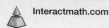 MyMathLab Interactmath.com

*As explained in this section, any values that would cause a denominator to equal 0 must be excluded from the domain and consequently as solutions of an equation that has variable expressions in the denominators. **(a)** Without actually solving the equation, list all possible numbers that would have to be rejected if they appeared as potential solutions. **(b)** Then give the domain using set notation. See Example 1.*

1. $\dfrac{1}{x+1} - \dfrac{1}{x-2} = 0$

2. $\dfrac{3}{x+4} - \dfrac{2}{x-9} = 0$

3. $\dfrac{5}{3x+5} - \dfrac{1}{x} = \dfrac{1}{2x+3}$

4. $\dfrac{6}{4x+7} - \dfrac{3}{x} = \dfrac{5}{6x-13}$

5. $\dfrac{1}{3x} + \dfrac{1}{2x} = \dfrac{x}{3}$

6. $\dfrac{5}{6x} - \dfrac{8}{2x} = \dfrac{x}{4}$

7. $\dfrac{3x+1}{x-4} = \dfrac{6x+5}{2x-7}$

8. $\dfrac{4x-1}{2x+3} = \dfrac{12x-25}{6x-2}$

9. $\dfrac{2}{x^2-x} + \dfrac{1}{x+3} = \dfrac{4}{x-2}$

10. Is it possible that any potential solutions to the equation

$$\dfrac{x+7}{4} - \dfrac{x+3}{3} = \dfrac{x}{12}$$

would have to be rejected? Explain.

Solve each equation. See Examples 2–5.

11. $\dfrac{-5}{2x} + \dfrac{3}{4x} = \dfrac{-7}{4}$

12. $\dfrac{6}{5x} - \dfrac{2}{3x} = \dfrac{-8}{45}$

13. $x - \dfrac{24}{x} = -2$

14. $p + \dfrac{15}{p} = -8$

15. $\dfrac{x-4}{x+6} = \dfrac{2x+3}{2x-1}$

16. $\dfrac{5x-8}{x+2} = \dfrac{5x-1}{x+3}$

17. $\dfrac{3x+1}{x-4} = \dfrac{6x+5}{2x-7}$

18. $\dfrac{4x-1}{2x+3} = \dfrac{12x-25}{6x-2}$

19. $\dfrac{1}{y-1} + \dfrac{5}{12} = \dfrac{-2}{3y-3}$

20. $\dfrac{4}{m+2} - \dfrac{11}{9} = \dfrac{1}{3m+6}$

21. $\dfrac{-2}{3t-6} - \dfrac{1}{36} = \dfrac{-3}{4t-8}$

22. $\dfrac{3}{4m+2} = \dfrac{17}{2} - \dfrac{7}{2m+1}$

23. $\dfrac{3}{k+2} - \dfrac{2}{k^2-4} = \dfrac{1}{k-2}$

24. $\dfrac{3}{x-2} + \dfrac{21}{x^2-4} = \dfrac{14}{x+2}$

25. $\dfrac{1}{y+2} + \dfrac{3}{y+7} = \dfrac{5}{y^2+9y+14}$

26. $\dfrac{1}{t+3} + \dfrac{4}{t+5} = \dfrac{2}{t^2+8t+15}$

27. $\dfrac{9}{x} + \dfrac{4}{6x-3} = \dfrac{2}{6x-3}$

28. $\dfrac{5}{n} + \dfrac{4}{6-3n} = \dfrac{2n}{6-3n}$

29. $\dfrac{6}{w+3} + \dfrac{-7}{w-5} = \dfrac{-48}{w^2-2w-15}$

30. $\dfrac{2}{r-5} + \dfrac{3}{2r+1} = \dfrac{22}{2r^2-9r-5}$

31. $\dfrac{x}{x-3} + \dfrac{4}{x+3} = \dfrac{18}{x^2-9}$

32. $\dfrac{2x}{x-3} + \dfrac{4}{x+3} = \dfrac{-24}{x^2-9}$

33. $\dfrac{6}{x-4} + \dfrac{5}{x} = \dfrac{-20}{x^2-4x}$

34. $\dfrac{7}{x-4} + \dfrac{3}{x} = \dfrac{-12}{x^2 - 4x}$

35. $\dfrac{2}{4x+7} + \dfrac{x}{3} = \dfrac{6}{12x+21}$

36. $\dfrac{5x+14}{x^2-9} = \dfrac{-2x^2 - 5x + 2}{x^2 - 9} + \dfrac{2x+4}{x-3}$

37. $\dfrac{4x-7}{4x^2-9} = \dfrac{-2x^2 + 5x - 4}{4x^2 - 9} + \dfrac{x+1}{2x+3}$

38. What is wrong with the following problem?   "Solve $\dfrac{2x+1}{3x-4} + \dfrac{1}{2x+3}$."

Graph each rational function. Give the equations of the vertical and horizontal asymptotes. See Objective 3 and Figures 2 and 3.

39. $f(x) = \dfrac{2}{x}$

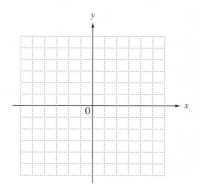

40. $f(x) = \dfrac{3}{x}$

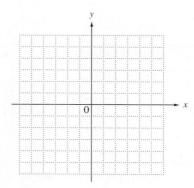

41. $f(x) = \dfrac{1}{x-2}$

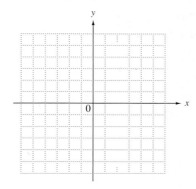

42. $f(x) = \dfrac{1}{x+2}$

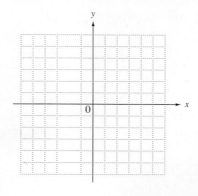

Solve each problem.

43. The average number of vehicles waiting in line to enter a sports arena parking area is modeled by the rational function defined by

$$w(x) = \frac{x^2}{2(1-x)},$$

where x is a quantity between 0 and 1 known as the **traffic intensity.** (*Source:* Mannering, F. and W. Kilareski, *Principles of Highway Engineering and Traffic Control,* John Wiley and Sons, 1990.) To the nearest tenth, find the average number of vehicles waiting for each traffic intensity.

(a) .1

(b) .8

(c) .9

(d) What happens to waiting time as traffic intensity increases?

44. The percent of deaths caused by smoking is modeled by the rational function defined by

$$p(x) = \frac{x-1}{x},$$

where x is the number of times a smoker is more likely to die of lung cancer than a nonsmoker. This is called the **incidence rate.** (*Source:* Walker, A., *Observation and Inference: An Introduction to the Methods of Epidemiology,* Epidemiology Resources Inc., 1991.) For example, $x = 10$ means that a smoker is 10 times more likely than a nonsmoker to die of lung cancer.

(a) Find $p(x)$ if x is 10.

(b) For what value of x is $p(x) = 80\%$? (*Hint:* Change 80% to a decimal.)

(c) Can the incidence rate equal 0? Explain.

RELATING CONCEPTS (EXERCISES 45–48)   For Individual or Group Work

An equation of the form

$$\frac{A}{x+B} + \frac{x}{x-B} = \frac{C}{x^2 - B^2}$$

will have one rejected solution if the relationship $C = -2AB$ holds true. (This can be proved using methods not covered in intermediate algebra.) For example, if $A = 1$ and $B = 2$, then $C = -2AB = -2(1)(2) = -4$, and the equation becomes

$$\frac{1}{x+2} + \frac{x}{x-2} = \frac{-4}{x^2-4}.$$

*This equation has solution set $\{-1\}$; the potential solution -2 must be rejected. To further understand this idea, **work Exercises 45–48 in order.***

45. Show that the second equation does indeed have solution set $\{-1\}$ and -2 must be rejected.

46. Let $A = 2$ and let $B = 1$. What is the corresponding value of C? Solve the equation determined by A, B, and C. What is the solution set? What value must be rejected?

47. Let $A = 4$ and let $B = -3$. What is the corresponding value of C? Solve the equation determined by A, B, and C. What is the solution set? What value must be rejected?

48. Choose two numbers of your own, letting one be A and the other be B. Repeat the process described in Exercises 46 and 47.

Summary Exercises on Rational Expressions and Equations

A common student error is to confuse an equation, *such as* $\frac{x}{2} + \frac{x}{3} = -5$, *with an* expression, *such as* $\frac{x}{2} + \frac{x}{3}$. *Look for the equals sign to distinguish between them. Equations are solved for a numerical answer, while expressions are simplified as shown below.*

Solving an Equation

Solve: $\dfrac{x}{2} + \dfrac{x}{3} = -5$.

Multiply each side by the LCD, 6.

$$6\left(\frac{x}{2} + \frac{x}{3}\right) = 6(-5)$$

$$6\left(\frac{x}{2}\right) + 6\left(\frac{x}{3}\right) = 6(-5)$$

$$3x + 2x = -30$$

$$5x = -30$$

$$x = -6$$

Check that the solution set is $\{-6\}$.

Simplifying an Expression

Add: $\dfrac{x}{2} + \dfrac{x}{3}$.

Write both fractions with the LCD, 6.

$$\frac{x}{2} + \frac{x}{3} = \frac{x \cdot 3}{2 \cdot 3} + \frac{x \cdot 2}{3 \cdot 2}$$

$$= \frac{3x}{6} + \frac{2x}{6}$$

$$= \frac{3x + 2x}{6}$$

$$= \frac{5x}{6}$$

Identify each exercise as an expression *or an* equation. *Then simplify the expression by performing the indicated operation, or solve the given equation, as appropriate.*

1. $\dfrac{x}{2} - \dfrac{x}{4} = 5$

2. $\dfrac{4x - 20}{x^2 - 25} \cdot \dfrac{(x + 5)^2}{10}$

3. $\dfrac{6}{7x} - \dfrac{4}{x}$

4. $\dfrac{\dfrac{1}{x} + \dfrac{1}{y}}{\dfrac{1}{x} - \dfrac{1}{y}}$

5. $\dfrac{5}{7t} = \dfrac{52}{7} - \dfrac{3}{t}$

6. $\dfrac{x - 5}{3} + \dfrac{1}{3} = \dfrac{x - 2}{5}$

7. $\dfrac{7}{6x} + \dfrac{5}{8x}$

8. $\dfrac{4}{x} - \dfrac{8}{x + 1} = 0$

9. $\dfrac{\dfrac{6}{x + 1} - \dfrac{1}{x}}{\dfrac{2}{x} - \dfrac{4}{x + 1}}$

10. $\dfrac{8}{r + 2} - \dfrac{7}{4r + 8}$

11. $\dfrac{x}{x + y} + \dfrac{2y}{x - y}$

12. $\dfrac{3p^2 - 6p}{p + 5} \div \dfrac{p^2 - 4}{8p + 40}$

13. $\dfrac{x-2}{9} \cdot \dfrac{5}{8-4x}$

14. $\dfrac{a-4}{3} + \dfrac{11}{6} = \dfrac{a+1}{2}$

15. $\dfrac{b^2 + b - 6}{b^2 + 2b - 8} \cdot \dfrac{b^2 + 8b + 16}{3b + 12}$

16. $\dfrac{10z^2 - 5z}{3z^3 - 6z^2} \div \dfrac{2z^2 + 5z - 3}{z^2 + z - 6}$

17. $\dfrac{5}{x^2 - 2x} - \dfrac{3}{x^2 - 4}$

18. $\dfrac{6}{t+1} + \dfrac{4}{5t+5} = \dfrac{34}{15}$

19. $\dfrac{\dfrac{5}{x} - \dfrac{3}{y}}{\dfrac{9x^2 - 25y^2}{x^2 y}}$

20. $\dfrac{-2}{a^2 + 2a - 3} - \dfrac{5}{3 - 3a} = \dfrac{4}{3a + 9}$

21. $\dfrac{4y^2 - 13y + 3}{2y^2 - 9y + 9} \div \dfrac{4y^2 + 11y - 3}{6y^2 - 5y - 6}$

22. $\dfrac{8}{3k+9} - \dfrac{8}{15} = \dfrac{2}{5k+15}$

23. $\dfrac{3r}{r-2} = 1 + \dfrac{6}{r-2}$

24. $\dfrac{6z^2 - 5z - 6}{6z^2 + 5z - 6} \cdot \dfrac{12z^2 - 17z + 6}{12z^2 - z - 6}$

25. $\dfrac{-1}{3-x} - \dfrac{2}{x-3}$

26. $\dfrac{\dfrac{t}{4} - \dfrac{1}{t}}{1 + \dfrac{t+4}{t}}$

27. $\dfrac{2}{y+1} - \dfrac{3}{y^2 - y - 2} = \dfrac{3}{y-2}$

28. $\dfrac{7}{2x^2 - 8x} + \dfrac{3}{x^2 - 16}$

29. $\dfrac{3}{y-3} - \dfrac{3}{y^2 - 5y + 6} = \dfrac{2}{y-2}$

30. $\dfrac{2k + \dfrac{5}{k-1}}{3k - \dfrac{2}{k}}$

8.5 Applications of Rational Expressions

OBJECTIVE 1 **Find the value of an unknown variable in a formula.**
In this section, we work with formulas that contain rational expressions.

EXAMPLE 1 Finding the Value of a Variable in a Formula

In physics, the focal length, f, of a lens is given by the formula

$$\frac{1}{f} = \frac{1}{p} + \frac{1}{q},$$

where p is the distance from the object to the lens and q is the distance from the lens to the image. See Figure 4. Find q if $p = 20$ cm and $f = 10$ cm.

Focal Length of Camera Lens
Figure 4

Replace f with 10 and p with 20.

$$\frac{1}{f} = \frac{1}{p} + \frac{1}{q}$$

$$\frac{1}{10} = \frac{1}{20} + \frac{1}{q} \qquad \text{Let } f = 10, p = 20.$$

$$20q \cdot \frac{1}{10} = 20q\left(\frac{1}{20} + \frac{1}{q}\right) \qquad \text{Multiply by the LCD, } 20q.$$

$$20q \cdot \frac{1}{10} = 20q\left(\frac{1}{20}\right) + 20q\left(\frac{1}{q}\right) \quad \text{Distributive property.}$$

$$2q = q + 20 \qquad \text{Multiply.}$$

$$q = 20 \qquad \text{Subtract } q.$$

The distance from the lens to the image is 20 cm.

Work Problem 1 at the Side. ▶▶▶

OBJECTIVE 2 **Solve a formula for a specified variable.** The goal in solving for a specified variable is to isolate it on one side of the equals sign.

EXAMPLE 2 Solving a Formula for a Specified Variable

Solve $\dfrac{1}{f} = \dfrac{1}{p} + \dfrac{1}{q}$ for p.

$$\frac{1}{f} = \frac{1}{p} + \frac{1}{q}$$

$$fpq \cdot \frac{1}{f} = fpq\left(\frac{1}{p} + \frac{1}{q}\right) \qquad \text{Multiply by the LCD, } fpq.$$

$$pq = fq + fp \qquad \text{Distributive property}$$

Continued on Next Page

OBJECTIVES

1 Find the value of an unknown variable in a formula.

2 Solve a formula for a specified variable.

3 Solve applications using proportions.

4 Solve applications about distance, rate, and time.

5 Solve applications about work rates.

❶ Use the formula given in Example 1 to answer each part.

(a) Find p if $f = 15$ and $q = 25$.

(b) Find f if $p = 6$ and $q = 9$.

(c) Find q if $f = 12$ and $p = 16$.

ANSWERS

1. **(a)** $\dfrac{75}{2}$ **(b)** $\dfrac{18}{5}$ **(c)** 48

2 Solve $\dfrac{3}{p} + \dfrac{3}{q} = \dfrac{5}{r}$ for q.

Transform the equation so that the terms with p (the specified variable) are on the same side. One way to do this is to subtract fp from each side.

$$pq = fq + fp$$
$$pq - fp = fq \qquad \text{Subtract } fp.$$
$$p(q - f) = fq \qquad \text{Factor out } p.$$
$$p = \frac{fq}{q - f} \qquad \text{Divide by } q - f.$$

Work Problem 2 at the Side.

EXAMPLE 3 Solving a Formula for a Specified Variable

Solve $I = \dfrac{nE}{R + nr}$ for n.

$$I = \frac{nE}{R + nr}$$
$$(R + nr)I = (R + nr)\frac{nE}{R + nr} \qquad \text{Multiply by } R + nr.$$
$$RI + nrI = nE$$
$$RI = nE - nrI \qquad \text{Subtract } nrI.$$
$$RI = n(E - rI) \qquad \text{Factor out } n.$$
$$\frac{RI}{E - rI} = n \qquad \text{Divide by } E - rI.$$

3 Solve $A = \dfrac{Rr}{R + r}$ for R.

CAUTION
Refer to the steps in Examples 2 and 3 that factor out the desired variable. This is a step that often gives students difficulty. *Remember that the variable for which you are solving must be a factor on only one side of the equation.* Then each side can be divided by the remaining factor in the last step.

Work Problem 3 at the Side.

We can now solve problems that translate into equations with rational expressions. To do so, we continue to use the six-step problem-solving method from **Section 2.3.**

OBJECTIVE 3 Solve applications using proportions. A **ratio** is a comparison of two quantities. The ratio of a to b may be written in any of the following ways:

$$a \text{ to } b, \quad a : b, \quad \text{or} \quad \frac{a}{b}. \qquad \text{Ratio of } a \text{ to } b$$

Ratios are usually written as quotients in algebra. A **proportion** is a statement that two ratios are equal, such as

$$\frac{a}{b} = \frac{c}{d}. \qquad \text{Proportion}$$

Proportions are a useful and important type of rational equation.

ANSWERS

2. $q = \dfrac{3rp}{5p - 3r}$ or $q = \dfrac{-3rp}{3r - 5p}$

3. $R = \dfrac{-Ar}{A - r}$ or $R = \dfrac{Ar}{r - A}$

EXAMPLE 4 Solving a Proportion

In 2002, about 15 of every 100 Americans had no health insurance coverage. The population at that time was about 288 million. How many million Americans had no health insurance? (*Source:* U.S. Bureau of the Census.)

Step 1 **Read** the problem.

Step 2 **Assign a variable.** Let x = the number (in millions) who had no health insurance.

Step 3 **Write an equation.** To get an equation, set up a proportion. The ratio 15 to 100 should equal the ratio x to 288.

$$\frac{15}{100} = \frac{x}{288} \qquad \text{Write a proportion.}$$

Step 4 **Solve.** $28{,}800\left(\frac{15}{100}\right) = 28{,}800\left(\frac{x}{288}\right)$ Multiply by a common denominator.

$$4320 = 100x \qquad \text{Simplify.}$$
$$x = 43.2 \qquad \text{Divide by 100.}$$

Step 5 **State the answer.** There were 43.2 million Americans with no health insurance in 2002.

Step 6 **Check** that the ratio of 43.2 million to 288 million equals $\frac{15}{100}$.

Work Problem 4 at the Side. ▶▶▶

EXAMPLE 5 Solving a Proportion Involving Rates

Marissa's car uses 10 gal of gas to travel 210 mi. She has 5 gal of gas in the car, and she still needs to drive 640 mi. If we assume the car continues to use gas at the same rate, how many more gallons will she need?

Step 1 **Read** the problem.

Step 2 **Assign a variable.** Let x = the additional number of gallons of gas.

Step 3 **Write an equation.** To get an equation, set up a proportion.

$$\text{gallons} \longrightarrow \frac{10}{210} = \frac{5 + x}{640} \longleftarrow \text{gallons}$$
$$\text{miles} \longrightarrow \qquad\qquad\qquad \longleftarrow \text{miles}$$

Step 4 **Solve.** The LCD is $10 \cdot 21 \cdot 64$.

$$10 \cdot 21 \cdot 64 \left(\frac{10}{210}\right) = 10 \cdot 21 \cdot 64 \left(\frac{5 + x}{640}\right)$$

$$64 \cdot 10 = 21(5 + x)$$
$$640 = 105 + 21x \qquad \text{Distributive property}$$
$$535 = 21x \qquad \text{Subtract 105.}$$
$$25.5 \approx x \qquad \text{Divide by 21; round to the nearest tenth.}$$

Step 5 **State the answer.** Marissa will need about 25.5 more gallons of gas.

Step 6 **Check.** The 25.5 gal plus the 5 gal equals 30.5 gal.

$$\frac{30.5}{640} \approx .047 \quad \text{and} \quad \frac{10}{210} \approx .047$$

Since the ratios are equal, the answer is correct.

Work Problem 5 at the Side. ▶▶▶

4 Solve the problem.

In 2002, approximately 11.6% (that is, 11.6 of every 100) of the 73,500,000 children under 18 yr of age in the United States had no health insurance. How many such children were uninsured? (*Source:* U.S. Bureau of the Census.)

5 Solve the problem.

In a recent year, the average American family spent 8.2 of every 100 dollars on health care. This amounted to $3665 per family. To the nearest dollar, what was the average family income at that time? (*Source:* U.S. Health Care Financing Administration, U.S. Bureau of the Census.)

ANSWERS
4. 8,526,000
5. $44,695

OBJECTIVE 4 **Solve applications about distance, rate, and time.** The next examples use the distance formula $d = rt$ introduced in **Section 2.2.** A familiar example of a rate is speed, which is the ratio of distance to time, or $r = \frac{d}{t}$.

EXAMPLE 6 **Solving a Problem about Distance, Rate, and Time**

A tour boat goes 10 mi against the current in a small river in the same time that it goes 15 mi with the current. If the speed of the current is 3 mph, find the speed of the boat in still water.

Step 1 **Read** the problem. We must find the speed of the boat in still water.

Step 2 **Assign a variable.**

Let x = the speed of the boat in still water.

When the boat is traveling *against* the current, the current slows the boat down, and the speed of the boat is the difference between its speed in still water and the speed of the current, that is, $x - 3$ mph.

When the boat is traveling *with* the current, the current speeds the boat up, and the speed of the boat is the sum of its speed in still water and the speed of the current, that is, $x + 3$ mph.

Thus, $x - 3$ = the speed of the boat *against* the current,

and $x + 3$ = the speed of the boat *with* the current.

Because the time is the same going against the current as with the current, find time in terms of distance and rate (speed) for each situation. Start with the distance formula,

$$d = rt,$$

and divide each side by r to get $t = \frac{d}{r}$. Going against the current, the distance is 10 mi and the rate is $x - 3$, giving

$$t = \frac{d}{r} = \frac{10}{x - 3}.$$

Going with the current, the distance is 15 mi and the rate is $x + 3$, so

$$t = \frac{d}{r} = \frac{15}{x + 3}.$$

This information is summarized in the following table.

	Distance	Rate	Time	
Against Current	10	$x - 3$	$\dfrac{10}{x - 3}$	Times are equal.
With Current	15	$x + 3$	$\dfrac{15}{x + 3}$	

Step 3 **Write an equation.** Because the times are equal,

$$\frac{10}{x - 3} = \frac{15}{x + 3}.$$

Continued on Next Page

Step 4 **Solve.** The LCD is $(x + 3)(x - 3)$.

$$\frac{10}{x - 3} = \frac{15}{x + 3}$$

$$(x + 3)(x - 3)\left(\frac{10}{x - 3}\right) = (x + 3)(x - 3)\left(\frac{15}{x + 3}\right) \quad \text{Multiply by the LCD.}$$

$$10(x + 3) = 15(x - 3) \qquad \text{Multiply.}$$

$$10x + 30 = 15x - 45 \qquad \text{Distributive property}$$

$$30 = 5x - 45 \qquad \text{Subtract } 10x.$$

$$75 = 5x \qquad \text{Add 45.}$$

$$15 = x \qquad \text{Divide by 5.}$$

Step 5 **State the answer.** The speed of the boat in still water is 15 mph.

Step 6 **Check** the answer: $\dfrac{10}{15 - 3} = \dfrac{15}{15 + 3}$ is true.

> **Work Problem 6 at the Side.** ▶▶▶

EXAMPLE 7 **Solving a Problem about Distance, Rate, and Time**

At O'Hare Airport, Cheryl and Bill are walking to the gate (at the same speed) to catch their flight to Akron, Ohio. Since Bill wants a window seat, he steps onto the moving sidewalk and continues to walk while Cheryl uses the stationary sidewalk. If the sidewalk moves at 1 m per sec and Bill saves 50 sec covering the 300-m distance, what is their walking speed?

Step 1 **Read** the problem. We must find their walking speed.

Step 2 **Assign a variable.** Let x represent their walking speed in meters per second. Thus Cheryl travels at x meters per second and Bill travels at $x + 1$ meters per second. Express their times in terms of the known distances and the variable rates. As in Example 6, start with $d = rt$ and divide each side by r to get $t = \frac{d}{r}$. For Cheryl, the distance is 300 m and the rate is x, so Cheryl's time is

$$t = \frac{d}{r} = \frac{300}{x}.$$

Bill travels 300 m at a rate of $x + 1$, so his time is

$$t = \frac{d}{r} = \frac{300}{x + 1}.$$

This information is summarized in the following table.

	Distance	Rate	Time
Cheryl	300	x	$\dfrac{300}{x}$
Bill	300	$x + 1$	$\dfrac{300}{x + 1}$

Step 3 **Write an equation** using the times from the table.

Bill's time is Cheryl's time less 50 seconds.

$$\frac{300}{x + 1} = \frac{300}{x} - 50$$

Continued on Next Page

---

6 Solve the problem.

A plane travels 100 mi against the wind in the same time that it takes to travel 120 mi with the wind. The wind speed is 20 mph.

(a) Complete this table.

	d	r	t
Against Wind	100	$x - 20$	
With Wind	120	$x + 20$	

(b) Find the speed of the plane in still air.

ANSWERS

6. (a) $\dfrac{100}{x - 20}; \dfrac{120}{x + 20}$ **(b)** 220 mph

7 Solve the problem.

Dona Kenly drove 300 mi north from San Antonio, mostly on the freeway. She usually averaged 55 mph, but an accident slowed her speed through Dallas to 15 mph. If her trip took 6 hr, how many miles did she drive at reduced speed?

	d	r	t
Normal Speed	$300 - x$	55	
Reduced Speed	x	15	

Step 4 **Solve.**

$$\frac{300}{x + 1} = \frac{300}{x} - 50$$

$$x(x + 1)\left(\frac{300}{x + 1}\right) = x(x + 1)\left(\frac{300}{x} - 50\right) \qquad \text{Multiply by the LCD, } x(x + 1).$$

$$x(x + 1)\left(\frac{300}{x + 1}\right) = x(x + 1)\left(\frac{300}{x}\right) - x(x + 1)(50)$$

Distributive property

$$300x = 300(x + 1) - 50x(x + 1) \qquad \text{Multiply.}$$

$$300x = 300x + 300 - 50x^2 - 50x \qquad \text{Distributive property}$$

$$50x^2 + 50x - 300 = 0 \qquad \text{Standard form}$$

$$x^2 + x - 6 = 0 \qquad \text{Divide by 50.}$$

$$(x + 3)(x - 2) = 0 \qquad \text{Factor.}$$

$$x + 3 = 0 \quad \text{or} \quad x - 2 = 0 \qquad \text{Zero-factor property}$$

$$x = -3 \quad \text{or} \qquad x = 2$$

Discard the negative answer, since speed cannot be negative.

Step 5 **State the answer.** Their walking speed is 2 m per sec.

Step 6 **Check** the answer in the words of the original problem.

◀◀◀ Work Problem 7 at the Side.

OBJECTIVE 5 **Solve applications about work rates.** Problems about work are closely related to distance problems.

PROBLEM-SOLVING HINT

People work at different rates. If the letters r, t, and A represent the rate at which the work is done, the time required, and the amount of work accomplished, respectively, then $A = rt$. Notice the similarity to the distance formula, $d = rt$.

Amount of work can be measured in terms of jobs accomplished. Thus, if 1 job is completed, $A = 1$, and the formula gives the rate as

$$1 = rt$$

$$r = \frac{1}{t}.$$

To solve a work problem, we begin by using the following fact to express all rates of work.

Rate of Work

If a job can be accomplished in t units of time, then the rate of work is

$$\frac{1}{t} \text{ job per unit of time.}$$

See if you can identify the six problem-solving steps in the next example.

ANSWERS

7. $11\frac{1}{4}$ mi

EXAMPLE 8 Solving a Problem about Work

Letitia and Kareem are working on a neighborhood cleanup. Kareem can clean up all the trash in the area in 7 hr, while Letitia can do the same job in 5 hr. How long will it take them if they work together?

Let $x =$ the number of hours it will take the two people working together. Just as we made a table for the distance formula, $d = rt$, make a table here for $A = rt$, with $A = 1$. Since $A = 1$, the rate for each person will be $\frac{1}{t}$, where t is the time it takes the person to complete the job alone. For example, since Kareem can clean up all the trash in 7 hr, his rate is $\frac{1}{7}$ of the job per hour. Similarly, Letitia's rate is $\frac{1}{5}$ of the job per hour.

	Rate	Time Working Together	Fractional Part of the Job Done
Kareem	$\frac{1}{7}$	x	$\frac{1}{7}x$
Letitia	$\frac{1}{5}$	x	$\frac{1}{5}x$

Since together they complete 1 job, the sum of the fractional parts accomplished by them should equal 1.

$$\begin{array}{ccccc} \text{Part done} & & \text{Part done} & & 1 \text{ whole} \\ \text{by Kareem} & + & \text{by Letitia} & \text{is} & \text{job.} \\ \frac{1}{7}x & + & \frac{1}{5}x & = & 1 \end{array}$$

$$35\left(\frac{1}{7}x + \frac{1}{5}x\right) = 35 \cdot 1 \quad \text{The LCD is 35.}$$

$$5x + 7x = 35$$

$$12x = 35$$

$$x = \frac{35}{12}$$

Working together, Kareem and Letitia can do the entire job in $\frac{35}{12}$ hr, or 2 hr and 55 min. Check this result in the original problem.

Work Problem 8 at the Side. ▶▶▶

There is another way to approach problems about work. For instance, in Example 8, x represents the number of hours it will take the two people working together to complete the entire job. In one hour, $\frac{1}{x}$ of the entire job will be completed. Kareem completes $\frac{1}{7}$ of the job in one hour, and Letitia completes $\frac{1}{5}$ of the job, so the sum of their rates should equal $\frac{1}{x}$. Thus,

$$\frac{1}{7} + \frac{1}{5} = \frac{1}{x}.$$

Multiplying each side of this equation by $35x$ gives $5x + 7x = 35$. This is the same equation we got in Example 8 in the third line from the bottom. Thus the solution of the equation is the same using either approach.

---

8 Solve each problem.

(a) Stan needs 45 min to do the dishes, while Deb can do them in 30 min. How long will it take them if they work together?

	Rate	Time Working Together	Fractional Part of the Job Done
Stan	$\frac{1}{45}$	x	
Deb	$\frac{1}{30}$	x	

(b) Suppose it takes Stan 35 min to do the dishes, and together they can do them in 15 min. How long will it take Deb to do them alone?

ANSWERS

8. (a) 18 min **(b)** $26\frac{1}{4}$ min

Real-Data Applications

It Depends on What You Mean by "Average"

Finding an average seems to be a simple process. Don't we just add the values and divide by the number of values? Well, for rational expressions, it all depends on what you mean by "average."

- To find the average of two fractions, say $\frac{1}{3}$ and $\frac{3}{4}$, add the two fractions and divide by 2.

$$\frac{\frac{1}{3} + \frac{3}{4}}{2} = \frac{\left(\frac{1}{3} + \frac{3}{4}\right) \cdot 12}{2 \cdot 12} = \frac{4 + 9}{24} = \frac{13}{24}$$

On a number line, the fraction $\frac{13}{24}$ is the **arithmetic mean,** which is exactly halfway between the fractions $\frac{1}{3}$ and $\frac{3}{4}$.

- Suppose you travel one direction at 60 mph and return at 30 mph. To find your average rate, you have to calculate the total distance divided by the total time. Recall that $d = rt$, so the total distance is $2d$, the time going is $\frac{d}{60}$, and the time returning is $\frac{d}{30}$. Since $r = \frac{d}{t}$,

$$\frac{2d}{\frac{d}{60} + \frac{d}{30}} = \frac{2d \cdot 60}{\left(\frac{d}{60} + \frac{d}{30}\right) \cdot 60} = \frac{120d}{d + 2d} = \frac{120d}{3d} = 40 \text{ mph.}$$

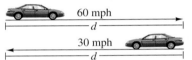

The average rate is 40 mph. This is the *harmonic mean* of 60 and 30. The **harmonic mean** of two numbers a and b is defined as $\frac{2ab}{a+b}$. Note that

$$\frac{2 \cdot 60 \cdot 30}{60 + 30} = \frac{3600}{90} = 40.$$

- To calculate a batting average, you find the **ratio** of the number of hits to the number of "at bats." Suppose a baseball player has 72 hits in 364 "at bats." His batting average would be $\frac{72}{364} \approx .198$. If the same player gets an additional 3 hits from 8 more "at bats" during the next week, then his revised batting average would be

$$\frac{72 + 3}{364 + 8} = \frac{75}{372} \approx .202.$$

For Group Discussion

A carpenter builds wine racks. For each situation, find the appropriate "average" quantity.

1. The carpenter told his helper to cut $\frac{1}{2}$ ft pieces from a dowel. The helper could not find a measuring tape, but he did recall that the distance from the tip of his middle finger to the tip of his thumb was approximately 6 in., so he estimated the lengths. When the carpenter checked his work, he found that the helper had actually cut two pieces that were $\frac{5}{12}$ ft and $\frac{1}{2}$ ft long. What was the average length of the two pieces?

2. Once the pieces are cut, the carpenter can assemble and finish a wine rack in 2 hr, working alone. His helper takes 4 hr to accomplish the same task, working alone. If the carpenter and the helper work together, what is their average time to assemble and finish a wine rack?

3. Of 115 wine racks built, 112 passed a quality control check. What was the acceptance rate? The carpenter built 35 additional wine racks, of which 28 were acceptable. What was the revised acceptance rate? Round answers to the nearest thousandth.

8.5 Exercises

FOR EXTRA HELP · Addison-Wesley Math Tutor Center · MathXL · Digital Video Tutor CD 4 Videotape 4 · Student's Solutions Manual · MyMathLab · Interactmath.com

In Exercises 1–4, a familiar formula is given. Give the letter of the choice that is an equivalent form of the given formula.

1. $p = br$ (percent)

   A. $b = \dfrac{p}{r}$ B. $r = \dfrac{b}{p}$

   C. $b = \dfrac{r}{p}$ D. $p = \dfrac{r}{b}$

2. $V = LWH$ (geometry)

   A. $H = \dfrac{LW}{V}$ B. $L = \dfrac{V}{WH}$

   C. $L = \dfrac{WH}{V}$ D. $W = \dfrac{H}{VL}$

3. $m = \dfrac{F}{a}$ (physics)

   A. $a = mF$ B. $F = \dfrac{m}{a}$

   C. $F = \dfrac{a}{m}$ D. $F = ma$

4. $I = \dfrac{E}{R}$ (electricity)

   A. $R = \dfrac{I}{E}$ B. $R = IE$

   C. $E = \dfrac{I}{R}$ D. $E = RI$

Solve each problem. See Example 1.

5. A gas law in chemistry says that
$$\frac{PV}{T} = \frac{pv}{t}.$$
Suppose that $T = 300$, $t = 350$, $V = 9$, $P = 50$, and $v = 8$. Find p.

6. In work with electric circuits, the formula
$$\frac{1}{a} = \frac{1}{b} + \frac{1}{c}$$
occurs. Find b if $a = 8$ and $c = 12$.

7. A formula from anthropology says that
$$c = \frac{100b}{L}.$$
Find L if $c = 80$ and $b = 5$.

8. The gravitational force between two masses is given by
$$F = \frac{GMm}{d^2}.$$
Find M if $F = 10$, $G = 6.67 \times 10^{-11}$, $m = 1$, and $d = 3 \times 10^{-6}$.

Solve each formula for the specified variable. See Examples 2 and 3.

9. $F = \dfrac{GMm}{d^2}$ for G (physics)

10. $F = \dfrac{GMm}{d^2}$ for M (physics)

11. $\dfrac{1}{a} = \dfrac{1}{b} + \dfrac{1}{c}$ for a (electricity)

12. $\dfrac{1}{a} = \dfrac{1}{b} + \dfrac{1}{c}$ for b (electricity)

13. $\dfrac{PV}{T} = \dfrac{pv}{t}$ for v (chemistry)

14. $\dfrac{PV}{T} = \dfrac{pv}{t}$ for T (chemistry)

15. $I = \dfrac{nE}{R + nr}$ for r (engineering)

16. $a = \dfrac{V - v}{t}$ for V (physics)

17. $A = \dfrac{1}{2}h(b + B)$ for b (mathematics)

18. $S = \dfrac{n}{2}(a + \ell)d$ for n (mathematics)

19. $\dfrac{E}{e} = \dfrac{R + r}{r}$ for r (engineering)

20. $y = \dfrac{x + z}{a - x}$ for x

21. To solve the equation $m = \dfrac{ab}{a - b}$ for a, what is the first step?

22. Suppose you are asked to solve the equation
$$rp - rq = p + q$$
for r. What is the first step?

Solve each problem mentally. Use proportions in Exercises 23 and 24.

23. In a mathematics class, 3 of every 4 students are girls. If there are 20 students in the class, how many are girls? How many are boys?

24. In a certain southern state, sales tax on a purchase of $1.50 is $.12. What is the sales tax on a purchase of $6.00?

25. If Marin can mow her yard in 2 hr, what is her rate (in job per hour)?

26. A van traveling from Atlanta to Detroit averages 50 mph and takes 14 hr to make the trip. How far is it from Atlanta to Detroit?

Use the bar graph to answer Exercises 27–30.

27. In which year was the ratio of truck accidents to car accidents the greatest?

28. In which year was the ratio of truck accidents to car accidents the least?

29. In which year was the ratio of car accidents to truck accidents closest to 3 to 1?

30. In which year(s) was the ratio of car accidents to truck accidents less than 2 to 1?

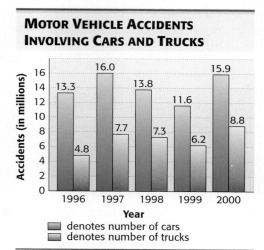

MOTOR VEHICLE ACCIDENTS INVOLVING CARS AND TRUCKS

Source: National Safety Council.

Solve each problem. See Examples 4 and 5.

31. The ratio of fast food restaurants to total restaurants in the United States in 2003 was approximately 1 to 5. If there were 844,000 total restaurants in 2003, how many fast food restaurants were there? (*Source:* National Restaurant Association.)

32. During 2000–2001, the ratio of teachers to students in private elementary and secondary schools was approximately 1 to 14. If a private school had 554 students, how many teachers would be at the school if this ratio was valid for that school? Round your answer to the nearest whole number. (*Source:* U.S. National Center for Education Statistics.)

33. Biologists tagged 500 fish in a lake on January 1. On February 1 they returned and collected a random sample of 400 fish, 8 of which had been previously tagged. Approximately how many fish does the lake have based on this experiment?

34. Suppose that in the experiment of Exercise 33, 10 of the previously tagged fish were collected on February 1. What would be the estimate of the fish population?

35. In a recent year, 50 shares of common stock in Merck Company earned $191.50. How much more would 75 shares of the stock have earned? (*Source:* Merck & Co., Inc.)

36. Seligman Communications and Information Fund, Inc. produced income of $22,950 on an investment of $100,000 in a recent year. If the investment had been increased to $260,000, how much more income would have been produced? (*Source:* Seligman Communications and Information Fund, Inc.)

Nurses use proportions to determine the amount of a drug to administer when the dose of the drug is measured in milligrams but the drug is packaged in a diluted form in milliliters. (Source: Hoyles, Celia, Richard Noss, and Stefano Pozzi, "Proportional Reasoning in Nursing Practice," *Journal for Research in Mathematics Education,* January 2001.) *For example, to find the number of milliliters of fluid needed to administer* 300 mg *of a drug that comes packaged as* 120 mg *in* 2 mL *of fluid, a nurse sets up the proportion*

$$\frac{120 \text{ mg}}{2 \text{ mL}} = \frac{300 \text{ mg}}{x \text{ mL}},$$

where x represents the amount to administer in milliliters. Use this method to find the correct dose for each prescription.

37. 120 mg of Amakacine packaged as 100 mg in 2-mL vials

38. 1.5 mg of morphine packaged as 20 mg ampules diluted in 10 mL of fluid

*In geometry, it is shown that two triangles with corresponding angle measures equal, called **similar triangles,** have corresponding sides proportional. For example, in the figure, angle A = angle D, angle B = angle E, and angle C = angle F, so the triangles are similar. Then the following ratios of corresponding sides are equal.*

$$\frac{4}{6} = \frac{6}{9} = \frac{2x + 1}{2x + 5}$$

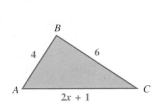
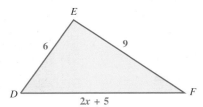

39. Solve for x using the given proportion to find the lengths of the third sides of the triangles.

40. Suppose the following triangles are similar. Find y and the lengths of the two longest sides of each triangle.

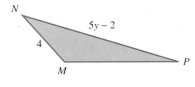
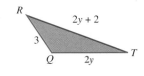

Solve each problem. See Examples 6 and 7.

41. Kellen's boat goes 12 mph. Find the rate of the current of the river if she can go 6 mi upstream in the same amount of time she can go 10 mi downstream.

	Distance	Rate	Time
Downstream	10	$12 + x$	
Upstream	6	$12 - x$	

42. Kasey can travel 8 mi upstream in the same time it takes her to go 12 mi downstream. Her boat goes 15 mph in still water. What is the rate of the current?

	Distance	Rate	Time
Downstream			
Upstream			

43. Driving from Tulsa to Detroit, Jeff averaged 50 mph. He figured that if he had averaged 60 mph, his driving time would have decreased 3 hr. How far is it from Tulsa to Detroit?

44. If Dr. Dawson rides his bike to his office, he averages 12 mph. If he drives his car, he averages 36 mph. His time driving is $\frac{1}{4}$ hr less than his time riding his bike. How far is his office from home?

45. A private plane traveled from San Francisco to a secret rendezvous. It averaged 200 mph. On the return trip, the average speed was 300 mph. If the total traveling time was 4 hr, how far from San Francisco was the secret rendezvous?

46. Johnny averages 30 mph when he drives on the old highway to his favorite fishing hole, and he averages 50 mph when most of his route is on the interstate. If both routes are the same length, and he saves 2 hr by traveling on the interstate, how far away is the fishing hole?

47. On the first part of a trip to Carmel traveling on the freeway, Marge averaged 60 mph. On the rest of the trip, which was 10 mi longer than the first part, she averaged 50 mph. Find the total distance to Carmel if the second part of the trip took 30 min more than the first part.

48. While on vacation, Jim and Annie decided to drive all day. During the first part of their trip on the highway, they averaged 60 mph. When they got to Houston, traffic caused them to average only 30 mph. The distance they drove in Houston was 100 mi less than their distance on the highway. What was their total driving distance if they spent 50 min more on the highway than they did in Houston?

Solve each problem. See Example 8.

49. Butch and Peggy want to pick up the mess that their grandson, Grant, has made in his playroom. Butch could do it in 15 min working alone. Peggy, working alone, could clean it in 12 min. How long will it take them if they work together?

	Rate	Time Working Together	Fractional Part of the Job Done
Butch	$\frac{1}{15}$	x	
Peggy	$\frac{1}{12}$	x	

50. Lou can groom Jay Beckenstein's dogs in 8 hr, but it takes his business partner, Janet, only 5 hr to groom the same dogs. How long will it take them to groom Jay's dogs if they work together?

	Rate	Time Working Together	Fractional Part of the Job Done
Lou	$\frac{1}{8}$	x	
Janet	$\frac{1}{5}$	x	

51. Ron Wood can paint a room in 6 hr working alone. If his son, Jason, helps him, the job takes 4 hr. How long would it take Jason to do the job if he worked alone?

52. Sandi and Cary Goldstein are refinishing a table. Working alone, Cary could do the job in 7 hr. If the two work together, the job takes 5 hr. How long will it take Sandi to refinish the table working alone?

53. If a vat of acid can be filled by an inlet pipe in 10 hr and emptied by an outlet pipe in 20 hr, how long will it take to fill the vat if both pipes are open?

54. A winery has a vat to hold chardonnay. An inlet pipe can fill the vat in 9 hr, while an outlet pipe can empty it in 12 hr. How long will it take to fill the vat if both the outlet and the inlet pipes are open?

55. Suppose that Hortense and Mort can clean their entire house in 7 hr, while their toddler, Mimi, just by being around, can completely mess it up in only 2 hr. If Hortense and Mort clean the house while Mimi is at her grandma's, and then start cleaning up after Mimi the minute she gets home, how long does it take from the time Mimi gets home until the whole place is a shambles?

56. An inlet pipe can fill an artificial lily pond in 60 min, while an outlet pipe can empty it in 80 min. Through an error, both pipes are left open. How long will it take for the pond to fill?

Chapter 8

SUMMARY

KEY TERMS

8.1	**rational expression**	A rational expression (algebraic fraction) is the quotient of two polynomials with denominator not 0.
	rational function	A rational function is a function that is defined by a rational expression in the form

$$f(x) = \frac{P(x)}{Q(x)},$$

where $Q(x) \neq 0$.

8.2	**least common denominator (LCD)**	The least common denominator in a group of denominators is the product of all different factors from each denominator, with each factor raised to the greatest power that occurs in any denominator.
8.3	**complex fraction**	A complex fraction is an expression having a fraction in the numerator, denominator, or both.
8.4	**domain of a rational equation**	The domain of a rational equation is the intersection (overlap) of the domains of the rational expressions in the equation.
	vertical asymptote	A rational function of the form $f(x) = \dfrac{P(x)}{x - a}$ has the line $x = a$ as a vertical asymptote; the graph approaches the line on each side but does not intersect it.
	horizontal asymptote	A horizontal line that a graph approaches as $|x|$ gets larger and larger without bound is called a horizontal asymptote.
8.5	**ratio**	A ratio is a comparison of two quantities using a quotient.
	proportion	A proportion is a statement that two ratios are equal.

TEST YOUR WORD POWER

See how well you have learned the vocabulary in this chapter. Answers, with examples, follow the Quick Review.

1. A **rational expression** is
   A. an algebraic expression made up of a term or the sum of a finite number of terms with real coefficients and integer exponents
   B. a polynomial equation of degree 2
   C. an expression with one or more fractions in the numerator, denominator, or both
   D. the quotient of two polynomials with denominator not zero.

2. In a given set of fractions, the **least common denominator** is
   A. the smallest denominator of all the denominators
   B. the smallest expression that is divisible by all the denominators

   C. the largest integer that evenly divides the numerator and denominator of all the fractions
   D. the largest denominator of all the denominators.

3. A **complex fraction** is
   A. an algebraic expression made up of a term or the sum of a finite number of terms with real coefficients and integer exponents
   B. a polynomial equation of degree 2
   C. an expression with one or more fractions in the numerator, denominator, or both
   D. the quotient of two polynomials with denominator not zero.

4. A **ratio**
   A. compares two quantities using a quotient
   B. says that two quotients are equal
   C. is a product of two quantities
   D. is a difference between two quantities.

5. A **proportion**
   A. compares two quantities using a quotient
   B. says that two quotients are equal
   C. is a product of two quantities
   D. is a difference between two quantities.

QUICK REVIEW

Concepts	Examples

8.1 *Rational Expressions and Functions;*
Multiplying and Dividing

Fundamental Property of Rational Numbers

If $\frac{a}{b}$ is a rational number and if c is any nonzero real number, then

$$\frac{a}{b} = \frac{ac}{bc}.$$

$$\frac{3}{4} = \frac{3 \cdot 5}{4 \cdot 5} = \frac{15}{20}$$

Writing a Rational Expression in Lowest Terms

Write in lowest terms.

Step 1 Factor the numerator and the denominator completely.

$$\frac{2x + 8}{x^2 - 16} = \frac{2(x + 4)}{(x - 4)(x + 4)}$$

Step 2 Apply the fundamental property.

$$= \frac{2}{x - 4}$$

Multiplying Rational Expressions

Multiply.

Step 1 Factor numerators and denominators.

Step 2 Apply the fundamental property.

$$\frac{x^2 + 2x + 1}{x^2 - 1} \cdot \frac{5}{3x + 3} = \frac{(x + 1)^2}{(x - 1)(x + 1)} \cdot \frac{5}{3(x + 1)}$$

Step 3 Multiply the remaining factors in the numerator and in the denominator.

$$= \frac{5}{3(x - 1)}$$

Step 4 Check that the product is in lowest terms.

Dividing Rational Expressions

Divide.

Multiply the first rational expression (dividend) by the reciprocal of the second (divisor).

$$\frac{2x + 5}{x - 3} \div \frac{2x^2 + 3x - 5}{x^2 - 9} = \frac{2x + 5}{x - 3} \cdot \frac{(x + 3)(x - 3)}{(2x + 5)(x - 1)}$$

$$= \frac{x + 3}{x - 1}$$

8.2 *Adding and Subtracting Rational Expressions*

Adding or Subtracting Rational Expressions

Subtract.

Step 1 If the denominators are the same, add or subtract the numerators. Place the result over the common denominator.

$$\frac{1}{x + 6} - \frac{3}{x + 2} = \frac{x + 2}{(x + 6)(x + 2)} - \frac{3(x + 6)}{(x + 6)(x + 2)}$$

If the denominators are different, write all rational expressions with the LCD. Then add or subtract the numerators, and place the result over the common denominator.

$$= \frac{x + 2 - 3(x + 6)}{(x + 6)(x + 2)}$$

$$= \frac{x + 2 - 3x - 18}{(x + 6)(x + 2)}$$

Step 2 Be sure the answer is in lowest terms.

$$= \frac{-2x - 16}{(x + 6)(x + 2)}$$

Concepts	**Examples**

8.3 *Complex Fractions*
Simplifying a Complex Fraction
Method 1   Simplify the numerator and denominator separately, as much as possible. Then multiply the numerator by the reciprocal of the denominator. Write the answer in lowest terms.

Simplify the complex fraction.

$$Method\ 1\quad \frac{\dfrac{1}{x^2} - \dfrac{1}{y^2}}{\dfrac{1}{x} + \dfrac{1}{y}} = \frac{\dfrac{y^2}{x^2y^2} - \dfrac{x^2}{x^2y^2}}{\dfrac{y}{xy} + \dfrac{x}{xy}}$$

$$= \frac{\dfrac{y^2 - x^2}{x^2y^2}}{\dfrac{y + x}{xy}} = \frac{y^2 - x^2}{x^2y^2} \div \frac{y + x}{xy}$$

$$= \frac{(y + x)(y - x)}{x^2y^2} \cdot \frac{xy}{y + x}$$

$$= \frac{y - x}{xy}$$

Method 2   Multiply the numerator and denominator of the complex fraction by the least common denominator of all fractions appearing in the complex fraction. Then simplify the result.

$$Method\ 2\quad \frac{\dfrac{1}{x^2} - \dfrac{1}{y^2}}{\dfrac{1}{x} + \dfrac{1}{y}} = \frac{x^2y^2\left(\dfrac{1}{x^2} - \dfrac{1}{y^2}\right)}{x^2y^2\left(\dfrac{1}{x} + \dfrac{1}{y}\right)}$$

$$= \frac{y^2 - x^2}{xy^2 + x^2y} = \frac{(y - x)(y + x)}{xy(y + x)}$$

$$= \frac{y - x}{xy}$$

8.4 *Equations with Rational Expressions and Graphs*
Solving an Equation with Rational Expressions
To solve an equation involving rational expressions, first determine the domain. Then multiply all the terms in the equation by the least common denominator. Solve the resulting equation. ***Each potential solution must be checked to see that it is in the domain of the equation.***

Solve.

$$\frac{1}{x} + x = \frac{26}{5}$$

Note that 0 is excluded from the domain.

$5 + 5x^2 = 26x$	Multiply by $5x$.
$5x^2 - 26x + 5 = 0$	Subtract $26x$.
$(5x - 1)(x - 5) = 0$	Factor.
$5x - 1 = 0 \quad or \quad x - 5 = 0$	Zero-factor property
$x = \dfrac{1}{5} \quad or \quad x = 5$	

Both check. The solution set is $\left\{\frac{1}{5}, 5\right\}$.

The graph of a rational function of the type covered in this section may have one or more breaks. At such points, the graph will approach an asymptote.

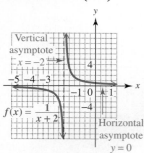

| *Concepts* | *Examples* |

8.5 *Applications of Rational Expressions*

To solve a formula for a particular variable, isolate that variable on one side.

Solve for L.

$$c = \frac{100b}{L}$$

$$cL = 100b \qquad \text{Multiply by } L.$$

$$L = \frac{100b}{c} \qquad \text{Divide by } c.$$

To solve a motion problem, use the formula

$$d = rt$$

or one of its equivalents,

$$t = \frac{d}{r} \quad \text{or} \quad r = \frac{d}{t}.$$

Solve.

A canal has a current of 2 mph. Find the speed of Amy's boat in still water if it goes 11 mi downstream in the same time that it goes 8 mi upstream.

Let x represent the speed of the boat in still water.

	Distance	Rate	Time
Downstream	11	$x + 2$	$\dfrac{11}{x + 2}$
Upstream	8	$x - 2$	$\dfrac{8}{x - 2}$

Because the times are the same, the equation is

$$\frac{11}{x + 2} = \frac{8}{x - 2}. \qquad \text{Use } t = \tfrac{d}{r}.$$

$$11(x - 2) = 8(x + 2) \qquad \text{Multiply by the LCD.}$$

$$11x - 22 = 8x + 16 \qquad \text{Distributive property}$$

$$3x = 38 \qquad \text{Subtract } 8x \text{ and add 22.}$$

$$x = 12\frac{2}{3} \qquad \text{Divide by 3.}$$

The speed in still water is $12\frac{2}{3}$ mph.

To solve a work problem, use the fact that if a complete job is done in t units of time, the rate of work is $\frac{1}{t}$ job per unit of time.

ANSWERS TO TEST YOUR WORD POWER

1. D; *Examples:* $-\dfrac{3}{4y^2}$, $\dfrac{5x^3}{x + 2}$, $\dfrac{a + 3}{a^2 - 4a - 5}$

2. B; *Example:* The LCD of $\dfrac{1}{x}$, $\dfrac{2}{3}$, and $\dfrac{5}{x + 1}$ is $3x(x + 1)$.

3. C; *Examples:* $\dfrac{\frac{2}{3}}{\frac{4}{7}}$, $\dfrac{x - \frac{1}{x}}{x + \frac{1}{y}}$, $\dfrac{\frac{2}{a + 1}}{a^2 - 1}$

4. A; *Example:* $\dfrac{7 \text{ in.}}{12 \text{ in.}}$ compares two quantities.

5. B; *Example:* The proportion $\dfrac{2}{3} = \dfrac{8}{12}$ states that the two ratios are equal.

Chapter **8**

REVIEW EXERCISES

[8.1] *(a) Find all real numbers that are excluded from the domain. (b) Give the domain using set notation.*

1. $f(x) = \dfrac{-7}{3x + 18}$

2. $f(x) = \dfrac{5x + 17}{x^2 - 7x + 10}$

3. $f(x) = \dfrac{9}{x^2 - 18x + 81}$

Write in lowest terms.

4. $\dfrac{12x^2 + 6x}{24x + 12}$

5. $\dfrac{25m^2 - n^2}{25m^2 - 10mn + n^2}$

6. $\dfrac{r - 2}{4 - r^2}$

7. What is meant by the reciprocal of a rational expression?

Multiply or divide. Write the answer in lowest terms.

8. $\dfrac{(2y + 3)^2}{5y} \cdot \dfrac{15y^3}{4y^2 - 9}$

9. $\dfrac{w^2 - 16}{w} \cdot \dfrac{3}{4 - w}$

10. $\dfrac{z^2 - z - 6}{z - 6} \div \dfrac{z^2 + 2z - 15}{z^2 - 6z}$

11. $\dfrac{m^3 - n^3}{m^2 - n^2} \div \dfrac{m^2 + mn + n^2}{m + n}$

[8.2] *Assume that each expression is the denominator of a rational expression. Find the least common denominator for each group.*

12. $32b^3, \quad 24b^5$

13. $9r^2, \quad 3r + 1$

14. $6x^2 + 13x - 5, \quad 9x^2 + 9x - 4$

Add or subtract as indicated.

15. $\dfrac{8}{z} - \dfrac{3}{2z^2}$

16. $\dfrac{5y + 13}{y + 1} - \dfrac{1 - 7y}{y + 1}$

17. $\dfrac{6}{5a + 10} + \dfrac{7}{6a + 12}$

18. $\dfrac{3r}{10r^2 - 3rs - s^2} + \dfrac{2r}{2r^2 + rs - s^2}$

[8.3] *Simplify each complex fraction.*

19. $\dfrac{\dfrac{3}{t} + 2}{\dfrac{4}{t} - 7}$

20. $\dfrac{\dfrac{2}{m - 3n}}{\dfrac{1}{3n - m}}$

21. $\dfrac{\dfrac{3}{p} - \dfrac{2}{q}}{\dfrac{9q^2 - 4p^2}{qp}}$

22. $\dfrac{x^{-2} - y^{-2}}{x^{-1} - y^{-1}}$

[8.4] *Solve each equation.*

23. $\dfrac{1}{t + 4} + \dfrac{1}{2} = \dfrac{3}{2t + 8}$

24. $\dfrac{-5m}{m + 1} + \dfrac{m}{3m + 3} = \dfrac{56}{6m + 6}$

25. $\dfrac{2}{k - 1} - \dfrac{4k + 1}{k^2 - 1} = \dfrac{-1}{k + 1}$

26. $\dfrac{5}{x + 2} + \dfrac{3}{x + 3} = \dfrac{x}{x^2 + 5x + 6}$

27. After solving the equation

$$\dfrac{3}{x - 3} - \dfrac{2}{x - 2} = \dfrac{3}{x^2 - 5x + 6},$$

a student got $x = 3$ as her final step. She could not understand why the answer in the back of the book was "∅," because she checked her algebra several times and was sure that all her algebraic work was correct. Was she wrong or was the answer in the back of the book wrong? Explain.

28. Explain the difference between simplifying the expression

$$\dfrac{4}{x} + \dfrac{1}{2} - \dfrac{1}{3}$$

and solving the equation

$$\dfrac{4}{x} + \dfrac{1}{2} = \dfrac{1}{3}.$$

29. Which is the graph of a rational function? Give the equations of its vertical and horizontal asymptotes?

A.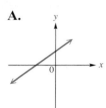

B.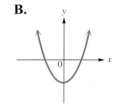

C.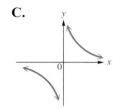

D.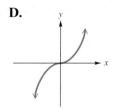

[8.5]

30. According to a law from physics, $\dfrac{1}{A} = \dfrac{1}{B} + \dfrac{1}{C}$. Find A if $B = 30$ and $C = 10$.

Solve each formula for the specified variable.

31. $F = \dfrac{GMm}{d^2}$ for m (physics)

32. $\mu = \dfrac{Mv}{M + m}$ for M (electronics)

Solve each problem.

33. An article in *Scientific American* predicts that 23,200 of the 58,000 passenger-km per day in North America will be provided by high-speed trains by the year 2050. If the traffic volume in a typical region of North America is 15,000, how many passenger-kilometers per day will high-speed trains provide there? (*Source:* Schafer, Andreas and David Victor, "The Past and Future of Global Mobility," *Scientific American*, October, 1997.)

34. A river has a current of 4 km per hr. Find the speed of Lynn McTernan's boat in still water if it goes 40 km downstream in the same time that it takes to go 24 km upstream.

	d	r	t
Upstream	24	$x - 4$	
Downstream	40		

35. A sink can be filled by a cold-water tap in 8 min, and filled by the hot-water tap in 12 min. How long would it take to fill the sink with both taps open?

36. Dona Kenly and Jay Jenkins need to sort a pile of bottles at the recycling center. Working alone, Dona could do the entire job in 9 hr, while Jay could do the entire job in 6 hr. How long will it take them if they work together?

MIXED REVIEW EXERCISES

Write in lowest terms.

37. $\dfrac{x + 2y}{x^2 - 4y^2}$

38. $\dfrac{x^2 + 2x - 15}{x^2 - x - 6}$

Perform the indicated operations.

39. $\dfrac{2}{m} + \dfrac{5}{3m^2}$

40. $\dfrac{k^2 - 6k + 9}{1 - 216k^3} \cdot \dfrac{6k^2 + 17k - 3}{9 - k^2}$

41. $\dfrac{\dfrac{-3}{x} + \dfrac{x}{2}}{1 + \dfrac{x+1}{x}}$

42. $\dfrac{9x^2 + 46x + 5}{3x^2 - 2x - 1} \div \dfrac{x^2 + 11x + 30}{x^3 + 5x^2 - 6x}$

43. $\dfrac{\dfrac{3}{x} - 5}{6 + \dfrac{1}{x}}$

44. $\dfrac{9}{3-x} - \dfrac{2}{x-3}$

45. $\dfrac{4y+16}{30} \div \dfrac{2y+8}{5}$

46. $\dfrac{t^{-2} + s^{-2}}{t^{-1} - s^{-1}}$

47. $\dfrac{4a}{a^2 - ab - 2b^2} - \dfrac{6b - a}{a^2 + 4ab + 3b^2}$

48. $\dfrac{a}{b} + \dfrac{b}{c} + \dfrac{c}{d}$

Solve.

49. $\dfrac{x+3}{x^2 - 5x + 4} - \dfrac{1}{x} = \dfrac{2}{x^2 - 4x}$

50. $A = \dfrac{Rr}{R+r}$ for r

51. $1 - \dfrac{5}{r} = \dfrac{-4}{r^2}$

52. $\dfrac{3x}{x-4} + \dfrac{2}{x} = \dfrac{48}{x^2 - 4x}$

53. The strength of a contact lens is given in units called diopters, and also in millimeters of arc. As the diopters increase, the millimeters of arc decrease. The rational function defined by

$$a(d) = \dfrac{337}{d}$$

relates the arc measurement a to the diopter measurement d. (*Source:* Bausch and Lomb.)

(a) What arc measurement will correspond to 40.5-diopter lenses?

(b) A lens with an arc measurement of 7.51 will provide what diopter strength?

54. The hot-water tap can fill a tub in 20 min. The cold-water tap takes 15 min to fill the tub. How long would it take to fill the tub with both taps open?

Chapter 8

T E S T

1. Find all real numbers excluded from the domain of
   $f(x) = \dfrac{x + 3}{3x^2 + 2x - 8}$. Then give the domain using set notation.

2. Write $\dfrac{6x^2 - 13x - 5}{9x^3 - x}$ in lowest terms.

Multiply or divide.

3. $\dfrac{(x + 3)^2}{4} \cdot \dfrac{6}{2x + 6}$

4. $\dfrac{y^2 - 16}{y^2 - 25} \cdot \dfrac{y^2 + 2y - 15}{y^2 - 7y + 12}$

5. $\dfrac{x^2 - 9}{x^3 + 3x^2} \div \dfrac{x^2 + x - 12}{x^3 + 9x^2 + 20x}$

6. Find the least common denominator for the following group of denominators: $t^2 + t - 6$, $t^2 + 3t$, t^2.

Add or subtract as indicated.

7. $\dfrac{7}{6t^2} - \dfrac{1}{3t}$

8. $\dfrac{9}{x - 7} + \dfrac{4}{x + 7}$

9. $\dfrac{6}{x + 4} + \dfrac{1}{x + 2} - \dfrac{3x}{x^2 + 6x + 8}$

Simplify each complex fraction.

10. $\dfrac{\dfrac{12}{r + 4}}{\dfrac{11}{6r + 24}}$

11. $\dfrac{\dfrac{1}{a} - \dfrac{1}{b}}{\dfrac{a}{b} - \dfrac{b}{a}}$

12. $\dfrac{\dfrac{2}{x^2} + \dfrac{1}{y^2}}{\dfrac{1}{x} - \dfrac{1}{y}}$

13. Identify each of the following as an *expression* to be simplified or an *equation* to be solved. Then simplify the one that is an expression, and solve the one that is an equation.

    (a) $\dfrac{2x}{3} + \dfrac{x}{4} - \dfrac{11}{2}$

    (b) $\dfrac{2x}{3} + \dfrac{x}{4} = \dfrac{11}{2}$

Solve each equation.

14. $\dfrac{1}{x} - \dfrac{4}{3x} = \dfrac{1}{x - 2}$

15. $\dfrac{y}{y + 2} - \dfrac{1}{y - 2} = \dfrac{8}{y^2 - 4}$

1. _____

2. _____

3. _____

4. _____

5. _____

6. _____

7. _____

8. _____

9. _____

10. _____

11. _____

12. _____

13. (a) _____

    (b) _____

14. _____

15. _____

16. _____

16. Checking the solution(s) of an equation in earlier chapters verified that the algebraic steps were performed correctly. When an equation includes a term with a variable denominator, what additional reason requires that the solutions be checked?

17. _____

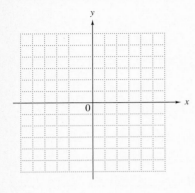

17. Sketch the graph of the function defined by $f(x) = \dfrac{-2}{x+1}$. Give the equations of its vertical and horizontal asymptotes.

18. _____

18. Solve for the variable ℓ in this formula from mathematics:

$$S = \frac{n}{2}(a + \ell).$$

Solve each problem.

19. _____

19. Wayne can do a job in 9 hr, while Susan can do the same job in 5 hr. How long would it take them to do the job if they worked together?

20. _____

20. The rate of the current in a stream is 3 mph. Nana's boat can go 36 mi downstream in the same time that it takes to go 24 mi upstream. Find the rate of her boat in still water.

21. _____

21. Biologists collected a sample of 600 fish from Lake Linda on May 1 and tagged each of them. When they returned on June 1, a new sample of 800 fish was collected, and 10 of these had been previously tagged. Use this experiment to determine the approximate fish population of Lake Linda.

22. **(a)** _____

(b) _____

22. In biology, the function defined by

$$g(x) = \frac{5x}{2 + x}$$

gives the growth rate g of a population for x units of available food. (*Source:* Smith, J. Maynard, *Models in Ecology,* Cambridge University Press, 1974.)

(a) What amount of food (in appropriate units) would produce a growth rate of 3 units of growth per unit of food?

(b) What is the growth rate if no food is available?

Cumulative Review Exercises

CHAPTERS 1–8

Evaluate if $x = -4$, $y = 3$, and $z = 6$.

1. $|2x| + 3y - z^3$

2. $\dfrac{x(2x-1)}{3y-z}$

Solve each equation.

3. $7(2x + 3) - 4(2x + 1) = 2(x + 1)$

4. $|6x - 8| - 4 = 0$

5. $ax + by = cx + d$ for x

Solve each inequality.

6. $\dfrac{2}{3}x + \dfrac{5}{12}x \le 20$

7. $|3x + 2| \ge 4$

Solve each problem.

8. Otis Taylor invested some money at 4% interest and twice as much at 3% interest. His interest for the first year was $400. How much did he invest at each rate?

9. A triangle has an area of 42 m². The base is 14 m long. Find the height of the triangle.

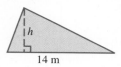

14 m

10. Graph $-4x + 2y = 8$ and give the intercepts.

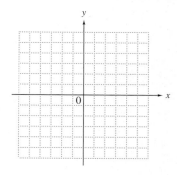

Find the slope of each line described in Exercises 11 and 12.

11. Through $(-5, 8)$ and $(-1, 2)$

12. Perpendicular to $4x - 3y = 12$

13. Write an equation of the line in Exercise 11 in the form $y = mx + b$.

Graph the solution set of each inequality.

14. $2x + 5y > 10$

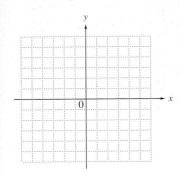

15. $x - y \geq 3$ and $3x + 4y \leq 12$

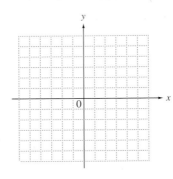

Decide whether each relation defined in Exercises 16–18 defines a function, and give its domain and range.

16. AVERAGE HOURLY WAGES IN MEXICO

Year	Wage (in dollars)
1990	1.25
1992	1.61
1994	1.80
1996	1.21
1998	1.94
2000	2.26

Source: John Christman, CIEMEX-WEFA.

17.

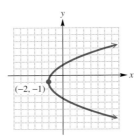

18. $y = -\sqrt{x + 2}$

19. Given the equation $5x - 3y = 8$,

  (a) write y as a function of x, using function notation $f(x)$;

  (b) find $f(1)$.

20. If $f(x) = 3x + 6$, what is $f(x + 3)$?

Solve each system.

21. $4x - y = -7$
    $5x + 2y = 1$

22. $x + y - 2z = -1$
    $2x - y + z = -6$
    $3x + 2y - 3z = -3$

23. $x + 2y + z = 5$
    $x - y + z = 3$
    $2x + 4y + 2z = 11$

24. As of June 2004, Lisa Leslie of the WNBA Los Angeles Sparks had 799 more career points than Sheryl Swoopes of the Houston Comets. Together the two players had 6737 career points. How many points did each of the two players have? (*Source:* WNBA.)

Simplify. Write each answer with only positive exponents. Assume all variables represent nonzero real numbers.

25. $\left(\dfrac{a^{-3}b^4}{a^2b^{-1}}\right)^{-2}$

26. $\left(\dfrac{m^{-4}n^2}{m^2n^{-3}}\right) \cdot \left(\dfrac{m^5n^{-1}}{m^{-2}n^5}\right)$

Perform the indicated operations.

27. $(3y^2 - 2y + 6) - (-y^2 + 5y + 12)$

28. $-6x^4(x^2 - 3x + 2)$

29. $(4f + 3)(3f - 1)$

30. $(7t^3 + 8)(7t^3 - 8)$

31. $\left(\dfrac{1}{4}x + 5\right)^2$

32. $(3x^3 + 13x^2 - 17x - 7) \div (3x + 1)$

33. For the polynomial functions defined by
$$f(x) = x^2 + 2x - 3 \quad \text{and} \quad g(x) = 2x^3 - 3x^2 + 4x - 1,$$
find each of the following.

(a) $(f + g)(x)$ **(b)** $(g - f)(x)$ **(c)** $(f + g)(-1)$

Factor each polynomial completely.

34. $2x^2 - 13x - 45$ **35.** $100t^4 - 25$ **36.** $8p^3 + 125$

37. Solve the equation $3x^2 + 4x = 7$.

Write each rational expression in lowest terms.

38. $\dfrac{y^2 - 16}{y^2 - 8y + 16}$

39. $\dfrac{8x^2 - 18}{8x^2 + 4x - 12}$

Perform the indicated operations. Express the answer in lowest terms.

40. $\dfrac{2a^2}{a + b} \cdot \dfrac{a - b}{4a}$

41. $\dfrac{x + 4}{x - 2} + \dfrac{2x - 10}{x - 2}$

42. $\dfrac{2x}{2x - 1} + \dfrac{4}{2x + 1} + \dfrac{8}{4x^2 - 1}$

43. Solve the equation

$$\dfrac{-3x}{x + 1} + \dfrac{4x + 1}{x} = \dfrac{-3}{x^2 + x}.$$

44. Solve the formula

$$\dfrac{1}{f} = \dfrac{1}{p} + \dfrac{1}{q} \quad \text{for } q.$$

Solve each problem.

45. Lucinda can fly her plane 200 mi against the wind in the same time it takes her to fly 300 mi with the wind. The wind blows at 30 mph. Find the speed of her plane in still air.

46. Machine A can complete a certain job in 2 hr. To speed up the work, Machine B, which could complete the job alone in 3 hr, is brought in to help. How long will it take the two machines to complete the job working together?

Roots, Radicals, and Root Functions

9

- **9.1** Radical Expressions and Graphs
- **9.2** Rational Exponents
- **9.3** Simplifying Radical Expressions
- **9.4** Adding and Subtracting Radical Expressions
- **9.5** Multiplying and Dividing Radical Expressions

Summary Exercises on Operations with Radicals and Rational Exponents

- **9.6** Solving Equations with Radicals
- **9.7** Complex Numbers

Tom Skilling is the chief meteorologist for the *Chicago Tribune*. He writes a column titled "Ask Tom Why," where readers question him on a variety of topics. Reader Ted Fleischaker wrote: "I cannot remember the formula to calculate the distance to the horizon. I have a stunning view from my 14th floor condo, 150 feet above the ground. How far can I see?" (See Exercise 125 in Section 9.3.)

In Skilling's answer, he explained the formula for finding the distance d to the horizon in miles,

$$d = 1.224\sqrt{h},$$

where h is the height in feet. Square roots such as this one are often found in formulas. This chapter deals with roots and radicals.

9.1 Radical Expressions and Graphs

OBJECTIVES

1. Find roots of numbers.
2. Find principal roots.
3. Graph functions defined by radical expressions.
4. Find nth roots of nth powers.
5. Use a calculator to find roots.

1 Simplify.

(a) $\sqrt[3]{8}$

(b) $\sqrt[3]{1000}$

(c) $\sqrt[4]{81}$

(d) $\sqrt[6]{64}$

OBJECTIVE 1 Find roots of numbers. In **Section 1.3** we found square roots of positive numbers such as

$$\sqrt{36} = 6, \text{ because } 6 \cdot 6 = 36 \quad \text{and} \quad \sqrt{144} = 12, \text{ because } 12 \cdot 12 = 144.$$

We now extend our discussion of roots to cube roots, fourth roots, and higher roots. In general, $\sqrt[n]{a}$ is a number whose nth power equals a. That is,

$$\sqrt[n]{a} = b \quad \text{means} \quad b^n = a.$$

The number a is the **radicand,** n is the **index** or **order,** and the expression $\sqrt[n]{a}$ is a **radical.**

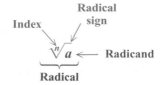

EXAMPLE 1 Simplifying Higher Roots

Simplify.

(a) $\sqrt[3]{27} = 3$, because $3^3 = 27$.

(b) $\sqrt[3]{125} = 5$, because $5^3 = 125$.

(c) $\sqrt[4]{16} = 2$, because $2^4 = 16$.

(d) $\sqrt[5]{32} = 2$, because $2^5 = 32$.

◀◀◀ Work Problem 1 at the Side.

OBJECTIVE 2 Find principal roots. If n is even, positive numbers have two nth roots. For example, both 4 and -4 are square roots of 16, and 2 and -2 are fourth roots of 16. In such cases, the notation $\sqrt[n]{a}$ represents the positive root, called the **principal root.**

nth Root

If n is *even* and a is *positive* or 0, then

$$\sqrt[n]{a} \text{ represents the principal } n\text{th root of } a, \text{ and}$$
$$-\sqrt[n]{a} \text{ represents the negative } n\text{th root of } a.$$

If n is *even* and a is *negative,* then

$$\sqrt[n]{a} \text{ is not a real number.}$$

If n is *odd,* then

$$\text{there is exactly one } n\text{th root of } a, \text{ written } \sqrt[n]{a}.$$

If n is even, then the two nth roots of a are often written together as $\pm \sqrt[n]{a}$, with $\pm$ read "positive or negative."

ANSWERS
1. (a) 2 (b) 10 (c) 3 (d) 2

EXAMPLE 2 **Finding Roots**

Find each root.

(a) $\sqrt{100} = 10$

Because the radicand is positive, there are two square roots, 10 and -10. We want the principal root, which is 10.

(b) $-\sqrt{100} = -10$

Here, we want the negative square root, -10.

(c) $\sqrt[4]{81} = 3$

(d) $\sqrt[6]{-64}$

The index is even and the radicand is negative, so this is not a real number.

(e) $\sqrt[3]{-8} = -2$, because $(-2)^3 = -8$.

Work Problem 2 at the Side. ▶▶▶

OBJECTIVE 3 **Graph functions defined by radical expressions.** A **radical expression** is an algebraic expression that contains radicals. For example,

$$3 - \sqrt{x}, \quad \sqrt[3]{x}, \quad \text{and} \quad \sqrt{2x - 1} \qquad \text{Radical expressions}$$

are radical expressions.

In earlier chapters we graphed functions defined by polynomial and rational expressions. Now we examine the graphs of functions defined by the radical expressions $f(x) = \sqrt{x}$ and $f(x) = \sqrt[3]{x}$.

Figure 1 shows the graph of the **square root function** with a table of selected points.

x	$f(x) = \sqrt{x}$
0	0
1	1
4	2
9	3

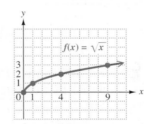

Figure 1

Only nonnegative values can be used for x, so the domain is $[0, \infty)$. Because $\sqrt{x}$ is the principal square root of x, it always has a nonnegative value, so the range is also $[0, \infty)$.

Figure 2 shows the graph of the **cube root function** and a table of selected points.

x	$f(x) = \sqrt[3]{x}$
-8	-2
-1	-1
0	0
1	1
8	2

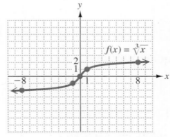

Figure 2

Since any real number (positive, negative, or 0) can be used for x in the cube root function, $\sqrt[3]{x}$ can be positive, negative, or 0. Thus both the domain and the range of the cube root function are $(-\infty, \infty)$.

2 Find each root.

(a) $\sqrt{4}$

(b) $\sqrt[3]{27}$

(c) $-\sqrt{36}$

(d) $\sqrt[4]{625}$

(e) $\sqrt[5]{-32}$

(f) $\sqrt[4]{-16}$

ANSWERS
2. (a) 2 (b) 3 (c) -6 (d) 5
   (e) -2 (f) not a real number

3 Graph each function by creating a table of values. Give the domain and range.

(a) $f(x) = \sqrt{x} + 2$

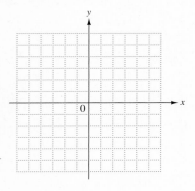

(b) $f(x) = \sqrt[3]{x} - 1$

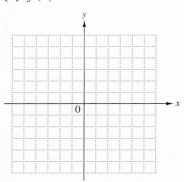

EXAMPLE 3 Graphing Functions Defined with Radicals

Graph each function by creating a table of values. Give the domain and the range.

(a) $f(x) = \sqrt{x - 3}$

A table of values is shown. The x-values were chosen in such a way that the function values are all integers. For the radicand to be nonnegative, we must have $x - 3 \geq 0$, or $x \geq 3$. Therefore, the domain is $[3, \infty)$. Again, function values are positive or 0, so the range is $[0, \infty)$. The graph is shown in Figure 3.

x	$f(x) = \sqrt{x - 3}$
3	$\sqrt{3 - 3} = 0$
4	$\sqrt{4 - 3} = 1$
7	$\sqrt{7 - 3} = 2$

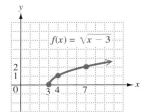

Figure 3

(b) $f(x) = \sqrt[3]{x} + 2$

See the table and Figure 4. Both the domain and the range are $(-\infty, \infty)$.

x	$f(x) = \sqrt[3]{x} + 2$
-8	$\sqrt[3]{-8} + 2 = 0$
-1	$\sqrt[3]{-1} + 2 = 1$
0	$\sqrt[3]{0} + 2 = 2$
1	$\sqrt[3]{1} + 2 = 3$
8	$\sqrt[3]{8} + 2 = 4$

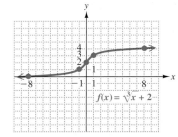

Figure 4

Work Problem 3 at the Side.

OBJECTIVE 4 Find nth roots of nth powers. A square root of a^2 (where $a \neq 0$) is a number that can be squared to give a^2. This number is either a or $-a$. Since the symbol $\sqrt{a^2}$ represents the *nonnegative* square root, we must write $\sqrt{a^2}$ with absolute value bars, as $|a|$, because a may be a negative number.

$\sqrt{a^2}$

For any real number a, $\quad \sqrt{a^2} = |a|$.

EXAMPLE 4 Simplifying Square Roots Using Absolute Value

Find each square root that is a real number.

(a) $\sqrt{7^2} = |7| = 7$

(b) $\sqrt{(-7)^2} = |-7| = 7$

(c) $\sqrt{k^2} = |k|$

(d) $\sqrt{(-k)^2} = |-k| = |k|$

ANSWERS

3. **(a)** domain: $[0, \infty)$; range: $[2, \infty)$

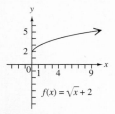

$f(x) = \sqrt{x} + 2$

(b) domain: $(-\infty, \infty)$; range: $(-\infty, \infty)$

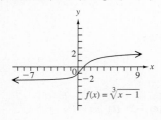

$f(x) = \sqrt[3]{x} - 1$

Work Problem 4 at the Side. ▶▶▶

We can generalize this idea to any nth root.

$$\sqrt[n]{a^n}$$

If n is an *even* positive integer, $\sqrt[n]{a^n} = |a|$,

and if n is an *odd* positive integer, $\sqrt[n]{a^n} = a$.

In words, use absolute value when n is even; do not use absolute value when n is odd.

EXAMPLE 5 **Simplifying Higher Roots Using Absolute Value**

Simplify each root.

(a) $\sqrt[6]{(-3)^6} = |-3| = 3$ n is even; use absolute value.

(b) $\sqrt[5]{(-4)^5} = -4$ n is odd.

(c) $-\sqrt[4]{(-9)^4} = -|-9| = -9$

(d) $\sqrt[3]{\dfrac{8}{27}} = \sqrt[3]{\left(\dfrac{2}{3}\right)^3} = \dfrac{2}{3}$

(e) $-\sqrt{m^4} = -|m^2| = -m^2$
No absolute value bars are needed here because m^2 is nonnegative for any real number value of m.

(f) $\sqrt[3]{a^{12}} = a^4$, because $a^{12} = (a^4)^3$.

(g) $\sqrt[4]{x^{12}} = |x^3|$
We use absolute value bars to guarantee that the result is not negative (because x^3 can be either positive or negative, depending on x). If desired, $|x^3|$ can be written as $x^2 \cdot |x|$.

Work Problem 5 at the Side. ▶▶▶

OBJECTIVE 5 **Use a calculator to find roots.** While numbers such as $\sqrt{9}$ and $\sqrt[3]{-8}$ are rational, radicals are often irrational numbers. To find approximations of roots such as $\sqrt{15}$, $\sqrt[3]{10}$, and $\sqrt[4]{2}$, we usually use scientific or graphing calculators. Using a calculator, we find

$$\sqrt{15} \approx 3.872983346, \quad \sqrt[3]{10} \approx 2.15443469, \quad \text{and} \quad \sqrt[4]{2} \approx 1.189207115,$$

where the symbol $\approx$ means "is approximately equal to." In this book we will usually show approximations rounded to three decimal places. Thus, we would write

$$\sqrt{15} \approx 3.873, \quad \sqrt[3]{10} \approx 2.154, \quad \text{and} \quad \sqrt[4]{2} \approx 1.189.$$

▦ **Calculator Tip** The methods for finding approximations differ among makes and models, and you should always consult your owner's manual for keystroke instructions. Be aware that graphing calculators often differ from scientific calculators in the order in which keystrokes are made.

4 Find each square root that is a real number.

(a) $\sqrt{49}$

(b) $-\sqrt{\dfrac{36}{25}}$

(c) $\sqrt{(-6)^2}$

(d) $\sqrt{r^2}$

5 Simplify.

(a) $\sqrt[6]{64}$

(b) $-\sqrt[4]{16}$

(c) $\sqrt[3]{\dfrac{216}{125}}$

(d) $\sqrt[5]{-243}$

(e) $\sqrt[6]{(-p)^6}$

(f) $-\sqrt[6]{y^{24}}$

ANSWERS

4. (a) 7   **(b)** $-\dfrac{6}{5}$ **(c)** 6   **(d)** $|r|$

5. (a) 2   **(b)** -2 **(c)** $\dfrac{6}{5}$

   (d) -3 **(e)** $|p|$ **(f)** $-y^4$

6 Use a calculator to approximate each radical to three decimal places.

(a) $\sqrt{17}$

Figure 5 shows how the preceding approximations are displayed on a TI-83 Plus or TI-84 graphing calculator. In Figure 5(a), eight or nine decimal places are shown, while in Figure 5(b), the number of decimal places is fixed at three.

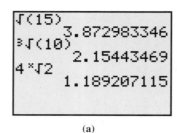
(a)

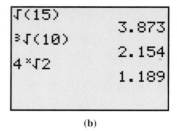
(b)

Figure 5

(b) $-\sqrt{362}$

There is a simple way to check that a calculator approximation is "in the ballpark." Because 16 is a little larger than 15, $\sqrt{16} = 4$ should be a little larger than $\sqrt{15}$. Thus, 3.873 is a reasonable approximation for $\sqrt{15}$.

EXAMPLE 6 Finding Approximations for Roots

Use a calculator to verify that each approximation is correct.

(a) $\sqrt{39} \approx 6.245$ (b) $-\sqrt{72} \approx -8.485$

(c) $\sqrt[3]{9482}$

(c) $\sqrt[3]{93} \approx 4.531$ (d) $\sqrt[4]{39} \approx 2.499$

◄◄◄ Work Problem 6 at the Side.

EXAMPLE 7 Using Roots to Calculate Resonant Frequency

In electronics, the resonant frequency f of a circuit may be found by the formula

$$f = \frac{1}{2\pi\sqrt{LC}},$$

(d) $\sqrt[4]{6825}$

where f is in cycles per second, L is in henrys, and C is in farads.* Find the resonant frequency f if $L = 5 \times 10^{-4}$ henrys and $C = 3 \times 10^{-10}$ farads. Give your answer to the nearest thousand.

Find the value of f when $L = 5 \times 10^{-4}$ and $C = 3 \times 10^{-10}$.

$$f = \frac{1}{2\pi\sqrt{LC}} \qquad \text{Given formula}$$

7 Use the formula in Example 7 to approximate f to the nearest thousand if

$$L = 6 \times 10^{-5}$$
and $$C = 4 \times 10^{-9}.$$

$$= \frac{1}{2\pi\sqrt{(5 \times 10^{-4})(3 \times 10^{-10})}} \qquad \text{Substitute for } L \text{ and } C.$$

$$\approx 411,000 \qquad \text{Use a calculator.}$$

The resonant frequency f is approximately 411,000 cycles per sec.

◄◄◄ Work Problem 7 at the Side.

ANSWERS
6. (a) 4.123   (b) -19.026
   (c) 21.166   (d) 9.089
7. 325,000 cycles per sec

*Henrys and farads are units of measure in electronics.

9.1 Exercises

FOR EXTRA HELP Tutor Center Addison-Wesley Math Tutor Center | MathXL MathXL | Digital Video Tutor CD 5 Videotape 5 | Student's Solutions Manual | MyMathLab MyMathLab | Interactmath.com

Match each expression with the equivalent choice from A–F. Answers may be used more than once.

1. $-\sqrt{16}$
2. $\sqrt{-16}$
3. $\sqrt[3]{-27}$
4. $\sqrt[5]{-32}$
5. $\sqrt[4]{81}$
6. $\sqrt[3]{8}$

A. 3 **B.** -2 **C.** 2 **D.** -3 **E.** -4 **F.** Not a real number

Choose the closest approximation of each square root.

7. $\sqrt{123.5}$
   A. 9 **B.** 10 **C.** 11 **D.** 12

8. $\sqrt{67.8}$
   A. 7 **B.** 8 **C.** 9 **D.** 10

Refer to the figure to answer the questions in Exercises 9–10.

$\sqrt{98}$
$\sqrt{26}$

9. Which one of the following is the best estimate of its area?
   A. 2500 **B.** 250 **C.** 50 **D.** 100

10. Which one of the following is the best estimate of its perimeter?
   A. 15 **B.** 250 **C.** 100 **D.** 30

11. Consider the expression $-\sqrt{-a}$. Decide whether it is positive, negative, 0, or not a real number if
   (a) $a > 0$, **(b)** $a < 0$, **(c)** $a = 0$.

12. If n is odd, under what conditions is $\sqrt[n]{a}$
   (a) positive, **(b)** negative, **(c)** 0?

Find each root that is a real number. Use a calculator as necessary. See Examples 1 and 2.

13. $-\sqrt{81}$
14. $-\sqrt{121}$
15. $\sqrt[3]{216}$
16. $\sqrt[3]{343}$

17. $\sqrt[3]{-64}$
18. $\sqrt[3]{-125}$
19. $-\sqrt[3]{512}$
20. $-\sqrt[3]{1000}$

21. $\sqrt[4]{1296}$
22. $\sqrt[4]{625}$
23. $-\sqrt[4]{81}$
24. $-\sqrt[4]{256}$

25. $\sqrt[4]{-16}$

26. $\sqrt[4]{-81}$

27. $\sqrt[6]{(-2)^6}$

28. $\sqrt[6]{(-4)^6}$

29. $\sqrt[5]{(-9)^5}$

30. $\sqrt[5]{(-8)^5}$

31. $\sqrt{\dfrac{64}{81}}$

32. $\sqrt{\dfrac{100}{9}}$

33. $\sqrt[3]{\dfrac{8}{27}}$

34. $\sqrt[4]{\dfrac{81}{16}}$

35. $\sqrt[6]{\dfrac{1}{64}}$

36. $\sqrt[5]{\dfrac{1}{32}}$

Graph each function and give its domain and range. See Example 3.

37. $f(x) = \sqrt{x + 3}$

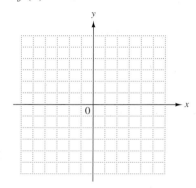

38. $f(x) = \sqrt{x - 5}$

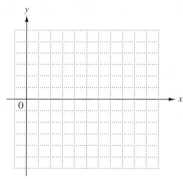

39. $f(x) = \sqrt{x} - 2$

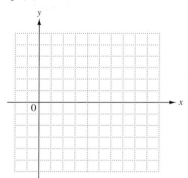

40. $f(x) = \sqrt{x} + 4$

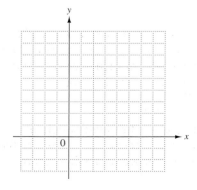

41. $f(x) = \sqrt[3]{x} - 3$

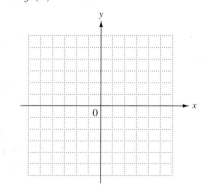

42. $f(x) = \sqrt[3]{x} + 1$

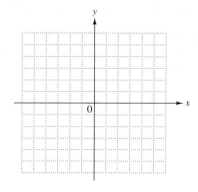

Simplify each root. See Examples 4 and 5.

43. $\sqrt{x^2}$

44. $-\sqrt{x^2}$

45. $\sqrt[3]{x^3}$

46. $-\sqrt[3]{x^3}$

47. $\sqrt[6]{x^{30}}$

48. $\sqrt[4]{k^{20}}$

49. $\sqrt{x^6}$

50. $\sqrt[4]{x^{12}}$

Use a calculator to find a decimal approximation for each radical. Round answers to three decimal places if necessary. See Example 6.

51. $\sqrt{9483}$

52. $\sqrt{6825}$

53. $\sqrt{284.361}$

54. $\sqrt{846.104}$

55. $\sqrt[4]{19.4481}$

56. $\sqrt[4]{39.0625}$

57. $\sqrt[3]{3.375}$

58. $\sqrt[3]{238.328}$

RELATING CONCEPTS (EXERCISES 59–64) For Individual or Group Work

Every positive number has two even nth roots, the principal (positive) root and a negative root. **Work Exercises 59–64 in order,** *to explore connections between these roots.*

59. Find the square roots of 16.

60. Find the principal square root of 16.

61. Find $\sqrt{16}$ and $-\sqrt{16}$.

62. What is the solution set of $x^2 = 16$?

63. Explain what is meant by $\pm\sqrt{16}$.

64. Explain why $\sqrt{x^2}$ is simplified as $|x|$.

Solve each problem. See Example 7.

65. Use the formula in Example 7 to calculate the resonant frequency of a circuit to the nearest thousand if $L = 7.237 \times 10^{-5}$ henrys and $C = 2.5 \times 10^{-10}$ farads.

66. The threshold weight T for a person is the weight above which the risk of death increases greatly. The threshold weight in pounds for men aged 40–49 is related to height in inches by the formula

$$h = 12.3\sqrt[3]{T}.$$

What height corresponds to a threshold weight of 216 lb for a 43-yr-old man? Round your answer to the nearest inch, and then to the nearest tenth of a foot.

67. According to an article in *The World Scanner Report,* the distance D, in miles, to the horizon from an observer's point of view over water or "flat" earth is given by

$$D = \sqrt{2H},$$

where H is the height of the point of view, in feet. If a person whose eyes are 6 ft above ground level is standing at the top of a hill 44 ft above "flat" earth, approximately how far to the horizon will she be able to see?

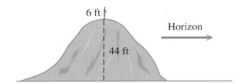

68. The time for one complete swing of a simple pendulum is

$$t = 2\pi\sqrt{\frac{L}{g}},$$

where t is time in seconds, L is the length of the pendulum in feet, and g, the force due to gravity, is about 32 ft per sec^2. Find the time of a complete swing of a 2-ft pendulum to the nearest tenth of a second.

69. Heron's formula gives a method of finding the area of a triangle if the lengths of its sides are known. Suppose that a, b, and c are the lengths of the sides. Let s denote one-half of the perimeter of the triangle (called the *semiperimeter*); that is,

$$s = \frac{1}{2}(a + b + c).$$

Then the area of the triangle is

$$A = \sqrt{s(s - a)(s - b)(s - c)}.$$

Find the area of the Bermuda Triangle, if the "sides" of this triangle measure approximately 850 mi, 925 mi, and 1300 mi. Give your answer to the nearest thousand square miles.

70. The Vietnam Veterans' Memorial in Washington, D.C., is in the shape of an unenclosed isosceles triangle with equal sides of length 246.75 ft. If the triangle were enclosed, the third side would have length 438.14 ft. Use Heron's formula from the previous exercise to find the area of this enclosure to the nearest hundred square feet. (*Source:* Information pamphlet obtained at the Vietnam Veterans' Memorial.)

The formula

$$I = \sqrt{\frac{2P}{L}}$$

relates the coefficient of self-induction L (in henrys), the energy P stored in an electronic circuit (in joules), and the current I (in amps). Round your answers in Exercises 71 and 72 to the nearest thousandth.

71. Find I if $P = 120$ and $L = 80$.

72. Find I if $P = 100$ and $L = 40$.

9.2 Rational Exponents

OBJECTIVE 1 Use exponential notation for nth roots. In mathematics we often formulate definitions so that previous rules remain valid. In **Section 6.1** we defined 0 as an exponent in such a way that the rules for products, quotients, and powers would still be valid. Now we look at exponents that are rational numbers of the form $\frac{1}{n}$, where n is a natural number.

For the rules of exponents to remain valid, the product $(3^{1/2})^2 = 3^{1/2} \cdot 3^{1/2}$ should be found by adding exponents.

$$(3^{1/2})^2 = 3^{1/2} \cdot 3^{1/2}$$
$$= 3^{1/2 + 1/2}$$
$$= 3^1$$
$$= 3$$

However, by definition $(\sqrt{3})^2 = \sqrt{3} \cdot \sqrt{3} = 3$. Since both $(3^{1/2})^2$ and $(\sqrt{3})^2$ are equal to 3, we must have

$$3^{1/2} = \sqrt{3}.$$

This suggests the following generalization.

$a^{1/n}$
If $\sqrt[n]{a}$ is a real number, then
$$a^{1/n} = \sqrt[n]{a}.$$

EXAMPLE 1 Evaluating Exponentials of the Form $a^{1/n}$

Evaluate each expression.

(a) $64^{1/3} = \sqrt[3]{64} = 4$

(b) $100^{1/2} = \sqrt{100} = 10$

(c) $-256^{1/4} = -\sqrt[4]{256} = -4$

(d) $(-256)^{1/4} = \sqrt[4]{-256}$ is not a real number because the radicand, -256, is negative and the index is even.

(e) $(-32)^{1/5} = \sqrt[5]{-32} = -2$

(f) $\left(\dfrac{1}{8}\right)^{1/3} = \sqrt[3]{\dfrac{1}{8}} = \dfrac{1}{2}$

CAUTION
Notice the difference between parts (c) and (d) in Example 1. The radical in part (c) is the *negative fourth root* of a positive number, while the radical in part (d) is the *principal fourth root of a negative number, which is not a real number.*

◀◀◀ **Work Problem 1 at the Side.**

OBJECTIVES

1 Use exponential notation for nth roots.

2 Define $a^{m/n}$.

3 Convert between radicals and rational exponents.

4 Use the rules for exponents with rational exponents.

1 Evaluate each exponential.

(a) $8^{1/3}$

(b) $9^{1/2}$

(c) $-81^{1/4}$

(d) $(-16)^{1/4}$

(e) $64^{1/3}$

(f) $\left(\dfrac{1}{32}\right)^{1/5}$

ANSWERS
1. (a) 2 (b) 3 (c) -3
(d) not a real number (e) 4 (f) $\dfrac{1}{2}$

2 Evaluate each exponential.

(a) $64^{2/3}$

(b) $100^{3/2}$

(c) $-16^{3/4}$

(d) $(-16)^{3/4}$

OBJECTIVE 2 Define $a^{m/n}$. We now define a number like $8^{2/3}$. For past rules of exponents to be valid,

$$8^{2/3} = 8^{(1/3)2} = (8^{1/3})^2.$$

Since $8^{1/3} = \sqrt[3]{8}$,

$$8^{2/3} = (\sqrt[3]{8})^2 = 2^2 = 4.$$

Generalizing from this example, we define $a^{m/n}$ as follows.

> $a^{m/n}$
>
> If m and n are positive integers with m/n in lowest terms, then
>
> $$a^{m/n} = (a^{1/n})^m,$$
>
> provided that $a^{1/n}$ is a real number. If $a^{1/n}$ is not a real number, then $a^{m/n}$ is not a real number.

EXAMPLE 2 Evaluating Exponentials of the Form $a^{m/n}$

Evaluate each exponential.

(a) $36^{3/2} = (36^{1/2})^3 = 6^3 = 216$

(b) $125^{2/3} = (125^{1/3})^2 = 5^2 = 25$

(c) $-4^{5/2} = -(4^{5/2}) = -(4^{1/2})^5 = -(2)^5 = -32$

(d) $(-27)^{2/3} = [(-27)^{1/3}]^2 = (-3)^2 = 9$

Notice how the $-$ sign is used in parts (c) and (d). In part (c), we first evaluate the exponential and then find its negative. In part (d), the $-$ sign is part of the base, -27.

(e) $(-100)^{3/2}$ is not a real number, since $(-100)^{1/2}$ is not a real number.

◀◀◀ **Work Problem 2 at the Side.**

EXAMPLE 3 Evaluating Exponentials with Negative Rational Exponents

Evaluate each exponential.

(a) $16^{-3/4}$

By the definition of a negative exponent,

$$16^{-3/4} = \frac{1}{16^{3/4}}.$$

Since $16^{3/4} = (\sqrt[4]{16})^3 = 2^3 = 8$,

$$16^{-3/4} = \frac{1}{16^{3/4}} = \frac{1}{8}.$$

(b) $25^{-3/2} = \frac{1}{25^{3/2}} = \frac{1}{(\sqrt{25})^3} = \frac{1}{5^3} = \frac{1}{125}$

Continued on Next Page

ANSWERS
2. **(a)** 16   **(b)** 1000   **(c)** -8
   (d) not a real number

(c) $\left(\dfrac{8}{27}\right)^{-2/3} = \dfrac{1}{\left(\dfrac{8}{27}\right)^{2/3}} = \dfrac{1}{\left(\sqrt[3]{\dfrac{8}{27}}\right)^{2}} = \dfrac{1}{\left(\dfrac{2}{3}\right)^{2}} = \dfrac{1}{\dfrac{4}{9}} = \dfrac{9}{4}$

We could also use the rule $\left(\dfrac{b}{a}\right)^{-m} = \left(\dfrac{a}{b}\right)^{m}$ here, as follows.

$$\left(\frac{8}{27}\right)^{-2/3} = \left(\frac{27}{8}\right)^{2/3} = \left(\sqrt[3]{\frac{27}{8}}\right)^{2} = \left(\frac{3}{2}\right)^{2} = \frac{9}{4}$$

CAUTION
When using the rule in Example 3(c), we take the reciprocal only of the base, *not* the exponent. Also, be careful to distinguish between exponential expressions like $-16^{1/4}$, $16^{-1/4}$, and $-16^{-1/4}$.

$$-16^{1/4} = -2, \quad 16^{-1/4} = \frac{1}{2}, \quad \text{and} \quad -16^{-1/4} = -\frac{1}{2}.$$

> **Work Problem 3 at the Side.** ▶▶▶

We get an alternative definition of $a^{m/n}$ by using the power rule for exponents a little differently than in the earlier definition. If all indicated roots are real numbers, then

$$a^{m/n} = a^{m(1/n)} = (a^{m})^{1/n},$$

so
$$a^{m/n} = (a^{m})^{1/n}.$$

$a^{m/n}$

If all indicated roots are real numbers, then
$$a^{m/n} = (a^{1/n})^{m} = (a^{m})^{1/n}.$$

We can now evaluate an expression such as $27^{2/3}$ in two ways:

$$27^{2/3} = (27^{1/3})^{2} = 3^{2} = 9$$

or
$$27^{2/3} = (27^{2})^{1/3} = 729^{1/3} = 9.$$

In most cases, it is easier to use $(a^{1/n})^{m}$.

This rule can also be expressed with radicals as follows.

Radical Form of $a^{m/n}$

If all indicated roots are real numbers, then
$$a^{m/n} = \sqrt[n]{a^{m}} = (\sqrt[n]{a})^{m}.$$

In words, we can raise to the power and then take the root, or take the root and then raise to the power.

For example,

$$8^{2/3} = \sqrt[3]{8^{2}} = \sqrt[3]{64} = 4, \quad \text{and} \quad 8^{2/3} = (\sqrt[3]{8})^{2} = 2^{2} = 4,$$

so
$$8^{2/3} = \sqrt[3]{8^{2}} = (\sqrt[3]{8})^{2}.$$

❸ Evaluate each exponential.

(a) $36^{-3/2}$

(b) $32^{-4/5}$

(c) $\left(\dfrac{4}{9}\right)^{-5/2}$

ANSWERS

3. (a) $\dfrac{1}{216}$ **(b)** $\dfrac{1}{16}$ **(c)** $\dfrac{243}{32}$

4 Write each exponential as a radical. Assume that all variables represent positive real numbers. Use the definition that takes the root first.

(a) $5^{2/3}$

(b) $4k^{3/5}$

(c) $(7r)^{4/3}$

(d) $(m^3 + n^3)^{1/3}$

OBJECTIVE 3 Convert between radicals and rational exponents. Using the definition of rational exponents, we can simplify many problems involving radicals by converting the radicals to numbers with rational exponents. After simplifying, we convert the answer back to radical form.

EXAMPLE 4 Converting between Rational Exponents and Radicals

Write each exponential as a radical. Assume that all variables represent positive real numbers. Use the definition that takes the root first.

(a) $13^{1/2} = \sqrt{13}$

(b) $6^{3/4} = (\sqrt[4]{6})^3$

(c) $9m^{5/8} = 9(\sqrt[8]{m})^5$

(d) $6x^{2/3} - (4x)^{3/5} = 6(\sqrt[3]{x})^2 - (\sqrt[5]{4x})^3$

(e) $r^{-2/3} = \dfrac{1}{r^{2/3}} = \dfrac{1}{(\sqrt[3]{r})^2}$

(f) $(a^2 + b^2)^{1/2} = \sqrt{a^2 + b^2}$ Note that $\sqrt{a^2 + b^2} \neq a + b$.

In (g)–(i), write each radical as an exponential. Simplify. Assume that all variables represent positive real numbers.

(g) $\sqrt{10} = 10^{1/2}$

(h) $\sqrt[4]{3^8} = 3^{8/4} = 3^2 = 9$

(i) $\sqrt[6]{z^6} = z$, since z is positive.

Work Problems 4 and 5 at the Side.

5 Write each radical as an exponential and simplify. Assume that all variables represent positive real numbers.

(a) $\sqrt{y^{10}}$

(b) $\sqrt[3]{27\,y^9}$

(c) $\sqrt[4]{t^4}$

OBJECTIVE 4 Use the rules for exponents with rational exponents. The definition of rational exponents allows us to apply the rules for exponents first introduced in **Section 6.1.**

Rules for Rational Exponents

Let r and s be rational numbers. For all real numbers a and b for which the indicated expressions exist:

$$a^r \cdot a^s = a^{r+s} \qquad a^{-r} = \frac{1}{a^r} \qquad \frac{a^r}{a^s} = a^{r-s} \qquad \left(\frac{a}{b}\right)^{-r} = \frac{b^r}{a^r}$$

$$(a^r)^s = a^{rs} \qquad (ab)^r = a^r b^r \qquad \left(\frac{a}{b}\right)^r = \frac{a^r}{b^r} \qquad a^{-r} = \left(\frac{1}{a}\right)^r.$$

EXAMPLE 5 Applying Rules for Rational Exponents

Write with only positive exponents. Assume that all variables represent positive real numbers.

(a) $2^{1/2} \cdot 2^{1/4} = 2^{1/2 + 1/4} = 2^{3/4}$ Product rule

(b) $\dfrac{5^{2/3}}{5^{7/3}} = 5^{2/3 - 7/3} = 5^{-5/3} = \dfrac{1}{5^{5/3}}$ Quotient rule

Continued on Next Page

ANSWERS

4. (a) $(\sqrt[3]{5})^2$ (b) $4(\sqrt[5]{k})^3$

   (c) $(\sqrt[3]{7r})^4$ (d) $\sqrt[3]{m^3 + n^3}$

5. (a) y^5 (b) $3y^3$ (c) t

(c) $\dfrac{(x^{1/2} y^{2/3})^4}{y} = \dfrac{(x^{1/2})^4 (y^{2/3})^4}{y}$ Power rule

$= \dfrac{x^2 y^{8/3}}{y^1}$ Power rule

$= x^2 y^{8/3 - 1}$ Quotient rule

$= x^2 y^{5/3}$

(d) $\left(\dfrac{x^4 y^{-6}}{x^{-2} y^{1/3}}\right)^{-2/3} = \dfrac{(x^4)^{-2/3}(y^{-6})^{-2/3}}{(x^{-2})^{-2/3}(y^{1/3})^{-2/3}}$

$= \dfrac{x^{-8/3} y^4}{x^{4/3} y^{-2/9}}$ Power rule

$= x^{-8/3 - 4/3} y^{4 - (-2/9)}$ Quotient rule

$= x^{-4} y^{38/9}$

$= \dfrac{y^{38/9}}{x^4}$ Definition of negative exponent

The same result is obtained if we simplify within the parentheses first, leading to $(x^6 y^{-19/3})^{-2/3}$. Then, apply the power rule. (Show that the result is the same.)

(e) $m^{3/4}(m^{5/4} - m^{1/4}) = m^{3/4} \cdot m^{5/4} - m^{3/4} \cdot m^{1/4}$ Distributive property

$= m^{3/4 + 5/4} - m^{3/4 + 1/4}$ Product rule

$= m^{8/4} - m^{4/4}$

$= m^2 - m$

Do not make the common mistake of multiplying exponents in the first step.

Work Problem 6 at the Side. ▶▶▶

CAUTION
Use the rules of exponents in problems like those in Example 5. Do not convert the expressions to radical form.

EXAMPLE 6 **Applying Rules for Rational Exponents**

Rewrite all radicals as exponentials, and then apply the rules for rational exponents. Leave answers in exponential form. Assume that all variables represent positive real numbers.

(a) $\sqrt[3]{x^2} \cdot \sqrt[4]{x} = x^{2/3} \cdot x^{1/4}$ Convert to rational exponents.

$= x^{2/3 + 1/4}$ Product rule

$= x^{8/12 + 3/12}$ Write exponents with a common denominator.

$= x^{11/12}$

(b) $\dfrac{\sqrt{x^3}}{\sqrt[3]{x^2}} = \dfrac{x^{3/2}}{x^{2/3}} = x^{3/2 - 2/3} = x^{5/6}$

(c) $\sqrt{\sqrt[4]{z}} = \sqrt{z^{1/4}} = (z^{1/4})^{1/2} = z^{1/8}$

Work Problem 7 at the Side. ▶▶▶

6 Write with only positive exponents. Assume that all variables represent positive real numbers.

(a) $11^{3/4} \cdot 11^{5/4}$

(b) $\dfrac{7^{3/4}}{7^{7/4}}$

(c) $\dfrac{9^{2/3}(x^{1/3})^4}{9^{-1/3}}$

(d) $\left(\dfrac{a^3 b^{-4}}{a^{-2} b^{1/5}}\right)^{-1/2}$

(e) $a^{2/3}(a^{7/3} + a^{1/3})$

7 Simplify using the rules for rational exponents. Assume that all variables represent positive real numbers. Leave answers in exponential form.

(a) $\sqrt[5]{m^3} \cdot \sqrt{m}$

(b) $\dfrac{\sqrt[3]{p^5}}{\sqrt{p^3}}$

(c) $\sqrt[4]{\sqrt[3]{x}}$

ANSWERS

6. (a) 11^2 or 121 **(b)** $\dfrac{1}{7}$ **(c)** $9x^{4/3}$

(d) $\dfrac{b^{21/10}}{a^{5/2}}$ **(e)** $a^3 + a$

7. (a) $m^{11/10}$ **(b)** $p^{1/6}$ **(c)** $x^{1/12}$

Real-Data Applications

Windchill—A Radical Idea

When the wind blows, the air feels much colder than the actual temperature. The **windchill factor** measures the cooling effect that the wind has on one's skin. The formula that the National Weather Service uses to compute windchill is $T_{wc} = .0817(3.71\sqrt{V} + 5.81 - .25V)(T - 91.4) + 91.4$, where T_{wc} is windchill, V is wind speed in miles per hour (mph), and T is air temperature in degrees Fahrenheit. The windchill for various wind speeds and temperatures is shown in the table.

WINDCHILL FACTOR

Wind Speed (mph)	Air Temperature (°Fahrenheit)														
	35	30	25	20	15	10	5	0	−5	−10	−15	−20	−25	−30	−35
4	35	30	25	20	15	10	5	0	−5	−10	−15	−20	−25	−30	−35
5	32	27	22	16	11	6	0	−5	−10	−15	−21	−26	−31	−36	−42
10	22	16	10	3	−3	−9	−15	−22	−27	−34	−40	−46	−52	−58	−64
15	16	9	2	−5	−11	−18	−25	−31	−38	−45	−51	−58	−65	−72	−78
20	12	4	−3	−10	−17	−24	−31	−39	−46	−53	−60	−67	−74	−81	−88
25	8	1	−7	−15	−22	−29	−36	−44	−51	−59	−66	−74	−81	−88	−96
30	6	−2	−10	−18	−25	−33	−41	−49	−56	−64	−71	−79	−86	−93	−101
35	4	−4	−12	−20	−27	−35	−43	−52	−58	−67	−74	−82	−89	−97	−105
40	3	−5	−13	−21	−29	−37	−45	−53	−60	−69	−76	−84	−92	−100	−107
45	2	−6	−14	−22	−30	−38	−46	−54	−62	−70	−78	−85	−93	−102	−109

Source: USA Today.

If you consider the vertical columns of numbers in the table, the data represents the relationships of windchill versus wind speed for a constant air temperature. For example, if you choose one measure of air temperature to keep constant, such as 10°F, then the following data gives wind speed (V) as the *input* and windchill T_{wc} as the *output*. Wind speed is measured in miles per hour (mph).

V	4	5	10	15	20	25	30	35	40	45
T_{wc}	10	6	−9	−18	−24	−29	−33	−35	−37	−38

For Group Discussion

1. Choose a temperature of 10°F. Use the formula to calculate the windchill for wind speeds of 4, 10, 25, and 40 mph. Round the results to the nearest degree. Do your results match those in the tables?

2. On a sheet of graph paper, sketch a graph of windchill, T_{wc}, versus wind speed, V, data for the temperature 10°F. Describe the resulting graph. Is the graph a line or a parabola, for example?

3. For four representative pairs of points, calculate $\dfrac{\text{change in windchill}}{\text{change in wind speed}}$. For example, using the ordered pairs (4, 10) and (5, 6), $\dfrac{\text{change in windchill}}{\text{change in wind speed}} = \dfrac{6 - 10}{5 - 4} = -4$.

   Recall that if this ratio is constant, then the data is linearly related. Is it approximately constant?

9.2 Exercises

FOR EXTRA HELP

Addison-Wesley Math Tutor Center MathXL Digital Video Tutor CD 5 Videotape 5 Student's Solutions Manual MyMathLab Interactmath.com

Match each expression from Column I with the equivalent choice from Column II.

	I			II	

1. $2^{1/2}$ **2.** $(-27)^{1/3}$ **A.** -4 **B.** 8

3. $-16^{1/2}$ **4.** $(-16)^{1/2}$ **C.** $\sqrt{2}$ **D.** $-\sqrt{6}$

5. $(-32)^{1/5}$ **6.** $(-32)^{2/5}$ **E.** -3 **F.** $\sqrt{6}$

7. $4^{3/2}$ **8.** $6^{2/4}$ **G.** 4 **H.** -2

9. $-6^{2/4}$ **10.** $36^{.5}$ **I.** 6 **J.** Not a real number

Simplify each expression. See Examples 1–3.

11. $169^{1/2}$ **12.** $121^{1/2}$ **13.** $729^{1/3}$ **14.** $512^{1/3}$

15. $16^{1/4}$ **16.** $625^{1/4}$ **17.** $\left(\dfrac{64}{81}\right)^{1/2}$ **18.** $\left(\dfrac{8}{27}\right)^{1/3}$

19. $(-27)^{1/3}$ **20.** $(-32)^{1/5}$ **21.** $100^{3/2}$ **22.** $64^{3/2}$

23. $-16^{5/2}$ **24.** $-32^{3/5}$ **25.** $(-144)^{1/2}$ **26.** $(-36)^{1/2}$

27. $64^{-3/2}$ **28.** $81^{-3/2}$ **29.** $\left(-\dfrac{8}{27}\right)^{-2/3}$ **30.** $\left(-\dfrac{64}{125}\right)^{-2/3}$

31. Explain why $(-64)^{1/2}$ is not a real number, while $-64^{1/2}$ is a real number.

32. Explain why $a^{1/n}$ is defined to be equal to $\sqrt[n]{a}$ when $\sqrt[n]{a}$ is real.

Write with radicals. Assume that all variables represent positive real numbers. See Example 4.

33. $12^{1/2}$

34. $3^{1/2}$

35. $8^{3/4}$

36. $7^{2/3}$

37. $(9q)^{5/8} - (2x)^{2/3}$

38. $(3p)^{3/4} + (4x)^{1/3}$

39. $(2m)^{-3/2}$

40. $(5y)^{-3/5}$

41. $(2y + x)^{2/3}$

42. $(r + 2z)^{3/2}$

43. $(3m^4 + 2k^2)^{-2/3}$

44. $(5x^2 + 3z^3)^{-5/6}$

45. Show that, in general, $\sqrt{a^2 + b^2} \neq a + b$ by replacing a with 3 and b with 4.

46. Suppose someone claims that $\sqrt[n]{a^n + b^n}$ must equal $a + b$, since when $a = 1$ and $b = 0$, a true statement results:

$$\sqrt[n]{a^n + b^n} = \sqrt[n]{1^n + 0^n} = \sqrt[n]{1^n} = 1 = 1 + 0 = a + b.$$

Explain why this is faulty reasoning.

Simplify by first converting to rational exponents. Assume that all variables represent positive real numbers. See Example 4.

47. $\sqrt{2^{12}}$

48. $\sqrt{5^{10}}$

49. $\sqrt[3]{4^9}$

50. $\sqrt[4]{6^8}$

51. $\sqrt{x^{20}}$

52. $\sqrt{r^{50}}$

53. $\sqrt[3]{x} \cdot \sqrt{x}$

54. $\sqrt[4]{y} \cdot \sqrt[5]{y^2}$

55. $\dfrac{\sqrt[3]{t^4}}{\sqrt[5]{t^4}}$

56. $\dfrac{\sqrt[4]{w^3}}{\sqrt[6]{w}}$

Simplify each expression. Write all answers with positive exponents. Assume that all variables represent positive real numbers. See Example 5.

57. $3^{1/2} \cdot 3^{3/2}$

58. $6^{4/3} \cdot 6^{2/3}$

59. $\dfrac{64^{5/3}}{64^{4/3}}$

60. $\dfrac{125^{7/3}}{125^{5/3}}$

61. $y^{7/3} \cdot y^{-4/3}$

62. $r^{-8/9} \cdot r^{17/9}$

63. $\dfrac{k^{1/3}}{k^{2/3} \cdot k^{-1}}$

64. $\dfrac{z^{3/4}}{z^{5/4} \cdot z^{-2}}$

65. $\dfrac{(x^{1/4} y^{2/5})^{20}}{x^2}$

66. $\dfrac{(r^{1/5}s^{2/3})^{15}}{r^2}$

67. $\dfrac{(x^{2/3})^2}{(x^2)^{7/3}}$

68. $\dfrac{(p^3)^{1/4}}{(p^{5/4})^2}$

69. $\dfrac{m^{3/4}n^{-1/4}}{(m^2n)^{1/2}}$

70. $\dfrac{(a^2b^5)^{-1/4}}{(a^{-3}b^2)^{1/6}}$

71. $\dfrac{p^{1/5}p^{7/10}p^{1/2}}{(p^3)^{-1/5}}$

72. $\dfrac{z^{1/3}z^{-2/3}z^{1/6}}{(z^{-1/6})^3}$

73. $\left(\dfrac{b^{-3/2}}{c^{-5/3}}\right)^2 (b^{-1/4}c^{-1/3})^{-1}$

74. $\left(\dfrac{m^{-2/3}}{a^{-3/4}}\right)^4 (m^{-3/8}a^{1/4})^{-2}$

75. $\left(\dfrac{p^{-1/4}q^{-3/2}}{3^{-1}p^{-2}q^{-2/3}}\right)^{-2}$

76. $\left(\dfrac{2^{-2}w^{-3/4}x^{-5/8}}{w^{3/4}x^{-1/2}}\right)^{-3}$

77. $p^{2/3}(p^{1/3} + 2p^{4/3})$

78. $z^{5/8}(3z^{5/8} + 5z^{11/8})$

79. $k^{1/4}(k^{3/2} - k^{1/2})$

80. $r^{3/5}(r^{1/2} + r^{3/4})$

81. $6a^{7/4}(a^{-7/4} + 3a^{-3/4})$

82. $4m^{5/3}(m^{-2/3} - 4m^{-5/3})$

83. $5m^{-2/3}(m^{2/3} + m^{-7/3})$

Write with rational exponents, and then apply the properties of exponents. Assume that all radicands represent positive real numbers. Give answers in exponential form. See Example 6.

84. $\sqrt[5]{x^3} \cdot \sqrt[4]{x}$

85. $\sqrt[6]{y^5} \cdot \sqrt[3]{y^2}$

86. $\dfrac{\sqrt{x^5}}{\sqrt{x^8}}$

87. $\dfrac{\sqrt[3]{k^5}}{\sqrt[3]{k^7}}$

88. $\sqrt{y} \cdot \sqrt[3]{yz}$

89. $\sqrt[3]{xz} \cdot \sqrt{z}$

90. $\sqrt[4]{\sqrt[3]{m}}$

91. $\sqrt[3]{\sqrt{k}}$

92. $\sqrt{\sqrt[3]{\sqrt[4]{x}}}$

93. $\sqrt[3]{\sqrt[5]{\sqrt{y}}}$

94. $\sqrt{y^{5/4}}$

95. $\sqrt[3]{x^{5/9}}$

Solve each problem.

96. Meteorologists can determine the duration of a storm by using the function defined by

$$T(D) = .07D^{3/2},$$

where D is the diameter of the storm in miles and T is the time in hours. Find the duration of a storm with a diameter of 16 mi. Round your answer to the nearest tenth of an hour.

97. The threshold weight T, in pounds, for a person is the weight above which the risk of death increases greatly. The threshold weight in pounds for men aged 40–49 is related to height in inches by the function defined by

$$h(T) = (1860.867T)^{1/3}.$$

What height corresponds to a threshold weight of 200 lb for a 46-yr-old man? Round your answer to the nearest inch, and then to the nearest tenth of a foot.

Earlier, we factored expressions like $x^4 - x^5$ by factoring out the greatest common factor to get $x^4 - x^5 = x^4(1 - x)$. We can adapt this approach to factor expressions with rational exponents. When one or more of the exponents is negative or a fraction, we use order on the number line discussed in **Section 1.1** *to decide on the common factor. In this type of factoring, we want the binomial factor to have only positive exponents, so we always factor out the variable with the least exponent. A positive exponent is greater than a negative exponent, so in $7z^{5/8} + z^{-3/4}$, we factor out $z^{-3/4}$, because $-\frac{3}{4}$ is less than $\frac{5}{8}$.*

Factor out the given common factor from each expression. Assume that all variables represent positive real numbers.

98. $3x^{-1/2} - 4x^{1/2}; \quad x^{-1/2}$

99. $m^3 - 3m^{5/2}; \quad m^{5/2}$

100. $4t^{-1/2} + 7t^{3/2}; \quad t^{-1/2}$

101. $8x^{2/3} + 5x^{-1/3}; \quad x^{-1/3}$

102. $4p - p^{3/4}; \quad p^{3/4}$

103. $2m^{1/8} - m^{5/8}; \quad m^{1/8}$

9.3 Simplifying Radical Expressions

OBJECTIVE 1 Use the product rule for radicals.   We now develop rules for multiplying and dividing radicals that have the same index. For example, is the product of two nth-root radicals equal to the nth root of the product of the radicands? For example, are $\sqrt{36 \cdot 4}$ and $\sqrt{36} \cdot \sqrt{4}$ equal?

$$\sqrt{36 \cdot 4} = \sqrt{144} = 12$$
$$\sqrt{36} \cdot \sqrt{4} = 6 \cdot 2 = 12$$

Notice that in both cases the result is the same. This is an example of the **product rule for radicals.**

Product Rule for Radicals

If $\sqrt[n]{a}$ and $\sqrt[n]{b}$ are real numbers and n is a natural number, then

$$\sqrt[n]{a} \cdot \sqrt[n]{b} = \sqrt[n]{ab}.$$

In words, the product of two radicals is the radical of the product.

We justify the product rule using the rules for rational exponents. Since $\sqrt[n]{a} = a^{1/n}$ and $\sqrt[n]{b} = b^{1/n}$,

$$\sqrt[n]{a} \cdot \sqrt[n]{b} = a^{1/n} \cdot b^{1/n} = (ab)^{1/n} = \sqrt[n]{ab}.$$

CAUTION
Use the product rule only when the radicals have the same indexes.

EXAMPLE 1 Using the Product Rule

Multiply. Assume that all variables represent positive real numbers.

(a) $\sqrt{5} \cdot \sqrt{7} = \sqrt{5 \cdot 7} = \sqrt{35}$

(b) $\sqrt{2} \cdot \sqrt{19} = \sqrt{2 \cdot 19} = \sqrt{38}$

(c) $\sqrt{11} \cdot \sqrt{p} = \sqrt{11p}$

(d) $\sqrt{7} \cdot \sqrt{11xyz} = \sqrt{77xyz}$

> **Work Problem 1 at the Side.** ▶▶▶

EXAMPLE 2 Using the Product Rule

Multiply. Assume that all variables represent positive real numbers.

(a) $\sqrt[3]{3} \cdot \sqrt[3]{12} = \sqrt[3]{3 \cdot 12} = \sqrt[3]{36}$

(b) $\sqrt[4]{8y} \cdot \sqrt[4]{3r^2} = \sqrt[4]{24yr^2}$

(c) $\sqrt[6]{10m^4} \cdot \sqrt[6]{5m} = \sqrt[6]{50m^5}$

(d) $\sqrt[4]{2} \cdot \sqrt[5]{2}$ cannot be simplified using the product rule for radicals, because the indexes (4 and 5) are different.

> **Work Problem 2 at the Side.** ▶▶▶

OBJECTIVES

1 Use the product rule for radicals.

2 Use the quotient rule for radicals.

3 Simplify radicals.

4 Simplify products and quotients of radicals with different indexes.

5 Use the Pythagorean formula.

6 Use the distance formula.

❶ Multiply. Assume that all variables represent positive real numbers.

(a) $\sqrt{5} \cdot \sqrt{13}$

(b) $\sqrt{10y} \cdot \sqrt{3k}$

(c) $\sqrt{\dfrac{5}{a}} \cdot \sqrt{\dfrac{11}{z}}$

❷ Multiply. Assume that all variables represent positive real numbers.

(a) $\sqrt[3]{2} \cdot \sqrt[3]{7}$

(b) $\sqrt[6]{8r^2} \cdot \sqrt[6]{2r^3}$

(c) $\sqrt[5]{9y^2x} \cdot \sqrt[5]{8xy^2}$

(d) $\sqrt{7} \cdot \sqrt[3]{5}$

ANSWERS

1. **(a)** $\sqrt{65}$ **(b)** $\sqrt{30yk}$ **(c)** $\sqrt{\dfrac{55}{az}}$

2. **(a)** $\sqrt[3]{14}$ **(b)** $\sqrt[6]{16r^5}$ **(c)** $\sqrt[5]{72y^4x^2}$
   (d) cannot be simplified using the product rule

3 Simplify. Assume that all variables represent positive real numbers.

(a) $\sqrt{\dfrac{100}{81}}$

(b) $\sqrt{\dfrac{11}{25}}$

(c) $\sqrt[3]{\dfrac{18}{125}}$

(d) $\sqrt{\dfrac{y^8}{16}}$

(e) $\sqrt[3]{\dfrac{x^2}{r^{12}}}$

OBJECTIVE 2 Use the quotient rule for radicals. The **quotient rule for radicals** is similar to the product rule.

Quotient Rule for Radicals

If $\sqrt[n]{a}$ and $\sqrt[n]{b}$ are real numbers, $b \neq 0$, and n is a natural number, then

$$\sqrt[n]{\dfrac{a}{b}} = \dfrac{\sqrt[n]{a}}{\sqrt[n]{b}}.$$

In words, the radical of a quotient is the quotient of the radicals.

EXAMPLE 3 Using the Quotient Rule

Simplify. Assume that all variables represent positive real numbers.

(a) $\sqrt{\dfrac{16}{25}} = \dfrac{\sqrt{16}}{\sqrt{25}} = \dfrac{4}{5}$ (b) $\sqrt{\dfrac{7}{36}} = \dfrac{\sqrt{7}}{\sqrt{36}} = \dfrac{\sqrt{7}}{6}$

(c) $\sqrt[3]{-\dfrac{8}{125}} = \sqrt[3]{\dfrac{-8}{125}} = \dfrac{\sqrt[3]{-8}}{\sqrt[3]{125}} = \dfrac{-2}{5} = -\dfrac{2}{5}$

(d) $\sqrt[3]{\dfrac{7}{216}} = \dfrac{\sqrt[3]{7}}{\sqrt[3]{216}} = \dfrac{\sqrt[3]{7}}{6}$

(e) $\sqrt[5]{\dfrac{x}{32}} = \dfrac{\sqrt[5]{x}}{\sqrt[5]{32}} = \dfrac{\sqrt[5]{x}}{2}$ (f) $\sqrt[3]{\dfrac{m^6}{125}} = \dfrac{\sqrt[3]{m^6}}{\sqrt[3]{125}} = \dfrac{m^2}{5}$

Work Problem 3 at the Side.

OBJECTIVE 3 Simplify radicals. We use the product and quotient rules to simplify radicals. A radical is **simplified** if the following four conditions are met.

Conditions for a Simplified Radical

1. The radicand has no factor raised to a power greater than or equal to the index.

2. The radicand has no fractions.

3. No denominator has a radical.

4. Exponents in the radicand and the index of the radical have no common factor (except 1).

EXAMPLE 4 Simplifying Roots of Numbers

Simplify.

(a) $\sqrt{24}$

Check to see whether 24 is divisible by a perfect square (the square of a natural number) such as 4, 9, ... Choose the largest perfect square that divides into 24. The largest such number is 4. Write 24 as the product of 4 and 6, and then use the product rule.

$$\sqrt{24} = \sqrt{4 \cdot 6} = \sqrt{4} \cdot \sqrt{6} = 2\sqrt{6}$$

Continued on Next Page

ANSWERS

3. (a) $\dfrac{10}{9}$ (b) $\dfrac{\sqrt{11}}{5}$ (c) $\dfrac{\sqrt[3]{18}}{5}$

  (d) $\dfrac{y^4}{4}$ (e) $\dfrac{\sqrt[3]{x^2}}{r^4}$

(b) $\sqrt{108}$

The number 108 is divisible by the perfect square 36: $\sqrt{108} = \sqrt{36 \cdot 3}$. If this is not obvious, try factoring 108 into its prime factors.

$$\sqrt{108} = \sqrt{2^2 \cdot 3^3}$$
$$= \sqrt{2^2 \cdot 3^2 \cdot 3}$$
$$= 2 \cdot 3 \cdot \sqrt{3} \qquad \text{Product rule}$$
$$= 6\sqrt{3}$$

(c) $\sqrt{10}$

No perfect square (other than 1) divides into 10, so $\sqrt{10}$ cannot be simplified further.

(d) $\sqrt[3]{16}$

Look for the largest perfect *cube* that divides into 16. The number 8 satisfies this condition, so write 16 as $8 \cdot 2$ (or factor 16 into prime factors).

$$\sqrt[3]{16} = \sqrt[3]{8 \cdot 2} = \sqrt[3]{8} \cdot \sqrt[3]{2} = 2\sqrt[3]{2}$$

(e) $-\sqrt[4]{162} = -\sqrt[4]{81 \cdot 2}$ 81 is a perfect 4th power.

$$= -\sqrt[4]{81} \cdot \sqrt[4]{2} \qquad \text{Product rule}$$
$$= -3\sqrt[4]{2}$$

CAUTION
Be careful with which factors belong outside the radical sign and which belong inside. Note in Example 4(b) how $2 \cdot 3$ is written outside because $\sqrt{2^2} = 2$ and $\sqrt{3^2} = 3$. The remaining 3 is left inside the radical.

Work Problem 4 at the Side. ▶▶▶

EXAMPLE 5 **Simplifying Radicals Involving Variables**

Simplify. Assume that all variables represent positive real numbers.

(a) $\sqrt{16m^3} = \sqrt{16m^2 \cdot m}$
$$= \sqrt{16m^2} \cdot \sqrt{m}$$
$$= 4m\sqrt{m}$$

No absolute value bars are needed around the m in color because of the assumption that all the variables represent *positive* real numbers.

(b) $\sqrt{200k^7q^8} = \sqrt{10^2 \cdot 2 \cdot (k^3)^2 \cdot k \cdot (q^4)^2}$ Factor.
$$= 10k^3q^4\sqrt{2k} \qquad \begin{array}{l}\text{Remove perfect square}\\\text{factors.}\end{array}$$

(c) $\sqrt[3]{8x^4y^5} = \sqrt[3]{(8x^3y^3)(xy^2)}$ $8x^3y^3$ is the largest perfect cube that divides $8x^4y^5$.
$$= \sqrt[3]{8x^3y^3} \cdot \sqrt[3]{xy^2}$$
$$= 2xy\sqrt[3]{xy^2}$$

(d) $-\sqrt[4]{32y^9} = -\sqrt[4]{(16y^8)(2y)}$ $16y^8$ is the largest 4th power that divides $32y^9$.
$$= -\sqrt[4]{16y^8} \cdot \sqrt[4]{2y}$$
$$= -2y^2\sqrt[4]{2y}$$

4 Simplify.

(a) $\sqrt{32}$

(b) $\sqrt{45}$

(c) $\sqrt{300}$

(d) $\sqrt{35}$

(e) $-\sqrt[3]{54}$

(f) $\sqrt[4]{243}$

ANSWERS
4. (a) $4\sqrt{2}$ **(b)** $3\sqrt{5}$ **(c)** $10\sqrt{3}$
    (d) cannot be simplified further
    (e) $-3\sqrt[3]{2}$ **(f)** $3\sqrt[4]{3}$

5 Simplify. Assume that all variables represent positive real numbers.

(a) $\sqrt{25p^7}$

(b) $\sqrt{72y^3x}$

(c) $\sqrt[3]{y^7x^5z^6}$

(d) $-\sqrt[4]{32a^5b^7}$

> **NOTE**
> From Example 5 we see that if a variable is raised to a power with an exponent divisible by 2, it is a perfect square. If it is raised to a power with an exponent divisible by 3, it is a perfect cube. In general, if it is raised to a power with an exponent divisible by n, it is a perfect nth power.

◀◀◀ **Work Problem 5 at the Side.**

The conditions for a simplified radical given earlier state that an exponent in the radicand and the index of the radical should have no common factor (except 1). The next example shows how to simplify radicals with such common factors.

EXAMPLE 6 Simplifying Radicals by Using Smaller Indexes

Simplify. Assume that all variables represent positive real numbers.

(a) $\sqrt[9]{5^6}$

We can write this radical using rational exponents and then write the exponent in lowest terms. We then express the answer as a radical.

$$\sqrt[9]{5^6} = 5^{6/9} = 5^{2/3} = \sqrt[3]{5^2} \quad \text{or} \quad \sqrt[3]{25}$$

(b) $\sqrt[4]{p^2} = p^{2/4} = p^{1/2} = \sqrt{p}$ (Recall the assumption that $p > 0$.)

These examples suggest the following rule.

6 Simplify. Assume that all variables represent positive real numbers.

(a) $\sqrt[12]{2^3}$

If m is an integer, n and k are natural numbers, and all indicated roots exist, then

$$\sqrt[kn]{a^{km}} = \sqrt[n]{a^m}.$$

◀◀◀ **Work Problem 6 at the Side.**

(b) $\sqrt[6]{t^2}$

OBJECTIVE 4 Simplify products and quotients of radicals with different indexes. Since the product and quotient rules for radicals apply only when they have the same index, we multiply and divide radicals with different indexes by using rational exponents.

EXAMPLE 7 Multiplying Radicals with Different Indexes

Simplify $\sqrt{7} \cdot \sqrt[3]{2}$.

7 Simplify $\sqrt{5} \cdot \sqrt[3]{4}$.

Because the different indexes, 2 and 3, have a least common index of 6, use rational exponents to write each radical as a sixth root.

$$\sqrt{7} = 7^{1/2} = 7^{3/6} = \sqrt[6]{7^3} = \sqrt[6]{343}$$
$$\sqrt[3]{2} = 2^{1/3} = 2^{2/6} = \sqrt[6]{2^2} = \sqrt[6]{4}$$

Therefore,

$$\sqrt{7} \cdot \sqrt[3]{2} = \sqrt[6]{343} \cdot \sqrt[6]{4} = \sqrt[6]{1372}. \quad \text{Product rule}$$

◀◀◀ **Work Problem 7 at the Side.**

ANSWERS
5. (a) $5p^3\sqrt{p}$ (b) $6y\sqrt{2yx}$ (c) $y^2xz^2\sqrt[3]{yx^2}$
   (d) $-2ab\sqrt[4]{2ab^3}$
6. (a) $\sqrt[4]{2}$ (b) $\sqrt[3]{t}$
7. $\sqrt[6]{2000}$

OBJECTIVE 5 Use the Pythagorean formula. The **Pythagorean formula** relates the lengths of the three sides of a right triangle.

Pythagorean Formula

If c is the length of the longest side of a right triangle and a and b are the lengths of the shorter sides, then

$$c^2 = a^2 + b^2.$$

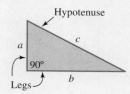

The longest side is the **hypotenuse** and the two shorter sides are the **legs** of the triangle. The hypotenuse is the side opposite the right angle.

In **Section 10.1** we will see that an equation such as $x^2 = 7$ has two solutions: $\sqrt{7}$ (the principal, or positive, square root of 7) and $-\sqrt{7}$. Similarly, $c^2 = 52$ has two solutions, $\pm\sqrt{52} = \pm 2\sqrt{13}$. In applications we often choose only the positive square root, as seen in the examples that follow.

EXAMPLE 8 Using the Pythagorean Formula

Use the Pythagorean formula to find the length of the hypotenuse in the triangle in Figure 6.
    To find the length of the hypotenuse c, let $a = 4$ and $b = 6$. Then, use the formula.

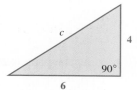

Figure 6

$$c^2 = a^2 + b^2$$
$$c^2 = 4^2 + 6^2 \qquad \text{Let } a = 4 \text{ and } b = 6.$$
$$c^2 = 52$$
$$c = \sqrt{52} \qquad \text{Choose the principal root.}$$
$$c = \sqrt{4 \cdot 13} \qquad \text{Factor.}$$
$$c = \sqrt{4} \cdot \sqrt{13} \qquad \text{Product rule}$$
$$c = 2\sqrt{13}$$

The length of the hypotenuse is $2\sqrt{13}$.

Work Problem 8 at the Side. ▶▶▶

OBJECTIVE 6 Use the distance formula. An important result in algebra is derived by using the Pythagorean formula. The *distance formula* allows us to find the distance between two points in the coordinate plane, or the length of the line segment joining those two points. Figure 7 shows the points $(3, -4)$ and $(-5, 3)$. The vertical line through $(-5, 3)$ and the horizontal line through $(3, -4)$ intersect at the point $(-5, -4)$. Thus, the point $(-5, -4)$ becomes the vertex of the right angle in a right triangle. By the Pythagorean formula, the square of the length of the hypotenuse, d, of the right triangle in Figure 7 is equal to the sum of the squares of the lengths of the two legs a and b:

$$d^2 = a^2 + b^2.$$

8 Find the length of the unknown side in each triangle.

(a)

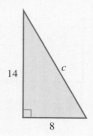

(b)

(*Hint:* Write the Pythagorean formula as $b^2 = c^2 - a^2$ here.)

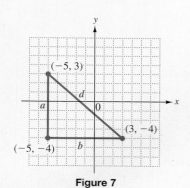

Figure 7

ANSWERS

8. (a) $2\sqrt{65}$ **(b)** $2\sqrt{5}$

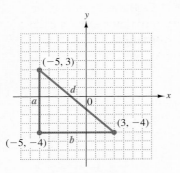

Figure 7 (repeat)

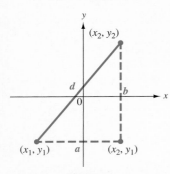

Figure 8

9 Find the distance between each pair of points.

(a) $(2, -1)$ and $(5, 3)$

(b) $(-3, 2)$ and $(0, -4)$

The length a is the difference between the y-coordinates of the endpoints. Since the x-coordinate of both points is -5, the side is vertical, and we can find a by finding the difference between the y-coordinates. We subtract -4 from 3 to get a positive value for a.

$$a = 3 - (-4) = 7$$

Similarly, we find b by subtracting -5 from 3.

$$b = 3 - (-5) = 8$$

Substituting these values into the formula, we have

$$d^2 = 7^2 + 8^2 \qquad \text{Let } a = 7 \text{ and } b = 8 \text{ in } d^2 = a^2 + b^2.$$
$$d^2 = 49 + 64$$
$$d^2 = 113$$
$$d = \sqrt{113}.$$

We choose the principal root since distance cannot be negative. Therefore, the distance between $(-5, 3)$ and $(3, -4)$ is $\sqrt{113}$.

This result can be generalized. Figure 8 shows the two points (x_1, y_1) and (x_2, y_2). Notice that the distance between (x_1, y_1) and (x_2, y_1) is given by

$$a = x_2 - x_1,$$

and the distance between (x_2, y_2) and (x_2, y_1) is given by

$$b = y_2 - y_1.$$

From the Pythagorean formula,

$$d^2 = a^2 + b^2$$
$$= (x_2 - x_1)^2 + (y_2 - y_1)^2.$$

Choosing the principal square root gives the **distance formula.**

Distance Formula

The distance between the points (x_1, y_1) and (x_2, y_2) is

$$d = \sqrt{(x_2 - x_1)^2 + (y_2 - y_1)^2}.$$

EXAMPLE 9 **Using the Distance Formula**

Find the distance between $(-3, 5)$ and $(6, 4)$.

When using the distance formula to find the distance between two points, designating the points as (x_1, y_1) and (x_2, y_2) is arbitrary. Let us choose $(x_1, y_1) = (-3, 5)$ and $(x_2, y_2) = (6, 4)$.

$$d = \sqrt{(x_2 - x_1)^2 + (y_2 - y_1)^2}$$
$$= \sqrt{(6 - (-3))^2 + (4 - 5)^2} \qquad x_2 = 6, y_2 = 4, x_1 = -3, y_1 = 5$$
$$= \sqrt{9^2 + (-1)^2}$$
$$= \sqrt{82} \qquad \text{Leave in radical form.}$$

Work Problem 9 at the Side.

ANSWERS
9. **(a)** 5 **(b)** $\sqrt{45}$ or $3\sqrt{5}$

9.3 Exercises

FOR EXTRA HELP Tutor Center Addison-Wesley Math Tutor Center   Math XL MathXL   Digital Video Tutor CD 5 Videotape 5   Student's Solutions Manual   *MyMathLab* MyMathLab   Interactmath.com

 Decide whether each statement is true *or* false *by using the product rule explained in this section. Then support your answer by finding a calculator approximation for each expression.*

1. $2\sqrt{12} = \sqrt{48}$

2. $\sqrt{72} = 2\sqrt{18}$

3. $3\sqrt{8} = 2\sqrt{18}$

4. $5\sqrt{72} = 6\sqrt{50}$

5. Explain why $\sqrt[3]{x} \cdot \sqrt[3]{x}$ is not equal to x. What is it equal to?

6. Explain why $\sqrt[4]{x} \cdot \sqrt[4]{x}$ is not equal to x, but *is* equal to $\sqrt{x}$, for $x \geq 0$.

7. Which one of the following is *not* equal to $\sqrt{\frac{1}{2}}$? (Do not use calculator approximations.)

   A. $\sqrt{.5}$ **B.** $\sqrt{\frac{2}{4}}$ **C.** $\sqrt{\frac{3}{6}}$ **D.** $\frac{\sqrt{4}}{\sqrt{16}}$

8. Use the π key on your calculator to get a value for π. Now find an approximation for $\sqrt[4]{\dfrac{2143}{22}}$. Does the result mean that π is actually equal to $\sqrt[4]{\dfrac{2143}{22}}$? Why or why not?

Multiply. Assume all variables represent positive real numbers. See Examples 1 and 2.

9. $\sqrt{5} \cdot \sqrt{6}$

10. $\sqrt{10} \cdot \sqrt{3}$

11. $\sqrt{14} \cdot \sqrt{x}$

12. $\sqrt{23} \cdot \sqrt{t}$

13. $\sqrt{14} \cdot \sqrt{3pqr}$

14. $\sqrt{7} \cdot \sqrt{5xt}$

15. $\sqrt[3]{7x} \cdot \sqrt[3]{2y}$

16. $\sqrt[3]{9x} \cdot \sqrt[3]{4y}$

17. $\sqrt[4]{11} \cdot \sqrt[4]{3}$

18. $\sqrt[4]{6} \cdot \sqrt[4]{9}$

19. $\sqrt[4]{2x} \cdot \sqrt[4]{3y^2}$

20. $\sqrt[4]{3y^2} \cdot \sqrt[4]{6yz}$

21. $\sqrt[3]{7} \cdot \sqrt[4]{3}$

22. $\sqrt[5]{8} \cdot \sqrt[6]{12}$

Simplify each radical. Assume that all variables represent positive real numbers. See Example 3.

23. $\sqrt{\dfrac{64}{121}}$

24. $\sqrt{\dfrac{16}{49}}$

25. $\sqrt{\dfrac{3}{25}}$

26. $\sqrt{\dfrac{13}{49}}$

27. $\sqrt{\dfrac{x}{25}}$

28. $\sqrt{\dfrac{k}{100}}$

29. $\sqrt{\dfrac{p^6}{81}}$

30. $\sqrt{\dfrac{w^{10}}{36}}$

31. $\sqrt[3]{\dfrac{27}{64}}$

32. $\sqrt[3]{\dfrac{216}{125}}$

33. $\sqrt[3]{-\dfrac{r^2}{8}}$

34. $\sqrt[3]{-\dfrac{t}{125}}$

35. $-\sqrt[4]{\dfrac{81}{x^4}}$

36. $-\sqrt[4]{\dfrac{625}{y^4}}$

37. $\sqrt[5]{\dfrac{1}{x^{15}}}$

38. $\sqrt[5]{\dfrac{32}{y^{20}}}$

Express each radical in simplified form. See Example 4.

39. $\sqrt{12}$

40. $\sqrt{18}$

41. $\sqrt{288}$

42. $\sqrt{72}$

43. $-\sqrt{32}$

44. $-\sqrt{48}$

45. $-\sqrt{28}$

46. $-\sqrt{24}$

47. $\sqrt{-300}$

48. $\sqrt{-150}$

49. $\sqrt[3]{128}$

50. $\sqrt[3]{24}$

51. $\sqrt[3]{-16}$

52. $\sqrt[3]{-250}$

53. $\sqrt[3]{40}$

54. $\sqrt[3]{375}$

55. $-\sqrt[4]{512}$

56. $-\sqrt[4]{1250}$

57. $\sqrt[5]{64}$

58. $\sqrt[5]{128}$

59. A student claimed that $\sqrt[3]{14}$ is not in simplified form, since $14 = 8 + 6$, and 8 is a perfect cube. Was his reasoning correct? Why or why not?

60. Explain in your own words why $\sqrt[3]{k^4}$ is not a simplified radical.

Express each radical in simplified form. Assume that all variables represent positive real numbers. See Example 5.

61. $\sqrt{72k^2}$

62. $\sqrt{18m^2}$

63. $\sqrt[3]{\dfrac{81}{64}}$

64. $\sqrt[3]{\dfrac{32}{216}}$

65. $\sqrt{121x^6}$

66. $\sqrt{256z^{12}}$

67. $-\sqrt[3]{27t^{12}}$

68. $-\sqrt[3]{64y^{18}}$

69. $-\sqrt{100m^8z^4}$

70. $-\sqrt{25t^6s^{20}}$

71. $-\sqrt[3]{-125a^6b^9c^{12}}$

72. $-\sqrt[3]{-216y^{15}x^6z^3}$

73. $\sqrt[4]{\dfrac{1}{16}r^8t^{20}}$

74. $\sqrt[4]{\dfrac{81}{256}t^{12}u^8}$

75. $\sqrt{50x^3}$

76. $\sqrt{300z^3}$

77. $-\sqrt{500r^{11}}$

78. $-\sqrt{200p^{13}}$

79. $\sqrt{13x^7y^8}$

80. $\sqrt{23k^9p^{14}}$

81. $\sqrt[3]{8z^6w^9}$

82. $\sqrt[3]{64a^{15}b^{12}}$

83. $\sqrt[3]{-16z^5t^7}$

84. $\sqrt[3]{-81m^4n^{10}}$

85. $\sqrt[4]{81x^{12}y^{16}}$

86. $\sqrt[4]{81t^8u^{28}}$

87. $-\sqrt[4]{162r^{15}s^{10}}$

88. $-\sqrt[4]{32k^5m^{10}}$

89. $\sqrt{\dfrac{y^{11}}{36}}$

90. $\sqrt{\dfrac{v^{13}}{49}}$

91. $\sqrt[3]{\dfrac{x^{16}}{27}}$

92. $\sqrt[3]{\dfrac{y^{17}}{125}}$

Simplify each radical. Assume that $x \geq 0$. See Example 6.

93. $\sqrt[4]{48^2}$

94. $\sqrt[4]{50^2}$

95. $\sqrt[4]{25}$

96. $\sqrt[6]{8}$

97. $\sqrt[10]{x^{25}}$

98. $\sqrt[12]{x^{44}}$

Simplify by first writing the radicals with the same index. Then multiply. Assume that $x \geq 0$. See Example 7.

99. $\sqrt[3]{4} \cdot \sqrt{3}$

100. $\sqrt[3]{5} \cdot \sqrt{6}$

101. $\sqrt[4]{3} \cdot \sqrt[3]{4}$

102. $\sqrt[5]{7} \cdot \sqrt[7]{5}$

103. $\sqrt{x} \cdot \sqrt[3]{x}$

104. $\sqrt[3]{x} \cdot \sqrt[4]{x}$

Find the unknown length in each right triangle. Simplify the answer if necessary.
See Example 8.

105.

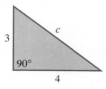

106.

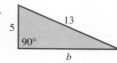

107.

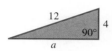

108.

Find the distance between each pair of points. See Example 9.

109. $(6, 13)$ and $(1, 1)$

110. $(8, 13)$ and $(2, 5)$

111. $(-6, 5)$ and $(3, -4)$

112. $(-1, 5)$ and $(-7, 7)$

113. $(-8, 2)$ and $(-4, 1)$

114. $(-1, 2)$ and $(5, 3)$

115. $(4.7, 2.3)$ and $(1.7, -1.7)$

116. $(-2.9, 18.2)$ and $(2.1, 6.2)$

117. $(\sqrt{2}, \sqrt{6})$ and $(-2\sqrt{2}, 4\sqrt{6})$

118. $(\sqrt{7}, 9\sqrt{3})$ and $(-\sqrt{7}, 4\sqrt{3})$

119. $(x + y, y)$ and $(x - y, x)$

120. $(c, c - d)$ and $(d, c + d)$

Solve each problem.

121. A Sanyo color television, model AVM-2755, has a rectangular screen with a 21.7-in. width. Its height is 16 in. What is the diagonal of the screen to the nearest tenth of an inch? (*Source:* Actual measurements of the author's television.)

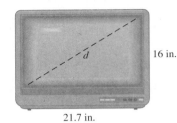

16 in.

21.7 in.

122. The length of the diagonal of a box is given by

$$D = \sqrt{L^2 + W^2 + H^2},$$

where L, W, and H are the length, width, and height of the box. Find the length of the diagonal, D, of a box that is 4 ft long, 3 ft high, and 2 ft wide. Give the exact value, then round to the nearest tenth of a foot.

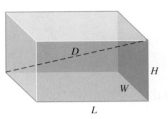

123. A formula from electronics dealing with impedance of parallel resonant circuits is

$$I = \frac{E}{\sqrt{R^2 + \omega^2 L^2}},$$

where the variables are in appropriate units. Find I if $E = 282$, $R = 100$, $L = 264$, and $\omega = 120\pi$. Give your answer to the nearest thousandth.

124. In the study of sound, one version of the law of tensions is

$$f_1 = f_2 \sqrt{\frac{F_1}{F_2}}.$$

If $F_1 = 300$, $F_2 = 60$, and $f_2 = 260$, find f_1 to the nearest unit.

125. The following letter appeared in the column "Ask Tom Why," written by Tom Skilling of the *Chicago Tribune.*

> *Dear Tom,*
> *I cannot remember the formula to calculate the distance to the horizon. I have a stunning view from my 14th floor condo, 150 feet above the ground. How far can I see?*
> *Ted Fleischaker; Indianapolis, Ind.*

Skilling's answer was as follows.

To find the distance to the horizon in miles, take the square root of the height of your view in feet and multiply that result by 1.224. Your answer will be the number of miles to the horizon. (*Source: Chicago Tribune,* August 17, 2002.)

Assuming Ted's eyes are 6 ft above the ground, the total height from the ground is $150 + 6 = 156$ ft. To the nearest tenth of a mile, how far can he see to the horizon?

9.4 Adding and Subtracting Radical Expressions

The examples in the preceding section discussed simplifying radical expressions that involve multiplication and division. Now we show how to simplify radical expressions that involve addition and subtraction.

OBJECTIVE 1 Simplify radical expressions involving addition and subtraction. An expression such as $4\sqrt{2} + 3\sqrt{2}$ can be simplified by using the distributive property.

$$4\sqrt{2} + 3\sqrt{2} = (4 + 3)\sqrt{2} = 7\sqrt{2}$$

As another example, $2\sqrt{3} - 5\sqrt{3} = (2 - 5)\sqrt{3} = -3\sqrt{3}$. This is similar to simplifying $2x + 3x$ to $5x$ or $5y - 8y$ to $-3y$.

> **CAUTION**
> *Only radical expressions with the same index and the same radicand may be combined.* Expressions such as $5\sqrt{3} + 2\sqrt{2}$ or $3\sqrt{3} + 2\sqrt[3]{3}$ cannot be simplified by combining terms.

EXAMPLE 1 Adding and Subtracting Radicals

Add or subtract to simplify each radical expression.

(a) $3\sqrt{24} + \sqrt{54}$

Begin by simplifying each radical; then use the distributive property to combine terms.

$$
\begin{aligned}
3\sqrt{24} + \sqrt{54} &= 3\sqrt{4} \cdot \sqrt{6} + \sqrt{9} \cdot \sqrt{6} && \text{Product rule} \\
&= 3 \cdot 2\sqrt{6} + 3\sqrt{6} \\
&= 6\sqrt{6} + 3\sqrt{6} \\
&= 9\sqrt{6} && \text{Combine terms.}
\end{aligned}
$$

(b)
$$
\begin{aligned}
2\sqrt{20x} - \sqrt{45x} &= 2\sqrt{4} \cdot \sqrt{5x} - \sqrt{9} \cdot \sqrt{5x} && \text{Product rule} \\
&= 2 \cdot 2\sqrt{5x} - 3\sqrt{5x} \\
&= 4\sqrt{5x} - 3\sqrt{5x} \\
&= \sqrt{5x}, \quad x \geq 0 && \text{Combine terms.}
\end{aligned}
$$

(c) $2\sqrt{3} - 4\sqrt{5}$

Here the radicals differ and are already simplified, so $2\sqrt{3} - 4\sqrt{5}$ cannot be simplified further.

Work Problem 1 at the Side. ▷▷▷

> **CAUTION**
> Do not confuse the product rule with combining like terms. *The root of a sum does not equal the sum of the roots.* For example,
> $$\sqrt{9 + 16} \neq \sqrt{9} + \sqrt{16}, \text{ since}$$
> $$\sqrt{9 + 16} = \sqrt{25} = 5, \quad \text{but} \quad \sqrt{9} + \sqrt{16} = 3 + 4 = 7.$$

OBJECTIVES

1 Simplify radical expressions involving addition and subtraction.

1 Add or subtract to simplify each radical expression.

(a) $3\sqrt{5} + 7\sqrt{5}$

(b) $2\sqrt{11} - \sqrt{11} + 3\sqrt{44}$

(c) $5\sqrt{12y} + 6\sqrt{75y}, \; y \geq 0$

(d) $3\sqrt{8} - 6\sqrt{50} + 2\sqrt{200}$

(e) $9\sqrt{5} - 4\sqrt{10}$

ANSWERS
1. **(a)** $10\sqrt{5}$ **(b)** $7\sqrt{11}$
   (c) $40\sqrt{3y}$ **(d)** $-4\sqrt{2}$
   (e) cannot be simplified further

2 Add or subtract to simplify each radical expression. Assume that all variables represent positive real numbers.

(a) $7\sqrt[3]{81} + 3\sqrt[3]{24}$

(b) $-2\sqrt[4]{32} - 7\sqrt[4]{162}$

(c) $\sqrt[3]{p^4q^7} - \sqrt[3]{64pq}$

3 Add. Assume that all variables represent positive real numbers.

$$\sqrt{\frac{80}{y^4}} + \sqrt{\frac{81}{y^{10}}}$$

EXAMPLE 2 Adding and Subtracting Radicals with Greater Indexes

Add or subtract to simplify each radical expression. Assume that all variables represent positive real numbers.

(a) $2\sqrt[3]{16} - 5\sqrt[3]{54} = 2\sqrt[3]{8 \cdot 2} - 5\sqrt[3]{27 \cdot 2}$ Factor.

$\phantom{(a) 2\sqrt[3]{16} - 5\sqrt[3]{54}} = 2\sqrt[3]{8} \cdot \sqrt[3]{2} - 5\sqrt[3]{27} \cdot \sqrt[3]{2}$ Product rule

$\phantom{(a) 2\sqrt[3]{16} - 5\sqrt[3]{54}} = 2 \cdot 2 \cdot \sqrt[3]{2} - 5 \cdot 3 \cdot \sqrt[3]{2}$

$\phantom{(a) 2\sqrt[3]{16} - 5\sqrt[3]{54}} = 4\sqrt[3]{2} - 15\sqrt[3]{2}$

$\phantom{(a) 2\sqrt[3]{16} - 5\sqrt[3]{54}} = -11\sqrt[3]{2}$ Combine terms.

(b) $2\sqrt[3]{x^2y} + \sqrt[3]{8x^5y^4} = 2\sqrt[3]{x^2y} + \sqrt[3]{(8x^3y^3)x^2y}$ Factor.

$\phantom{(b) 2\sqrt[3]{x^2y} + \sqrt[3]{8x^5y^4}} = 2\sqrt[3]{x^2y} + 2xy\sqrt[3]{x^2y}$ Product rule

$\phantom{(b) 2\sqrt[3]{x^2y} + \sqrt[3]{8x^5y^4}} = (2 + 2xy)\sqrt[3]{x^2y}$ Distributive property

CAUTION
Remember to write the index when working with cube roots, fourth roots, and so on.

Work Problem 2 at the Side.

EXAMPLE 3 Adding and Subtracting Radicals with Fractions

Perform the indicated operations. Assume that all variables represent positive real numbers.

(a) $2\sqrt{\frac{75}{16}} + 4\frac{\sqrt{8}}{\sqrt{32}} = 2\frac{\sqrt{25 \cdot 3}}{\sqrt{16}} + 4\frac{\sqrt{4 \cdot 2}}{\sqrt{16 \cdot 2}}$ Quotient rule

$ = 2\left(\frac{5\sqrt{3}}{4}\right) + 4\left(\frac{2\sqrt{2}}{4\sqrt{2}}\right)$ Product rule

$ = \frac{5\sqrt{3}}{2} + 2$ Multiply; $\frac{\sqrt{2}}{\sqrt{2}} = 1.$

$ = \frac{5\sqrt{3}}{2} + \frac{4}{2}$ Write with a common denominator.

$ = \frac{5\sqrt{3} + 4}{2}$

(b) $10\sqrt[3]{\frac{5}{x^6}} - 3\sqrt[3]{\frac{4}{x^9}} = 10\frac{\sqrt[3]{5}}{\sqrt[3]{x^6}} - 3\frac{\sqrt[3]{4}}{\sqrt[3]{x^9}}$ Quotient rule

$ = \frac{10\sqrt[3]{5}}{x^2} - \frac{3\sqrt[3]{4}}{x^3}$

$ = \frac{10x\sqrt[3]{5}}{x^3} - \frac{3\sqrt[3]{4}}{x^3}$ Write with a common denominator.

$ = \frac{10x\sqrt[3]{5} - 3\sqrt[3]{4}}{x^3}$

Work Problem 3 at the Side.

ANSWERS
2. **(a)** $27\sqrt[3]{3}$ **(b)** $-25\sqrt[4]{2}$
    (c) $(pq^2 - 4)\sqrt[3]{pq}$

3. $\dfrac{4y^3\sqrt{5} + 9}{y^5}$

9.4 Exercises

FOR EXTRA HELP Tutor Center — Addition-Wesley Math Tutor Center MathXL — MathXL Digital Video Tutor CD 5 / Videotape 5 Student's Solutions Manual MyMathLab — MyMathLab Interactmath.com

1. Which one of the following sums could be simplified without first simplifying the individual radical expressions?

   A. $\sqrt{50} + \sqrt{32}$ **B.** $3\sqrt{6} + 9\sqrt{6}$ **C.** $\sqrt[3]{32} - \sqrt[3]{108}$ **D.** $\sqrt[5]{6} - \sqrt[5]{192}$

2. Let $a = 1$ and $b = 64$.

   (a) Evaluate $\sqrt{a} + \sqrt{b}$. Then find $\sqrt{a+b}$. Are they equal?

   (b) Evaluate $\sqrt[3]{a} + \sqrt[3]{b}$. Then find $\sqrt[3]{a+b}$. Are they equal?

   (c) Complete the following: In general, $\sqrt[n]{a} + \sqrt[n]{b} \neq$ _____, based on the observations in parts (a) and (b) of this exercise.

3. Even though the indexes of the terms are not equal, the sum $\sqrt{64} + \sqrt[3]{125} + \sqrt[4]{16}$ can be simplified quite easily. What is this sum? Why can these terms be combined so easily?

4. Explain why $28 - 4\sqrt{2}$ *is not equal to* $24\sqrt{2}$. (This is a common error among algebra students.)

Simplify. Assume that all variables represent positive real numbers. See Examples 1 and 2.

5. $\sqrt{36} - \sqrt{100}$ 6. $\sqrt{25} - \sqrt{81}$ 7. $-2\sqrt{48} + 3\sqrt{75}$ 8. $4\sqrt{32} - 2\sqrt{8}$

9. $\sqrt[3]{16} + 4\sqrt[3]{54}$ 10. $3\sqrt[3]{24} - 2\sqrt[3]{192}$ 11. $\sqrt[4]{32} + 3\sqrt[4]{2}$ 12. $\sqrt[4]{405} - 2\sqrt[4]{5}$

13. $6\sqrt{18} - \sqrt{32} + 2\sqrt{50}$ 14. $5\sqrt{8} + 3\sqrt{72} - 3\sqrt{50}$ 15. $5\sqrt{6} + 2\sqrt{10}$

16. $3\sqrt{11} - 5\sqrt{13}$ 17. $2\sqrt{5} + 3\sqrt{20} + 4\sqrt{45}$ 18. $5\sqrt{54} - 2\sqrt{24} - 2\sqrt{96}$

19. $8\sqrt{2x} - \sqrt{8x} + \sqrt{72x}$ 20. $4\sqrt{18k} - \sqrt{72k} + \sqrt{50k}$ 21. $3\sqrt{72m^2} - 5\sqrt{32m^2}$

22. $9\sqrt{27p^2} - 14\sqrt{108p^2}$ 23. $-\sqrt[3]{54} + 2\sqrt[3]{16}$ 24. $15\sqrt[3]{81} - 4\sqrt[3]{24}$

25. $2\sqrt[3]{27x} - 2\sqrt[3]{8x}$ 26. $6\sqrt[3]{128m} + 3\sqrt[3]{16m}$ 27. $\sqrt[3]{x^2y} - \sqrt[3]{8x^2y}$

28. $3\sqrt[3]{x^2y^2} - 2\sqrt[3]{64x^2y^2}$ 29. $3x\sqrt[3]{xy^2} - 2\sqrt[3]{8x^4y^2}$ 30. $6q^2\sqrt[3]{5q} - 2q\sqrt[3]{40q^4}$

31. $5\sqrt[4]{32} + 3\sqrt[4]{162}$

32. $2\sqrt[4]{512} + 4\sqrt[4]{32}$

33. $3\sqrt[4]{x^5 y} - 2x\sqrt[4]{xy}$

34. $2\sqrt[4]{m^9 p^6} - 3m^2 p\sqrt[4]{mp^2}$

35. $2\sqrt[4]{32a^3} + 5\sqrt[4]{2a^3}$

36. $-\sqrt[4]{16r} + 5\sqrt[4]{r}$

Simplify. Assume all variables represent positive real numbers. See Example 3.

37. $\dfrac{2\sqrt{5}}{3} + \dfrac{\sqrt{5}}{6}$

38. $\dfrac{4\sqrt{3}}{3} + \dfrac{2\sqrt{3}}{9}$

39. $\sqrt{\dfrac{8}{9}} + \sqrt{\dfrac{18}{36}}$

40. $\sqrt{\dfrac{12}{16}} + \sqrt{\dfrac{48}{64}}$

41. $\dfrac{\sqrt{32}}{3} + \dfrac{2\sqrt{2}}{3} - \dfrac{\sqrt{2}}{\sqrt{9}}$

42. $\dfrac{\sqrt{27}}{2} - \dfrac{3\sqrt{3}}{2} + \dfrac{\sqrt{3}}{\sqrt{4}}$

43. $3\sqrt{\dfrac{50}{9}} + 8\dfrac{\sqrt{2}}{\sqrt{8}}$

44. $9\sqrt{\dfrac{48}{25}} - 2\dfrac{\sqrt{2}}{\sqrt{98}}$

45. $\sqrt{\dfrac{25}{x^8}} - \sqrt{\dfrac{9}{x^6}}$

46. $\sqrt{\dfrac{100}{y^4}} + \sqrt{\dfrac{81}{y^{10}}}$

47. $3\sqrt[3]{\dfrac{m^5}{27}} - 2m\sqrt[3]{\dfrac{m^2}{64}}$

48. $2a\sqrt[4]{\dfrac{a}{16}} - 5a\sqrt[4]{\dfrac{a}{81}}$

49. $3\sqrt[3]{\dfrac{2}{x^6}} - 4\sqrt[3]{\dfrac{5}{x^9}}$

50. $-4\sqrt[3]{\dfrac{4}{t^9}} + 3\sqrt[3]{\dfrac{9}{t^{12}}}$

Solve each problem. Give answers as simplified radical expressions.

51. Find the perimeter of the triangle.

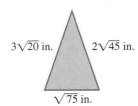

$3\sqrt{20}$ in. $2\sqrt{45}$ in. $\sqrt{75}$ in.

52. Find the perimeter of the rectangle.

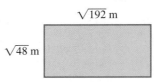

$\sqrt{192}$ m $\sqrt{48}$ m

53. What is the perimeter of the computer graphic?

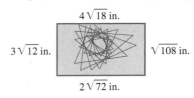

$4\sqrt{18}$ in. $3\sqrt{12}$ in. $\sqrt{108}$ in. $2\sqrt{72}$ in.

54. Find the area of the trapezoid.

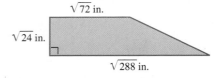

$\sqrt{72}$ in. $\sqrt{24}$ in. $\sqrt{288}$ in.

9.5 Multiplying and Dividing Radical Expressions

OBJECTIVE 1 Multiply radical expressions.   We multiply binomial expressions involving radicals by using the FOIL (First, Outer, Inner, Last) method from **Section 6.4.** For example, we find the product of the binomials $\sqrt{5} + 3$ and $\sqrt{6} + 1$ as follows.

$$
\begin{array}{c}
\overbrace{}^{First} \quad \overbrace{}^{Outer} \quad \overbrace{}^{Inner} \quad \overbrace{}^{Last}
\end{array}
$$

$$(\sqrt{5} + 3)(\sqrt{6} + 1) = \sqrt{5} \cdot \sqrt{6} + \sqrt{5} \cdot 1 + 3 \cdot \sqrt{6} + 3 \cdot 1$$

$$= \sqrt{30} + \sqrt{5} + 3\sqrt{6} + 3$$

This result cannot be simplified further.

OBJECTIVES

1 **Multiply radical expressions.**

2 **Rationalize denominators with one radical term.**

3 **Rationalize denominators with binomials involving radicals.**

4 **Write radical quotients in lowest terms.**

EXAMPLE 1 **Multiplying Binomials Involving Radical Expressions**

Multiply using FOIL.

$$
 \text{F} \text{O} \text{I} \text{L}
$$

(a) $(7 - \sqrt{3})(\sqrt{5} + \sqrt{2}) = 7\sqrt{5} + 7\sqrt{2} - \sqrt{3} \cdot \sqrt{5} - \sqrt{3} \cdot \sqrt{2}$

$$= 7\sqrt{5} + 7\sqrt{2} - \sqrt{15} - \sqrt{6}$$

(b) $(\sqrt{10} + \sqrt{3})(\sqrt{10} - \sqrt{3})$

$$= \sqrt{10} \cdot \sqrt{10} - \sqrt{10} \cdot \sqrt{3} + \sqrt{10} \cdot \sqrt{3} - \sqrt{3} \cdot \sqrt{3}$$

$$= 10 - 3$$

$$= 7$$

Notice that this is the kind of product that results in the difference of squares:

$$(x + y)(x - y) = x^2 - y^2.$$

Here, $x = \sqrt{10}$ and $y = \sqrt{3}$.

(c) $(\sqrt{7} - 3)^2 = (\sqrt{7} - 3)(\sqrt{7} - 3)$

$$= \sqrt{7} \cdot \sqrt{7} - 3\sqrt{7} - 3\sqrt{7} + 3 \cdot 3$$

$$= 7 - 6\sqrt{7} + 9$$

$$= 16 - 6\sqrt{7}$$

(d) $(5 - \sqrt[3]{3})(5 + \sqrt[3]{3}) = 5 \cdot 5 + 5\sqrt[3]{3} - 5\sqrt[3]{3} - \sqrt[3]{3} \cdot \sqrt[3]{3}$

$$= 25 - \sqrt[3]{3^2}$$

$$= 25 - \sqrt[3]{9}$$

(e) $(\sqrt{k} + \sqrt{y})(\sqrt{k} - \sqrt{y}) = (\sqrt{k})^2 - (\sqrt{y})^2$

$$= k - y, \quad k \geq 0 \text{ and } y \geq 0$$

❶ Multiply using FOIL.

(a) $(2 + \sqrt{3})(1 + \sqrt{5})$

(b) $(2\sqrt{3} + \sqrt{5})(\sqrt{6} - 3\sqrt{5})$

(c) $(4 + \sqrt{3})(4 - \sqrt{3})$

(d) $(\sqrt{6} - \sqrt{5})^2$

(e) $(4 + \sqrt[3]{7})(4 - \sqrt[3]{7})$

(f) $(\sqrt{p} + \sqrt{2})(\sqrt{p} - \sqrt{2})$

> **NOTE**
> In Example 1(c) we could have used the formula for the square of a binomial,
> $$(x - y)^2 = x^2 - 2xy + y^2,$$
> to get the same result.
> $$(\sqrt{7} - 3)^2 = (\sqrt{7})^2 - 2(\sqrt{7})(3) + 3^2$$
> $$= 7 - 6\sqrt{7} + 9$$
> $$= 16 - 6\sqrt{7}$$

◀◀◀ **Work Problem 1 at the Side.**

OBJECTIVE ❷ Rationalize denominators with one radical term. As defined earlier, a simplified radical expression will have no radical in the denominator. The origin of this agreement no doubt occurred before the days of high-speed calculation, when computation was a tedious process performed by hand. To see this, consider the radical expression $\frac{1}{\sqrt{2}}$. To find a decimal approximation by hand, it would be necessary to divide 1 by a decimal approximation for $\sqrt{2}$, such as 1.414. It would be much easier if the divisor were a whole number. This can be accomplished by multiplying $\frac{1}{\sqrt{2}}$ by 1 in the form $\frac{\sqrt{2}}{\sqrt{2}}$:

$$\frac{1}{\sqrt{2}} \cdot \frac{\sqrt{2}}{\sqrt{2}} = \frac{\sqrt{2}}{2}.$$

Now the computation would require dividing 1.414 by 2 to obtain .707, a much easier task.

With current technology, either form of this fraction can be approximated with the same number of keystrokes. See Figure 9, which shows how a calculator gives the same approximation for both forms of the expression.

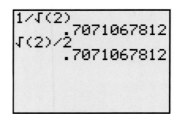

Figure 9

A common way of "standardizing" the form of a radical expression is to have the denominator contain no radicals. The process of removing radicals from a denominator so that the denominator contains only rational numbers is called **rationalizing the denominator.**

EXAMPLE 2 **Rationalizing Denominators with Square Roots**

Rationalize each denominator.

(a) $\dfrac{3}{\sqrt{7}}$

Multiply the numerator and denominator by $\sqrt{7}$. This is, in effect, multiplying by 1.

Continued on Next Page

ANSWERS
1. **(a)** $2 + 2\sqrt{5} + \sqrt{3} + \sqrt{15}$
   (b) $6\sqrt{2} - 6\sqrt{15} + \sqrt{30} - 15$
   (c) 13   **(d)** $11 - 2\sqrt{30}$
   (e) $16 - \sqrt[3]{49}$ **(f)** $p - 2, p \geq 0$

$$\frac{3}{\sqrt{7}} = \frac{3 \cdot \sqrt{7}}{\sqrt{7} \cdot \sqrt{7}}$$

In the denominator, $\sqrt{7} \cdot \sqrt{7} = \sqrt{7 \cdot 7} = \sqrt{49} = 7$, so

$$\frac{3}{\sqrt{7}} = \frac{3\sqrt{7}}{7}.$$

The denominator is now a rational number.

(b) $\dfrac{5\sqrt{2}}{\sqrt{5}} = \dfrac{5\sqrt{2} \cdot \sqrt{5}}{\sqrt{5} \cdot \sqrt{5}} = \dfrac{5\sqrt{10}}{5} = \sqrt{10}$

(c) $\dfrac{6}{\sqrt{12}}$

Less work is involved if the radical in the denominator is simplified first.

$$\frac{6}{\sqrt{12}} = \frac{6}{\sqrt{4 \cdot 3}} = \frac{6}{2\sqrt{3}} = \frac{3}{\sqrt{3}}$$

Now rationalize the denominator by multiplying the numerator and denominator by $\sqrt{3}$.

$$\frac{3}{\sqrt{3}} = \frac{3 \cdot \sqrt{3}}{\sqrt{3} \cdot \sqrt{3}} = \frac{3\sqrt{3}}{3} = \sqrt{3}$$

Work Problem 2 at the Side. ▶▶▶

EXAMPLE 3 **Rationalizing Denominators in Roots of Fractions**

Simplify each radical. Assume that all variables represent positive real numbers.

(a) $\sqrt{\dfrac{18}{125}} = \dfrac{\sqrt{18}}{\sqrt{125}}$ Quotient rule

$$= \frac{\sqrt{9 \cdot 2}}{\sqrt{25 \cdot 5}} \qquad \text{Factor.}$$

$$= \frac{3\sqrt{2}}{5\sqrt{5}} \qquad \text{Product rule}$$

$$= \frac{3\sqrt{2} \cdot \sqrt{5}}{5\sqrt{5} \cdot \sqrt{5}} \qquad \text{Multiply by } \tfrac{\sqrt{5}}{\sqrt{5}}.$$

$$= \frac{3\sqrt{10}}{5 \cdot 5} \qquad \text{Product rule}$$

$$= \frac{3\sqrt{10}}{25}$$

Continued on Next Page

2 Rationalize each denominator.

(a) $\dfrac{8}{\sqrt{3}}$

(b) $\dfrac{\sqrt{3}}{\sqrt{7}}$

(c) $\dfrac{3}{\sqrt{48}}$

(d) $\dfrac{-16}{\sqrt{32}}$

ANSWERS

2. (a) $\dfrac{8\sqrt{3}}{3}$ **(b)** $\dfrac{\sqrt{21}}{7}$

    (c) $\dfrac{\sqrt{3}}{4}$ **(d)** $-2\sqrt{2}$

3 Simplify. Assume that all variables represent positive real numbers.

(a) $\sqrt{\dfrac{8}{45}}$

(b) $\sqrt{\dfrac{72}{y}}$

(c) $\sqrt{\dfrac{200k^6}{y^7}}$

4 Simplify.

(a) $\sqrt[3]{\dfrac{15}{32}}$

(b) $\sqrt[3]{\dfrac{m^{12}}{n}}$, $\quad n \neq 0$

(c) $\sqrt[4]{\dfrac{6y}{w^2}}$, $\quad y \geq 0, w \neq 0$

(b) $\sqrt{\dfrac{50m^4}{p^5}}$, $\quad p > 0$

$$\sqrt{\dfrac{50m^4}{p^5}} = \dfrac{\sqrt{50m^4}}{\sqrt{p^5}} \qquad \text{Quotient rule}$$

$$= \dfrac{5m^2\sqrt{2}}{p^2\sqrt{p}} \qquad \text{Product rule}$$

$$= \dfrac{5m^2\sqrt{2} \cdot \sqrt{p}}{p^2\sqrt{p} \cdot \sqrt{p}} \qquad \text{Multiply by } \tfrac{\sqrt{p}}{\sqrt{p}}.$$

$$= \dfrac{5m^2\sqrt{2p}}{p^2 \cdot p} \qquad \text{Product rule}$$

$$= \dfrac{5m^2\sqrt{2p}}{p^3}$$

◀◀◀ **Work Problem 3 at the Side.**

EXAMPLE 4 **Rationalizing Denominators with Cube Roots**

Simplify.

(a) $\sqrt[3]{\dfrac{27}{16}}$

Use the quotient rule and simplify the numerator and denominator.

$$\sqrt[3]{\dfrac{27}{16}} = \dfrac{\sqrt[3]{27}}{\sqrt[3]{16}} = \dfrac{3}{\sqrt[3]{8} \cdot \sqrt[3]{2}} = \dfrac{3}{2\sqrt[3]{2}}$$

To get a rational denominator, multiply the numerator and denominator by a number that will result in a perfect cube in the radicand in the denominator. Since $2 \cdot 4 = 8$, a perfect cube, multiply the numerator and denominator by $\sqrt[3]{4}$.

$$\sqrt[3]{\dfrac{27}{16}} = \dfrac{3}{2\sqrt[3]{2}} = \dfrac{3 \cdot \sqrt[3]{4}}{2\sqrt[3]{2} \cdot \sqrt[3]{4}} = \dfrac{3\sqrt[3]{4}}{2\sqrt[3]{8}} = \dfrac{3\sqrt[3]{4}}{2 \cdot 2} = \dfrac{3\sqrt[3]{4}}{4}$$

(b) $\sqrt[4]{\dfrac{5x}{z}} = \dfrac{\sqrt[4]{5x}}{\sqrt[4]{z}} \cdot \dfrac{\sqrt[4]{z^3}}{\sqrt[4]{z^3}} = \dfrac{\sqrt[4]{5xz^3}}{\sqrt[4]{z^4}} = \dfrac{\sqrt[4]{5xz^3}}{z}$, $\quad x \geq 0, z > 0$

CAUTION

In problems like the one in Example 4(a), a typical error is to multiply the numerator and denominator by $\sqrt[3]{2}$, forgetting that

$$\sqrt[3]{2} \cdot \sqrt[3]{2} \neq 2.$$

We need *three* factors of 2 to get 2^3 under the radical. As implied in Example 4(a),

$$\sqrt[3]{2} \cdot \sqrt[3]{2} \cdot \sqrt[3]{2} = 2.$$

ANSWERS

3. (a) $\dfrac{2\sqrt{10}}{15}$ (b) $\dfrac{6\sqrt{2y}}{y}$ (c) $\dfrac{10k^3\sqrt{2y}}{y^4}$

4. (a) $\dfrac{\sqrt[3]{30}}{4}$ (b) $\dfrac{m^4\sqrt[3]{n^2}}{n}$ (c) $\dfrac{\sqrt[4]{6yw^2}}{w}$

◀◀◀ **Work Problem 4 at the Side.**

OBJECTIVE 3 Rationalize denominators with binomials involving radicals. Recall the special product

$$(x + y)(x - y) = x^2 - y^2.$$

To rationalize a denominator that contains a binomial expression (one that contains exactly two terms) involving radicals, such as

$$\frac{3}{1 + \sqrt{2}},$$

we must use *conjugates*. The conjugate of $1 + \sqrt{2}$ is $1 - \sqrt{2}$. In general, $x + y$ and $x - y$ are **conjugates.**

Rationalizing a Binomial Denominator

If a radical expression has a sum or difference with square root radicals in the denominator, rationalize the denominator by multiplying both the numerator and denominator by the conjugate of the denominator.

For the expression $\dfrac{3}{1 + \sqrt{2}}$, we rationalize the denominator by multiplying both the numerator and denominator by $1 - \sqrt{2}$, the conjugate of the denominator.

$$\frac{3}{1 + \sqrt{2}} = \frac{3(1 - \sqrt{2})}{(1 + \sqrt{2})(1 - \sqrt{2})}$$

$$= \frac{3(1 - \sqrt{2})}{-1} \qquad \begin{aligned} (1 + \sqrt{2})(1 - \sqrt{2}) \\ = 1^2 - (\sqrt{2})^2 \\ = 1 - 2 = -1 \end{aligned}$$

$$= \frac{3}{-1}(1 - \sqrt{2})$$

$$= -3(1 - \sqrt{2}) \quad \text{or} \quad -3 + 3\sqrt{2}$$

EXAMPLE 5 **Rationalizing Binomial Denominators**

Rationalize each denominator.

(a) $\dfrac{5}{4 - \sqrt{3}}$

To rationalize the denominator, multiply both the numerator and denominator by the conjugate of the denominator, $4 + \sqrt{3}$.

$$\frac{5}{4 - \sqrt{3}} = \frac{5(4 + \sqrt{3})}{(4 - \sqrt{3})(4 + \sqrt{3})}$$

$$= \frac{5(4 + \sqrt{3})}{16 - 3}$$

$$= \frac{5(4 + \sqrt{3})}{13}$$

Notice that the numerator is left in factored form. This makes it easier to determine whether the expression is written in lowest terms.

Continued on Next Page

5 Rationalize each denominator.

(a) $\dfrac{-4}{\sqrt{5} + 2}$

(b) $\dfrac{15}{\sqrt{7} + \sqrt{2}}$

(c) $\dfrac{\sqrt{3} + \sqrt{5}}{\sqrt{2} - \sqrt{7}}$

(d) $\dfrac{2}{\sqrt{k} + \sqrt{z}}$, $k \neq z, k > 0, z > 0$

(b) $\dfrac{\sqrt{2} - \sqrt{3}}{\sqrt{5} + \sqrt{3}} = \dfrac{(\sqrt{2} - \sqrt{3})(\sqrt{5} - \sqrt{3})}{(\sqrt{5} + \sqrt{3})(\sqrt{5} - \sqrt{3})}$ Multiply the numerator and denominator by $\sqrt{5} - \sqrt{3}$.

$= \dfrac{\sqrt{10} - \sqrt{6} - \sqrt{15} + 3}{5 - 3}$

$= \dfrac{\sqrt{10} - \sqrt{6} - \sqrt{15} + 3}{2}$

(c) $\dfrac{3}{\sqrt{5m} - \sqrt{p}} = \dfrac{3(\sqrt{5m} + \sqrt{p})}{(\sqrt{5m} - \sqrt{p})(\sqrt{5m} + \sqrt{p})}$

$= \dfrac{3(\sqrt{5m} + \sqrt{p})}{5m - p}$, $5m \neq p, m > 0, p > 0$

Work Problem 5 at the Side.

OBJECTIVE 4 Write radical quotients in lowest terms.

EXAMPLE 6 Writing Radical Quotients in Lowest Terms

Write each quotient in lowest terms.

(a) $\dfrac{6 + 2\sqrt{5}}{4}$

Factor the numerator and denominator, then write in lowest terms.

$$\dfrac{6 + 2\sqrt{5}}{4} = \dfrac{2(3 + \sqrt{5})}{2 \cdot 2} = \dfrac{3 + \sqrt{5}}{2}$$

Here is an alternative method for writing this expression in lowest terms.

$$\dfrac{6 + 2\sqrt{5}}{4} = \dfrac{6}{4} + \dfrac{2\sqrt{5}}{4} = \dfrac{3}{2} + \dfrac{\sqrt{5}}{2} = \dfrac{3 + \sqrt{5}}{2}$$

6 Write each quotient in lowest terms.

(a) $\dfrac{15 - 5\sqrt{3}}{5}$

(b) $\dfrac{24 - 36\sqrt{7}}{16}$

(b) $\dfrac{5y - \sqrt{8y^2}}{6y} = \dfrac{5y - 2y\sqrt{2}}{6y}$, $y > 0$ Product rule

$= \dfrac{y(5 - 2\sqrt{2})}{6y}$ Factor the numerator.

$= \dfrac{5 - 2\sqrt{2}}{6}$ Lowest terms

Note that the final fraction cannot be simplified further because there is no common factor of 2 in the numerator.

CAUTION
Be careful to factor before writing a quotient in lowest terms.

Work Problem 6 at the Side.

ANSWERS

5. (a) $-4(\sqrt{5} - 2)$ (b) $3(\sqrt{7} - \sqrt{2})$

(c) $\dfrac{-(\sqrt{6} + \sqrt{21} + \sqrt{10} + \sqrt{35})}{5}$

(d) $\dfrac{2(\sqrt{k} - \sqrt{z})}{k - z}$

6. (a) $3 - \sqrt{3}$ (b) $\dfrac{6 - 9\sqrt{7}}{4}$

9.5 Exercises

FOR EXTRA HELP   Tutor Center Addison-Wesley Math Tutor Center   MathXL MathXL   Digital Video Tutor CD 5 Videotape 5   Student's Solutions Manual   MyMathLab MyMathLab   Interactmath.com

Match each part of a rule for a special product in Column I with the part it equals in Column II.

I

1. $(x + \sqrt{y})(x - \sqrt{y})$

2. $(\sqrt{x} + y)(\sqrt{x} - y)$

3. $(\sqrt{x} + \sqrt{y})(\sqrt{x} - \sqrt{y})$

4. $(\sqrt{x} + \sqrt{y})^2$

5. $(\sqrt{x} - \sqrt{y})^2$

6. $(\sqrt{x} + y)^2$

II

A. $x - y$

B. $x + 2y\sqrt{x} + y^2$

C. $x - y^2$

D. $x - 2\sqrt{xy} + y$

E. $x^2 - y$

F. $x + 2\sqrt{xy} + y$

Multiply, then simplify each product. Assume that all variables represent positive real numbers. See Example 1.

7. $\sqrt{3}(\sqrt{12} - 4)$

8. $\sqrt{5}(\sqrt{125} - 6)$

9. $\sqrt{2}(\sqrt{18} - \sqrt{3})$

10. $\sqrt{5}(\sqrt{15} + \sqrt{5})$

11. $(\sqrt{6} + 2)(\sqrt{6} - 2)$

12. $(\sqrt{7} + 8)(\sqrt{7} - 8)$

13. $(\sqrt{12} - \sqrt{3})(\sqrt{12} + \sqrt{3})$

14. $(\sqrt{18} + \sqrt{8})(\sqrt{18} - \sqrt{8})$

15. $(\sqrt{3} + 2)(\sqrt{6} - 5)$

16. $(\sqrt{7} + 1)(\sqrt{2} - 4)$

17. $(\sqrt{3x} + 2)(\sqrt{3x} - 2)$

18. $(\sqrt{6y} - 4)(\sqrt{6y} + 4)$

19. $(2\sqrt{x} + \sqrt{y})(2\sqrt{x} - \sqrt{y})$

20. $(\sqrt{p} + 5\sqrt{s})(\sqrt{p} - 5\sqrt{s})$

21. $(4\sqrt{x} + 3)^2$

22. $(5\sqrt{p} - 6)^2$

23. $(9 - \sqrt[3]{2})(9 + \sqrt[3]{2})$

24. $(7 + \sqrt[3]{6})(7 - \sqrt[3]{6})$

25. The correct answer to Exercise 7 is $6 - 4\sqrt{3}$. Explain why this is not equal to $2\sqrt{3}$.

26. When we rationalize the denominator in the radical expression $\frac{1}{\sqrt{2}}$, we multiply both the numerator and denominator by $\sqrt{2}$. What property of real numbers covered in **Section 1.4** justifies this procedure?

Rationalize the denominator in each expression. Assume that all variables represent positive real numbers. See Example 2.

27. $\dfrac{7}{\sqrt{7}}$

28. $\dfrac{11}{\sqrt{11}}$

29. $\dfrac{15}{\sqrt{3}}$

30. $\dfrac{12}{\sqrt{6}}$

31. $\dfrac{\sqrt{3}}{\sqrt{2}}$

32. $\dfrac{\sqrt{7}}{\sqrt{6}}$

33. $\dfrac{9\sqrt{3}}{\sqrt{5}}$

34. $\dfrac{3\sqrt{2}}{\sqrt{11}}$

35. $\dfrac{-6}{\sqrt{18}}$

36. $\dfrac{-5}{\sqrt{24}}$

37. $\dfrac{-8\sqrt{3}}{\sqrt{k}}$

38. $\dfrac{-4\sqrt{13}}{\sqrt{m}}$

39. $\dfrac{6\sqrt{3y}}{\sqrt{y^3}}$

40. $\dfrac{-8\sqrt{5y}}{\sqrt{y^5}}$

41. Look again at the expression in Exercise 39. Start by multiplying both the numerator and the denominator by $\sqrt{y}$, to obtain the final answer. Then start over, multiplying both the numerator and denominator by $\sqrt{y^3}$, to obtain the same answer. Which method do you prefer? Why?

42. Explain why $\dfrac{1}{\sqrt[3]{2}}$ would not be written with the denominator rationalized if you begin by multiplying both the numerator and denominator by $\sqrt[3]{2}$. By what should you multiply them both to achieve the desired result?

Simplify. Assume that all variables represent positive real numbers. See Examples 3 and 4.

43. $\sqrt{\dfrac{7}{2}}$

44. $\sqrt{\dfrac{10}{3}}$

45. $-\sqrt{\dfrac{7}{50}}$

46. $-\sqrt{\dfrac{13}{75}}$

47. $\sqrt{\dfrac{24}{x}}$

48. $\sqrt{\dfrac{52}{y}}$

49. $-\sqrt{\dfrac{98r^3}{s}}$

50. $-\sqrt{\dfrac{150m^5}{n}}$

51. $\sqrt{\dfrac{288x^7}{y^9}}$

52. $\sqrt{\dfrac{242t^9}{u^{11}}}$

53. $\sqrt[3]{\dfrac{2}{3}}$

54. $\sqrt[3]{\dfrac{4}{5}}$

55. $\sqrt[3]{\dfrac{4}{9}}$

56. $\sqrt[3]{\dfrac{5}{16}}$

57. $-\sqrt[3]{\dfrac{2p}{r^2}}$

58. $-\sqrt[3]{\dfrac{6x}{y^2}}$

59. $\sqrt[4]{\dfrac{16}{x}}$

60. $\sqrt[4]{\dfrac{81}{y}}$

61. Explain the procedure you will use to rationalize the denominator of the expression in Exercise 63:
$$\frac{2}{4 + \sqrt{3}}.$$

62. Would multiplying both the numerator and denominator of $\dfrac{2}{4 + \sqrt{3}}$ by $4 + \sqrt{3}$ lead to a rationalized denominator? Why or why not?

Rationalize the denominator in each expression. Assume that all variables represent positive real numbers and that no denominators are 0. See Example 5.

63. $\dfrac{2}{4 + \sqrt{3}}$

64. $\dfrac{6}{5 + \sqrt{2}}$

65. $\dfrac{6}{\sqrt{5} + \sqrt{3}}$

66. $\dfrac{12}{\sqrt{6} + \sqrt{3}}$

67. $\dfrac{-4}{\sqrt{3} - \sqrt{7}}$

68. $\dfrac{-3}{\sqrt{2} + \sqrt{5}}$

69. $\dfrac{1 - \sqrt{2}}{\sqrt{7} + \sqrt{6}}$

70. $\dfrac{-1 - \sqrt{3}}{\sqrt{6} + \sqrt{5}}$

71. $\dfrac{4\sqrt{x}}{\sqrt{x} - 2\sqrt{y}}$

72. $\dfrac{5\sqrt{r}}{3\sqrt{r} + \sqrt{s}}$

73. $\dfrac{\sqrt{x} - \sqrt{y}}{\sqrt{2x} + \sqrt{3y}}$

74. $\dfrac{\sqrt{a} + \sqrt{b}}{\sqrt{5a} - \sqrt{2b}}$

75. If a and b are both positive numbers and $a^2 = b^2$, then $a = b$. Use this fact to show that

$$\frac{\sqrt{6} - \sqrt{2}}{4} = \frac{\sqrt{2 - \sqrt{3}}}{2}.$$

76. Use a calculator approximation to support the result in Exercise 75.

Write each quotient in lowest terms. Assume that all variables represent positive real numbers. See Example 6.

77. $\dfrac{25 + 10\sqrt{6}}{20}$

78. $\dfrac{12 - 6\sqrt{2}}{24}$

79. $\dfrac{16 + 4\sqrt{8}}{12}$

80. $\dfrac{12 + 9\sqrt{72}}{18}$

81. $\dfrac{6x + \sqrt{24x^3}}{3x}$

82. $\dfrac{11y + \sqrt{242y^5}}{22y}$

RELATING CONCEPTS (EXERCISES 83–86) For Individual or Group Work

Sometimes it is desirable to rationalize the numerator *in an expression. The procedure is similar to rationalizing the denominator. For example, to rationalize the numerator of*

$$\frac{6 - \sqrt{2}}{3},$$

we multiply both the numerator and denominator by the conjugate of the numerator, $6 + \sqrt{2}$.

$$\frac{6 - \sqrt{2}}{3} = \frac{(6 - \sqrt{2})(6 + \sqrt{2})}{3(6 + \sqrt{2})} = \frac{36 - 2}{3(6 + \sqrt{2})} = \frac{34}{3(6 + \sqrt{2})}$$

In the final expression, the numerator is rationalized. **Work Exercises 83–86 in order.**

83. Rationalize the numerator of $\dfrac{8\sqrt{5} - 1}{6}$.

84. Rationalize the numerator of $\dfrac{3\sqrt{a} + \sqrt{b}}{\sqrt{b} - \sqrt{a}}$. Assume a and b are positive and $a \neq b$.

85. Rationalize the denominator of the expression in Exercise 84.

86. Describe the difference in the procedures used in Exercises 84 and 85.

Summary Exercises on Operations with Radicals and Rational Exponents

Recall that a simplified radical satisfies the following conditions.

Conditions for a Simplified Radical

1. The radicand has no factor raised to a power greater than or equal to the index.

2. The radicand has no fractions.

3. No denominator has a radical.

4. Exponents in the radicand and the index of the radical have no common factor (except 1).

Perform all indicated operations and express each answer in simplest form with positive exponents. Assume that all variables represent positive real numbers.

1. $6\sqrt{10} - 12\sqrt{10}$

2. $\sqrt{7}(\sqrt{7} - \sqrt{2})$

3. $(1 - \sqrt{3})(2 + \sqrt{6})$

4. $\sqrt{50} - \sqrt{98} + \sqrt{72}$

5. $(3\sqrt{5} + 2\sqrt{7})^2$

6. $\dfrac{-3}{\sqrt{6}}$

7. $\dfrac{8}{\sqrt{7} + \sqrt{5}}$

8. $\sqrt[3]{16x^2} - \sqrt[3]{54x^2} + \sqrt[3]{128x^2}$

9. $\dfrac{1 - \sqrt{2}}{1 + \sqrt{2}}$

10. $(1 - \sqrt[3]{3})(1 + \sqrt[3]{3} + \sqrt[3]{9})$

11. $(\sqrt{5} + 7)(\sqrt{5} - 7)$

12. $\dfrac{1}{\sqrt{x} - \sqrt{5}}, \quad x \neq 5$

13. $\sqrt[3]{8a^3 b^5 c^9}$

14. $\dfrac{15}{\sqrt[3]{9}}$

15. $\dfrac{3}{\sqrt{5} + 2}$

16. $\sqrt{\dfrac{3}{5x}}$

17. $\dfrac{16\sqrt{3}}{5\sqrt{12}}$

18. $\dfrac{2\sqrt{25}}{8\sqrt{50}}$

19. $\dfrac{-10}{\sqrt[3]{10}}$

20. $\dfrac{\sqrt{6}+\sqrt{5}}{\sqrt{6}-\sqrt{5}}$

21. $\sqrt{12x}-\sqrt{75x}$

22. $(5-3\sqrt{3})^2$

23. $(\sqrt{74}-\sqrt{73})(\sqrt{74}+\sqrt{73})$

24. $\sqrt[3]{\dfrac{13}{81}}$

25. $-t^2\sqrt[4]{t}+3\sqrt[4]{t^9}-t\sqrt[4]{t^5}$

26. $\dfrac{\sqrt{3}+\sqrt{7}}{\sqrt{6}-\sqrt{5}}$

27. $\dfrac{6}{\sqrt[4]{3}}$

28. $\dfrac{1}{1-\sqrt[3]{3}}$

29. $\sqrt[3]{\dfrac{x^2y}{x^{-3}y^4}}$

30. $\sqrt{12}-\sqrt{108}-\sqrt[3]{27}$

31. $\dfrac{x^{-2/3}y^{4/5}}{x^{-5/3}y^{-2/5}}$

32. $\left(\dfrac{x^{3/4}y^{2/3}}{x^{1/3}y^{5/8}}\right)^{24}$

33. $(125x^3)^{-2/3}$

34. $(3x^{-2/3}y^{1/2})(-2x^{5/8}y^{-1/3})$

35. $\dfrac{4^{1/2}+3^{1/2}}{4^{1/2}-3^{1/2}}$

36. $(\sqrt{6}-\sqrt{5})^2(\sqrt{6}+\sqrt{5})^2$

9.6 Solving Equations with Radicals

An equation that includes one or more radical expressions with a variable is called a **radical equation.** Some examples of radical equations are

$$\sqrt{x - 4} = 8, \quad \sqrt{5x + 12} = 3\sqrt{2x - 1}, \quad \text{and} \quad \sqrt[3]{6 + x} = 27.$$

OBJECTIVE 1 Solve radical equations using the power rule. The equation $x = 1$ has only one solution. Its solution set is $\{1\}$. If we square both sides of this equation, we get $x^2 = 1$. This new equation has two solutions: -1 and 1. Notice that the solution of the original equation is also a solution of the squared equation. However, the squared equation has another solution, -1, that is *not* a solution of the original equation. When solving equations with radicals, we use this idea of raising both sides to a power. It is an application of the **power rule.**

> **Power Rule for Solving Equations with Radicals**
>
> If both sides of an equation are raised to the same power, all solutions of the original equation are also solutions of the new equation.

Read the power rule carefully; it does *not* say that all solutions of the new equation are solutions of the original equation. They may or may not be. Solutions that do not satisfy the original equation are called **extraneous solutions;** they must be discarded.

> **CAUTION**
> When the power rule is used to solve an equation, *every solution of the new equation* **must** *be checked in the original equation.*

EXAMPLE 1 Using the Power Rule

Solve $\sqrt{3x + 4} = 8$.

Use the power rule and square both sides to get

$$(\sqrt{3x + 4})^2 = 8^2$$

$$3x + 4 = 64$$

$$3x = 60 \qquad \text{Subtract 4.}$$

$$x = 20. \qquad \text{Divide by 3.}$$

To check, substitute the potential solution in the *original* equation.

$$\sqrt{3x + 4} = 8$$

$$\sqrt{3 \cdot 20 + 4} = 8 \quad ? \qquad \text{Let } x = 20.$$

$$\sqrt{64} = 8 \quad ?$$

$$8 = 8 \qquad \text{True}$$

Since 20 satisfies the *original* equation, the solution set is $\{20\}$.

Work Problem 1 at the Side. ▶▶▶

The solution of the equation in Example 1 can be generalized to give a method for solving equations with radicals.

OBJECTIVES

1 Solve radical equations using the power rule.

2 Solve radical equations that require additional steps.

3 Solve radical equations with indexes greater than 2.

1 Solve.

(a) $\sqrt{r} = 3$

(b) $\sqrt{5x + 1} = 4$

ANSWERS
1. (a) $\{9\}$ (b) $\{3\}$

2 Solve.

(a) $\sqrt{x} + 4 = -3$

Solving an Equation with Radicals

Step 1 **Isolate the radical.** Make sure that one radical term is alone on one side of the equation.

Step 2 **Apply the power rule.** Raise both sides of the equation to a power that is the same as the index of the radical.

Step 3 **Solve.** Solve the resulting equation; if it still contains a radical, repeat Steps 1 and 2.

Step 4 **Check** all potential solutions in the original equation.

CAUTION
Remember Step 4 or you may get an incorrect solution set.

EXAMPLE 2 Using the Power Rule

Solve $\sqrt{5x - 1} + 3 = 0$.

Step 1 To isolate the radical on one side, subtract 3 from each side.

$$\sqrt{5x - 1} = -3$$

Step 2 Now square both sides.

$$(\sqrt{5x - 1})^2 = (-3)^2$$

Step 3
$$5x - 1 = 9$$
$$5x = 10$$

(b) $\sqrt{x - 9} - 3 = 0$
$$x = 2$$

Step 4 Check the potential solution, 2, by substituting it in the original equation.

$$\sqrt{5x - 1} + 3 = 0$$
$$\sqrt{5 \cdot 2 - 1} + 3 = 0 \quad ? \quad \text{Let } x = 2.$$
$$3 + 3 = 0 \quad \text{False}$$

This false result shows that 2 is *not* a solution of the original equation; it is extraneous. The solution set is $\emptyset$.

NOTE
We could have determined after Step 1 that the equation in Example 2 has no solution because the expression on the left cannot be negative.

Work Problem 2 at the Side.

OBJECTIVE 2 Solve radical equations that require additional steps.
The next examples involve finding the square of a binomial. Recall that

$$(x + y)^2 = x^2 + 2xy + y^2.$$

ANSWERS
2. (a) $\emptyset$ (b) $\{18\}$

EXAMPLE 3 **Using the Power Rule; Squaring a Binomial**

Solve $\sqrt{4 - x} = x + 2$.

Step 1 The radical is alone on the left side of the equation.

Step 2 Square both sides; on the right, $(x + 2)^2 = x^2 + 2(x)(2) + 2^2$.

$$(\sqrt{4 - x})^2 = (x + 2)^2$$
$$4 - x = x^2 + 4x + 4$$

Twice the product of 2 and x

Step 3 The new equation is quadratic, so get 0 on one side.

$$0 = x^2 + 5x \qquad \text{Subtract 4 and add } x.$$
$$0 = x(x + 5) \qquad \text{Factor.}$$
$$x = 0 \quad \text{or} \quad x + 5 = 0 \qquad \text{Zero-factor property}$$
$$x = -5$$

Step 4 Check each potential solution in the original equation.

If $x = 0$, then

$$\sqrt{4 - x} = x + 2$$
$$\sqrt{4 - 0} = 0 + 2 \quad ?$$
$$\sqrt{4} = 2 \quad ?$$
$$2 = 2. \qquad \text{True}$$

If $x = -5$, then

$$\sqrt{4 - x} = x + 2$$
$$\sqrt{4 - (-5)} = -5 + 2 \quad ?$$
$$\sqrt{9} = -3 \quad ?$$
$$3 = -3. \qquad \text{False}$$

The solution set is $\{0\}$. The other potential solution, -5, is extraneous.

CAUTION

When a radical equation requires squaring a binomial as in Example 3, *remember to include the middle term.*

$$(x + 2)^2 = x^2 + 4x + 4$$

Work Problem 3 at the Side.

EXAMPLE 4 **Using the Power Rule; Squaring a Binomial**

Solve $\sqrt{x^2 - 4x + 9} = x - 1$.

Square both sides; $(x - 1)^2 = x^2 - 2(x)(1) + 1^2$ on the right.

$$(\sqrt{x^2 - 4x + 9})^2 = (x - 1)^2$$
$$x^2 - 4x + 9 = x^2 - 2x + 1$$

Twice the product of x and -1

$$-2x = -8 \qquad \text{Subtract } x^2 \text{ and 9; add } 2x.$$
$$x = 4 \qquad \text{Divide by } -2.$$

Check:
$$\sqrt{x^2 - 4x + 9} = x - 1$$
$$\sqrt{4^2 - 4 \cdot 4 + 9} = 4 - 1 \quad ? \qquad \text{Let } x = 4.$$
$$3 = 3 \qquad \text{True}$$

The solution set of the original equation is $\{4\}$.

Work Problem 4 at the Side.

3 Solve.

(a) $\sqrt{3x - 5} = x - 1$

(b) $x + 1 = \sqrt{-2x - 2}$

4 Solve.

$$\sqrt{4x^2 + 2x - 3} = 2x + 7$$

ANSWERS
3. (a) $\{2, 3\}$ (b) $\{-1\}$
4. $\{-2\}$

5 (a) Verify that 15 is not a solution of the equation in Example 5.

EXAMPLE 5 Using the Power Rule; Squaring Twice

Solve $\sqrt{5x + 6} + \sqrt{3x + 4} = 2$.

Start by isolating one radical on one side of the equation by subtracting $\sqrt{3x + 4}$ from each side. Then square both sides.

$$\sqrt{5x + 6} = 2 - \sqrt{3x + 4}$$
$$(\sqrt{5x + 6})^2 = (2 - \sqrt{3x + 4})^2$$
$$5x + 6 = 4 - 4\sqrt{3x + 4} + (3x + 4)$$

Twice the product of 2 and $-\sqrt{3x + 4}$

This equation still contains a radical, so square both sides again. Before doing this, isolate the radical term on the right.

$$5x + 6 = 8 + 3x - 4\sqrt{3x + 4}$$

(b) Solve.

$\sqrt{x + 1} - \sqrt{x - 4} = 1$

$$2x - 2 = -4\sqrt{3x + 4} \qquad \text{Subtract 8 and } 3x.$$
$$x - 1 = -2\sqrt{3x + 4} \qquad \text{Divide by 2.}$$
$$(x - 1)^2 = (-2\sqrt{3x + 4})^2 \qquad \text{Square both sides again.}$$
$$x^2 - 2x + 1 = (-2)^2 (\sqrt{3x + 4})^2 \qquad (ab)^2 = a^2b^2$$
$$x^2 - 2x + 1 = 4(3x + 4)$$
$$x^2 - 2x + 1 = 12x + 16 \qquad \text{Distributive property}$$
$$x^2 - 14x - 15 = 0 \qquad \text{Standard form}$$
$$(x + 1)(x - 15) = 0 \qquad \text{Factor.}$$

6 Solve each equation.

(a) $\sqrt[3]{x^2 + 3x + 12} = \sqrt[3]{x^2}$

$$x + 1 = 0 \quad \text{or} \quad x - 15 = 0 \qquad \text{Zero-factor property}$$
$$x = -1 \quad \text{or} \qquad x = 15$$

Check each of these potential solutions in the original equation. Only -1 satisfies the equation, so the solution set, $\{-1\}$, has only one element.

Work Problem 5 at the Side.

OBJECTIVE 3 Solve radical equations with indexes greater than 2. The power rule also works for powers greater than 2.

EXAMPLE 6 Using the Power Rule for a Power Greater than 2

(b) $\sqrt[4]{2x + 5} + 1 = 0$

Solve $\sqrt[3]{x + 5} = \sqrt[3]{2x - 6}$.

Raise both sides to the third power.

$$(\sqrt[3]{x + 5})^3 = (\sqrt[3]{2x - 6})^3$$
$$x + 5 = 2x - 6$$
$$11 = x \qquad \text{Subtract } x; \text{ add 6.}$$

Check this result in the original equation.

$$\sqrt[3]{x + 5} = \sqrt[3]{2x - 6}$$
$$\sqrt[3]{11 + 5} = \sqrt[3]{2 \cdot 11 - 6} \qquad ? \qquad \text{Let } x = 11.$$
$$\sqrt[3]{16} = \sqrt[3]{16} \qquad \text{True}$$

The solution set is $\{11\}$.

ANSWERS
5. (a) The final step in the check leads to
    $16 = 2$, which is false.
   (b) $\{8\}$
6. (a) $\{-4\}$ (b) $\emptyset$

Work Problem 6 at the Side.

9.6 Exercises

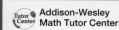 Addison-Wesley Math Tutor Center

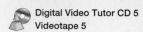 MathXL

 Digital Video Tutor CD 5 Videotape 5

Student's Solutions Manual

MyMathLab MyMathLab

Interactmath.com

Check each equation to see if the given value for x is a solution.

1. $\sqrt{3x + 18} = x$

  (a) 6   **(b)** -3

2. $\sqrt{3x - 3} = x - 1$

  (a) 1   **(b)** 4

3. $\sqrt{x + 2} = \sqrt{9x - 2} - 2\sqrt{x - 1}$

  (a) 2   **(b)** 7

4. $\sqrt{8x - 3} = 2x$

  (a) $\dfrac{3}{2}$ **(b)** $\dfrac{1}{2}$

5. Is 9 a solution of the equation $\sqrt{x} = -3$? If not, what is the solution of this equation? Explain.

6. Before even attempting to solve $\sqrt{3x + 18} = x$, how can you be sure that the equation cannot have a negative solution?

Solve each equation. See Examples 1–4.

7. $\sqrt{x - 2} = 3$

8. $\sqrt{x + 1} = 7$

9. $\sqrt{6x - 1} = 1$

10. $\sqrt{7x - 3} = 5$

11. $\sqrt{4x + 3} + 1 = 0$

12. $\sqrt{5x - 3} + 2 = 0$

13. $\sqrt{3k + 1} - 4 = 0$

14. $\sqrt{5z + 1} - 11 = 0$

15. $4 - \sqrt{x - 2} = 0$

16. $9 - \sqrt{4k + 1} = 0$

17. $\sqrt{9a - 4} = \sqrt{8a + 1}$

18. $\sqrt{4p - 2} = \sqrt{3p + 5}$

19. $2\sqrt{x} = \sqrt{3x + 4}$

20. $2\sqrt{m} = \sqrt{5m - 16}$

21. $3\sqrt{z - 1} = 2\sqrt{2z + 2}$

22. $5\sqrt{4x + 1} = 3\sqrt{10x + 25}$

23. $k = \sqrt{k^2 + 4k - 20}$

24. $p = \sqrt{p^2 - 3p + 18}$

25. $x = \sqrt{x^2 + 3x + 9}$

26. $z = \sqrt{z^2 - 4z - 8}$

27. $\sqrt{9 - x} = x + 3$

28. $\sqrt{5 - x} = x + 1$

29. $\sqrt{k^2 + 2k + 9} = k + 3$

30. $\sqrt{x^2 - 3x + 3} = x - 1$

31. $\sqrt{r^2 + 9r + 3} = -r$

32. $\sqrt{p^2 - 15p + 15} = p - 5$

33. $\sqrt{z^2 + 12z - 4} + 4 - z = 0$

34. $\sqrt{m^2 + 3m + 12} - m - 2 = 0$

35. What is *wrong* with this first step in the solution process for $\sqrt{3x + 4} = 8 - x$. Solve it correctly.
$$3x + 4 = 64 + x^2$$

36. Explain what is *wrong* with this first step in the solution process for $\sqrt{5x + 6} - \sqrt{x + 3} = 3$. Solve it correctly.
$$(5x + 6) + (x + 3) = 9$$

Solve each equation. See Examples 5 and 6.

37. $\sqrt[3]{2x + 5} = \sqrt[3]{6x + 1}$

38. $\sqrt[3]{p - 1} = 2$

39. $\sqrt[3]{a^2 + 5a + 1} = \sqrt[3]{a^2 + 4a}$

40. $\sqrt[3]{r^2 + 2r + 8} = \sqrt[3]{r^2}$

41. $\sqrt[3]{2m - 1} = \sqrt[3]{m + 13}$

42. $\sqrt[3]{2k - 11} - \sqrt[3]{5k + 1} = 0$

43. $\sqrt[4]{a + 8} = \sqrt[4]{2a}$

44. $\sqrt[4]{z + 11} = \sqrt[4]{2z + 6}$

45. $\sqrt[3]{x - 8} + 2 = 0$

46. $\sqrt[3]{r + 1} + 1 = 0$

47. $\sqrt[4]{2k - 5} + 4 = 0$

48. $\sqrt[4]{8z - 3} + 2 = 0$

49. $\sqrt{k + 2} - \sqrt{k - 3} = 1$

50. $\sqrt{r + 6} - \sqrt{r - 2} = 2$

51. $\sqrt{2r + 11} - \sqrt{5r + 1} = -1$

52. $\sqrt{3x - 2} - \sqrt{x + 3} = 1$

53. $\sqrt{3p + 4} - \sqrt{2p - 4} = 2$

54. $\sqrt{4x + 5} - \sqrt{2x + 2} = 1$

55. $\sqrt{3 - 3p} - 3 = \sqrt{3p + 2}$

56. $\sqrt{4x + 7} - 4 = \sqrt{4x - 1}$

57. $\sqrt{2\sqrt{x + 11}} = \sqrt{4x + 2}$

58. $\sqrt{1 + \sqrt{24 - 10x}} = \sqrt{3x + 5}$

For each equation, rewrite the expressions with rational exponents as radical expressions, and then solve using the procedures explained in this section.

59. $(2x - 9)^{1/2} = 2 + (x - 8)^{1/2}$

60. $(3w + 7)^{1/2} = 1 + (w + 2)^{1/2}$

61. $(2w - 1)^{2/3} - w^{1/3} = 0$

62. $(x^2 - 2x)^{1/3} - x^{1/3} = 0$

Solve each formula from electricity and radio for the indicated variable. (Source: Cooke, Nelson M., and Joseph B. Orleans, Mathematics Essential to Electricity and Radio, *McGraw-Hill, 1943.)*

63. $V = \sqrt{\dfrac{2K}{m}}$ for K

64. $V = \sqrt{\dfrac{2K}{m}}$ for m

65. $f = \dfrac{1}{2\pi\sqrt{LC}}$ for L

66. $r = \sqrt{\dfrac{Mm}{F}}$ for F

A number of useful formulas involve radicals or radical expressions. Many occur in the mathematics needed for working with objects in space. The formula

$$N = \frac{1}{2\pi}\sqrt{\frac{a}{r}}$$

is used to find the rotational rate N of a space station. Here a is the acceleration and r represents the radius of the space station in meters. To find the value of r that will make N simulate the effect of gravity on Earth, the equation must be solved for r, using the required value of N. (Source: Kastner, Bernice, Space Mathematics, *NASA, 1972.)*

67. Solve the equation for r.

68. Find the value of r that makes $N = .063$ rotation per sec, if $a = 9.8$ m per sec^2.

9.7 Complex Numbers

As we saw in **Section 1.1,** the set of real numbers includes many other number sets (the rational numbers, integers, and natural numbers, for example). In this section a new set of numbers is introduced that includes the set of real numbers, as well as numbers that are even roots of negative numbers, like $\sqrt{-2}$.

OBJECTIVE 1 Simplify numbers of the form $\sqrt{-b}$, where $b > 0$. The equation $x^2 + 1 = 0$ has no real number solution since any solution must be a number whose square is -1. In the set of real numbers, all squares are nonnegative numbers because the product of two positive numbers or two negative numbers is positive and $0^2 = 0$. To provide a solution for the equation $x^2 + 1 = 0$, a new number i, the **imaginary unit,** is defined so that

$$i^2 = -1.$$

That is, i is a number whose square is -1, so $i = \sqrt{-1}$. This definition of i makes it possible to define any square root of a negative number as follows.

$\sqrt{-b}$

For any positive number b,

$$\sqrt{-b} = i\sqrt{b}.$$

EXAMPLE 1 Simplifying Square Roots of Negative Numbers

Write each number as a product of a real number and i.

(a) $\sqrt{-100} = i\sqrt{100} = 10i$ **(b)** $-\sqrt{-36} = -i\sqrt{36} = -6i$

(c) $\sqrt{-2} = i\sqrt{2}$

(d) $\sqrt{-8} = \sqrt{-4 \cdot 2} = \sqrt{-4} \cdot \sqrt{2} = 2i\sqrt{2}$

CAUTION
It is easy to mistake $\sqrt{2i}$ for $\sqrt{2}i$, with the i under the radical. For this reason, we usually write $\sqrt{2}i$ as $i\sqrt{2}$, as in the definition of $\sqrt{-b}$.

Work Problem 1 at the Side. ▶▶▶

When finding a product such as $\sqrt{-4} \cdot \sqrt{-9}$, we cannot use the product rule for radicals because it applies only to nonnegative radicands. For this reason, we change $\sqrt{-b}$ to the form $i\sqrt{b}$ before performing any multiplications or divisions. For example,

$$\sqrt{-4} \cdot \sqrt{-9} = i\sqrt{4} \cdot i\sqrt{9}$$
$$= i \cdot 2 \cdot i \cdot 3$$
$$= 6i^2$$
$$= 6(-1) \qquad \text{Substitute: } i^2 = -1.$$
$$= -6.$$

OBJECTIVES

1 Simplify numbers of the form $\sqrt{-b}$, where $b > 0$.

2 Recognize subsets of the complex numbers.

3 Add and subtract complex numbers.

4 Multiply complex numbers.

5 Divide complex numbers.

6 Find powers of i.

1 Write each number as a product of a real number and i.

(a) $\sqrt{-16}$

(b) $-\sqrt{-81}$

(c) $\sqrt{-7}$

(d) $\sqrt{-32}$

ANSWERS
1. **(a)** $4i$ **(b)** $-9i$ **(c)** $i\sqrt{7}$ **(d)** $4i\sqrt{2}$

2 Multiply.

(a) $\sqrt{-7} \cdot \sqrt{-7}$

CAUTION

Using the product rule for radicals *before* using the definition of $\sqrt{-b}$ gives a *wrong* answer. The preceding example shows that

$$\sqrt{-4} \cdot \sqrt{-9} = -6, \text{ but}$$
$$\sqrt{-4(-9)} = \sqrt{36} = 6,$$

so

$$\sqrt{-4} \cdot \sqrt{-9} \neq \sqrt{-4(-9)}.$$

(b) $\sqrt{-5} \cdot \sqrt{-10}$

EXAMPLE 2 Multiplying Square Roots of Negative Numbers

Multiply.

(a) $\sqrt{-3} \cdot \sqrt{-7} = i\sqrt{3} \cdot i\sqrt{7}$
$$= i^2\sqrt{3 \cdot 7}$$
$$= (-1)\sqrt{21} \quad \text{Substitute: } i^2 = -1.$$
$$= -\sqrt{21}$$

(c) $\sqrt{-15} \cdot \sqrt{2}$

(b) $\sqrt{-2} \cdot \sqrt{-8} = i\sqrt{2} \cdot i\sqrt{8}$
$$= i^2\sqrt{2 \cdot 8}$$
$$= (-1)\sqrt{16}$$
$$= (-1)4$$
$$= -4$$

3 Divide.

(a) $\dfrac{\sqrt{-32}}{\sqrt{-2}}$

(c) $\sqrt{-5} \cdot \sqrt{6} = i\sqrt{5} \cdot \sqrt{6} = i\sqrt{30}$

◀◀◀ Work Problem 2 at the Side.

The methods used to find products also apply to quotients.

EXAMPLE 3 Dividing Square Roots of Negative Numbers

Divide.

(b) $\dfrac{\sqrt{-27}}{\sqrt{-3}}$

(a) $\dfrac{\sqrt{-75}}{\sqrt{-3}} = \dfrac{i\sqrt{75}}{i\sqrt{3}} = \sqrt{\dfrac{75}{3}} = \sqrt{25} = 5$

(b) $\dfrac{\sqrt{-32}}{\sqrt{8}} = \dfrac{i\sqrt{32}}{\sqrt{8}} = i\sqrt{\dfrac{32}{8}} = i\sqrt{4} = 2i$

◀◀◀ Work Problem 3 at the Side.

(c) $\dfrac{\sqrt{-40}}{\sqrt{10}}$

OBJECTIVE 2 Recognize subsets of the complex numbers. With the imaginary unit i and the real numbers, a new set of numbers can be formed that includes the real numbers as a subset. The *complex numbers* are defined as follows.

Complex Number

If a and b are real numbers, then any number of the form $a + bi$ is called a **complex number.**

ANSWERS
2. (a) -7 (b) $-5\sqrt{2}$ (c) $i\sqrt{30}$
3. (a) 4 (b) 3 (c) $2i$

In the complex number $a + bi$, the number a is called the **real part** and b is called the **imaginary part.*** When $b = 0$, $a + bi$ is a real number, so the real numbers are a subset of the complex numbers. Complex numbers with $a = 0$ and $b \neq 0$ are called **pure imaginary numbers.** In spite of their name, these numbers are very useful in applications, particularly in work with electricity.

The relationships among the sets of numbers are shown in Figure 10.

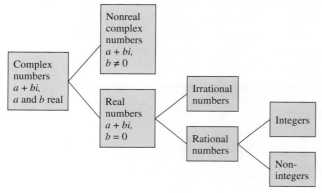

Figure 10

OBJECTIVE 3 **Add and subtract complex numbers.** The commutative, associative, and distributive properties for real numbers are also valid for complex numbers. Thus, to add complex numbers, we add their real parts and add their imaginary parts.

EXAMPLE 4 **Adding Complex Numbers**

Add.

(a) $(2 + 3i) + (6 + 4i)$
$= (2 + 6) + (3 + 4)i$ Commutative, associative, and distributive properties
$= 8 + 7i$

(b) $5 + (9 - 3i) = (5 + 9) - 3i$
$= 14 - 3i$

Work Problem 4 at the Side.

We subtract complex numbers by subtracting their real parts and subtracting their imaginary parts.

EXAMPLE 5 **Subtracting Complex Numbers**

Subtract.

(a) $(6 + 5i) - (3 + 2i) = (6 - 3) + (5 - 2)i$
$= 3 + 3i$

(b) $(7 - 3i) - (8 - 6i) = (7 - 8) + [-3 - (-6)]i$
$= -1 + 3i$

(c) $(-9 + 4i) - (-9 + 8i) = (-9 + 9) + (4 - 8)i$
$= 0 - 4i$
$= -4i$

Work Problem 5 at the Side.

*Some texts define bi as the imaginary part of the complex number $a + bi$.

4 Add.

(a) $(4 + 6i) + (-3 + 5i)$

(b) $(-1 + 8i) + (9 - 3i)$

5 Subtract.

(a) $(7 + 3i) - (4 + 2i)$

(b) $(-6 - i) - (-5 - 4i)$

(c) $8 - (3 - 2i)$

ANSWERS
4. (a) $1 + 11i$ **(b)** $8 + 5i$
5. (a) $3 + i$ **(b)** $-1 + 3i$ **(c)** $5 + 2i$

6 Multiply.

(a) $6i(4 + 3i)$

(b) $(6 - 4i)(2 + 4i)$

(c) $(3 - 2i)(3 + 2i)$

In Example 5(c), the answer was written as $0 - 4i$ and then as just $-4i$. A complex number written in the form $a + bi$, like $0 - 4i$, is in **standard form.** In this section, most answers will be given in standard form, but if a or b is 0, we consider answers such as a or bi to be in standard form.

OBJECTIVE 4 Multiply complex numbers. We multiply complex numbers as we multiply polynomials. Complex numbers of the form $a + bi$ have the same form as binomials, so we multiply two complex numbers in standard form by using the FOIL method for multiplying binomials. (Recall from **Section 6.4** that FOIL stands for *First, Outer, Inner, Last.*)

EXAMPLE 6 **Multiplying Complex Numbers**

Multiply.

(a) $4i(2 + 3i)$
Use the distributive property.

$$4i(2 + 3i) = 4i(2) + 4i(3i)$$
$$= 8i + 12i^2$$
$$= 8i + 12(-1) \qquad \text{Substitute: } i^2 = -1.$$
$$= -12 + 8i$$

(b) $(3 + 5i)(4 - 2i)$
Use the FOIL method.

$$(3 + 5i)(4 - 2i) = \underbrace{3(4)}_{\text{First}} + \underbrace{3(-2i)}_{\text{Outer}} + \underbrace{5i(4)}_{\text{Inner}} + \underbrace{5i(-2i)}_{\text{Last}}$$
$$= 12 - 6i + 20i - 10i^2$$
$$= 12 + 14i - 10(-1) \qquad \text{Substitute: } i^2 = -1.$$
$$= 12 + 14i + 10$$
$$= 22 + 14i$$

(c) $(2 + 3i)(1 - 5i) = 2(1) + 2(-5i) + 3i(1) + 3i(-5i) \qquad \text{FOIL}$$
$$= 2 - 10i + 3i - 15i^2$$
$$= 2 - 7i - 15(-1)$$
$$= 2 - 7i + 15$$
$$= 17 - 7i$$

◄◄◄ Work Problem 6 at the Side.

The two complex numbers $a + bi$ and $a - bi$ are called *complex conjugates* of each other. The product of a complex number and its conjugate is always a real number, as shown here.

$$(a + bi)(a - bi) = a^2 - abi + abi - b^2i^2$$
$$= a^2 - b^2(-1)$$
$$(a + bi)(a - bi) = a^2 + b^2$$

For example, $(3 + 7i)(3 - 7i) = 3^2 + 7^2 = 9 + 49 = 58.$

OBJECTIVE 5 Divide complex numbers. The quotient of two complex numbers should be a complex number. To write the quotient as a complex number, we need to eliminate i in the denominator. We use conjugates to do this.

ANSWERS
6. (a) $-18 + 24i$ (b) $28 + 16i$ (c) 13

EXAMPLE 7 **Dividing Complex Numbers**

Find each quotient.

(a) $\dfrac{8 + 9i}{5 + 2i}$

Multiply both the numerator and denominator by the conjugate of the denominator. The conjugate of $5 + 2i$ is $5 - 2i$.

$$\frac{8 + 9i}{5 + 2i} = \frac{(8 + 9i)(5 - 2i)}{(5 + 2i)(5 - 2i)} \qquad \tfrac{5-2i}{5-2i} = 1$$

$$= \frac{40 - 16i + 45i - 18i^2}{5^2 + 2^2}$$

$$= \frac{58 + 29i}{29} \qquad \text{Substitute: } i^2 = -1; \text{ combine terms.}$$

$$= \frac{29(2 + i)}{29} \qquad \text{Factor the numerator.}$$

$$= 2 + i \qquad \text{Lowest terms}$$

Notice that this is just like rationalizing a denominator. The final result is in standard form.

(b) $\dfrac{1 + i}{i}$

The conjugate of i is $-i$. Multiply both the numerator and denominator by $-i$.

$$\frac{1 + i}{i} = \frac{(1 + i)(-i)}{i(-i)}$$

$$= \frac{-i - i^2}{-i^2}$$

$$= \frac{-i - (-1)}{-(-1)} \qquad \text{Substitute: } i^2 = -1.$$

$$= \frac{-i + 1}{1}$$

$$= 1 - i$$

Work Problem 7 at the Side. ▶▶▶

Calculator Tip   In Examples 4–7, we showed how complex numbers can be added, subtracted, multiplied, and divided algebraically. Many current models of graphing calculators can perform these operations. Figure 11 shows how the computations in parts of Examples 4–7 are displayed on a TI-83 Plus or TI-84 calculator. Be sure to use parentheses as shown.

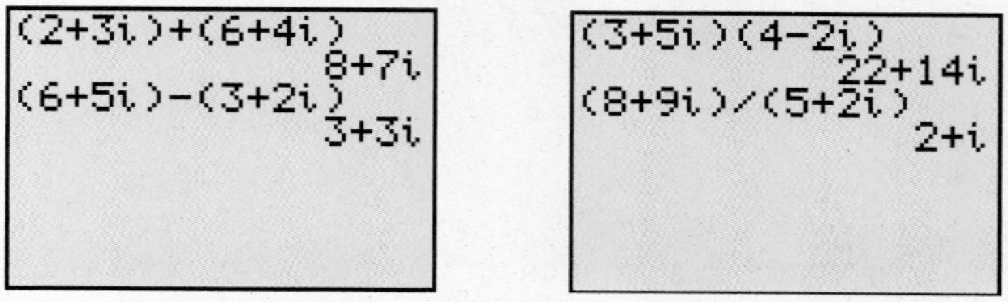

```
(2+3i)+(6+4i)
             8+7i
(6+5i)-(3+2i)
             3+3i
```

```
(3+5i)(4-2i)
          22+14i
(8+9i)/(5+2i)
             2+i
```

Figure 11

7 Find each quotient.

(a) $\dfrac{2 + i}{3 - i}$

(b) $\dfrac{8 - 4i}{1 - i}$

(c) $\dfrac{5}{3 - 2i}$

(d) $\dfrac{5 - i}{i}$

ANSWERS

7. (a) $\dfrac{1}{2} + \dfrac{1}{2}i$ (b) $6 + 2i$

(c) $\dfrac{15}{13} + \dfrac{10}{13}i$ (d) $-1 - 5i$

8 Find each power of i.

(a) i^{21}

OBJECTIVE 6 Find powers of i. Because i^2 is defined to be -1, we can find higher powers of i as shown in the following examples.

$$i^3 = i \cdot i^2 = i(-1) = -i \qquad i^6 = i^2 \cdot i^4 = (-1) \cdot 1 = -1$$
$$i^4 = i^2 \cdot i^2 = (-1)(-1) = 1 \qquad i^7 = i^3 \cdot i^4 = (-i) \cdot 1 = -i$$
$$i^5 = i \cdot i^4 = i \cdot 1 = i \qquad i^8 = i^4 \cdot i^4 = 1 \cdot 1 = 1$$

As these examples suggest, the powers of i rotate through the four numbers i, -1, $-i$, and 1. Larger powers of i can be simplified by using the fact that $i^4 = 1$. For example,

$$i^{75} = (i^4)^{18} \cdot i^3 = 1^{18} \cdot i^3 = 1 \cdot i^3 = i^3 = -i.$$

This example suggests a quick method for simplifying larger powers of i.

EXAMPLE 8 Simplifying Powers of i

(b) i^{36}

Find each power of i.

(a) $i^{12} = (i^4)^3 = 1^3 = 1$

(b) $i^{39} = i^{36} \cdot i^3 = (i^4)^9 \cdot i^3 = 1^9 \cdot (-i) = -i$

(c) $i^{-2} = \dfrac{1}{i^2} = \dfrac{1}{-1} = -1$

(d) $i^{-1} = \dfrac{1}{i}$

To simplify this quotient, multiply both the numerator and denominator by $-i$, the conjugate of i.

$$\frac{1}{i} = \frac{1(-i)}{i(-i)} = \frac{-i}{-i^2} = \frac{-i}{-(-1)} = \frac{-i}{1} = -i$$

(c) i^{50}

Work Problem 8 at the Side.

(d) i^{-9}

ANSWERS
8. (a) i (b) 1 (c) -1 (d) $-i$

9.7 Exercises

FOR EXTRA HELP

Tutor Center — Addison-Wesley Math Tutor Center

Math XL — MathXL

Digital Video Tutor CD 5 Videotape 5

Student's Solutions Manual

MyMathLab — MyMathLab

Interactmath.com

Decide whether each expression is equal to 1, −1, i, or −i.

1. $\sqrt{-1}$

2. $-i^2$

3. $\dfrac{1}{i}$

4. $(-i)^2$

5. Every real number is a complex number. Explain why this is so.

6. Not every complex number is a real number. Give an example of this, and explain why this statement is true.

Write each number as a product of a real number and i. Simplify all radical expressions. See Example 1.

7. $\sqrt{-169}$

8. $\sqrt{-225}$

9. $-\sqrt{-144}$

10. $-\sqrt{-196}$

11. $\sqrt{-5}$

12. $\sqrt{-21}$

13. $\sqrt{-48}$

14. $\sqrt{-96}$

Multiply or divide as indicated. See Examples 2 and 3.

15. $\sqrt{-15} \cdot \sqrt{-15}$

16. $\sqrt{-19} \cdot \sqrt{-19}$

17. $\sqrt{-4} \cdot \sqrt{-25}$

18. $\sqrt{-9} \cdot \sqrt{-81}$

19. $\dfrac{\sqrt{-300}}{\sqrt{-100}}$

20. $\dfrac{\sqrt{-40}}{\sqrt{-10}}$

21. $\dfrac{\sqrt{-75}}{\sqrt{3}}$

22. $\dfrac{\sqrt{-160}}{\sqrt{10}}$

Add or subtract as indicated. Write your answers in standard form. See Examples 4 and 5.

23. $(3 + 2i) + (-4 + 5i)$

24. $(7 + 15i) + (-11 + 14i)$

25. $(5 - i) + (-5 + i)$

26. $(-2 + 6i) + (2 - 6i)$

27. $(4 + i) - (-3 - 2i)$

28. $(9 + i) - (3 + 2i)$

29. $(-3 - 4i) - (-1 - 4i)$

30. $(-2 - 3i) - (-5 - 3i)$

31. $(-4 + 11i) + (-2 - 4i) + (7 + 6i)$

32. $(-1 + i) + (2 + 5i) + (3 + 2i)$

33. $[(7 + 3i) - (4 - 2i)] + (3 + i)$

34. $[(7 + 2i) + (-4 - i)] - (2 + 5i)$

35. Fill in the blank with the correct response: Because $(4 + 2i) - (3 + i) = 1 + i$, using the definition of subtraction we can check this to find that $(1 + i) + (3 + i) =$ _____ .

36. Fill in the blank with the correct response: Because $\frac{-5}{2-i} = -2 - i$, using the definition of division we can check this to find that $(-2 - i)(2 - i) =$ _____ .

Multiply. See Example 6.

37. $(3i)(27i)$

38. $(5i)(125i)$

39. $(-8i)(-2i)$

40. $(-32i)(-2i)$

41. $5i(-6 + 2i)$

42. $3i(4 + 9i)$

43. $(4 + 3i)(1 - 2i)$

44. $(7 - 2i)(3 + i)$

45. $(4 + 5i)^2$

46. $(3 + 2i)^2$

47. $(12 + 3i)(12 - 3i)$

48. $(6 + 7i)(6 - 7i)$

49. (a) What is the conjugate of $a + bi$?

(b) If we multiply $a + bi$ by its conjugate, we get _____ + _____ , which is always a real number.

50. Explain the procedure you would use to find the quotient

$$\frac{-1 + 5i}{3 + 2i}.$$

Write each quotient in the form a + bi. See Example 7.

51. $\dfrac{2}{1-i}$

52. $\dfrac{29}{5+2i}$

53. $\dfrac{-7+4i}{3+2i}$

54. $\dfrac{-38-8i}{7+3i}$

55. $\dfrac{8i}{2+2i}$

56. $\dfrac{-8i}{1+i}$

57. $\dfrac{2-3i}{2+3i}$

58. $\dfrac{-1+5i}{3+2i}$

RELATING CONCEPTS (EXERCISES 59–64)   For Individual or Group Work

Consider these expressions:

Binomials	**Complex Numbers**
$x+2, \quad 3x-1$	$1+2i, \quad 3-i.$

When we add, subtract, or multiply complex numbers in standard form, the rules are the same as those for the corresponding operations on binomials. That is, we add or subtract like terms, and we use FOIL to multiply. Division, however, is comparable to division by the sum or difference of radicals, where we multiply by the conjugate of the denominator to get a rational denominator. To express the quotient of two complex numbers in standard form, we also multiply by the conjugate of the denominator. **Work Exercises 59–64 in order,** *to better understand these ideas.*

59. (a) Add the two binomials.

(b) Add the two complex numbers.

60. (a) Subtract the second binomial from the first.

(b) Subtract the second complex number from the first.

61. (a) Multiply the two binomials.

(b) Multiply the two complex numbers.

62. (a) Rationalize the denominator: $\dfrac{\sqrt{3}-1}{1+\sqrt{2}}$.

(b) Write in standard form: $\dfrac{3-i}{1+2i}$.

63. Explain why the answers for parts (a) and (b) in Exercise 61 do not correspond as the answers in Exercises 59 and 60 do.

64. Explain why the answers for parts (a) and (b) in Exercise 62 do not correspond as the answers in Exercises 59 and 60 do.

65. Recall that if $a \neq 0$, $\frac{1}{a}$ is called the reciprocal of a. Use this definition to express the reciprocal of $5 - 4i$ in the form $a + bi$.

66. Recall that if $a \neq 0$, a^{-1} is defined to be $\frac{1}{a}$. Use this definition to express $(4 - 3i)^{-1}$ in the form $a + bi$.

Find each power of i. See Example 8.

67. i^{18}

68. i^{26}

69. i^{89}

70. i^{45}

71. i^{96}

72. i^{48}

73. i^{-5}

74. i^{-17}

75. A student simplified i^{-18} as follows:

$$i^{-18} = i^{-18} \cdot i^{20} = i^{-18 + 20} = i^2 = -1.$$

Explain the mathematical justification for this correct work.

76. Explain why

$$(46 + 25i)(3 - 6i) \quad \text{and} \quad (46 + 25i)(3 - 6i)i^{12}$$

must be equal. (Do not actually perform the computation.)

Ohm's law for the current I in a circuit with voltage E, resistance R, capacitance reactance X_c, and inductive reactance X_L is

$$I = \frac{E}{R + (X_L - X_c)i}.$$

Use this law to work Exercises 77 and 78.

77. Find I if $E = 2 + 3i$, $R = 5$, $X_L = 4$, and $X_c = 3$.

78. Find E if $I = 1 - i$, $R = 2$, $X_L = 3$, and $X_c = 1$.

79. Show that $1 + 5i$ is a solution of

$$x^2 - 2x + 26 = 0.$$

80. Show that $3 + 2i$ is a solution of

$$x^2 - 6x + 13 = 0.$$

Chapter 9

SUMMARY

9.1   **radicand, index**   In the expression $\sqrt[n]{a}$, a is the radicand and n is the index (order).

radical   The expression $\sqrt[n]{a}$ is a radical.

principal root   If a is positive and n is even, the principal nth root of a is the positive root.

radical expression   A radical expression is an algebraic expression that contains radicals.

9.5   **rationalizing the denominator**   The process of removing radicals from the denominator so that the denominator contains only rational quantities is called rationalizing the denominator.

conjugate   The conjugate of $a + b$ is $a - b$.

9.6   **radical equation**   A radical equation is an equation that includes one or more radical expressions with variables.

extraneous solution   An extraneous solution of a radical equation is a solution of $x = a^2$ that is not a solution of $\sqrt{x} = a$.

9.7   **complex number**   A complex number is a number that can be written in the form $a + bi$, where a and b are real numbers.

real part   The real part of $a + bi$ is a.

imaginary part   The imaginary part of $a + bi$ is b.

pure imaginary number   A complex number $a + bi$ with $a = 0$ and $b \neq 0$ is called a pure imaginary number.

standard form (of a complex number)   A complex number is in standard form if it is written in the form $a + bi$.

complex conjugates   The complex conjugate of $a + bi$ is $a - bi$.

NEW SYMBOLS

$\sqrt{}$ radical sign

$\sqrt[n]{a}$ radical; principal nth root of a

$\pm$ positive or negative

$\approx$ is approximately equal to

$a^{1/n}$ a to the power $\dfrac{1}{n}$

$a^{m/n}$ a to the power $\dfrac{m}{n}$

i a number whose square is -1

TEST YOUR WORD POWER

See how well you have learned the vocabulary in this chapter. Answers, with examples, follow the Quick Review.

1. A **radicand** is
   A. the index of a radical
   B. the number or expression under the radical sign
   C. the positive root of a number
   D. the radical sign.

2. The **Pythagorean formula** states that, in a right triangle,
   A. the sum of the measures of the angles is 180°
   B. the sum of the lengths of the two shorter sides equals the length of the longest side
   C. the longest side is opposite the right angle
   D. the square of the length of the longest side equals the sum of the squares of the lengths of the two shorter sides.

3. A **hypotenuse** is
   A. either of the two shorter sides of a triangle
   B. the shortest side of a triangle
   C. the side opposite the right angle in a triangle
   D. the longest side in any triangle.

4. **Rationalizing the denominator** is the process of
   A. eliminating fractions from a radical expression
   B. changing the denominator of a fraction from a radical to a rational number
   C. clearing a radical expression of radicals
   D. multiplying radical expressions.

5. An **extraneous solution** is a solution
   A. that does not satisfy the original equation
   B. that makes an equation true
   C. that makes an expression equal 0
   D. that checks in the original equation.

6. A **complex number** is
   A. a real number that includes a complex fraction
   B. a zero multiple of i
   C. a number of the form $a + bi$, where a and b are real numbers
   D. the square root of -1.

QUICK REVIEW

Concepts	*Examples*

9.1 *Radical Expressions and Graphs*

$\sqrt[n]{a} = b$ means $b^n = a$.

$\sqrt[n]{a}$ is the principal nth root of a.

$\sqrt[n]{a^n} = |a|$ if n is even.

$\sqrt[n]{a^n} = a$ if n is odd.

Functions Defined by Radical Expressions

The square root function with $f(x) = \sqrt{x}$ and the cube root function with $f(x) = \sqrt[3]{x}$ are two important functions defined by radical expressions.

The two square roots of 64 are $\sqrt{64} = 8$, the principal square root, and $-\sqrt{64} = -8$.

$$\sqrt[3]{-27} = -3 \qquad \sqrt[4]{(-2)^4} = |-2| = 2$$

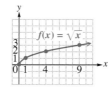

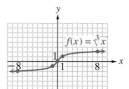

9.2 *Rational Exponents*

$a^{1/n} = \sqrt[n]{a}$ whenever $\sqrt[n]{a}$ exists.

If m and n are positive integers with m/n in lowest terms, then $a^{m/n} = (a^{1/n})^m$, provided that $a^{1/n}$ is a real number.

All of the usual definitions and rules for exponents are valid for rational exponents.

$$81^{1/2} = \sqrt{81} = 9 \qquad -64^{1/3} = -\sqrt[3]{64} = -4$$

$$8^{5/3} = (8^{1/3})^5 = 2^5 = 32$$

$$5^{-1/2} \cdot 5^{1/4} = 5^{-1/2 + 1/4} = 5^{-1/4} = \frac{1}{5^{1/4}} \qquad (y^{2/5})^{10} = y^4$$

$$\frac{x^{-1/3}}{x^{-1/2}} = x^{-1/3 - (-1/2)} = x^{-1/3 + 1/2} = x^{1/6}, \quad x > 0$$

Concepts	*Examples*
9.3 *Simplifying Radical Expressions* **Product and Quotient Rules for Radicals** If $\sqrt[n]{a}$ and $\sqrt[n]{b}$ are real numbers and n is a natural number, $$\sqrt[n]{a} \cdot \sqrt[n]{b} = \sqrt[n]{ab}$$ and $$\sqrt[n]{\frac{a}{b}} = \frac{\sqrt[n]{a}}{\sqrt[n]{b}}, \quad b \neq 0.$$	$$\sqrt{3} \cdot \sqrt{7} = \sqrt{21}$$ $$\sqrt[5]{x^3 y} \cdot \sqrt[5]{xy^2} = \sqrt[5]{x^4 y^3}$$ $$\frac{\sqrt{x^5}}{\sqrt{x^4}} = \sqrt{\frac{x^5}{x^4}} = \sqrt{x}, \quad x > 0$$
Conditions for a Simplified Radical **1.** The radicand has no factor raised to a power greater than or equal to the index. **2.** The radicand has no fractions. **3.** No denominator has a radical. **4.** Exponents in the radicand and the index of the radical have no common factors (except 1).	$$\sqrt{18} = \sqrt{9 \cdot 2} = 3\sqrt{2}$$ $$\sqrt[3]{54x^5 y^3} = \sqrt[3]{27x^3 y^3 \cdot 2x^2} = 3xy\sqrt[3]{2x^2}$$ $$\sqrt{\frac{7}{4}} = \frac{\sqrt{7}}{\sqrt{4}} = \frac{\sqrt{7}}{2}$$ $$\sqrt[9]{x^3} = x^{3/9} = x^{1/3} \quad \text{or} \quad \sqrt[3]{x}$$
Pythagorean Formula If c is the length of the longest side of a right triangle and a and b are the lengths of the shorter sides, then $$c^2 = a^2 + b^2.$$ The longest side is the hypotenuse and the two shorter sides are the legs of the triangle. The hypotenuse is opposite the right angle.	Find b for the triangle in the figure. $$10^2 + b^2 = (2\sqrt{61})^2$$ $$b^2 = 4(61) - 100$$ $$b^2 = 144$$ $$b = 12$$
Distance Formula The distance between (x_1, y_1) and (x_2, y_2) is $$d = \sqrt{(x_2 - x_1)^2 + (y_2 - y_1)^2}.$$	The distance between $(3, -2)$ and $(-1, 1)$ is $$\sqrt{(-1 - 3)^2 + [1 - (-2)]^2}$$ $$= \sqrt{(-4)^2 + 3^2} = \sqrt{16 + 9} = \sqrt{25} = 5.$$
9.4 *Adding and Subtracting Radical Expressions* Only radical expressions with the same index and the same radicand may be combined.	$$3\sqrt{17} + 2\sqrt{17} - 8\sqrt{17} = (3 + 2 - 8)\sqrt{17}$$ $$= -3\sqrt{17}$$ $$\sqrt[3]{2} - \sqrt[3]{250} = \sqrt[3]{2} - 5\sqrt[3]{2} = -4\sqrt[3]{2}$$ $$\left. \begin{array}{l} \sqrt{15} + \sqrt{30} \\ \sqrt{3} + \sqrt[3]{9} \end{array} \right\} \begin{array}{l} \text{cannot be} \\ \text{simplified further} \end{array}$$
9.5 *Multiplying and Dividing Radical Expressions* Multiply binomial radical expressions by using the FOIL method. Special products from **Section 6.4** may apply.	$$(\sqrt{2} + \sqrt{7})(\sqrt{3} - \sqrt{6})$$ $$= \sqrt{6} - 2\sqrt{3} + \sqrt{21} - \sqrt{42} \quad \sqrt{12} = 2\sqrt{3}$$ $$(\sqrt{5} - \sqrt{10})(\sqrt{5} + \sqrt{10}) = 5 - 10 = -5$$ $$(\sqrt{3} - \sqrt{2})^2 = 3 - 2\sqrt{3} \cdot \sqrt{2} + 2 = 5 - 2\sqrt{6}$$
Rationalize the denominator by multiplying both the numerator and denominator by the same expression.	$$\frac{\sqrt{7}}{\sqrt{5}} = \frac{\sqrt{7} \cdot \sqrt{5}}{\sqrt{5} \cdot \sqrt{5}} = \frac{\sqrt{35}}{5}$$ $$\frac{4}{\sqrt{5} - \sqrt{2}} = \frac{4(\sqrt{5} + \sqrt{2})}{(\sqrt{5} - \sqrt{2})(\sqrt{5} + \sqrt{2})}$$ $$= \frac{4(\sqrt{5} + \sqrt{2})}{5 - 2} = \frac{4(\sqrt{5} + \sqrt{2})}{3}$$

Concepts	Examples

9.6 *Solving Equations with Radicals*

Solving an Equation with Radicals

Step 1 Isolate one radical on one side of the equation.

Step 2 Raise each side of the equation to a power that is the same as the index of the radical.

Step 3 Solve the resulting equation; if it still contains a radical, repeat Steps 1 and 2.

Step 4 Check all potential solutions in the *original* equation.

Potential solutions that do not check are extraneous; they are not part of the solution set.

Solve $\sqrt{2x + 3} - x = 0$.

$$\sqrt{2x + 3} = x$$
$$(\sqrt{2x + 3})^2 = x^2$$
$$2x + 3 = x^2$$
$$x^2 - 2x - 3 = 0$$
$$(x + 1)(x - 3) = 0$$
$$x + 1 = 0 \quad \text{or} \quad x - 3 = 0$$
$$x = -1 \quad \text{or} \quad x = 3$$

A check shows that 3 is a solution, but -1 is extraneous. The solution set is $\{3\}$.

9.7 *Complex Numbers*

$i^2 = -1$, so $i = \sqrt{-1}$.

For any positive number b, $\sqrt{-b} = i\sqrt{b}$.

To multiply radicals with negative radicands, first change each factor to the form $i\sqrt{b}$, then multiply. The same procedure applies to quotients.

$$\sqrt{-25} = i\sqrt{25} = 5i$$
$$\sqrt{-3} \cdot \sqrt{-27} = i\sqrt{3} \cdot i\sqrt{27}$$
$$= i^2\sqrt{81}$$
$$= -1 \cdot 9$$
$$= -9$$

$$\frac{\sqrt{-18}}{\sqrt{-2}} = \frac{i\sqrt{18}}{i\sqrt{2}} = \sqrt{\frac{18}{2}} = \sqrt{9} = 3$$

Adding and Subtracting Complex Numbers

Add (or subtract) the real parts and add (or subtract) the imaginary parts.

$$(5 + 3i) + (8 - 7i) = 13 - 4i$$
$$(5 + 3i) - (8 - 7i) = -3 + 10i$$

Multiplying and Dividing Complex Numbers

Multiply complex numbers by using the FOIL method.

$$(2 + i)(5 - 3i) = 10 - 6i + 5i - 3i^2$$
$$= 10 - i - 3(-1)$$
$$= 10 - i + 3$$
$$= 13 - i$$

Divide complex numbers by multiplying the numerator and the denominator by the conjugate of the denominator.

$$\frac{2}{3 + i} = \frac{2(3 - i)}{(3 + i)(3 - i)} = \frac{2(3 - i)}{9 - i^2}$$
$$= \frac{2(3 - i)}{10} = \frac{3 - i}{5}$$

ANSWERS TO TEST YOUR WORD POWER

1. B; *Example:* In $\sqrt{3xy}$, $3xy$ is the radicand.
2. D; *Example:* In a right triangle where $a = 6$, $b = 8$, and $c = 10$, $6^2 + 8^2 = 10^2$.
3. C; *Example:* In a right triangle where the sides measure 9, 12, and 15 units, the hypotenuse is the side with measure 15 units.
4. B; *Example:* To rationalize the denominator of $\dfrac{5}{\sqrt{3} + 1}$, multiply both the numerator and denominator by $\sqrt{3} - 1$ to get $\dfrac{5(\sqrt{3} - 1)}{2}$.
5. A; *Example:* The potential solution 2 is extraneous in $\sqrt{5x - 1} + 3 = 0$.
6. C; *Examples:* -5 (or $-5 + 0i$), $7i$ (or $0 + 7i$), and $\sqrt{2} - 4i$.

Chapter 9
REVIEW EXERCISES

[9.1] *Find each real number root. Use a calculator as necessary.*

1. $\sqrt{1764}$

2. $-\sqrt{289}$

3. $-\sqrt{-841}$

4. $\sqrt[3]{216}$

5. $\sqrt[5]{-32}$

6. $\sqrt{x^2}$

7. $\sqrt[3]{x^3}$

8. $\sqrt[4]{x^{20}}$

Graph each function. Give the domain and the range.

9. $f(x) = \sqrt{x - 1}$

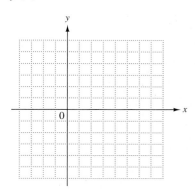

10. $f(x) = \sqrt[3]{x} + 4$

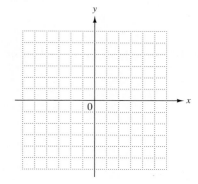

11. Under what conditions is $\sqrt[n]{a}$ not a real number?

12. If a is negative and n is even, what can be said about $a^{1/n}$?

Use a calculator to find a decimal approximation for each radical. Round to the nearest thousandth.

13. $\sqrt{40}$

14. $\sqrt{77}$

15. $\sqrt{310}$

16. Use the formula for the time for one complete swing of a pendulum from Exercise 68 in **Section 9.1,** $t = 2\pi\sqrt{\dfrac{L}{g}}$, to find the time of a complete swing if the pendulum is 3 ft long and g is 32 ft per sec^2.

17. Use Heron's formula from Exercise 69 in **Section 9.1,** $A = \sqrt{s(s - a)(s - b)(s - c)}$, where $s = \frac{1}{2}(a + b + c)$, to find the area of a triangle with sides of lengths 11, 13, and 20 in.

[9.2] *Find each real number root.*

18. $49^{1/2}$

19. $-8^{1/3}$

20. $(-16)^{1/4}$

21. Explain the relationship between the expressions $a^{m/n}$ and $\sqrt[n]{a^m}$.

Use the definitions and rules for exponents to simplify each expression. Assume that all variables represent positive real numbers.

22. $16^{5/4}$

23. $-8^{2/3}$

24. $-\left(\dfrac{36}{25}\right)^{3/2}$

25. $\left(-\dfrac{1}{8}\right)^{-5/3}$

26. $\left(\dfrac{81}{10,000}\right)^{-3/4}$

27. $7^{1/3} \cdot 7^{5/3}$

28. $\dfrac{96^{2/3}}{96^{-1/3}}$

29. $\dfrac{k^{2/3} k^{-1/2} k^{3/4}}{2(k^2)^{-1/4}}$

30. Write $2^{4/5}$ as a radical.

Simplify each expression. Write answers in radical form. Assume that all variables represent positive real numbers.

31. $\sqrt{3^{18}}$

32. $\sqrt{7^9}$

33. $\sqrt[3]{m^5} \cdot \sqrt[3]{m^8}$

34. $\sqrt[4]{k^2} \cdot \sqrt[4]{k^7}$

35. $\sqrt[3]{\sqrt{m}}$

36. $\sqrt[4]{16y^5}$

37. $\sqrt[5]{y} \cdot \sqrt[3]{y}$

38. $\dfrac{\sqrt[3]{y^2}}{\sqrt[4]{y}}$

[9.3] *Simplify each expression. Assume that all variables represent positive real numbers.*

39. $\sqrt{6} \cdot \sqrt{11}$

40. $\sqrt{5} \cdot \sqrt{r}$

41. $\sqrt[3]{6} \cdot \sqrt[3]{5}$

42. $\sqrt[4]{7} \cdot \sqrt[4]{3}$

43. $\sqrt{20}$

44. $-\sqrt{125}$

45. $\sqrt[3]{-108x^4y}$

46. $\sqrt[3]{64p^4q^6}$

47. $\sqrt{\dfrac{49}{81}}$

48. $\sqrt{\dfrac{y^3}{144}}$

49. $\sqrt[3]{\dfrac{m^{15}}{27}}$

50. $\sqrt[3]{\dfrac{r^2}{8}}$

51. $\dfrac{\sqrt[3]{2^4}}{\sqrt[4]{32}}$

52. $\dfrac{\sqrt{x}}{\sqrt[5]{x}}$

Find the distance between each pair of points.

53. $(2, 7)$ and $(-1, -4)$

54. $(-3, -5)$ and $(4, -3)$

[9.4] *Perform the indicated operations. Assume that all variables represent positive real numbers.*

55. $2\sqrt{8} - 3\sqrt{50}$

56. $8\sqrt{80} - 3\sqrt{45}$

57. $-\sqrt{27y} + 2\sqrt{75y}$

58. $2\sqrt{54m^3} + 5\sqrt{96m^3}$

59. $3\sqrt[3]{54} + 5\sqrt[3]{16}$

60. $-6\sqrt[4]{32} + \sqrt[4]{512}$

[9.5] *Multiply, then simplify the products.*

61. $(\sqrt{3} + 1)(\sqrt{3} - 2)$

62. $(\sqrt{7} + \sqrt{5})(\sqrt{7} - \sqrt{5})$

63. $(3\sqrt{2} + 1)(2\sqrt{2} - 3)$

64. $(\sqrt{11} + 3\sqrt{5})(\sqrt{11} + 5\sqrt{5})$

65. $(\sqrt{13} - \sqrt{2})^2$

66. $(\sqrt{5} - \sqrt{7})^2$

Rationalize each denominator. Assume that all variables represent positive real numbers.

67. $\dfrac{-6\sqrt{3}}{\sqrt{2}}$

68. $\dfrac{3\sqrt{7p}}{\sqrt{y}}$

69. $-\sqrt[3]{\dfrac{9}{25}}$

70. $\sqrt[3]{\dfrac{108m^3}{n^5}}$

71. $\dfrac{1}{\sqrt{2} + \sqrt{7}}$

72. $\dfrac{-5}{\sqrt{6} - \sqrt{3}}$

[9.6] *Solve each equation.*

73. $\sqrt{8x + 9} = 5$

74. $\sqrt{2z - 3} - 3 = 0$

75. $\sqrt{3m + 1} = -1$

76. $\sqrt{7z + 1} = z + 1$

77. $3\sqrt{m} = \sqrt{10m - 9}$

78. $\sqrt{p^2 + 3p + 7} = p + 2$

79. $\sqrt{x + 2} - \sqrt{x - 3} = 1$

80. $\sqrt[3]{5m - 1} = \sqrt[3]{3m - 2}$

81. $\sqrt[4]{x + 6} = \sqrt[4]{2x}$

[9.7] *Write as a product of a real number and i.*

82. $\sqrt{-25}$

83. $\sqrt{-200}$

84. $\sqrt{-160}$

Perform the indicated operations. Write answers in standard form.

85. $(-2 + 5i) + (-8 - 7i)$

86. $(5 + 4i) - (-9 - 3i)$

87. $\sqrt{-5} \cdot \sqrt{-7}$

88. $\sqrt{-25} \cdot \sqrt{-81}$

89. $\dfrac{\sqrt{-72}}{\sqrt{-8}}$

90. $(2 + 3i)(1 - i)$

91. $(6 - 2i)^2$

92. $\dfrac{3 - i}{2 + i}$

93. $\dfrac{5 + 14i}{2 + 3i}$

Find each power of i.

94. i^{11}

95. i^{52}

96. i^{-13}

MIXED REVIEW EXERCISES

Simplify. Assume that all variables represent positive real numbers.

97. $-\sqrt{169a^2b^4}$

98. $1000^{-2/3}$

99. $\dfrac{y^{-1/3} \cdot y^{5/6}}{y}$

100. $\dfrac{z^{-1/4} x^{1/2}}{z^{1/2} x^{-1/4}}$

101. $\sqrt[4]{k^{24}}$

102. $\sqrt[3]{54z^9 t^8}$

103. $-5\sqrt{18} + 12\sqrt{72}$

104. $8\sqrt[3]{x^3 y^2} - 2x\sqrt[3]{y^2}$

105. $(\sqrt{5} - \sqrt{3})(\sqrt{7} + \sqrt{3})$

106. $\dfrac{-1}{\sqrt{12}}$

107. $\sqrt[3]{\dfrac{12}{25}}$

108. $\dfrac{2\sqrt{z}}{\sqrt{z} - 2}$

109. $\sqrt{-49}$

110. $(4 - 9i) + (-1 + 2i)$

111. $\dfrac{\sqrt{50}}{\sqrt{-2}}$

Solve each equation.

112. $\sqrt{x + 4} = x - 2$

113. $\sqrt{6 + 2x} - 1 = \sqrt{7 - 2x}$

Solve each problem.

114. Carpenters stabilize wall frames with a diagonal brace as shown in the figure. The length of the brace is given by $L = \sqrt{H^2 + W^2}$. If the bottom of the brace is attached 9 ft from the corner and the brace is 12 ft long, how far up the corner post should it be nailed (to the nearest tenth of a foot)?

115. Find the perimeter of a triangular electronic highway road sign having the dimensions shown in the figure.

All
Traffic
Must Exit
Highway 59

$\sqrt{108}$ ft $\qquad$ $2\sqrt{27}$ ft

$\sqrt{50}$ ft

Chapter 9

TEST

Find each root. Use a calculator as necessary.

1. $-\sqrt{841}$

2. $125^{1/3}$

3. For $\sqrt{146.25}$, which choice gives the best estimate?

    A. 10   **B.** 11   **C.** 12   **D.** 13

4. Give a calculator approximation of $\sqrt{146.25}$ to the nearest hundredth.

5. Graph the function defined by $f(x) = \sqrt{x + 6}$, and give the domain and the range.

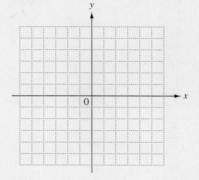

Simplify each expression. Assume that all variables represent positive real numbers.

6. $(-64)^{-4/3}$

7. $\dfrac{3^{2/5}x^{-1/4}y^{2/5}}{3^{-8/5}x^{7/4}y^{1/10}}$

8. $\sqrt{54x^5y^6}$

9. $\sqrt[4]{32a^7b^{13}}$

10. $\sqrt{2} \cdot \sqrt[3]{5}$

11. $3\sqrt{20} - 5\sqrt{80} + 4\sqrt{500}$

12. $(7\sqrt{5} + 4)(2\sqrt{5} - 1)$

1. _____

2. _____

3. _____

4. _____

5. _____

6. _____

7. _____

8. _____

9. _____

10. _____

11. _____

12. _____

13. _____

13. $\dfrac{-4}{\sqrt{7} + \sqrt{5}}$

14. _____

14. $\dfrac{-5}{\sqrt{40}}$

15. _____

15. $\dfrac{2}{\sqrt[3]{5}}$

16. _____

16. Find the distance between the points $(-3, 8)$ and $(2, 7)$.

17. _____

17. Use the Pythagorean formula to find the exact length of side b in the figure.

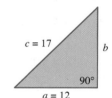

Solve each equation.

18. _____

18. $\sqrt[3]{5x} = \sqrt[3]{2x - 3}$

19. _____

19. $\sqrt{7 - x} + 5 = x$

20. _____

20. $\sqrt{x + 4} - \sqrt{1 - x} = -1$

Perform the indicated operations. Express answers in the form $a + bi$.

21. _____

21. $(-2 + 5i) - (3 + 6i) - 7i$ 22. $(-4 + 2i)(3 - i)$

22. _____

23. _____

23. $\dfrac{7 + i}{1 - i}$ 24. Simplify i^{35}.

24. _____

Cumulative Review Exercises

CHAPTERS 1–9

Solve each equation.

1. $7 - (4 + 3t) + 2t = -6(t - 2) - 5$

2. $|6x - 9| = |-4x + 2|$

Solve each inequality.

3. $-5 - 3(x - 2) < 11 - 2(x + 2)$ **4.** $1 + 4x > 5$ and $-2x > -6$ **5.** $-2 < 1 - 3x < 7$

6. Write an equation of the line through the points $(-4, 6)$ and $(7, -6)$.

7. Choose the correct response: The lines with equations $2x + 3y = 8$ and $6y = 4x + 16$ are

  A. parallel,   **B.** perpendicular,   **C.** neither.

8. For the graph of $f(x) = -3x + 6$,

  (a) what is the y-intercept?

  (b) what is the x-intercept?

9. For some items, the cost per item to manufacture it varies inversely as the number made. Widgets are this type of item. It costs $200 each to manufacture 1500 widgets. How much will it cost per widget to make 2500 widgets?

10. Graph the inequality $-2x + y < -6$.

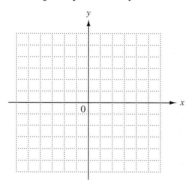

11. Find the measures of the marked angles.

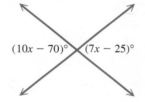

$(10x - 70)°$ $(7x - 25)°$

Solve each system.

12. $3x - y = 23$
$2x + 3y = 8$

13. $5x + 2y = 7$
$10x + 4y = 12$

14. $2x + y - z = 5$
$3x + 2y + z = 8$
$4x + 2y - 2z = 10$

15. In 2004, if you had sent five 2-oz letters and three 3-oz letters using first-class mail, it would have cost $5.49. Sending three 2-oz letters and five 3-oz letters would have cost $5.95. What was the 2004 postage rate for one 2-oz letter and for one 3-oz letter? (*Source:* U.S. Postal Service.)

Perform the indicated operations.

16. $(3k^3 - 5k^2 + 8k - 2) - (4k^3 + 11k + 7) + (2k^2 - 5k)$

17. $(8x - 7)(x + 3)$

18. $\dfrac{8z^3 - 16z^2 + 24z}{8z^2}$

19. $\dfrac{6y^4 - 3y^3 + 5y^2 + 6y - 9}{2y + 1}$

Factor each polynomial completely.

20. $2p^2 - 5pq + 3q^2$

21. $18k^4 + 9k^2 - 20$

22. $x^3 + 512$

Perform each operation and express answers in lowest terms.

23. $\dfrac{y^2 + y - 12}{y^3 + 9y^2 + 20y} \div \dfrac{y^2 - 9}{y^3 + 3y^2}$

24. $\dfrac{1}{x + y} + \dfrac{3}{x - y}$

Simplify each complex fraction.

25. $\dfrac{\dfrac{-6}{x - 2}}{\dfrac{8}{3x - 6}}$

26. $\dfrac{\dfrac{1}{a} - \dfrac{1}{b}}{\dfrac{a}{b} - \dfrac{b}{a}}$

Solve by factoring.

27. $2x^2 + 11x + 15 = 0$

28. $5t(t - 1) = 2(1 - t)$

Simplify.

29. $27^{-5/3}$

30. $\dfrac{x^{-2/3}}{x^{-3/4}}, \; x \neq 0$

31. $8\sqrt{20} + 3\sqrt{80} - 2\sqrt{500}$

32. $\dfrac{-9}{\sqrt{80}}$

33. $\dfrac{4}{\sqrt{6} - \sqrt{5}}$

34. $\dfrac{12}{\sqrt[3]{2}}$

35. Find the distance between the points $(-4, 4)$ and $(-2, 9)$.

36. Solve $\sqrt{8x - 4} - \sqrt{7x + 2} = 0$.

Solve each problem.

37. The current of a river runs at 3 mph. Brent's boat can go 36 mi downstream in the same time that it takes to go 24 mi upstream. Find the speed of the boat in still water.

38. How many liters of pure alcohol must be mixed with 40 L of 18% alcohol to obtain a 22% alcohol solution?

39. A jar containing only dimes and quarters has 29 coins with a face value of $4.70. How many of each denomination are there?

40. Brenda rides her bike 4 mph faster than her husband, Chuck. If Brenda can ride 48 mi in the same time that Chuck can ride 24 mi, what are their speeds?

Quadratic Equations, Inequalities, and Functions

10

10.1 **The Square Root Property and Completing the Square**

10.2 **The Quadratic Formula**

10.3 **Equations Quadratic in Form**

Summary Exercises on Solving Quadratic Equations

10.4 **Formulas and Further Applications**

10.5 **Graphs of Quadratic Functions**

10.6 **More about Parabolas; Applications**

10.7 **Quadratic and Rational Inequalities**

Since 1980, the number of multiple births in the United States has increased 59%, primarily due to greater use of fertility drugs and greater numbers of births to women over age 40. The number of higher-order multiple births—that is, births involving triplets or more—has increased over 400%. One of the most publicized higher-order multiple births occurred November 19, 1997, with the birth of the McCaughey septuplets in Des Moines, Iowa. All seven premature babies survived, a first in medical history. (*Source:* American College of Obstetricians and Gynecologists; *The Gazette,* November 19, 2003.)

In Example 6 of Section 10.5, we determine a *quadratic function* that models the number of higher-order multiple births in the United States.

10.1 The Square Root Property and Completing the Square

OBJECTIVES

1. Learn the square root property.
2. Solve quadratic equations of the form $(ax + b)^2 = c$ by using the square root property.
3. Solve quadratic equations by completing the square.
4. Solve quadratic equations with nonreal complex solutions.

1. (a) Which of the following are quadratic equations?

   A. $x + 2y = 0$
   B. $x^2 - 8x + 16 = 0$
   C. $2t^2 - 5t = 3$
   D. $x^3 + x^2 + 4 = 0$

(b) Which quadratic equation identified in part (a) is in standard form?

2. Solve each equation by factoring.

(a) $x^2 + 3x + 2 = 0$

(b) $3m^2 = 3 - 8m$

   (*Hint:* Remember to write the equation in standard form first.)

We introduced quadratic equations in **Section 7.4.** Recall that a *quadratic equation* is defined as follows.

> **Quadratic Equation**
>
> An equation that can be written in the form
> $$ax^2 + bx + c = 0,$$
> where a, b, and c are real numbers, with $a \neq 0$, is a **quadratic equation.** The given form is called **standard form.**

A quadratic equation is a *second-degree equation,* that is, an equation with a squared term and no terms of higher degree. For example,

$$4m^2 + 4m - 5 = 0 \quad \text{and} \quad 3x^2 = 4x - 8 \qquad \text{Quadratic equations}$$

are quadratic equations, with the first equation in standard form.

◄◄◄ Work Problem 1 at the Side.

In **Section 7.4** we used factoring and the zero-factor property to solve quadratic equations.

> **Zero-Factor Property**
>
> If two numbers have a product of 0, then at least one of the numbers must be 0. That is, if $ab = 0$, then $a = 0$ or $b = 0$.

We solved a quadratic equation such as $3x^2 - 5x - 28 = 0$ using the zero-factor property as follows.

$$3x^2 - 5x - 28 = 0$$
$$(3x + 7)(x - 4) = 0 \qquad \text{Factor.}$$
$$3x + 7 = 0 \quad \text{or} \quad x - 4 = 0 \qquad \text{Zero-factor property}$$
$$3x = -7 \quad \text{or} \qquad x = 4 \qquad \text{Solve each equation.}$$
$$x = -\frac{7}{3}$$

The solution set is $\left\{-\frac{7}{3}, 4\right\}$.

◄◄◄ Work Problem 2 at the Side.

OBJECTIVE 1 Learn the square root property. Although factoring is the simplest way to solve quadratic equations, not every quadratic equation can be solved easily by factoring. In this section and the next, we develop other methods of solving quadratic equations based on the following property.

> **Square Root Property**
>
> If x and k are complex numbers and $x^2 = k$, then
> $$x = \sqrt{k} \quad \text{or} \quad x = -\sqrt{k}.$$

ANSWERS
1. (a) B, C   (b) B
2. (a) $\{-2, -1\}$ (b) $\left\{-3, \frac{1}{3}\right\}$

The following steps justify the square root property.

$$x^2 = k$$
$$x^2 - k = 0 \qquad \text{Subtract } k.$$
$$(x - \sqrt{k})(x + \sqrt{k}) = 0 \qquad \text{Factor.}$$
$$x - \sqrt{k} = 0 \quad \text{or} \quad x + \sqrt{k} = 0 \qquad \text{Zero-factor property}$$
$$\boldsymbol{x = \sqrt{k}} \quad \text{or} \quad \boldsymbol{x = -\sqrt{k}} \qquad \text{Solve each equation.}$$

Thus, the solutions of the equation $x^2 = k$ are $x = \sqrt{k}$ or $x = -\sqrt{k}$.

3 Solve each equation.

(a) $m^2 = 64$

> **CAUTION**
> If $k \neq 0$, then using the square root property always produces *two* square roots, one positive and one negative.

EXAMPLE 1 Using the Square Root Property

Solve each equation.

(a) $r^2 = 5$

By the square root property, if $r^2 = 5$, then

$$r = \sqrt{5} \quad \text{or} \quad r = -\sqrt{5},$$

and the solution set is $\{\sqrt{5}, -\sqrt{5}\}$.

(b) $p^2 = 7$

(b) $4x^2 - 48 = 0$

Solve for x^2.

$$4x^2 - 48 = 0$$
$$4x^2 = 48 \qquad \text{Add 48.}$$
$$x^2 = 12 \qquad \text{Divide by 4.}$$
$$x = \sqrt{12} \quad \text{or} \quad x = -\sqrt{12} \qquad \text{Square root property}$$
$$x = 2\sqrt{3} \quad \text{or} \quad x = -2\sqrt{3} \qquad \sqrt{12} = \sqrt{4} \cdot \sqrt{3} = 2\sqrt{3}$$

Check: $\qquad 4x^2 - 48 = 0 \qquad$ Original equation

$4(2\sqrt{3})^2 - 48 = 0$?	$4(-2\sqrt{3})^2 - 48 = 0$?
$4(12) - 48 = 0$?	$4(12) - 48 = 0$?
$48 - 48 = 0$?	$48 - 48 = 0$?
$0 = 0 \qquad$ True	$0 = 0 \qquad$ True

(c) $3x^2 - 54 = 0$

The solution set is $\{2\sqrt{3}, -2\sqrt{3}\}$.

Work Problem 3 at the Side. ▶▶▶

> **NOTE**
> Recall that solutions such as those in Example 1 are sometimes abbreviated with the symbol $\pm$ (read "positive or negative"); with this symbol the solutions in Example 1 would be written $\pm\sqrt{5}$ and $\pm2\sqrt{3}$.

ANSWERS
3. (a) $\{8, -8\}$ (b) $\{\sqrt{7}, -\sqrt{7}\}$
   (c) $\{3\sqrt{2}, -3\sqrt{2}\}$

4 Solve the problem.

An expert marksman can hold a silver dollar at forehead level, drop it, draw his gun, and shoot the coin as it passes waist level. If the coin falls about 4 ft, use the formula in Example 2 to find the time that elapses between the dropping of the coin and the shot.

EXAMPLE 2 Using the Square Root Property in an Application

Galileo Galilei (1564–1642) developed a formula for freely falling objects described by

$$d = 16t^2,$$

where d is the distance in feet that an object falls (disregarding air resistance) in t seconds, regardless of weight. Galileo dropped objects from the Leaning Tower of Pisa to develop this formula. If the Leaning Tower is about 180 ft tall, use Galileo's formula to determine how long it would take an object dropped from the tower to fall to the ground. (*Source: Microsoft Encarta Encyclopedia 2002.*)

We substitute 180 for d in Galileo's formula.

$$d = 16t^2$$
$$\mathbf{180} = 16t^2 \qquad \text{Let } d = 180.$$
$$11.25 = t^2 \qquad \text{Divide by 16.}$$
$$t = \sqrt{11.25} \quad \text{or} \quad t = -\sqrt{11.25} \qquad \text{Square root property}$$

Since time cannot be negative, we discard the negative solution. In applied problems, we usually prefer approximations to exact values. Using a calculator, $\sqrt{11.25} \approx 3.4$ so $t \approx 3.4$. The object would fall to the ground in about 3.4 sec.

Work Problem 4 at the Side.

OBJECTIVE 2 Solve quadratic equations of the form $(ax + b)^2 = c$ by using the square root property. To solve more complicated equations using the square root property, such as

$$(x - 5)^2 = 36,$$

substitute $(x - 5)^2$ for x^2 and 36 for k, to get

$$x - 5 = \sqrt{36} \quad \text{or} \quad x - 5 = -\sqrt{36}$$
$$x - 5 = 6 \quad \text{or} \quad x - 5 = -6$$
$$x = 11 \quad \text{or} \quad x = -1.$$

Check: $(x - 5)^2 = 36$ Original equation

$$(11 - 5)^2 = 36 \quad ? \qquad\qquad (-1 - 5)^2 = 36 \quad ?$$
$$6^2 = 36 \quad ? \qquad\qquad (-6)^2 = 36 \quad ?$$
$$36 = 36 \qquad \text{True} \qquad\qquad 36 = 36 \qquad \text{True}$$

EXAMPLE 3 Using the Square Root Property

Solve $(2x - 3)^2 = 18$.

$$2x - 3 = \sqrt{18} \qquad \text{or} \quad 2x - 3 = -\sqrt{18} \qquad \text{Square root property}$$
$$2x = 3 + \sqrt{18} \quad \text{or} \qquad 2x = 3 - \sqrt{18} \qquad \text{Add 3.}$$
$$x = \frac{3 + \sqrt{18}}{2} \quad \text{or} \qquad x = \frac{3 - \sqrt{18}}{2} \qquad \text{Divide by 2.}$$
$$x = \frac{3 + 3\sqrt{2}}{2} \quad \text{or} \qquad x = \frac{3 - 3\sqrt{2}}{2} \qquad \sqrt{18} = \sqrt{9} \cdot \sqrt{2} = 3\sqrt{2}$$

Continued on Next Page

ANSWERS
4. .5 sec

We show the check for the first solution. The check for the second solution is similar.

Check:
$$(2x - 3)^2 = 18 \qquad \text{Original equation}$$

$$\left[2\left(\frac{3 + 3\sqrt{2}}{2}\right) - 3\right]^2 = 18 \quad ?$$

$$(3 + 3\sqrt{2} - 3)^2 = 18 \quad ?$$

$$(3\sqrt{2})^2 = 18 \quad ?$$

$$18 = 18 \qquad \text{True}$$

The solution set is $\left\{ \dfrac{3 + 3\sqrt{2}}{2}, \dfrac{3 - 3\sqrt{2}}{2} \right\}$.

Work Problem 5 at the Side. ▶▶▶

OBJECTIVE 3 Solve quadratic equations by completing the square.
We can use the square root property to solve *any* quadratic equation by writing it in the form $(x + k)^2 = n$. That is, we must write the left side of the equation as a perfect square trinomial that can be factored as $(x + k)^2$, the square of a binomial, and the right side must be a constant. Rewriting a quadratic equation in this form is called **completing the square.**

Recall that the perfect square trinomial

$$x^2 + 10x + 25$$

can be factored as $(x + 5)^2$. In the trinomial, the coefficient of x (the first-degree term) is 10 and the constant term is 25. Notice that if we take half of 10 and square it, we get the constant term, 25.

$$\underset{\text{Coefficient of } x}{\left[\frac{1}{2}(10)\right]^2} = 5^2 = \underset{\text{Constant}}{25}$$

Similarly, in

$$x^2 + 12x + 36, \qquad \left[\frac{1}{2}(12)\right]^2 = 6^2 = 36,$$

and in

$$m^2 - 6m + 9, \qquad \left[\frac{1}{2}(-6)\right]^2 = (-3)^2 = 9.$$

This relationship is true in general and is the idea behind completing the square.

Work Problem 6 at the Side. ▶▶▶

EXAMPLE 4 Solving a Quadratic Equation by Completing the Square

Solve $x^2 + 8x + 10 = 0$.

This quadratic equation cannot be solved easily by factoring, and it is not in the correct form to solve using the square root property. To solve it by completing the square, we need a perfect square trinomial on the left side of the equation. To get this form, we first subtract 10 from each side.

Continued on Next Page

---

5 Solve each equation.

(a) $(x - 3)^2 = 25$

(b) $(3k + 1)^2 = 2$

(c) $(2r + 3)^2 = 8$

6 Find the constant to be added to get a perfect square trinomial. In each case, take half the coefficient of the first-degree term and square the result.

(a) $x^2 + 4x +$ ____

(b) $t^2 - 2t +$ ____

(c) $m^2 + 5m +$ ____

(d) $x^2 - \dfrac{2}{3}x +$ ____

ANSWERS
5. (a) $\{-2, 8\}$

(b) $\left\{ \dfrac{-1 + \sqrt{2}}{3}, \dfrac{-1 - \sqrt{2}}{3} \right\}$

(c) $\left\{ \dfrac{-3 + 2\sqrt{2}}{2}, \dfrac{-3 - 2\sqrt{2}}{2} \right\}$

6. (a) 4 (b) 1 (c) $\dfrac{25}{4}$ (d) $\dfrac{1}{9}$

7 Solve $n^2 + 6n + 4 = 0$ by completing the square.

$$x^2 + 8x + 10 = 0 \qquad \text{Original equation}$$
$$x^2 + 8x = -10 \qquad \text{Subtract 10.}$$

We must add a constant to get a perfect square trinomial on the left.

$$\underbrace{x^2 + 8x + \underline{\quad}}_{\substack{\text{Needs to be a perfect} \\ \text{square trinomial}}}$$

To find this constant, we apply the ideas preceding this example—we take half the coefficient of the first-degree term and square the result.

$$\left[\frac{1}{2}(8)\right]^2 = 4^2 = 16 \leftarrow \text{Desired constant}$$

Now we add 16 to *each* side of the equation. (Why?)

$$x^2 + 8x + 16 = -10 + 16$$

Next we factor on the left side and add on the right.

$$(x + 4)^2 = 6$$

We can now use the square root property.

$$x + 4 = \sqrt{6} \qquad \text{or} \qquad x + 4 = -\sqrt{6}$$
$$x = -4 + \sqrt{6} \qquad \text{or} \qquad x = -4 - \sqrt{6}$$

Check:
$$x^2 + 8x + 10 = 0 \qquad \text{Original equation}$$
$$(-4 + \sqrt{6})^2 + 8(-4 + \sqrt{6}) + 10 = 0 \qquad ? \quad \text{Let } x = -4 + \sqrt{6}.$$
$$16 - 8\sqrt{6} + 6 - 32 + 8\sqrt{6} + 10 = 0 \qquad ?$$
$$0 = 0 \qquad \text{True}$$

The check of the other solution is similar. Thus, $\{-4 + \sqrt{6}, -4 - \sqrt{6}\}$ is the solution set.

Work Problem 7 at the Side.

The procedure from Example 4 can be generalized.

Completing the Square

To solve $ax^2 + bx + c = 0$ $(a \neq 0)$ by completing the square, use these steps.

Step 1   **Be sure the squared term has coefficient 1.** If the coefficient of the squared term is some other nonzero number a, divide each side of the equation by a.

Step 2   **Write the equation in correct form** so that terms with variables are on one side of the equals sign and the constant is on the other side.

Step 3   **Square half the coefficient of the first-degree term.**

Step 4   **Add the square to each side.**

Step 5   **Factor the perfect square trinomial.** One side should now be a perfect square trinomial. Factor it as the square of a binomial. Simplify the other side.

Step 6   **Solve the equation.** Apply the square root property to complete the solution.

ANSWERS
7. $\{-3 + \sqrt{5}, -3 - \sqrt{5}\}$

EXAMPLE 5 **Solving a Quadratic Equation with $a = 1$ by Completing the Square**

Solve $k^2 + 5k - 1 = 0$.

Since the coefficient of the squared term is 1, begin with Step 2.

Step 2 $k^2 + 5k = 1$ Add 1 to each side.

Step 3 Take half the coefficient of the first-degree term and square the result.

$$\left[\frac{1}{2}(5)\right]^2 = \left(\frac{5}{2}\right)^2 = \frac{25}{4}$$

Step 4 $k^2 + 5k + \frac{25}{4} = 1 + \frac{25}{4}$ Add the square to each side of the equation.

Step 5 $\left(k + \frac{5}{2}\right)^2 = \frac{29}{4}$ Factor on the left; add on the right.

Step 6 $k + \frac{5}{2} = \sqrt{\frac{29}{4}}$ or $k + \frac{5}{2} = -\sqrt{\frac{29}{4}}$ Square root property

$k + \frac{5}{2} = \frac{\sqrt{29}}{2}$ or $k + \frac{5}{2} = -\frac{\sqrt{29}}{2}$

$k = -\frac{5}{2} + \frac{\sqrt{29}}{2}$ or $k = -\frac{5}{2} - \frac{\sqrt{29}}{2}$

$k = \frac{-5 + \sqrt{29}}{2}$ or $k = \frac{-5 - \sqrt{29}}{2}$

Check that the solution set is $\left\{\frac{-5 + \sqrt{29}}{2}, \frac{-5 - \sqrt{29}}{2}\right\}$.

> **Work Problem 8 at the Side.** ▶▶▶

EXAMPLE 6 **Solving a Quadratic Equation with $a \neq 1$ by Completing the Square**

Solve $2x^2 - 4x - 5 = 0$.

First divide each side of the equation by 2 to get 1 as the coefficient of the squared term.

$$x^2 - 2x - \frac{5}{2} = 0 \qquad\qquad \text{Step 1}$$

$$x^2 - 2x = \frac{5}{2} \qquad\qquad \text{Step 2}$$

$$\left[\frac{1}{2}(-2)\right]^2 = (-1)^2 = 1 \qquad\qquad \text{Step 3}$$

$$x^2 - 2x + 1 = \frac{5}{2} + 1 \qquad\qquad \text{Step 4}$$

$$(x - 1)^2 = \frac{7}{2} \qquad\qquad \text{Step 5}$$

$$x - 1 = \sqrt{\frac{7}{2}} \quad \text{or} \quad x - 1 = -\sqrt{\frac{7}{2}} \qquad \text{Step 6}$$

Continued on Next Page

8 Solve each equation by completing the square.

(a) $x^2 + 2x - 10 = 0$

(b) $r^2 + 3r - 1 = 0$

ANSWERS

8. (a) $\{-1 + \sqrt{11}, -1 - \sqrt{11}\}$

(b) $\left\{\frac{-3 + \sqrt{13}}{2}, \frac{-3 - \sqrt{13}}{2}\right\}$

9 Solve each equation by completing the square.

(a) $2r^2 - 4r + 1 = 0$

(b) $3z^2 - 6z - 2 = 0$

(c) $8x^2 - 4x - 2 = 0$

10 Solve each equation.

(a) $x^2 = -17$

(b) $(k + 5)^2 = -100$

(c) $5t^2 - 15t + 12 = 0$

ANSWERS

9. (a) $\left\{ \dfrac{2 + \sqrt{2}}{2}, \dfrac{2 - \sqrt{2}}{2} \right\}$

(b) $\left\{ \dfrac{3 + \sqrt{15}}{3}, \dfrac{3 - \sqrt{15}}{3} \right\}$

(c) $\left\{ \dfrac{1 + \sqrt{5}}{4}, \dfrac{1 - \sqrt{5}}{4} \right\}$

10. (a) $\{ i\sqrt{17}, -i\sqrt{17} \}$

(b) $\{ -5 + 10i, -5 - 10i \}$

(c) $\left\{ \dfrac{15 + i\sqrt{15}}{10}, \dfrac{15 - i\sqrt{15}}{10} \right\}$

$$x = 1 + \sqrt{\frac{7}{2}} \quad \text{or} \quad x = 1 - \sqrt{\frac{7}{2}} \qquad \text{Add 1.}$$

$$x = 1 + \frac{\sqrt{14}}{2} \quad \text{or} \quad x = 1 - \frac{\sqrt{14}}{2} \qquad \text{Rationalize denominators.}$$

Since $1 = \frac{2}{2}$, add the two terms in each solution as follows:

$$1 + \frac{\sqrt{14}}{2} = \frac{2}{2} + \frac{\sqrt{14}}{2} = \frac{2 + \sqrt{14}}{2}$$

$$1 - \frac{\sqrt{14}}{2} = \frac{2}{2} - \frac{\sqrt{14}}{2} = \frac{2 - \sqrt{14}}{2}.$$

Check that the solution set is $\left\{ \dfrac{2 + \sqrt{14}}{2}, \dfrac{2 - \sqrt{14}}{2} \right\}$.

◀◀◀ Work Problem 9 at the Side.

OBJECTIVE 4 **Solve quadratic equations with nonreal complex solutions.** In the equation $x^2 = k$, if $k < 0$, there will be two nonreal complex solutions.

EXAMPLE 7 **Solving Quadratic Equations with Nonreal Complex Solutions**

Solve each equation.

(a) $x^2 = -15$

$$x = \sqrt{-15} \quad \text{or} \quad x = -\sqrt{-15} \qquad \text{Square root property}$$
$$x = i\sqrt{15} \quad \text{or} \quad x = -i\sqrt{15} \qquad \sqrt{-1} = i$$

The solution set is $\{ i\sqrt{15}, -i\sqrt{15} \}$.

(b) $(t + 2)^2 = -16$

$$t + 2 = \sqrt{-16} \quad \text{or} \quad t + 2 = -\sqrt{-16} \qquad \text{Square root property}$$
$$t + 2 = 4i \quad \text{or} \quad t + 2 = -4i \qquad \sqrt{-16} = 4i$$
$$t = -2 + 4i \quad \text{or} \quad t = -2 - 4i$$

The solution set is $\{ -2 + 4i, -2 - 4i \}$.

(c) $x^2 + 2x + 7 = 0$

$$x^2 + 2x = -7 \qquad \text{Subtract 7.}$$
$$x^2 + 2x + 1 = -7 + 1 \qquad [\tfrac{1}{2}(2)]^2 = 1; \text{ add 1 to each side.}$$
$$(x + 1)^2 = -6 \qquad \text{Factor on the left; add on the right.}$$
$$x + 1 = \pm\, i\sqrt{6} \qquad \text{Square root property}$$
$$x = -1 \pm i\sqrt{6} \qquad \text{Subtract 1.}$$

The solution set is $\{ -1 + i\sqrt{6}, -1 - i\sqrt{6} \}$.

◀◀◀ Work Problem 10 at the Side.

NOTE
We will use completing the square in **Section 10.6** when we graph quadratic equations and in **Section 12.2** when we work with circles.

10.1 Exercises

FOR EXTRA HELP Addison-Wesley Math Tutor Center 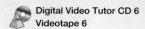 MathXL Digital Video Tutor CD 6 Videotape 6 Student's Solutions Manual MyMathLab Interactmath.com

1. A student was asked to solve the quadratic equation $x^2 = 16$ and did not get full credit for the solution set $\{4\}$. Why?

2. Why can't the zero-factor property be used to solve every quadratic equation?

3. Give a one-sentence description or explanation of each of the following.

   (a) Quadratic equation in standard form

4. What is wrong with the following "solution"?
$$x^2 - x - 2 = 5$$
$$(x - 2)(x + 1) = 5$$
$$x - 2 = 5 \quad \text{or} \quad x + 1 = 5 \quad \text{Zero-factor property}$$
$$x = 7 \quad \text{or} \quad \quad x = 4$$

   (b) Zero-factor property

   (c) Square root property

Use the square root property to solve each equation. See Examples 1 and 3.

5. $x^2 = 81$

6. $z^2 = 225$

7. $t^2 = 17$

8. $k^2 = 19$

9. $m^2 = 32$

10. $x^2 = 54$

11. $t^2 - 20 = 0$

12. $p^2 - 50 = 0$

13. $3n^2 - 72 = 0$

14. $5z^2 - 200 = 0$

15. $(x + 2)^2 = 25$

16. $(t + 8)^2 = 9$

17. $(x - 4)^2 = 3$

18. $(x + 3)^2 = 11$

19. $(t + 5)^2 = 48$

20. $(m - 6)^2 = 27$

21. $(3k - 1)^2 = 7$

22. $(2x + 4)^2 = 10$

23. $(4p + 1)^2 = 24$

24. $(5k - 2)^2 = 12$

Solve Exercises 25 and 26 using Galileo's formula, $d = 16t^2$. Round answers to the nearest tenth. See Example 2.

25. Mount Rushmore National Memorial in South Dakota features a sculpture of four of America's favorite presidents carved into the rim of the mountain, 500 ft above the valley floor. How long would it take a rock dropped from the top of the sculpture to fall to the ground? (*Source: Microsoft Encarta Encyclopedia 2002.*)

26. The Gateway Arch in St. Louis, Missouri, is 630 ft tall. How long would it take an object dropped from the top of it to fall to the ground? (*Source: Home & Away*, November/December 2000.)

27. Of the two equations

$$(2x + 1)^2 = 5 \quad \text{and} \quad x^2 + 4x = 12,$$

one is more suitable for solving by the square root property, and the other is more suitable for solving by completing the square. Which method do you think most students would use for each equation?

28. Why would most students find the equation $x^2 + 4x = 20$ easier to solve by completing the square than the equation $5x^2 + 2x = 3$?

29. Decide what number must be added to make each expression a perfect square trinomial.

(a) $x^2 + 6x + $ _____

(b) $x^2 + 14x + $ _____

(c) $p^2 - 12p + $ _____

(d) $x^2 + 3x + $ _____

(e) $q^2 - 9q + $ _____

(f) $t^2 - \frac{1}{2}t + $ _____

30. What would be the first step in solving $2x^2 + 8x = 9$ by completing the square?

Determine the number that will complete the square to solve each equation after the constant term has been written on the right side. Do not actually solve. See Examples 4–6.

31. $x^2 + 4x - 2 = 0$

32. $t^2 + 2t - 1 = 0$

33. $x^2 + 10x + 18 = 0$

34. $x^2 + 8x + 11 = 0$

35. $3w^2 - w - 24 = 0$

36. $4z^2 - z - 39 = 0$

Solve each equation by completing the square. Use the results of Exercises 31–36 to solve Exercises 39–44. See Examples 4–6.

37. $x^2 - 2x - 24 = 0$

38. $m^2 - 4m - 32 = 0$

39. $x^2 + 4x - 2 = 0$

40. $t^2 + 2t - 1 = 0$

41. $x^2 + 10x + 18 = 0$

42. $x^2 + 8x + 11 = 0$

43. $3w^2 - w = 24$

44. $4z^2 - z = 39$

45. $2k^2 + 5k - 2 = 0$

46. $3r^2 + 2r - 2 = 0$

47. $5x^2 - 10x + 2 = 0$

48. $2x^2 - 16x + 25 = 0$

49. $9x^2 - 24x = -13$

50. $25n^2 - 20n = 1$

51. $z^2 - \dfrac{4}{3}z = -\dfrac{1}{9}$

52. $p^2 - \dfrac{8}{3}p = -1$

53. $.1x^2 - .2x - .1 = 0$
(*Hint:* First clear the decimals.)

54. $.1p^2 - .4p + .1 = 0$
(*Hint:* First clear the decimals.)

Find the all complex solutions of each equation. See Example 7.

55. $x^2 = -12$

56. $x^2 = -18$

57. $(r - 5)^2 = -3$

58. $(t + 6)^2 = -5$

59. $(6k - 1)^2 = -8$

60. $(4m - 7)^2 = -27$

61. $m^2 + 4m + 13 = 0$

62. $t^2 + 6t + 10 = 0$

63. $3r^2 + 4r + 4 = 0$

64. $4x^2 + 5x + 5 = 0$

65. $-m^2 - 6m - 12 = 0$

66. $-k^2 - 5k - 10 = 0$

RELATING CONCEPTS (EXERCISES 67–72)   For Individual or Group Work

The Greeks had a method of completing the square geometrically in which they literally changed a figure into a square. For example, to complete the square for $x^2 + 6x$, we begin with a square of side x, as in the figure. We add three rectangles of width 1 to the right side and the bottom to get a region with area $x^2 + 6x$. To fill in the corner (complete the square), we must add 9 1-by-1 squares as shown.

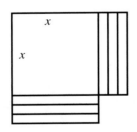 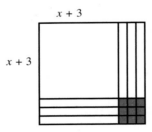

Work Exercises 67–72 in order.

67. What is the area of the original square?

68. What is the area of each strip?

69. What is the total area of the six strips?

70. What is the area of each small square in the corner of the second figure?

71. What is the total area of the small squares?

72. What is the area of the new, larger square?

10.2 The Quadratic Formula

The examples in the previous section showed that any quadratic equation can be solved by completing the square; however, completing the square can be tedious and time consuming. In this section, we complete the square to solve the general quadratic equation

$$ax^2 + bx + c = 0,$$

where a, b, and c are complex numbers and $a \neq 0$. The solution of this general equation gives a formula for finding the solution of any specific quadratic equation.

OBJECTIVES

1. Derive the quadratic formula.
2. Solve quadratic equations using the quadratic formula.
3. Use the discriminant to determine the number and type of solutions.

OBJECTIVE 1 Derive the quadratic formula. To solve $ax^2 + bx + c = 0$ by completing the square (assuming $a > 0$), we follow the steps given in **Section 10.1.**

$$ax^2 + bx + c = 0$$

$$x^2 + \frac{b}{a}x + \frac{c}{a} = 0 \qquad \text{Divide by } a. \text{ (Step 1)}$$

$$x^2 + \frac{b}{a}x = -\frac{c}{a} \qquad \text{Subtract } \tfrac{c}{a}. \text{ (Step 2)}$$

$$\left[\frac{1}{2}\left(\frac{b}{a}\right)\right]^2 = \left(\frac{b}{2a}\right)^2 = \frac{b^2}{4a^2} \qquad \text{(Step 3)}$$

$$x^2 + \frac{b}{a}x + \frac{b^2}{4a^2} = -\frac{c}{a} + \frac{b^2}{4a^2} \qquad \text{Add } \frac{b^2}{4a^2} \text{ to each side. (Step 4)}$$

Write the left side as a perfect square, and rearrange the right side.

$$\left(x + \frac{b}{2a}\right)^2 = \frac{b^2}{4a^2} + \frac{-c}{a} \qquad \text{(Step 5)}$$

$$\left(x + \frac{b}{2a}\right)^2 = \frac{b^2}{4a^2} + \frac{-4ac}{4a^2} \qquad \text{Write with a common denominator.}$$

$$\left(x + \frac{b}{2a}\right)^2 = \frac{b^2 - 4ac}{4a^2} \qquad \text{Add fractions.}$$

$$x + \frac{b}{2a} = \sqrt{\frac{b^2 - 4ac}{4a^2}} \quad \text{or} \quad x + \frac{b}{2a} = -\sqrt{\frac{b^2 - 4ac}{4a^2}} \qquad \begin{array}{l}\text{Square root} \\ \text{property} \\ \text{(Step 6)}\end{array}$$

Since

$$\sqrt{\frac{b^2 - 4ac}{4a^2}} = \frac{\sqrt{b^2 - 4ac}}{\sqrt{4a^2}} = \frac{\sqrt{b^2 - 4ac}}{2a},$$

the right sides of these equations can be expressed as

$$x + \frac{b}{2a} = \frac{\sqrt{b^2 - 4ac}}{2a} \quad \text{or} \quad x + \frac{b}{2a} = \frac{-\sqrt{b^2 - 4ac}}{2a}$$

$$x = \frac{-b}{2a} + \frac{\sqrt{b^2 - 4ac}}{2a} \quad \text{or} \quad x = \frac{-b}{2a} - \frac{\sqrt{b^2 - 4ac}}{2a}$$

$$x = \frac{-b + \sqrt{b^2 - 4ac}}{2a} \quad \text{or} \quad x = \frac{-b - \sqrt{b^2 - 4ac}}{2a}.$$

If $a < 0$, the same two solutions are obtained. The result is the **quadratic formula,** which is abbreviated as shown on the next page.

1 Identify the values of a, b, and c. (*Hint:* If necessary, first write the equation in standard form with 0 on the right side.) *Do not actually solve.*

(a) $-3q^2 + 9q - 4 = 0$

Quadratic Formula

The solutions of $ax^2 + bx + c = 0$ ($a \neq 0$) are given by

$$x = \frac{-b \pm \sqrt{b^2 - 4ac}}{2a}.$$

CAUTION

In the quadratic formula, ***the square root is added to or subtracted from the value of $-b$ BEFORE dividing by $2a$.***

OBJECTIVE 2 Solve quadratic equations using the quadratic formula.
To use the quadratic formula, first write the given equation in standard form $ax^2 + bx + c = 0$; then identify the values of a, b, and c and substitute them into the quadratic formula, as shown in the next examples.

Work Problem 1 at the Side.

EXAMPLE 1 Using the Quadratic Formula (Rational Solutions)

Solve $6x^2 - 5x - 4 = 0$.

(b) $3x^2 = 6x + 2$

Here a, the coefficient of the second-degree term, is 6, while b, the coefficient of the first-degree term, is -5, and the constant c is -4. Substitute these values into the quadratic formula.

$$x = \frac{-b \pm \sqrt{b^2 - 4ac}}{2a} \qquad \text{Quadratic formula}$$

$$x = \frac{-(-5) \pm \sqrt{(-5)^2 - 4(6)(-4)}}{2(6)} \qquad a = 6, b = -5, c = -4$$

$$x = \frac{5 \pm \sqrt{25 + 96}}{12}$$

$$x = \frac{5 \pm \sqrt{121}}{12}$$

2 Solve $4x^2 - 11x - 3 = 0$ using the quadratic formula.

$$x = \frac{5 \pm 11}{12}$$

This last statement leads to two solutions, one from $+$ and one from $-$.

$$x = \frac{5 + 11}{12} = \frac{16}{12} = \frac{4}{3} \quad \text{or} \quad x = \frac{5 - 11}{12} = \frac{-6}{12} = -\frac{1}{2}$$

Check each solution in the original equation. The solution set is $\left\{-\frac{1}{2}, \frac{4}{3}\right\}$.

Work Problem 2 at the Side.

We could have used factoring to solve the equation in Example 1.

$$6x^2 - 5x - 4 = 0$$

$$(3x - 4)(2x + 1) = 0 \qquad \text{Factor.}$$

$$3x - 4 = 0 \quad \text{or} \quad 2x + 1 = 0 \qquad \text{Zero-factor property}$$

$$3x = 4 \quad \text{or} \qquad 2x = -1 \qquad \text{Solve each equation.}$$

ANSWERS
1. (a) $-3; 9; -4$ (b) $3; -6; -2$
2. $\left\{-\frac{1}{4}, 3\right\}$

$$x = \frac{4}{3} \quad \text{or} \qquad x = -\frac{1}{2} \qquad \text{Same solutions as in Example 1}$$

When solving quadratic equations, it is a good idea to try factoring first. If the equation cannot be factored or if factoring is difficult, then use the quadratic formula. Later in this section, we will show a way to determine whether factoring can be used to solve a quadratic equation.

EXAMPLE 2 Using the Quadratic Formula (Irrational Solutions)

Solve $4r^2 = 8r - 1$.

Write the equation in standard form as $4r^2 - 8r + 1 = 0$.

$$r = \frac{-b \pm \sqrt{b^2 - 4ac}}{2a}$$ Quadratic formula

$$r = \frac{-(-8) \pm \sqrt{(-8)^2 - 4(4)(1)}}{2(4)}$$ $a = 4, b = -8, c = 1$

$$= \frac{8 \pm \sqrt{64 - 16}}{8}$$

$$= \frac{8 \pm \sqrt{48}}{8}$$

$$= \frac{8 \pm 4\sqrt{3}}{8}$$ $\sqrt{48} = \sqrt{16} \cdot \sqrt{3} = 4\sqrt{3}$

$$= \frac{4(2 \pm \sqrt{3})}{4(2)}$$ Factor.

$$= \frac{2 \pm \sqrt{3}}{2}$$ Lowest terms

The solution set is $\left\{ \dfrac{2 + \sqrt{3}}{2}, \dfrac{2 - \sqrt{3}}{2} \right\}$.

CAUTION

1. *Every quadratic equation must be written in standard form $ax^2 + bx + c = 0$ before we begin to solve it,* whether we use factoring or the quadratic formula.
2. *When writing solutions in lowest terms, be sure to factor first; then divide out the common factor,* as shown in the last two steps in Example 2.

Work Problem 3 at the Side. ▶▶▶

EXAMPLE 3 Using the Quadratic Formula (Nonreal Complex Solutions)

Solve $(9q + 3)(q - 1) = -8$.

To write this equation in standard form, we first multiply and collect all nonzero terms on the left.

$$(9q + 3)(q - 1) = -8$$
$$9q^2 - 6q - 3 = -8$$
$$9q^2 - 6q + 5 = 0 \qquad \text{Standard form}$$

Continued on Next Page

3 Solve each equation using the quadratic formula.

(a) $6x^2 + 4x - 1 = 0$

(b) $2k^2 + 19 = 14k$

ANSWERS

3. (a) $\left\{ \dfrac{-2 + \sqrt{10}}{6}, \dfrac{-2 - \sqrt{10}}{6} \right\}$

(b) $\left\{ \dfrac{7 + \sqrt{11}}{2}, \dfrac{7 - \sqrt{11}}{2} \right\}$

4 Solve each equation using the quadratic formula.

(a) $x^2 + x + 1 = 0$

From the equation $9q^2 - 6q + 5 = 0$, we identify $a = 9$, $b = -6$, and $c = 5$.

$$q = \frac{-(-6) \pm \sqrt{(-6)^2 - 4(9)(5)}}{2(9)} \quad \text{Substitute in the quadratic formula.}$$

$$= \frac{6 \pm \sqrt{-144}}{18}$$

$$= \frac{6 \pm 12i}{18} \quad \sqrt{-144} = 12i$$

$$= \frac{6(1 \pm 2i)}{6(3)} \quad \text{Factor.}$$

$$= \frac{1 \pm 2i}{3} \quad \text{Lowest terms}$$

The solution set is $\left\{ \dfrac{1 + 2i}{3}, \dfrac{1 - 2i}{3} \right\}$.

NOTE

We could have written the solutions in Example 3 in the form $a + bi$, the standard form for complex numbers, as follows:

$$\frac{1 \pm 2i}{3} = \frac{1}{3} \pm \frac{2}{3}i. \quad \text{Standard form}$$

◄◄◄ Work Problem 4 at the Side.

(b) $(z + 2)(z - 6) = -17$

OBJECTIVE 3 Use the discriminant to determine the number and type of solutions. The solutions of the quadratic equation $ax^2 + bx + c = 0$ are given by

$$x = \frac{-b \pm \sqrt{b^2 - 4ac}}{2a}. \quad \leftarrow \text{Discriminant}$$

If a, b, and c are integers, the type of solutions of a quadratic equation—that is, rational, irrational, or nonreal complex—is determined by the expression under the radical sign, $b^2 - 4ac$. Because it distinguishes among the three types of solutions, $b^2 - 4ac$ is called the *discriminant*. By calculating the discriminant before solving a quadratic equation, we can predict whether the solutions will be rational numbers, irrational numbers, or nonreal complex numbers. (This can be useful in an applied problem, for example, where irrational or nonreal complex solutions are not acceptable.)

Discriminant

The **discriminant** of $ax^2 + bx + c = 0$ is $b^2 - 4ac$. If a, b, and c are integers, then the number and type of solutions are determined as follows.

Discriminant	Number and Type of Solutions
Positive, and the square of an integer	Two rational solutions
Positive, but not the square of an integer	Two irrational solutions
Zero	One rational solution
Negative	Two nonreal complex solutions

ANSWERS

4. **(a)** $\left\{ \dfrac{-1 + i\sqrt{3}}{2}, \dfrac{-1 - i\sqrt{3}}{2} \right\}$

  (b) $\{2 + i, 2 - i\}$

Calculating the discriminant can also help you decide whether to solve a quadratic equation by factoring or by using the quadratic formula. *If the discriminant is a perfect square (including 0), then the equation can be solved by factoring. Otherwise, the quadratic formula should be used.*

EXAMPLE 4 Using the Discriminant

Find the discriminant. Use it to predict the number and type of solutions for each equation. Tell whether the equation can be solved by factoring or whether the quadratic formula should be used.

(a) $6x^2 - x - 15 = 0$

We find the discriminant by evaluating $b^2 - 4ac$.

$$b^2 - 4ac = (-1)^2 - 4(6)(-15) \qquad a = 6, b = -1, c = -15$$
$$= 1 + 360$$
$$= 361$$

A calculator shows that $361 = 19^2$, a perfect square. Since a, b, and c are integers and the discriminant is a perfect square, there will be two rational solutions and the equation can be solved by factoring.

(b) $3m^2 - 4m = 5$

Write the equation in standard form as $3m^2 - 4m - 5 = 0$ to find $a = 3$, $b = -4$, and $c = -5$.

$$b^2 - 4ac = (-4)^2 - 4(3)(-5)$$
$$= 16 + 60$$
$$= 76$$

Because 76 is positive but not the square of an integer and a, b, and c are integers, the equation will have two irrational solutions and is best solved using the quadratic formula.

(c) $4x^2 + x + 1 = 0$

Since $a = 4$, $b = 1$, and $c = 1$, the discriminant is

$$1^2 - 4(4)(1) = -15.$$

Since the discriminant is negative and a, b, and c are integers, this quadratic equation will have two nonreal complex solutions. The quadratic formula should be used to solve it.

(d) $4t^2 + 9 = 12t$

Write the equation as $4t^2 - 12t + 9 = 0$ to find $a = 4$, $b = -12$, and $c = 9$. The discriminant is

$$b^2 - 4ac = (-12)^2 - 4(4)(9)$$
$$= 144 - 144$$
$$= 0.$$

Because the discriminant is 0, the quantity under the radical in the quadratic formula is 0, and there is only one rational solution. Again, the equation can be solved by factoring.

Work Problem 5 at the Side. ▶▶▶

5 Find the discriminant. Use it to predict the number and type of solutions for each equation.

(a) $2x^2 + 3x = 4$

(b) $2x^2 + 3x + 4 = 0$

(c) $x^2 + 20x + 100 = 0$

(d) $15k^2 + 11k = 14$

(e) Which of the equations in parts (a)–(d) can be solved by factoring?

ANSWERS
5. **(a)** 41; two; irrational
   (b) −23; two; nonreal complex
   (c) 0; one; rational
   (d) 961; two; rational   **(e)** (c) and (d)

Real-Data Applications

Almost, but Not Quite Right

Algebra is a precise language in which the correct order of operations must be followed and details must be watched. A slight change to a formula makes a big difference in the results.

- The Cadillac Bar in Houston, Texas, encourages patrons to write (tasteful) messages on their wall. Markers are provided, and the customers' creative juices flow. One person attempted to write the quadratic formula, which is shown in (1) below. Instead, that person wrote the formula shown in (2). It was not quite right.

$$(1)\ \textit{Quadratic Formula:}\quad \frac{-b \pm \sqrt{b^2 - 4ac}}{2a}$$

$$(2)\ \textit{Formula on Wall of Cadillac Bar:}\quad \frac{-b\sqrt{b^2 - 4ac}}{2a}$$

- An early version of Microsoft Word for Windows included the 1.0 edition of the Equation Editor. The documentation that explained how to use the Equation Editor used the following formula in the sample explanation. That author probably intended to use the quadratic formula, but again, it was not quite right.

$$(3)\ \textit{Equation Editor Sample Formula:}\quad -b \pm \frac{\sqrt{b^2 - 4ac}}{2a}$$

For Group Discussion

1. Evaluate the quadratic formula for $a = 2$, $b = 7$, and $c = -15$. Note that you have to evaluate the formula twice: first use the $+$ sign, then use the $-$ sign.

2. Using the order of operations, describe the steps to correctly evaluate the quadratic formula.

3. Explain how the incorrect Cadillac Bar formula used the order of operations differently than the correct quadratic formula.

4. Explain how the incorrect Microsoft Equation Editor formula used the order of operations differently than the correct quadratic formula.

10.2 Exercises

FOR EXTRA HELP

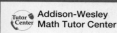 Addison-Wesley Math Tutor Center

Math XP MathXL

Digital Video Tutor CD 6 Videotape 6

Student's Solutions Manual

MyMathLab MyMathLab.com

Interactmath.com

1. A student wrote the following as the quadratic formula for solving $ax^2 + bx + c = 0$, $a \neq 0$:

$$x = -b \pm \frac{\sqrt{b^2 - 4ac}}{2a}.$$

Was this correct? Explain.

2. What is wrong with the following "solution" of $5x^2 - 5x + 1 = 0$?

$$x = \frac{5 \pm \sqrt{25 - 4(5)(1)}}{2(5)} \qquad a = 5, b = -5, c = 1$$

$$x = \frac{5 \pm \sqrt{5}}{10}$$

$$x = \frac{1}{2} \pm \sqrt{5}$$

Use the quadratic formula to solve each equation. (All solutions for these equations are real numbers.) See Examples 1 and 2.

3. $m^2 - 8m + 15 = 0$

4. $x^2 + 3x - 28 = 0$

5. $2k^2 + 4k + 1 = 0$

6. $2w^2 + 3w - 1 = 0$

7. $2x^2 - 2x = 1$

8. $9t^2 + 6t = 1$

9. $x^2 + 18 = 10x$

10. $x^2 - 4 = 2x$

11. $4k^2 + 4k - 1 = 0$

12. $4r^2 - 4r - 19 = 0$

13. $2 - 2x = 3x^2$

14. $26r - 2 = 3r^2$

15. $\dfrac{x^2}{4} - \dfrac{x}{2} = 1$
(*Hint:* First clear the fractions.)

16. $p^2 + \dfrac{p}{3} = \dfrac{1}{6}$
(*Hint:* First clear the fractions.)

17. $-2t(t + 2) = -3$

18. $-3x(x + 2) = -4$

19. $(r - 3)(r + 5) = 2$

20. $(k + 1)(k - 7) = 1$

Use the quadratic formula to solve each equation. (All solutions for these equations are nonreal complex numbers.) See Example 3.

21. $x^2 - 3x + 17 = 0$

22. $x^2 - 5x + 20 = 0$

23. $r^2 - 6r + 14 = 0$

24. $t^2 + 4t + 11 = 0$

25. $4x^2 - 4x = -7$

26. $9x^2 - 6x = -7$

27. $x(3x + 4) = -2$

28. $p(2p + 3) = -2$

Use the discriminant to determine whether the solutions for each equation are
    A. *two rational numbers,*   **B.** *one rational number,*
    C. *two irrational numbers,*   **D.** *two nonreal complex numbers.*
Do not actually solve. See Example 4.

29. $25x^2 + 70x + 49 = 0$

30. $4k^2 - 28k + 49 = 0$

31. $x^2 + 4x + 2 = 0$

32. $9x^2 - 12x - 1 = 0$

33. $3x^2 = 5x + 2$

34. $4x^2 = 4x + 3$

35. $3m^2 - 10m + 15 = 0$

36. $18x^2 + 60x + 82 = 0$

37. Using the discriminant, which equations in Exercises 29–36 can be solved by factoring?

38. Based on your answer in Exercise 37, solve the equation given in each exercise.
    (a) Exercise 29   **(b)** Exercise 33

10.3 Equations Quadratic in Form

OBJECTIVE 1 Solve an equation with fractions by writing it in quadratic form. A variety of nonquadratic equations can be written in the form of a quadratic equation and solved by using one of the methods from **Sections 10.1 and 10.2.**

OBJECTIVES

1 Solve an equation with fractions by writing it in quadratic form.

2 Use quadratic equations to solve applied problems.

3 Solve an equation with radicals by writing it in quadratic form.

4 Solve an equation that is quadratic in form by substitution.

EXAMPLE 1 Solving an Equation with Fractions That Leads to a Quadratic Equation

Solve $\dfrac{1}{x} + \dfrac{1}{x-1} = \dfrac{7}{12}$.

Clear fractions by multiplying each term by the least common denominator, $12x(x-1)$. (Note that the domain must be restricted to $x \neq 0$ and $x \neq 1$.)

$$12x(x-1)\frac{1}{x} + 12x(x-1)\frac{1}{x-1} = 12x(x-1)\frac{7}{12}$$

$$12(x-1) + 12x = 7x(x-1)$$

$$12x - 12 + 12x = 7x^2 - 7x \qquad \text{Distributive property}$$

$$24x - 12 = 7x^2 - 7x \qquad \text{Combine terms.}$$

Combine and rearrange terms so that the quadratic equation is in standard form. Then factor to solve the resulting equation.

$$7x^2 - 31x + 12 = 0 \qquad \text{Standard form}$$

$$(7x - 3)(x - 4) = 0 \qquad \text{Factor.}$$

$$7x - 3 = 0 \quad \text{or} \quad x - 4 = 0 \qquad \text{Zero-factor property}$$

$$x = \frac{3}{7} \quad \text{or} \qquad x = 4 \qquad \text{Solve each equation.}$$

Check by substituting these solutions in the original equation. The solution set is $\left\{\frac{3}{7}, 4\right\}$.

Work Problem 1 at the Side. ▶▶▶

1 Solve each equation. Check your solutions.

(a) $\dfrac{5}{m} + \dfrac{12}{m^2} = 2$

(b) $\dfrac{2}{x} + \dfrac{1}{x-2} = \dfrac{5}{3}$

(c) $\dfrac{4}{m-1} + 9 = -\dfrac{7}{m}$

OBJECTIVE 2 Use quadratic equations to solve applied problems. In **Sections 2.4 and 8.5** we solved distance-rate-time (or motion) problems that led to linear equations or rational equations. Now we can extend that work to motion problems that lead to quadratic equations. We continue to use the six-step problem-solving method from **Section 2.3.**

EXAMPLE 2 Solving a Motion Problem

A riverboat for tourists averages 12 mph in still water. It takes the boat 1 hr, 4 min to go 6 mi upstream and return. Find the speed of the current.

Step 1 **Read** the problem carefully.

Step 2 **Assign a variable.** Let $x =$ the speed of the current. The current slows down the boat when it is going upstream, so the rate (or speed) upstream is the speed of the boat in still water less the speed of the current, or $12 - x$. See Figure 1 on the next page.

Continued on Next Page

ANSWERS

1. (a) $\left\{-\frac{3}{2}, 4\right\}$ (b) $\left\{\frac{4}{5}, 3\right\}$

(c) $\left\{\frac{7}{9}, -1\right\}$

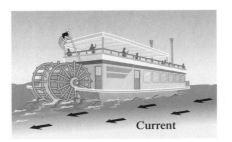

Riverboat traveling *upstream*—the current slows it down.

Figure 1

Similarly, the current speeds up the boat as it travels downstream, so its speed downstream is $12 + x$. Thus,

$$12 - x = \text{the rate upstream;}$$
$$12 + x = \text{the rate downstream.}$$

This information can be used to complete a table. We use the distance formula, $d = rt$, solved for time t, $t = \frac{d}{r}$, to write expressions for t.

	d	r	t
Upstream	6	$12 - x$	$\dfrac{6}{12 - x}$
Downstream	6	$12 + x$	$\dfrac{6}{12 + x}$

Times in hours

Step 3 **Write an equation.** The total time, 1 hr and 4 min, can be written as

$$1 + \frac{4}{60} = 1 + \frac{1}{15} = \frac{16}{15} \text{ hr.}$$

Because the time upstream plus the time downstream equals $\frac{16}{15}$ hr,

$$\underset{\downarrow}{\text{Time upstream}} \quad + \quad \underset{\downarrow}{\text{Time downstream}} \quad = \quad \underset{\downarrow}{\text{Total time}}$$

$$\frac{6}{12 - x} \quad + \quad \frac{6}{12 + x} \quad = \quad \frac{16}{15}.$$

Step 4 **Solve** the equation. Multiply each side by $15(12 - x)(12 + x)$, the LCD, and solve the resulting quadratic equation.

$$15(12 + x)6 + 15(12 - x)6 = 16(12 - x)(12 + x)$$
$$90(12 + x) + 90(12 - x) = 16(144 - x^2)$$

$$1080 + 90x + 1080 - 90x = 2304 - 16x^2 \qquad \text{Distributive property}$$
$$2160 = 2304 - 16x^2 \qquad \text{Combine terms.}$$
$$16x^2 = 144$$
$$x^2 = 9 \qquad \text{Divide by 16.}$$
$$x = 3 \quad \text{or} \quad x = -3 \qquad \text{Square root property}$$

Step 5 **State the answer.** The speed of the current cannot be -3, so the answer is 3 mph.

Step 6 **Check** that this value satisfies the original problem.

CAUTION
As shown in Example 2, when a quadratic equation is used to solve an applied problem, sometimes only *one* answer satisfies the application. ***Always check each answer in the words of the original problem.***

Work Problem 2 at the Side. ▶▶

In **Section 8.5** we solved problems about work rates. Recall that a person's work rate is $\frac{1}{t}$ part of the job per hour, where t is the time in hours required to do the complete job. Thus, the part of the job the person will do in x hours is $\frac{1}{t}x$.

EXAMPLE 3 **Solving a Work Problem**

It takes two carpet layers 4 hr to carpet a room. If each worked alone, one of them could do the job in 1 hr less time than the other. How long would it take each carpet layer to complete the job alone?

Step 1 **Read** the problem again. There will be two answers.

Step 2 **Assign a variable.** Let x represent the number of hours for the slower carpet layer to complete the job alone. Then the faster carpet layer could do the entire job in $(x - 1)$ hours. The slower person's rate is $\frac{1}{x}$, and the faster person's rate is $\frac{1}{x-1}$. Together, they can do the job in 4 hr. Complete a table as shown.

	Rate	Time Working Together	Fractional Part of the Job Done
Slower Worker	$\frac{1}{x}$	4	$\frac{1}{x}(4)$
Faster Worker	$\frac{1}{x-1}$	4	$\frac{1}{x-1}(4)$

← Sum is 1 whole job.

Step 3 **Write an equation.** The sum of the fractional parts done by the workers should equal 1 (the whole job).

Part done by slower worker + part done by faster worker = 1 whole job.

$$\frac{4}{x} + \frac{4}{x-1} = 1$$

Continued on Next Page

---

2 Solve each problem.

(a) In 4 hr, Kerrie can go 15 mi upriver and come back. The speed of the current is 5 mph. Complete this table.

	d	r	t
Up			
Down			

(b) Find the speed of the boat from part (a) in still water.

(c) In $1\frac{3}{4}$ hr, Ken rows his boat 5 mi upriver and comes back. The speed of the current is 3 mph. How fast does Ken row?

ANSWERS

2. (a) row 1: 15; $x - 5$; $\dfrac{15}{x-5}$

row 2: 15; $x + 5$; $\dfrac{15}{x+5}$

(b) 10 mph **(c)** 7 mph

❸ Solve each problem. Round answers to the nearest tenth.

(a) Carlos can complete a certain lab test in 2 hr less time than Jaime can. If they can finish the job together in 2 hr, how long would it take each of them working alone?

	Rate	Time Working Together	Fractional Part of the Job Done
Carlos			
Jaime			

Step 4 **Solve** the equation from Step 3.

$$\frac{4}{x} + \frac{4}{x-1} = 1$$

$$x(x-1)\left(\frac{4}{x} + \frac{4}{x-1}\right) = x(x-1)(1) \qquad \text{Multiply by the LCD.}$$

$$4(x-1) + 4x = x(x-1) \qquad \text{Distributive property}$$

$$4x - 4 + 4x = x^2 - x \qquad \text{Distributive property}$$

$$x^2 - 9x + 4 = 0 \qquad \text{Standard form}$$

This equation cannot be solved by factoring, so use the quadratic formula.

$$x = \frac{9 \pm \sqrt{81 - 16}}{2} = \frac{9 \pm \sqrt{65}}{2} \qquad a = 1, b = -9, c = 4$$

$$x = \frac{9 + \sqrt{65}}{2} \approx 8.5 \quad \text{or} \quad x = \frac{9 - \sqrt{65}}{2} \approx .5 \qquad \text{Use a calculator.}$$

Step 5 **State the answer.** Only the solution 8.5 makes sense in the original problem. (Why?) Thus, the slower worker can do the job in about 8.5 hr and the faster in about $8.5 - 1 = 7.5$ hr.

Step 6 **Check** that these results satisfy the original problem.

◀◀◀ Work Problem 3 at the Side.

OBJECTIVE ❸ **Solve an equation with radicals by writing it in quadratic form.**

EXAMPLE 4 **Solving Radical Equations That Lead to Quadratic Equations**

Solve each equation.

(a) $k = \sqrt{6k - 8}$

This equation is not quadratic. However, squaring both sides of the equation gives a quadratic equation that can be solved by factoring.

$$k^2 = 6k - 8 \qquad \text{Square both sides.}$$

$$k^2 - 6k + 8 = 0 \qquad \text{Standard form}$$

$$(k - 4)(k - 2) = 0 \qquad \text{Factor.}$$

$$k - 4 = 0 \quad \text{or} \quad k - 2 = 0 \qquad \text{Zero-factor property}$$

$$k = 4 \quad \text{or} \quad k = 2 \qquad \text{Potential solutions}$$

Recall from **Section 9.6** that squaring both sides of a radical equation can introduce extraneous solutions that do not satisfy the original equation. *All potential solutions must be checked in the original (not the squared) equation.*

(b) Two chefs are preparing a banquet. One chef could prepare the banquet in 2 hr less time than the other. Together, they complete the job in 5 hr. How long would it take the faster chef working alone?

Check: If $k = 4$, then

$$k = \sqrt{6k - 8}$$
$$4 = \sqrt{6(4) - 8} \qquad ?$$
$$4 = \sqrt{16} \qquad ?$$
$$4 = 4. \qquad \text{True}$$

If $k = 2$, then

$$k = \sqrt{6k - 8}$$
$$2 = \sqrt{6(2) - 8} \qquad ?$$
$$2 = \sqrt{4} \qquad ?$$
$$2 = 2. \qquad \text{True}$$

Both solutions check, so the solution set is $\{2, 4\}$.

ANSWERS
3. (a) Jaime: 5.2 hr; Carlos: 3.2 hr **(b)** 9.1 hr

Continued on Next Page

(b) $x + \sqrt{x} = 6$

$$\sqrt{x} = 6 - x \qquad \text{Isolate the radical on one side.}$$
$$x = 36 - 12x + x^2 \qquad \text{Square both sides.}$$
$$0 = x^2 - 13x + 36 \qquad \text{Standard form}$$
$$0 = (x - 4)(x - 9) \qquad \text{Factor.}$$
$$x - 4 = 0 \quad \text{or} \quad x - 9 = 0 \qquad \text{Zero-factor property}$$
$$x = 4 \quad \text{or} \qquad x = 9 \qquad \text{Potential solutions}$$

Check both potential solutions in the *original* equation.

If $x = 4$, then

$$x + \sqrt{x} = 6$$
$$4 + \sqrt{4} = 6 \quad ?$$
$$6 = 6. \qquad \text{True}$$

If $x = 9$, then

$$x + \sqrt{x} = 6$$
$$9 + \sqrt{9} = 6 \quad ?$$
$$12 = 6. \qquad \text{False}$$

Only the solution 4 checks, so the solution set is {4}.

Work Problem 4 at the Side. ▷▷▷

OBJECTIVE 4 Solve an equation that is quadratic in form by substitution. A nonquadratic equation that can be written in the form

$$au^2 + bu + c = 0,$$

for $a \neq 0$ and an algebraic expression u, is called **quadratic in form.**

EXAMPLE 5 Solving Equations That Are Quadratic in Form

Solve each equation.

(a) $x^4 - 13x^2 + 36 = 0$

Because $x^4 = (x^2)^2$, we can write this equation in quadratic form with $u = x^2$ and $u^2 = x^4$. (Instead of u, any letter other than x could be used.)

$$x^4 - 13x^2 + 36 = 0$$
$$(x^2)^2 - 13x^2 + 36 = 0 \qquad x^4 = (x^2)^2$$
$$u^2 - 13u + 36 = 0 \qquad \text{Let } u = x^2.$$
$$(u - 4)(u - 9) = 0 \qquad \text{Factor.}$$
$$u - 4 = 0 \quad \text{or} \quad u - 9 = 0 \qquad \text{Zero-factor property}$$
$$u = 4 \quad \text{or} \qquad u = 9 \qquad \text{Solve.}$$

To find x, we substitute x^2 for u.

$$x^2 = 4 \quad \text{or} \qquad x^2 = 9$$
$$x = \pm 2 \quad \text{or} \qquad x = \pm 3 \qquad \text{Square root property}$$

The equation $x^4 - 13x^2 + 36 = 0$, a fourth-degree equation, has four solutions.* The solution set is $\{-3, -2, 2, 3\}$. Check by substitution.

(b)
$$4x^4 + 1 = 5x^2$$
$$4(x^2)^2 + 1 = 5x^2 \qquad x^4 = (x^2)^2$$
$$4u^2 + 1 = 5u \qquad \text{Let } u = x^2.$$

Continued on Next Page

*In general, an equation in which an nth-degree polynomial equals 0 has n solutions, although some of them may be repeated.

4 Solve each equation. Check your solutions.

(a) $x = \sqrt{7x - 10}$

(b) $2x = \sqrt{x} + 1$

ANSWERS
4. **(a)** $\{2, 5\}$ **(b)** $\{1\}$

5 Solve each equation. Check your solutions.

(a) $m^4 - 10m^2 + 9 = 0$

(b) $9k^4 - 37k^2 + 4 = 0$

(c) $x^4 - 4x^2 = -2$

$4u^2 - 5u + 1 = 0$		Standard form
$(4u - 1)(u - 1) = 0$		Factor.
$4u - 1 = 0$ or $u - 1 = 0$		Zero-factor property
$u = \dfrac{1}{4}$ or $u = 1$		Solve.
$x^2 = \dfrac{1}{4}$ or $x^2 = 1$		Substitute x^2 for u.
$x = \pm\dfrac{1}{2}$ or $x = \pm 1$		Square root property

Check that the solution set is $\{-1, -\frac{1}{2}, \frac{1}{2}, 1\}$.

(c) $x^4 = 6x^2 - 3$

First write the equation as

$$x^4 - 6x^2 + 3 = 0 \quad \text{or} \quad (x^2)^2 - 6x^2 + 3 = 0,$$

which is quadratic in form with $u = x^2$. Substitute u for x^2 and u^2 for x^4 to get

$$u^2 - 6u + 3 = 0.$$

Since this equation cannot be solved by factoring, use the quadratic formula.

$$u = \frac{6 \pm \sqrt{36 - 12}}{2} \qquad a = 1, b = -6, c = 3$$

$$u = \frac{6 \pm \sqrt{24}}{2}$$

$$u = \frac{6 \pm 2\sqrt{6}}{2} \qquad \sqrt{24} = \sqrt{4} \cdot \sqrt{6} = 2\sqrt{6}$$

$$u = \frac{2(3 \pm \sqrt{6})}{2} \qquad \text{Factor.}$$

$$u = 3 \pm \sqrt{6} \qquad \text{Lowest terms}$$

$$x^2 = 3 + \sqrt{6} \quad \text{or} \quad x^2 = 3 - \sqrt{6} \qquad \text{Substitute } x^2 \text{ for } u.$$

$$x = \pm\sqrt{3 + \sqrt{6}} \quad \text{or} \quad x = \pm\sqrt{3 - \sqrt{6}} \qquad \text{Square root property}$$

The solution set contains four numbers:

$$\left\{\sqrt{3 + \sqrt{6}}, -\sqrt{3 + \sqrt{6}}, \sqrt{3 - \sqrt{6}}, -\sqrt{3 - \sqrt{6}}\right\}.$$

NOTE

Some students prefer to solve equations like those in Examples 5(a) and (b) by factoring directly. For example,

$$x^4 - 13x^2 + 36 = 0 \qquad \text{Example 5(a) equation}$$
$$(x^2 - 9)(x^2 - 4) = 0 \qquad \text{Factor.}$$
$$(x + 3)(x - 3)(x + 2)(x - 2) = 0. \qquad \text{Factor again.}$$

Using the zero-factor property gives the same solutions obtained in Example 5(a). Equations that cannot be solved by factoring (as in Example 5(c)) must be solved by substitution and the quadratic formula.

ANSWERS

5. (a) $\{-3, -1, 1, 3\}$ **(b)** $\left\{-2, -\dfrac{1}{3}, \dfrac{1}{3}, 2\right\}$

(c) $\{\sqrt{2 + \sqrt{2}}, -\sqrt{2 + \sqrt{2}}, \sqrt{2 - \sqrt{2}}, -\sqrt{2 - \sqrt{2}}\}$

◄◄◄ Work Problem 5 at the Side.

EXAMPLE 6 **Solving Equations That Are Quadratic in Form**

Solve each equation.

(a) $2(4m - 3)^2 + 7(4m - 3) + 5 = 0$

Because of the repeated quantity $4m - 3$, this equation is quadratic in form with $u = 4m - 3$.

$$2(4m - 3)^2 + 7(4m - 3) + 5 = 0$$
$$2u^2 + 7u + 5 = 0 \qquad \text{Let } 4m - 3 = u.$$
$$(2u + 5)(u + 1) = 0 \qquad \text{Factor.}$$
$$2u + 5 = 0 \quad \text{or} \quad u + 1 = 0 \qquad \text{Zero-factor property}$$
$$u = -\frac{5}{2} \quad \text{or} \quad u = -1$$
$$4m - 3 = -\frac{5}{2} \quad \text{or} \quad 4m - 3 = -1 \qquad \text{Substitute } 4m - 3 \text{ for } u.$$
$$4m = \frac{1}{2} \quad \text{or} \quad 4m = 2 \qquad \text{Solve for } m.$$
$$m = \frac{1}{8} \quad \text{or} \quad m = \frac{1}{2}$$

Check that the solution set of the original equation is $\{\frac{1}{8}, \frac{1}{2}\}$.

(b) $2a^{2/3} - 11a^{1/3} + 12 = 0$

Let $a^{1/3} = u$; then $a^{2/3} = (a^{1/3})^2 = u^2$. Substitute into the given equation.

$$2u^2 - 11u + 12 = 0 \qquad \text{Let } a^{1/3} = u; a^{2/3} = u^2.$$
$$(2u - 3)(u - 4) = 0 \qquad \text{Factor.}$$
$$2u - 3 = 0 \quad \text{or} \quad u - 4 = 0 \qquad \text{Zero-factor property}$$
$$u = \frac{3}{2} \quad \text{or} \quad u = 4$$
$$a^{1/3} = \frac{3}{2} \quad \text{or} \quad a^{1/3} = 4 \qquad u = a^{1/3}$$
$$(a^{1/3})^3 = \left(\frac{3}{2}\right)^3 \quad \text{or} \quad (a^{1/3})^3 = 4^3 \qquad \text{Cube each side.}$$
$$a = \frac{27}{8} \quad \text{or} \quad a = 64$$

Check that the solution set is $\{\frac{27}{8}, 64\}$.

CAUTION
A common error when solving problems like those in Examples 5 and 6 is to stop too soon. *Once you have solved for u, remember to substitute and solve for the values of the original variable.*

Work Problem 6 at the Side.

6 Solve each equation. Check your solutions.

(a) $5(r + 3)^2 + 9(r + 3) = 2$

(b) $4m^{2/3} = 3m^{1/3} + 1$

ANSWERS
6. (a) $\left\{-5, -\frac{14}{5}\right\}$ (b) $\left\{-\frac{1}{64}, 1\right\}$

Real-Data Applications

Smile! You're on Golden Ratio!

The **Golden Ratio** is the number phi, $\phi = \frac{1 + \sqrt{5}}{2}$. The Rhind Papyrus, dated 1600 B.C., referred to the **sacred ratio** used in building the Great Pyramids at Giza, Egypt. The ancient Greeks used ϕ in art and architecture, striving for a proportion that was the most pleasing to the eye. The Parthenon is the classic illustration of the use of ϕ in achieving that goal.

In a segment of length 1 that is divided into two parts, the Golden Ratio is defined as the proportion that equates the ratio of the whole segment to the larger segment and the ratio of the larger segment to the smaller segment. In the diagram, segment AC has length 1. Point B divides the segment so that AB has length x and BC has length $1 - x$. Since x represents the length of the larger segment AB, it follows that ϕ is the ratio $\frac{1}{x}$ and the Golden Proportion is

$$\frac{\text{whole}}{\text{larger}} = \frac{\text{larger}}{\text{smaller}} \quad \text{or} \quad \frac{1}{x} = \frac{x}{1 - x}.$$

The Golden Ratio is used in dentistry and medicine today. Eddy Levin, an English dentist, became interested in applications of the Golden Ratio, or Golden Proportion, to orthodontia and dentistry in 1978. His work is now a compulsory topic of study in U.S. dental schools. Viewed from the front, the "four front teeth, from central incisor to the premolar are the most significant part of the smile and they are in Golden Proportion to each other." He invented the Golden Mean Gauge, which is a tool that measures the Golden Proportion. (*Source:* www.goldenmeangauge.co.uk)

For Group Discussion

1. Write the Golden Proportion as a quadratic equation.

   (a) Use the quadratic formula to solve this quadratic equation for x. Note that x must be a positive number since it is the length of AB.

   (b) The Golden Ratio is $\phi = \frac{1}{x}$. Rationalize the denominator to write ϕ in exact form (using radicals).

   (c) Write an approximate value for ϕ, rounded to 6 decimal places.

2. The Golden Ratio is a mathematically curious number. The reciprocal of ϕ is one less than ϕ, and the square of ϕ is one more than ϕ.

   (a) Find $\frac{1}{\phi}$ and $\phi - 1$. (*Hint:* You have previously found the quantity $\frac{1}{\phi}$.)

   (b) Find ϕ^2 and $\phi + 1$ in exact form.

10.3 Exercises

FOR EXTRA HELP

Tutor Center — Addison-Wesley Math Tutor Center

Math XP MathXL

Digital Video Tutor CD 6 Videotape 6

Student's Solutions Manual

MyMathLab MyMathLab

Interactmath.com

Based on the discussion and examples of this section, write a sentence describing the first step you would take to solve each equation. Do not actually solve.

1. $\dfrac{14}{x} = x - 5$

2. $\sqrt{1 + x} + x = 5$

3. $(r^2 + r)^2 - 8(r^2 + r) + 12 = 0$

4. $3t = \sqrt{16 - 10t}$

5. What is wrong with the following "solution"?

$$x = \sqrt{3x + 4}$$
$$x^2 = 3x + 4 \qquad \text{Square both sides.}$$
$$x^2 - 3x - 4 = 0$$
$$(x - 4)(x + 1) = 0$$
$$x - 4 = 0 \quad \text{or} \quad x + 1 = 0$$
$$x = 4 \quad \text{or} \quad x = -1$$

Solution set: $\{4, -1\}$

6. What is wrong with the following "solution"?

$$2(m - 1)^2 - 3(m - 1) + 1 = 0$$
$$2u^2 - 3u + 1 = 0 \qquad \text{Let } u = m - 1.$$
$$(2u - 1)(u - 1) = 0$$
$$2u - 1 = 0 \quad \text{or} \quad u - 1 = 0$$
$$u = \frac{1}{2} \quad \text{or} \qquad u = 1$$

Solution set: $\{\frac{1}{2}, 1\}$

Solve each equation. Check your solutions. See Example 1.

7. $1 - \dfrac{3}{x} - \dfrac{28}{x^2} = 0$

8. $4 - \dfrac{7}{r} - \dfrac{2}{r^2} = 0$

9. $3 - \dfrac{1}{t} = \dfrac{2}{t^2}$

10. $1 + \dfrac{2}{k} = \dfrac{3}{k^2}$

11. $\dfrac{1}{x} + \dfrac{2}{x + 2} = \dfrac{17}{35}$

12. $\dfrac{2}{m} + \dfrac{3}{m + 9} = \dfrac{11}{4}$

13. $\dfrac{2}{x + 1} + \dfrac{3}{x + 2} = \dfrac{7}{2}$

14. $\dfrac{4}{3 - p} + \dfrac{2}{5 - p} = \dfrac{26}{15}$

15. $\dfrac{3}{2x} - \dfrac{1}{2(x + 2)} = 1$

16. $\dfrac{4}{3x} - \dfrac{1}{2(x + 1)} = 1$

17. $\dfrac{6}{p} = 2 + \dfrac{p}{p + 1}$

18. $\dfrac{k}{2 - k} + \dfrac{2}{k} = 5$

19. A boat goes 20 mph in still water, and the rate of the current is t mph.

(a) What is the rate of the boat when it travels upstream?

(b) What is the rate of the boat when it travels downstream?

20. If it takes m hours to grade a set of papers, what is the grader's rate (in job per hour)?

Solve each problem. See Examples 2 and 3.

21. On a windy day Yoshiaki found that he could go 16 mi downstream and then 4 mi back upstream at top speed in a total of 48 min. What was the top speed of Yoshiaki's boat if the current was 15 mph?

	d	*r*	*t*
Upstream	4	$x - 15$	
Downstream	16		

22. Lekesha flew her plane for 6 hr at a constant speed. She traveled 810 mi with the wind, then turned around and traveled 720 mi against the wind. The wind speed was a constant 15 mph. Find the speed of the plane.

	d	*r*	*t*
With Wind	810		
Against Wind	720		

23. In Canada, Medicine Hat and Cranbrook are 300 km apart. Harry rides his Honda 20 km per hr faster than Yoshi rides his Yamaha. Find Harry's average speed if he travels from Cranbrook to Medicine Hat in $1\frac{1}{4}$ hr less time than Yoshi. (*Source: State Farm Road Atlas.*)

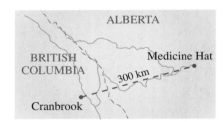

24. In California, the distance from Jackson to Lodi is about 40 mi, as is the distance from Lodi to Manteca. Rico drove from Jackson to Lodi during the rush hour, stopped in Lodi for a root beer, and then drove on to Manteca at 10 mph faster. Driving time for the entire trip was 88 min. Find his speed from Jackson to Lodi. (*Source: State Farm Road Atlas.*)

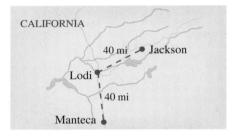

25. Working together, two people can cut a large lawn in 2 hr. One person can do the job alone in 1 hr less time than the other. How long (to the nearest tenth) would it take the faster person to do the job? (*Hint:* x is the time of the faster person.)

	Rate	Time Working Together	Fractional Part of the Job Done
Faster Worker	$\dfrac{1}{x}$	2	
Slower Worker		2	

26. A janitorial service provides two people to clean an office building. Working together, the two can clean the building in 5 hr. One person is new to the job and would take 2 hr longer than the other person to clean the building alone. How long (to the nearest tenth) would it take the new worker to clean the building alone?

	Rate	Time Working Together	Fractional Part of the Job Done
Faster Worker			
Slower Worker			

27. A washing machine can be filled in 6 min if both the hot and cold water taps are fully opened. Filling the washer with hot water alone takes 9 min longer than filling it with cold water alone. How long does it take to fill the washer with cold water?

28. Two pipes together can fill a large tank in 2 hr. One of the pipes, used alone, takes 3 hr longer than the other to fill the tank. How long would each pipe take to fill the tank alone?

Solve each equation. Check your solutions. See Example 4.

29. $z = \sqrt{5z - 4}$

30. $x = \sqrt{9x - 14}$

31. $2x = \sqrt{11x + 3}$

32. $4x = \sqrt{6x + 1}$

33. $3x = \sqrt{16 - 10x}$

34. $4t = \sqrt{8t + 3}$

35. $p - 2\sqrt{p} = 8$

36. $k + \sqrt{k} = 12$

37. $m = \sqrt{\dfrac{6 - 13m}{5}}$

38. $r = \sqrt{\dfrac{20 - 19r}{6}}$

Solve each equation. Check your solutions. See Examples 5 and 6.

39. $t^4 - 18t^2 + 81 = 0$

40. $x^4 - 8x^2 + 16 = 0$

41. $4k^4 - 13k^2 + 9 = 0$

42. $9x^4 - 25x^2 + 16 = 0$

43. $x^4 + 48 = 16x^2$

44. $z^4 = 17z^2 - 72$

45. $2x^4 - 9x^2 = -2$

46. $8x^4 + 1 = 11x^2$

47. $(x + 3)^2 + 5(x + 3) + 6 = 0$

48. $(k - 4)^2 + (k - 4) - 20 = 0$

49. $(t + 5)^2 + 6 = 7(t + 5)$

50. $3(m + 4)^2 - 8 = 2(m + 4)$

51. $2 + \dfrac{5}{3k - 1} = \dfrac{-2}{(3k - 1)^2}$

52. $3 - \dfrac{7}{2p + 2} = \dfrac{6}{(2p + 2)^2}$

53. $x^{2/3} + x^{1/3} - 2 = 0$

54. $x^{2/3} - 2x^{1/3} - 3 = 0$ **55.** $r^{2/3} + r^{1/3} - 12 = 0$ **56.** $3x^{2/3} - x^{1/3} - 24 = 0$

57. $2(1 + \sqrt{r})^2 = 13(1 + \sqrt{r}) - 6$ **58.** $(k^2 + k)^2 + 12 = 8(k^2 + k)$

RELATING CONCEPTS (EXERCISES 59–64)   For Individual or Group Work

Consider the following equation, which contains variable expressions in the denominators.
Work Exercises 59–64 in order.

$$\frac{x^2}{(x - 3)^2} + \frac{3x}{x - 3} - 4 = 0$$

59. Why must 3 be excluded from the domain of this equation?

60. Multiply each side of the equation by the LCD, $(x - 3)^2$, and solve. There is only one solution—what is it?

61. Write the equation in a different manner so that it is quadratic in form using the expression $\frac{x}{x - 3}$.

62. In your own words, explain why the expression $\frac{x}{x - 3}$ cannot equal 1.

63. Solve the equation from Exercise 61 by making the substitution $t = \frac{x}{x - 3}$. You should get two values for t. Why is one of them impossible for this equation?

64. Solve the equation $x^2(x - 3)^{-2} + 3x(x - 3)^{-1} - 4 = 0$ by letting $s = (x - 3)^{-1}$. You should get two values for s. Why is this impossible for this equation?

Summary Exercises on Solving Quadratic Equations

We have introduced four methods for solving quadratic equations written in standard form $ax^2 + bx + c = 0$. *The following table lists some advantages and disadvantages of each method.*

METHODS FOR SOLVING QUADRATIC EQUATIONS

Method	Advantages	Disadvantages
Factoring	This is usually the fastest method.	Not all polynomials are factorable; some factorable polynomials are hard to factor.
Square root property	This is the simplest method for solving equations of the form $(ax + b)^2 = c$.	Few equations are given in this form.
Completing the square	This method can always be used, although most people prefer the quadratic formula.	It requires more steps than other methods.
Quadratic formula	This method can always be used.	It is more difficult than factoring because of the square root, although calculators can simplify its use.

Refer to the preceding box. Decide whether factoring, the square root property, *or the* quadratic formula *is most appropriate for solving each quadratic equation. Do not actually solve the equations.*

1. $(2x + 3)^2 = 4$

2. $4x^2 - 3x = 1$

3. $z^2 + 5z - 8 = 0$

4. $2k^2 + 3k = 1$

5. $3m^2 = 2 - 5m$

6. $p^2 = 5$

Solve each quadratic equation by the method of your choice. Check your solutions.

7. $p^2 = 47$

8. $6x^2 - x - 15 = 0$

9. $n^2 + 8n + 6 = 0$

10. $(x - 4)^2 = 49$

11. $\dfrac{9}{m} + \dfrac{5}{m^2} = 2$

12. $3m^2 = 3 - 8m$

13. $3x^2 - 9x + 4 = 0$

*14. $x^2 = -12$

15. $x\sqrt{2} = \sqrt{5x - 2}$

16. $12x^4 - 11x^2 + 2 = 0$

17. $(2k + 5)^2 = 12$

18. $\dfrac{2}{x} + \dfrac{1}{x - 2} - \dfrac{5}{3} = 0$

19. $t^4 + 14 = 9t^2$

20. $2x^2 + 4x = 5$

*21. $z^2 + z + 2 = 0$

22. $x^4 - 8x^2 = -1$

23. $4t^2 - 12t + 9 = 0$

24. $x\sqrt{3} = \sqrt{2 - x}$

25. $r^2 - 72 = 0$

26. $-3x^2 + 4x = -4$

27. $x^2 - 5x - 36 = 0$

28. $w^2 = 169$

*29. $3p^2 = 6p - 4$

30. $z = \sqrt{\dfrac{5z + 3}{2}}$

31. $2(3k - 1)^2 + 5(3k - 1) = -2$

*32. $\dfrac{4}{r^2} + 3 = \dfrac{1}{r}$

33. $x - \sqrt{15 - 2x} = 0$

34. $3 = \dfrac{1}{t + 2} + \dfrac{2}{(t + 2)^2}$

*35. $4k^4 + 5k^2 + 1 = 0$

36. $(x + 1)^{2/3} - (x + 1)^{1/3} = 2$

---

* This exercise requires knowledge of complex numbers.

10.4 Exercises

FOR EXTRA HELP

Tutor Center Addison-Wesley Math Tutor Center

MathXL MathXL

Digital Video Tutor CD 6 Videotape 6

Student's Solutions Manual

MyMathLab MyMathLab

Interactmath.com

1. What is the first step in solving a formula like $gw^2 = 2r$ for w?

2. What is the first step in solving a formula like $gw^2 = kw + 24$ for w?

In Exercises 3 and 4, solve for m in terms of the other variables ($m > 0$).

3.

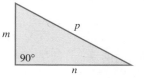

4.

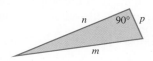

Solve each equation for the indicated variable. (Leave $\pm$ in your answers.) See Examples 1 and 2.

5. $d = kt^2$ for t

6. $s = kwd^2$ for d

7. $I = \dfrac{ks}{d^2}$ for d

8. $R = \dfrac{k}{d^2}$ for d

9. $F = \dfrac{kA}{v^2}$ for v

10. $L = \dfrac{kd^4}{h^2}$ for h

11. $V = \dfrac{1}{3}\pi r^2 h$ for r

12. $V = \pi(r^2 + R^2)h$ for r

13. $At^2 + Bt = -C$ for t

14. $S = 2\pi rh + \pi r^2$ for r

15. $D = \sqrt{kh}$ for h

16. $F = \dfrac{k}{\sqrt{d}}$ for d

17. $p = \sqrt{\dfrac{k\ell}{g}}$ for ℓ

18. $p = \sqrt{\dfrac{k\ell}{g}}$ for g

Solve each problem. When appropriate, round answers to the nearest tenth. See Example 3.

19. Find the lengths of the sides of the triangle.

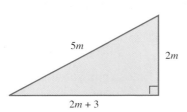

20. Find the lengths of the sides of the triangle.

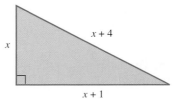

21. Two ships leave port at the same time, one heading due south and the other heading due east. Several hours later, they are 170 mi apart. If the ship traveling south traveled 70 mi farther than the other, how many miles did they each travel?

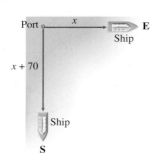

22. Anna Sudak is flying a kite that is 30 ft farther above her hand than its horizontal distance from her. The string from her hand to the kite is 150 ft long. How high is the kite?

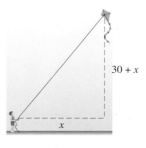

Solve each problem. See Example 4.

23. A couple wants to buy a rug for a room that is 20 ft long and 15 ft wide. They want to leave an even strip of flooring uncovered around the edges of the room. How wide a strip will they have if they buy a rug with an area of 234 ft^2?

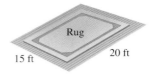

24. A club swimming pool is 30 ft wide and 40 ft long. The club members want an exposed aggregate border in a strip of uniform width around the pool. They have enough material for 296 ft^2. How wide can the strip be?

25. A rectangular piece of sheet metal has a length that is 4 in. less than twice the width. A square piece 2 in. on a side is cut from each corner. The sides are then turned up to form an uncovered box of volume 256 in.^{3}. Find the length and width of the original piece of metal.

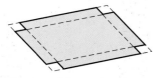

26. Another rectangular piece of sheet metal is 2 in. longer than it is wide. A square piece 3 in. on a side is cut from each corner. The sides are then turned up to form an uncovered box of volume 765 in.^{3}. Find the dimensions of the original piece of metal.

Solve each problem. Round answers to the nearest tenth. See Example 5.

27. A ball is projected upward from the ground. Its distance in feet from the ground in t seconds is given by

$$s(t) = -16t^2 + 128t.$$

At what times will the ball be 213 ft from the ground?

28. A toy rocket is launched from ground level. Its distance in feet from the ground in t seconds is given by

$$s(t) = -16t^2 + 208t.$$

At what times will the rocket be 550 ft from the ground?

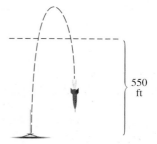

29. The function defined by

$$D(t) = 13t^2 - 100t$$

gives the distance in feet a car going approximately 68 mph will skid in t seconds. Find the time it would take for the car to skid 180 ft.

30. The function given in Exercise 29 becomes

$$D(t) = 13t^2 - 73t$$

for a car going 50 mph. Find the time for this car to skid 218 ft.

A rock is projected upward from ground level, and its distance in feet from the ground in t seconds is given by $s(t) = -16t^2 + 160t$. Use algebra and a short explanation to answer Exercises 31 and 32.

31. After how many seconds does it reach a height of 400 ft? How would you describe in words its position at this height?

32. After how many seconds does it reach a height of 425 ft? How would you interpret the mathematical result here?

Solve each problem using a quadratic equation.

33. A certain bakery has found that the daily demand for blueberry muffins is $\frac{3200}{p}$, where p is the price of a muffin in cents. The daily supply is $3p - 200$. Find the price at which supply and demand are equal.

34. In one area the demand for compact discs is $\frac{700}{P}$ per day, where P is the price in dollars per disc. The supply is $5P - 1$ per day. At what price does supply equal demand?

Sales of SUVs (sport-utility vehicles) in the United States (in millions) for the years 1990–1999 are shown in the bar graph and can be modeled by the quadratic function defined by

$$f(x) = .016x^2 + .124x + .787.$$

Here, $x = 0$ represents 1990, $x = 1$ represents 1991, and so on. Use the graph and the model to work Exercises 35–38. See Example 6.

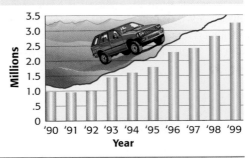

SALES OF SUVS IN THE UNITED STATES (IN MILLIONS)

Source: CNW Marketing Research of Bandon, OR, based on automakers' reported sales.

35. **(a)** Use the graph to estimate sales in 1997 to the nearest tenth.

   (b) Use the model to approximate sales in 1997 to the nearest tenth. How does this result compare to your estimate from part (a)?

36. **(a)** Use the model to estimate sales in 2000 to the nearest tenth.

   (b) Sales through October 2000 were about 2.9 million. Based on this, is the sales estimate for 2000 from part (a) reasonable? Explain.

37. Based on the model, in what year did sales reach 2 million? (Round down to the nearest year.) How does this result compare to the sales shown in the graph?

38. Based on the model, in what year did sales reach 3 million? (Round down to the nearest year.) How does this result compare to the sales shown in the graph?

William Froude was a 19th-century naval architect who used the expression

$$\frac{v^2}{g\ell}$$

in shipbuilding. This expression, known as the Froude number, was also used by R. McNeill Alexander in his research on dinosaurs. (Source: "How Dinosaurs Ran," Scientific American, April 1991.) In Exercises 39 and 40, find the value of v (in meters per second), given that $g = 9.8$ m per sec^2.

39. Rhinoceros: $\ell = 1.2$; Froude number $= 2.57$

40. Triceratops: $\ell = 2.8$; Froude number $= .16$

*Recall from the **Section 8.5** exercises that corresponding sides of similar triangles are proportional. Use this fact to find the lengths of the indicated sides of each pair of similar triangles. Check all possible solutions in both triangles. Sides of a triangle cannot be negative (and are not drawn to scale here).*

41. Side AC

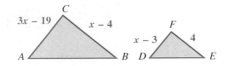

42. Side RQ

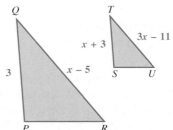

10.5 Graphs of Quadratic Functions

OBJECTIVE 1 Graph a quadratic function. Figure 4 gives a graph of the simplest *quadratic function,* defined by $y = x^2$.

OBJECTIVES

1 Graph a quadratic function.

2 Graph parabolas with horizontal and vertical shifts.

3 Predict the shape and direction of a parabola from the coefficient of x^2.

4 Find a quadratic function to model data.

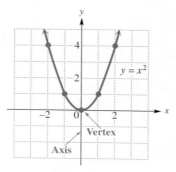

x	y
-2	4
-1	1
0	0
1	1
2	4

Figure 4

As mentioned in **Section 6.3,** this graph is called a **parabola.** The point $(0, 0)$, the lowest point on the curve, is the **vertex** of this parabola. The vertical line through the vertex is the **axis** of the parabola, here $x = 0$. A parabola is **symmetric about its axis;** that is, if the graph were folded along the axis, the two portions of the curve would coincide. As Figure 4 suggests, x can be any real number, so the domain of the function defined by $y = x^2$ is $(-\infty, \infty)$. Since y is always nonnegative, the range is $[0, \infty)$.

In **Section 10.4,** we solved applications modeled by quadratic functions. We now consider graphs of general quadratic functions as defined here.

> **Quadratic Function**
>
> A function that can be written in the form
>
> $$f(x) = ax^2 + bx + c$$
>
> for real numbers a, b, and c, with $a \neq 0$, is a **quadratic function.**

The graph of any quadratic function is a parabola with a vertical axis.

> **NOTE**
> We use the variable y and function notation $f(x)$ interchangeably. Although we use the letter f most often to name quadratic functions, other letters can be used. We use the capital letter F to distinguish between different parabolas graphed on the same coordinate axes.

Parabolas, which are a type of *conic section* (**Chapter 12**), have many applications. Cross sections of satellite dishes and automobile headlights form parabolas, as do the cables that support suspension bridges.

OBJECTIVE 2 Graph parabolas with horizontal and vertical shifts. Parabolas need not have their vertices at the origin, as does the graph of $f(x) = x^2$. For example, to graph a parabola of the form $F(x) = x^2 + k$, start by selecting sample values of x like those that were used to graph $f(x) = x^2$. The corresponding values of $F(x)$ in $F(x) = x^2 + k$ differ by k from those of $f(x) = x^2$. For this reason, the graph of $F(x) = x^2 + k$ is *shifted,* or *translated,* k units vertically compared with that of $f(x) = x^2$.

1 Graph each parabola. Give the vertex, domain, and range.

(a) $f(x) = x^2 + 3$

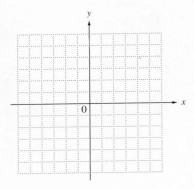

(b) $f(x) = x^2 - 1$

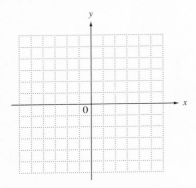

EXAMPLE 1 **Graphing a Parabola with a Vertical Shift**

Graph $F(x) = x^2 - 2$.

This graph has the same shape as that of $f(x) = x^2$, but since k here is -2, the graph is shifted 2 units down, with vertex $(0, -2)$. Every function value is 2 less than the corresponding function value of $f(x) = x^2$. Plotting points on both sides of the vertex gives the graph in Figure 5.

Notice that since the parabola is symmetric about its axis $x = 0$, the plotted points are "mirror images" of each other. Since x can be any real number, the domain is still $(-\infty, \infty)$; the value of y (or $F(x)$) is always greater than or equal to -2, so the range is $[-2, \infty)$. The graph of $f(x) = x^2$ is shown for comparison.

x	$f(x) = x^2$	$F(x) = x^2 - 2$
-2	4	2
-1	1	-1
0	0	-2
1	1	-1
2	4	2

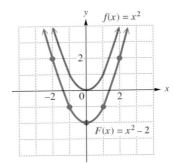

Figure 5

Vertical Shift

The graph of $F(x) = x^2 + k$ is a parabola with the same shape as the graph of $f(x) = x^2$. The parabola is shifted vertically: k units up if $k > 0$, and $|k|$ units down if $k < 0$. The vertex is $(0, k)$.

Work Problem 1 at the Side.

The graph of $F(x) = (x - h)^2$ is also a parabola with the same shape as that of $f(x) = x^2$. Because $(x - h)^2 \geq 0$ for all x, the vertex of $F(x) = (x - h)^2$ is the lowest point on the parabola. The lowest point occurs here when $F(x)$ is 0. To get $F(x)$ equal to 0, let $x = h$ so the vertex of $F(x) = (x - h)^2$ is $(h, 0)$. Based on this, the graph of $F(x) = (x - h)^2$ is shifted h units horizontally compared with that of $f(x) = x^2$.

EXAMPLE 2 **Graphing a Parabola with a Horizontal Shift**

Graph $F(x) = (x - 2)^2$.

When $x = 2$, then $F(x) = 0$, giving the vertex $(2, 0)$. The graph of $F(x) = (x - 2)^2$ has the same shape as that of $f(x) = x^2$ but is shifted 2 units to the right. Plotting several points on one side of the vertex and using symmetry about the axis $x = 2$ to find corresponding points on the other side of the vertex gives the graph in Figure 6. Again, the domain is $(-\infty, \infty)$; the range is $[0, \infty)$.

Continued on Next Page

ANSWERS
1. **(a)**

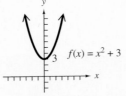

vertex: $(0, 3)$; domain: $(-\infty, \infty)$; range: $[3, \infty)$

(b)

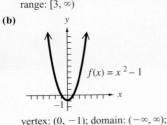

vertex: $(0, -1)$; domain: $(-\infty, \infty)$; range: $[-1, \infty)$

x	$F(x) = (x-2)^2$
0	4
1	1
2	0
3	1
4	4

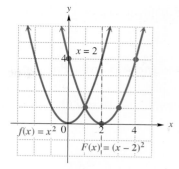

Figure 6

Horizontal Shift

The graph of $F(x) = (x - h)^2$ is a parabola with the same shape as the graph of $f(x) = x^2$. The parabola is shifted h units horizontally: h units to the right if $h > 0$, and $|h|$ units to the left if $h < 0$. The vertex is $(h, 0)$.

CAUTION

Errors frequently occur when horizontal shifts are involved. To determine the direction and magnitude of a horizontal shift, find the value that would cause the expression $x - h$ to equal 0. For example, the graph of $F(x) = (x - 5)^2$ would be shifted 5 units to the *right,* because $+5$ would cause $x - 5$ to equal 0. On the other hand, the graph of $F(x) = (x + 5)^2$ would be shifted 5 units to the *left,* because -5 would cause $x + 5$ to equal 0.

Work Problem 2 at the Side. ▶▶▶

A parabola can have both horizontal and vertical shifts.

EXAMPLE 3   Graphing a Parabola with Horizontal and Vertical Shifts

Graph $F(x) = (x + 3)^2 - 2$.

This graph has the same shape as that of $f(x) = x^2$, but is shifted 3 units to the left (since $x + 3 = 0$ if $x = -3$) and 2 units down (because of the -2). As shown in Figure 7, the vertex is $(-3, -2)$, with axis $x = -3$. This function has domain $(-\infty, \infty)$ and range $[-2, \infty)$.

x	$F(x)$
-5	2
-4	-1
-3	-2
-2	-1
-1	2

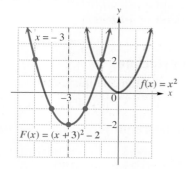

Figure 7

2 Graph each parabola. Give the vertex, axis, domain, and range.

(a) $f(x) = (x - 3)^2$

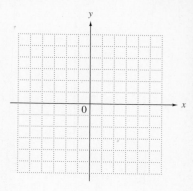

(b) $f(x) = (x + 2)^2$

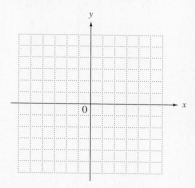

ANSWERS

2. **(a)**

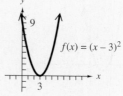

vertex: $(3, 0)$; axis: $x = 3$;
domain: $(-\infty, \infty)$; range: $[0, \infty)$

(b)

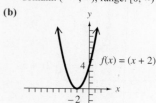

vertex: $(-2, 0)$; axis: $x = -2$;
domain: $(-\infty, \infty)$; range: $[0, \infty)$

3 Graph each parabola. Give the vertex, axis, domain, and range.

(a) $f(x) = (x + 2)^2 - 1$

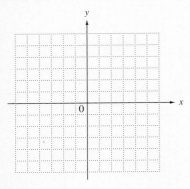

(b) $f(x) = (x - 2)^2 + 5$

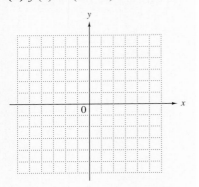

The characteristics of the graph of a parabola of the form $F(x) = (x - h)^2 + k$ are summarized as follows.

Vertex and Axis of a Parabola

The graph of $F(x) = (x - h)^2 + k$ is a parabola with the same shape as the graph of $f(x) = x^2$ with vertex (h, k). The axis is the vertical line $x = h$.

◀◀◀ **Work Problem 3 at the Side.**

OBJECTIVE 3 **Predict the shape and direction of a parabola from the coefficient of x^2.** Not all parabolas open up, and not all parabolas have the same shape as the graph of $f(x) = x^2$.

EXAMPLE 4 **Graphing a Parabola That Opens Down**

Graph $f(x) = -\dfrac{1}{2}x^2$.

This parabola is shown in Figure 8. The coefficient $-\frac{1}{2}$ affects the shape of the graph; the $\frac{1}{2}$ makes the parabola wider (since the values of $\frac{1}{2}x^2$ increase more slowly than those of x^2), and the negative sign makes the parabola open down. The graph is not shifted in any direction; the vertex is still $(0, 0)$. Unlike the parabolas graphed in Examples 1–3, the vertex here has the *greatest* function value of any point on the graph. The domain is $(-\infty, \infty)$; the range is $(-\infty, 0]$.

x	$f(x)$
-2	-2
-1	$-\frac{1}{2}$
0	0
1	$-\frac{1}{2}$
2	-2

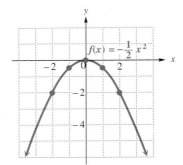

Figure 8

Some general principles concerning the graph of $F(x) = a(x - h)^2 + k$ are summarized as follows.

General Principles

1. The graph of the quadratic function defined by
$$F(x) = a(x - h)^2 + k, \quad a \neq 0,$$
is a parabola with vertex (h, k) and the vertical line $x = h$ as axis.

2. The graph opens up if a is positive and down if a is negative.

3. The graph is wider than that of $f(x) = x^2$ if $0 < |a| < 1$. The graph is narrower than that of $f(x) = x^2$ if $|a| > 1$.

ANSWERS

3. (a)

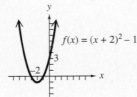

vertex: $(-2, -1)$; axis: $x = -2$;
domain: $(-\infty, \infty)$; range: $[-1, \infty)$

(b)

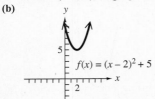

vertex: $(2, 5)$; axis: $x = 2$;
domain: $(-\infty, \infty)$; range: $[5, \infty)$

Work Problems 4 and 5 at the Side. ▶▶▶

EXAMPLE 5 **Using the General Principles to Graph a Parabola**

Graph $F(x) = -2(x + 3)^2 + 4$.

The parabola opens down (because $a < 0$), and is narrower than the graph of $f(x) = x^2$, since $|-2| = 2 > 1$, causing values of $F(x)$ to decrease more quickly than those of $f(x) = -x^2$. This parabola has vertex $(-3, 4)$ as shown in Figure 9. To complete the graph, we plotted the ordered pairs $(-4, 2)$ and, by symmetry, $(-2, 2)$. Symmetry can be used to find additional ordered pairs that satisfy the equation, if desired.

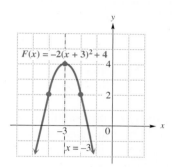

Figure 9

Work Problem 6 at the Side. ▶▶▶

OBJECTIVE **4** **Find a quadratic function to model data.**

EXAMPLE 6 **Finding a Quadratic Function to Model the Rise in Multiple Births**

The number of higher-order multiple births in the United States is rising. Let x represent the number of years since 1970 and y represent the rate of higher-order multiples born per 100,000 births since 1971. The data are shown in the following table. Find a quadratic function that models the data.

U.S. HIGHER-ORDER
MULTIPLE BIRTHS

Year	x	y
1971	1	29.1
1976	6	35.0
1981	11	40.0
1986	16	47.0
1991	21	100.0
1996	26	152.6
2001	31	185.6

Source: National Center for Health Statistics.

A scatter diagram of the ordered pairs (x, y) is shown in Figure 10 on the next page. The general shape suggested by the scatter diagram indicates that a parabola should approximate these points, as shown by the dashed curve in Figure 11. The equation for such a parabola would have a positive coefficient for x^2 since the graph opens up.

——— **Continued on Next Page**

4 Decide whether each parabola opens up or down.

(a) $f(x) = -\dfrac{2}{3}x^2$

(b) $f(x) = \dfrac{3}{4}x^2 + 1$

(c) $f(x) = -2x^2 - 3$

(d) $f(x) = 3x^2 + 2$

5 Decide whether each parabola in Problem 4 is wider or narrower than the graph of $f(x) = x^2$.

6 Graph
$$f(x) = \frac{1}{2}(x - 2)^2 + 1.$$

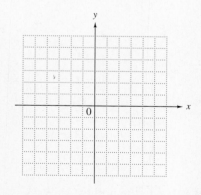

ANSWERS
4. (a) down **(b)** up **(c)** down **(d)** up
5. (a) wider **(b)** wider **(c)** narrower
   (d) narrower
6.

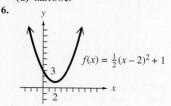

$f(x) = \frac{1}{2}(x - 2)^2 + 1$

7 Tell whether a linear or quadratic function would be a more appropriate model for each set of graphed data. If linear, tell whether the slope should be positive or negative. If quadratic, tell whether the coefficient a of x^2 should be positive or negative.

(a) AVERAGE DAILY E-MAIL VOLUME

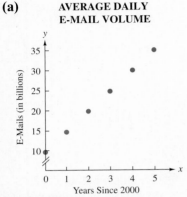

Source: General Accounting Office.

(b) INCREASES IN WHOLESALE DRUG PRICES

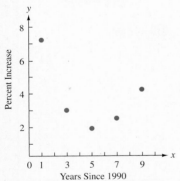

Source: IMS Health, Retail and Provider Perspective.

8 Using the points $(1, 29.1)$, $(6, 35)$, and $(26, 152.6)$, find another quadratic model for the data on higher-order multiple births in Example 6.

U.S. HIGHER-ORDER MULTIPLE BIRTHS

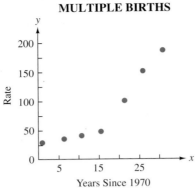

Figure 10

U.S. HIGHER-ORDER MULTIPLE BIRTHS

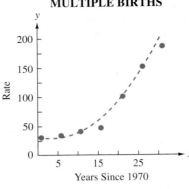

Figure 11

To find a quadratic function of the form

$$y = ax^2 + bx + c$$

that models, or *fits*, these data, we choose three representative ordered pairs and use them to write a system of three equations. Using $(1, 29.1)$, $(11, 40)$, and $(21, 100)$, we substitute the x- and y-values from the ordered pairs into the quadratic form $y = ax^2 + bx + c$ to get the three equations

$$a(1)^2 + b(1) + c = 29.1 \quad \text{or} \quad a + b + c = 29.1 \quad (1)$$
$$a(11)^2 + b(11) + c = 40 \quad \text{or} \quad 121a + 11b + c = 40 \quad (2)$$
$$a(21)^2 + b(21) + c = 100 \quad \text{or} \quad 441a + 21b + c = 100. \quad (3)$$

We can find the values of a, b, and c by solving this system of three equations in three variables using the methods of **Section 5.2.** Multiplying equation (1) by -1 and adding the result to equation (2) gives

$$120a + 10b = 10.9. \quad (4)$$

Multiplying equation (2) by -1 and adding the result to equation (3) gives

$$320a + 10b = 60. \quad (5)$$

We can eliminate b from this system of equations in two variables by multiplying equation (4) by -1 and adding the result to equation (5) to get

$$200a = 49.1$$
$$a = .2455. \quad \text{Use a calculator.}$$

We substitute .2455 for a in equation (4) or (5) to find that $b = -1.856$. Substituting the values of a and b into equation (1) gives $c = 30.7105$. Using these values of a, b, and c, our model is defined by

$$y = .2455x^2 - 1.856x + 30.7105.$$

>>> **Work Problems 7 and 8 at the Side.**

NOTE
If we had chosen three different ordered pairs of data in Example 6, a slightly different model would have resulted.

Calculator Tip The *quadratic regression* feature on a graphing calculator can be used to generate a quadratic model that fits given data. See your owner's manual for details on how to do this.

ANSWERS
7. **(a)** linear; positive **(b)** quadratic; positive
8. $y = .188x^2 - .136x + 29.05$

10.5 Exercises

FOR EXTRA HELP · Tutor Center Addison-Wesley Math Tutor Center · MathXL · Digital Video Tutor CD 6 Videotape 6 · Student's Solutions Manual · MyMathLab · Interactmath.com

1. Match each quadratic function with its graph from choices A–D.

(a) $f(x) = (x + 2)^2 - 1$
(b) $f(x) = (x + 2)^2 + 1$
(c) $f(x) = (x - 2)^2 - 1$
(d) $f(x) = (x - 2)^2 + 1$

A.
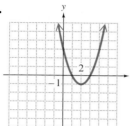

B.
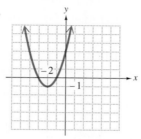

C.
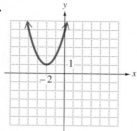

D.
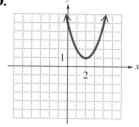

2. Match each quadratic function with its graph from choices A–D.

(a) $f(x) = -x^2 + 2$
(b) $f(x) = -x^2 - 2$
(c) $f(x) = -(x + 2)^2$
(d) $f(x) = -(x - 2)^2$

A.
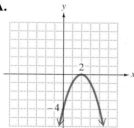

B.
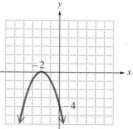

C.
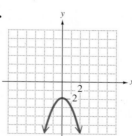

D.
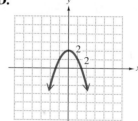

Identify the vertex of each parabola. See Examples 1–4.

3. $f(x) = -3x^2$

4. $f(x) = \dfrac{1}{2}x^2$

5. $f(x) = x^2 + 4$

6. $f(x) = x^2 - 4$

7. $f(x) = (x - 1)^2$

8. $f(x) = (x + 3)^2$

9. $f(x) = (x + 3)^2 - 4$

10. $f(x) = (x - 5)^2 - 8$

11. Describe how each of the parabolas in Exercises 9 and 10 is shifted compared to the graph of $f(x) = x^2$.

12. What does the value of a in $F(x) = a(x - h)^2 + k$ tell you about the graph of the equation compared to the graph of $f(x) = x^2$?

For each quadratic function, tell whether the graph opens up or down and whether the graph is wider, narrower, or the same shape as the graph of $f(x) = x^2$. See Examples 4 and 5.

13. $f(x) = -\dfrac{2}{5}x^2$
14. $f(x) = -2x^2$
15. $f(x) = 3x^2 + 1$
16. $f(x) = \dfrac{2}{3}x^2 - 4$

17. For $f(x) = a(x - h)^2 + k$, in what quadrant is the vertex if

(a) $h > 0, k > 0$; (b) $h > 0, k < 0$;
(c) $h < 0, k > 0$; (d) $h < 0, k < 0$?

18. Match each quadratic function with the description of the parabola that is its graph.

(a) $f(x) = (x - 4)^2 - 2$ **A.** Vertex $(2, -4)$, opens down

(b) $f(x) = (x - 2)^2 - 4$ **B.** Vertex $(2, -4)$, opens up

(c) $f(x) = -(x - 4)^2 - 2$ **C.** Vertex $(4, -2)$, opens down

(d) $f(x) = -(x - 2)^2 - 4$ **D.** Vertex $(4, -2)$, opens up

Sketch the graph of each parabola. Plot at least two points in addition to the vertex. In Exercises 25–32, give the vertex, axis, domain, and range of the parabola.

19. $f(x) = -2x^2$
20. $f(x) = \dfrac{1}{3}x^2$
21. $f(x) = x^2 - 1$

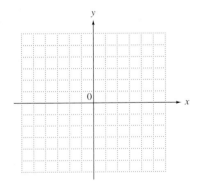
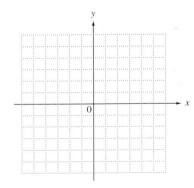
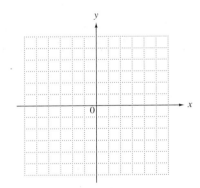

22. $f(x) = x^2 + 3$
23. $f(x) = -x^2 + 2$
24. $f(x) = 2x^2 - 2$

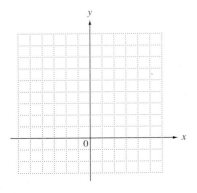
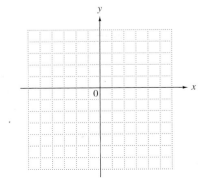
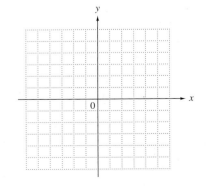

25. $f(x) = \dfrac{1}{2}(x - 4)^2$

vertex:
axis:
domain:
range:

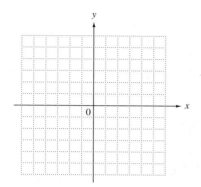

26. $f(x) = -2(x + 1)^2$

vertex:
axis:
domain:
range:

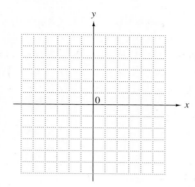

27. $f(x) = (x + 2)^2 - 1$

vertex:
axis:
domain:
range:

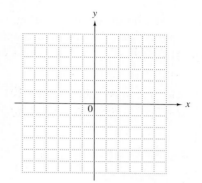

28. $f(x) = (x - 1)^2 + 2$

vertex:
axis:
domain:
range:

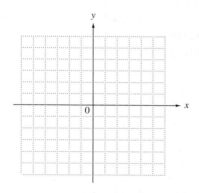

29. $f(x) = -2(x + 3)^2 + 4$

vertex:
axis:
domain:
range:

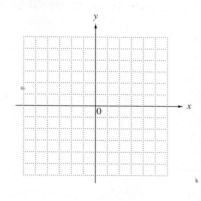

30. $f(x) = 2(x - 2)^2 - 3$

vertex:
axis:
domain:
range:

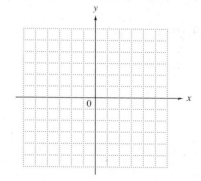

31. $f(x) = -\dfrac{2}{3}(x + 2)^2 + 1$

vertex:
axis:
domain:
range:

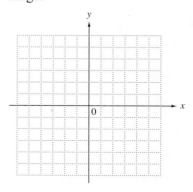

32. $f(x) = -\dfrac{1}{2}(x + 1)^2 + 2$

vertex:
axis:
domain:
range:

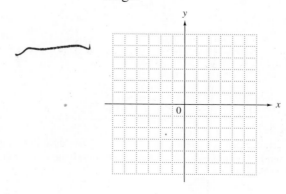

RELATING CONCEPTS (EXERCISES 33–38)   For Individual or Group Work

The procedures described in this section that allow the graph of $f(x) = x^2$ to be shifted vertically and horizontally are applicable to other types of functions. In **Section 4.5** *we introduced linear functions of the form $g(x) = ax + b$. Consider the graph of the simplest linear function defined by $g(x) = x$, shown here, and then* **work Exercises 33–38 in order.**

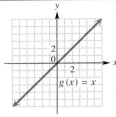

33. Based on the concepts of this section, how does the graph of $F(x) = x^2 + 6$ compare to the graph of $f(x) = x^2$ if a *vertical* shift is considered?

34. Graph the linear function defined by $G(x) = x + 6$.

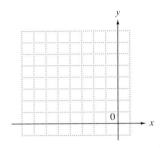

35. Based on the concepts of **Chapter 4,** how does the graph of $G(x) = x + 6$ compare to the graph of $g(x) = x$ if a *vertical* shift is considered? (*Hint:* Look at the y-intercept.)

36. Based on the concepts of this section, how does the graph of $F(x) = (x - 6)^2$ compare to the graph of $f(x) = x^2$ if a *horizontal* shift is considered?

37. Graph the linear function defined by $G(x) = x - 6$.

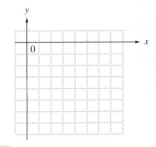

38. Based on the concepts of **Chapter 4,** how does the graph of $G(x) = x - 6$ compare to the graph of $g(x) = x$ if a *horizontal* shift is considered? (*Hint:* Look at the x-intercept.)

In Exercises 39–44, tell whether a linear or quadratic function would be a more appropriate model for each set of graphed data. If linear, tell whether the slope should be positive or negative. If quadratic, tell whether the coefficient of x^2 should be positive or negative. See Example 6.

39.

U.S. TRADE DEFICIT

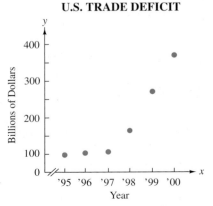

Source: U.S. Department of Commerce.

40.   **AVERAGE DAILY VOLUME OF FIRST-CLASS MAIL**

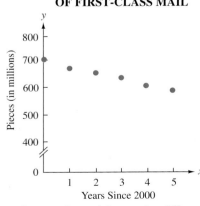

Source: General Accounting Office.

41.   **SOCIAL SECURITY ASSETS***

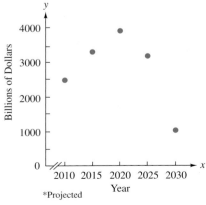

*Projected

Source: Social Security Administration.

42. **CEDAR RAPIDS SCHOOLS— GENERAL RESERVE FUND**

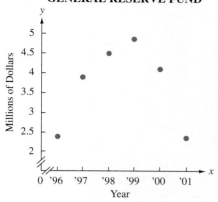

Year

Source: Cedar Rapids School District.

43. **CONSUMER DEMAND FOR ELECTRICITY**

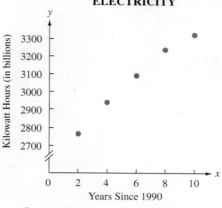

Years Since 1990

Source: U.S. Department of Energy.

44. **U.S. COMMERCIAL BANK FAILURES**

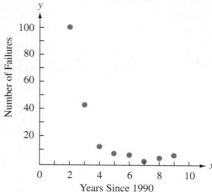

Years Since 1990

Source: www.ABA.com

Solve each problem. See Example 6.

45. The number of publicly traded companies filing for bankruptcy for selected years between 1990 and 2000 are shown in the table. In the year column, 0 represents 1990, 2 represents 1992, and so on.

COMPANY BANKRUPTCY FILINGS

Year	Number of Bankruptcies
0	115
2	91
4	70
6	84
8	120
10	176

Source: www.BankruptcyData.com

(a) Use the ordered pairs (year, number of bankruptcies) to make a scatter diagram of the data.

COMPANY BANKRUPTCY FILINGS

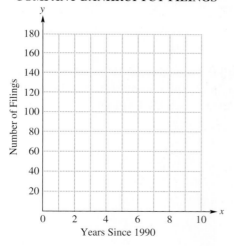

Years Since 1990

(b) Use the scatter diagram to decide whether a linear or quadratic function would better model the data. If quadratic, should the coefficient a of x^2 be positive or negative?

(c) Use the ordered pairs (0, 115), (4, 70), and (8, 120) to find a quadratic function that models the data. Round the values of a, b, and c in your model to three decimal places, as necessary.

(d) Use your model from part (c) to approximate the number of company bankruptcy filings in 2002. Round your answer to the nearest whole number.

(e) The number of company bankruptcy filings through August 16, 2002, was 129. Based on this, is your estimate from part (d) reasonable? Explain.

46. The percent of U.S. high school students in grades 9–12 who smoke is shown in the table for selected years. In the year column, 1 represents 1991, 3 represents 1993, and so on.

HIGH SCHOOL STUDENTS WHO SMOKE

Year	Percent of Students
1	28
3	31
5	35
7	36
9	35
11	29
13	22

Source: National Center for Health Statistics.

(a) Use the ordered pairs (year, percent of students) to make a scatter diagram of the data.

PERCENT OF HIGH SCHOOL STUDENTS WHO SMOKE

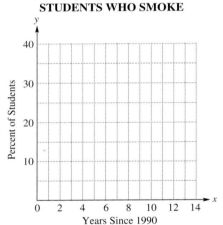

(b) Would a linear or quadratic function better model the data?

(c) Should the coefficient a of x^2 in a quadratic model be positive or negative?

(d) Use the ordered pairs (1, 28), (7, 36), and (11, 29) to find a quadratic function that models the data. Round the values of a, b, and c in your model to the nearest tenth, as necessary.

(e) Use your model from part (d) to approximate the percent of high school students who smoked during 1995 and 2003 to the nearest percent. How well does the model approximate the actual data from the table?

47. In Example 6, we determined that the quadratic function defined by

$$y = .2455x^2 - 1.856x + 30.7105$$

modeled the rate (per 100,000) of higher-order multiple births, where x represents the number of years since 1970.

(a) Use this model to approximate the rate of higher-order births in 2002 to the nearest tenth.

(b) The actual rate of higher-order births in 2002 was 184.0. (*Source:* National Center for Health Statistics.) How does the approximation using the model compare to the actual rate for 2002?

48. Should the model from Exercise 47 be used to approximate the rate of higher-order multiple births in years after 2002? Explain.

10.6 More about Parabolas; Applications

OBJECTIVE 1 Find the vertex of a vertical parabola. When the equation of a parabola is given in the form $f(x) = ax^2 + bx + c$, we need to locate the vertex in order to sketch an accurate graph. There are two ways to do this:

1. Complete the square, as shown in Examples 1 and 2, or

2. Use a formula derived by completing the square, as shown in Example 3.

EXAMPLE 1 Completing the Square to Find the Vertex

Find the vertex of the graph of $f(x) = x^2 - 4x + 5$.

To find the vertex, we need to write the expression $x^2 - 4x + 5$ in the form $(x - h)^2 + k$. We do this by completing the square on $x^2 - 4x$, as in **Section 10.1**. The process is a little different here because we want to keep $f(x)$ alone on one side of the equation. Instead of adding the appropriate number to each side, we *add and subtract* it on the right. This is equivalent to adding 0.

$$f(x) = x^2 - 4x + 5$$
$$= (x^2 - 4x \quad) + 5 \qquad \text{Group the variable terms.}$$
$$\left[\frac{1}{2}(-4)\right]^2 = (-2)^2 = 4$$
$$= (x^2 - 4x + 4 - 4) + 5 \qquad \text{Add and subtract 4.}$$
$$= (x^2 - 4x + 4) - 4 + 5 \qquad \text{Bring } -4 \text{ outside the parentheses.}$$
$$f(x) = (x - 2)^2 + 1 \qquad \text{Factor; combine terms.}$$

The vertex of this parabola is $(2, 1)$.

> **Work Problem 1 at the Side.** ▶▶▶

EXAMPLE 2 Completing the Square to Find the Vertex When $a \neq 1$

Find the vertex of the graph of $f(x) = -3x^2 + 6x - 1$.

We must complete the square on $-3x^2 + 6x$. Because the x^2-term has a coefficient other than 1, we factor that coefficient out of the first two terms and then proceed as in Example 1.

$$f(x) = -3x^2 + 6x - 1$$
$$= -3(x^2 - 2x) - 1 \qquad \text{Factor out } -3.$$
$$\left[\frac{1}{2}(-2)\right]^2 = (-1)^2 = 1$$
$$= -3(x^2 - 2x + 1 - 1) - 1 \qquad \text{Add and subtract 1.}$$

Bring -1 outside the parentheses; be sure to multiply it by -3.

$$= -3(x^2 - 2x + 1) + (-3)(-1) - 1 \qquad \text{Distributive property}$$
$$= -3(x^2 - 2x + 1) + 3 - 1$$
$$f(x) = -3(x - 1)^2 + 2 \qquad \text{Factor; combine terms.}$$

The vertex is $(1, 2)$.

> **Work Problem 2 at the Side.** ▶▶▶

OBJECTIVES

1 Find the vertex of a vertical parabola.

2 Graph a quadratic function.

3 Use the discriminant to find the number of x-intercepts of a vertical parabola.

4 Use quadratic functions to solve problems involving maximum or minimum value.

5 Graph horizontal parabolas.

---

❶ Find the vertex of the graph of each quadratic function.

(a) $f(x) = x^2 - 6x + 7$

(b) $f(x) = x^2 + 4x - 9$

❷ Find the vertex of the graph of each quadratic function.

(a) $f(x) = 2x^2 - 4x + 1$

(b) $f(x) = -\dfrac{1}{2}x^2 + 2x - 3$

ANSWERS
1. (a) $(3, -2)$ (b) $(-2, -13)$
2. (a) $(1, -1)$ (b) $(2, -1)$

3 Use the formula to find the vertex of the graph of each quadratic function.

(a) $f(x) = -2x^2 + 3x - 1$

To derive a formula for the vertex of the graph of the quadratic function defined by $f(x) = ax^2 + bx + c$, complete the square.

$$f(x) = ax^2 + bx + c \quad (a \neq 0) \qquad \text{Standard form}$$

$$= a\left(x^2 + \frac{b}{a}x\right) + c \qquad \begin{array}{l}\text{Factor } a \text{ from the first} \\ \text{two terms.}\end{array}$$

$$\left[\frac{1}{2}\left(\frac{b}{a}\right)\right]^2 = \left(\frac{b}{2a}\right)^2 = \frac{b^2}{4a^2}$$

$$= a\left(x^2 + \frac{b}{a}x + \frac{b^2}{4a^2} - \frac{b^2}{4a^2}\right) + c \qquad \text{Add and subtract } \frac{b^2}{4a^2}.$$

$$= a\left(x^2 + \frac{b}{a}x + \frac{b^2}{4a^2}\right) + a\left(-\frac{b^2}{4a^2}\right) + c \qquad \text{Distributive property}$$

$$= a\left(x^2 + \frac{b}{a}x + \frac{b^2}{4a^2}\right) - \frac{b^2}{4a} + c$$

$$= a\left(x + \frac{b}{2a}\right)^2 + \frac{4ac - b^2}{4a} \qquad \text{Factor; combine terms.}$$

$$f(x) = a\left[x - \left(\frac{-b}{2a}\right)\right]^2 + \frac{4ac - b^2}{4a} \qquad f(x) = (x - h)^2 + k$$

$$\underbrace{\phantom{x - \left(\frac{-b}{2a}\right)}}_{h} \qquad \underbrace{\phantom{\frac{4ac - b^2}{4a}}}_{k}$$

Thus, the vertex (h, k) can be expressed in terms of a, b, and c. It is not necessary to remember the expression for k, since it can be found by replacing x with $\frac{-b}{2a}$. Using function notation, if $y = f(x)$, then the y-value of the vertex is $f\left(\frac{-b}{2a}\right)$.

(b) $f(x) = 4x^2 - x + 5$

Vertex Formula

The graph of the quadratic function defined by $f(x) = ax^2 + bx + c$ has vertex

$$\left(\frac{-b}{2a}, \ f\left(\frac{-b}{2a}\right)\right),$$

and the axis of the parabola is the line

$$x = \frac{-b}{2a}.$$

EXAMPLE 3 Using the Formula to Find the Vertex

Use the vertex formula to find the vertex of the graph of $f(x) = x^2 - x - 6$.
For this function, $a = 1$, $b = -1$, and $c = -6$. The x-coordinate of the vertex of the parabola is given by

$$\frac{-b}{2a} = \frac{-(-1)}{2(1)} = \frac{1}{2}.$$

The y-coordinate is $f\left(\frac{-b}{2a}\right) = f\left(\frac{1}{2}\right)$.

$$f\left(\frac{1}{2}\right) = \left(\frac{1}{2}\right)^2 - \frac{1}{2} - 6 = \frac{1}{4} - \frac{1}{2} - 6 = -\frac{25}{4}$$

The vertex is $\left(\frac{1}{2}, -\frac{25}{4}\right)$.

Work Problem 3 at the Side.

ANSWERS

3. (a) $\left(\frac{3}{4}, \frac{1}{8}\right)$ **(b)** $\left(\frac{1}{8}, \frac{79}{16}\right)$

OBJECTIVE 2 Graph a quadratic function. We give a general approach for graphing any quadratic function here.

Graphing a Quadratic Function f

Step 1 **Determine whether the graph opens up or down.** If $a > 0$, the parabola opens up; if $a < 0$, it opens down.

Step 2 **Find the vertex.** Use either the vertex formula or completing the square.

Step 3 **Find any intercepts.** To find the x-intercepts (if any), solve $f(x) = 0$. To find the y-intercept, evaluate $f(0)$.

Step 4 **Complete the graph.** Plot the points found so far. Find and plot additional points as needed, using symmetry about the axis.

EXAMPLE 4 Using the Steps to Graph a Quadratic Function

Graph the quadratic function defined by $f(x) = x^2 - x - 6$.

Step 1 From the equation, $a = 1$, so the graph of the function opens up.

Step 2 The vertex, $\left(\frac{1}{2}, -\frac{25}{4}\right)$, was found in Example 3 by substituting the values $a = 1$, $b = -1$, and $c = -6$ in the vertex formula.

Step 3 Now find any intercepts. Since the vertex, $\left(\frac{1}{2}, -\frac{25}{4}\right)$, is in quadrant IV and the graph opens up, there will be two x-intercepts. To find them, let $f(x) = 0$ and solve the equation.

$$f(x) = x^2 - x - 6$$
$$0 = x^2 - x - 6 \qquad \text{Let } f(x) = 0.$$
$$0 = (x - 3)(x + 2) \qquad \text{Factor.}$$
$$x - 3 = 0 \quad \text{or} \quad x + 2 = 0 \qquad \text{Zero-factor property}$$
$$x = 3 \quad \text{or} \qquad x = -2$$

The x-intercepts are $(3, 0)$ and $(-2, 0)$. Find the y-intercept.

$$f(x) = x^2 - x - 6$$
$$f(0) = 0^2 - 0 - 6 \qquad \text{Let } x = 0.$$
$$f(0) = -6$$

The y-intercept is $(0, -6)$.

Step 4 Plot the points found so far and additional points as needed using symmetry about the axis $x = \frac{1}{2}$. The graph is shown in Figure 12. The domain is $(-\infty, \infty)$, and the range is $\left[-\frac{25}{4}, \infty\right)$.

x	y
-2	0
-1	-4
0	-6
$\frac{1}{2}$	$-\frac{25}{4}$
2	-4
3	0

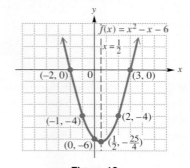

Figure 12

Work Problem 4 at the Side. ▶▶▶

4 Graph the quadratic function defined by

$$f(x) = x^2 - 6x + 5.$$

Give the axis, domain, and range.

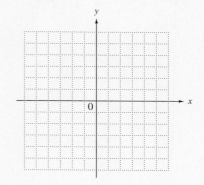

ANSWERS

4.

$f(x) = x^2 - 6x + 5$

axis: $x = 3$; domain: $(-\infty, \infty)$; range: $[-4, \infty)$

5 Use the discriminant to determine the number of x-intercepts of the graph of each quadratic function.

(a) $f(x) = 4x^2 - 20x + 25$

(b) $f(x) = 2x^2 + 3x + 5$

(c) $f(x) = -3x^2 - x + 2$

OBJECTIVE 3 **Use the discriminant to find the number of x-intercepts of a vertical parabola.** Recall from **Section 10.2** that the expression $b^2 - 4ac$ is called the discriminant of the quadratic *equation* $ax^2 + bx + c = 0$ and that we can use it to determine the number of real solutions of a quadratic equation. In a similar way, we can use the discriminant of a quadratic *function* to determine the number of x-intercepts of its graph. See Figure 13. If the discriminant is positive, the parabola will have two x-intercepts. If the discriminant is 0, there will be only one x-intercept, and it will be the vertex of the parabola. If the discriminant is negative, the graph will have no x-intercepts.

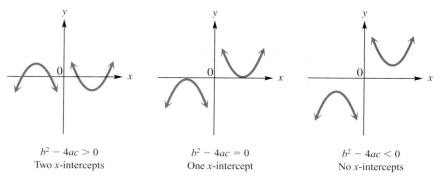

$b^2 - 4ac > 0$
Two x-intercepts

$b^2 - 4ac = 0$
One x-intercept

$b^2 - 4ac < 0$
No x-intercepts

Figure 13

EXAMPLE 5 **Using the Discriminant to Determine the Number of x-Intercepts**

Use the discriminant to determine the number of x-intercepts of the graph of each quadratic function.

(a) $f(x) = 2x^2 + 3x - 5$
The discriminant is $b^2 - 4ac$. Here $a = 2$, $b = 3$, and $c = -5$, so
$$b^2 - 4ac = 9 - 4(2)(-5) = \mathbf{49}.$$

Since the discriminant is positive, the parabola has two x-intercepts.

(b) $f(x) = -3x^2 - 1$
In this equation, $a = -3$, $b = 0$, and $c = -1$. The discriminant is
$$b^2 - 4ac = 0 - 4(-3)(-1) = \mathbf{-12}.$$

The discriminant is negative, so the graph has no x-intercepts.

(c) $f(x) = 9x^2 + 6x + 1$
Here, $a = 9$, $b = 6$, and $c = 1$. The discriminant is
$$b^2 - 4ac = 36 - 4(9)(1) = \mathbf{0}.$$

The parabola has only one x-intercept (its vertex) because the value of the discriminant is 0.

Work Problem 5 at the Side.

OBJECTIVE 4 **Use quadratic functions to solve problems involving maximum or minimum value.** The vertex of a parabola is either the highest or the lowest point on the parabola. The y-value of the vertex gives the maximum or minimum value of y, while the x-value tells where that maximum or minimum occurs.

ANSWERS
5. **(a)** discriminant is 0; one x-intercept
   (b) discriminant is -31; no x-intercepts
   (c) discriminant is 25; two x-intercepts

PROBLEM-SOLVING HINT
In many applied problems we must find the largest or smallest value of some quantity. When we can express that quantity as a quadratic function, the value of k in the vertex (h, k) gives that optimum value.

6 Solve Example 6 if the farmer has only 100 ft of fencing.

EXAMPLE 6 Finding the Maximum Area of a Rectangular Region

A farmer has 120 ft of fencing to enclose a rectangular area next to a building. See Figure 14. Find the maximum area he can enclose.

Figure 14

Let x represent the width of the rectangle. Since he has 120 ft of fencing,

$$x + x + \text{length} = 120 \qquad \text{Sum of the sides is 120 ft.}$$
$$2x + \text{length} = 120 \qquad \text{Combine terms.}$$
$$\text{length} = 120 - 2x. \qquad \text{Subtract } 2x.$$

The area $A(x)$ is given by the product of the width and length, so

$$A(x) = x(120 - 2x)$$
$$= 120x - 2x^2.$$

To determine the maximum area, find the vertex of the parabola given by $A(x) = 120x - 2x^2$ using the vertex formula. Writing the equation in standard form as $A(x) = -2x^2 + 120x$ gives $a = -2$, $b = 120$, and $c = 0$, so

$$h = \frac{-b}{2a} = \frac{-120}{2(-2)} = \frac{-120}{-4} = 30;$$
$$A(30) = -2(30)^2 + 120(30) = -2(900) + 3600 = \mathbf{1800}.$$

The graph is a parabola that opens down, and its vertex is $(30, 1800)$. Thus, the maximum area will be 1800 ft^2. This area will occur if x, the width of the rectangle, is 30 ft.

CAUTION
Be careful when interpreting the meanings of the coordinates of the vertex. The first coordinate, x, gives the value for which the *function value* is a maximum or a minimum. Be sure to read the problem carefully to determine whether you are asked to find the value of the independent variable, the function value, or both.

Work Problem 6 at the Side. ▶▶▶

ANSWERS
6. The rectangle should be 25 ft by 50 ft with a maximum area of 1250 ft^2.

7 Solve the problem.

A toy rocket is launched from the ground so that its distance in feet above the ground after t seconds is

$$s(t) = -16t^2 + 208t.$$

Find the maximum height it reaches and the number of seconds it takes to reach that height.

8 Graph $x = (y + 1)^2 - 4$. Give the vertex, axis, domain, and range.

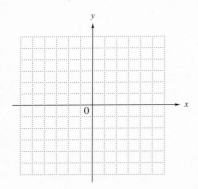

EXAMPLE 7 Finding the Maximum Height Attained by a Projectile

If air resistance is neglected, a projectile on Earth shot straight upward with an initial velocity of 40 m per sec will be at a height s in meters given by

$$s(t) = -4.9t^2 + 40t,$$

where t is the number of seconds elapsed after projection. After how many seconds will it reach its maximum height, and what is this maximum height?

For this function, $a = -4.9$, $b = 40$, and $c = 0$. Use the vertex formula.

$$h = \frac{-b}{2a} = \frac{-40}{2(-4.9)} \approx 4.1 \qquad \text{Use a calculator.}$$

Thus, the maximum height is attained at 4.1 sec. To find this maximum height, calculate $s(4.1)$.

$$s(4.1) = -4.9(4.1)^2 + 40(4.1) \approx 81.6 \qquad \text{Use a calculator.}$$

The projectile will attain a maximum height of approximately 81.6 m.

Work Problem 7 at the Side.

OBJECTIVE 5 Graph horizontal parabolas. If x and y are interchanged in the equation $y = ax^2 + bx + c$, the equation becomes $x = ay^2 + by + c$. Because of the interchange of the roles of x and y, these parabolas are horizontal (with horizontal lines as axes).

Graph of a Horizontal Parabola

The graph of

$$x = ay^2 + by + c \quad \text{or} \quad x = a(y - k)^2 + h$$

is a parabola with vertex (h, k) and the horizontal line $y = k$ as axis. The graph opens to the right if $a > 0$ and to the left if $a < 0$.

EXAMPLE 8 Graphing a Horizontal Parabola

Graph $x = (y - 2)^2 - 3$.

This graph has its vertex at $(-3, 2)$, since the roles of x and y are reversed. It opens to the right, the positive x-direction, and has the same shape as $y = x^2$. Plotting a few additional points gives the graph shown in Figure 15. Note that the graph is symmetric about its axis, $y = 2$. The domain is $[-3, \infty)$, and the range is $(-\infty, \infty)$.

x	y
-3	2
-2	3
-2	1
1	4
1	0

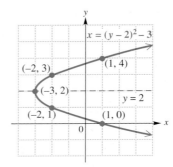

Figure 15

Work Problem 8 at the Side.

ANSWERS
7. 676 ft; 6.5 sec
8.

$x = (y + 1)^2 - 4$

vertex: $(-4, -1)$; axis: $y = -1$; domain: $[-4, \infty)$; range: $(-\infty, \infty)$

When a quadratic equation is given in the form $x = ay^2 + by + c$, completing the square on y will allow us to find the vertex.

EXAMPLE 9 **Completing the Square to Graph a Horizontal Parabola**

Graph $x = -2y^2 + 4y - 3$. Give the domain and range of the relation.

$$x = -2y^2 + 4y - 3$$
$$= -2(y^2 - 2y) - 3 \qquad \text{Factor out } -2.$$
$$= -2(y^2 - 2y + 1 - 1) - 3 \qquad \begin{array}{l}\text{Complete the square;}\\ \text{add and subtract 1.}\end{array}$$
$$= -2(y^2 - 2y + 1) + (-2)(-1) - 3 \qquad \text{Distributive property}$$
$$x = -2(y - 1)^2 - 1 \qquad \text{Factor; simplify.}$$

Because of the negative coefficient (-2), the graph opens to the left (the negative x-direction) and is narrower than the graph of $y = x^2$. As shown in Figure 16, the vertex is $(-1, 1)$. The domain is $(-\infty, -1]$, and the range is $(-\infty, \infty)$.

x	y
-3	2
-3	0
-1	1

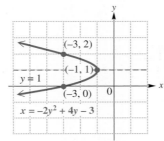

Figure 16

◄◄ Work Problems 9 and 10 at the Side.

In summary, the graphs of parabolas studied in **Sections 10.5 and 10.6** fall into the following categories.

GRAPHS OF PARABOLAS

Equation	Graph	
$y = ax^2 + bx + c$ $y = a(x - h)^2 + k$	(h, k) $a > 0$ — These graphs represent functions.	(h, k) $a < 0$
$x = ay^2 + by + c$ $x = a(y - k)^2 + h$	(h, k) $a > 0$ — These graphs are not graphs of functions.	(h, k) $a < 0$

9 Find the vertex of each parabola. Tell whether the graph opens to the right or to the left. Give the domain and range.

(a) $x = 2y^2 - 6y + 5$

(b) $x = -y^2 + 2y + 5$

10 Graph $x = -y^2 + 2y + 5$. Give the vertex, axis, domain, and range.

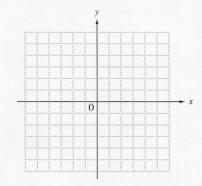

ANSWERS

9. **(a)** $\left(\frac{1}{2}, \frac{3}{2}\right)$; right; domain: $\left[\frac{1}{2}, \infty\right)$; range: $(-\infty, \infty)$

   (b) $(6, 1)$; left; domain; $(-\infty, 6]$; range: $(-\infty, \infty)$

10.

vertex: $(6, 1)$; axis: $y = 1$; domain: $(-\infty, 6]$; range: $(-\infty, \infty)$

11 **(a)** Tell whether each of the following equations has a vertical or horizontal parabola as its graph.

A. $y = -x^2 + 20x + 80$

B. $x = 2y^2 + 6y + 5$

C. $x + 1 = (y + 2)^2$

D. $f(x) = (x - 4)^2$

CAUTION

Only quadratic equations solved for y (whose graphs are vertical parabolas) are examples of functions. The horizontal parabolas in Examples 8 and 9 are *not* graphs of functions, because they do not satisfy the vertical line test. Furthermore, the vertex formula given earlier does not apply to parabolas with horizontal axes.

◀◀◀ **Work Problem 11 at the Side.**

(b) Which of the equations in part (a) represent functions?

ANSWERS

11. (a) A, D are vertical parabolas; B, C are horizontal parabolas.

(b) A, D

10.6 Exercises

FOR EXTRA HELP Addition-Wesley Math Tutor Center MathXL Digital Video Tutor CD 6 Videotape 6 Student's Solutions Manual MyMathLab Interactmath.com

1. How can you determine just by looking at the equation of a parabola whether it has a vertical or a horizontal axis?

2. Why can't the graph of a quadratic function be a horizontal parabola?

3. How can you determine the number of x-intercepts of the graph of a quadratic function without graphing the function?

4. If the vertex of the graph of a quadratic function is $(1, -3)$ and the graph opens down, how many x-intercepts does the graph have?

Find the vertex of each parabola. For each equation, decide whether the graph opens up, down, to the left, or to the right, and whether it is wider, narrower, or the same shape as the graph of $y = x^2$. If it is a vertical parabola, use the discriminant to determine the number of x-intercepts. See Examples 1–3, 5, 8, and 9.

5. $y = 2x^2 + 4x + 5$

6. $y = 3x^2 - 6x + 4$

7. $y = -x^2 + 5x + 3$

8. $x = -y^2 + 7y - 2$

9. $x = \dfrac{1}{3}y^2 + 6y + 24$

10. $x = \dfrac{1}{2}y^2 + 10y - 5$

Graph each parabola. Give the vertex, axis, domain, and range. See Examples 4, 8, and 9.

11. $f(x) = x^2 + 4x + 3$
   vertex:
   axis:
   domain:
   range:

12. $f(x) = x^2 + 2x - 2$
   vertex:
   axis:
   domain:
   range:

13. $f(x) = -2x^2 + 4x - 5$
   vertex:
   axis:
   domain:
   range:

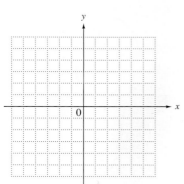

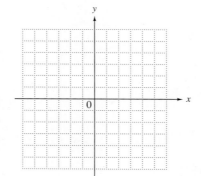

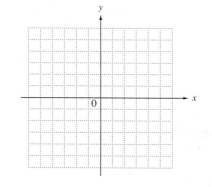

14. $f(x) = -3x^2 + 12x - 8$
vertex:
axis:
domain:
range:

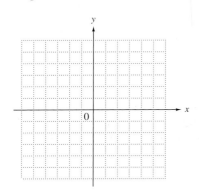

15. $x = -\frac{1}{5}y^2 + 2y - 4$
vertex:
axis:
domain:
range:

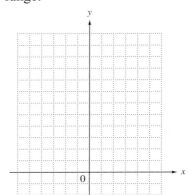

16. $x = -\frac{1}{2}y^2 - 4y - 6$
vertex:
axis:
domain:
range:

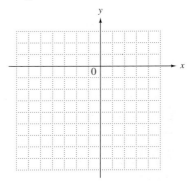

17. $x = 3y^2 + 12y + 5$
vertex:
axis:
domain:
range:

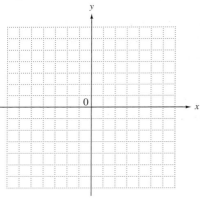

18. $x = 4y^2 + 16y + 11$
vertex:
axis:
domain:
range:

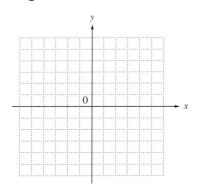

Use the concepts of this section to match each equation with its graph.

19. $y = 2x^2 + 4x - 3$

20. $y = -x^2 + 3x + 5$

21. $y = -\frac{1}{2}x^2 - x + 1$

22. $x = y^2 + 6y + 3$

23. $x = -y^2 - 2y + 4$

24. $x = 3y^2 + 6y + 5$

A.

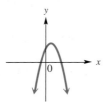

B.

C.

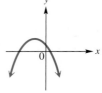

D.

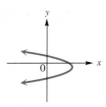

E.

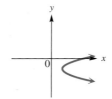

F.

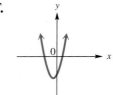

Solve each problem. See Examples 6 and 7.

25. Find the pair of numbers whose sum is 60 and whose product is a maximum. (*Hint:* Let x and $60 - x$ represent the two numbers.)

26. Find the pair of numbers whose sum is 10 and whose product is a maximum.

27. Palo Alto College is planning to construct a rectangular parking lot on land bordered on one side by a highway. The plan is to use 640 ft of fencing to fence off the other three sides. What should the dimensions of the lot be if the enclosed area is to be a maximum?

28. Keisha Hughes has 100 m of fencing material to enclose a rectangular exercise run for her dog. What width will give the enclosure the maximum area?

29. If an object on Earth is propelled upward with an initial velocity of 32 ft per sec, then its height (in feet) after t seconds is given by

$$h(t) = 32t - 16t^2.$$

Find the maximum height attained by the object and the number of seconds it takes to hit the ground.

30. A projectile on Earth is fired straight upward so that its distance (in feet) above the ground t seconds after firing is given by

$$s(t) = -16t^2 + 400t.$$

Find the maximum height it reaches and the number of seconds it takes to reach that height.

31. A charter flight charges a fare of $200 per person, plus $4 per person for each unsold seat on the plane. If the plane holds 100 passengers and if x represents the number of unsold seats, find the following.

    (a) A function defined by $R(x)$ that describes the total revenue received for the flight (*Hint:* Multiply the number of people flying, $100 - x$, by the price per ticket, $200 + 4x$.)

    (b) The number of unsold seats that will produce the maximum revenue

    (c) The maximum revenue

32. For a trip, a charter bus company charges a fare of $48 per person, plus $2 per person for each unsold seat on the bus. If the bus has 42 seats and x represents the number of unsold seats, find the following.

    (a) A function defined by $R(x)$ that describes the total revenue from the trip (*Hint:* Multiply the total number riding, $42 - x$, by the price per ticket, $48 + 2x$.)

    (b) The number of unsold seats that produces the maximum revenue

    (c) The maximum revenue

33. The annual percent increase in the amount pharmacies paid wholesalers for drugs in the years 1990–1999 can be modeled by the quadratic function defined by

$$f(x) = .228x^2 - 2.57x + 8.97,$$

where $x = 0$ represents 1990, $x = 1$ represents 1991, and so on. (*Source: IMS Health*, Retail and Provider Perspective.)

(a) Since the coefficient of x^2 in the model is positive, the graph of this quadratic function is a parabola that opens up. Will the y-value of the vertex of this graph be a maximum or minimum?

(b) In what year was the minimum percent increase? (Round down to the nearest year.) Use the actual x-value of the vertex, to the nearest tenth, to find this increase.

34. The number of tickets sold (in millions) to the top 50 rock concerts from 1998–2002 is modeled by the quadratic function defined by

$$f(x) = -.386x^2 + 1.28x + 11.3,$$

where $x = 0$ represents 1998, $x = 1$ represents 1999, and so on. (*Source:* Pollstar Online.)

(a) Since the coefficient of x^2 in the model is negative, the graph of this quadratic function is a parabola that opens down. Will the y-value of the vertex of this graph be a maximum or minimum?

(b) In what year was the maximum number of tickets sold? (Round down to the nearest year.) Use the actual x-value of the vertex, to the nearest tenth, to find this number.

35. The graph shows how Social Security assets are expected to change as the number of retirees receiving benefits increases.

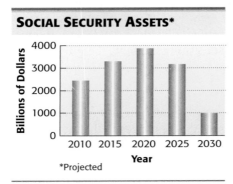

Source: Social Security Administration.

The graph suggests that a quadratic function would be a good fit to the data. The data are approximated by the function defined by

$$f(x) = -20.57x^2 + 758.9x - 3140.$$

In the model, $x = 10$ represents 2010, $x = 15$ represents 2015, and so on, and $f(x)$ is in billions of dollars.

(a) Explain why the coefficient of x^2 in the model is negative, based on the graph.

(b) Algebraically determine the vertex of the graph, with coordinates to four significant digits.

(c) Interpret the answer to part (b) as it applies to the application.

36. The graph shows the performance of investment portfolios with different mixtures of U.S. and foreign investments over a 25-yr period.

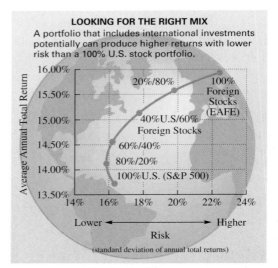

Source: Financial Ink Newsletter, Investment Management and Research, Inc., Feb. 1998. Thanks to David Van Geffen for this information.

(a) Is this the graph of a function? Explain.

(b) What investment mixture shown on the graph appears to represent the vertex? What relative amount of risk does this point represent? What return on investment does it provide?

(c) Which point on the graph represents the riskiest investment mixture? What return on investment does it provide?

10.7 Quadratic and Rational Inequalities

We combine methods of solving linear inequalities and methods of solving quadratic equations to solve *quadratic inequalities*.

Quadratic Inequality

A **quadratic inequality** can be written in the form

$$ax^2 + bx + c < 0 \quad \text{or} \quad ax^2 + bx + c > 0,$$

where a, b, and c are real numbers, with $a \neq 0$.

As before, $<$ and $>$ may be replaced with $\leq$ and $\geq$.

OBJECTIVE 1 Solve quadratic inequalities. One method for solving a quadratic inequality is by graphing the related quadratic function.

EXAMPLE 1 Solving Quadratic Inequalities by Graphing

Solve each inequality.

(a) $x^2 - x - 12 > 0$

To solve the inequality, we graph the related quadratic function defined by $f(x) = x^2 - x - 12$. We are particularly interested in the x-intercepts, which are found as in **Section 10.6** by letting $f(x) = 0$ and solving the quadratic equation

$$x^2 - x - 12 = 0.$$
$$(x - 4)(x + 3) = 0 \qquad \text{Factor.}$$
$$x - 4 = 0 \quad \text{or} \quad x + 3 = 0 \qquad \text{Zero-factor property}$$
$$x = 4 \quad \text{or} \qquad x = -3$$

Thus, the x-intercepts are $(4, 0)$ and $(-3, 0)$. The graph, which opens up since the coefficient of x^2 is positive, is shown in Figure 17(a). Notice from this graph that x-values less than -3 or greater than 4 result in y-values *greater than* 0. Therefore, the solution set of $x^2 - x - 12 > 0$, written in interval notation, is $(-\infty, -3) \cup (4, \infty)$.

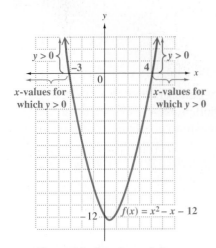

The graph is *above* the x-axis for $(-\infty, -3) \cup (4, \infty)$.

(a)

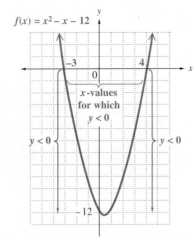

The graph is *below* the x-axis for $(-3, 4)$.

(b)

Figure 17

Continued on Next Page

1 Use the graph to solve each quadratic inequality.

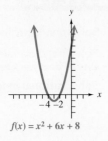

$f(x) = x^2 + 6x + 8$

(a) $x^2 + 6x + 8 > 0$

(b) $x^2 + 6x + 8 < 0$

2 Graph $f(x) = x^2 + 3x - 4$ and use the graph to solve each quadratic inequality.

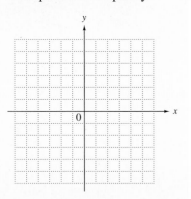

(a) $x^2 + 3x - 4 \geq 0$

(b) $x^2 + 3x - 4 \leq 0$

ANSWERS
1. **(a)** $(-\infty, -4) \cup (-2, \infty)$ **(b)** $(-4, -2)$
2. **(a)** $(-\infty, -4] \cup [1, \infty)$ **(b)** $[-4, 1]$

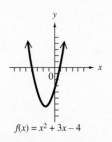

$f(x) = x^2 + 3x - 4$

(b) $x^2 - x - 12 < 0$

Here we want values of y that are *less than* 0. Referring to Figure 17(b) on the previous page, we notice from the graph that x-values between -3 and 4 result in y-values less than 0. Therefore, the solution set of the inequality $x^2 - x - 12 < 0$, written in interval notation, is $(-3, 4)$.

NOTE
If the inequalities in Example 1 had used $\geq$ and $\leq$, the solution sets would have included the x-values of the intercepts and been written in interval notation as $(-\infty, -3] \cup [4, \infty)$ for Example 1(a) and $[-3, 4]$ for Example 1(b).

◀◀◀ **Work Problems 1 and 2 at the Side.**

In Example 1, we used graphing to divide the x-axis into intervals. Then using the graphs in Figure 17, we determined which x-values resulted in y-values that were either greater than or less than 0. Another method for solving a quadratic inequality uses these basic ideas without actually graphing the related quadratic function.

EXAMPLE 2 Solving a Quadratic Inequality Using Test Numbers

Solve $x^2 - x - 12 > 0$.

First solve the quadratic equation $x^2 - x - 12 = 0$ by factoring, as in Example 1(a).

$$(x - 4)(x + 3) = 0$$
$$x - 4 = 0 \quad \text{or} \quad x + 3 = 0$$
$$x = 4 \quad \text{or} \quad x = -3$$

The numbers 4 and -3 divide the number line into the three intervals shown in Figure 18. *Be careful to put the lesser number on the left.* (Notice the similarity between Figure 18 and the x-axis with intercepts $(-3, 0)$ and $(4, 0)$ in Figure 17(a).)

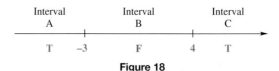

Figure 18

The numbers 4 and -3 are the only numbers that make the expression $x^2 - x - 12$ equal to 0. All other numbers make the expression either positive or negative. The sign of the expression can change from positive to negative or from negative to positive only at a number that makes it 0. Therefore, if one number in an interval satisfies the inequality, then all the numbers in that interval will satisfy the inequality.

To see if the numbers in Interval A satisfy the inequality, choose any number from Interval A in Figure 18 (that is, any number less than -3). Substitute this test number for x in the original inequality $x^2 - x - 12 > 0$. If the result is *true*, then all numbers in Interval A satisfy the inequality.

Continued on Next Page

We choose -5 from Interval A. Substitute -5 for x.

$$x^2 - x - 12 > 0 \qquad \text{Original inequality}$$
$$(-5)^2 - (-5) - 12 > 0 \qquad ?$$
$$25 + 5 - 12 > 0 \qquad ?$$
$$18 > 0 \qquad \text{True}$$

Because -5 from Interval A satisfies the inequality, all numbers from Interval A are solutions.

Try 0 from Interval B. If $x = 0$, then

$$0^2 - 0 - 12 > 0 \qquad ?$$
$$-12 > 0. \qquad \text{False}$$

The numbers in Interval B are *not* solutions.

Work Problem 3 at the Side. ▶▶▶

In Problem 3 at the side, the test number 5 satisfies the inequality, so the numbers in Interval C are also solutions.

Based on these results (shown by the colored letters in Figure 18), the solution set includes the numbers in Intervals A and C, as shown on the graph in Figure 19. The solution set is written in interval notation as

$$(-\infty, -3) \cup (4, \infty).$$

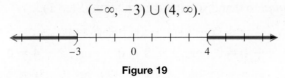

Figure 19

This agrees with the solution set we found by graphing the related quadratic function in Example 1(a).

In summary, a quadratic inequality is solved by following these steps.

Solving a Quadratic Inequality

Step 1 **Write the inequality as an equation and solve it.**

Step 2 **Use the solutions from Step 1 to determine intervals.** Graph the numbers found in Step 1 on a number line. These numbers divide the number line into intervals.

Step 3 **Find the intervals that satisfy the inequality.** Substitute a test number from each interval into the original inequality to determine the intervals that satisfy the inequality. All numbers in those intervals are in the solution set. A graph of the solution set will usually look like one of these. (Square brackets might be used instead of parentheses.)

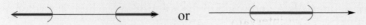

Step 4 **Consider the endpoints separately.** The numbers from Step 1 are included in the solution set if the inequality is $\leq$ or $\geq$; they are not included if it is $<$ or $>$.

Work Problem 4 at the Side. ▶▶▶

3 Does the number 5 from Interval C satisfy $x^2 - x - 12 > 0$?

4 Solve each inequality, and graph the solution set.

(a) $x^2 + x - 6 > 0$

(b) $3m^2 - 13m - 10 \leq 0$

ANSWERS

3. yes

4. (a) $(-\infty, -3) \cup (2, \infty)$

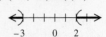

(b) $\left[-\dfrac{2}{3}, 5\right]$

5 Solve each inequality.

(a) $(3k - 2)^2 > -2$

(b) $(5z + 3)^2 < -3$

6 Solve each inequality, and graph the solution set.

(a) $(x - 3)(x + 2)(x + 1) > 0$

(b) $(k - 5)(k + 1)(k - 3) \le 0$

ANSWERS
5. (a) $(-\infty, \infty)$ (b) $\emptyset$
6. (a) $(-2, -1) \cup (3, \infty)$

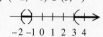

$-2\ -1\ 0\ 1\ 2\ 3\ 4$

(b) $(-\infty, -1] \cup [3, 5]$

$-1\ 0\ 1\ \ \ 3\ \ \ 5$

Special cases of quadratic inequalities may occur, as in the next example.

EXAMPLE 3 Solving Special Cases

Solve $(2t - 3)^2 > -1$. Then solve $(2t - 3)^2 < -1$.

Because $(2t - 3)^2$ is never negative, it is always greater than -1. Thus, the solution set for $(2t - 3)^2 > -1$ is the set of all real numbers, $(-\infty, \infty)$. In the same way, there is no solution for $(2t - 3)^2 < -1$ and the solution set is $\emptyset$.

Work Problem 5 at the Side.

OBJECTIVE 2 Solve polynomial inequalities of degree 3 or more. Higher-degree polynomial inequalities that can be factored are solved in the same way as quadratic inequalities.

EXAMPLE 4 Solving a Third-Degree Polynomial Inequality

Solve $(x - 1)(x + 2)(x - 4) \le 0$.

This is a *cubic* (third-degree) inequality rather than a quadratic inequality, but it can be solved using the method shown in the box by extending the zero-factor property to more than two factors. Begin by setting the factored polynomial *equal* to 0 and solving the equation (Step 1).

$$(x - 1)(x + 2)(x - 4) = 0$$
$$x - 1 = 0 \quad \text{or} \quad x + 2 = 0 \quad \text{or} \quad x - 4 = 0$$
$$x = 1 \quad \text{or} \quad x = -2 \quad \text{or} \quad x = 4$$

Locate the numbers -2, 1, and 4 on a number line, as in Figure 20, to determine the Intervals A, B, C, and D (Step 2).

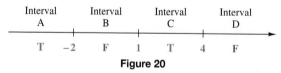

Figure 20

Substitute a test number from each interval in the *original* inequality to determine which intervals satisfy the inequality (Step 3). It is helpful to organize this information in a table.

Interval	Test Number	Test of Inequality	True or False?
A	-3	$-28 \le 0$	T
B	0	$8 \le 0$	F
C	2	$-8 \le 0$	T
D	5	$28 \le 0$	F

Verify the information given in the table and graphed in Figure 21. The numbers in Intervals A and C are in the solution set, which is written as

$$(-\infty, -2] \cup [1, 4].$$

The three endpoints are included since the inequality symbol is $\le$ (Step 4).

Figure 21

Work Problem 6 at the Side.

OBJECTIVE 3 Solve rational inequalities.   Inequalities that involve rational expressions, called **rational inequalities**, are solved similarly using the following steps.

Solving a Rational Inequality

Step 1   **Write the inequality** so that 0 is on one side and there is a single fraction on the other side.

Step 2   **Determine the numbers that make the numerator and denominator equal to 0.**

Step 3   **Divide a number line into intervals.** Use the numbers from Step 2.

Step 4   **Find the intervals that satisfy the inequality.** Test a number from each interval by substituting it into the *original* inequality.

Step 5   **Consider the endpoints separately.** Exclude any values that make the denominator 0.

EXAMPLE 5   Solving a Rational Inequality

Solve $\dfrac{-1}{p-3} > 1$.

Write the inequality so that 0 is on one side (Step 1).

$$\frac{-1}{p-3} - 1 > 0 \qquad \text{Subtract 1.}$$

$$\frac{-1}{p-3} - \frac{p-3}{p-3} > 0 \qquad \text{Use } p-3 \text{ as the common denominator.}$$

$$\frac{-1-p+3}{p-3} > 0 \qquad \begin{array}{l}\text{Write the left side as a single fraction;}\\ \text{be careful with signs in the numerator.}\end{array}$$

$$\frac{-p+2}{p-3} > 0 \qquad \text{Combine terms in the numerator.}$$

The sign of the rational expression $\frac{-p+2}{p-3}$ will change from positive to negative or negative to positive only at those numbers that make the numerator or denominator 0. The number 2 makes the numerator 0, and 3 makes the denominator 0 (Step 2). These two numbers, 2 and 3, divide a number line into three intervals. See Figure 22 (Step 3).

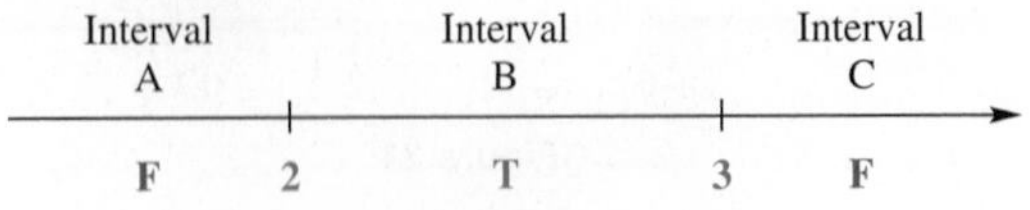

Figure 22

Testing a number from each interval in the *original* inequality, $\frac{-1}{p-3} > 1$, gives the results shown in the table (Step 4).

Interval	Test Number	Test of Inequality	True or False?
A	0	$\frac{1}{3} > 1$	F
B	2.5	$2 > 1$	T
C	4	$-1 > 1$	F

Continued on Next Page

7 Solve each inequality, and graph the solution set.

(a) $\dfrac{2}{x-4} < 3$

(b) $\dfrac{5}{z+1} > 4$

8 Solve $\dfrac{k+2}{k-1} \le 5$, and graph the solution set.

The solution set of $\dfrac{-1}{p-3} > 1$ is the interval $(2, 3)$. This interval does not include 3 since it would make the denominator of the original inequality 0; 2 is not included either since the inequality symbol is $>$ (Step 5). A graph of the solution set is given in Figure 23.

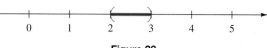

Figure 23

▶◀◀ **Work Problem 7 at the Side.**

CAUTION
When solving a rational inequality, *any number that makes the denominator 0 must be excluded from the solution set.*

EXAMPLE 6 Solving a Rational Inequality

Solve $\dfrac{m-2}{m+2} \le 2$.

Write the inequality so that 0 is on one side (Step 1).

$$\dfrac{m-2}{m+2} - 2 \le 0 \qquad \text{Subtract 2.}$$

$$\dfrac{m-2}{m+2} - \dfrac{2(m+2)}{m+2} \le 0 \qquad \text{Use } m+2 \text{ as the common denominator.}$$

$$\dfrac{m-2-2m-4}{m+2} \le 0 \qquad \text{Write as a single fraction.}$$

$$\dfrac{-m-6}{m+2} \le 0 \qquad \text{Combine terms in the numerator.}$$

The number -6 makes the numerator 0, and -2 makes the denominator 0 (Step 2). These two numbers determine three intervals (Step 3). Test one number from each interval (Step 4) to see that the solution set is the interval

$$(-\infty, -6] \cup (-2, \infty).$$

The number -6 satisfies the original inequality, but -2 cannot be used as a solution since it makes the denominator 0 (Step 5). A graph of the solution set is shown in Figure 24.

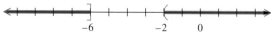

Figure 24

▶◀◀ **Work Problem 8 at the Side.**

ANSWERS

7. (a) $(-\infty, 4) \cup \left(\dfrac{14}{3}, \infty\right)$

(b) $\left(-1, \dfrac{1}{4}\right)$

8. $(-\infty, 1) \cup \left[\dfrac{7}{4}, \infty\right)$

10.7 Exercises

FOR EXTRA HELP

Tutor Center — Addison-Wesley Math Tutor Center

Math XL — MathXL

Digital Video Tutor CD 6 — Videotape 6

Student's Solutions Manual

MyMathLab — MyMathLab

Interactmath.com

In Example 1, we determined the solution sets of the quadratic inequalities $x^2 - x - 12 > 0$ and $x^2 - x - 12 < 0$ by graphing $f(x) = x^2 - x - 12$. The x-intercepts of this graph indicated the solutions of the equation $x^2 - x - 12 = 0$. The x-values of the points on the graph that were **above** *the x-axis formed the solution set of $x^2 - x - 12 > 0$, and the x-values of the points on the graph that were* **below** *the x-axis formed the solution set of $x^2 - x - 12 < 0$.*

In Exercises 1–4, the graph of a quadratic function f is given. Use the graph to find the solution set of each equation or inequality. See Example 1.

1. (a) $x^2 - 4x + 3 = 0$

(b) $x^2 - 4x + 3 > 0$

(c) $x^2 - 4x + 3 < 0$

2. (a) $3x^2 + 10x - 8 = 0$

(b) $3x^2 + 10x - 8 \geq 0$

(c) $3x^2 + 10x - 8 < 0$

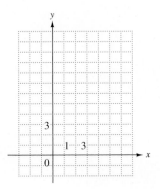

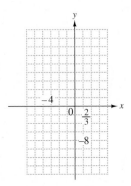

3. (a) $-2x^2 - x + 15 = 0$

(b) $-2x^2 - x + 15 \geq 0$

(c) $-2x^2 - x + 15 \leq 0$

4. (a) $-x^2 + 3x + 10 = 0$

(b) $-x^2 + 3x + 10 \geq 0$

(c) $-x^2 + 3x + 10 \leq 0$

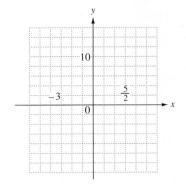

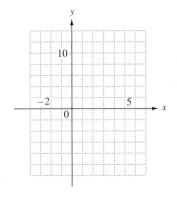

5. Explain how to determine whether to include or exclude endpoints when solving a quadratic or higher-degree inequality.

6. The solution set of the inequality $x^2 + x - 12 < 0$ is the interval $(-4, 3)$. Without actually performing any work, give the solution set of the inequality $x^2 + x - 12 \geq 0$.

Solve each inequality, and graph the solution set. See Example 2.

7. $(x + 1)(x - 5) > 0$

8. $(m + 6)(m - 2) > 0$

9. $(r + 4)(r - 6) < 0$

10. $(x + 4)(x - 8) < 0$

11. $x^2 - 4x + 3 \geq 0$

12. $m^2 - 3m - 10 \geq 0$

13. $10t^2 + 9t \geq 9$

14. $3r^2 + 10r \geq 8$

15. $9p^2 + 3p < 2$

16. $2x^2 + x < 15$

17. $6x^2 + x \geq 1$

18. $4m^2 + 7m \geq -3$

19. $x^2 - 6x + 6 \geq 0$
    (*Hint:* Use the quadratic formula.)

20. $3k^2 - 6k + 2 \leq 0$
    (*Hint:* Use the quadratic formula.)

Solve each inequality. See Example 3.

21. $(4 - 3x)^2 \geq -2$ **22.** $(6p + 7)^2 \geq -1$ **23.** $(3x + 5)^2 \leq -4$ **24.** $(8t + 5)^2 \leq -5$

Solve each inequality, and graph the solution set. See Example 4.

25. $(p - 1)(p - 2)(p - 4) < 0$

26. $(2r + 1)(3r - 2)(4r + 7) < 0$

27. $(x - 4)(2x + 3)(3x - 1) \geq 0$

28. $(z + 2)(4z - 3)(2z + 7) \geq 0$

Solve each inequality, and graph the solution set. See Examples 5 and 6.

29. $\dfrac{x - 1}{x - 4} > 0$

30. $\dfrac{x + 1}{x - 5} > 0$

31. $\dfrac{2n + 3}{n - 5} \leq 0$

32. $\dfrac{3t + 7}{t - 3} \leq 0$

33. $\dfrac{8}{x - 2} \geq 2$

34. $\dfrac{20}{x - 1} \geq 1$

35. $\dfrac{3}{2t - 1} < 2$

36. $\dfrac{6}{m - 1} < 1$

37. $\dfrac{w}{w + 2} \geq 2$

38. $\dfrac{m}{m+5} \geq 2$

39. $\dfrac{4k}{2k-1} < k$

40. $\dfrac{r}{r+2} < 2r$

41. $\dfrac{x-8}{x-4} \leq 3$

42. $\dfrac{2t-3}{t+1} \geq 4$

RELATING CONCEPTS (EXERCISES 43–46)   For Individual or Group Work

A rock is projected vertically upward from the ground. Its distance s in feet above the ground after t seconds is given by the quadratic function defined by

$$s(t) = -16t^2 + 256t.$$

Work Exercises 43–46 in order, *to see how quadratic equations and inequalities are related.*

43. At what times will the rock be 624 ft above the ground? (*Hint:* Let $s(t) = 624$ and solve the quadratic *equation*.)

44. At what times will the rock be more than 624 ft above the ground? (*Hint:* Set $s(t) > 624$ and solve the quadratic *inequality*.)

45. At what times will the rock be at ground level? (*Hint:* Let $s(t) = 0$ and solve the quadratic *equation*.)

46. At what times will the rock be less than 624 ft above the ground? (*Hint:* Set $s(t) < 624$, solve the quadratic *inequality,* and observe the solutions in Exercises 44 and 45 to determine the least and greatest possible values of t.)

Chapter 10

SUMMARY

KEY TERMS

10.1	**quadratic equation**	A quadratic equation is an equation that can be written in the form $ax^2 + bx + c = 0$, where a, b, and c are real numbers, with $a \neq 0$. This form is called standard form.
10.2	**quadratic formula**	The quadratic formula is a formula for solving quadratic equations.
	discriminant	The discriminant is the expression under the radical in the quadratic formula.
10.3	**quadratic in form**	A nonquadratic equation that can be written as a quadratic equation is called quadratic in form.
10.5	**parabola**	The graph of a quadratic function is a parabola.
	vertex	The point on a parabola that has the least y-value (if the parabola opens up) or the greatest y-value (if the parabola opens down) is called the vertex of the parabola.
	axis	The vertical (or horizontal) line through the vertex of a vertical (or horizontal) parabola is its axis.
	quadratic function	A function defined by $f(x) = ax^2 + bx + c$, for real numbers a, b, and c, with $a \neq 0$, is a quadratic function.
10.7	**quadratic inequality**	A quadratic inequality is an inequality that can be written in the form $ax^2 + bx + c < 0$ or $ax^2 + bx + c > 0$ (or with $\leq$ or $\geq$) where a, b, and c are real numbers, with $a \neq 0$.
	rational inequality	An inequality that involves a rational expression is a rational inequality.

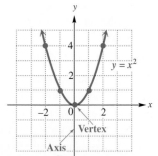

TEST YOUR WORD POWER

See how well you have learned the vocabulary in this chapter. Answers, with examples, follow the Quick Review.

1. The **quadratic formula** is
   A. a formula to find the number of solutions of a quadratic equation
   B. a formula to find the type of solutions of a quadratic equation
   C. the standard form of a quadratic equation
   D. a general formula for solving any quadratic equation.

2. A **quadratic function** is a function that can be written in the form
   A. $f(x) = mx + b$ for real numbers m and b
   B. $f(x) = \frac{P(x)}{Q(x)}$, where $Q(x) \neq 0$
   C. $f(x) = ax^2 + bx + c$ for real numbers a, b, and c ($a \neq 0$)
   D. $f(x) = \sqrt{x}$ for $x \geq 0$.

3. A **parabola** is the graph of
   A. any equation in two variables
   B. a linear equation
   C. an equation of degree 3
   D. a quadratic equation in 2 variables.

4. The **vertex** of a parabola is
   A. the point where the graph intersects the y-axis
   B. the point where the graph intersects the x-axis
   C. the lowest point on a parabola that opens up or the highest point on a parabola that opens down
   D. the origin.

5. The **axis** of a parabola is
   A. either the x-axis or the y-axis
   B. the vertical line (of a vertical parabola) or the horizontal line (of a horizontal parabola) through the vertex
   C. the lowest or highest point on the graph of a parabola
   D. a line through the origin.

6. A parabola is **symmetric about its axis** since
   A. its graph is near the axis
   B. its graph is a mirror image on each side of the axis
   C. its graph looks different on each side of the axis
   D. its graph intersects the axis.

QUICK REVIEW

Concepts	Examples

10.1 *The Square Root Property and Completing the Square*

Square Root Property

If x and k are complex numbers and $x^2 = k$, then

$$x = \sqrt{k} \quad \text{or} \quad x = -\sqrt{k}.$$

Solve $(x - 1)^2 = 8$.

$$x - 1 = \sqrt{8} \qquad \text{or} \quad x - 1 = -\sqrt{8}$$
$$x = 1 + 2\sqrt{2} \quad \text{or} \qquad x = 1 - 2\sqrt{2}$$

Solution set: $\{1 + 2\sqrt{2}, 1 - 2\sqrt{2}\}$

Completing the Square

To solve $ax^2 + bx + c = 0 \ (a \neq 0)$:

Step 1 If $a \neq 1$, divide each side by a.

Step 2 Write the equation with the variable terms on one side and the constant on the other.

Step 3 Take half the coefficient of x and square it.

Step 4 Add the square to each side.

Step 5 Factor the perfect square trinomial, and write it as the square of a binomial. Simplify the other side.

Step 6 Use the square root property to complete the solution.

Solve $2x^2 - 4x - 18 = 0$.

$$x^2 - 2x - 9 = 0 \qquad \text{Divide by 2.}$$
$$x^2 - 2x = 9 \qquad \text{Add 9.}$$

$$\left[\frac{1}{2}(-2)\right]^2 = (-1)^2 = 1$$

$$x^2 - 2x + 1 = 9 + 1 \qquad \text{Add 1.}$$
$$(x - 1)^2 = 10 \qquad \text{Factor; add.}$$

$$x - 1 = \sqrt{10} \qquad \text{or} \quad x - 1 = -\sqrt{10}$$
$$x = 1 + \sqrt{10} \quad \text{or} \qquad x = 1 - \sqrt{10}$$

Solution set: $\{1 + \sqrt{10}, 1 - \sqrt{10}\}$

10.2 *The Quadratic Formula*

Quadratic Formula

The solutions of $ax^2 + bx + c = 0 \ (a \neq 0)$ are given by

$$x = \frac{-b \pm \sqrt{b^2 - 4ac}}{2a}.$$

Solve $3x^2 + 5x + 2 = 0$.

$$x = \frac{-5 \pm \sqrt{5^2 - 4(3)(2)}}{2(3)} = \frac{-5 \pm 1}{6}$$

$$x = -1 \text{ or } x = -\frac{2}{3}$$

Solution set: $\{-1, -\frac{2}{3}\}$

The Discriminant

If a, b, and c are integers, then the discriminant, $b^2 - 4ac$, of $ax^2 + bx + c = 0$ determines the number and type of solutions as follows.

Discriminant	Number and Type of Solutions
Positive, the square of an integer	Two rational solutions
Positive, not the square of an integer	Two irrational solutions
Zero	One rational solution
Negative	Two nonreal complex solutions

For $x^2 + 3x - 10 = 0$, the discriminant is

$$3^2 - 4(1)(-10) = 49. \qquad \text{Two rational solutions}$$

For $4x^2 + x + 1 = 0$, the discriminant is

$$1^2 - 4(4)(1) = -15. \qquad \text{Two nonreal complex solutions}$$

Concepts	*Examples*

10.3 *Equations Quadratic in Form*

A nonquadratic equation that can be written in the form

$$au^2 + bu + c = 0,$$

for $a \neq 0$ and an algebraic expression u, is called quadratic in form. Substitute u for the expression, solve for u, and then solve for the variable in the expression.

Solve $3(x + 5)^2 + 7(x + 5) + 2 = 0$.

$$3u^2 + 7u + 2 = 0 \qquad \text{Let } u = x + 5.$$
$$(3u + 1)(u + 2) = 0$$
$$u = -\frac{1}{3} \quad \text{or} \quad u = -2$$
$$x + 5 = -\frac{1}{3} \quad \text{or} \quad x + 5 = -2 \qquad x + 5 = u$$
$$x = -\frac{16}{3} \quad \text{or} \quad x = -7$$

Solution set: $\{-7, -\frac{16}{3}\}$

10.4 *Formulas and Further Applications*

To solve a formula for a squared variable, proceed as follows.

(a) If the variable appears only to the second power: Isolate the squared variable on one side of the equation, and then use the square root property.

Solve $A = \dfrac{2mp}{r^2}$ for r.

$$r^2 A = 2mp \qquad \text{Multiply by } r^2.$$
$$r^2 = \frac{2mp}{A} \qquad \text{Divide by } A.$$
$$r = \pm\sqrt{\frac{2mp}{A}} = \frac{\pm\sqrt{2mpA}}{A}$$

(b) If the variable appears to the first and second powers: Write the equation in standard form, and then use the quadratic formula.

Solve $m^2 + rm = t$ for m.

$$m^2 + rm - t = 0 \qquad \text{Standard form}$$
$$m = \frac{-r \pm \sqrt{r^2 - 4(1)(-t)}}{2(1)} = \frac{-r \pm \sqrt{r^2 + 4t}}{2}$$
$$a = 1, b = r, c = -t$$

10.5 *Graphs of Quadratic Functions*

1. The graph of the quadratic function defined by $F(x) = a(x - h)^2 + k$, $a \neq 0$, is a parabola with vertex at (h, k) and the vertical line $x = h$ as axis.

2. The graph opens up if a is positive and down if a is negative.

3. The graph is wider than the graph of $f(x) = x^2$ if $0 < |a| < 1$ and narrower if $|a| > 1$.

Graph $f(x) = -(x + 3)^2 + 1$.

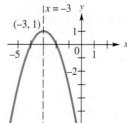

The graph opens down since $a < 0$. It is shifted 3 units left and 1 unit up, so the vertex is $(-3, 1)$, with axis $x = -3$. The domain is $(-\infty, \infty)$; the range is $(-\infty, 1]$.

10.6 *More about Parabolas; Applications*

The vertex of the graph of $f(x) = ax^2 + bx + c$, $a \neq 0$, may be found by completing the square. The vertex has coordinates

$$\left(\frac{-b}{2a}, f\left(\frac{-b}{2a}\right)\right).$$

Graphing a Quadratic Function

Step 1 Determine whether the graph opens up or down.

Step 2 Find the vertex.

Step 3 Find the x-intercepts (if any). Find the y-intercept.

Step 4 Find and plot additional points as needed.

Graph $f(x) = x^2 + 4x + 3$.

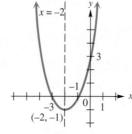

The graph opens up since $a > 0$. The vertex is $(-2, -1)$. The solutions of $x^2 + 4x + 3 = 0$ are -1 and -3, so the x-intercepts are $(-1, 0)$ and $(-3, 0)$. Since $f(0) = 3$, the y-intercept is $(0, 3)$. The domain is $(-\infty, \infty)$; the range is $[-1, \infty)$.

Concepts	Examples

10.6 *More about Parabolas; Applications (continued)*

The graph of

$$x = ay^2 + by + c \quad \text{or} \quad x = a(y - k)^2 + h$$

is a horizontal parabola with vertex (h, k) and the horizontal line $y = k$ as axis. The graph opens to the right if $a > 0$ and to the left if $a < 0$.

Horizontal parabolas do not represent functions.

Graph $x = 2y^2 + 6y + 5$.

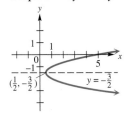

The graph opens to the right since $a > 0$. The vertex is $(\frac{1}{2}, -\frac{3}{2})$. The axis is $y = -\frac{3}{2}$. The domain is $[\frac{1}{2}, \infty)$; the range is $(-\infty, \infty)$.

10.7 *Quadratic and Rational Inequalities*

Solving a Quadratic (or Higher-Degree Polynomial) Inequality

Step 1 Write the inequality as an equation and solve.

Solve $2x^2 + 5x + 2 < 0$.

$$2x^2 + 5x + 2 = 0$$

$$x = -\frac{1}{2} \quad \text{or} \quad x = -2$$

Step 2 Use the numbers found in Step 1 to divide a number line into intervals.

$x = -3$ makes the original inequality false; $x = -1$ makes it true; $x = 0$ makes it false.

Step 3 Substitute a number from each interval into the original inequality to determine the intervals that belong in the solution set.

Step 4 Consider the endpoints separately.

The solution set is the interval $(-2, -\frac{1}{2})$.

Solving a Rational Inequality

Solve $\dfrac{x}{x + 2} \geq 4$.

Step 1 Write the inequality so that 0 is on one side and there is a single fraction on the other side.

$$\frac{x}{x + 2} - 4 \geq 0 \qquad \text{Subtract 4.}$$

$$\frac{x}{x + 2} - \frac{4(x + 2)}{x + 2} \geq 0 \qquad \begin{array}{l}\text{Get a common}\\\text{denominator.}\end{array}$$

$$\frac{-3x - 8}{x + 2} \geq 0$$

Step 2 Determine the numbers that make the numerator and denominator 0.

$-\frac{8}{3}$ makes the numerator 0; -2 makes the denominator 0.

Step 3 Use the numbers from Step 2 to divide a number line into intervals.

Step 4 Substitute a number from each interval into the original inequality to determine the intervals that belong in the solution set.

-4 makes the original inequality false; $-\frac{7}{3}$ makes it true; 0 makes it false.

Step 5 Consider the endpoints separately.

The solution set is the interval $[-\frac{8}{3}, -2)$; -2 is excluded since it makes the denominator 0.

ANSWERS TO TEST YOUR WORD POWER

1. D; *Example:* The solutions of $ax^2 + bx + c = 0$ $(a \neq 0)$ are given by $x = \dfrac{-b \pm \sqrt{b^2 - 4ac}}{2a}$.

2. C; *Examples:* $f(x) = x^2 - 2, f(x) = (x + 4)^2 + 1, f(x) = x^2 - 4x + 5$ **3.** D; *Examples:* See the figures in the Quick Review for **Sections 10.5 and 10.6.**   **4.** C; *Example:* The graph of $y = (x + 3)^2$ has vertex $(-3, 0)$, which is the lowest point on the graph. **5.** B; *Example:* The axis of $y = (x + 3)^2$ is the vertical line $x = -3$.   **6.** B; *Example:* Since the graph of $y = (x + 3)^2$ is symmetric about its axis $x = -3$, the points $(-2, 1)$ and $(-4, 1)$ are on the graph.

Chapter **10**

REVIEW EXERCISES

[10.1] *Solve each equation by using the square root property or completing the square.*

1. $t^2 = 121$

2. $p^2 = 3$

3. $(2x + 5)^2 = 100$

***4.** $(3k - 2)^2 = -25$

5. $x^2 + 4x = 15$

6. $2m^2 - 3m = -1$

7. A student gave the following "solution" to the equation $x^2 = 12$.

$$x^2 = 12$$
$$x = \sqrt{12} \quad \text{Square root property}$$
$$x = 2\sqrt{3}$$

What is wrong with this solution?

8. Navy Pier Center in Chicago, Illinois, features a 150-ft tall Ferris wheel. Use Galileo's formula $d = 16t^2$ to find how long it would take a wallet dropped from the top of the Ferris wheel to fall to the ground. Round your answer to the nearest tenth of a second. (*Source: Microsoft Encarta Encyclopedia 2002.*)

[10.2] *Solve each equation using the quadratic formula.*

9. $2x^2 + x - 21 = 0$

10. $k^2 + 5k = 7$

11. $(t + 3)(t - 4) = -2$

***12.** $2x^2 + 3x + 4 = 0$

***13.** $3p^2 = 2(2p - 1)$

14. $m(2m - 7) = 3m^2 + 3$

Use the discriminant to predict whether the solutions to each equation are
A. *two rational numbers;*   **B.** *one rational number;*
C. *two irrational numbers;*   **D.** *two nonreal complex numbers.*

15. $x^2 + 5x + 2 = 0$

16. $4t^2 = 3 - 4t$

17. $4x^2 = 6x - 8$

18. $9z^2 + 30z + 25 = 0$

---

* This exercise requires knowledge of complex numbers.

[10.3] *Solve each equation.*

19. $\dfrac{15}{x} = 2x - 1$

20. $\dfrac{1}{n} + \dfrac{2}{n+1} = 2$

21. $-2r = \sqrt{\dfrac{48 - 20r}{2}}$

22. $8(3x+5)^2 + 2(3x+5) - 1 = 0$

23. $2x^{2/3} - x^{1/3} - 28 = 0$

24. $p^4 - 5p^2 + 4 = 0$

Solve each problem. Round answers to the nearest tenth, as necessary.

25. Phong paddled his canoe 20 mi upstream, then paddled back. If the speed of the current was 3 mph and the total trip took 7 hr, what was Phong's speed?

26. Maureen O'Connor drove 8 mi to pick up her friend Laurie, and then drove 11 mi to a mall at a speed 15 mph faster. If Maureen's total travel time was 24 min, what was her speed on the trip to pick up Laurie?

27. An old machine processes a batch of checks in 1 hr more time than a new one. How long would it take the old machine to process a batch of checks that the two machines together process in 2 hr?

28. Greg Tobin can process a stack of invoices 1 hr faster than Carter Fenton can. Working together, they take 1.5 hr. How long would it take each person working alone?

[10.4] *Solve each formula for the indicated variable. (Give answers with $\pm$ when applicable.)*

29. $k = \dfrac{rF}{wv^2}$ for v

30. $p = \sqrt{\dfrac{yz}{6}}$ for y

31. $mt^2 = 3mt + 6$ for t

Solve each problem. Round answers to the nearest tenth, as necessary.

32. A large machine requires a part in the shape of a right triangle with a hypotenuse 9 ft less than twice the length of the longer leg. The shorter leg must be $\frac{3}{4}$ the length of the longer leg. Find the lengths of the three sides of the part.

33. A square has an area of 256 cm^2. If the same amount is removed from one dimension and added to the other, the resulting rectangle has an area 16 cm^2 less. Find the dimensions of the rectangle.

34. Nancy wants to buy a mat for a photograph that measures 14 in. by 20 in. She wants to have an even border around the picture when it is mounted on the mat. If the area of the mat she chooses is 352 in.^{2}, how wide will the border be?

35. A search light moves horizontally back and forth along a wall with the distance of the light from a starting point at t minutes given by the quadratic function defined by

$$f(t) = 100t^2 - 300t.$$

How long will it take before the light returns to the starting point?

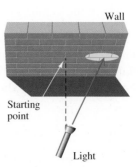

36. Lewis Tower, built in Philadelphia, Pennsylvania, in 1929, is 400 ft high. Suppose that a ball is projected upward from the top of the Tower, and its position in feet above the ground is given by the quadratic function defined by

$$f(t) = -16t^2 + 45t + 400,$$

where t is the number of seconds elapsed. How long will it take for the ball to reach a height of 200 ft above the ground? (*Source: World Almanac and Book of Facts,* 2004.)

37. The Alberta Stock Exchange in Calgary, Alberta, is 407 ft high. Suppose that a ball is projected upward from the top of the Exchange, and its position in feet above the ground is given by the quadratic function defined by

$$s(t) = -16t^2 + 75t + 407,$$

where t is the number of seconds elapsed. How long will it take for the ball to reach a height of 450 ft above the ground? (*Source: World Almanac and Book of Facts*, 2004.)

38. The manager of a fast food outlet has determined that the demand for frozen yogurt is $\frac{25}{p}$ units per day, where p is the price (in dollars) per unit. The supply is $70p + 15$ units per day. Find the price at which supply and demand are equal.

39. Use the formula $A = P(1 + r)^2$ to find the interest rate r at which a principal P of $10,000 will increase to $10,920.25 in 2 yr.

40. The numbers of e-mail boxes in North America (in millions) for the years 1995–2001 are shown in the graph and can be modeled by the quadratic function defined by

$$f(x) = 3.29x^2 - 10.4x + 21.6.$$

In the model, $x = 5$ represents 1995, $x = 10$ represents 2000, and so on.

(a) Use the model to approximate the number of e-mail boxes in 2001 to the nearest whole number. How does this result compare to the number shown in the graph?

(b) Based on the model, in what year did the number of e-mail boxes reach 200 million? (Round down to the nearest year.) How does this result compare to the number shown in the graph?

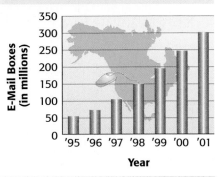

GROWTH OF E-MAIL BOXES IN NORTH AMERICA

Source: IDC research.

[10.5–10.6] *Identify the vertex of the graph of each parabola.*

41. $f(x) = -(x - 1)^2$ **42.** $f(x) = (x - 3)^2 + 7$ **43.** $y = -3x^2 + 4x - 2$ **44.** $x = (y - 3)^2 - 4$

Graph each parabola. Give the vertex, axis, domain, and range.

45. $y = 2(x - 2)^2 - 3$

    vertex: domain:
    axis: range:

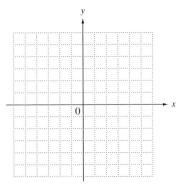

46. $f(x) = -2x^2 + 8x - 5$

    vertex: domain:
    axis: range:

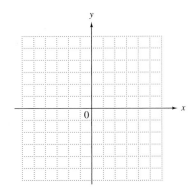

47. $x = 2(y + 3)^2 - 4$

    vertex: domain:
    axis: range:

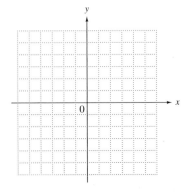

48. $x = -\dfrac{1}{2}y^2 + 6y - 14$

    vertex: domain:
    axis: range:

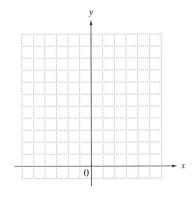

Solve each problem.

⊞ **49.** Consumer spending for home video games in dollars per person per year is given in the table. Let $x = 0$ represent 1995, $x = 1$ represent 1996, and so on.

CONSUMER SPENDING FOR HOME VIDEO GAMES

Year	Dollars
1995	10.54
1996	11.47
1997	16.45
1998	18.49
1999	24.45
2000	24.65
2001	27.69

Source: Communications Industry Forecast, (annual).

(a) Use the data for 1995, 1998, and 2001 in the quadratic form $ax^2 + bx + c = y$ to write a system of three equations.

(b) Solve the system from part (a) to get a quadratic function f that models the data.

(c) Use the model found in part (b) to approximate consumer spending for home video games in 2000 to the nearest cent. How does your answer compare to the actual data from the table?

50. The height (in feet) of a projectile t seconds after being fired from Earth into the air is given by

$$f(t) = -16t^2 + 160t.$$

Find the number of seconds required for the projectile to reach maximum height. What is the maximum height?

51. Find the length and width of a rectangle having a perimeter of 200 m if the area is to be a maximum.

[10.7] *Solve each inequality, and graph the solution set.*

52. $(x - 4)(2x + 3) > 0$

53. $x^2 + x \leq 12$

54. $(x + 2)(x - 3)(x + 5) \leq 0$

55. $(4m + 3)^2 \leq -4$

56. $\dfrac{6}{2z - 1} < 2$

57. $\dfrac{3t + 4}{t - 2} \leq 1$

MIXED REVIEW EXERCISES

Solve.

58. $V = r^2 + R^2h$ for R

*59. $3t^2 - 6t = -4$

60. $(x^2 - 2x)^2 = 11(x^2 - 2x) - 24$

61. $(r - 1)(2r + 3)(r + 6) < 0$

62. $(3k + 11)^2 = 7$

63. $S = \dfrac{Id^2}{k}$ for d

64. $2x - \sqrt{x} = 6$

65. $6 + \dfrac{15}{s^2} = -\dfrac{19}{s}$

66. $\dfrac{-2}{x + 5} \le -5$

67. Graph $f(x) = 4x^2 + 4x - 2$. Give the vertex, axis, domain, and range.

vertex: domain:

axis: range:

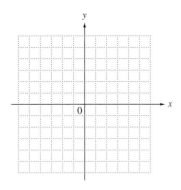

68. Natural gas use in the United States in trillions of cubic feet (ft^3) from 1970 through 1999 can be modeled by the quadratic function defined by

$$f(x) = .014x^2 - .396x + 21.2,$$

where $x = 0$ represents 1970, $x = 5$ represents 1975, and so on. (*Source:* Energy Information Administration.)

(a) Use the model to approximate natural gas use in 2000.

(b) Actual natural gas use in 2000 was 23.33 trillion ft^3. How does this compare to the answer found in part (a) using the model?

(c) Based on the model, in what year will natural gas use reach 25 trillion ft^3? (Round down to the nearest year.)

* This exercise requires knowledge of complex numbers.

Chapter **10**

TEST

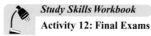

Study Skills Workbook
Activity 12: Final Exams

Solve by using either the square root property or completing the square.

1. $t^2 = 54$

1. _____

2. $(7x + 3)^2 = 25$

2. _____

3. $x^2 + 2x = 1$

3. _____

Solve using the quadratic formula.

4. $2x^2 - 3x - 1 = 0$

4. _____

*5. $3t^2 - 4t = -5$

5. _____

6. $3x = \sqrt{\dfrac{9x + 2}{2}}$

6. _____

*7. If k is a negative number, then which one of the following equations will have two nonreal complex solutions?

A. $x^2 = 4k$ **B.** $x^2 = -4k$
C. $(x + 2)^2 = -k$ **D.** $x^2 + k = 0$

7. _____

8. What is the discriminant for $2x^2 - 8x - 3 = 0$? How many and what type of solutions does this equation have? (Do not actually solve.)

8. _____

Solve by any method.

9. $3 - \dfrac{16}{x} - \dfrac{12}{x^2} = 0$

9. _____

10. $4x^2 + 7x - 3 = 0$

10. _____

_____

* This exercise requires knowledge of complex numbers.

11. _____

11. $9x^4 + 4 = 37x^2$

12. _____

12. $12 = (2n + 1)^2 + (2n + 1)$

13. _____

13. Solve for r: $S = 4\pi r^2$. (Leave $\pm$ in your answer.)

Solve each problem.

14. _____

14. Andrew and Kent do word processing. Kent can prepare a certain prospectus 2 hr faster than Andrew. If they work together, they can do the entire prospectus in 5 hr. How long will it take each of them working alone to prepare the prospectus? Round your answers to the nearest tenth of an hour.

15. _____

15. Abby Tanenbaum paddled her canoe 10 mi upstream, and then paddled back to her starting point. If the rate of the current was 3 mph and the entire trip took $3\frac{1}{2}$ hr, what was Abby's rate?

16. _____

16. Tyler McGinnis has a pool 24 ft long and 10 ft wide. He wants to construct a concrete walk around the pool. If he plans for the walk to be of uniform width and cover 152 ft^2, what will the width of the walk be?

17. _____

17. At a point 30 m from the base of a tower, the distance to the top of the tower is 2 m more than twice the height of the tower. Find the height of the tower.

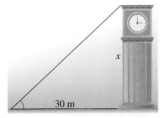

18. The percent increase for in-state tuition at Iowa public universities during the years 1992–2002 can be modeled by the quadratic function defined by

$$f(x) = .156x^2 - 2.05x + 10.2,$$

where $x = 2$ represents 1992, $x = 3$ represents 1993, and so on. (*Source:* Iowa Board of Regents.)

(a) Based on this model, by what percent (to the nearest tenth) did tuition increase in 2001?

(b) In what year was the mini-mum tuition increase? (Round down to the nearest year.) To the nearest tenth, by what percent did tuition increase that year?

18. (a) _____

(b) _____

19. Which one of the following most closely resembles the graph of $f(x) = a(x - h)^2 + k$ if $a < 0$, $h > 0$, and $k < 0$?

19. _____

A.

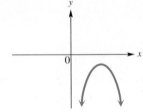

B.

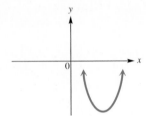

C.

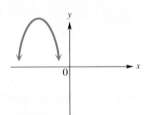

D.

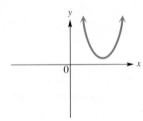

Graph each parabola. Give the vertex, axis, domain, and range.

20. $f(x) = \dfrac{1}{2}x^2 - 2$

20. _____

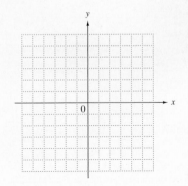

21. _____

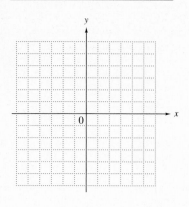

21. $f(x) = -x^2 + 4x - 1$

22. _____

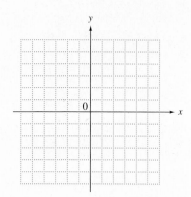

22. $x = 2y^2 + 8y + 3$

23. _____

23. Morgan's Department Store wants to construct a rectangular parking lot on land bordered on one side by a highway. The store has 280 ft of fencing that is to be used to fence off the other three sides. What should be the dimensions of the lot if the enclosed area is to be a maximum? What is the maximum area?

Solve. Graph each solution set.

24. _____→

24. $2x^2 + 7x > 15$

25. _____→

25. $\dfrac{5}{t - 4} \le 1$

Cumulative Review Exercises

CHAPTERS 1–10

1. Let $S = \{-\frac{7}{3}, -2, -\sqrt{3}, 0, .7, \sqrt{12}, \sqrt{-8}, 7, \frac{32}{3}\}$. List the elements of S that are elements of each set.

   (a) Integers   (b) Rational numbers   (c) Real numbers   (d) Complex numbers

Simplify each expression.

2. $|-3| + 8 - |-9| - (-7 + 3)$

3. $2(-3)^2 + (-8)(-5) + (-17)$

In this day of Automated Teller Machines (ATMs), people often find themselves doing what they have done for years when faced with a soft drink machine that won't respond: They talk to it. According to one report, the following are percentages of people in the United States, the United Kingdom (UK), and Germany who talk to ATMs and what they say.

	United States	UK	Germany
Thanking the ATM	22%	24%	14%
Cursing the ATM	31%	41%	53%
Telling the ATM to Hurry Up	47%	36%	33%

Source: BMRB International for NCR.

In a random sample of 3000 people, how many would there be in each category?

4. People in the United States who curse the ATM

5. People in the UK who thank the ATM

6. People in Germany who tell the ATM to hurry up

7. How many more German cursers would there be than United States thankers?

Solve each equation or inequality.

8. $-2x + 4 = 5(x - 4) + 17$

9. $-2x + 4 \leq -x + 3$

10. $|3x - 7| \leq 1$

11. Find the slope and y-intercept of the line with equation $2x - 4y = 7$.

12. Write the equation in standard form of the line through $(2, -1)$ and perpendicular to $-3x + y = 5$.

Graph each relation. Tell whether or not each is a function, and if it is, give its domain and range.

13. $4x - 5y = 15$

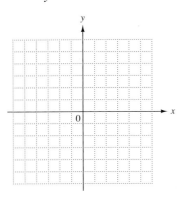

14. $4x - 5y < 15$

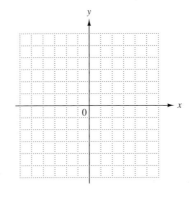

15. $f(x) = -2(x - 1)^2 + 3$

16. Sales of NASCAR-licensed merchandise for recent years are given in the table and can be modeled by a linear equation. Let $x = 0$ represent 1995, $x = 2$ represent 1997, and so on.

Year	Sales (in millions of dollars)
1995	600
1997	800
1999	1130
2001	1340
2002	1400

Source: NASCAR.

(a) Use the ordered pairs (0, 600) and (6, 1340) to write a linear equation that models these data. Round the slope to the nearest whole number.

(b) Use your model to approximate sales of NASCAR-licensed merchandise in 2002. How does it compare to the actual value from the table?

17. Does the relation $x = 5$ define a function? Explain why or why not.

Solve each system of equations.

18. $2x - 4y = 10$
$9x + 3y = 3$

19. $x + y + 2z = 3$
$-x + y + z = -5$
$2x + 3y - z = -8$

20. When America Online and Time Warner merged, the two companies had combined sales of $34.2 billion. Sales for AOL were $.3 billion less than 4 times the sales of Time Warner. What were the sales for each company? (*Source:* Company reports.)

(a) Write a system of equations to solve the problem.

(b) Solve the problem.

Write with positive exponents only. Assume variables represent positive real numbers.

21. $\left(\dfrac{x^{-3}y^2}{x^5y^{-2}}\right)^{-1}$

22. $\dfrac{(4x^{-2})^2\,(2y^3)}{8x^{-3}y^5}$

Perform the indicated operations.

23. $\left(\dfrac{2}{3}t + 9\right)^2$

24. $(3t^3 + 5t^2 - 8t + 7) - (6t^3 + 4t - 8)$

25. Divide $4x^3 + 2x^2 - x + 26$ by $x + 2$.

26. Scientists worldwide are analyzing fragments of a meteorite that shattered over Tagish Lake in British Columbia in January 1999. The meteorite represents the most pristine specimen ever recovered, remaining virtually unchanged since the solar system formed some 4.6×10^9 yr ago.
(*Source: The Gazette*, October 13, 2000.)
Write the age of the meteorite without scientific notation.

Factor completely.

27. $16x - x^3$

28. $24m^2 + 2m - 15$

29. $9x^2 - 30xy + 25y^2$

Perform the operations, and express answers in lowest terms. Assume denominators represent nonzero real numbers.

30. $\dfrac{5t + 2}{-6} \div \dfrac{15t + 6}{5}$

31. $\dfrac{3}{2 - k} - \dfrac{5}{k} + \dfrac{6}{k^2 - 2k}$

32. $\dfrac{\dfrac{r}{s} - \dfrac{s}{r}}{\dfrac{r}{s} + 1}$

Simplify each radical expression.

33. $\sqrt[3]{\dfrac{27}{16}}$

34. $\dfrac{2}{\sqrt{7} - \sqrt{5}}$

Solve each equation.

35. $2x = \sqrt{\dfrac{5x + 2}{3}}$

36. $2x^2 - 4x - 3 = 0$

37. $z^2 - 2z = 15$

38. $\dfrac{3}{x - 3} - \dfrac{2}{x - 2} = \dfrac{3}{x^2 - 5x + 6}$

39. $p^4 - 10p^2 + 9 = 0$

40. Two cars left an intersection at the same time, one heading due south and the other due east. Later they were exactly 95 mi apart. The car heading east had gone 38 mi less than twice as far as the car heading south. How far had each car traveled?

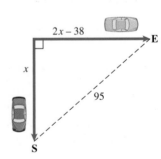

Exponential and Logarithmic Functions

11

11.1 **Inverse Functions**

11.2 **Exponential Functions**

11.3 **Logarithmic Functions**

11.4 **Properties of Logarithms**

11.5 **Common and Natural Logarithms**

11.6 **Exponential and Logarithmic Equations; Further Applications**

With the many advances made in electronics over the past decade, home theater is now a reality. The operating instructions for the Pioneer PD-F1009 compact disc player, a component in one author's system, includes a warning that loud noises can cause hearing damage, and a list of sound levels, measured in *decibels* and shown in the accompanying table, that can be dangerous under constant exposure.

In Exercise 39 of Section 11.5 we examine the meaning of decibel, which is based on *logarithmic functions,* one topic covered in this chapter.

Decibel Level	Example
90	Subway, motorcycle, truck traffic, lawn mower
100	Garbage truck, chain saw, pneumatic drill
120	Rock concert in front of speakers, thunderclap
140	Gunshot blast, jet plane
180	Rocket launching pad

Source: Deafness Research Foundation.

11.1 Inverse Functions

OBJECTIVES

1 Decide whether a function is one-to-one and, if it is, find its inverse.

2 Use the horizontal line test to determine whether a function is one-to-one.

3 Find the equation of the inverse of a function.

4 Graph f^{-1} from the graph of f.

In this chapter we will study two important types of functions, *exponential* and *logarithmic*. These functions are related in a special way: They are *inverses* of one another. We begin by discussing inverse functions in general.

▦ **Calculator Tip** A calculator with the following keys will be essential in this chapter.

$\boxed{y^x}$, $\boxed{10^x}$ or $\boxed{\text{LOG}}$, $\boxed{e^x}$ or $\boxed{\ln x}$

We will explain how these keys are used at appropriate places in the chapter.

OBJECTIVE 1 Decide whether a function is one-to-one and, if it is, find its inverse. Suppose we define the function

$$G = \{(-2, 2), (-1, 1), (0, 0), (1, 3), (2, 5)\}.$$

We can form another set of ordered pairs from G by interchanging the x- and y-values of each pair in G. Call this set F, with

$$F = \{(2, -2), (1, -1), (0, 0), (3, 1), (5, 2)\}.$$

To show that these two sets are related, F is called the *inverse* of G. For a function f to have an inverse, f must be *one-to-one*.

One-to-One Function

In a **one-to-one function,** each x-value corresponds to only one y-value, and each y-value corresponds to just one x-value.

The function shown in Figure 1(a) is not one-to-one because the y-value 7 corresponds to *two x*-values, 2 and 3. That is, the ordered pairs (2, 7) and (3, 7) both appear in the function. The function in Figure 1(b) is one-to-one.

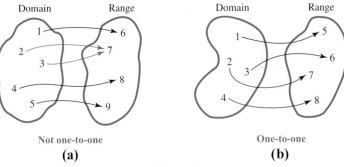

Not one-to-one
(a)

One-to-one
(b)

Figure 1

The *inverse* of any one-to-one function f is found by interchanging the components of the ordered pairs of f. The inverse of f is written f^{-1}. Read f^{-1} as "the inverse of f" or "f-inverse."

CAUTION

The symbol $f^{-1}(x)$ does not represent $\dfrac{1}{f(x)}$.

The definition of the inverse of a function follows.

Inverse of a Function

The **inverse** of a one-to-one function f, written f^{-1}, is the set of all ordered pairs of the form (y, x), where (x, y) belongs to f. Since the inverse is formed by interchanging x and y, the domain of f becomes the range of f^{-1} and the range of f becomes the domain of f^{-1}.

For inverses f and f^{-1}, it follows that

$$f(f^{-1}(x)) = x \quad \text{and} \quad f^{-1}(f(x)) = x.$$

EXAMPLE 1 Finding the Inverses of One-to-One Functions

Find the inverse of each function that is one-to-one.

(a) $F = \{(-2, 1), (-1, 0), (0, 1), (1, 2), (2, 2)\}$
    Each x-value in F corresponds to just one y-value. However, the y-value 2 corresponds to two x-values, 1 and 2. Also, the y-value 1 corresponds to both -2 and 0. Because some y-values correspond to more than one x-value, F is not one-to-one and does not have an inverse.

(b) $G = \{(3, 1), (0, 2), (2, 3), (4, 0)\}$
    Every x-value in G corresponds to only one y-value, and every y-value corresponds to only one x-value, so G is a one-to-one function. The inverse function is found by interchanging the x- and y-values in each ordered pair.

$$G^{-1} = \{(1, 3), (2, 0), (3, 2), (0, 4)\}$$

Notice how the domain and range of G become the range and domain, respectively, of G^{-1}.

(c) The U.S. Environmental Protection Agency has developed an indicator of air quality called the Pollutant Standard Index (PSI). If the PSI exceeds 100 on a particular day, that day is classified as unhealthy. The table shows the number of unhealthy days in Chicago for the years 1991–2002, based on new standards set in 1998.

Year	Number of Unhealthy Days	Year	Number of Unhealthy Days
1991	24	1997	10
1992	5	1998	12
1993	4	1999	19
1994	13	2000	2
1995	24	2001	22
1996	7	2002	21

Source: U.S. Environmental Protection Agency, Office of Air Quality Planning and Standards.

Let f be the function defined in the table, with the years forming the domain and the numbers of unhealthy days forming the range. Then f is not one-to-one, because in two different years (1991 and 1995), the number of unhealthy days was the same, 24.

Work Problem 1 at the Side.

1 Find the inverse of each function that is one-to-one.

(a) $\{(1, 2), (2, 4), (3, 3), (4, 5)\}$

(b) $\{(0, 3), (-1, 2), (1, 3)\}$

(c) A Norwegian physiologist has developed a rule for predicting running times based on the time to run 5 km (5K). An example for one runner is shown here. (*Source:* Stephen Seiler, Agder College, Kristiansand, Norway.)

Distance	Time
1.5K	4:22
3K	9:18
5K	16:00
10K	33:40

ANSWERS
1. (a) $\{(2, 1), (4, 2), (3, 3), (5, 4)\}$
   (b) not a one-to-one function
   (c)

Time	Distance
4:22	1.5K
9:18	3K
16:00	5K
33:40	10K

2 Use the horizontal line test to determine whether each graph is the graph of a one-to-one function.

(a)

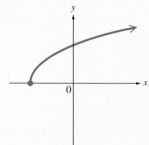

OBJECTIVE 2 Use the horizontal line test to determine whether a function is one-to-one. It may be difficult to decide whether a function is one-to-one just by looking at the equation that defines the function. However, by graphing the function and observing the graph, we can use the *horizontal line test* to tell whether the function is one-to-one.

Horizontal Line Test

A function is one-to-one if every horizontal line intersects the graph of the function at most once.

The horizontal line test follows from the definition of a one-to-one function. Any two points that lie on the same horizontal line have the same y-coordinate. No two ordered pairs that belong to a one-to-one function may have the same y-coordinate, and therefore no horizontal line will intersect the graph of a one-to-one function more than once.

EXAMPLE 2 Using the Horizontal Line Test

Use the horizontal line test to determine whether the graphs in Figures 2 and 3 are graphs of one-to-one functions.

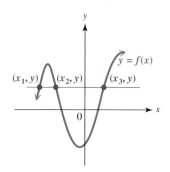

Figure 2

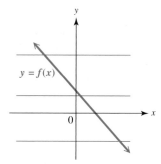

Figure 3

(b)

Because the horizontal line shown in Figure 2 intersects the graph in more than one point (actually three points), the function is not one-to-one.

Every horizontal line will intersect the graph in Figure 3 in exactly one point. This function is one-to-one.

Work Problem 2 at the Side.

OBJECTIVE 3 Find the equation of the inverse of a function. By definition, the inverse of a function is found by interchanging the x- and y-values of each of its ordered pairs. The equation of the inverse of a function defined by $y = f(x)$ is found in the same way.

Finding the Equation of the Inverse of $y = f(x)$

For a one-to-one function f defined by an equation $y = f(x)$, find the defining equation of the inverse as follows.

Step 1   Interchange x and y.

Step 2   Solve for y.

Step 3   Replace y with $f^{-1}(x)$.

ANSWERS
2. (a) one-to-one   **(b)** not one-to-one

EXAMPLE 3 **Finding Equations of Inverses**

Decide whether each equation defines a one-to-one function. If so, find the equation of the inverse.

(a) $f(x) = 2x + 5$

The graph of $y = 2x + 5$ is a nonvertical line, so by the horizontal line test, f is a one-to-one function. To find the inverse, let $y = f(x)$ so that

$$y = 2x + 5$$
$$x = 2y + 5 \qquad \text{Interchange } x \text{ and } y. \text{ (Step 1)}$$
$$2y = x - 5 \qquad \text{Solve for } y. \text{ (Step 2)}$$
$$y = \frac{x - 5}{2}$$
$$f^{-1}(x) = \frac{x - 5}{2}. \qquad \text{Replace } y \text{ with } f^{-1}(x). \text{ (Step 3)}$$

Thus, f^{-1} is a linear function. In the function with $y = 2x + 5$, the value of y is found by starting with a value of x, multiplying by 2, and adding 5. The equation for the inverse has us *subtract* 5, and then *divide* by 2. This shows how an inverse is used to "undo" what a function does to the variable x.

(b) $y = x^2 + 2$

This equation has a vertical parabola as its graph, so some horizontal lines will intersect the graph at two points. For example, both $x = 3$ and $x = -3$ correspond to $y = 11$. Because of the x^2-term, there are many pairs of x-values that correspond to the same y-value. This means that the function defined by $y = x^2 + 2$ is not one-to-one and does not have an inverse.

If this is not noticed, following the steps for finding the equation of an inverse leads to

$$y = x^2 + 2$$
$$x = y^2 + 2 \qquad \text{Interchange } x \text{ and } y.$$
$$x - 2 = y^2 \qquad \text{Solve for } y.$$
$$\pm\sqrt{x - 2} = y. \qquad \text{Square root property}$$

The last step shows that there are two y-values for each choice of $x > 2$, so the given function is not one-to-one and cannot have an inverse.

(c) $f(x) = (x - 2)^3$

Refer to **Section 6.3** to see from its graph that a cubing function like this is a one-to-one function.

$$y = (x - 2)^3 \qquad \text{Replace } f(x) \text{ with } y.$$
$$x = (y - 2)^3 \qquad \text{Interchange } x \text{ and } y.$$
$$\sqrt[3]{x} = \sqrt[3]{(y - 2)^3} \qquad \text{Take the cube root on each side.}$$
$$\sqrt[3]{x} = y - 2$$
$$y = \sqrt[3]{x} + 2 \qquad \text{Solve for } y.$$
$$f^{-1}(x) = \sqrt[3]{x} + 2 \qquad \text{Replace } y \text{ with } f^{-1}(x).$$

Work Problem 3 at the Side. ▶▶▶

3 Decide whether each equation defines a one-to-one function. If so, find the equation that defines the inverse.

(a) $f(x) = 3x - 4$

(b) $f(x) = x^3 + 1$

(c) $f(x) = (x - 3)^2$

ANSWERS

3. **(a)** one-to-one function; $f^{-1}(x) = \dfrac{x + 4}{3}$

   (b) one-to-one function; $f^{-1}(x) = \sqrt[3]{x - 1}$

   (c) not a one-to-one function

4 Use the given graphs to graph each inverse.

(a)

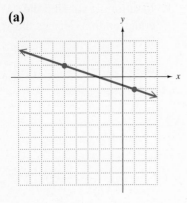

(b)

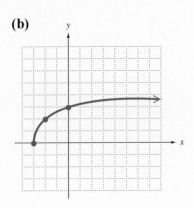

(c)

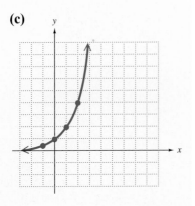

OBJECTIVE 4 Graph f^{-1} from the graph of f. One way to graph the inverse of a function f whose equation is known is to find some ordered pairs that belong to f, interchange x and y to get ordered pairs that belong to f^{-1}, plot those points, and sketch the graph of f^{-1} through the points. A simpler way is to select points on the graph of f and use symmetry to find corresponding points on the graph of f^{-1}.

For example, suppose the point (a, b) shown in Figure 4 belongs to a one-to-one function f. Then the point (b, a) belongs to f^{-1}. The line segment connecting (a, b) and (b, a) is perpendicular to, and cut in half by, the line $y = x$. The points (a, b) and (b, a) are "mirror images" of each other with respect to $y = x$. For this reason we can find the graph of f^{-1} from the graph of f by locating the mirror image of each point in f with respect to the line $y = x$.

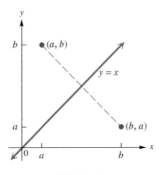

Figure 4

EXAMPLE 4 Graphing the Inverse

Graph the inverses of the functions shown in Figure 5.

In Figure 5 the graphs of two functions are shown in blue. Their inverses are shown in red. In each case, the graph of f^{-1} is symmetric to the graph of f with respect to the line $y = x$.

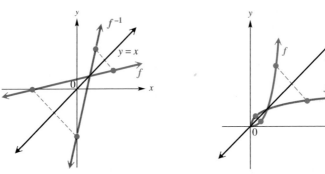

Figure 5

Work Problem 4 at the Side.

ANSWERS
4. (a) (b)

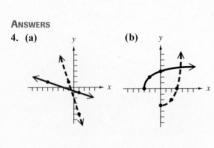

(c)

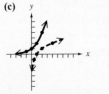

11.1 Exercises

FOR EXTRA HELP

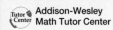 Addison-Wesley Math Tutor Center

 MathXL

Digital Video Tutor CD 7 Videotape 7

Student's Solutions Manual

 MyMathLab

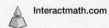 Interactmath.com

1. The table shows the number of uncontrolled hazardous waste sites that require further investigation to determine whether remedies are needed under the Superfund program. The seven states listed are ranked in the top ten in the United States.

   If this correspondence is considered to be a function that pairs each state with its number of uncontrolled waste sites, is it one-to-one? If not, explain why.

State	Number of Sites
New Jersey	116
California	99
Pennsylvania	96
New York	91
Florida	52
Illinois	45
Wisconsin	40

   Source: U.S. Environmental Protection Agency.

2. The table shows emissions of a major air pollutant, carbon monoxide, in the United States for the years 1992–1998.

   If this correspondence is considered to be a function that pairs each year with its emissions amount, is it one-to-one? If not, explain why.

Year	Amount of Emissions (in thousands of tons)
1992	97,630
1993	98,160
1994	102,643
1995	93,353
1996	95,479
1997	94,410
1998	89,454

   Source: U.S. Environmental Protection Agency.

3. Suppose you consider the set of ordered pairs (x, y) such that x represents a person in your mathematics class and y represents that person's mother. Explain how this function might not be a one-to-one function.

4. The road mileage between Denver, Colorado, and several selected U.S. cities is shown in the table below.

City	Distance to Denver (in miles)
Atlanta	1398
Dallas	781
Indianapolis	1058
Kansas City, MO	600
Los Angeles	1059
San Francisco	1235

   If we consider this as a function that pairs each city with a distance, is it a one-to-one function? How could we change the answer to this question by adding 1 mile to one of the distances shown?

1e correct response from the given list.

5. function is made up of ordered pairs in such a way that the same y-value appears in a correspondence with two different x-values, then

  A. the function is one-to-one

  B. the function is not one-to-one

  C. its graph does not pass the vertical line test

  D. it has an inverse function associated with it.

6. Which equation defines a one-to-one function? Explain why the others are not, using specific examples.

  A. $f(x) = x$ B. $f(x) = x^2$
  C. $f(x) = |x|$ D. $f(x) = -x^2 + 2x - 1$

7. Only one of the graphs illustrates a one-to-one function. Which one is it?

A. 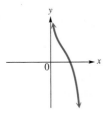   B.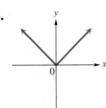

C. 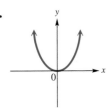   D.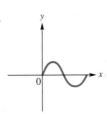

8. If a function f is one-to-one and the point (p, q) lies on the graph of f, then which point *must* lie on the graph of f^{-1}?

  A. $(-p, q)$ B. $(-q, -p)$
  C. $(p, -q)$ D. (q, p)

If the function is one-to-one, find its inverse. See Examples 1–3.

9. $\{(3, 6), (2, 10), (5, 12)\}$

10. $\left\{(-1, 3), (0, 5), (5, 0), \left(7, -\dfrac{1}{2}\right)\right\}$

11. $\{(-1, 3), (2, 7), (4, 3), (5, 8)\}$

12. $\{(-8, 6), (-4, 3), (0, 6), (5, 10)\}$

13. $f(x) = 2x + 4$

14. $f(x) = 3x + 1$

15. $g(x) = \sqrt{x - 3}, x \geq 3$

16. $g(x) = \sqrt{x + 2}, x \geq -2$

17. $f(x) = 3x^2 + 2$

18. $f(x) = -4x^2 - 1$

19. $f(x) = x^3 - 4$

20. $f(x) = x^3 - 3$

Let $f(x) = 2^x$. We will see in the next section that the function f is one-to-one. Find each value, always working part (a) before part (b).

21. (a) $f(3)$ **22. (a)** $f(4)$ **23. (a)** $f(0)$ **24. (a)** $f(-2)$

     (b) $f^{-1}(8)$ **(b)** $f^{-1}(16)$ **(b)** $f^{-1}(1)$ **(b)** $f^{-1}\left(\dfrac{1}{4}\right)$

The graphs of some functions are given in Exercises 25–30. **(a)** *Use the horizontal line test to determine whether each function is one-to-one.* **(b)** *If the function is one-to-one, graph the inverse of the function with a dashed line (or curve) on the same set of axes. (Remember that if f is one-to-one and $f(a) = b$, then $f^{-1}(b) = a$.) See Example 4.*

25.

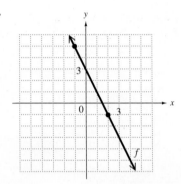

26.

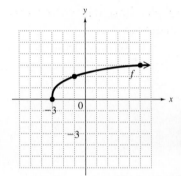

27.

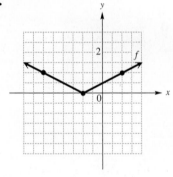

28.

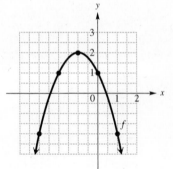

29.

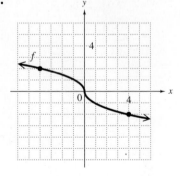

30.

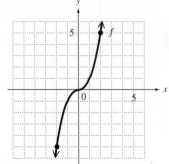

...tion defined in Exercises 31–38 is a one-to-one function. Graph the function as ...e (or curve), and then graph its inverse on the same set of axes as a dashed line (or cu... ...). In Exercises 37 and 38 you are given a table to complete so that graphing the function will be easier. See Example 4.

31. $f(x) = 2x - 1$

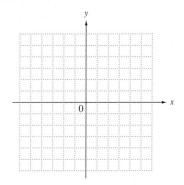

32. $f(x) = 2x + 3$

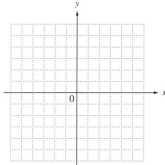

33. $g(x) = -4x$

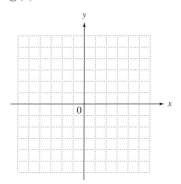

34. $g(x) = -2x$

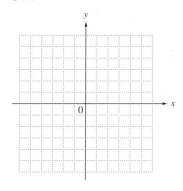

35. $f(x) = \sqrt{x}, x \geq 0$

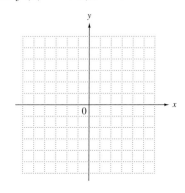

36. $f(x) = -\sqrt{x}, x \geq 0$

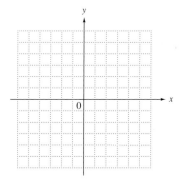

37. $y = x^3 - 2$

x	y
−1	
0	
1	
2	

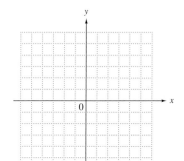

38. $y = x^3 + 3$

x	y
−2	
−1	
0	
1	

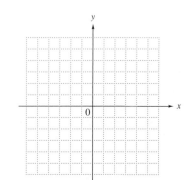

11.2 Exponential Functions

OBJECTIVE 1 Define exponential functions. In **Section 9.2** we showed how to evaluate 2^x for rational values of x. For example,

$$2^3 = 8, \qquad 2^{-1} = \frac{1}{2}, \qquad 2^{1/2} = \sqrt{2}, \qquad 2^{3/4} = \sqrt[4]{2^3} = \sqrt[4]{8}.$$

In more advanced courses it is shown that 2^x exists for all real number values of x, both rational and irrational. (Later in this chapter, we will see how to approximate the value of 2^x for irrational x.) The following definition of an exponential function assumes that a^x exists for all real numbers x.

OBJECTIVES

1 Define exponential functions.

2 Graph exponential functions.

3 Solve exponential equations of the form $a^x = a^k$ for x.

4 Use exponential functions in applications involving growth or decay.

Exponential Function

For $a > 0$, $a \neq 1$, and all real numbers x,

$$f(x) = a^x$$

defines the **exponential function with base a.**

NOTE
The two restrictions on a in the definition of an exponential function are important. The restriction that a must be positive is necessary so that the function can be defined for all real numbers x. For example, letting a be negative ($a = -2$, for instance) and letting $x = \frac{1}{2}$ would give the expression $(-2)^{1/2}$, which is not real. The other restriction, $a \neq 1$, is necessary because 1 raised to any power is equal to 1, and the function would then be the linear function defined by $f(x) = 1$.

OBJECTIVE 2 Graph exponential functions. We can graph an exponential function by finding several ordered pairs that belong to the function, plotting these points, and connecting them with a smooth curve.

EXAMPLE 1 Graphing an Exponential Function with $a > 1$

Graph $f(x) = 2^x$.

Choose some values of x, and find the corresponding values of $f(x)$.

x	-3	-2	-1	0	1	2	3	4
$f(x) = 2^x$	$\frac{1}{8}$	$\frac{1}{4}$	$\frac{1}{2}$	1	2	4	8	16

Plotting these points and drawing a smooth curve through them gives the graph shown in Figure 6 on the next page. This graph is typical of the graphs of exponential functions of the form $F(x) = a^x$, where $a > 1$. The larger the value of a, the faster the graph rises. To see this, compare the graph of $F(x) = 5^x$ with the graph of $f(x) = 2^x$ in Figure 6.

Continued on Next Page

1.

(a) $f(x) = 10^x$

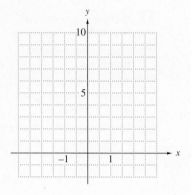

(b) $g(x) = \left(\dfrac{1}{4}\right)^x$

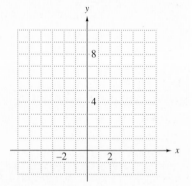

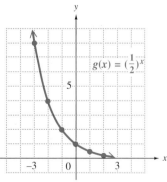

Figure 6

By the vertical line test, the graphs in Figure 6 represent functions. As these graphs suggest, the domain of an exponential function includes all real numbers. Because y is always positive, the range is $(0, \infty)$. Figure 6 also shows an important characteristic of exponential functions where $a > 1$: as x gets larger, y increases at a faster and faster rate.

CAUTION
Be sure to plot a sufficient number of points to see how rapidly the graph rises.

EXAMPLE 2 Graphing an Exponential Function with $a < 1$

Graph $g(x) = \left(\dfrac{1}{2}\right)^x$.

Again, find some points on the graph.

x	-3	-2	-1	0	1	2	3
$g(x) = (\frac{1}{2})^x$	8	4	2	1	$\frac{1}{2}$	$\frac{1}{4}$	$\frac{1}{8}$

The graph, shown in Figure 7, is very similar to that of $f(x) = 2^x$ (Figure 6) with the same domain and range, except that here as x gets larger, y *decreases*. This graph is typical of the graph of a function of the form $F(x) = a^x$, where $0 < a < 1$.

Answers
1. **(a)**

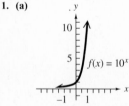

(b)

Figure 7

Work Problem 1 at the Side.

Based on Examples 1 and 2, we make the following generalizations about the graphs of exponential functions of the form $F(x) = a^x$.

Characteristics of the Graph of $F(x) = a^x$

1. The graph contains the point $(0, 1)$.

2. When $a > 1$, the graph *rises* from left to right. When $0 < a < 1$, the graph *falls* from left to right. In both cases, the graph goes from the second quadrant to the first.

3. The graph approaches the x-axis, but never touches it. (Recall from **Section 8.4** that such a line is called an *asymptote*.)

4. The domain is $(-\infty, \infty)$, and the range is $(0, \infty)$.

EXAMPLE 3 **Graphing a More Complicated Exponential Function**

Graph $f(x) = 3^{2x-4}$.
  Find some ordered pairs.

$$\text{If } x = 0, \text{ then } y = 3^{2(0)-4} = 3^{-4} = \frac{1}{81}.$$

$$\text{If } x = 2, \text{ then } y = 3^{2(2)-4} = 3^0 = 1.$$

These ordered pairs, $(0, \frac{1}{81})$ and $(2, 1)$, along with the other ordered pairs shown in the table, lead to the graph in Figure 8. The graph is similar to the graph of $f(x) = 3^x$ except that it is shifted to the right and rises more rapidly.

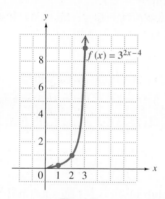

x	y
0	$\frac{1}{81}$
1	$\frac{1}{9}$
2	1
3	9

Figure 8

Work Problem 2 at the Side.)))

OBJECTIVE 3 **Solve exponential equations of the form $a^x = a^k$ for x.** Until this chapter, we have solved only equations that had the variable as a base, like $x^2 = 8$; all exponents have been constants. An **exponential equation** is an equation that has a variable in an exponent, such as

$$9^x = 27.$$

By the horizontal line test, the exponential function defined by $F(x) = a^x$ is a one-to-one function, so we can use the following property to solve many exponential equations.

Property for Solving an Exponential Equation
  For $a > 0$ and $a \neq 1$, if $a^x = a^y$ then $x = y.$

This property would not necessarily be true if $a = 1$.

2 Graph $y = 2^{4x-3}$.

ANSWERS
2.
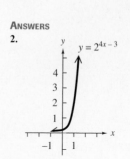

3 Solve each equation and check the solution.

(a) $25^x = 125$

To solve an exponential equation using this property, follow these steps.

Solving an Exponential Equation

Step 1 **Each side must have the same base.** If the two sides of the equation do not have the same base, express each as a power of the same base.

Step 2 **Simplify exponents,** if necessary, using the rules of exponents.

Step 3 **Set exponents equal** using the property given in this section.

Step 4 **Solve** the equation obtained in Step 3.

NOTE

These steps cannot be applied to an exponential equation like

$$3^x = 12$$

because Step 1 cannot easily be done. A method for solving such equations is given in **Section 11.6.**

(b) $4^x = 32$

EXAMPLE 4 Solving an Exponential Equation

Solve the equation $9^x = 27$.

We can use the property given in the box if both sides are written with the same base. Since $9 = 3^2$ and $27 = 3^3$,

$$9^x = 27$$
$$(3^2)^x = 3^3 \qquad \text{Write with the same base. (Step 1)}$$
$$3^{2x} = 3^3 \qquad \text{Power rule for exponents (Step 2)}$$
$$2x = 3 \qquad \text{If } a^x = a^y, \text{ then } x = y. \text{ (Step 3)}$$
$$x = \frac{3}{2}. \qquad \text{Solve for } x. \text{ (Step 4)}$$

Check that the solution set is $\left\{\frac{3}{2}\right\}$ by substituting $\frac{3}{2}$ for x in the original equation.

(c) $81^p = 27$

Work Problem 3 at the Side.

EXAMPLE 5 Solving Exponential Equations

Solve each equation.

(a) $4^{3x-1} = 16^{x+2}$
Since $4 = 2^2$ and $16 = 2^4$,

$$(2^2)^{3x-1} = (2^4)^{x+2} \qquad \text{Write with the same base.}$$
$$2^{6x-2} = 2^{4x+8} \qquad \text{Power rule for exponents}$$
$$6x - 2 = 4x + 8 \qquad \text{Set exponents equal.}$$
$$2x = 10 \qquad \text{Subtract } 4x; \text{ add 2.}$$
$$x = 5. \qquad \text{Divide by 2.}$$

Verify that the solution set is $\{5\}$.

Continued on Next Page

ANSWERS

3. **(a)** $\left\{\frac{3}{2}\right\}$ **(b)** $\left\{\frac{5}{2}\right\}$ **(c)** $\left\{\frac{3}{4}\right\}$

(b) $6^x = \dfrac{1}{216}$

$6^x = \dfrac{1}{6^3}$ $216 = 6^3$

$6^x = 6^{-3}$ Write with the same base; $\dfrac{1}{6^3} = 6^{-3}$.

$x = -3$ Set exponents equal.

Verify that the solution set is $\{-3\}$.

(c) $\left(\dfrac{2}{3}\right)^x = \dfrac{9}{4}$

$\left(\dfrac{2}{3}\right)^x = \left(\dfrac{4}{9}\right)^{-1}$ $\dfrac{9}{4} = \left(\dfrac{4}{9}\right)^{-1}$

$\left(\dfrac{2}{3}\right)^x = \left[\left(\dfrac{2}{3}\right)^2\right]^{-1}$ Write with the same base.

$\left(\dfrac{2}{3}\right)^x = \left(\dfrac{2}{3}\right)^{-2}$ Power rule for exponents

$x = -2$ Set exponents equal.

Check that the solution set is $\{-2\}$.

> **Work Problem 4 at the Side.** ▶▶▶

OBJECTIVE 4 Use exponential functions in applications involving growth or decay.

EXAMPLE 6 Solving an Application Involving Exponential Growth

One result of the rapidly increasing world population is an increase of carbon dioxide in the air, which scientists believe may be contributing to global warming. Both population and carbon dioxide in the air are increasing exponentially. This means that the growth rate is continually increasing. The graph in Figure 9 shows the concentration of carbon dioxide (in parts per million) in the air.

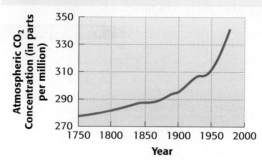

CARBON DIOXIDE IN THE AIR

Source: Sacramento Bee, Monday, September 13, 1993.

Figure 9

Continued on Next Page

---

4 Solve each equation and check the solution.

(a) $25^{x-2} = 125^x$

(b) $4^x = \dfrac{1}{32}$

(c) $\left(\dfrac{3}{4}\right)^x = \dfrac{16}{9}$

ANSWERS

4. **(a)** $\{-4\}$ **(b)** $\left\{-\dfrac{5}{2}\right\}$ **(c)** $\{-2\}$

5 Solve each problem.

(a) Use the function in Example 6 to approximate the carbon dioxide concentration in 1925.

The data are approximated by the function defined by

$$f(x) = 278(1.00084)^x,$$

where x is the number of years since 1750. Use this function and a calculator to approximate the concentration of carbon dioxide in parts per million for each year.

(a) 1900

Since x represents the number of years since 1750, in this case $x = 1900 - 1750 = 150$. Thus, evaluate $f(150)$.

$$f(150) = 278(1.00084)^{150} \qquad \text{Let } x = 150.$$
$$\approx 315 \text{ parts per million} \qquad \text{Use a calculator.}$$

(b) 1950

Use $x = 1950 - 1750 = 200$.

$$f(200) = 278(1.00084)^{200}$$
$$\approx 329 \text{ parts per million}$$

EXAMPLE 7 Applying an Exponential Decay Function

The atmospheric pressure (in millibars) at a given altitude x, in meters, can be approximated by the function defined by

$$f(x) = 1038(1.000134)^{-x},$$

for values of x between 0 and 10,000. Because the base is greater than 1 and the coefficient of x in the exponent is negative, the function values decrease as x increases. This means that as the altitude increases, the atmospheric pressure decreases. (*Source:* Miller, A. and J. Thompson, *Elements of Meteorology,* Fourth Edition, Charles E. Merrill Publishing Company, 1993.)

(b) Use the function in Example 7 to find the pressure at 8000 m.

(a) According to this function, what is the pressure at ground level?

At ground level, $x = 0$, so

$$f(0) = 1038(1.000134)^{-0} = 1038(1) = 1038.$$

The pressure is 1038 millibars.

(b) What is the pressure at 5000 m?

Use a calculator to find $f(5000)$.

$$f(5000) = 1038(1.000134)^{-5000} \approx 531$$

The pressure is approximately 531 millibars.

◀◀◀ Work Problem 5 at the Side.

ANSWERS
5. (a) 322 parts per million
   (b) approximately 355 millibars

11.2 Exercises

FOR EXTRA HELP

Tutor Center · Addison-Wesley Math Tutor Center

MathXL · MathXL

Digital Video Tutor CD 7 · Videotape 7

Student's Solutions Manual

MyMathLab · MyMathLab

Interactmath.com

Choose the correct response in Exercises 1–4.

1. Which point lies on the graph of $f(x) = 2^x$?

   A. $(1, 0)$ **B.** $(2, 1)$

   C. $(0, 1)$ **D.** $\left(\sqrt{2}, \dfrac{1}{2}\right)$

2. Which statement is true?

   A. The y-intercept of the graph of $f(x) = 10^x$ is $(0, 10)$.

   B. For any $a > 1$, the graph of $f(x) = a^x$ falls from left to right.

   C. The point $(\frac{1}{2}, \sqrt{5})$ lies on the graph of $f(x) = 5^x$.

   D. The graph of $y = 4^x$ rises at a faster rate than the graph of $y = 10^x$.

3. The asymptote of the graph of $F(x) = a^x$

   A. is the x-axis.   **B.** is the y-axis.

   C. has equation $x = 1$.   **D.** has equation $y = 1$.

4. Which equation is graphed here?

   A. $y = 1000\left(\dfrac{1}{2}\right)^{.3x}$

   B. $y = 1000\left(\dfrac{1}{2}\right)^{x}$

   C. $y = 1000(2)^{.3x}$

   D. $y = 1000^{x}$

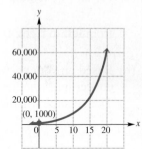

Graph each exponential function. See Examples 1–3.

5. $f(x) = 3^x$

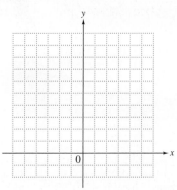

6. $f(x) = 5^x$

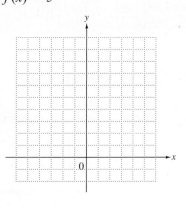

7. $g(x) = \left(\dfrac{1}{3}\right)^{x}$

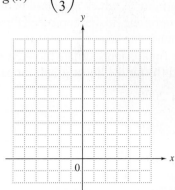

8. $g(x) = \left(\dfrac{1}{5}\right)^{x}$

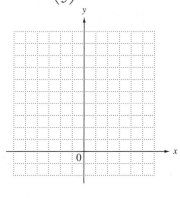

9. $y = 2^{2x-2}$

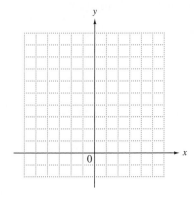

10. $y = 2^{2x+1}$

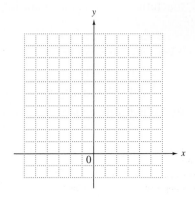

Solve each equation. See Examples 4 and 5.

11. $6^x = 36$

12. $8^x = 64$

13. $100^x = 1000$

14. $8^x = 4$

15. $16^{2x+1} = 64^{x+3}$

16. $9^{2x-8} = 27^{x-4}$

17. $5^x = \dfrac{1}{125}$

18. $3^x = \dfrac{1}{81}$

19. $5^x = .2$

20. $10^x = .1$

21. $\left(\dfrac{3}{2}\right)^x = \dfrac{8}{27}$

22. $\left(\dfrac{4}{3}\right)^x = \dfrac{27}{64}$

23. (a) For an exponential function defined by $f(x) = a^x$, if $a > 1$, then the graph _____ (rises/falls) from left to right. If $0 < a < 1$, then the graph _____ from left to right. (rises/falls)

(b) Based on your answers in part (a), make a conjecture (an educated guess) concerning whether an exponential function defined by $f(x) = a^x$ is one-to-one. Then decide whether it has an inverse based on the concepts of **Section 11.1.**

Solve each problem. See Examples 6 and 7.

The figure shown here accompanied the article "Is Our World Warming?", which appeared in the October 1990 issue of *National Geographic*. It shows projected temperature increases using two graphs: one an exponential-type curve and the other linear. From the figure, approximate the increase **(a)** for the exponential curve, and **(b)** for the linear graph for each of the following years.

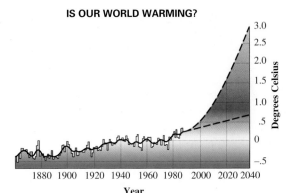

IS OUR WORLD WARMING?

Graph, "Zero Equals Average Global Temperature for the Period 1950–1979." Dale D. Glasgow, © National Geographic Society. Reprinted by permission.

24. 2000

25. 2010

26. 2020

27. 2040

28. A small business estimates that the value $V(t)$ of a copy machine is decreasing according to the function defined by

$$V(t) = 5000(2)^{-.15t},$$

where t is the number of years that have elapsed since the machine was purchased and $V(t)$ is in dollars.

(a) What was the original value of the machine?

(b) What is the value of the machine 5 yr after purchase? Give your answer to the nearest dollar.

(c) What is the value of the machine 10 yr after purchase? Give your answer to the nearest dollar.

(d) Graph the function.

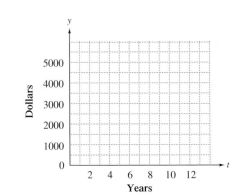

11.3 Logarithmic Functions

The graph of $y = 2^x$ is the curve shown in blue in Figure 10. Because $y = 2^x$ defines a one-to-one function, it has an inverse. Interchanging x and y gives

$$x = 2^y, \quad \text{the inverse of} \quad y = 2^x.$$

As we saw in **Section 11.1,** the graph of the inverse is found by reflecting the graph of $y = 2^x$ about the line $y = x$. The graph of $x = 2^y$ is shown as a red curve in Figure 10.

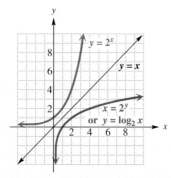

Figure 10

OBJECTIVES

1 Define a logarithm.

2 Convert between exponential and logarithmic forms.

3 Solve logarithmic equations of the form $\log_a b = k$ for a, b, or k.

4 Define and graph logarithmic functions.

5 Use logarithmic functions in applications of growth or decay.

OBJECTIVE 1 Define a logarithm. We cannot solve the equation $x = 2^y$ for the dependent variable y with the methods presented up to now. The following definition is used to solve $x = 2^y$ for y.

Logarithm

For all positive numbers a, with $a \neq 1$, and all positive numbers x,

$$y = \log_a x \quad \text{means the same as} \quad x = a^y.$$

This key statement should be memorized. The abbreviation **log** is used for **logarithm.** Read $\log_a x$ as "the logarithm of x to the base a." To remember the location of the base and the exponent in each form, refer to the following diagrams.

Exponent $\downarrow$

Logarithmic form: $y = \log_a x$

$\uparrow$ Base

Exponent $\downarrow$

Exponential form: $x = a^y$

$\uparrow$ Base

In working with logarithmic form and exponential form, remember the following.

Meaning of $\log_a x$

A logarithm is an exponent. The expression $\log_a x$ represents the exponent to which the base a must be raised to obtain x.

1 Complete the table.

Exponential Form	Logarithmic Form
$2^5 = 32$	
$100^{1/2} = 10$	
	$\log_8 4 = \dfrac{2}{3}$
	$\log_6 \dfrac{1}{1296} = -4$

OBJECTIVE 2 Convert between exponential and logarithmic forms. We can use the definition of logarithm to write exponential statements in logarithmic form and logarithmic statements in exponential form. The following table shows several pairs of equivalent statements.

Exponential Form	Logarithmic Form
$3^2 = 9$	$\log_3 9 = 2$
$\left(\dfrac{1}{5}\right)^{-2} = 25$	$\log_{1/5} 25 = -2$
$10^5 = 100,000$	$\log_{10} 100,000 = 5$
$4^{-3} = \dfrac{1}{64}$	$\log_4 \dfrac{1}{64} = -3$

◀◀◀ Work Problem 1 at the Side.

OBJECTIVE 3 Solve logarithmic equations of the form $\log_a b = k$ for a, b, or k. A **logarithmic equation** is an equation with a logarithm in at least one term. We solve logarithmic equations of the form $\log_a b = k$ for any of the three variables by first writing the equation in exponential form.

EXAMPLE 1 Solving Logarithmic Equations

Solve each equation.

(a) $\log_4 x = -2$

By the definition of logarithm, $\log_4 x = -2$ is equivalent to $x = 4^{-2}$. Solve this exponential equation.

$$x = 4^{-2} = \frac{1}{16}$$

The solution set is $\left\{\frac{1}{16}\right\}$.

(b) $\log_{1/2} (3x + 1) = 2$

$$3x + 1 = \left(\frac{1}{2}\right)^2 \qquad \text{Write in exponential form.}$$

$$3x + 1 = \frac{1}{4}$$

$$12x + 4 = 1 \qquad \text{Multiply by 4.}$$

$$12x = -3 \qquad \text{Subtract 4.}$$

$$x = -\frac{1}{4} \qquad \text{Divide by 12; lowest terms}$$

The solution set is $\left\{-\frac{1}{4}\right\}$.

(c) $\log_x 3 = 2$

$$x^2 = 3 \qquad \text{Write in exponential form.}$$

$$x = \sqrt{3} \qquad \text{Take the } \textit{principal} \text{ square root.}$$

Notice that only the principal square root satisfies the equation, since the base must be a positive number. The solution set is $\{\sqrt{3}\}$.

Continued on Next Page

ANSWERS

1. $\log_2 32 = 5$; $\log_{100} 10 = \dfrac{1}{2}$;

$8^{2/3} = 4$; $6^{-4} = \dfrac{1}{1296}$

(d) $\log_{49} \sqrt[3]{7} = x$

$$49^x = \sqrt[3]{7} \qquad \text{Write in exponential form.}$$
$$(7^2)^x = 7^{1/3} \qquad \text{Write with the same base.}$$
$$7^{2x} = 7^{1/3} \qquad \text{Power rule for exponents.}$$
$$2x = \frac{1}{3} \qquad \text{Set exponents equal.}$$
$$x = \frac{1}{6} \qquad \text{Divide by 2.}$$

The solution set is $\{\frac{1}{6}\}$.

Work Problem 2 at the Side.

For any real number b, we know that $b^1 = b$ and $b^0 = 1$. Writing these two statements in logarithmic form gives the following two properties of logarithms.

Properties of Logarithms

For any positive real number b, with $b \neq 1$,
$$\log_b b = 1 \quad \text{and} \quad \log_b 1 = 0.$$

EXAMPLE 2 Using Properties of Logarithms

Use the preceding two properties of logarithms to evaluate each logarithm.

(a) $\log_7 7 = 1$ **(b)** $\log_{\sqrt{2}} \sqrt{2} = 1$

(c) $\log_9 1 = 0$ **(d)** $\log_{.2} 1 = 0$

Work Problem 3 at the Side.

OBJECTIVE 4 Define and graph logarithmic functions. Now we define the logarithmic function with base a.

Logarithmic Function

If a and x are positive numbers, with $a \neq 1$, then
$$G(x) = \log_a x$$
defines the **logarithmic function with base** a.

To graph a logarithmic function, it is helpful to write it in exponential form first. Then plot selected ordered pairs to determine the graph.

EXAMPLE 3 Graphing a Logarithmic Function

Graph $y = \log_{1/2} x$.

By writing $y = \log_{1/2} x$ in exponential form as $x = (\frac{1}{2})^y$, we can identify ordered pairs that satisfy the equation. Here it is easier to choose values for y and find the corresponding values of x. See the table of ordered pairs on the next page.

——— Continued on Next Page

2 Solve each equation.

(a) $\log_3 27 = x$

(b) $\log_5 p = 2$

(c) $\log_m \frac{1}{16} = -4$

(d) $\log_x 12 = 3$

3 Evaluate each logarithm.

(a) $\log_{2/5} \frac{2}{5}$

(b) $\log_\pi \pi$

(c) $\log_4 1$

(d) $\log_6 1$

ANSWERS
2. **(a)** {3} **(b)** {25} **(c)** {2} **(d)** $\{\sqrt[3]{12}\}$
3. **(a)** 1 **(b)** 1 **(c)** 0 **(d)** 0

4 Graph.

(a) $y = \log_3 x$

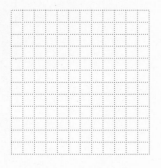

(b) $y = \log_{1/10} x$

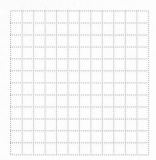

x	y
$\frac{1}{4}$	2
$\frac{1}{2}$	1
1	0
2	−1
4	−2

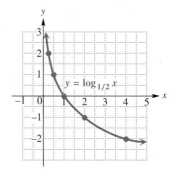

Figure 11

Plotting these points (be careful to get the values of x and y in the right order) and connecting them with a smooth curve gives the graph in Figure 11. This graph is typical of logarithmic functions with $0 < a < 1$. The graph of $x = 2^y$ in Figure 10, which is equivalent to $y = \log_2 x$, is typical of graphs of logarithmic functions with base $a > 1$.

◀◀◀ Work Problem 4 at the Side.

Based on the graphs of the functions defined by $y = \log_2 x$ in Figure 10 and $y = \log_{1/2} x$ in Figure 11, we make the following generalizations about the graphs of logarithmic functions of the form $G(x) = \log_a x$.

Characteristics of the Graph of $G(x) = \log_a x$

1. The graph contains the point $(1, 0)$.

2. When $a > 1$, the graph *rises* from left to right, from the fourth quadrant to the first. When $0 < a < 1$, the graph *falls* from left to right, from the first quadrant to the fourth.

3. The graph approaches the y-axis, but never touches it. (The y-axis is an asymptote.)

4. The domain is $(0, \infty)$, and the range is $(-\infty, \infty)$.

Compare these characteristics to the analogous ones for exponential functions in **Section 11.2.**

OBJECTIVE 5 Use logarithmic functions in applications of growth or decay. Logarithmic functions, like exponential functions, can be applied to growth or decay of real-world phenomena.

EXAMPLE 4 Solving an Application of a Logarithmic Function

The function defined by

$$f(x) = 27 + 1.105 \log_{10}(x + 1)$$

approximates the barometric pressure in inches of mercury at a distance of x miles from the eye of a typical hurricane. (*Source:* Miller, A. and R. Anthes, *Meteorology,* Fifth Edition, Charles E. Merrill Publishing Company, 1985.)

Continued on Next Page

ANSWERS

4. (a)

(b)

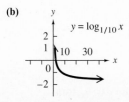

(a) Approximate the pressure 9 mi from the eye of the hurricane. Let $x = 9$, and find $f(9)$.

$$\begin{aligned} f(9) &= 27 + 1.105 \log_{10}(9 + 1) &&\text{Let } x = 9. \\ &= 27 + 1.105 \log_{10} 10 &&\text{Add inside parentheses.} \\ &= 27 + 1.105(1) &&\log_{10} 10 = 1 \\ &= 28.105 &&\text{Add.} \end{aligned}$$

The pressure 9 mi from the eye of the hurricane is 28.105 in.

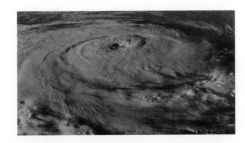

(b) Approximate the pressure 99 mi from the eye of the hurricane.

$$\begin{aligned} f(99) &= 27 + 1.105 \log_{10}(99 + 1) &&\text{Let } x = 99. \\ &= 27 + 1.105 \log_{10} 100 &&\text{Add inside parentheses.} \\ &= 27 + 1.105(2) &&\log_{10} 100 = 2 \\ &= 29.21 \end{aligned}$$

The pressure 99 mi from the eye of the hurricane is 29.21 in.

Work Problem 5 at the Side. ▶▶▶

5 Solve the problem.

A population of mites in a laboratory is growing according to the function defined by

$$P(t) = 80 \log_{10}(t + 10),$$

where t is the number of days after a study is begun.

(a) Find the number of mites at the beginning of the study.

(b) Find the number present after 90 days.

(c) Find the number present after 990 days.

ANSWERS
5. (a) 80 (b) 160 (c) 240

m&m's and Exponential Decay

Exponential functions are important for modeling decay patterns, including the life of a lightbulb and radioactive decaying elements, such as carbon-14. You can simulate an exponential decay problem with an m&m experiment.

Use a fun-size or small package of regular m&m's. Before you begin the simulation, check that each candy has the logo "m" stamped on one side—you may eat the candies with no logo. Place the m&m's in a cup, shake, and toss them onto a napkin. In the table, record the number of m&m's showing the logo. Discard (or eat) all the candies for which the m&m logo is not showing. Repeat until there are 1 or no candies left. The data from one such simulation using 64 m&m's are shown in the table.

EXPERIMENTAL DATA

Toss	Number of m&m Logos Showing	Your Results
0	64	
1	29	
2	17	
3	6	
4	4	
5	3	
6	2	
7	1	

For Group Discussion

1. In a perfect world, you might expect the number of candies left after each toss in our simulation using 64 m&m's to follow the pattern 64, 32, 16, 8, 4, 2, 1. An exponential model has an equation of the form $y = ab^x$. The constant a represents the initial quantity (value when $x = 0$), and the constant b represents the growth rate factor ($b > 1$) or decay rate factor ($0 < b < 1$). For our perfect-world data, what would be the values of a and b? What would be the model exponential equation?

2. On the grid, plot the points for the experimental data, your results, and the points for the theoretical (perfect-world) model. Use a different color to plot each set of data. How well do the graphs match?

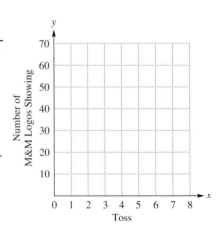

3. You can use a graphing calculator to find the statistical model for the sample experimental data, $y = 50.4(.56)^x$. Develop a table of values for $x = 0, 1, 2, \ldots, 7$ and superimpose the plot of the statistical model on your graph from Problem 2.

   (a) Does the theoretical or the statistical equation better model your experimental data?

   (b) Based on the statistical equation, give an estimate of the initial number of m&m's.

   (c) Based on the statistical equation, estimate the decay rate factor.

11.3 Exercises

FOR EXTRA HELP

Tutor Center Addison-Wesley Math Tutor Center

Math XL MathXL

Digital Video Tutor CD 7 Videotape 7

Student's Solutions Manual

MyMathLab MyMathLab

Interactmath.com

1. By definition, $\log_a x$ is the exponent to which the base a must be raised in order to obtain x. Use this definition to match the logarithm in Column I with its value in Column II. (*Example:* $\log_3 9$ is equal to 2 because 2 is the exponent to which 3 must be raised in order to obtain 9.)

I	II
(a) $\log_4 16$	**A.** -2
(b) $\log_3 81$	**B.** -1
(c) $\log_3\left(\dfrac{1}{3}\right)$	**C.** 2
(d) $\log_{10} .01$	**D.** 0
(e) $\log_5 \sqrt{5}$	**E.** $\dfrac{1}{2}$
(f) $\log_{13} 1$	**F.** 4

2. Match the logarithmic equation in Column I with the corresponding exponential equation from Column II.

I	II
(a) $\log_{1/3} 3 = -1$	**A.** $8^{1/3} = \sqrt[3]{8}$
(b) $\log_5 1 = 0$	**B.** $\left(\dfrac{1}{3}\right)^{-1} = 3$
(c) $\log_2 \sqrt{2} = \dfrac{1}{2}$	**C.** $4^1 = 4$
(d) $\log_{10} 1000 = 3$	**D.** $2^{1/2} = \sqrt{2}$
(e) $\log_8 \sqrt[3]{8} = \dfrac{1}{3}$	**E.** $5^0 = 1$
(f) $\log_4 4 = 1$	**F.** $10^3 = 1000$

Write in logarithmic form. See the table in Objective 2.

3. $4^5 = 1024$

4. $3^6 = 729$

5. $\left(\dfrac{1}{2}\right)^{-3} = 8$

6. $\left(\dfrac{1}{6}\right)^{-3} = 216$

7. $10^{-3} = .001$

8. $36^{1/2} = 6$

9. $\sqrt[4]{625} = 5$

10. $\sqrt[3]{343} = 7$

Write in exponential form. See the table in Objective 2.

11. $\log_4 64 = 3$

12. $\log_2 512 = 9$

13. $\log_{10} \dfrac{1}{10,000} = -4$

14. $\log_{100} 100 = 1$

15. $\log_6 1 = 0$

16. $\log_\pi 1 = 0$

17. $\log_9 3 = \dfrac{1}{2}$

18. $\log_{64} 2 = \dfrac{1}{6}$

19. When a student asked his teacher to explain to him how to evaluate $\log_9 3$ without showing any work, his teacher told him, "Think radically." Explain what the teacher meant by this hint.

20. A student told her teacher, "I know that $\log_2 1$ is the exponent to which 2 must be raised in order to obtain 1, but I can't think of any such number." How would you explain to the student that the value of $\log_2 1$ is 0?

Solve each equation for x. See Examples 1 and 2.

21. $x = \log_{27} 3$

22. $x = \log_{125} 5$

23. $\log_x 9 = \dfrac{1}{2}$

24. $\log_x 5 = \dfrac{1}{2}$

25. $\log_x 125 = -3$

26. $\log_x 64 = -6$

27. $\log_{12} x = 0$

28. $\log_4 x = 0$

29. $\log_x x = 1$

30. $\log_x 1 = 0$

31. $\log_x \dfrac{1}{25} = -2$

32. $\log_x \dfrac{1}{10} = -1$

33. $\log_8 32 = x$

34. $\log_{81} 27 = x$

35. $\log_\pi \pi^4 = x$

36. $\log_{\sqrt{2}} \sqrt{2}^9 = x$

37. $\log_6 \sqrt{216} = x$

38. $\log_4 \sqrt{64} = x$

*If the point (p, q) is on the graph of $f(x) = a^x$ (for $a > 0$ and $a \neq 1$), then the point (q, p) is on the graph of $f^{-1}(x) = \log_a x$. Use this fact and refer to the graphs required in Exercises 5–8 in **Section 11.2** to graph each logarithmic function. See Example 3.*

39. $y = \log_3 x$

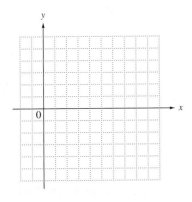

40. $y = \log_5 x$

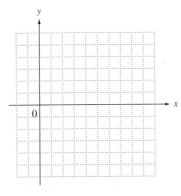

41. $y = \log_{1/3} x$

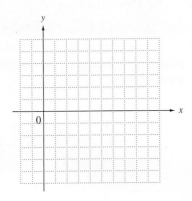

42. $y = \log_{1/5} x$

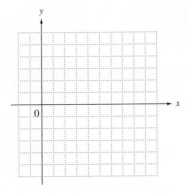

43. Compare the summary of characteristics of the graph of $F(x) = a^x$ in **Section 11.2** with the similar summary of characteristics of the graph of $G(x) = \log_a x$ in this section. Make a list of the characteristics that reinforce the concept that F and G are inverse functions.

44. The domain of $F(x) = a^x$ is $(-\infty, \infty)$, while the range is $(0, \infty)$. Therefore, since $G(x) = \log_a x$ defines the inverse of F, the domain of G is _____, while the range of G is _____.

Use the graph to predict the value of $f(t)$ for each value of t.

45. $t = 0$

46. $t = 10$

47. $t = 60$

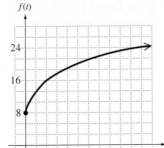

48. Show that the points determined in Exercises 45–47 lie on the graph of $f(t) = 8 \log_5(2t + 5)$.

49. Explain why 1 is not allowed as a base for a logarithmic function.

50. Explain why $\log_a 1$ is 0 for any value of a that is allowed as the base of a logarithm. Use a rule of exponents introduced earlier in your explanation.

51. The graphs of both $f(x) = 3^x$ and $g(x) = \log_3 x$ rise from left to right. Which one rises at a faster rate?

52. Use the exponential key of your calculator to find approximations for the expression $(1 + \frac{1}{x})^x$, using x values of 1, 10, 100, 1000, and 10,000. Explain what seems to be happening as x gets larger and larger.

Solve each application of a logarithmic function. See Example 4.

53. According to selected figures from 1981 through 2003, the number of billion cubic feet of natural gas gross withdrawals from crude oil wells in the United States can be approximated by the function defined by

$$f(x) = 3800 + 585 \log_2 x,$$

where $x = 1$ represents 1981, $x = 2$ represents 1982, and so on. (*Source:* Energy Information Administration, Annual Energy Review 2003.) Use this function to approximate the number of cubic feet withdrawn in each of the following years.

(a) 1982

(b) 1988

(c) 1996

54. According to selected figures from the last two decades of the twentieth century, the number of trillion cubic feet of dry natural gas consumed worldwide was approximated by the function defined by

$$f(x) = 51.47 + 6.044 \log_2 x,$$

where $x = 1$ corresponds to 1980, $x = 2$ to 1981, and so on. (*Source:* Energy Information Administration.) Use the function to approximate consumption in each of the following years.

(a) 1980

(b) 1987

(c) 1995

*In the United States, the intensity of an earthquake is rated using the **Richter scale**. The Richter scale rating of an earthquake of intensity x is given by*

$$R = \log_{10} \frac{x}{x_0},$$

where x_0 is the intensity of an earthquake of a certain (small) size. The figure shows Richter scale ratings for major southern California earthquakes since 1920. As the figure indicates, earthquakes "come in bunches" and the 1990s were an especially busy time.

55. The 1994 Northridge earthquake had a Richter scale rating of 6.7; the Landers earthquake had a rating of 7.3. How much more powerful was the Landers earthquake than the Northridge earthquake?

56. Compare the smallest rated earthquake in the figure (at 4.8) with the Landers quake. How much more powerful was the Landers quake?

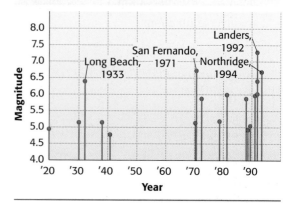

MAJOR SOUTHERN CALIFORNIA EARTHQUAKES

Earthquakes with magnitudes greater than 4.8

Source: Caltech; U.S. Geological Survey.

11.4 Properties of Logarithms

Logarithms have been used as an aid to numerical calculation for several hundred years. Today the widespread use of calculators has made the use of logarithms for calculation obsolete. However, logarithms are still very important in applications and in further work in mathematics.

OBJECTIVE 1 **Use the product rule for logarithms.** One way in which logarithms simplify problems is by changing a problem of multiplication into one of addition. We know that $\log_2 4 = 2$, $\log_2 8 = 3$, and $\log_2 32 = 5$. Since $2 + 3 = 5$,

$$\log_2 32 = \log_2 4 + \log_2 8$$
$$\log_2(4 \cdot 8) = \log_2 4 + \log_2 8.$$

This is true in general.

OBJECTIVES

1. Use the product rule for logarithms.
2. Use the quotient rule for logarithms.
3. Use the power rule for logarithms.
4. Use properties to write alternative forms of logarithmic expressions.

Product Rule for Logarithms

If x, y, and b are positive real numbers, where $b \neq 1$, then

$$\log_b xy = \log_b x + \log_b y.$$

In words, the logarithm of a product is the sum of the logarithms of the factors.

NOTE
The word statement of the product rule can be restated by replacing "logarithm" with "exponent." The rule then becomes the familiar rule for multiplying exponential expressions: The *exponent* of a product is equal to the sum of the *exponents* of the factors.

To prove this rule, let $m = \log_b x$ and $n = \log_b y$, and recall that

$$\log_b x = m \quad \text{means} \quad b^m = x.$$
$$\log_b y = n \quad \text{means} \quad b^n = y.$$

Now consider the product xy.

$$xy = b^m \cdot b^n \qquad \text{Substitute.}$$
$$xy = b^{m+n} \qquad \text{Product rule for exponents}$$
$$\log_b xy = m + n \qquad \text{Convert to logarithmic form.}$$
$$\log_b xy = \log_b x + \log_b y \qquad \text{Substitute.}$$

The last statement is the result we wished to prove.

EXAMPLE 1 **Using the Product Rule**

Use the product rule to rewrite each expression. Assume $x > 0$.

(a) $\log_5(6 \cdot 9)$

By the product rule,

$$\log_5(6 \cdot 9) = \log_5 6 + \log_5 9.$$

(b) $\log_7 8 + \log_7 12 = \log_7(8 \cdot 12) = \log_7 96$

Continued on Next Page

1 Use the product rule to rewrite each expression.

(a) $\log_6(5 \cdot 8)$

(c) $\log_3(3x) = \log_3 3 + \log_3 x$

$= 1 + \log_3 x \qquad \log_3 3 = 1$

(d) $\log_4 x^3 = \log_4(x \cdot x \cdot x) \qquad x^3 = x \cdot x \cdot x$

$= \log_4 x + \log_4 x + \log_4 x \quad$ Product rule

$= 3 \log_4 x$

Work Problem 1 at the Side.

(b) $\log_4 3 + \log_4 7$

OBJECTIVE 2 Use the quotient rule for logarithms. The rule for division is similar to the rule for multiplication.

Quotient Rule for Logarithms

If x, y, and b are positive real numbers, where $b \neq 1$, then

(c) $\log_8 8k, \quad k > 0$

$$\log_b \frac{x}{y} = \log_b x - \log_b y.$$

In words, the logarithm of a quotient is the difference between the logarithm of the numerator and the logarithm of the denominator.

(d) $\log_5 m^2, \quad m > 0$

The proof of this rule is very similar to the proof of the product rule.

EXAMPLE 2 Using the Quotient Rule

2 Use the quotient rule to rewrite each expression.

Use the quotient rule to rewrite each logarithm.

(a) $\log_7 \frac{9}{4}$

(a) $\log_4 \frac{7}{9} = \log_4 7 - \log_4 9$

(b) $\log_5 6 - \log_5 x = \log_5 \frac{6}{x}, \quad x > 0$

(c) $\log_3 \frac{27}{5} = \log_3 27 - \log_3 5$

$= 3 - \log_3 5 \qquad \log_3 27 = 3$

(b) $\log_3 p - \log_3 q,$
$p > 0, \quad q > 0$

CAUTION

There is no property of logarithms to rewrite the logarithm of a sum or difference. For example, we *cannot* write $\log_b(x + y)$ in terms of $\log_b x$ and $\log_b y$. Also,

(c) $\log_4 \frac{3}{16}$

$$\log_b \frac{x}{y} \neq \frac{\log_b x}{\log_b y}.$$

Work Problem 2 at the Side.

ANSWERS

1. (a) $\log_6 5 + \log_6 8$ (b) $\log_4 21$
   (c) $1 + \log_8 k$ (d) $2 \log_5 m$
2. (a) $\log_7 9 - \log_7 4$ (b) $\log_3 \frac{p}{q}$
   (c) $\log_4 3 - 2$

OBJECTIVE 3 Use the power rule for logarithms. The next rule gives a method for evaluating powers and roots such as

$$2^{\sqrt{2}}, \quad (\sqrt{2})^{3/4}, \quad (.032)^{5/8}, \quad \text{and} \quad \sqrt[5]{12}.$$

This rule makes it possible to find approximations for numbers that could not be evaluated before. By the product rule for logarithms,

$$\begin{aligned}
\log_5 2^3 &= \log_5(2 \cdot 2 \cdot 2) \\
&= \log_5 2 + \log_5 2 + \log_5 2 \\
&= 3 \log_5 2.
\end{aligned}$$

Also,

$$\begin{aligned}
\log_2 7^4 &= \log_2(7 \cdot 7 \cdot 7 \cdot 7) \\
&= \log_2 7 + \log_2 7 + \log_2 7 + \log_2 7 \\
&= 4 \log_2 7.
\end{aligned}$$

Furthermore, we saw in Example 1(d) that $\log_4 x^3 = 3 \log_4 x$. These examples suggest the following rule.

Power Rule for Logarithms

If x and b are positive real numbers, where $b \neq 1$, and if r is any real number, then

$$\log_b x^r = r \log_b x.$$

In words, the logarithm of a number to a power equals the exponent times the logarithm of the number.

As examples of this result,

$$\log_b m^5 = 5 \log_b m \quad \text{and} \quad \log_3 5^4 = 4 \log_3 5.$$

To prove the power rule, let

$$\log_b x = m.$$

$b^m = x$	Convert to exponential form.
$(b^m)^r = x^r$	Raise to the power r.
$b^{mr} = x^r$	Power rule for exponents
$\log_b x^r = mr$	Convert to logarithmic form.
$\log_b x^r = rm$	Commutative property
$\log_b x^r = r \log_b x$	$m = \log_b x$

This is the statement to be proved.

As a special case of the power rule, let $r = \frac{1}{p}$, so

$$\log_b \sqrt[p]{x} = \log_b x^{1/p} = \frac{1}{p} \log_b x.$$

For example, using this result, with $x > 0$,

$$\log_b \sqrt[5]{x} = \log_b x^{1/5} = \frac{1}{5} \log_b x \quad \text{and} \quad \log_b \sqrt[3]{x^4} = \log_b x^{4/3} = \frac{4}{3} \log_b x.$$

Another special case is

$$\log_b \frac{1}{x} = \log_b x^{-1} = -\log_b x.$$

❸ Use the power rule to rewrite each logarithm. Assume $a > 0, b > 0, x > 0, a \neq 1$, and $b \neq 1$.

(a) $\log_3 5^2$

(b) $\log_a x^4$

(c) $\log_b \sqrt{8}$

(d) $\log_2 \sqrt[3]{2}$

NOTE
For a review of rational exponents, refer to **Section 9.2.**

EXAMPLE 3 Using the Power Rule

Use the power rule to rewrite each logarithm. Assume $b > 0, x > 0$, and $b \neq 1$.

(a) $\log_5 4^2 = 2 \log_5 4$

(b) $\log_b x^5 = 5 \log_b x$

(c) $\log_b \sqrt{7}$

When using the power rule with logarithms of expressions involving radicals, begin by rewriting the radical expression with a rational exponent.

$$\log_b \sqrt{7} = \log_b 7^{1/2} \qquad \sqrt{x} = x^{1/2}$$

$$= \frac{1}{2} \log_b 7 \qquad \text{Power rule}$$

(d) $\log_2 \sqrt[5]{x^2} = \log_2 x^{2/5} \qquad \sqrt[5]{x^2} = x^{2/5}$

$$= \frac{2}{5} \log_2 x \qquad \text{Power rule}$$

◀◀◀ Work Problem 3 at the Side.

Two special properties involving both exponential and logarithmic expressions come directly from the fact that logarithmic and exponential functions are inverses of each other.

❹ Find the value of each logarithmic expression.

(a) $\log_{10} 10^3$

(b) $\log_2 8$

(c) $5^{\log_5 3}$

Special Properties
If $b > 0$ and $b \neq 1$, then

$$b^{\log_b x} = x, \; x > 0 \quad \text{and} \quad \log_b b^x = x.$$

To prove the first statement, let

$$y = \log_b x.$$
$$b^y = x \qquad \qquad \text{Convert to exponential form.}$$
$$b^{\log_b x} = x \qquad \qquad \text{Replace } y \text{ with } \log_b x.$$

The proof of the second statement is similar.

EXAMPLE 4 Using the Special Properties

Find the value of each logarithmic expression.

(a) $\log_5 5^4$
Since $\log_b b^x = x$,

$$\log_5 5^4 = 4.$$

(b) $\log_3 9 = \log_3 3^2 = 2$

(c) $4^{\log_4 10} = 10$

◀◀◀ Work Problem 4 at the Side.

ANSWERS
3. (a) $2 \log_3 5$ **(b)** $4 \log_a x$
(c) $\frac{1}{2} \log_b 8$ **(d)** $\frac{1}{3}$
4. (a) 3 **(b)** 3 **(c)** 3

Here is a summary of the properties of logarithms.

Properties of Logarithms

If x, y, and b are positive real numbers, where $b \neq 1$, and r is any real number, then

Product Rule $\log_b xy = \log_b x + \log_b y$

Quotient Rule $\log_b \dfrac{x}{y} = \log_b x - \log_b y$

Power Rule $\log_b x^r = r \log_b x$

Special Properties $b^{\log_b x} = x$ and $\log_b b^x = x.$

OBJECTIVE 4 Use properties to write alternative forms of logarithmic expressions.   Applying the properties of logarithms is important for solving equations with logarithms and in calculus.

EXAMPLE 5 Writing Logarithms in Alternative Forms

Use the properties of logarithms to rewrite each expression. Assume all variables represent positive real numbers.

(a) $\log_4 4x^3 = \log_4 4 + \log_4 x^3$ Product rule

$\qquad\qquad = 1 + 3 \log_4 x$ $\log_4 4 = 1$; power rule

(b) $\log_7 \sqrt{\dfrac{m}{n}} = \log_7 \left(\dfrac{m}{n}\right)^{1/2}$

$\qquad\qquad = \dfrac{1}{2} \log_7 \dfrac{m}{n}$ Power rule

$\qquad\qquad = \dfrac{1}{2}(\log_7 m - \log_7 n)$ Quotient rule

(c) $\log_5 \dfrac{a^2}{bc} = \log_5 a^2 - \log_5 bc$ Quotient rule

$\qquad\qquad = 2 \log_5 a - \log_5 bc$ Power rule

$\qquad\qquad = 2 \log_5 a - (\log_5 b + \log_5 c)$ Product rule

$\qquad\qquad = 2 \log_5 a - \log_5 b - \log_5 c$ Distributive property

Notice the careful use of parentheses in the third step. Since we are subtracting the logarithm of a product and rewriting it as a sum of two terms, we must place parentheses around the sum.

(d) $4 \log_b m - \log_b n = \log_b m^4 - \log_b n$ Power rule

$\qquad\qquad = \log_b \dfrac{m^4}{n}$ Quotient rule

Continued on Next Page

5 Use the properties of logarithms to rewrite each expression. Assume all variables represent positive real numbers.

(a) $\log_6 36m^5$

(e) $\log_b(x + 1) + \log_b(2x - 1) - \dfrac{2}{3}\log_b x$

$= \log_b(x + 1) + \log_b(2x - 1) - \log_b x^{2/3}$ Power rule

$= \log_b \dfrac{(x + 1)(2x - 1)}{x^{2/3}}$ Product and quotient rules

$= \log_b \dfrac{2x^2 + x - 1}{x^{2/3}}$

(f) $\log_8(2p + 3r)$ cannot be rewritten using the properties of logarithms. There is no property of logarithms to rewrite the logarithm of a sum.

◀◀◀ **Work Problem 5 at the Side.**

(b) $\log_2 \sqrt{9z}$

(c) $\log_q \dfrac{8r^2}{m - 1}, m > 1, q \neq 1$

(d) $2\log_a x + 3\log_a y, a \neq 1$

(e) $\log_4(3x + y)$

ANSWERS

5. **(a)** $2 + 5\log_6 m$ **(b)** $\log_2 3 + \dfrac{1}{2}\log_2 z$

(c) $\log_q 8 + 2\log_q r - \log_q(m - 1)$
(d) $\log_a x^2 y^3$ **(e)** cannot be rewritten

11.4 Exercises

FOR EXTRA HELP

Tutor Center Addison-Wesley Math Tutor Center

Math XL MathXL

Digital Video Tutor CD 7 Videotape 7

Student's Solutions Manual

MyMathLab MyMathLab

Interactmath.com

Decide whether each statement of a logarithmic property is true *or* false. *If it is* false, *correct it by changing the right side of the equation.*

1. $\log_b x + \log_b y = \log_b(x + y)$

2. $\log_b \dfrac{x}{y} = \log_b x - \log_b y$

3. $\log_b b^x = x$

4. $\log_b x^r = \log_b rx$

Use the properties of logarithms introduced in this section to express each logarithm as a sum or difference of logarithms, or as a single number if possible. Assume all variables represent positive real numbers. See Examples 1–5.

5. $\log_7 \dfrac{4}{5}$

6. $\log_8 \dfrac{9}{11}$

7. $\log_2 8^{1/4}$

8. $\log_3 9^{3/4}$

9. $\log_4 \dfrac{3\sqrt{x}}{y}$

10. $\log_5 \dfrac{6\sqrt{z}}{w}$

11. $\log_3 \dfrac{\sqrt[3]{4}}{x^2 y}$

12. $\log_7 \dfrac{\sqrt[3]{13}}{pq^2}$

13. $\log_3 \sqrt{\dfrac{xy}{5}}$

14. $\log_6 \sqrt{\dfrac{pq}{7}}$

15. $\log_2 \dfrac{\sqrt[3]{x} \cdot \sqrt[5]{y}}{r^2}$

16. $\log_4 \dfrac{\sqrt[4]{z} \cdot \sqrt[5]{w}}{s^2}$

17. A student erroneously wrote

$$\log_a(x + y) = \log_a x + \log_a y.$$

When his teacher explained that this was wrong, the student claimed he had used the distributive property. Write a few sentences explaining why the distributive property does not apply in this case.

18. Write a few sentences explaining how the rules for multiplying and dividing powers of the same base are similar to the rules for finding logarithms of products and quotients.

Use the properties of logarithms introduced in this section to rewrite each expression as a single logarithm. Assume all variables are defined in such a way that the variable expressions are positive, and bases are positive numbers not equal to 1. See Examples 1–5.

19. $\log_b x + \log_b y$

20. $\log_b 2 + \log_b z$

21. $3 \log_a m - \log_a n$

22. $5 \log_b x - \log_b y$

23. $(\log_a r - \log_a s) + 3 \log_a t$

24. $(\log_a p - \log_a q) + 2 \log_a r$

25. $3 \log_a 5 - 4 \log_a 3$

26. $3 \log_a 5 + \dfrac{1}{2} \log_a 9$

27. $\log_{10}(x + 3) + \log_{10}(x - 3)$

28. $\log_{10}(y + 4) + \log_{10}(y - 4)$

29. $3 \log_p x + \dfrac{1}{2} \log_p y - \dfrac{3}{2} \log_p z - 3 \log_p a$

30. $\dfrac{1}{3} \log_b x + \dfrac{2}{3} \log_b y - \dfrac{3}{4} \log_b s - \dfrac{2}{3} \log_b t$

31. Explain why the statement for the power rule for logarithms requires that x be a positive real number.

32. What is wrong with the following "proof" that $\log_2 16$ does not exist?

$$\log_2 16 = \log_2 (-4)(-4)$$
$$= \log_2(-4) + \log_2(-4)$$

Since the logarithm of a negative number is not defined, the final step cannot be evaluated, and so $\log_2 16$ does not exist.

RELATING CONCEPTS (EXERCISES 33–38)   For Individual or Group Work

Work Exercises 33–38 in order.

33. Evaluate $\log_3 81$.

34. Write the *meaning* of the expression $\log_3 81$.

35. Evaluate $3^{\log_3 81}$.

36. Write the *meaning* of the expression $\log_2 19$.

37. Evaluate $2^{\log_2 19}$.

38. Keeping in mind that a logarithm is an exponent, and using the results from Exercises 33–37, what is the simplest form of the expression $k^{\log_k m}$?

11.5 Common and Natural Logarithms

As mentioned earlier, logarithms are important in many applications of mathematics to everyday problems, particularly in biology, engineering, economics, and social science. In this section we find numerical approximations for logarithms. Traditionally, base 10 logarithms were used most often because our number system is base 10. Logarithms to base 10 are called **common logarithms,** and $\log_{10} x$ is abbreviated as simply **log x,** where the base is understood to be 10.

OBJECTIVE 1 Evaluate common logarithms using a calculator. We use calculators to evaluate common logarithms. In the next example we give the results of evaluating some common logarithms using a calculator with a LOG key. (This may be a second function key on some calculators.) For simple scientific calculators, just enter the number, then press the LOG key. For graphing calculators, these steps are reversed. We will give all approximations for logarithms to four decimal places.

EXAMPLE 1 Evaluating Common Logarithms

Evaluate each logarithm using a calculator.

(a) $\log 327.1 \approx 2.5147$ **(b)** $\log 437,000 \approx 5.6405$

(c) $\log .0615 \approx -1.2111$

Notice that $\log .0615 \approx -1.2111$, a negative result. *The common logarithm of a number between 0 and 1 is always negative* because the logarithm is the exponent on 10 that produces the number. For example,

$$10^{-1.2111} \approx .0615.$$

If the exponent (the logarithm) were positive, the result would be greater than 1 because $10^0 = 1$. See Figure 12.

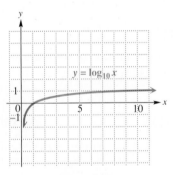

Figure 12

Work Problem 1 at the Side.

OBJECTIVE 2 Use common logarithms in applications. In chemistry, pH is a measure of the acidity or alkalinity of a solution; water, for example, has pH 7. In general, acids have pH numbers less than 7, and alkaline solutions have pH values greater than 7. The **pH** of a solution is defined as

$$pH = -\log[H_3O^+],$$

where $[H_3O^+]$ is the hydronium ion concentration in moles per liter. It is customary to round pH values to the nearest tenth.

OBJECTIVES

1 Evaluate common logarithms using a calculator.
2 Use common logarithms in applications.
3 Evaluate natural logarithms using a calculator.
4 Use natural logarithms in applications.

1 Evaluate each logarithm to four decimal places using a calculator.

(a) log 41,600

(b) log 43.5

(c) log .442

ANSWERS
1. (a) 4.6191   **(b)** 1.6385   **(c)** −.3546

2 Solve the problem.

Find the pH of water with a hydronium ion concentration of 1.2×10^{-3}. If this water had been taken from a wetland, is the wetland a rich fen, a poor fen, or a bog?

3 Find the hydronium ion concentrations of solutions with the following pH values.

(a) 4.6

(b) 7.5

EXAMPLE 2 Using pH in an Application

Wetlands are classified as *bogs, fens, marshes,* and *swamps.* These classifications are based on pH values. A pH value between 6.0 and 7.5, such as that of Summerby Swamp in Michigan's Hiawatha National Forest, indicates that the wetland is a "rich fen." When the pH is between 4.0 and 6.0, the wetland is a "poor fen," and if the pH falls to 3.0 or less, it is a "bog." (*Source:* Mohlenbrock, R., "Summerby Swamp, Michigan," *Natural History,* March 1994.)

Suppose that the hydronium ion concentration of a sample of water from a wetland is 6.3×10^{-3}. How would this wetland be classified?

Use the definition of pH.

$$
\begin{aligned}
\text{pH} &= -\log(6.3 \times 10^{-3}) \\
&= -(\log 6.3 + \log 10^{-3}) && \text{Product rule} \\
&= -[.7993 - 3(1)] && \text{Use a calculator.} \\
&= -.7993 + 3 \\
&\approx 2.2
\end{aligned}
$$

Since the pH is less than 3.0, the wetland is a bog.

◀◀◀ Work Problem 2 at the Side.

EXAMPLE 3 Finding Hydronium Ion Concentration

Find the hydronium ion concentration of drinking water with pH 6.5.

$$
\begin{aligned}
\textbf{pH} &= -\log[\text{H}_3\text{O}^+] \\
\textbf{6.5} &= -\log[\text{H}_3\text{O}^+] && \text{Let pH} = 6.5. \\
\log[\text{H}_3\text{O}^+] &= -6.5 && \text{Multiply by } -1.
\end{aligned}
$$

Solve for $[\text{H}_3\text{O}^+]$ by writing the equation in exponential form, remembering that the base is 10.

$$
\begin{aligned}
[\text{H}_3\text{O}^+] &= 10^{-6.5} \\
[\text{H}_3\text{O}^+] &\approx 3.2 \times 10^{-7} && \text{Use a calculator.}
\end{aligned}
$$

◀◀◀ Work Problem 3 at the Side.

OBJECTIVE 3 **Evaluate natural logarithms using a calculator.** The most important logarithms used in applications are **natural logarithms,** which have as base the number e. The number e is a fundamental number in our universe. For this reason e, like π, is called a *universal constant.* The letter e is used to honor Leonhard Euler, who published extensive results on the number in 1748. Since it is an irrational number, its decimal expansion never terminates and never repeats.

ANSWERS
2. 2.9; bog
3. (a) 2.5×10^{-5} (b) 3.2×10^{-8}

The first few digits of the decimal value of e are 2.718281828. A calculator key e^x or the two keys INV and lnx are used to approximate powers of e. For example, a calculator gives

$$e^2 \approx 7.389056099, \quad e^3 \approx 20.08553692, \quad \text{and} \quad e^{.6} \approx 1.8221188.$$

Logarithms to base e are called natural logarithms because they occur in biology and the social sciences in natural situations that involve growth or decay. The base e logarithm of x is written **ln x** (read "el en x"). A graph of $y = \ln x$, the equation that defines the natural logarithmic function, is given in Figure 13.

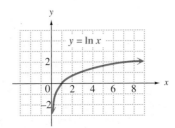

Figure 13

A calculator key labeled lnx is used to evaluate natural logarithms. If your calculator has an e^x key, but not a key labeled lnx, find natural logarithms by entering the number, pressing the INV key, and then pressing the e^x key. This works because $y = e^x$ defines the inverse function of $y = \ln x$ (or $y = \log_e x$).

EXAMPLE 4 **Finding Natural Logarithms**

Find each logarithm to four decimal places.

(a) $\ln .5841 \approx -.5377$

As with common logarithms, *a number between 0 and 1 has a negative natural logarithm.*

(b) $\ln 192.7 \approx 5.2611$ **(c)** $\ln 10.84 \approx 2.3832$

Work Problem 4 at the Side.)))

OBJECTIVE 4 **Use natural logarithms in applications.** A common application of natural logarithmic functions is to express growth or decay of a quantity, as in the next example.

EXAMPLE 5 **Applying Natural Logarithms**

The altitude in meters that corresponds to an atmospheric pressure of x millibars is given by the logarithmic function defined by

$$f(x) = 51{,}600 - 7457 \ln x.$$

(*Source:* Miller, A. and J. Thompson, *Elements of Meteorology,* Fourth Edition, Charles E. Merrill Publishing Company, 1993.) Use this function to find the altitude when atmospheric pressure is 400 millibars.

Let $x = 400$ and substitute in the expression for $f(x)$.

$$f(\mathbf{400}) = 51{,}600 - 7457 \ln \mathbf{400}$$

$$\approx 6900 \text{ (to the nearest hundred)}$$

Atmospheric pressure is 400 millibars at approximately 6900 m.

4️⃣ Find each logarithm to four decimal places.

(a) ln .01

(b) ln 27

(c) ln 529

ANSWERS
4. **(a)** -4.6052 **(b)** 3.2958 **(c)** 6.2710

5 Use the logarithmic function in Example 5 to approximate the altitude at 700 millibars of pressure.

Calculator Tip In Example 5, the final answer was obtained using a calculator *without* rounding the intermediate values. In general, it is best to wait until the final step to round the answer; otherwise, a build-up of round-off error may cause the final answer to have an incorrect final decimal place digit.

◀◀◀ **Work Problem 5 at the Side.**

ANSWERS
5. approximately 2700 m

11.5 Exercises

FOR EXTRA HELP   Addison-Wesley Math Tutor Center   MathXL   Digital Video Tutor CD 7 Videotape 7   Student's Solutions Manual   MyMathLab   Interactmath.com

Choose the correct response in Exercises 1–4.

1. What is the base in the expression log x?

   A. e **B.** 1   **C.** 10   **D.** x

2. What is the base in the expression ln x?

   A. e **B.** 1   **C.** 10   **D.** x

3. Since $10^0 = 1$ and $10^1 = 10$, between what two consecutive integers is the value of log 5.6?

   A. 5 and 6   **B.** 10 and 11   **C.** 0 and 1   **D.** -1 and 0

4. Since $e^1 \approx 2.718$ and $e^2 \approx 7.389$, between what two consecutive integers is the value of ln 5.6?

   A. 5 and 6   **B.** 2 and 3   **C.** 1 and 2   **D.** 0 and 1

5. Without using a calculator, give the value of log $10^{19.2}$.

6. Without using a calculator, give the value of ln $e^{\sqrt{2}}$.

You will need a calculator for the remaining exercises in this set.

Find each logarithm. Give an approximation to four decimal places. See Examples 1 and 4.

7. log 43

8. log 98

9. log 328.4

10. log 457.2

11. log .0326

12. log .1741

13. $\log(4.76 \times 10^9)$

14. $\log(2.13 \times 10^4)$

15. ln 7.84

16. ln 8.32

17. ln .0556

18. ln .0217

19. ln 388.1

20. ln 942.6

21. $\ln(8.59 \times e^2)$

22. $\ln(7.46 \times e^3)$

23. ln 10

24. log e

25. Let m be the number of letters in your first name, and let n be the number of letters in your last name.

(a) In your own words, explain what $\log_m n$ means.

(b) In the next section, we introduce the rule $\log_m n = \dfrac{\log n}{\log m}$. Use your calculator to find $\log_m n$.

(c) Raise m to the power indicated by the number you found in part (b). What is your result?

26. Use your calculator to find approximations of the following logarithms.

(a) $\log 356.8$

(b) $\log 35.68$

(c) $\log 3.568$

(d) Observe your answers and make a conjecture concerning the decimal values of the common logarithms of numbers greater than 1 that have the same digits.

27. Try to find $\log(-1)$ using a calculator. (If you have a graphing calculator, it should be in real number mode.) What happens? Explain why this happens.

Refer to Example 2. In Exercises 28 and 29, suppose that water from a wetland area is sampled and found to have the given hydronium ion concentration. Determine whether the wetland is a rich fen, *a* poor fen, *or a* bog.

28. 2.5×10^{-5}

29. 2.5×10^{-2}

Find the pH *of the substance with the given hydronium ion concentration. See Example 2.*

30. Ammonia, 2.5×10^{-12}

31. Tuna, 1.3×10^{-6}

Use the formula for pH *to find the hydronium ion concentration of the substance with the given* pH. *See Example 3.*

32. Human blood plasma, 7.4

33. Human gastric contents, 2.0

34. Spinach, 5.4

35. Bananas, 4.6

Solve each problem. See Example 5.

36. The number of years, $N(r)$, since two independently evolving languages split off from a common ancestral language is approximated by

$$N(r) = -5000 \ln r,$$

where r is the percent of words (in decimal form) from the ancestral language common to both languages now. Find the number of years since the split for each percent of common words.

(a) 85% (or .85)

(b) 35% (or .35)

(c) 10% (or .10)

37. The time t in years for an amount increasing at a rate of r (in decimal form) to double is given by

$$t = \frac{\ln 2}{\ln(1 + r)}.$$

This is called **doubling time.** Find the doubling time to the nearest tenth for an investment at each interest rate.

(a) 2% = .02

(b) 5% = .05

(c) 8% = .08

38. The concentration of a drug injected into the bloodstream decreases with time. The intervals of time T when the drug should be administered are given by

$$T = \frac{1}{k} \ln \frac{C_2}{C_1},$$

where k is a constant determined by the drug in use, C_2 is the concentration at which the drug is harmful, and C_1 is the concentration below which the drug is ineffective. (*Source:* Horelick, Brindell and Sinan Koont, "Applications of Calculus to Medicine: Prescribing Safe and Effective Dosage," *UMAP Module 202,* 1977.) Thus, if $T = 4$, the drug should be administered every 4 hr. For a certain drug, $k = \frac{1}{3}$, $C_2 = 5$, and $C_1 = 2$. How often should the drug be administered? (*Hint:* Round down.)

39. The loudness of sounds is measured in a unit called a **decibel,** abbreviated dB. A very faint sound, called the **threshold sound,** is assigned an intensity I_0. If a particular sound has intensity I, then the decibel level of this louder sound is

$$D = 10 \log\left(\frac{I}{I_0}\right).$$

Consumers can now enjoy movies at home in elaborate home-theater systems. Find the average decibel level for each popular movie with the given intensity I.

(a) *Spider-Man 2;* $5.012 \times 10^{10} I_0$

(b) *Finding Nemo;* $10^{10} I_0$

(c) *National Treasure;* $6,310,000,000 I_0$

40. The growth of outpatient surgery as a percent of total surgeries at hospitals is approximated by

$$f(x) = -1317 + 304 \ln x,$$

where x represents the number of years since 1900. (*Source:* American Hospital Association.)

(a) What does this function predict for the percent of outpatient surgeries in 1998?

(b) When did outpatient surgeries reach 50%? (*Hint:* Substitute for y, then write the equation in exponential form to solve it.)

41. In the central Sierra Nevada of California, the percent of moisture p that falls as snow rather than rain is approximated reasonably well by

$$p(h) = 86.3 \ln h - 680,$$

where h is the altitude in feet.

(a) What percent of the moisture at 5000 ft falls as snow?

(b) What percent at 7500 ft falls as snow?

42. The **cost-benefit equation**

$$T = -.642 - 189 \ln(1 - p)$$

describes the approximate tax T, in dollars per ton, that would result in a $p\%$ (in decimal form) reduction in carbon dioxide emissions.

(a) What tax will reduce emissions 25%?

(b) Explain why the equation is not valid for $p = 0$ or $p = 1$.

43. The age in years of a female blue whale is approximated by

$$t = -2.57 \ln\left(\frac{87 - L}{63}\right),$$

where L is its length in feet.

(a) How old is a female blue whale that measures 80 ft?

(b) The equation that defines t has domain $24 < L < 87$. Explain why.

11.6 Exponential and Logarithmic Equations; Further Applications

As mentioned earlier, exponential and logarithmic functions are important in many applications of mathematics. Using these functions in applications requires solving exponential and logarithmic equations. Some simple equations were solved in **Sections 11.2** and **11.3**. More general methods for solving these equations depend on the following properties.

OBJECTIVES

1 Solve equations involving variables in the exponents.

2 Solve equations involving logarithms.

3 Solve applications of compound interest.

4 Solve applications involving base e exponential growth and decay.

5 Use the change-of-base rule.

Properties for Solving Exponential and Logarithmic Equations

For all real numbers $b > 0$, $b \neq 1$, and any real numbers x and y:

1. If $x = y$, then $b^x = b^y$.
2. If $b^x = b^y$, then $x = y$.
3. If $x = y$, and $x > 0$, $y > 0$, then $\log_b x = \log_b y$.
4. If $x > 0$, $y > 0$, and $\log_b x = \log_b y$, then $x = y$.

We used Property 2 to solve exponential equations in **Section 11.2**.

OBJECTIVE 1 Solve equations involving variables in the exponents. The first examples illustrate a general method for solving exponential equations using Property 3.

1 Solve each equation and give the decimal approximation to three places.

(a) $2^x = 9$

EXAMPLE 1 Solving an Exponential Equation

Solve $3^x = 12$.

$$3^x = 12$$
$$\log 3^x = \log 12 \qquad \text{Property 3}$$
$$x \log 3 = \log 12 \qquad \text{Power rule}$$
$$x = \frac{\log 12}{\log 3} \qquad \text{Divide by log 3.}$$

This quotient is the exact solution. To find a decimal approximation for the solution, use a calculator.

$$x \approx 2.262$$

The solution set is $\{2.262\}$. Check that $3^{2.262} \approx 12$.

(b) $10^x = 4$

CAUTION
Be careful: $\dfrac{\log 12}{\log 3}$ is *not* equal to log 4. Note that $\log 4 \approx .6021$, but $\dfrac{\log 12}{\log 3} \approx 2.262$.

Work Problem 1 at the Side.

ANSWERS
1. (a) {3.170} (b) {.602}

2 Solve $e^{-.01t} = .38$.

When an exponential equation has e as the base, it is appropriate to use base e logarithms.

EXAMPLE 2 Solving an Exponential Equation with Base e

Solve $e^{.003x} = 40$.

Take base e logarithms on both sides.

$$\ln e^{.003x} = \ln 40$$

$$.003x \ln e = \ln 40 \qquad \text{Power rule}$$

$$.003x = \ln 40 \qquad \ln e = \ln e^1 = 1$$

$$x = \frac{\ln 40}{.003} \qquad \text{Divide by .003.}$$

$$x \approx 1230 \qquad \text{Use a calculator.}$$

The solution set is $\{1230\}$. Check that $e^{.003(1230)} \approx 40$.

Work Problem 2 at the Side.

General Method for Solving an Exponential Equation

Take logarithms to the same base on both sides and then use the power rule of logarithms or the special property $\log_b b^x = x$. (See Examples 1 and 2.)

As a special case, if both sides can be written as exponentials with the same base, do so, and set the exponents equal. (See **Section 11.2.**)

3 Solve $\log_3 (x + 1)^5 = 3$. Give the exact solution.

OBJECTIVE 2 Solve equations involving logarithms. The properties of logarithms from **Section 11.4** are useful here, as is using the definition of a logarithm to change the equation to exponential form.

EXAMPLE 3 Solving a Logarithmic Equation

Solve $\log_2 (x + 5)^3 = 4$. Give the exact solution.

$$(x + 5)^3 = 2^4 \qquad \text{Convert to exponential form.}$$

$$(x + 5)^3 = 16$$

$$x + 5 = \sqrt[3]{16} \qquad \text{Take the cube root on each side.}$$

$$x = -5 + \sqrt[3]{16} \qquad \text{Add } -5.$$

$$x = -5 + 2\sqrt[3]{2} \qquad \text{Simplify the radical.}$$

Verify that the solution satisfies the equation, so the solution set is $\{-5 + 2\sqrt[3]{2}\}$.

CAUTION
Recall that the domain of $y = \log_b x$ is $(0, \infty)$. For this reason, *it is always necessary to check that the solution of an equation with logarithms yields only logarithms of positive numbers in the original equation.*

Work Problem 3 at the Side.

ANSWERS
2. $\{96.8\}$
3. $\{-1 + \sqrt[5]{27}\}$

EXAMPLE 4 Solving a Logarithmic Equation

Solve $\log_2(x + 1) - \log_2 x = \log_2 7$.

$$\log_2(x + 1) - \log_2 x = \log_2 7$$

$$\log_2 \frac{x + 1}{x} = \log_2 7 \qquad \text{Quotient rule}$$

$$\frac{x + 1}{x} = 7 \qquad \text{Property 4}$$

$$x + 1 = 7x \qquad \text{Multiply by } x.$$

$$\frac{1}{6} = x \qquad \text{Subtract } x; \text{ divide by 6.}$$

Check this solution by substituting in the original equation. Here, both $x + 1$ and x must be positive. If $x = \frac{1}{6}$, this condition is satisfied, so the solution set is $\left\{\frac{1}{6}\right\}$.

Work Problem 4 at the Side. ▶▶▶

④ Solve
$\log_8(2x + 5) + \log_8 3 = \log_8 33.$

EXAMPLE 5 Solving a Logarithmic Equation

Solve $\log x + \log(x - 21) = 2$.

For this equation, write the left side as a single logarithm. Then write in exponential form and solve the equation.

$$\log x + \log(x - 21) = 2$$

$$\log x(x - 21) = 2 \qquad \text{Product rule}$$

$$x(x - 21) = 10^2 \qquad \log x = \log_{10} x; \text{ Write in exponential form.}$$

$$x^2 - 21x = 100 \qquad \text{Distributive property; multiply.}$$

$$x^2 - 21x - 100 = 0 \qquad \text{Standard form}$$

$$(x - 25)(x + 4) = 0 \qquad \text{Factor.}$$

$$x - 25 = 0 \quad \text{or} \quad x + 4 = 0 \qquad \text{Zero-factor property}$$

$$x = 25 \quad \text{or} \qquad x = -4 \qquad \text{Solve each equation.}$$

The value -4 must be rejected as a solution since it leads to the logarithm of at least one negative number in the original equation.

$$\log(-4) + \log(-4 - 21) = 2 \qquad \text{The left side is undefined.}$$

The only solution, therefore, is 25, and the solution set is $\{25\}$.

⑤ Solve
$\log_3 2x - \log_3(3x + 15) = -2.$

> **CAUTION**
> *Do not reject a potential solution just because it is nonpositive. Reject any value that leads to the logarithm of a nonpositive number.*

Work Problem 5 at the Side. ▶▶▶

ANSWERS
4. $\{3\}$
5. $\{1\}$

6 Find the value of $2000 deposited at 5% compounded annually for 10 yr.

In summary, we use the following steps to solve a logarithmic equation.

Solving a Logarithmic Equation

Step 1 **Transform the equation so that a single logarithm appears on one side.** Use the product rule or quotient rule of logarithms to do this.

Step 2 **(a) Use Property 4.** If $\log_b x = \log_b y$, then $x = y$. (See Example 4.)

(b) Write the equation in exponential form. If $\log_b x = k$, then $x = b^k$. (See Examples 3 and 5.)

OBJECTIVE 3 Solve applications of compound interest. So far in this book, problems involving applications of interest have been limited to simple interest using the formula $I = prt$. In most cases, interest paid or charged is **compound interest** (interest paid on both principal and interest). The formula for compound interest is an important application of exponential functions.

Compound Interest Formula (for a Finite Number of Periods)

If a principal of P dollars is deposited at an annual rate of interest r compounded (paid) n times per year, the account will contain

$$A = P\left(1 + \frac{r}{n}\right)^{nt}$$

dollars after t years. (In this formula, r is expressed as a decimal.)

EXAMPLE 6 Solving a Compound Interest Problem for A

How much money will there be in an account at the end of 5 yr if $1000 is deposited at 6% compounded quarterly? (Assume no withdrawals are made.)

Because interest is compounded quarterly, $n = 4$. The other values given in the problem are $P = 1000$, $r = .06$ (because $6\% = .06$), and $t = 5$. Substitute into the compound interest formula to get the value of A.

$$A = 1000\left(1 + \frac{.06}{4}\right)^{4 \cdot 5}$$

$$A = 1000(1.015)^{20}$$

Now use the y^x key on a calculator, and round the answer to the nearest cent.

$$A = 1346.86$$

The account will contain $1346.86. (The actual amount of interest earned is $1346.86 - $1000 = $346.86. Why?)

Work Problem 6 at the Side.

EXAMPLE 7 Solving a Compound Interest Problem for t

Suppose inflation is averaging 3% per year. How many years will it take for prices to double?

We want to find the number of years t for $1 to grow to $2 at a rate of 3% per year. In the compound interest formula, we let $A = 2$, $P = 1$, $r = .03$, and $n = 1$.

Continued on Next Page

ANSWERS
6. about $3257.79

$$2 = 1\left(1 + \frac{.03}{1}\right)^{1t} \qquad \text{Substitute in the compound interest formula.}$$

$$2 = (1.03)^t \qquad \text{Simplify.}$$

$$\log 2 = \log(1.03)^t \qquad \text{Property 3}$$

$$\log 2 = t \log 1.03 \qquad \text{Power rule}$$

$$t = \frac{\log 2}{\log 1.03} \qquad \text{Divide by log 1.03.}$$

$$t \approx 23.45 \qquad \text{Use a calculator.}$$

Prices will double in about 23 yr. (This is called the **doubling time** of the money.) To check, verify that $1.03^{23.45} \approx 2$.

Work Problem 7 at the Side. ▶▶▶

7 Find the number of years it will take for $500 to increase to $750 in an account paying 4% interest compounded semiannually.

Interest can be compounded annually, semiannually, quarterly, daily, and so on. The number of compounding periods can get larger and larger. If the value of n is allowed to approach infinity, we have an example of **continuous compounding.** However, the compound interest formula above cannot be used for continuous compounding since there is no finite value for n. The formula for continuous compounding is an example of exponential growth involving the number e.

8 (a) How much will $2500 grow to at 4% interest compounded continuously for 3 yr?

Continuous Compound Interest Formula

If a principal of P dollars is deposited at an annual rate of interest r compounded continuously for t years, the final amount on deposit is

$$A = Pe^{rt}.$$

EXAMPLE 8 Solving a Continuous Interest Problem

(a) In Example 6, we found that $1000 invested for 5 yr at 6% interest compounded quarterly would grow to $1346.86. How much would this same investment grow to if compounded continuously?

$$A = Pe^{rt} \qquad \text{Continuous compound interest formula}$$

$$A = 1000e^{(.06)5} \qquad \text{Let } P = 1000, r = .06, \text{ and } t = 5.$$

$$A \approx 1349.86 \qquad \text{Use a calculator; round to two decimal places.}$$

The account will grow to $1349.86.

(b) How long would it take for the initial investment amount to double? We must find the value of t that will cause A to be $2(\$1000) = \$2000$.

$$A = Pe^{rt}$$

$$2000 = 1000e^{.06t} \qquad \text{Let } A = 2P = 2000.$$

$$2 = e^{.06t} \qquad \text{Divide by 1000.}$$

$$\ln 2 = .06t \qquad \text{Take natural logarithms; } \ln e^k = k.$$

$$t = \frac{\ln 2}{.06} \qquad \text{Divide by .06.}$$

$$t \approx 11.55 \qquad \text{Use a calculator.}$$

It would take about 11.55 yr for the original investment to double.

(b) How long would it take for the initial investment in part (a) to double?

Work Problem 8 at the Side. ▶▶▶

ANSWERS
7. about 10.24 yr
8. (a) $2818.74 (b) about 17.33 yr

9 Radioactive strontium decays according to the function defined by

$$y = y_0 e^{-.0239t},$$

where t is time in years.

(a) If an initial sample contains $y_0 = 12$ g of radioactive strontium, how many grams will be present after 35 yr?

OBJECTIVE 4 Solve applications involving base e exponential growth and decay. You may have heard of the carbon-14 dating process used to determine the age of fossils. The method used is based on a base e exponential decay function.

EXAMPLE 9 Solving an Exponential Decay Application

Carbon-14 is a radioactive form of carbon that is found in all living plants and animals. After a plant or animal dies, the radioactive carbon-14 disintegrates according to the function defined by

$$y = y_0 e^{-.000121t},$$

where t is time in years, y is the amount of the sample at time t, and y_0 is the initial amount present at $t = 0$.

(a) If an initial sample contains $y_0 = 10$ g of carbon-14, how many grams will be present after 3000 yr?

Let $y_0 = 10$ and $t = 3000$ in the formula, and use a calculator.

$$y = 10e^{-.000121(3000)} \approx 6.96 \text{ g}$$

(b) How long would it take for the initial sample to decay to half of its original amount? (This is called the **half-life.**)

Let $y = \frac{1}{2}(10) = 5$, and solve for t.

$$5 = 10e^{-.000121t} \qquad \text{Substitute.}$$

$$\frac{1}{2} = e^{-.000121t} \qquad \text{Divide by 10.}$$

$$\ln \frac{1}{2} = -.000121t \qquad \text{Take natural logarithms; } \ln e^k = k.$$

$$t = \frac{\ln \frac{1}{2}}{-.000121} \qquad \text{Divide by } -.000121.$$

$$t \approx 5728 \qquad \text{Use a calculator.}$$

(b) What is the half-life of radioactive strontium?

The half-life is just over 5700 yr.

Work Problem 9 at the Side.

OBJECTIVE 5 Use the change-of-base rule. In **Section 11.5** we used a calculator to approximate the values of common logarithms (base 10) or natural logarithms (base e). However, some applications involve logarithms to other bases. For example, for the years 1980–1996, the percentage of women who had a baby in the last year and returned to work is given by

$$f(x) = 38.83 + 4.208 \log_2 x,$$

for year x. (*Source:* U.S. Bureau of the Census.) To use this function, we need to find a base 2 logarithm. The following rule is used to convert logarithms from one base to another.

Change-of-Base Rule

If $a > 0$, $a \neq 1$, $b > 0$, $b \neq 1$, and $x > 0$, then

$$\log_a x = \frac{\log_b x}{\log_b a}.$$

ANSWERS
9. (a) 5.20 g   (b) 29 yr

NOTE
Any positive number other than 1 can be used for base b in the change-of-base rule, but usually the only practical bases are e and 10 because calculators give logarithms only for these two bases.

10 (a) Find $\log_3 17$ using common logarithms.

To derive the change-of-base rule, let $\log_a x = m$.

$$\log_a x = m$$
$$a^m = x \qquad \text{Change to exponential form.}$$
$$\log_b(a^m) = \log_b x \qquad \text{Property 3}$$
$$m \log_b a = \log_b x \qquad \text{Power rule}$$
$$(\log_a x)(\log_b a) = \log_b x \qquad \text{Substitute for } m.$$
$$\log_a x = \frac{\log_b x}{\log_b a} \qquad \text{Divide by } \log_b a.$$

The last step gives the change-of-base rule.

(b) Find $\log_3 17$ using natural logarithms.

EXAMPLE 10 **Using the Change-of-Base Rule**

Find $\log_5 12$.
   Use common logarithms and the change-of-base rule.

$$\log_5 12 = \frac{\log 12}{\log 5}$$
$$\approx 1.5440 \qquad \text{Use a calculator.}$$

NOTE
Either common or natural logarithms can be used when applying the change-of-base rule. Verify that the same value is found in Example 10 if natural logarithms are used.

11 In Example 11, what percent of women returned to work after having a baby in 1990?

Work Problem 10 at the Side. ▶▶▶

EXAMPLE 11 **Using the Change-of-Base Rule in an Application**

Use natural logarithms in the change-of-base rule and the function defined by

$$f(x) = 38.83 + 4.208 \log_2 x$$

(given earlier) to find the percent of women who returned to work after having a baby in 1995. In the function, $x = 0$ represents 1980.
   Substitute $1995 - 1980 = 15$ for x.

$$f(15) = 38.83 + 4.208 \log_2 15$$
$$= 38.83 + 4.208 \left(\frac{\ln 15}{\ln 2}\right) \qquad \text{Change-of-base rule}$$
$$\approx 55.3\% \qquad \text{Use a calculator.}$$

This is very close to the actual value of 55%.

Work Problem 11 at the Side. ▶▶▶

ANSWERS
10. (a) 2.5789 **(b)** 2.5789
11. 52.8%; This is very close to the actual value of 53%.

Real-Data Applications

Evaluating Investments: The Rule of 72

The Rule of 72 gives an estimate of the doubling time of an investment. It is a useful tool in evaluating and comparing investments.

- The Rule of 72 is $t = \frac{72}{100r}$, where r is the annual interest rate and t is the doubling time. (Since r is the interest rate as a *decimal*, $100r$ is the interest rate as a *percent*.)

- The compound interest formula is $A = P(1 + \frac{r}{n})^{nt}$, for $\$P$ invested at interest rate r (in decimal form), compounded n times per year, that accumulates to $\$A$ after t years.

- The continuous interest formula is $A = Pe^{rt}$, for $\$P$ invested at interest rate r (in decimal form) that accumulates to $\$A$ after t years.

For Group Discussion

1. To investigate how the Rule of 72 works, we will use the Rule of 72 to estimate the doubling time for money invested at 10%.

   (a) What is the estimated doubling time, to the nearest year, for the investment?

   (b) If $2000 is invested at 10% compounded quarterly, what is its accumulated value after the predicted doubling time? Did the Rule of 72 give a good estimate?

   (c) If $2000 is invested at 10% compounded continuously, what is its accumulated value after the predicted doubling time? Did the Rule of 72 give a good estimate?

2. If money is invested at 8%, the Rule of 72 predicts a 9-year doubling time. Sketch a graph to illustrate the doubling effect of an investment of $2000 over time. The x-axis represents time in years with 0, 9, 18, 27, 36, and 45 representing five doubling-time periods. The y-axis represents the value of the investment in dollars.

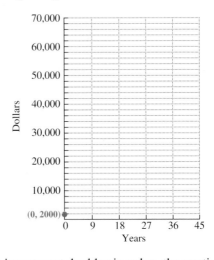

3. Now investigate why the Rule of 72 works. If an investment doubles in value, the continuous interest formula has the form $2P = Pe^{rt}$. Since P is not 0, divide each side of the equation by P to get $2 = e^{rt}$. To solve this equation for t, take the natural logarithm on each side, $\ln 2 = \ln e^{rt}$. Using the power rule, this simplifies to $\ln 2 = rt$. Therefore, $t = \frac{\ln 2}{r}$. Since $\ln 2 \approx .69$, this formula becomes $t = \frac{100 \ln 2}{100r} \approx \frac{69}{100r}$. The number 69 is less useful than 72, which has more factors (i.e., 2, 3, 4, 6, 8, 9, 12, 18, 24, 36), and the doubling time for compound interest will be slightly longer anyway. So, the Rule of 72 estimates this formula as $t = \frac{72}{100r}$. Does the Rule of 72 underestimate or overestimate the true doubling time for continuously compounded investments? Explain your answer.

11.6 Exercises

FOR EXTRA HELP

Tutor Center Addison-Wesley Math Tutor Center

Math XL MathXL

Digital Video Tutor CD 7 Videotape 7

Student's Solutions Manual

MyMathLab MyMathLab

Interactmath.com

RELATING CONCEPTS (EXERCISES 1–4)   For Individual or Group Work

*In **Section 11.2** we solved an equation such as $5^x = 125$ by writing each side as a power of the same base, setting exponents equal, and then solving the resulting equation. The equation is solved as follows.*

$$5^x = 125 \qquad \text{Original equation}$$
$$5^x = 5^3 \qquad 125 = 5^3$$
$$x = 3 \qquad \text{Set exponents equal.}$$

Solution set: {3}

The method described in this section can also be used to solve this equation.
Work Exercises 1–4 in order, *to see how this is done.*

1. Take common logarithms on both sides, and write this equation.

2. Apply the power rule for logarithms on the left.

3. Write the equation so that x is alone on the left.

4. Use a calculator to find the decimal form of the solution. What is the solution set?

Many of the problems in the remaining exercises require a scientific calculator.

Solve each equation. Give solutions to three decimal places. See Example 1.

5. $7^x = 5$

6. $4^x = 3$

7. $9^{-x+2} = 13$

8. $6^{-t+1} = 22$

9. $3^{2x} = 14$

10. $5^{.3x} = 11$

11. $2^{y+3} = 5^y$

12. $6^{m+3} = 4^m$

13. $2^{x+3} = 3^{x-4}$

Solve each equation. Use natural logarithms. Give solutions to three decimal places. See Example 2.

14. $e^{.006x} = 30$

15. $e^{.012x} = 23$

16. $e^{-.103x} = 7$

17. $e^{-.205x} = 9$

18. $\ln e^x = 4$

19. $\ln e^{3x} = 9$

20. $\ln e^{.04x} = \sqrt{3}$

21. $\ln e^{.45x} = \sqrt{7}$

22. $\ln e^{2x} = \pi$

23. Try solving one of the equations in Exercises 14–17 using common logarithms rather than natural logarithms. (You should get the same solution.) Explain why using natural logarithms is a better choice.

24. If you were asked to solve $10^{.0025x} = 75$, would natural or common logarithms be a better choice? Explain.

Solve each equation. Give the exact solution. See Example 3.

25. $\log_3(6x + 5) = 2$

26. $\log_5(12x - 8) = 3$

27. $\log_2(2x - 1) = 5$

28. $\log_6(4x + 2) = 2$

29. $\log_7(x + 1)^3 = 2$

30. $\log_4(x - 3)^3 = 4$

31. Suppose that in solving a logarithmic equation having the term $\log(x - 3)$ you obtain an apparent solution of 2. All algebraic work is correct. Explain why you must reject 2 as a solution of the equation.

32. Suppose that in solving a logarithmic equation having the term $\log(3 - x)$ you obtain an apparent solution of -4. All algebraic work is correct. Should you reject -4 as a solution of the equation? Explain why or why not.

Solve each equation. Give exact solutions. See Examples 4 and 5.

33. $\log(6x + 1) = \log 3$

34. $\log(7 - x) = \log 12$

35. $\log_5(3t + 2) - \log_5 t = \log_5 4$

36. $\log_2(x + 5) - \log_2(x - 1) = \log_2 3$

37. $\log 4x - \log(x - 3) = \log 2$

38. $\log(-x) + \log 3 = \log(2x - 15)$

39. $\log_2 x + \log_2(x - 7) = 3$

40. $\log(2x - 1) + \log 10x = \log 10$

41. $\log 5x - \log(2x - 1) = \log 4$

42. $\log_3 x + \log_3(2x + 5) = 1$

43. $\log_2 x + \log_2(x - 6) = 4$

44. $\log_2 x + \log_2(x + 4) = 5$

Solve each problem. See Examples 6–8.

45. (a) How much money will there be in an account at the end of 6 yr if \$2000 is deposited at 4% compounded quarterly? (Assume no withdrawals are made.)

(b) To one decimal place, how long will it take for the account to grow to \$3000?

46. (a) How much money will there be in an account at the end of 7 yr if \$3000 is deposited at 3.5% compounded quarterly? (Assume no withdrawals are made.)

(b) To one decimal place, how long will it take for the account to grow to \$5000?

47. What will be the amount A in an account with initial principal \$4000 if interest is compounded continuously at an annual rate of 3.5% for 6 yr?

48. Refer to Exercise 46. Does the money grow to a larger value under those conditions, or when invested for 7 yr at 3% compounded continuously?

49. How long would it take an initial principal P to double if it is invested at 4.5% compounded continuously?

50. How long would it take $4000 to double at 3.25% compounded continuously?

Solve each problem. See Example 9.

51. A sample of 400 g of lead-210 decays to polonium-210 according to the function defined by

$$A(t) = 400e^{-.032t},$$

where t is time in years. How much lead will be left in the sample after 25 yr?

52. How long will it take the initial sample of lead in Exercise 51 to decay to half of its original amount?

Use the change-of-base rule (with either common or natural logarithms) to find each logarithm. Give approximations to four decimal places. See Example 10.

53. $\log_6 13$

54. $\log_7 19$

55. $\log_{\sqrt{2}} \pi$

56. $\log_\pi \sqrt{2}$

57. $\log_{21} .7496$

58. $\log_{19} .8325$

59. $\log_{1/2} 5$

60. $\log_{1/3} 7$

61. $\log_{.3} 12$

One measure of the diversity of the species in an ecological community is the **index of diversity,** *the logarithmic expression*

$$-(p_1 \ln p_1 + p_2 \ln p_2 + \ldots + p_n \ln p_n),$$

where $p_1, p_2, \ldots, p_n$ are the proportions of a sample belonging to each of n species in the sample. (Source: Ludwig, John and James Reynolds, Statistical Ecology: A Primer on Methods and Computing, *New York, Wiley, 1988.) Find the index of diversity to three decimal places if a sample of 100 from a community produces the following numbers.*

62. 90 of one species, 10 of another

63. 60 of one species, 40 of another

Chapter 11

SUMMARY

KEY TERMS

11.1 one-to-one function
A one-to-one function is a function in which each x-value corresponds to just one y-value and each y-value corresponds to just one x-value.

inverse of a function f
If f is a one-to-one function, the inverse of f is the set of all ordered pairs of the form (y, x), where (x, y) belongs to f.

11.2 exponential equation
An equation involving an exponential, where the variable is in the exponent, is an exponential equation.

11.3 logarithm
A logarithm is an exponent. The expression $\log_a x$ represents the exponent on the base a that gives the number x.

logarithmic equation
A logarithmic equation is an equation with a logarithm in at least one term.

11.5 common logarithm
A common logarithm is a logarithm to the base 10.

natural logarithm
A natural logarithm is a logarithm to the base e.

NEW SYMBOLS

f^{-1} inverse of f

$\log_a x$ logarithm of x to the base a

$\log x$ common (base 10) logarithm of x

$\ln x$ natural (base e) logarithm of x

e a constant, approximately 2.718281828

TEST YOUR WORD POWER

See how well you have learned the vocabulary in this chapter. Answers, with examples, follow the Quick Review.

1. In a **one-to-one function**
  A. each x-value corresponds to only one y-value
  B. each x-value corresponds to one or more y-values
  C. each x-value is the same as each y-value
  D. each x-value corresponds to only one y-value and each y-value corresponds to only one x-value.

2. If f is a one-to-one function, then the **inverse** of f is
  A. the set of all solutions of f
  B. the set of all ordered pairs formed by interchanging the coordinates of the ordered pairs of f

  C. an equation involving an exponential expression
  D. the set of all ordered pairs that are the opposite (negative) of the coordinates of the ordered pairs of f.

3. An **exponential function** is a function defined by an expression of the form
  A. $f(x) = ax^2 + bx + c$ for real numbers a, b, c ($a \neq 0$)
  B. $f(x) = \log_a x$, for a and x positive numbers ($a \neq 1$)
  C. $f(x) = a^x$ for all real numbers x ($a > 0, a \neq 1$)
  D. $f(x) = \sqrt{x}$ for $x \geq 0$.

4. A **logarithm** is
  A. an exponent
  B. a base
  C. an equation
  D. a radical expression.

5. A **logarithmic function** is a function defined by an expression of the form
  A. $f(x) = ax^2 + bx + c$ for real numbers a, b, c ($a \neq 0$)
  B. $f(x) = \log_a x$, for a and x positive numbers ($a \neq 1$)
  C. $f(x) = a^x$ for all real numbers x ($a > 0, a \neq 1$)
  D. $f(x) = \sqrt{x}$ for $x \geq 0$.

QUICK REVIEW

| *Concepts* | *Examples* |

11.1 *Inverse Functions*

Horizontal Line Test

If a horizontal line intersects the graph of a function in no more than one point, then the function is one-to-one.

Inverse Functions

For a one-to-one function f defined by an equation $y = f(x)$, the equation that defines the inverse function f^{-1} is found by interchanging x and y, solving for y, and replacing y with $f^{-1}(x)$.

In general, the graph of f^{-1} is the mirror image of the graph of f with respect to the line $y = x$.

Find f^{-1} if $f(x) = 2x - 3$. The graph of f is a straight line, so f is one-to-one by the horizontal line test.

Interchange x and y in the equation $y = 2x - 3$.

$$x = 2y - 3$$

Solve for y to get $\qquad y = \dfrac{x + 3}{2}.$

Therefore, $\qquad f^{-1}(x) = \dfrac{x + 3}{2}.$

The graphs of a nonlinear function f and its inverse f^{-1} are shown here.

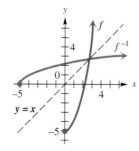

11.2 *Exponential Functions*

For $a > 0$, $a \neq 1$, $f(x) = a^x$ defines an exponential function with base a.

Characteristics of the Graph of $F(x) = a^x$
1. The graph contains the point $(0, 1)$.
2. When $a > 1$, the graph rises from left to right.
   When $0 < a < 1$, the graph falls from left to right.
3. The x-axis is an asymptote.
4. The domain is $(-\infty, \infty)$; the range is $(0, \infty)$.

$F(x) = 3^x$ defines the exponential function with base 3.

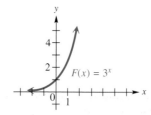

11.3 *Logarithmic Functions*

$y = \log_a x$ means $x = a^y$.

For $b > 0$, $b \neq 1$, $\log_b b = 1$ and $\log_b 1 = 0$.

For $a > 0$, $a \neq 1$, $x > 0$, $G(x) = \log_a x$ defines the logarithmic function with base a.

Characteristics of the Graph of $G(x) = \log_a x$
1. The graph contains the point $(1, 0)$.
2. When $a > 1$, the graph rises from left to right.
   When $0 < a < 1$, the graph falls from left to right.
3. The y-axis is an asymptote.
4. The domain is $(0, \infty)$; the range is $(-\infty, \infty)$.

$y = \log_2 x$ means $x = 2^y$.

$$\log_3 3 = 1 \qquad \log_5 1 = 0$$

$G(x) = \log_3 x$ defines the logarithmic function with base 3.

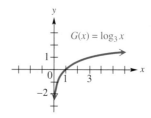

Concepts	Examples

11.4 *Properties of Logarithms*

Product Rule

$$\log_a xy = \log_a x + \log_a y$$

Quotient Rule

$$\log_a \frac{x}{y} = \log_a x - \log_a y$$

Power Rule

$$\log_a x^r = r \log_a x$$

Special Properties

$$b^{\log_b x} = x \quad \text{and} \quad \log_b b^x = x$$

$$\log_2 (3 \cdot m) = \log_2 3 + \log_2 m$$

$$\log_5 \frac{9}{4} = \log_5 9 - \log_5 4$$

$$\log_{10} 2^3 = 3 \log_{10} 2$$

$$6^{\log_6 10} = 10 \quad \log_3 3^4 = 4$$

11.5 *Common and Natural Logarithms*

Common logarithms (base 10) are used in applications such as pH, sound level, and intensity of an earthquake. Use the ⬭LOG key of a calculator to evaluate common logarithms.

Use the formula $pH = -\log [H_3O^+]$ to find the pH (to one decimal place) of grapes with hydronium ion concentration 5.0×10^{-5}.

$$
\begin{aligned}
pH &= -\log (5.0 \times 10^{-5}) &&\text{Substitute.}\\
&= -(\log 5.0 + \log 10^{-5}) &&\text{Property of logarithms}\\
&\approx 4.3 &&\text{Evaluate.}
\end{aligned}
$$

Natural logarithms (base e) are most often used in applications of growth and decay, such as time for money invested to double, decay of chemical compounds, and biological growth. Use the ⬭ln x key or both the ⬭INV and ⬭e^x keys to evaluate natural logarithms.

Use the formula for doubling time (in years) $t = \dfrac{\ln 2}{\ln (1 + r)}$ to find the doubling time to the nearest tenth at an interest rate of 4% compounded annually.

$$
\begin{aligned}
t &= \frac{\ln 2}{\ln (1 + .04)} &&\text{Substitute.}\\
&\approx 17.7 &&\text{Evaluate.}
\end{aligned}
$$

The doubling time is about 17.7 yr.

11.6 *Exponential and Logarithmic Equations; Further Applications*

To solve exponential equations, use these properties $(b > 0, b \neq 1)$.

1. If $b^x = b^y$, then $x = y$.

2. If $x = y \ (x > 0, y > 0)$, then $\log_b x = \log_b y$.

Solve $\qquad\qquad 2^{3x} = 2^5$.

$$3x = 5$$

$$x = \frac{5}{3}$$

The solution set is $\left\{ \frac{5}{3} \right\}$.

Solve $\qquad\qquad 5^x = 8$.

$$\log 5^x = \log 8$$

$$x \log 5 = \log 8$$

$$x = \frac{\log 8}{\log 5} \approx 1.2920$$

The solution set is $\{1.2920\}$. *(continued)*

Concepts	Examples
11.6 *Exponential and Logarithmic Equations; Further Applications* (*continued*)	

To solve logarithmic equations, use these properties, where $b > 0$, $b \neq 1$, $x > 0$, $y > 0$. First use the properties of **Section 11.4,** if necessary, to write the equation in the proper form.

1. If $\log_b x = \log_b y$, then $x = y$.

Solve $\log_3 2x = \log_3 (x + 1)$.

$$2x = x + 1$$
$$x = 1$$

The solution set is $\{1\}$.

2. If $\log_b x = y$, then $b^y = x$.

Solve $\log_2 (3x - 1) = 4$.

$$3x - 1 = 2^4$$
$$3x - 1 = 16$$
$$3x = 17$$
$$x = \frac{17}{3}$$

The solution set is $\left\{\frac{17}{3}\right\}$.

Change-of-Base Rule

If $a > 0$, $a \neq 1$, $b > 0$, $b \neq 1$, $x > 0$, then

$$\log_a x = \frac{\log_b x}{\log_b a}.$$

Approximate $\log_3 37$.

$$\log_3 37 = \frac{\ln 37}{\ln 3} = \frac{\log 37}{\log 3} \approx 3.2868$$

ANSWERS TO TEST YOUR WORD POWER

1. D; *Example:* The function $f = \{(0, 2), (1, -1), (3, 5), (-2, 3)\}$ is one-to-one.
2. B; *Example:* The inverse of the one-to-one function f defined in Answer 1 is $f^{-1} = \{(2, 0), (-1, 1), (5, 3), (3, -2)\}$.
3. C; *Examples:* $f(x) = 4^x$, $g(x) = \left(\frac{1}{2}\right)^x$
4. A; *Example:* $\log_a x$ is the exponent to which a must be raised to obtain x; $\log_3 9 = 2$ since $3^2 = 9$.
5. B; *Examples:* $y = \log_3 x$, $y = \log_{1/3} x$

Chapter **11**

R E V I E W E X E R C I S E S

[11.1] *Determine whether each graph is the graph of a one-to-one function.*

1.

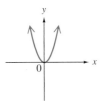

2.

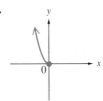

3. The table lists caffeine amounts in several popular 12-oz sodas. If the set of sodas is the domain and the set of caffeine amounts is the range of the function consisting of the six pairs listed, is it a one-to-one function? Why or why not?

Soda	Caffeine (mg)
Mountain Dew	55
Diet Coke	45
Dr. Pepper	41
Sunkist Orange Soda	41
Diet Pepsi-Cola	36
Coca-Cola Classic	34

Source: National Soft Drink Association.

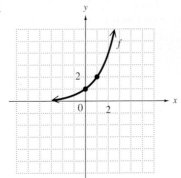

Determine whether each function is one-to-one. If it is, find its inverse.

4. $f(x) = -3x + 7$

5. $f(x) = \sqrt[3]{6x - 4}$

6. $f(x) = -x^2 + 3$

Each function graphed is one-to-one. Graph its inverse.

7.

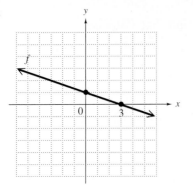

8.

[11.2] *Graph each function.*

9. $f(x) = 3^x$

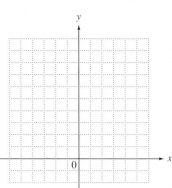

10. $f(x) = \left(\dfrac{1}{3}\right)^x$

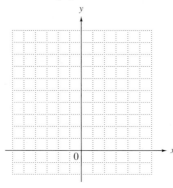

11. $y = 3^{x+1}$

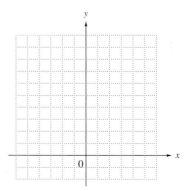

Solve each equation.

12. $4^{3x} = 8^{x+4}$

13. $\left(\dfrac{1}{27}\right)^{x-1} = 9^{2x}$

14. $5^x = 1$

▦ *In the remainder of the Chapter Review, many exercises will require a scientific calculator. We do not mark each such exercise.*

15. A recent report predicts that the U.S. Hispanic population will increase from 26.7 million in 1995 to 96.5 million in 2050.(*Source:* U.S. Bureau of the Census.) Assuming an exponential growth pattern, the population is approximated by

$$f(x) = 26.7\,(2)^{.0332x}$$

where x represents the number of years since 1995. Use this function to estimate the Hispanic population in each year.

(a) 2010 **(b)** 2015

[11.3]

16. (a) Write in exponential form: $\log_5 625 = 4$.

(b) Write in logarithmic form: $5^{-2} = .04$.

17. (a) In your own words, explain the meaning of $\log_b a$.

(b) Based on the meaning of $\log_b a$, what is the simplest form of $b^{\log_b a}$?

Graph each function.

18. $g(x) = \log_3 x$ (*Hint:* See Exercise 9.)

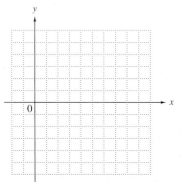

19. $g(x) = \log_{1/3} x$ (*Hint:* See Exercise 10.)

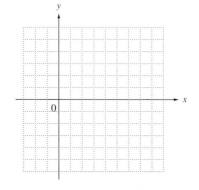

Solve each equation.

20. $\log_8 64 = x$

21. $\log_7\left(\dfrac{1}{49}\right) = x$

22. $\log_4 x = \dfrac{3}{2}$

23. $\log_b b^2 = 2$

[11.4] *Apply the properties of logarithms to express each logarithm as a sum or difference of logarithms. Assume that all variables represent positive real numbers.*

24. $\log_4 3x^2$

25. $\log_2 \dfrac{p^2 r}{\sqrt{z}}$

Use the properties of logarithms to write each expression as a single logarithm. Assume that all variables represent positive real numbers, $b \neq 1$.

26. $\log_b 3 + \log_b x - 2 \log_b y$

27. $\log_3(x + 7) - \log_3(4x + 6)$

[11.5] *Evaluate each logarithm. Give approximations to four decimal places.*

28. $\log 28.9$

29. $\log .257$

30. $\ln 28.9$

31. $\ln .257$

Find the pH of each substance with the given hydronium ion concentration.

32. Milk, 4.0×10^{-7}

33. Crackers, 3.8×10^{-9}

34. If orange juice has pH 4.6, what is its hydronium ion concentration?

Solve each problem.

35. Section 11.5 Exercise 37 introduced the formula for doubling time,

$$t = \frac{\ln 2}{\ln(1 + r)},$$

which gives the number of years required to double your money when it is invested at interest rate r (in decimal form) compounded annually. How long does it take to double your money at each rate? Round answers to the nearest year.

(a) 4%   **(b)** 6%

(c) 10%   **(d)** 12%

(e) Compare each answer in parts (a)–(d) with these numbers:

$$\frac{72}{4}, \frac{72}{6}, \frac{72}{10}, \frac{72}{12}.$$

What do you find?

36. The graph shows the percent change in commercial rents in California from 1992 through 1999.

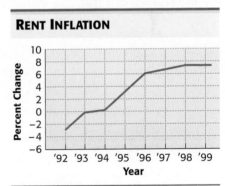

RENT INFLATION

Source: CB Commercial/Torto. Wheaton Research.

The percent change in rents is approximated by the logarithmic function defined by

$$g(x) = -650 + 143 \ln x,$$

where x represents the number of years since 1900.

(a) Find $g(92)$ and $g(99)$.

(b) Compare your results with the corresponding values in the graph.

[11.6] *Solve each equation. Give solutions to three decimal places.*

37. $3^x = 9.42$

38. $2^{x-1} = 15$

39. $e^{.06x} = 3$

Solve each equation. Give exact solutions.

40. $\log_3(9x + 8) = 2$

41. $\log_5(x + 6)^3 = 2$

42. $\log_3(p + 2) - \log_3 p = \log_3 2$

43. $\log(2x + 3) - \log x = 1$

44. $\log_4 x + \log_4(8 - x) = 2$

45. $\log_2 x + \log_2(x + 15) = 4$

Solve each problem.

46. How much would be in an account after 3 yr if $6500.00 was invested at 3% annual interest, compounded daily (use $n = 365$)?

47. Which is a better plan?

Plan A: Invest $1000.00 at 4% compounded quarterly for 3 yr

Plan B: Invest $1000.00 at 3.9% compounded monthly for 3 yr

A machine purchased for business use depreciates, or loses value, over a period of years. The value of the machine at the end of its useful life is called its scrap value. By one method of depreciation (where it is assumed a constant percentage of the value depreciates annually), the scrap value, S, is given by

$$S = C(1 - r)^n,$$

where C is the original cost, n is the useful life in years, and r is the constant percent of depreciation.

48. Find the scrap value of a machine costing $30,000, having a useful life of 12 yr and a constant annual rate of depreciation of 15%.

49. A machine has a "half-life" of 6 yr. Find the constant annual rate of depreciation.

Use the change-of-base rule (with either common or natural logarithms) to find each logarithm. Give approximations to four decimal places.

50. $\log_{16} 13$

51. $\log_4 12$

52. $\log_{\sqrt{6}} \sqrt{13}$

MIXED REVIEW EXERCISES

Solve.

53. $\log_3(x + 9) = 4$

54. $\log_2 32 = x$

55. $\log_x \dfrac{1}{81} = 2$

56. $27^x = 81$

57. $2^{2x-3} = 8$

58. $\log_3(x + 1) - \log_3 x = 2$

59. $\log(3x - 1) = \log 10$

60. Find the value of n in the equation for Exercise 48 if the scrap value is $10,000, the cost is $30,000, and the depreciation rate is 15%.

Chapter **11**

T E S T

 Study Skills Workbook
Activity 12: Final Exams

1. Decide whether each function is one-to-one.

(a) $f(x) = x^2 + 9$ **(b)**

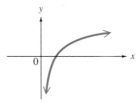

1. (a) _____

(b) _____

2. Find $f^{-1}(x)$ for the one-to-one function defined by $f(x) = \sqrt[3]{x + 7}$.

2. _____

3. Graph the inverse of f, given the graph of f here.

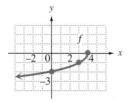

3.

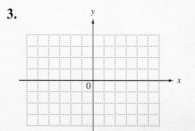

Graph each function.

4. $y = 6^x$

4.

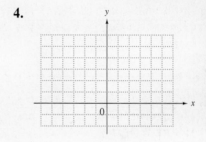

5. $y = \log_6 x$

5.

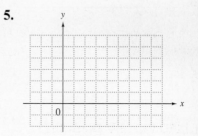

6. Explain how the graph of the function in Exercise 5 can be obtained from the graph of the function in Exercise 4.

6. _____

Solve each equation. Give the exact solution.

7. $5^x = \dfrac{1}{625}$ **8.** $2^{3x-7} = 8^{2x+2}$

7. _____

8. _____

9. (a) _____

(b) _____

9. The atmospheric pressure (in millibars) at a given altitude x (in meters) is approximated by

$$f(x) = 1013e^{-.0001341x}.$$

Use this function to approximate the atmospheric pressure at

(a) 2000 m　　　　**(b)** 10,000 m.

10. _____

10. Write in logarithmic form: $4^{-2} = .0625$.

11. _____

11. Write in exponential form: $\log_7 49 = 2$.

Solve each equation.

12. _____

12. $\log_{1/2} x = -5$

13. _____

13. $x = \log_9 3$

14. _____

14. $\log_x 16 = 4$

15. _____

15. Use properties of logarithms to write $\log_3 x^2 y$ as a sum or difference of logarithms. Assume the variables represent positive real numbers.

16. _____

16. Use properties of logarithms to write $\dfrac{1}{4}\log_b r + 2\log_b s - \dfrac{2}{3}\log_b t$ as a single logarithm. Assume the variables represent positive real numbers, $b \neq 1$.

17. (a) _____

(b) _____

(c) _____

17. Use a calculator to find an approximation to four decimal places for each logarithm.

(a) $\log 21.3$　　　　**(b)** $\ln .43$　　　　**(c)** $\log_6 45$

18. _____

18. Solve $3^x = 78$, giving the solution to four decimal places.

19. _____

19. Solve $\log_8(x + 5) + \log_8(x - 2) = \log_8 8$.

20. (a) _____

(b) _____

20. Suppose that $10,000 is invested at 4.5% annual interest, compounded quarterly.

(a) How much will be in the account in 5 yr if no money is withdrawn?

(b) How long will it take for the initial principal to double?

Cumulative Review Exercises

CHAPTERS 1–11

Let $S = \left\{-\frac{9}{4}, -2, -\sqrt{2}, 0, .6, \sqrt{11}, \sqrt{-8}, 6, \frac{30}{3}\right\}$. *List the elements of S that are elements of each set.*

1. Integers

2. Rational numbers

3. Irrational numbers

Simplify each expression.

4. $|-8| + 6 - |-2| - (-6 + 2)$

5. $2(-5) + (-8)(4) - (-3)$

Solve each equation or inequality.

6. $7 - (3 + 4x) + 2x = -5(x - 1) - 3$

7. $2x + 2 \leq 5x - 1$

8. $|2x - 5| = 9$

9. $|3x| - 4 = 12$

10. $|3x - 8| \leq 1$

11. $|4x + 2| > 10$

Graph each equation or inequality.

12. $y = -2.5x + 5$

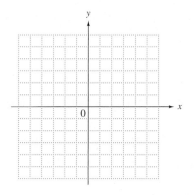

13. $-4x + y \leq 5$

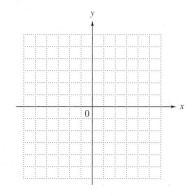

14. The graph indicates that the number of U.S. travelers to foreign countries increased from 44,623 thousand in 1990 to 60,816 thousand in 2000.

(a) Is this the graph of a function?

(b) What is the slope of the line in the graph? Interpret the slope in the context of U.S. travelers to foreign countries.

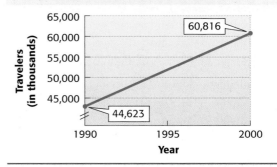

U.S. TRAVELERS TO FOREIGN COUNTRIES

Source: U.S. Department of Commerce.

15. Find the slope-intercept form of the equation of the line through $(5, -1)$ and parallel to the line with equation $3x - 4y = 12$.

Solve each system of equations.

16. $5x - 3y = 14$
$2x + 5y = 18$

17. $x + 2y + 3z = 11$
$3x - y + z = 8$
$2x + 2y - 3z = -12$

18. Candy worth $1.00 per lb is to be mixed with candy worth $1.96 per lb to get 16 lb of a mixture that will be sold for $1.60 per lb. How many pounds of each candy should be used?

Price per Pound	Number of Pounds	Value
$1.00	x	$1x$
	y	
$1.60		

Perform the indicated operations.

19. $(2p + 3)(3p - 1)$

20. $(4k - 3)^2$

21. $(3m^3 + 2m^2 - 5m) - (8m^3 + 2m - 4)$

22. Divide $6t^4 + 17t^3 - 4t^2 + 9t + 4$ by $3t + 1$.

Factor completely.

23. $8x + x^3$

24. $24y^2 - 7y - 6$

25. $5z^3 - 19z^2 - 4z$

26. $16a^2 - 25b^4$

27. $8c^3 + d^3$

28. $16r^2 + 56rq + 49q^2$

Perform the indicated operations.

29. $\dfrac{(5p^3)^4(-3p^7)}{2p^2(4p^4)}$

30. $\dfrac{x^2 - 9}{x^2 + 7x + 12} \div \dfrac{x - 3}{x + 5}$

31. $\dfrac{2}{k + 3} - \dfrac{5}{k - 2}$

32. $\dfrac{3}{p^2 - 4p} - \dfrac{4}{p^2 + 2p}$

33. Solve $\dfrac{1}{x} - \dfrac{3}{2x} = \dfrac{1}{x + 1}$.

Simplify.

34. $\sqrt{288}$

35. $\dfrac{-8^{4/3}}{8^2}$

36. $2\sqrt{32} - 5\sqrt{98}$

37. Solve $\sqrt{2x + 1} - \sqrt{x} = 1$.

38. Multiply $(5 + 4i)(5 - 4i)$.

Solve each equation or inequality.

39. $3x^2 = x + 1$

40. $x^2 + 2x - 8 > 0$

41. $x^4 - 5x^2 + 4 = 0$

Graph.

42. $y = \dfrac{1}{3}(x - 1)^2 + 2$

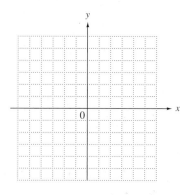

43. $f(x) = 2^x$

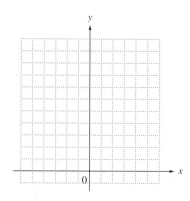

44. $f(x) = \log_3 x$

Solve.

45. $5^{x+3} = \left(\dfrac{1}{25}\right)^{3x+2}$

46. $\log_5 x + \log_5(x + 4) = 1$

47. Write $\log_5 125 = 3$ in exponential form.

48. Rewrite the following using the product, quotient, and power rules for logarithms:

$$\log \frac{x^3 \sqrt{y}}{z}.$$

49. We used the formula for continuous compounding

$$A = Pe^{rt}$$

in **Section 11.6.** To three decimal places, what growth rate r will triple the value of P in 10 yr?

50. An advertisement for a Pioneer 43 in. plasma HDTV states that its image processor "produces a staggering amount of 1,073,741,824 colors." (*Source:* Circuit City advertisement.) This number is an integer power of 2. Determine this exponent by solving the equation

$$2^x = 1,073,741,824$$

using the power rule for logarithms and the change-of-base rule.

Nonlinear Functions, Conic Sections, and Nonlinear Systems

12

12.1 Additional Graphs of Functions; Composition

12.2 The Circle and the Ellipse

12.3 The Hyperbola and Other Functions Defined by Radicals

12.4 Nonlinear Systems of Equations

12.5 Second-Degree Inequalities and Systems of Inequalities

In this chapter, we study a group of curves known as *conic sections*. One conic section, the *ellipse,* has a special reflecting property responsible for "whispering galleries." In a whispering gallery, a person whispering at a certain point in the room can be heard clearly at another point across the room.

The Old House Chamber of the U.S. Capitol, now called Statuary Hall, is a whispering gallery. History has it that John Quincy Adams, whose desk was positioned at exactly the right point beneath the ellipsoidal ceiling, often pretended to sleep there as he listened to political opponents whispering strategies across the room. (*Source: We, the People, The Story of the United States Capitol,* 1991.)

In Section 12.2, we investigate ellipses.

721

12.1 Additional Graphs of Functions; Composition

OBJECTIVES

1 Recognize the graphs of the elementary functions defined by $|x|, \frac{1}{x}$, and $\sqrt{x}$, and graph their translations.

2 Recognize and graph step functions.

3 Find the composition of functions.

In earlier chapters we introduced the function defined by $f(x) = x^2$, sometimes called the **squaring function.** This is one of the most important elementary functions in algebra.

OBJECTIVE 1 Recognize the graphs of the elementary functions defined by $|x|, \frac{1}{x}$, and $\sqrt{x}$, and graph their translations. Another one of the elementary functions, defined by $f(x) = |x|$, is called the **absolute value function.** Its graph, along with a table of selected ordered pairs, is shown in Figure 1. Its domain is $(-\infty, \infty)$, and its range is $[0, \infty)$.

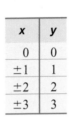

x	y
0	0
±1	1
±2	2
±3	3

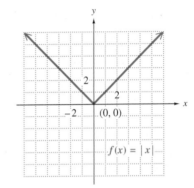

Figure 1

The **reciprocal function,** defined by $f(x) = \frac{1}{x}$, was introduced in **Section 8.4.** Its graph is shown in Figure 2, along with a table of selected ordered pairs. Notice that x can never equal 0 for this function, and as a result, as x gets closer and closer to 0, the graph approaches either ∞ or $-\infty$. Also, $\frac{1}{x}$ can never equal 0, and as x approaches ∞ or $-\infty$, $\frac{1}{x}$ approaches 0. The axes are called **asymptotes** for the function. (Asymptotes are studied in more detail in college algebra courses.) For the reciprocal function, the domain and the range are both $(-\infty, 0) \cup (0, \infty)$.

x	y
$\frac{1}{3}$	3
$\frac{1}{2}$	2
1	1
2	$\frac{1}{2}$
3	$\frac{1}{3}$

x	y
$-\frac{1}{3}$	-3
$-\frac{1}{2}$	-2
-1	-1
-2	$-\frac{1}{2}$
-3	$-\frac{1}{3}$

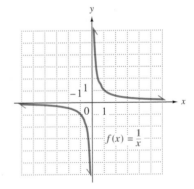

Figure 2

The **square root function,** defined by $f(x) = \sqrt{x}$, was introduced in **Section 9.1.** Its graph is shown in Figure 3 on the next page. Notice that since we restrict function values to be real numbers, x cannot take on negative values. Thus, the domain of the square root function is $[0, \infty)$. Because the principal square root is always nonnegative, the range is also $[0, \infty)$. A table of values is shown along with the graph.

x	y
0	0
1	1
4	2

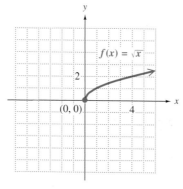

Figure 3

1 Graph $f(x) = \sqrt{x + 4}$.
Give the domain and range.

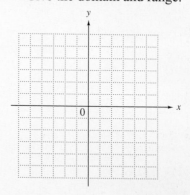

Just as the graph of $f(x) = x^2$ can be shifted, or translated, as we saw in **Section 10.5,** so can the graphs of these other elementary functions.

EXAMPLE 1 **Applying a Horizontal Shift**

Graph $f(x) = |x - 2|$.

The graph of $y = (x - 2)^2$ is obtained by shifting the graph of $y = x^2$ two units to the right. In a similar manner, the graph of $f(x) = |x - 2|$ is found by shifting the graph of $y = |x|$ two units to the right, as shown in Figure 4. The table of ordered pairs accompanying the graph supports this, as can be seen by comparing it to the table with Figure 1. The domain of this function is $(-\infty, \infty)$, and its range is $[0, \infty)$.

x	y
0	2
1	1
2	0
3	1
4	2

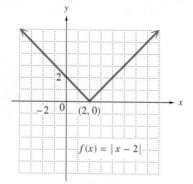

Figure 4

Work Problem 1 at the Side. ▶▶▶

EXAMPLE 2 **Applying a Vertical Shift**

Graph $f(x) = \dfrac{1}{x} + 3$.

The graph of this function is found by shifting the graph of $y = \frac{1}{x}$ three units up. See Figure 5 on the next page. The domain is $(-\infty, 0) \cup (0, \infty)$, and the range is $(-\infty, 3) \cup (3, \infty)$.

—— **Continued on Next Page**

ANSWERS

1.

graph

$[-4, \infty); [0, \infty)$

2 Graph $f(x) = \dfrac{1}{x} - 2$.

Give the domain and range.

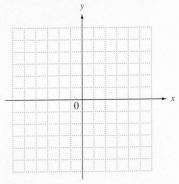

x	y
$\frac{1}{3}$	6
$\frac{1}{2}$	5
1	4
2	3.5

x	y
$-\frac{1}{3}$	0
$-\frac{1}{2}$	1
-1	2
-2	2.5

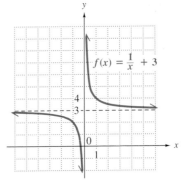

Figure 5

Work Problem 2 at the Side.

3 Graph $f(x) = |x + 2| + 1$.
Give the domain and range.

EXAMPLE 3 Applying Both Horizontal and Vertical Shifts

Graph $f(x) = \sqrt{x + 1} - 4$.

The graph of $y = (x + 1)^2 - 4$ is obtained by shifting the graph of $y = x^2$ one unit to the left and four units down. Following this pattern here, we shift the graph of $y = \sqrt{x}$ one unit to the left and four units down to get the graph of $f(x) = \sqrt{x + 1} - 4$. See Figure 6. The domain is $[-1, \infty)$, and the range is $[-4, \infty)$.

x	y
-1	-4
0	-3
3	-2

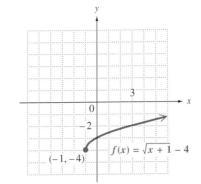

Figure 6

4 Find each of the following.

(a) $[\![18]\!]$ (b) $[\![8.7]\!]$

(c) $[\![-5]\!]$ (d) $[\![-6.9]\!]$

(e) $\left[\!\left[\dfrac{1}{2}\right]\!\right]$

Work Problem 3 at the Side.

OBJECTIVE 2 Recognize and graph step functions. The **greatest integer function,** usually written $f(x) = [\![x]\!]$, is defined as follows:

$[\![x]\!]$ denotes the largest integer that is less than or equal to x.

For example,

$$[\![8]\!] = 8, \quad [\![7.45]\!] = 7, \quad [\![\pi]\!] = 3, \quad [\![-1]\!] = -1, \quad [\![-2.6]\!] = -3,$$

and so on.

Work Problem 4 at the Side.

ANSWERS

2.

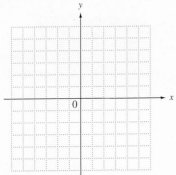

$f(x) = \frac{1}{x} - 2$

$(-\infty, 0) \cup (0, \infty); (-\infty, -2) \cup (-2, \infty)$

3.

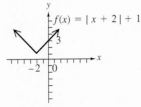

$f(x) = |x + 2| + 1$

$(-\infty, \infty); [1, \infty)$

4. (a) 18   (b) 8   (c) -5 (d) -7 (e) 0

EXAMPLE 4 **Graphing the Greatest Integer Function**

Graph $f(x) = [\![x]\!]$.
   For $[\![x]\!]$,

$$\text{if } -1 \le x < 0, \quad \text{then } [\![x]\!] = -1;$$
$$\text{if } \quad 0 \le x < 1, \quad \text{then } [\![x]\!] = 0;$$
$$\text{if } \quad 1 \le x < 2, \quad \text{then } [\![x]\!] = 1,$$

and so on. Thus, the graph, as shown in Figure 7, consists of a series of horizontal line segments. In each one, the left endpoint is included and the right endpoint is excluded. These segments continue infinitely following this pattern to the left and right. Since x can take any real number value, the domain is $(-\infty, \infty)$. The range is the set of integers $\{\ldots, -4, -3, -2, -1, 0, 1, 2, 3, 4, \ldots\}$. The appearance of the graph is the reason that this function is called a **step function.**

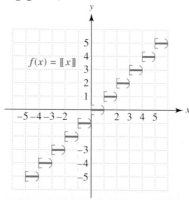

Figure 7

The graph of a step function also may be shifted. For example, the graph of $h(x) = [\![x - 2]\!]$ is the same as the graph of $f(x) = [\![x]\!]$ shifted two units to the right. Similarly, the graph of $g(x) = [\![x]\!] + 2$ is the graph of $f(x)$ shifted two units up.

Work Problem 5 at the Side. ▶▶▶

EXAMPLE 5 **Applying a Greatest Integer Function**

An overnight delivery service charges $25 for a package weighing up to 2 lb. For each additional pound or fraction of a pound there is an additional charge of $3. Let $D(x)$ represent the cost to send a package weighing x pounds. Graph $D(x)$ for x in the interval $(0, 6]$.

For x in the interval $(0, 2]$, $y = 25$.
For x in the interval $(2, 3]$, $y = 25 + 3 = 28$.
For x in the interval $(3, 4]$, $y = 28 + 3 = 31$, and so on.

The graph, which is that of a step function, is shown in Figure 8.

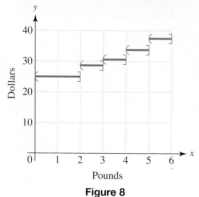

Figure 8

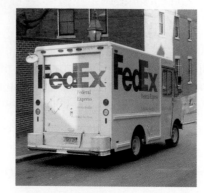

Work Problem 6 at the Side. ▶▶▶

5 Graph $f(x) = [\![x + 1]\!]$. Give the domain and range.

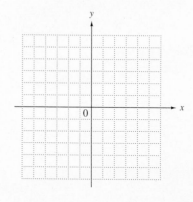

6 Assume that the post office charges 80¢ per oz (or fraction of an ounce) to mail a letter to Europe. Graph the ordered pairs (ounces, cost) for x in the interval $(0, 4]$.

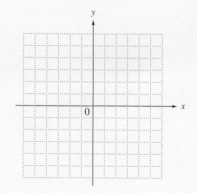

ANSWERS
5.

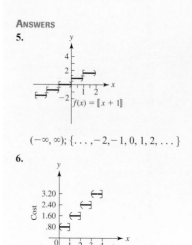

$(-\infty, \infty); \{\ldots, -2, -1, 0, 1, 2, \ldots\}$

6.

OBJECTIVE 3 Find the composition of functions. The diagram in Figure 9 shows a function f that assigns to each element x of set X some element y of set Y. Suppose that a function g takes each element of set Y and assigns a value z of set Z. Using both f and g, then, an element x in X is assigned to an element z in Z. The result of this process is a new function h, which takes an element x in X and assigns an element z in Z.

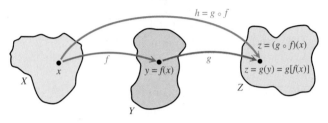

Figure 9

This function h is called the *composition* of functions g and f, written $g \circ f$, and is defined as follows.

Composition of Functions

If f and g are functions, then the **composite function,** or **composition,** of g and f is defined by

$$(g \circ f)(x) = g[f(x)]$$

for all x in the domain of f such that $f(x)$ is in the domain of g.

Read $g \circ f$ as "g of f."

As a real-life example of how composite functions occur, consider the following retail situation.

A $40 pair of blue jeans is on sale for 25% off. If you purchase the jeans before noon, the retailer offers an additional 10% off. What is the final sale price of the blue jeans?

You might be tempted to say that the jeans are 35% off and calculate $\$40(.35) = \$14$, giving a final sale price of $\$40 - \$14 = \$26$ for the jeans. ***This is not correct.*** To find the final sale price, we must first find the price after taking 25% off, and then take an additional 10% off that price.

$\$40(.25) = \$10$, giving a sale price of $\$40 - \$10 = \mathbf{\$30}$.   Take 25% off original price.

$\mathbf{\$30}(.10) = \$3$, giving a ***final sale price*** of $\$30 - \$3 = \mathbf{\$27}$.   Take additional 10% off.

This is the idea behind composition of functions.

As another example, suppose an oil well off the California coast is leaking, with the leak spreading oil in a circular layer over the surface. See Figure 10.

Figure 10

At any time t, in minutes, after the beginning of the leak, the radius of the circular oil slick is given by $r(t) = 5t$ feet. Since $A(r) = \pi r^2$ gives the area of a circle of radius r, the area can be expressed as a function of time by substituting $5t$ for r in $A(r) = \pi r^2$ to get

$$A(r) = \pi r^2$$
$$A[r(t)] = \pi(5t)^2 = 25\pi t^2.$$

The function $A[r(t)]$ is a composite function of the functions A and r.

EXAMPLE 6 **Evaluating a Composite Function**

Let $f(x) = x^2$ and $g(x) = x + 3$. Find $(f \circ g)(4)$.

$$
\begin{aligned}
(f \circ g)(4) &= f[g(4)] &&\text{Definition}\\
&= f(4 + 3) &&\text{Use the rule for } g(x); g(4) = 4 + 3.\\
&= f(7) &&\text{Add.}\\
&= 7^2 &&\text{Use the rule for } f(x); f(7) = 7^2.\\
&= 49
\end{aligned}
$$

Notice in Example 6 that if we reverse the order of the functions, the composition of g and f is defined by $g[f(x)]$. Once again, letting $x = 4$, we have

$$
\begin{aligned}
(g \circ f)(4) &= g[f(4)] &&\text{Definition}\\
&= g(4^2) &&\text{Use the rule for } f(x); f(4) = 4^2.\\
&= g(16) &&\text{Square 4.}\\
&= 16 + 3 &&\text{Use the rule for } g(x); g(16) = 16 + 3.\\
&= 19.
\end{aligned}
$$

Here we see that $(f \circ g)(4) \neq (g \circ f)(4)$ because $49 \neq 19$. In general,

$$(f \circ g)(x) \neq (g \circ f)(x).$$

7 Let $f(x) = 3x + 6$ and $g(x) = x^3$. Find each of the following.

(a) $(f \circ g)(2)$

EXAMPLE 7 Finding Composite Functions

Let $f(x) = 4x - 1$ and $g(x) = x^2 + 5$. Find each of the following.

(a) $(f \circ g)(2)$

$$(f \circ g)(2) = f[g(2)]$$
$$= f(2^2 + 5) \quad g(x) = x^2 + 5$$
$$= f(9)$$
$$= 4(9) - 1 \quad f(x) = 4x - 1$$
$$= 35$$

(b) $(f \circ g)(x)$

Here, use $g(x)$ as the input for the function f.

$$(f \circ g)(x) = f[g(x)]$$
$$= 4(g(x)) - 1 \quad \text{Use the rule for } f(x); f(x) = 4x - 1.$$
$$= 4(x^2 + 5) - 1 \quad g(x) = x^2 + 5$$
$$= 4x^2 + 20 - 1 \quad \text{Distributive property}$$
$$= 4x^2 + 19 \quad \text{Combine terms.}$$

(b) $(g \circ f)(2)$

(c) Find $(f \circ g)(2)$ again, this time using the rule obtained in part (b).

$$(f \circ g)(x) = 4x^2 + 19 \quad \text{From part (b)}$$
$$(f \circ g)(2) = 4(2)^2 + 19 \quad \text{Let } x = 2.$$
$$= 4(4) + 19$$
$$= 16 + 19$$
$$= 35$$

The result, 35, is the same as the result in part (a).

(c) $(f \circ g)(x)$

(d) $(g \circ f)(x)$

Here, use $f(x)$ as the input for the function g.

$$(g \circ f)(x) = g[f(x)]$$
$$= [f(x)]^2 + 5 \quad \text{Use the rule for } g(x); g(x) = x^2 + 5.$$
$$= (4x - 1)^2 + 5 \quad f(x) = 4x - 1$$
$$= 16x^2 - 8x + 1 + 5 \quad (x - y)^2 = x^2 - 2xy + y^2$$
$$= 16x^2 - 8x + 6 \quad \text{Combine terms.}$$

Compare this result to that in part (b). Again, $(f \circ g)(x) \neq (g \circ f)(x)$.

◄◄◄ **Work Problem 7 at the Side.**

(d) $(g \circ f)(x)$

ANSWERS
7. **(a)** 30 **(b)** 1728 **(c)** $3x^3 + 6$
   (d) $(3x + 6)^3$

12.1 Exercises

FOR EXTRA HELP

Addison-Wesley Math Tutor Center | MathXL | Digital Video Tutor CD 8 Videotape 8 | Student's Solutions Manual | MyMathLab | Interactmath.com

Fill in each blank with the correct response.

1. For the reciprocal function defined by $f(x) = \frac{1}{x}$, _____ is the only real number not in the domain.

2. The range of the square root function, given by $f(x) = \sqrt{x}$, is _____.

3. The lowest point on the graph of $f(x) = |x|$ has coordinates (_____, _____).

4. The range of $f(x) = [\![x]\!]$, the greatest integer function, is _____.

Without actually plotting points, match each function defined by the absolute value expression with its graph. See Example 1.

5. $f(x) = |x - 2| + 2$ **6.** $f(x) = |x + 2| + 2$ **7.** $f(x) = |x - 2| - 2$ **8.** $f(x) = |x + 2| - 2$

A.
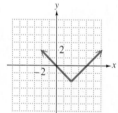

B.
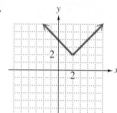

C.
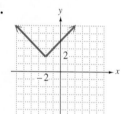

D.
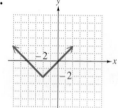

Graph each function. Give the domain and range. See Examples 1–3.

9. $f(x) = |x + 1|$

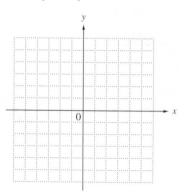

10. $f(x) = |x - 1|$

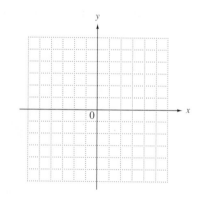

11. $f(x) = \frac{1}{x} + 1$

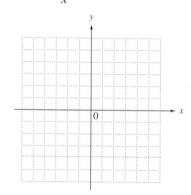

12. $f(x) = \dfrac{1}{x} - 1$

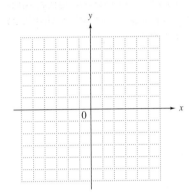

13. $f(x) = \sqrt{x - 2}$

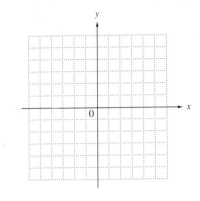

14. $f(x) = \sqrt{x + 5}$

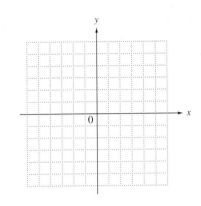

15. $f(x) = \dfrac{1}{x - 2}$

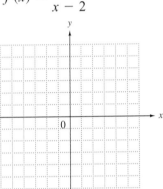

16. $f(x) = \dfrac{1}{x + 2}$

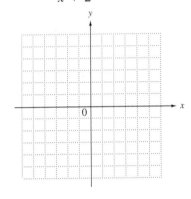

17. $f(x) = \sqrt{x + 3} - 3$

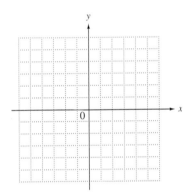

18. Explain how the graph of $f(x) = \dfrac{1}{x - 3} + 2$ is obtained from the graph of $g(x) = \dfrac{1}{x}$.

Graph each step function. See Examples 4 and 5.

19. $f(x) = [\![x - 3]\!]$

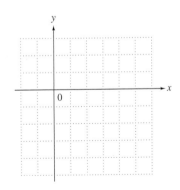

20. $g(x) = [\![x + 2]\!]$

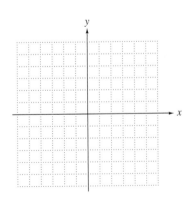

21. Assume that postage rates are 37¢ for the first ounce, plus 23¢ for each additional ounce, and that each letter carries one 37¢ stamp and as many 23¢ stamps as necessary. Graph the function defined by $p(x)$ = the number of stamps on a letter weighing x ounces. Use the interval $(0, 5]$.

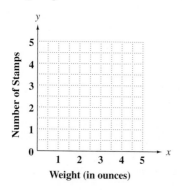

22. The cost of parking a car at an airport hourly parking lot is $3 for the first half-hour and $2 for each additional half-hour or fraction thereof. Graph the function defined by $f(x)$ = the cost of parking a car for x hours. Use the interval $(0, 2]$.

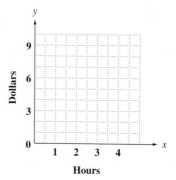

Let $f(x) = x^2 + 4,$ $g(x) = 2x + 3,$ *and* $h(x) = x + 5.$ *Find each value or expression. See Examples 6 and 7.*

23. $(h \circ g)(4)$ **24.** $(f \circ g)(4)$ **25.** $(g \circ f)(6)$ **26.** $(h \circ f)(6)$

27. $(f \circ h)(-2)$ **28.** $(h \circ g)(-2)$ **29.** $(f \circ g)(x)$ **30.** $(g \circ h)(x)$

31. $(f \circ h)(x)$ **32.** $(g \circ f)(x)$ **33.** $(h \circ g)(x)$ **34.** $(h \circ f)(x)$

Solve each problem.

35. The function defined by $f(x) = 12x$ computes the number of inches in x feet and the function defined by $g(x) = 5280x$ computes the number of feet in x miles. What is $(f \circ g)(x)$ and what does it compute?

36. The perimeter x of a square with sides of length s is given by the formula $x = 4s$.

(a) Solve for s in terms of x.

(b) If y represents the area of this square, write y as a function of the perimeter x.

(c) Use the composite function of part (b) to find the area of a square with perimeter 6.

37. When a thermal inversion layer is over a city (as happens often in Los Angeles), pollutants cannot rise vertically but are trapped below the layer and must disperse horizontally. Assume that a factory smokestack begins emitting a pollutant at 8 A.M. Assume that the pollutant disperses horizontally over a circular area. Suppose that t represents the time, in hours, since the factory began emitting pollutants ($t = 0$ represents 8 A.M.), and assume that the radius of the circle of pollution is $r(t) = 2t$ miles. Let $A(r) = \pi r^2$ represent the area of a circle of radius r. Find and interpret $(A \circ r)(t)$.

38. An oil well off the Gulf Coast is leaking, with the leak spreading oil over the surface as a circle. At any time t, in minutes, after the beginning of the leak, the radius of the circular oil slick on the surface is $r(t) = 4t$ feet. Let $A(r) = \pi r^2$ represent the area of a circle of radius r. Find and interpret $(A \circ r)(t)$.

12.2 The Circle and the Ellipse

When an infinite cone is intersected by a plane, the resulting figure is called a **conic section.** The parabola is one example of a conic section; circles, ellipses, and hyperbolas may also result. See Figure 11.

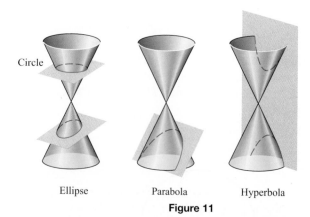

Ellipse   Parabola   Hyperbola

Figure 11

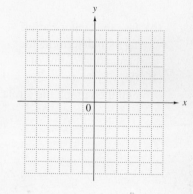

OBJECTIVES

1. Find the equation of a circle given the center and radius.

2. Determine the center and radius of a circle given its equation.

3. Recognize the equation of an ellipse.

4. Graph ellipses.

OBJECTIVE 1 Find an equation of a circle given the center and radius.
A **circle** is the set of all points in a plane that lie a fixed distance from a fixed point. The fixed point is called the **center,** and the fixed distance is called the **radius.** We use the distance formula from **Section 9.3** to find an equation of a circle.

EXAMPLE 1 Finding an Equation of a Circle and Graphing It

Find an equation of the circle with radius 3 and center at $(0, 0)$, and graph it.

If the point (x, y) is on the circle, the distance from (x, y) to the center $(0, 0)$ is 3. By the distance formula,

$$\sqrt{(x_2 - x_1)^2 + (y_2 - y_1)^2} = d \qquad \text{Distance formula}$$

$$\sqrt{(x - 0)^2 + (y - 0)^2} = 3$$

$$x^2 + y^2 = 9. \qquad \text{Square both sides.}$$

An equation of this circle is $x^2 + y^2 = 9$. The graph is shown in Figure 12.

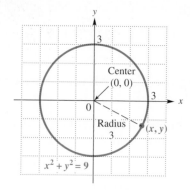

Figure 12

Work Problem 1 at the Side.

1. Find an equation of the circle with radius 4 and center $(0, 0)$. Sketch its graph.

ANSWERS
1. $x^2 + y^2 = 16$

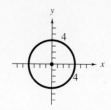

2 **(a)** Find an equation of the circle with center at $(3, -2)$ and radius 4. Graph the circle.

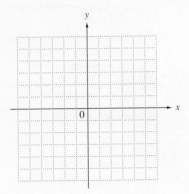

(b) Use the center-radius form to determine the center and radius of $(x - 5)^2 + (y + 2)^2 = 9$, and then graph the circle.

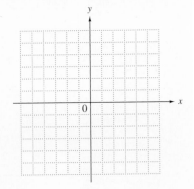

ANSWERS
2. (a) $(x - 3)^2 + (y + 2)^2 = 16$

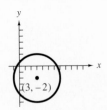

(b) center at $(5, -2)$; radius 3

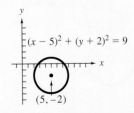
$(x - 5)^2 + (y + 2)^2 = 9$
$(5, -2)$

A circle may not be centered at the origin, as seen in the next example.

EXAMPLE 2 Finding an Equation of a Circle and Graphing It

Find an equation of the circle with center at $(4, -3)$ and radius 5, and graph it. Use the distance formula again.

$$\sqrt{(x - 4)^2 + [y - (-3)]^2} = 5$$
$$(x - 4)^2 + (y + 3)^2 = 25 \quad \text{Square both sides.}$$

To graph the circle, plot the center $(4, -3)$, then move 5 units right, left, up, and down from the center. Draw a smooth curve through these four points, sketching one quarter of the circle at a time. The graph of this circle is shown in Figure 13.

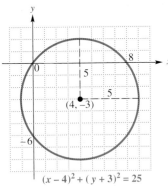

$(x - 4)^2 + (y + 3)^2 = 25$

Figure 13

Examples 1 and 2 suggest the form of an equation of a circle with radius r and center at (h, k). If (x, y) is a point on the circle, then the distance from the center (h, k) to the point (x, y) is r. By the distance formula,

$$\sqrt{(x - h)^2 + (y - k)^2} = r.$$

Squaring both sides gives us the following **center-radius form** of the equation of a circle.

Equation of a Circle (Center-Radius Form)

An equation of a circle of radius r with center at (h, k) is

$$(x - h)^2 + (y - k)^2 = r^2.$$

EXAMPLE 3 Using the Center-Radius Form of the Equation of a Circle

Find an equation of the circle with center at $(-1, 2)$ and radius 4. Use the center-radius form, with $h = -1, k = 2,$ and $r = 4$.

$$(x - h)^2 + (y - k)^2 = r^2$$
$$[x - (-1)]^2 + (y - 2)^2 = 4^2$$
$$(x + 1)^2 + (y - 2)^2 = 16$$

◀◀◀ Work Problem 2 at the Side.

OBJECTIVE 2 Determine the center and radius of a circle given its equation. In the equation found in Example 2, multiplying out $(x - 4)^2$ and $(y + 3)^2$ and then combining like terms gives

$$(x - 4)^2 + (y + 3)^2 = 25$$
$$x^2 - 8x + 16 + y^2 + 6y + 9 = 25$$
$$x^2 + y^2 - 8x + 6y = 0.$$

This general form suggests that an equation with both x^2- and y^2-terms with equal coefficients may represent a circle. The next example shows how to tell, by completing the square. This procedure was introduced in **Section 10.1.**

EXAMPLE 4 Completing the Square to Find the Center and Radius

Graph $x^2 + y^2 + 2x + 6y - 15 = 0$.

Since the equation has x^2- and y^2-terms with equal coefficients, its graph might be that of a circle. To find the center and radius, complete the squares on x and y.

$$x^2 + y^2 + 2x + 6y = 15$$ Get the constant on the right.

$$(x^2 + 2x \quad) + (y^2 + 6y \quad) = 15$$ Rewrite in anticipation of completing the square.

$$\left[\frac{1}{2}(2)\right]^2 = 1 \qquad \left[\frac{1}{2}(6)\right]^2 = 9$$ Square half the coefficient of each middle term.

$$(x^2 + 2x + 1) + (y^2 + 6y + 9) = 15 + 1 + 9$$ Complete the squares on both x and y.

$$(x + 1)^2 + (y + 3)^2 = 25$$ Factor on the left; add on the right.

$$[x - (-1)]^2 + [y - (-3)]^2 = 5^2$$ Center-radius form

The last equation shows that the graph is a circle with center at $(-1, -3)$ and radius 5. The graph is shown in Figure 14.

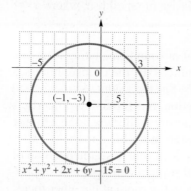

Figure 14

3 Find the center and radius of the circle with equation

$$x^2 + y^2 - 6x + 8y - 11 = 0.$$

NOTE
If the procedure of Example 4 leads to an equation of the form $(x - h)^2 + (y - k)^2 = 0$, then the graph is the single point (h, k). If the constant on the right side is negative, then the equation has no graph.

Work Problem 3 at the Side. ▶▶▶

ANSWERS

3. center at $(3, -4)$; radius 6

OBJECTIVE 3 Recognize the equation of an ellipse. An **ellipse** is the set of all points in a plane the *sum* of whose distances from two fixed points is constant. These fixed points are called **foci** (singular: *focus*). Figure 15 shows an ellipse whose foci are $(c, 0)$ and $(-c, 0)$, with x-intercepts $(a, 0)$ and $(-a, 0)$ and y-intercepts $(0, b)$ and $(0, -b)$. It can be shown that $c^2 = a^2 - b^2$ for an ellipse of this type. The origin is the **center** of the ellipse.

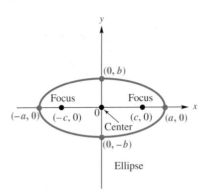

Figure 15

From the preceding definition, it can be shown by the distance formula that an ellipse has the following equation.

Equation of an Ellipse

The ellipse whose x-intercepts are $(a, 0)$ and $(-a, 0)$ and whose y-intercepts are $(0, b)$ and $(0, -b)$ has an equation of the form

$$\frac{x^2}{a^2} + \frac{y^2}{b^2} = 1.$$

NOTE
A circle is a special case of an ellipse, where $a^2 = b^2$.

When a ray of light or sound emanating from one focus of an ellipse bounces off the ellipse, it passes through the other focus. See the figure. As mentioned in the chapter introduction, this reflecting property is responsible for whispering galleries. John Quincy Adams was able to listen in on his opponents' conversations because his desk was positioned at one of the foci beneath the ellipsoidal ceiling and his opponents were located across the room at the other focus.

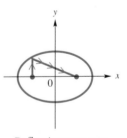

Reflecting property of an ellipse

The paths of Earth and other planets around the sun are approximately ellipses; the sun is at one focus and a point in space is at the other. The orbits of communication satellites and other space vehicles are also elliptical.

Elliptical bicycle gears are designed to respond to the legs' natural strengths and weaknesses. At the top and bottom of the powerstroke, where the legs have the least leverage, the gear offers little resistance, but as the gear rotates, the resistance increases. This allows the legs to apply more power where it is most naturally available. See Figure 16.

Figure 16

OBJECTIVE 4 Graph ellipses.   To graph an ellipse centered at the origin, we plot the four intercepts and then sketch the ellipse through those points.

EXAMPLE 5 Graphing Ellipses

Graph each ellipse.

(a) $\dfrac{x^2}{49} + \dfrac{y^2}{36} = 1$

Here, $a^2 = 49$, so $a = 7$, and the x-intercepts for this ellipse are $(7, 0)$ and $(-7, 0)$. Similarly, $b^2 = 36$, so $b = 6$, and the y-intercepts are $(0, 6)$ and $(0, -6)$. Plotting the intercepts and sketching the ellipse through them gives the graph in Figure 17.

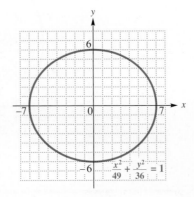

Figure 17

(b) $\dfrac{x^2}{36} + \dfrac{y^2}{121} = 1$

The x-intercepts for this ellipse are $(6, 0)$ and $(-6, 0)$, and the y-intercepts are $(0, 11)$ and $(0, -11)$. Join these intercepts with the smooth curve of an ellipse. The graph has been sketched in Figure 18.

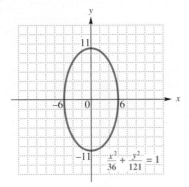

Figure 18

Work Problem 4 at the Side.

4 Graph each ellipse.

(a) $\dfrac{x^2}{4} + \dfrac{y^2}{25} = 1$

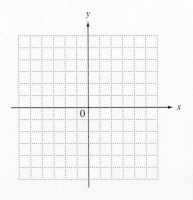

(b) $\dfrac{x^2}{64} + \dfrac{y^2}{49} = 1$

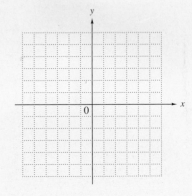

ANSWERS
4. (a)   (b)

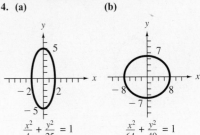

5 Graph

$$\frac{(x + 4)^2}{16} + \frac{(y - 1)^2}{36} = 1.$$

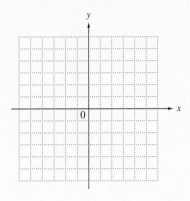

As with the graphs of parabolas and circles, the graph of an ellipse may be shifted horizontally and vertically, as in the next example.

EXAMPLE 6 Graphing an Ellipse Shifted Horizontally and Vertically

Graph $\dfrac{(x - 2)^2}{25} + \dfrac{(y + 3)^2}{49} = 1$.

Just as $(x - 2)^2$ and $(y + 3)^2$ would indicate that the center of a circle would be $(2, -3)$, so it is with this ellipse. Figure 19 shows that the graph goes through the four points $(2, 4)$, $(7, -3)$, $(2, -10)$, and $(-3, -3)$. The x-values of these points are found by adding $\pm a = \pm 5$ to 2, and the y-values come from adding $\pm b = \pm 7$ to -3.

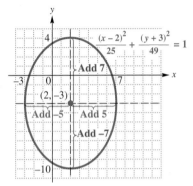

Figure 19

Work Problem 5 at the Side.

NOTE
The graphs in this section are not graphs of functions. The only conic section whose graph is a function is the vertical parabola with equation $f(x) = ax^2 + bx + c$.

ANSWERS

5.

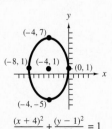

$$\frac{(x + 4)^2}{16} + \frac{(y - 1)^2}{36} = 1$$

12.2 Exercises

FOR EXTRA HELP

Tutor Center — Addison-Wesley Math Tutor Center

MathXL — MathXL

Digital Video Tutor CD 8 — Videotape 8

Student's Solutions Manual

MyMathLab — MyMathLab

Interactmath.com

1. See Example 1. Consider the circle whose equation is $x^2 + y^2 = 25$.

   (a) What are the coordinates of its center?

   (b) What is its radius?

   (c) Sketch its graph.

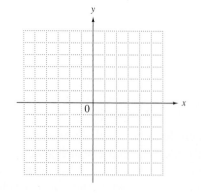

2. Explain why a set of points defined by a circle does not satisfy the definition of a function.

Match each equation with the correct graph. See Examples 1–3.

3. $(x - 3)^2 + (y - 2)^2 = 25$

   A.
   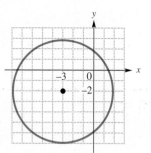

   B.
   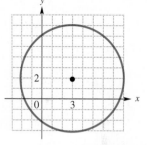

4. $(x - 3)^2 + (y + 2)^2 = 25$

5. $(x + 3)^2 + (y - 2)^2 = 25$

   C.
   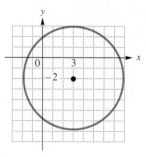

   D.
   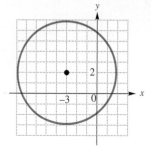

6. $(x + 3)^2 + (y + 2)^2 = 25$

Find the equation of a circle satisfying the given conditions. See Examples 2 and 3.

7. Center: $(-4, 3)$; radius: 2

8. Center: $(5, -2)$; radius: 4

9. Center: $(-8, -5)$; radius: $\sqrt{5}$

10. Center: $(-12, 13)$; radius: $\sqrt{7}$

Find the center and radius of each circle. (Hint: In Exercises 15 and 16, divide each side by a common factor.) See Example 4.

11. $x^2 + y^2 + 4x + 6y + 9 = 0$

12. $x^2 + y^2 - 8x - 12y + 3 = 0$

13. $x^2 + y^2 + 10x - 14y - 7 = 0$

14. $x^2 + y^2 - 2x + 4y - 4 = 0$

15. $3x^2 + 3y^2 - 12x - 24y + 12 = 0$

16. $2x^2 + 2y^2 + 20x + 16y + 10 = 0$

17. A circle can be drawn on a piece of posterboard by fastening one end of a string with a thumbtack, pulling the string taut with a pencil, and tracing a curve, as shown in the figure. Explain why this method works.

Graph each circle. See Examples 1–4.

18. $x^2 + y^2 = 9$

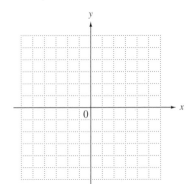

19. $x^2 + y^2 = 4$

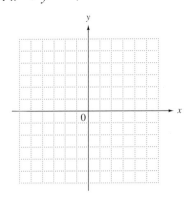

20. $2y^2 = 10 - 2x^2$

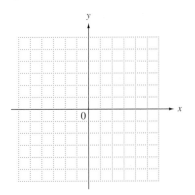

21. $3x^2 = 48 - 3y^2$

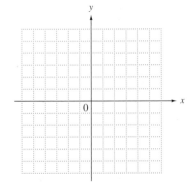

22. $(x + 3)^2 + (y - 2)^2 = 9$

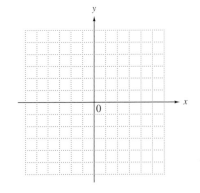

23. $(x - 1)^2 + (y + 3)^2 = 16$

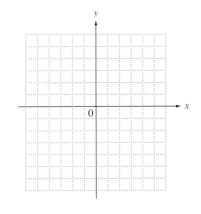

24. $x^2 + y^2 - 4x - 6y + 9 = 0$

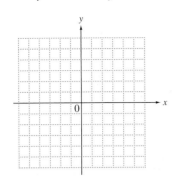

25. $x^2 + y^2 + 8x + 2y - 8 = 0$

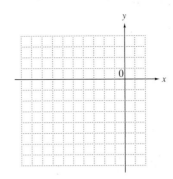

26. An ellipse can be drawn on a piece of posterboard by fastening two ends of a length of string with thumbtacks, pulling the string taut with a pencil, and tracing a curve, as shown in the figure. Explain why this method works.

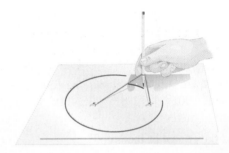

Graph each ellipse. See Examples 5 and 6.

27. $\dfrac{x^2}{9} + \dfrac{y^2}{25} = 1$

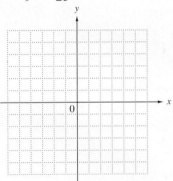

28. $\dfrac{x^2}{9} + \dfrac{y^2}{16} = 1$

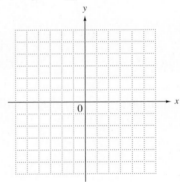

29. $\dfrac{x^2}{36} + \dfrac{y^2}{16} = 1$

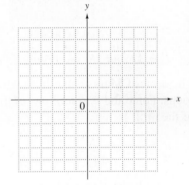

30. $\dfrac{x^2}{9} + \dfrac{y^2}{4} = 1$

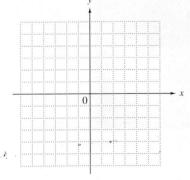

31. $\dfrac{x^2}{49} + \dfrac{y^2}{25} = 1$

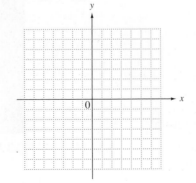

32. $\dfrac{x^2}{16} + \dfrac{y^2}{9} = 1$

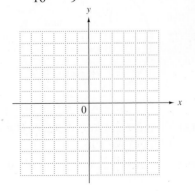

33. $\dfrac{(x-2)^2}{16} + \dfrac{(y-1)^2}{9} = 1$

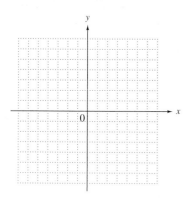

34. $\dfrac{(x-4)^2}{9} + \dfrac{(y+2)^2}{4} = 1$

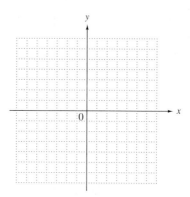

Solve each problem.

35. An arch has the shape of half an ellipse. The equation of the ellipse is $100x^2 + 324y^2 = 32{,}400$, where x and y are in meters.

  (a) How high is the center of the arch?

  (b) How wide is the arch across the bottom?

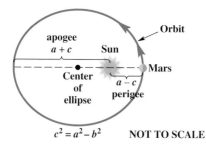

NOT TO SCALE

36. A one-way street passes under an overpass, which is in the form of the top half of an ellipse, as shown in the figure. Suppose that a truck 12 ft wide passes directly under the overpass. What is the maximum possible height of this truck?

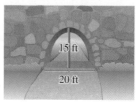

NOT TO SCALE

In Exercises 37 and 38, see Figure 15 and use the fact that $c^2 = a^2 - b^2$, where $a^2 > b^2$.

37. The orbit of Mars is an ellipse with the sun at one focus. For x and y in millions of miles, the equation of the orbit is

$$\frac{x^2}{141.7^2} + \frac{y^2}{141.1^2} = 1.$$

(*Source:* Kaler, James B., *Astronomy!*, Addison-Wesley, 1997.)

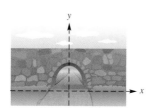

$c^2 = a^2 - b^2$ NOT TO SCALE

  (a) Find the greatest distance (the **apogee**) from Mars to the sun.

  (b) Find the least distance (the **perigee**) from Mars to the sun.

38. The orbit of Venus around the sun (one of the foci) is an ellipse with equation

$$\frac{x^2}{5013} + \frac{y^2}{4970} = 1,$$

where x and y are measured in millions of miles.

(*Source:* Kaler, James B., *Astronomy!*, Addison-Wesley, 1997.)

  (a) Find the greatest distance between Venus and the sun.

  (b) Find the least distance between Venus and the sun.

12.3 The Hyperbola and Other Functions Defined by Radicals

OBJECTIVE 1 **Recognize the equation of a hyperbola.** A hyperbola is the set of all points in a plane such that the absolute value of the *difference* of the distances from two fixed points (called *foci*) is constant. Figure 20 shows a hyperbola. Using the distance formula and the definition above, we can show that this hyperbola has equation $\dfrac{x^2}{16} - \dfrac{y^2}{12} = 1$.

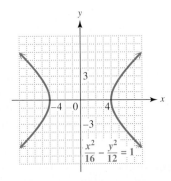

Figure 20

To graph hyperbolas centered at the origin, we need to find their intercepts. For the hyperbola in Figure 20, we proceed as follows.

x-Intercepts	y-Intercepts

x-Intercepts

Let $y = 0$.

$\dfrac{x^2}{16} - \dfrac{0^2}{12} = 1$ Let $y = 0$.

$\dfrac{x^2}{16} = 1$

$x^2 = 16$ Multiply by 16.

$x = \pm 4$

The x-intercepts are $(4, 0)$ and $(-4, 0)$.

y-Intercepts

Let $x = 0$.

$\dfrac{0^2}{16} - \dfrac{y^2}{12} = 1$ Let $x = 0$.

$-\dfrac{y^2}{12} = 1$

$y^2 = -12$ Multiply by -12.

Because there are no *real* solutions to $y^2 = -12$, the graph has no y-intercepts.

The graph of $\dfrac{x^2}{16} - \dfrac{y^2}{12} = 1$ in Figure 20 has no y-intercepts. On the other hand, the hyperbola in Figure 21 has no x-intercepts. Its equation is $\dfrac{y^2}{25} - \dfrac{x^2}{9} = 1$, with y-intercepts $(0, 5)$ and $(0, -5)$.

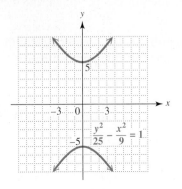

Figure 21

Equations of Hyperbolas

A hyperbola with x-intercepts $(a, 0)$ and $(-a, 0)$ has an equation of the form

$$\frac{x^2}{a^2} - \frac{y^2}{b^2} = 1,$$

and a hyperbola with y-intercepts $(0, b)$ and $(0, -b)$ has an equation of the form

$$\frac{y^2}{b^2} - \frac{x^2}{a^2} = 1.$$

OBJECTIVE 2 Graph hyperbolas by using asymptotes. The two branches of the graph of a hyperbola approach a pair of intersecting straight lines, which are its *asymptotes*. (See Figure 22 on the next page.) The asymptotes are useful for sketching the graph of the hyperbola.

Asymptotes of Hyperbolas

The extended diagonals of the rectangle with vertices (corners) at the points (a, b), $(-a, b)$, $(-a, -b)$, and $(a, -b)$ are the **asymptotes** of the hyperbolas

$$\frac{x^2}{a^2} - \frac{y^2}{b^2} = 1 \quad \text{and} \quad \frac{y^2}{b^2} - \frac{x^2}{a^2} = 1.$$

This rectangle is called the **fundamental rectangle.** Using the methods of **Chapter 4,** we could show that the equations of these asymptotes are

$$y = \frac{b}{a}x \quad \text{and} \quad y = -\frac{b}{a}x.$$

To graph hyperbolas, follow these steps.

Graphing a Hyperbola

Step 1 **Find the intercepts.** Locate the intercepts at $(a, 0)$ and $(-a, 0)$ if the x^2-term has a positive coefficient, or at $(0, b)$ and $(0, -b)$ if the y^2-term has a positive coefficient.

Step 2 **Find the fundamental rectangle.** Locate the vertices of the fundamental rectangle at (a, b), $(-a, b)$, $(-a, -b)$, and $(a, -b)$.

Step 3 **Sketch the asymptotes.** The extended diagonals of the rectangle are the asymptotes of the hyperbola, and they have equations $y = \pm \frac{b}{a}x$.

Step 4 **Draw the graph.** Sketch each branch of the hyperbola through an intercept and approaching (but not touching) the asymptotes.

EXAMPLE 1 **Graphing a Horizontal Hyperbola**

Graph $\dfrac{x^2}{16} - \dfrac{y^2}{25} = 1$.

Step 1 Here $a = 4$ and $b = 5$. The x-intercepts are $(4, 0)$ and $(-4, 0)$.

Step 2 The four points $(4, 5)$, $(-4, 5)$, $(-4, -5)$, and $(4, -5)$ are the vertices of the fundamental rectangle, as shown in Figure 22.

Steps 3 and 4 The equations of the asymptotes are $y = \pm\frac{5}{4}x$. The hyperbola approaches these lines as x and y get larger in absolute value.

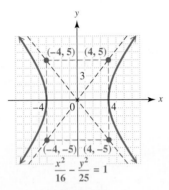

Figure 22

1 Graph $\dfrac{x^2}{4} - \dfrac{y^2}{25} = 1$.

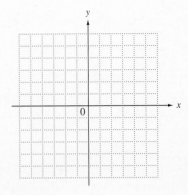

CAUTION
When sketching the graph of a hyperbola, be sure that the branches do not touch the asymptotes.

Work Problem 1 at the Side. ▶▶▶

2 Graph $\dfrac{y^2}{81} - \dfrac{x^2}{64} = 1$.

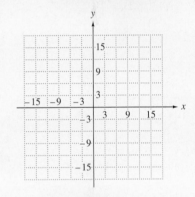

EXAMPLE 2 **Graphing a Vertical Hyperbola**

Graph $\dfrac{y^2}{49} - \dfrac{x^2}{16} = 1$.

This hyperbola has y-intercepts $(0, 7)$ and $(0, -7)$. The asymptotes are the extended diagonals of the rectangle with vertices at $(4, 7)$, $(-4, 7)$, $(-4, -7)$, and $(4, -7)$. Their equations are $y = \pm\frac{7}{4}x$. See Figure 23.

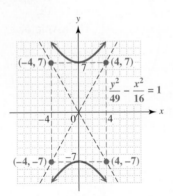

Figure 23

Work Problem 2 at the Side. ▶▶▶

ANSWERS

1.

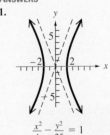

$$\frac{x^2}{4} - \frac{y^2}{25} = 1$$

2.

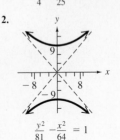

$$\frac{y^2}{81} - \frac{x^2}{64} = 1$$

OBJECTIVE 3 Identify conic sections by their equations. Rewriting a second-degree equation in one of the forms given for ellipses, hyperbolas, circles, or parabolas makes it possible to identify the graph of the equation.

SUMMARY OF CONIC SECTIONS

Equation	Graph	Description	Identification
$y = ax^2 + bx + c$ or $y = a(x - h)^2 + k$	Parabola	It opens up if $a > 0$, down if $a < 0$. The vertex is (h, k).	It has an x^2-term. y is not squared.
$x = ay^2 + by + c$ or $x = a(y - k)^2 + h$	Parabola	It opens to the right if $a > 0$, to the left if $a < 0$. The vertex is (h, k).	It has a y^2-term. x is not squared.
$(x - h)^2 + (y - k)^2 = r^2$	Circle	The center is (h, k), and the radius is r.	x^2- and y^2-terms have the same positive coefficient.
$\dfrac{x^2}{a^2} + \dfrac{y^2}{b^2} = 1$	Ellipse	The x-intercepts are $(a, 0)$ and $(-a, 0)$. The y-intercepts are $(0, b)$ and $(0, -b)$.	x^2- and y^2-terms have different positive coefficients.
$\dfrac{x^2}{a^2} - \dfrac{y^2}{b^2} = 1$	Hyperbola	The x-intercepts are $(a, 0)$ and $(-a, 0)$. The asymptotes are found from $(a, b), (a, -b), (-a, -b),$ and $(-a, b)$.	x^2 has a positive coefficient. y^2 has a negative coefficient.
$\dfrac{y^2}{b^2} - \dfrac{x^2}{a^2} = 1$	Hyperbola	The y-intercepts are $(0, b)$ and $(0, -b)$. The asymptotes are found from $(a, b), (a, -b), (-a, -b),$ and $(-a, b)$.	y^2 has a positive coefficient. x^2 has a negative coefficient.

EXAMPLE 3 **Identifying the Graphs of Equations**

Identify the graph of each equation.

(a) $9x^2 = 108 + 12y^2$

Both variables are squared, so the graph is either an ellipse or a hyperbola. (This situation also occurs for a circle, which is a special case of the ellipse.) To see which one it is, rewrite the equation so that the x^2- and y^2-terms are on one side of the equation and 1 is on the other.

$$9x^2 - 12y^2 = 108 \quad \text{Subtract } 12y^2.$$

$$\frac{x^2}{12} - \frac{y^2}{9} = 1 \quad \text{Divide by 108.}$$

Because of the minus sign, the graph of this equation is a hyperbola.

(b) $x^2 = y - 3$

Only one of the two variables, x, is squared, so this is the vertical parabola $y = x^2 + 3$.

(c) $x^2 = 9 - y^2$

Write the variable terms on the same side of the equation.

$$x^2 + y^2 = 9 \quad \text{Add } y^2.$$

The graph of this equation is a circle with center at the origin and radius 3.

Work Problem 3 at the Side. ▶▶▶

OBJECTIVE 4 Graph certain square root functions. Recall from **Section 4.5** that no vertical line will intersect the graph of a function in more than one point. Thus, horizontal parabolas and all circles, ellipses, and hyperbolas are examples of graphs that do not satisfy the conditions of a function. However, by considering only a part of the graph of each of these we have the graph of a function, as seen in Figure 24.

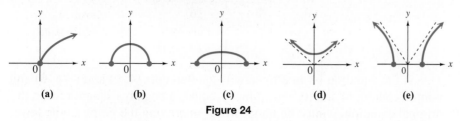

(a)   (b)   (c)   (d)   (e)

Figure 24

In parts (a)–(d) of Figure 24, the top portion of a conic section is shown (parabola, circle, ellipse, and hyperbola, respectively). In part (e), the top two portions of a hyperbola are shown. In each case, the graph is that of a function since the graph satisfies the conditions of the vertical line test.

In **Sections 9.1** and **12.1** we observed the square root function defined by $f(x) = \sqrt{x}$. To find equations for the types of graphs shown in Figure 24, we extend its definition.

Square Root Function

For an algebraic expression u, with $u \geq 0$, a function of the form

$$f(x) = \sqrt{u}$$

is called a **square root function.**

3 Identify the graph of each equation.

(a) $3x^2 = 27 - 4y^2$

(b) $6x^2 = 100 + 2y^2$

(c) $3x^2 = 27 - 4y$

(d) $3x^2 = 27 - 3y^2$

ANSWERS
3. (a) ellipse **(b)** hyperbola **(c)** parabola
(d) circle

④ Graph $f(x) = \sqrt{36 - x^2}$.
Give the domain and range.

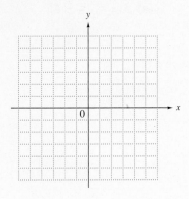

⑤ Graph

$$\frac{y}{3} = -\sqrt{1 - \frac{x^2}{4}}.$$

Give the domain and range.

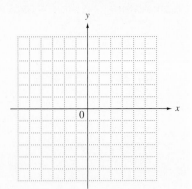

ANSWERS

4.

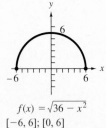

$f(x) = \sqrt{36 - x^2}$
$[-6, 6]; [0, 6]$

5.

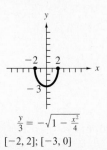

$\frac{y}{3} = -\sqrt{1 - \frac{x^2}{4}}$

$[-2, 2]; [-3, 0]$

EXAMPLE 4 Graphing a Semicircle

Graph $f(x) = \sqrt{25 - x^2}$. Give the domain and range.

Replace $f(x)$ with y and square both sides to get the equation

$$y^2 = 25 - x^2, \quad \text{or} \quad x^2 + y^2 = 25.$$

This is the graph of a circle with center at $(0, 0)$ and radius 5. Since $f(x)$, or y, represents a principal square root in the original equation, $f(x)$ must be nonnegative. This restricts the graph to the upper half of the circle, as shown in Figure 25. Use the graph and the vertical line test to verify that it is indeed a function. The domain is $[-5, 5]$, and the range is $[0, 5]$.

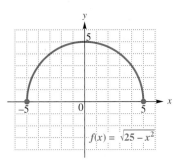

$f(x) = \sqrt{25 - x^2}$

Figure 25

◄◄◄ **Work Problem 4 at the Side.**

EXAMPLE 5 Graphing a Portion of an Ellipse

Graph $\frac{y}{6} = -\sqrt{1 - \frac{x^2}{16}}$. Give the domain and range.

Square both sides to get an equation whose form is known.

$$\frac{y^2}{36} = 1 - \frac{x^2}{16}$$

$$\frac{x^2}{16} + \frac{y^2}{36} = 1 \qquad \text{Add } \frac{x^2}{16}.$$

This is the equation of an ellipse with x-intercepts $(4, 0)$ and $(-4, 0)$ and y-intercepts $(0, 6)$ and $(0, -6)$. Since $\frac{y}{6}$ equals a negative square root in the original equation, y must be nonpositive, restricting the graph to the lower half of the ellipse, as shown in Figure 26. Verify that this is the graph of a function, using the vertical line test. The domain is $[-4, 4]$, and the range is $[-6, 0]$.

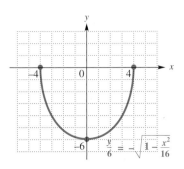

$\frac{y}{6} = -\sqrt{1 - \frac{x^2}{16}}$

Figure 26

◄◄◄ **Work Problem 5 at the Side.**

12.3 Exercises

FOR EXTRA HELP

Tutor Center Addison-Wesley Math Tutor Center

Math XL MathXL

Digital Video Tutor CD 8 Videotape 8

Student's Solutions Manual

MyMathLab MyMathLab

Interactmath.com

*Based on the discussions of ellipses in **Section 12.2** and of hyperbolas in this section, match each equation with its graph.*

1. $\dfrac{x^2}{25} + \dfrac{y^2}{9} = 1$

2. $\dfrac{x^2}{9} + \dfrac{y^2}{25} = 1$

3. $\dfrac{x^2}{9} - \dfrac{y^2}{25} = 1$

4. $\dfrac{x^2}{25} - \dfrac{y^2}{9} = 1$

A.
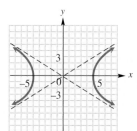

B.
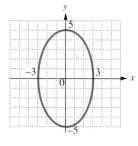

C.
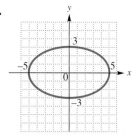

D.
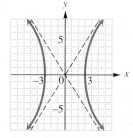

5. Write an explanation of how you can tell from the equation whether the branches of a hyperbola open up and down or left and right.

6. Describe how the fundamental rectangle is used to sketch a hyperbola.

Graph each hyperbola. See Examples 1 and 2.

7. $\dfrac{x^2}{16} - \dfrac{y^2}{9} = 1$

8. $\dfrac{y^2}{4} - \dfrac{x^2}{25} = 1$

9. $\dfrac{y^2}{9} - \dfrac{x^2}{9} = 1$

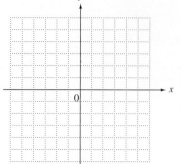

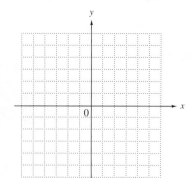

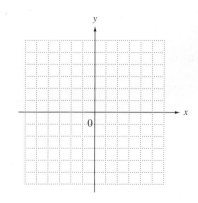

10. $\dfrac{x^2}{49} - \dfrac{y^2}{16} = 1$

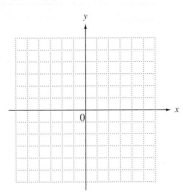

11. $\dfrac{x^2}{25} - \dfrac{y^2}{36} = 1$

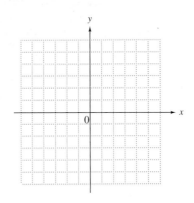

12. $\dfrac{y^2}{9} - \dfrac{x^2}{4} = 1$

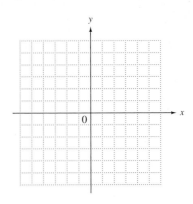

Identify the graph of each equation as a parabola, circle, ellipse, *or* hyperbola, *and sketch it. See Example 3.*

13. $x^2 - y^2 = 16$

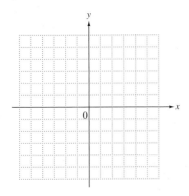

14. $x^2 + y^2 = 16$

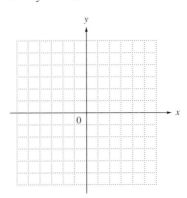

15. $4x^2 + y^2 = 16$

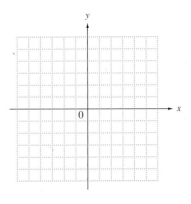

16. $x^2 - 2y = 0$

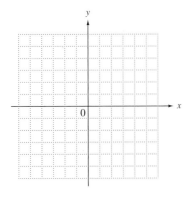

17. $y^2 = 36 - x^2$

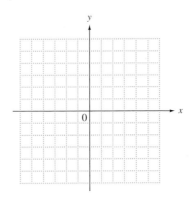

18. $9x^2 + 25y^2 = 225$

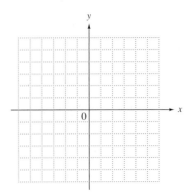

19. $9x^2 = 144 + 16y^2$

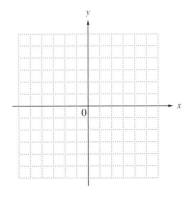

20. $y^2 = 4 + x^2$

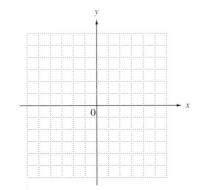

21. $x^2 + 9y^2 = 9$

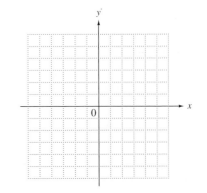

Graph each function defined by a radical expression. Give the domain and range. See Examples 4 and 5.

22. $f(x) = \sqrt{16 - x^2}$

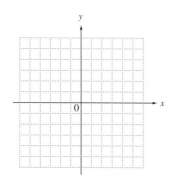

23. $f(x) = \sqrt{9 - x^2}$

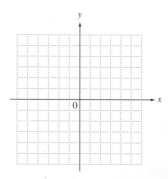

24. $f(x) = -\sqrt{36 - x^2}$

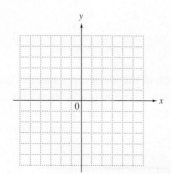

25. $f(x) = -\sqrt{25 - x^2}$

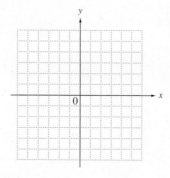

26. $\dfrac{y}{3} = \sqrt{1 + \dfrac{x^2}{9}}$

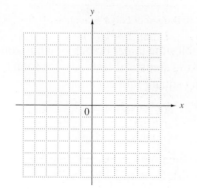

27. $y = \sqrt{\dfrac{x + 4}{2}}$

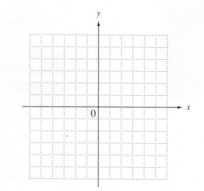

28. $y = -2\sqrt{\dfrac{9 - x^2}{9}}$

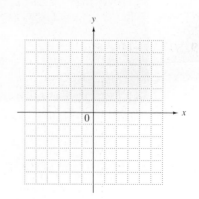

Solve each problem.

29. Two buildings in a sports complex are shaped and positioned like a portion of the branches of the hyperbola with equation

$$400x^2 - 625y^2 = 250,000,$$

where x and y are in meters.

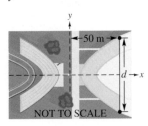

NOT TO SCALE

(a) How far apart are the buildings at their closest point?

(b) Find the distance d in the figure.

30. In rugby, after a *try* (similar to a touchdown in American football) the scoring team attempts a kick for extra points. The ball must be kicked from directly behind the point where the try was scored. The kicker can choose the distance but cannot move the ball sideways. It can be shown that the kicker's best choice is on the hyperbola with equation

$$\frac{x^2}{g^2} - \frac{y^2}{g^2} = 1,$$

where $2g$ is the distance between the goal posts. Since the hyperbola approaches its asymptotes, it is easier for the kicker to estimate points on the asymptotes instead of on the hyperbola. What are the equations of the asymptotes of this hyperbola? Why is it relatively easy to estimate them? (*Source:* Isaksen, Daniel C., "How to Kick a Field Goal," *The College Mathematics Journal,* September 1996.)

31. When a satellite is launched into orbit, the shape of its trajectory is determined by its velocity. The trajectory will be hyperbolic if the velocity V, in meters per second, satisfies the inequality

$$V > \frac{2.82 \times 10^7}{\sqrt{D}},$$

where D is the distance, in meters, from the center of Earth. For what values of V will the trajectory be hyperbolic if $D = 4.25 \times 10^7$ m? (*Source:* Kaler, James B., *Astronomy!,* Addison-Wesley, 1997.)

32. The percent of women in the work force has increased steadily for many years. The line graph shows the change for the period from 1975 to 1999, where $x = 75$ represents 1975, $x = 80$ represents 1980, and so on.

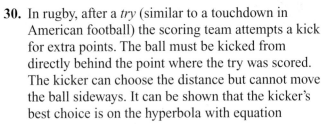

WOMEN IN THE WORK FORCE

Source: U.S. Bureau of Labor Statistics.

The graph resembles the upper branch of a horizontal hyperbola. Using statistical methods, we found the corresponding square root equation

$$y = .607\sqrt{383.9 + x^2},$$

which closely approximates the graph.

(a) According to the graph, what percent of women were in the work force in 1999?

(b) According to the equation, what percent of women worked in 1999? (Round to the nearest percent.)

12.4 Nonlinear Systems of Equations

An equation in which some terms have more than one variable or a variable of degree 2 or greater is called a **nonlinear equation**. A **nonlinear system of equations** includes at least one nonlinear equation.

When solving a nonlinear system, it helps to visualize the types of graphs of the equations of the system to determine the possible number of points of intersection. For example, if a system includes two equations where the graph of one is a parabola and the graph of the other is a line, then there may be 0, 1, or 2 points of intersection, as illustrated in Figure 27.

OBJECTIVES

1 Solve a nonlinear system by substitution.

2 Use the elimination method to solve a system with two second-degree equations.

3 Solve a system that requires a combination of methods.

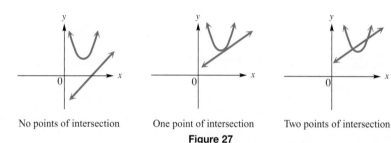

No points of intersection   One point of intersection   Two points of intersection
Figure 27

OBJECTIVE 1 Solve a nonlinear system by substitution. We solve nonlinear systems by the substitution method, the elimination method, or a combination of the two. The substitution method **(Section 5.1)** is usually appropriate when one of the equations is linear.

EXAMPLE 1 Solving a Nonlinear System by Substitution

Solve the system.

$$x^2 + y^2 = 9 \quad (1)$$
$$2x - y = 3 \quad (2)$$

The graph of (1) is a circle and the graph of (2) is a line. Visualizing the possible ways the graphs could intersect indicates that there may be 0, 1, or 2 points of intersection. See Figure 28. First solve the linear equation for one of the two variables, and then substitute the resulting expression into the nonlinear equation to obtain an equation in one variable.

$$2x - y = 3 \quad (2)$$
$$y = 2x - 3 \quad (3)$$

Substitute $2x - 3$ for y in equation (1).

$$x^2 + y^2 = 9 \quad (1)$$
$$x^2 + (2x - 3)^2 = 9$$
$$x^2 + 4x^2 - 12x + 9 = 9$$
$$5x^2 - 12x = 0 \qquad \text{Subtract 9; combine terms.}$$
$$x(5x - 12) = 0 \qquad \text{Factor; GCF is } x.$$
$$x = 0 \quad \text{or} \quad x = \frac{12}{5} \qquad \text{Zero-factor property}$$

Let $x = 0$ in equation (3) to get $y = -3$. If $x = \frac{12}{5}$, then $y = \frac{9}{5}$. The solution set of the system is $\{(0, -3), (\frac{12}{5}, \frac{9}{5})\}$. The graph in Figure 29 on the next page confirms the two points of intersection.

Continued on Next Page

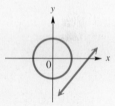

No points of intersection

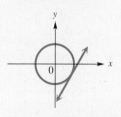

One point of intersection

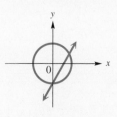

Two points of intersection
Figure 28

1 Solve each system.

(a) $x^2 + y^2 = 10$
$x = y + 2$

(b) $x^2 - 2y^2 = 8$
$y + x = 6$

2 Solve each system.

(a) $xy = 8$
$x + y = 6$

(b) $xy + 10 = 0$
$4x + 9y = -2$

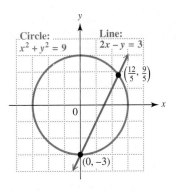

Figure 29

Work Problem 1 at the Side.

EXAMPLE 2 **Solving a Nonlinear System by Substitution**

Solve the system.

$$6x - y = 5 \quad (1)$$
$$xy = 4 \quad (2)$$

The graph of (1) is a line. We have not specifically mentioned equations like (2); however, it can be shown by plotting points that its graph is a hyperbola. Visualizing a line and a hyperbola indicates that there may be 0, 1, or 2 points of intersection. Since neither equation has a squared term, we can solve either equation for one of the variables and then substitute the result into the other equation. Solving $xy = 4$ for x gives $x = \frac{4}{y}$. Substitute $\frac{4}{y}$ for x in equation (1).

$$6\left(\frac{4}{y}\right) - y = 5 \qquad \text{Let } x = \frac{4}{y}.$$

$$\frac{24}{y} - y = 5 \qquad \text{Multiply.}$$

$$24 - y^2 = 5y \qquad \text{Multiply by } y, y \neq 0.$$

$$0 = y^2 + 5y - 24 \qquad \text{Standard form}$$

$$0 = (y - 3)(y + 8) \qquad \text{Factor.}$$

$$y = 3 \quad \text{or} \quad y = -8 \qquad \text{Zero-factor property}$$

We substitute these results into $x = \frac{4}{y}$ to obtain the corresponding values of x.

If $y = 3$, then $x = \dfrac{4}{3}$. \qquad If $y = -8$, then $x = -\dfrac{1}{2}$.

The solution set of the system is $\{(\frac{4}{3}, 3), (-\frac{1}{2}, -8)\}$. See Figure 30.

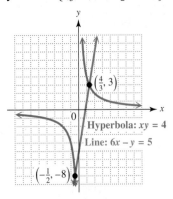

Figure 30

Work Problem 2 at the Side.

ANSWERS
1. (a) $\{(3, 1), (-1, -3)\}$
   (b) $\{(4, 2), (20, -14)\}$
2. (a) $\{(4, 2), (2, 4)\}$
   (b) $\left\{(-5, 2), \left(\frac{9}{2}, -\frac{20}{9}\right)\right\}$

OBJECTIVE 2 Use the elimination method to solve a system with two second-degree equations. The elimination method **(Section 5.1)** is often used when both equations are second degree.

EXAMPLE 3 Solving a Nonlinear System by Elimination

Solve the system.

$$x^2 + y^2 = 9 \qquad (1)$$
$$2x^2 - y^2 = -6 \qquad (2)$$

The graph of (1) is a circle, while the graph of (2) is a hyperbola. By analyzing the possibilities we conclude that there may be 0, 1, 2, 3, or 4 points of intersection. Adding the two equations will eliminate y, leaving an equation that can be solved for x.

$$
\begin{array}{rl}
x^2 + y^2 = & 9 \\
\underline{2x^2 - y^2 = -6} & \\
3x^2 = & 3 \\
\end{array}
$$

$$x^2 = 1 \qquad \text{Divide by 3.}$$
$$x = 1 \quad \text{or} \quad x = -1 \qquad \text{Square root property}$$

Each value of x gives corresponding values for y when substituted into one of the original equations. Using equation (1) is easier since the coefficients of the x^2- and y^2-terms are 1.

If $x = 1$, then

$$1^2 + y^2 = 9$$
$$y^2 = 8$$
$$y = \sqrt{8} \quad \text{or} \quad y = -\sqrt{8}$$
$$y = 2\sqrt{2} \quad \text{or} \quad y = -2\sqrt{2}.$$

If $x = -1$, then

$$(-1)^2 + y^2 = 9$$
$$y^2 = 8$$
$$y = \sqrt{8} \quad \text{or} \quad y = -\sqrt{8}$$
$$y = 2\sqrt{2} \quad \text{or} \quad y = -2\sqrt{2}.$$

The solution set is

$$\{(1, 2\sqrt{2}), (1, -2\sqrt{2}), (-1, 2\sqrt{2}), (-1, -2\sqrt{2})\}.$$

Figure 31 shows the four points of intersection.

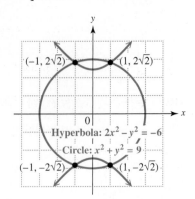

Figure 31

Work Problem 3 at the Side.

3 Solve each system.

(a) $x^2 + y^2 = 41$
$x^2 - y^2 = 9$

(b) $x^2 + 3y^2 = 40$
$4x^2 - y^2 = 4$

ANSWERS
3. (a) $\{(5, 4), (5, -4), (-5, 4), (-5, -4)\}$
 **(b)** $\{(2, 2\sqrt{3}), (2, -2\sqrt{3}),$
$\phantom{3. (b) \{}(-2, 2\sqrt{3}), (-2, -2\sqrt{3})\}$

OBJECTIVE 3 Solve a system that requires a combination of methods. Solving a system of second-degree equations may require a combination of methods.

EXAMPLE 4 Solving a Nonlinear System by a Combination of Methods

Solve the system.

$$x^2 + 2xy - y^2 = 7 \quad (1)$$
$$x^2 - y^2 = 3 \quad (2)$$

While we have not graphed equations like (1), its graph is a hyperbola. The graph of (2) is also a hyperbola. Two hyperbolas may have 0, 1, 2, 3, or 4 points of intersection. We use the elimination method here in combination with the substitution method. We begin by eliminating the squared terms by multiplying each side of equation (2) by -1 and then adding the result to equation (1).

$$
\begin{array}{rcl}
x^2 + 2xy - y^2 &=& 7 \\
-x^2 + y^2 &=& -3 \\
\hline
2xy &=& 4
\end{array}
$$

Next, we solve $2xy = 4$ for y. (Either variable would do.)

$$2xy = 4$$
$$y = \frac{2}{x} \quad (3)$$

Now, we substitute $y = \frac{2}{x}$ into one of the original equations. It is easier to do this with equation (2).

$$x^2 - y^2 = 3 \quad (2)$$
$$x^2 - \left(\frac{2}{x}\right)^2 = 3$$
$$x^2 - \frac{4}{x^2} = 3$$

$$x^4 - 4 = 3x^2 \qquad \text{Multiply by } x^2, x \neq 0.$$
$$x^4 - 3x^2 - 4 = 0 \qquad \text{Subtract } 3x^2.$$
$$(x^2 - 4)(x^2 + 1) = 0 \qquad \text{Factor.}$$
$$x^2 - 4 = 0 \quad \text{or} \quad x^2 + 1 = 0$$
$$x^2 = 4 \quad \text{or} \quad x^2 = -1$$
$$x = 2 \quad \text{or} \quad x = -2 \qquad x = i \quad \text{or} \quad x = -i$$

Substituting these four values of x into equation (3) gives the corresponding values for y.

If $x = 2$, then $y = 1$. \qquad If $x = i$, then $y = -2i$.

If $x = -2$, then $y = -1$. \qquad If $x = -i$, then $y = 2i$.

Note that if we substitute the x-values we found into equation (1) or (2) instead of into equation (3), we get extraneous solutions. ***It is always wise to check all solutions in both of the given equations.*** There are four ordered pairs in the solution set, two with real values and two with nonreal complex values. The solution set is

$$\{(2, 1), (-2, -1), (i, -2i), (-i, 2i)\}.$$

Continued on Next Page

The graph of the system, shown in Figure 32, shows only the two real intersection points because the graph is in the real number plane. The two ordered pairs with nonreal complex components are solutions of the system, but do not appear on the graph.

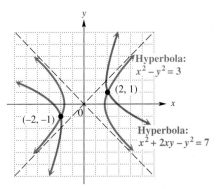

Figure 32

Work Problem 4 at the Side. ▶▶▶

NOTE
In the examples of this section, we analyzed the possible number of points of intersection of the graphs in each system. However, in Examples 2 and 4, we worked with equations whose graphs had not been studied. Keep in mind that it is not absolutely essential to visualize the number of points of intersection in order to solve the system. Furthermore, as in Example 4, there are sometimes nonreal complex solutions to nonlinear systems that do not appear as points of intersection in the real plane. Visualizing the geometry of the graphs is only an aid to solving these systems.

4️⃣ Solve each system.

(a) $x^2 + xy + y^2 = 3$
    $x^2 + y^2 = 5$

(b) $x^2 + 7xy - 2y^2 = -8$
    $-2x^2 + 4y^2 = 16$

ANSWERS
4. (a) $\{(1, -2), (-1, 2), (2, -1), (-2, 1)\}$
   (b) $\{(0, 2), (0, -2), (2i\sqrt{2}, 0),$
       $(-2i\sqrt{2}, 0)\}$

Real-Data Applications

Who Arrived First?

Suppose Vivian, Tommy, Carmen, and Manuel all leave home at 9:00 A.M. to drive to Atlanta, Georgia, along U.S. Interstate 75. The distance between Valdosta, Georgia, and Atlanta is 245 mi.

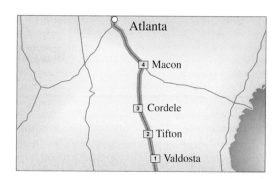

1. Vivian leaves from Valdosta, Georgia. She drives a Miata convertible and cruises at 85 mph.

2. Tommy leaves from Tifton, Georgia, which is 50 mi north of Valdosta. He drives a 1984 Toyota truck averaging 65 mph.

3. Carmen leaves from Cordele, Georgia, which is 40 mi north of Tifton. She drives a 1995 Honda Accord averaging 60 mph.

4. Manuel leaves from Macon, Georgia, which is 65 mi north of Cordele. He is riding a twenty-speed bike at 25 mph.

For Group Discussion

The independent variable is the number of hours traveled after 9:00 A.M. The dependent variable is distance, relative to Valdosta, after t hours. To compare the four trips, distances are measured from a common starting point in Valdosta. Recall that distance equals the product of rate and time. Tommy will have driven a distance of $65t$ miles after t hours, and since he starts 50 mi north of Valdosta, an equation that represents Tommy's distance relative to Valdosta is $d = 65t + 50$.

1. Write equations to represent the distances after t hours (relative to Valdosta) of Vivian, Carmen, and Manuel.

2. On a sheet of graph paper, sketch graphs of the four distance equations. Graph the horizontal line $d = 245$ to represent the distance between Valdosta and Atlanta. Based on your graphs, list the order in which the drivers reach Atlanta.

3. Based on your equations from Problem 1, at what time (rounded to the nearest minute) does each driver reach Atlanta? Are the results consistent with your conclusions based on your graphs?

4. Use a system of equations to find each time and location (distance from Valdosta). Round times to the nearest minute and distances to the nearest tenth of a mile, as necessary.

   (a) Find the time and location at which Vivian passes Tommy.

   (b) Find the time and location at which Carmen passes Manuel.

   (c) Does Carmen pass any other traveler before reaching Atlanta?

12.4 **Exercises**

FOR EXTRA HELP

Tutor Center Addison-Wesley Math Tutor Center

MathXL MathXL

Digital Video Tutor CD 8 Videotape 8

Student's Solutions Manual

MyMathLab MyMathLab

Interactmath.com

1. Write an explanation of the steps you would use to solve the system

$$x^2 + y^2 = 25$$
$$y = x - 1$$

   by the substitution method. Why would the elimination method not be appropriate for this system?

2. Write an explanation of the steps you would use to solve the system

$$x^2 + y^2 = 12$$
$$x^2 - y^2 = 13$$

   by the elimination method.

Each sketch represents the graphs of a pair of equations in a system. How many ordered pairs of real numbers are in each solution set?

3.
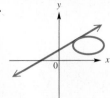

4.
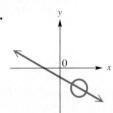

5.
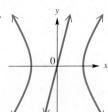

6.
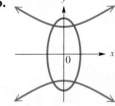

Suppose that a nonlinear system is composed of equations whose graphs are those described, and the number of points of intersection of the two graphs is as given. Make a sketch satisfying these conditions. (There may be more than one way to do this.)

7. A line and a circle; no points

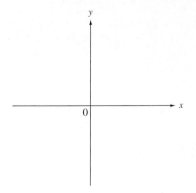

8. A line and a circle; one point

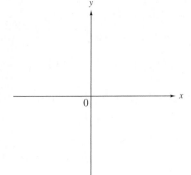

9. A line and an ellipse; two points

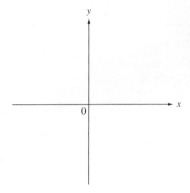

10. A line and a hyperbola; no points

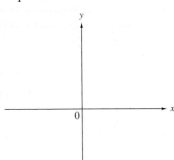

11. A circle and an ellipse; four points

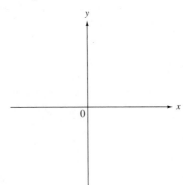

12. A parabola and an ellipse; one point

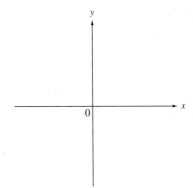

Solve each system by the substitution method. See Examples 1 and 2.

13. $y = 4x^2 - x$
$y = x$

14. $y = x^2 + 6x$
$3y = 12x$

15. $y = x^2 + 6x + 9$
$x + y = 3$

16. $y = x^2 + 8x + 16$
$x - y = -4$

17. $x^2 + y^2 = 2$
$2x + y = 1$

18. $2x^2 + 4y^2 = 4$
$x = 4y$

19. $xy = 4$
$3x + 2y = -10$

20. $xy = -5$
$2x + y = 3$

21. $xy = -3$
$x + y = -2$

22. $xy = 12$
$x + y = 8$

23. $y = 3x^2 + 6x$
$y = x^2 - x - 6$

24. $y = 2x^2 + 1$
$y = 5x^2 + 2x - 7$

25. $2x^2 - y^2 = 6$
$y = x^2 - 3$

26. $x^2 + y^2 = 4$
$y = x^2 - 2$

Solve each system using the elimination method or a combination of the elimination and substitution methods. See Examples 3 and 4.

27. $3x^2 + 2y^2 = 12$
$\quad x^2 + 2y^2 = 4$

28. $2x^2 + \ y^2 = 28$
$\quad 4x^2 - 5y^2 = 28$

29. $xy = 6$
$\quad 3x^2 - y^2 = 12$

30. $xy = 5$
$\quad 2y^2 - x^2 = 5$

31. $2x^2 + 2y^2 = 8$
$\quad 3x^2 + 4y^2 = 24$

32. $5x^2 + 5y^2 = 20$
$\quad x^2 + 2y^2 = 2$

33. $x^2 + xy + y^2 = 15$
$\quad x^2 + y^2 = 10$

34. $2x^2 + 3xy + 2y^2 = 21$
$\quad x^2 + y^2 = 6$

35. $3x^2 + 2xy - 3y^2 = 5$
$\quad -x^2 - 3xy + y^2 = 3$

36. $-2x^2 + 7xy - 3y^2 = 4$
$\quad 2x^2 - 3xy + 3y^2 = 4$

Solve each problem by using a nonlinear system.

37. The area of a rectangular rug is 84 ft² and its perimeter is 38 ft. Find the length and width of the rug.

38. Find the length and width of a rectangular room whose perimeter is 50 m and whose area is 100 m².

39. A company has found that the price p (in dollars) of its scientific calculator is related to the supply x (in thousands) by the equation

$$px = 16.$$

The price is related to the demand x (in thousands) for the calculator by the equation

$$p = 10x + 12.$$

The **equilibrium price** is the value of p where demand equals supply. Find the equilibrium price and the supply/demand at that price by solving a system of equations. (*Hint:* Demand, price, and supply must all be positive.)

40. The calculator company in Exercise 39 has also determined that the cost y to make x (thousand) calculators is

$$y = 4x^2 + 36x + 20,$$

while the revenue y from the sale of x (thousand) calculators is

$$36x^2 - 3y = 0.$$

Find the **break-even point,** where cost equals revenue, by solving a system of equations.

41. In the 1970s, the number of bachelor's degrees earned by men began to decrease. It stayed fairly constant in the 1980s, and then in the 1990s slowly began to increase again. Meanwhile, the number of bachelor's degrees earned by women continued to rise steadily throughout this period. Functions that model the situation are defined by the following equations, where y is the number of degrees (in thousands) granted in year x, with $x = 0$ corresponding to 1970.

Men: $\quad y = .138x^2 + .064x + 451$

Women: $\quad y = 12.1x + 334$

Solve this system of equations to find the year when the same number of bachelor's degrees was awarded to men and women. How many bachelor's degrees were awarded in that year? Give answer to the nearest ten thousand. (*Source:* U.S. National Center for Education Statistics, *Digest of Education Statistics,* annual.)

42. Andy Grove, chairman and co-founder of chip maker Intel Corp., recently noted that decreasing prices for computers and stable prices for Internet access implied that the trend lines for these costs either have crossed or soon will. He predicted that the time is not far away when computers, like cell phones, may be given away to sell on-line time. To see this, assume a price of $1000 for a computer, and let x represent the number of months it will be used. (*Source:* Corcoran, Elizabeth, "Can Free Computers Be That Far Away?", *Washington Post,* from *Sacramento Bee,* February 3, 1999.)

(a) Write an equation for the monthly cost y of the computer over this period.

(b) The average monthly on-line cost is about $20. Assume this will remain constant and write an equation to express this cost.

(c) Solve the system of equations from parts (a) and (b). Interpret your answer in relation to the situation.

12.5 Second-Degree Inequalities and Systems of Inequalities

OBJECTIVE 1 Graph second-degree inequalities. The linear inequality $3x + 2y \le 5$ is graphed by first graphing the boundary line $3x + 2y = 5$. A **second-degree inequality** is an inequality with at least one variable of degree 2 and no variable with degree greater than 2. An example is $x^2 + y^2 \le 36$. Such inequalities are graphed in the same way. The boundary of the inequality $x^2 + y^2 \le 36$ is the graph of the equation $x^2 + y^2 = 36$, a circle with radius 6 and center at the origin, as shown in Figure 33.

The graph of the inequality $x^2 + y^2 \le 36$ will include either the points outside the circle or the points inside the circle, as well as the boundary. We decide which region to shade by substituting any test point not on the circle, such as $(0, 0)$, into the original inequality. Since $0^2 + 0^2 \le 36$ is a true statement, the original inequality includes the points inside the circle, the shaded region in Figure 33, and the boundary.

OBJECTIVES

1 Graph second-degree inequalities.

2 Graph the solution set of a system of inequalities.

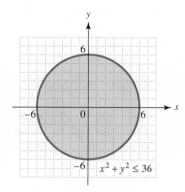

Figure 33

EXAMPLE 1 Graphing a Second-Degree Inequality

Graph $y < -2(x - 4)^2 - 3$.

The boundary, $y = -2(x - 4)^2 - 3$, is a parabola that opens down with vertex at $(4, -3)$. Using $(0, 0)$ as a test point gives

$$0 < -2(0 - 4)^2 - 3 \quad ?$$
$$0 < -32 - 3 \quad ?$$
$$0 < -35. \qquad \text{False}$$

Because the final inequality is a false statement, the points in the region containing $(0, 0)$ do not satisfy the inequality. Figure 34 shows the final graph. The parabola is drawn as a dashed curve since the points of the parabola itself do not satisfy the inequality, and the region inside (or below) the parabola is shaded.

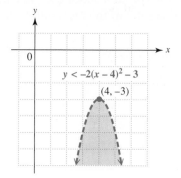

Figure 34

1 Graph $y \geq (x + 1)^2 - 5$.

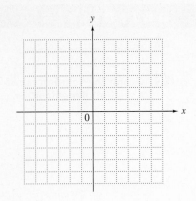

> **NOTE**
> Since the substitution is easy, the origin is the test point of choice unless the graph actually passes through $(0, 0)$.

Work Problem 1 at the Side.

EXAMPLE 2 Graphing a Second-Degree Inequality

Graph $16y^2 \leq 144 + 9x^2$.

First rewrite the inequality as follows.

$$16y^2 - 9x^2 \leq 144 \qquad \text{Subtract } 9x^2.$$

$$\frac{y^2}{9} - \frac{x^2}{16} \leq 1 \qquad \text{Divide by 144.}$$

This form shows that the boundary is the hyperbola given by

$$\frac{y^2}{9} - \frac{x^2}{16} = 1.$$

Since the graph is a vertical hyperbola, the desired region will be either the region between the branches or the regions above the top branch and below the bottom branch. Choose $(0, 0)$ as a test point. Substituting into the original inequality leads to $0 \leq 144$, a true statement, so the region between the branches containing $(0, 0)$ is shaded, as shown in Figure 35.

2 Graph $x^2 + 4y^2 > 36$.

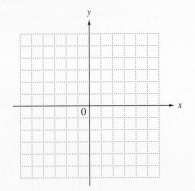

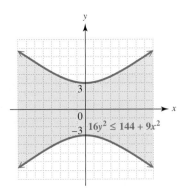

Figure 35

Work Problem 2 at the Side.

OBJECTIVE 2 Graph the solution set of a system of inequalities. If two or more inequalities are considered at the same time, we have a **system of inequalities.** To find the solution set of the system, we find the intersection of the graphs (solution sets) of the inequalities in the system.

EXAMPLE 3 Graphing a System of Two Inequalities

Graph the solution set of the system.

$$2x + 3y > 6$$
$$x^2 + y^2 < 16$$

Begin by graphing the solution set of $2x + 3y > 6$. The boundary line is the graph of $2x + 3y = 6$ and is a dashed line because of the symbol $>$. The test point $(0, 0)$ leads to a false statement in the inequality $2x + 3y > 6$,

ANSWERS

1.

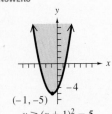

$y \geq (x + 1)^2 - 5$

2.

$x^2 + 4y^2 > 36$

— **Continued on Next Page**

so shade the region above the line, as shown in Figure 36. The graph of $x^2 + y^2 < 16$ is the interior of a dashed circle centered at the origin with radius 4. This is shown in Figure 37.

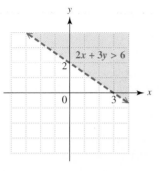

Figure 36 **Figure 37**

Finally, to show the graph of the solution set of the system, determine the intersection of the graphs of the two inequalities. The overlapping region in Figure 38 is the solution set.

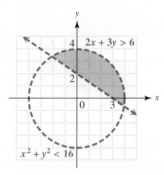

Figure 38

> **Work Problem 3 at the Side.** ⟩⟩⟩

EXAMPLE 4 **Graphing a Linear System with Three Inequalities**

Graph the solution set of the system.

$$x + y < 1$$
$$y \le 2x + 3$$
$$y \ge -2$$

Graph each inequality separately, on the same axes. The graph of $x + y < 1$ consists of all points below the dashed line $x + y = 1$. The graph of $y \le 2x + 3$ is the region that lies below the solid line $y = 2x + 3$. Finally, the graph of $y \ge -2$ is the region above the solid horizontal line $y = -2$. The graph of the system, the intersection of these three graphs, is the triangular region enclosed by the three boundary lines in Figure 39, including two of its boundaries.

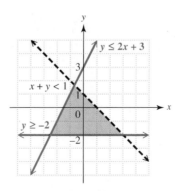

Figure 39

> **Work Problem 4 at the Side.** ⟩⟩⟩

❸ Graph the solution set of the system.

$$x^2 + y^2 \le 25$$
$$x + y \le 3$$

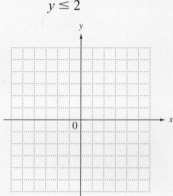

❹ Graph the solution set of the system.

$$3x - 4y \ge 12$$
$$x + 3y \ge 6$$
$$y \le 2$$

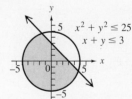

ANSWERS

3.

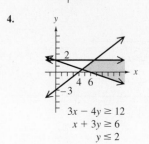

4.

5 Graph the solution set of the system.

$$y \geq x^2 + 1$$
$$\frac{x^2}{9} + \frac{y^2}{4} \geq 1$$
$$y \leq 5$$

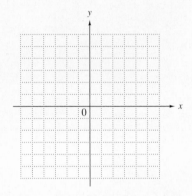

EXAMPLE 5 **Graphing a System with Three Inequalities**

Graph the solution set of the system.

$$y \geq x^2 - 2x + 1$$
$$2x^2 + y^2 > 4$$
$$y < 4$$

The graph of $y = x^2 - 2x + 1$ is a parabola with vertex at $(1, 0)$. Those points above (or in the interior of) the parabola satisfy the condition $y > x^2 - 2x + 1$. Thus, points on the parabola or in the interior are in the solution set of $y \geq x^2 - 2x + 1$.

The graph of the equation $2x^2 + y^2 = 4$ is an ellipse. We draw it as a dashed curve. To satisfy the inequality $2x^2 + y^2 > 4$, a point must lie outside the ellipse.

The graph of $y < 4$ includes all points below the dashed line $y = 4$. Finally, the graph of the system is the shaded region in Figure 40 that lies outside the ellipse, inside or on the boundary of the parabola, and below the line $y = 4$.

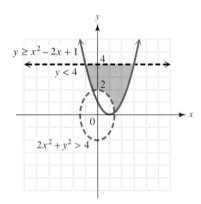

Figure 40

Work Problem 5 at the Side.

ANSWERS
5.

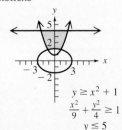

12.5 Exercises

FOR EXTRA HELP

Tutor Center — Addison-Wesley Math Tutor Center

MathXL — MathXL

Digital Video Tutor CD 8 — Videotape 8

Student's Solutions Manual

MyMathLab — MyMathLab

Interactmath.com

1. Which one of the following is a description of the graph of the solution set of this system?

$$x^2 + y^2 < 25$$
$$y > -2$$

   A. All points outside the circle $x^2 + y^2 = 25$ and above the line $y = -2$

   B. All points outside the circle $x^2 + y^2 = 25$ and below the line $y = -2$

   C. All points inside the circle $x^2 + y^2 = 25$ and above the line $y = -2$

   D. All points inside the circle $x^2 + y^2 = 25$ and below the line $y = -2$

2. Fill in each blank with the appropriate response. The graph of the system

$$y > x^2 + 1$$
$$\frac{x^2}{9} + \frac{y^2}{4} > 1$$
$$y < 5$$

   consists of all points _____ the parabola
   (above/below)

   $y = x^2 + 1$, _____ the ellipse
   (inside/outside)

   $\frac{x^2}{9} + \frac{y^2}{4} = 1$, and _____ the line $y = 5$.
   (above/below)

3. Explain how to graph the solution set of a nonlinear inequality.

4. Explain how to graph the solution set of a system of inequalities.

Graph each inequality. See Examples 1 and 2.

5. $y > x^2 - 1$

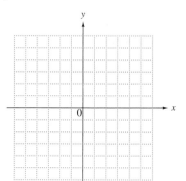

6. $y^2 > 4 + x^2$

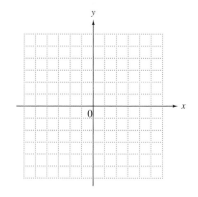

7. $y^2 \leq 4 - 2x^2$

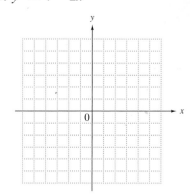

8. $y + 2 \geq x^2$

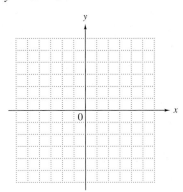

9. $x^2 \leq 16 - y^2$

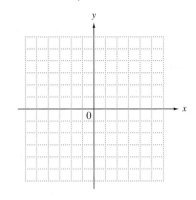

10. $2y^2 \geq 8 - x^2$

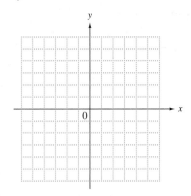

11. $x^2 \leq 16 + 4y^2$

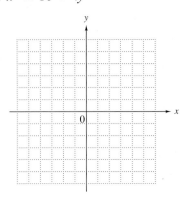

12. $y \leq x^2 + 4x + 2$

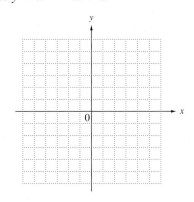

13. $9x^2 < 16y^2 - 144$

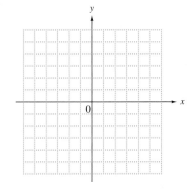

14. $9x^2 > 16y^2 + 144$

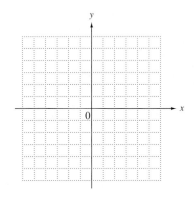

15. $4y^2 \leq 36 - 9x^2$

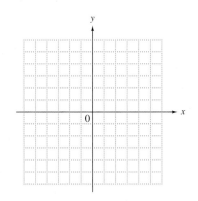

16. $x^2 - 4 \geq -4y^2$

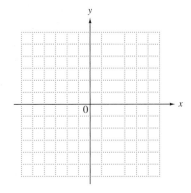

17. $x \geq y^2 - 8y + 14$

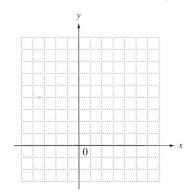

18. $x \leq -y^2 + 6y - 7$

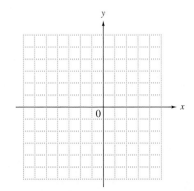

19. $25x^2 \leq 9y^2 + 225$

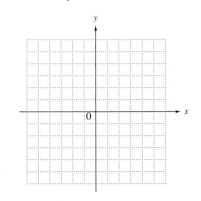

Graph each system of inequalities. See Examples 3–5.

20. $2x + 5y < 10$
$\quad x - 2y < 4$

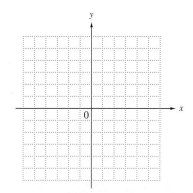

21. $3x - y > -6$
$\quad 4x + 3y > 12$

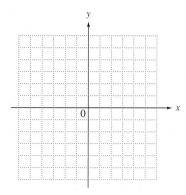

22. $5x - 3y \leq 15$
$\quad 4x + y \geq 4$

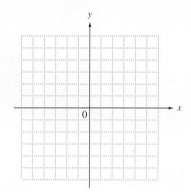

23. $4x - 3y \leq 0$
$\quad x + y \leq 5$

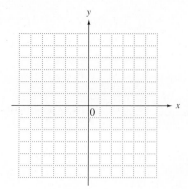

24. $x \leq 5$
$\quad y \leq 4$

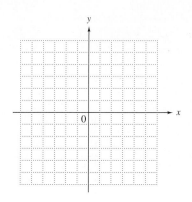

25. $x \geq -2$
$\quad y \leq 4$

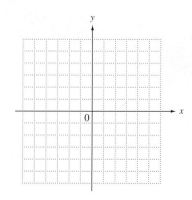

26. $y > x^2 - 4$
   $y < -x^2 + 3$

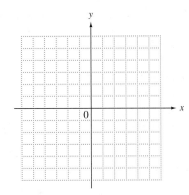

27. $x^2 - y^2 \geq 9$
   $\dfrac{x^2}{16} + \dfrac{y^2}{9} \leq 1$

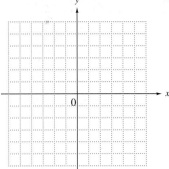

28. $y^2 - x^2 \geq 4$
   $-5 \leq y \leq 5$

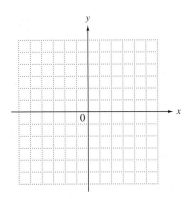

29. $x \geq 0$
   $y \geq 0$
   $x^2 + y^2 \geq 4$
   $x + y \leq 5$

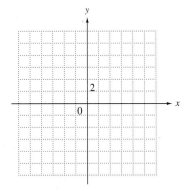

30. $y \leq -x^2$
   $y \geq x - 3$
   $y \leq -1$
   $x < 1$

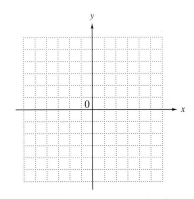

31. $y < x^2$
   $y > -2$
   $x + y < 3$
   $3x - 2y > -6$

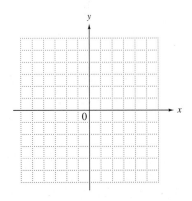

Chapter 12

SUMMARY

12.1 asymptotes
Lines that a graph approaches, such as the x- and y-axes for the graph of the reciprocal function, are called asymptotes of the graph.

greatest integer function
The function defined by $f(x) = [\![x]\!]$, where the symbol $[\![x]\!]$ represents the greatest integer less than or equal to x, is called the greatest integer function.

step function
A step function is a function with a graph that looks like a series of steps.

composition (composite function)
If f and g are functions, then the composition of g and f is defined by $(g \circ f)(x) = g[f(x)]$ for all x in the domain of f such that $f(x)$ is in the domain of g.

12.2 conic section
When a plane intersects an infinite cone at different angles, the figures formed by the intersections are called conic sections.

circle
A circle is the set of all points in a plane that lie a fixed distance from a fixed point.

center
The fixed point discussed in the definition of a circle is the center of the circle.

radius
The radius of a circle is the fixed distance between the center and any point on the circle.

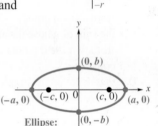
Circle: $x^2 + y^2 = r^2$

ellipse
An ellipse is the set of all points in a plane the sum of whose distances from two fixed points **(foci)** is constant.

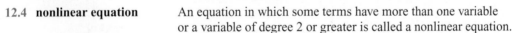
Ellipse: $\dfrac{x^2}{a^2} + \dfrac{y^2}{b^2} = 1$

12.3 hyperbola
A hyperbola is the set of all points in a plane such that the absolute value of the difference of the distances from two fixed points (foci) is constant.

asymptotes of a hyperbola
The two intersecting lines that the branches of a hyperbola approach are called asymptotes of the hyperbola.

fundamental rectangle
The asymptotes of a hyperbola are the extended diagonals of its fundamental rectangle.

Hyperbola: $\dfrac{x^2}{a^2} - \dfrac{y^2}{b^2} = 1$

12.4 nonlinear equation
An equation in which some terms have more than one variable or a variable of degree 2 or greater is called a nonlinear equation.

nonlinear system of equations
A nonlinear system of equations is a system with at least one nonlinear equation.

12.5 second-degree inequality
A second-degree inequality is an inequality with at least one variable of degree 2 and no variable with degree greater than 2.

system of inequalities
A system of inequalities consists of two or more inequalities to be solved at the same time.

NEW SYMBOLS

$[\![x]\!]$ greatest integer less than or equal to x

$(f \circ g)(x) = f[g(x)]$ composite function

TEST YOUR WORD POWER

See how well you have learned the vocabulary in this chapter. Answers, with examples, follow the Quick Review.

1. **Conic sections** are
   A. graphs of first-degree equations
   B. the result of two or more intersecting planes
   C. graphs of first-degree inequalities
   D. figures that result from the intersection of an infinite cone with a plane.

2. A **circle** is the set of all points in a plane
   A. such that the absolute value of the difference of the distances from two fixed points is constant
   B. that lie a fixed distance from a fixed point
   C. the sum of whose distances from two fixed points is constant
   D. that make up the graph of any second-degree equation.

3. An **ellipse** is the set of all points in a plane
   A. such that the absolute value of the difference of the distances from two fixed points is constant
   B. that lie a fixed distance from a fixed point
   C. the sum of whose distances from two fixed points is constant
   D. that make up the graph of any second-degree equation.

4. A **hyperbola** is the set of all points in a plane
   A. such that the absolute value of the difference of the distances from two fixed points is constant
   B. that lie a fixed distance from a fixed point
   C. the sum of whose distances from two fixed points is constant
   D. that make up the graph of any second-degree equation.

5. A **nonlinear equation** is an equation
   A. in which some terms have more than one variable or a variable of degree 2 or greater
   B. in which the terms have only one variable
   C. of degree 1
   D. of a linear function.

6. A **nonlinear system of equations** is a system
   A. with at least one linear equation
   B. with two or more inequalities
   C. with at least one nonlinear equation
   D. with at least two linear equations.

QUICK REVIEW

Concepts	*Examples*

12.1 *Additional Graphs of Functions; Composition*
Other Functions
In addition to the squaring function, some other important elementary functions in algebra are the absolute value function, defined by $f(x) = |x|$; the reciprocal function, defined by $f(x) = \frac{1}{x}$; the square root function, defined by $f(x) = \sqrt{x}$; and step functions, such as the greatest integer function defined by $f(x) = [\![x]\!]$.

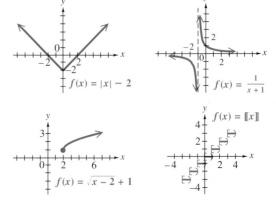

Composition of *f* and *g*

$$(f \circ g)(x) = f[g(x)]$$

If $f(x) = x^2$ and $g(x) = 2x + 1$, then

$$(f \circ g)(x) = f[g(x)]$$
$$= (2x + 1)^2 = 4x^2 + 4x + 1$$

and $\quad (g \circ f)(x) = g[f(x)]$
$$= 2x^2 + 1.$$

Concepts	*Examples*

12.2 *The Circle and the Ellipse*

Circle

The circle with radius r and center at (h, k) has an equation of the form

$$(x - h)^2 + (y - k)^2 = r^2.$$

The circle with equation $(x + 2)^2 + (y - 3)^2 = 25$, which can be written $[x - (-2)]^2 + (y - 3)^2 = 5^2$, has center $(-2, 3)$ and radius **5**.

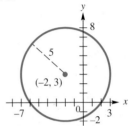

Ellipse

The ellipse whose x-intercepts are $(a, 0)$ and $(-a, 0)$ and whose y-intercepts are $(0, b)$ and $(0, -b)$ has an equation of the form

$$\frac{x^2}{a^2} + \frac{y^2}{b^2} = 1.$$

Graph $\dfrac{x^2}{9} + \dfrac{y^2}{4} = 1.$

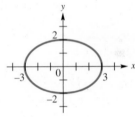

12.3 *The Hyperbola and Other Functions Defined by Radicals*

Hyperbola

A hyperbola with x-intercepts $(a, 0)$ and $(-a, 0)$ has an equation of the form

$$\frac{x^2}{a^2} - \frac{y^2}{b^2} = 1,$$

and a hyperbola with y-intercepts $(0, b)$ and $(0, -b)$ has an equation of the form

$$\frac{y^2}{b^2} - \frac{x^2}{a^2} = 1.$$

The extended diagonals of the fundamental rectangle with vertices at the points (a, b), $(-a, b)$, $(-a, -b)$, and $(a, -b)$ are the asymptotes of these hyperbolas.

Graph $\dfrac{x^2}{4} - \dfrac{y^2}{4} = 1.$

The graph has x-intercepts $(2, 0)$ and $(-2, 0)$.

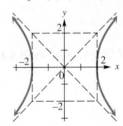

The fundamental rectangle has vertices at $(2, 2)$, $(-2, 2)$, $(-2, -2)$, and $(2, -2)$.

Graphing a Square Root Function

To graph a square root function defined by $f(x) = \sqrt{u}$ for an algebraic expression u, with $u > 0$, square both sides so that the equation can be easily recognized. Then graph only the part indicated by the original equation.

Graph $y = -\sqrt{4 - x^2}.$

Square both sides and rearrange terms to get

$$x^2 + y^2 = 4.$$

This equation has a circle as its graph. However, graph only the lower half of the circle, since the original equation indicates that y cannot be positive.

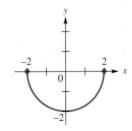

Concepts	*Examples*

12.4 *Nonlinear Systems of Equations*
Solving a Nonlinear System
A nonlinear system can be solved by the substitution method, the elimination method, or a combination of the two.

Solve the system.

$$x^2 + 2xy - y^2 = 14 \qquad (1)$$
$$x^2 - y^2 = -16 \qquad (2)$$

Multiply equation (2) by -1 and use elimination.

$$
\begin{array}{rr}
x^2 + 2xy - y^2 &= 14 \\
-x^2 \qquad\quad + y^2 &= 16 \\
\hline
2xy \qquad\quad &= 30 \\
xy &= 15
\end{array}
$$

Solve for y to obtain $y = \frac{15}{x}$, and substitute into equation (2).

$$x^2 - \left(\frac{15}{x}\right)^2 = -16$$

$$x^2 - \frac{225}{x^2} = -16$$

$$x^4 + 16x^2 - 225 = 0 \qquad \text{Multiply by } x^2 \text{; add } 16x^2.$$

$$(x^2 - 9)(x^2 + 25) = 0 \qquad \text{Factor.}$$

$$x = \pm 3 \quad \text{or} \quad x = \pm 5i \qquad \text{Zero-factor property}$$

Find corresponding y-values to obtain the solution set

$$\{(3, 5), (-3, -5), (5i, -3i), (-5i, 3i)\}.$$

12.5 *Second-Degree Inequalities and Systems of Inequalities*
Graphing a Second-Degree Inequality
To graph a second-degree inequality, graph the corresponding equation as a boundary and use test points to determine which region(s) form the solution set. Shade the appropriate region(s).

Graphing a System of Inequalities
The solution set of a system of inequalities is the intersection of the solution sets of the individual inequalities.

Graph $y \geq x^2 - 2x + 3$.

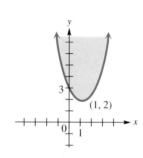

Graph the solution set of the system.

$$3x - 5y > -15$$
$$x^2 + y^2 \leq 25$$

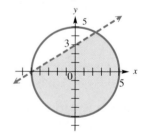

ANSWERS TO TEST YOUR WORD POWER

1. D; *Example:* Parabolas, circles, ellipses, and hyperbolas are conic sections.
2. B; *Example:* See the graph of $x^2 + y^2 = 9$ in Figure 12 of **Section 12.2.**
3. C; *Example:* See the graph of $\dfrac{x^2}{49} + \dfrac{y^2}{36} = 1$ in Figure 17 of **Section 12.2.**
4. A; *Example:* See the graph of $\dfrac{x^2}{16} - \dfrac{y^2}{12} = 1$ in Figure 20 of **Section 12.3.**
5. A; *Examples:* $y = x^2 + 8x + 16$, $xy = 5$, $2x^2 - y^2 = 6$
6. C; *Example:* $x^2 + y^2 = 2$
   $\qquad 2x + y = 1$

Chapter **12**

REVIEW EXERCISES

[12.1] *Graph each function.*

1. $f(x) = |x + 4|$

2. $f(x) = \dfrac{1}{x - 4}$

3. $f(x) = \sqrt{x} + 3$

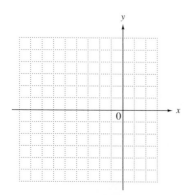
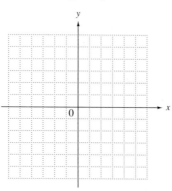
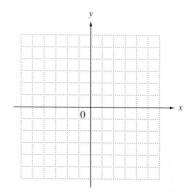

Find each of the following.

4. $[12]$

5. $\left[2\dfrac{1}{4}\right]$

6. $[-4.75]$

Let $f(x) = 3x^2 + 2x - 1$ and $g(x) = 5x + 7$. Find each of the following.

7. **(a)** $(g \circ f)(3)$ **(b)** $(f \circ g)(3)$

8. **(a)** $(f \circ g)(-2)$ **(b)** $(g \circ f)(-2)$

9. **(a)** $(f \circ g)(x)$ **(b)** $(g \circ f)(x)$

10. Based on your answers to Exercises 7–9, discuss whether composition of functions is a commutative operation.

[12.2] *Write an equation for each circle.*

11. Center $(-2, 4)$, $r = 3$

12. Center $(-1, -3)$, $r = 5$

13. Center $(4, 2)$, $r = 6$

Find the center and radius of each circle.

14. $x^2 + y^2 + 6x - 4y - 3 = 0$

15. $x^2 + y^2 - 8x - 2y + 13 = 0$

16. $2x^2 + 2y^2 + 4x + 20y = -34$

17. $4x^2 + 4y^2 - 24x + 16y = 48$

Graph each equation.

18. $x^2 + y^2 = 16$

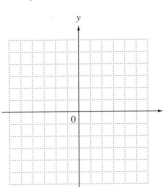

19. $\dfrac{x^2}{16} + \dfrac{y^2}{9} = 1$

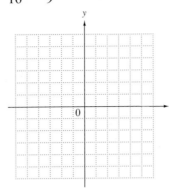

20. $\dfrac{x^2}{49} + \dfrac{y^2}{25} = 1$

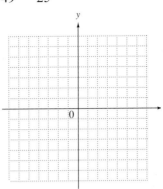

21. A satellite is in an elliptical orbit around Earth with perigee altitude of 160 km and apogee altitude of 16,000 km. See the figure. (*Source:* Kastner, Bernice, *Space Mathematics,* NASA, 1985.) Find the equation of the ellipse.

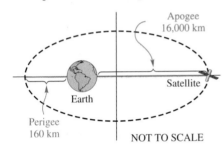

22. (a) The Roman Colosseum is an ellipse with $a = 310$ ft and $b = \frac{513}{2}$ ft. Find the distance between the foci of this ellipse.

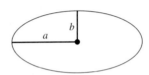

(b) A formula for the approximate circumference of an ellipse is

$$C \approx 2\pi\sqrt{\dfrac{a^2 + b^2}{2}},$$

where a and b are the lengths given in part (a). Use this formula to find the approximate circumference of the Roman Colosseum.

[12.3] *Graph each equation.*

23. $\dfrac{x^2}{16} - \dfrac{y^2}{25} = 1$

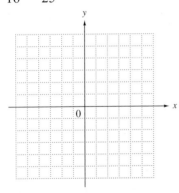

24. $\dfrac{y^2}{25} - \dfrac{x^2}{4} = 1$

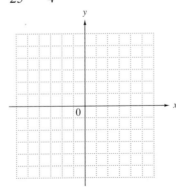

25. $f(x) = -\sqrt{16 - x^2}$

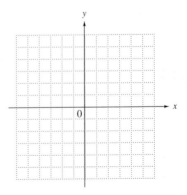

Identify the graph of each equation as a parabola, circle, ellipse, *or* hyperbola.

26. $x^2 + y^2 = 64$

27. $y = 2x^2 - 3$

28. $y^2 = 2x^2 - 8$

29. $y^2 = 8 - 2x^2$

30. $x = y^2 + 4$

31. $x^2 - y^2 = 64$

32. Ships and planes often use a location-finding system called LORAN. With this system, a radio transmitter at M sends out a series of pulses. (See the figure.) When each pulse is received at transmitter S, it then sends out a pulse. A ship at P receives pulses from both M and S. A receiver on the ship measures the difference in the arrival times of the pulses. A special map gives hyperbolas that correspond to the differences in arrival times (which give the distances d_1 and d_2 in the figure). The ship can then be located as lying on a branch of a particular hyperbola. Suppose $d_1 = 80$ mi and $d_2 = 30$ mi, and the distance between transmitters M and S is 100 mi. Use the definition to find an equation of the hyperbola on which the ship is located.

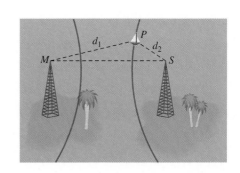

[12.4] *Solve each system.*

33. $2y = 3x - x^2$
$x + 2y = -12$

34. $y + 1 = x^2 + 2x$
$y + 2x = 4$

35. $x^2 + 3y^2 = 28$
$y - x = -2$

36. $xy = 8$
$x - 2y = 6$

37. $x^2 + y^2 = 6$
$x^2 - 2y^2 = -6$

38. $3x^2 - 2y^2 = 12$
$x^2 + 4y^2 = 18$

39. How many solutions are possible for a system of two equations whose graphs are a circle and a line?

40. How many solutions are possible for a system of two equations whose graphs are a parabola and a hyperbola?

[12.5] *Graph each inequality.*

41. $9x^2 \geq 16y^2 + 144$

42. $4x^2 + y^2 \geq 16$

43. $y < -(x + 2)^2 + 1$

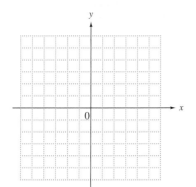

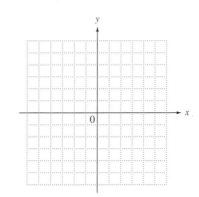

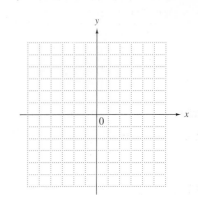

Graph each system of inequalities.

44. $2x + 5y \le 10$
$3x - y \le 6$

45. $|x| \le 2$
$|y| > 1$
$4x^2 + 9y^2 \le 36$

46. $9x^2 \le 4y^2 + 36$
$x^2 + y^2 \le 16$

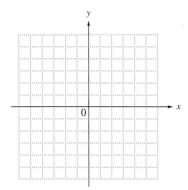
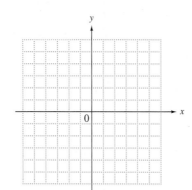
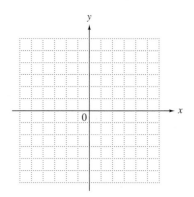

MIXED REVIEW EXERCISES

Graph.

47. $x^2 + y^2 = 25$

48. $x^2 + 9y^2 = 9$

49. $x^2 - 9y^2 = 9$

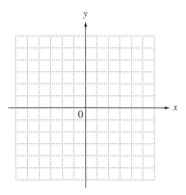
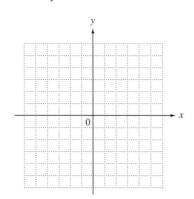

50. $f(x) = \sqrt{4 - x}$

51. $f(x) = [\![x]\!] - 1$

52. $4y > 3x - 12$
$x^2 < 16 - y^2$

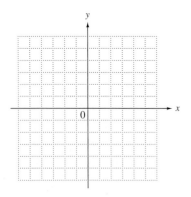
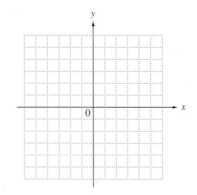
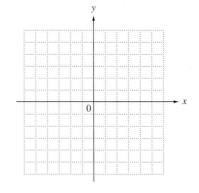

53. Explain why a set of points that form an ellipse does not satisfy the definition of a function.

Chapter **12**

T E S T

Study Skills Workbook
Activity 12: Final Exams

Match each function with its graph from choices A, B, C, and D.

1. $f(x) = \sqrt{x-2}$ **A.**   **B.**

1. _____

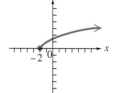

2. $f(x) = \sqrt{x} + 2$

2. _____

3. $f(x) = \sqrt{x} + 2$ **C.**   **D.**

3. _____

4. $f(x) = \sqrt{x} - 2$

4. _____

5. Sketch the graph of $f(x) = |x - 3| + 4$.

5.

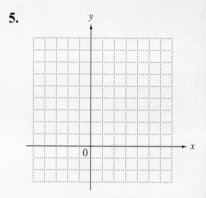

6. For $f(x) = 3x + 5$ and $g(x) = x^2 + 2$, find each of the following.

6. (a) _____

   (a) $(f \circ g)(-2)$

   (b) $(f \circ g)(x)$

(b) _____

   (c) $(g \circ f)(x)$

(c) _____

7. Find the center and radius of the circle whose equation is
$(x - 2)^2 + (y + 3)^2 = 16$. Sketch the graph.

7. _____

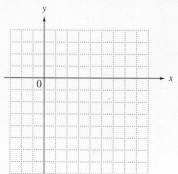

8. _____

8. Find the center and radius of the circle whose equation is $x^2 + y^2 + 8x - 2y = 8$.

Graph.

9.

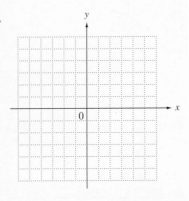

9. $f(x) = \sqrt{9 - x^2}$

10.

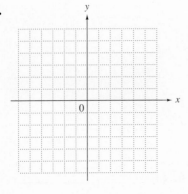

10. $4x^2 + 9y^2 = 36$

11.

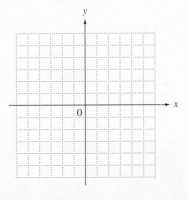

11. $16y^2 - 4x^2 = 64$

12. $\dfrac{y}{2} = -\sqrt{1 - \dfrac{x^2}{9}}$

12.

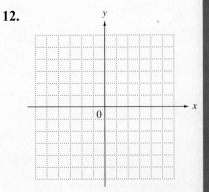

Identify the graph of each equation as a parabola, hyperbola, ellipse, *or* circle.

13. $6x^2 + 4y^2 = 12$

13. _____

14. $16x^2 = 144 + 9y^2$

14. _____

15. $4y^2 + 4x = 9$

15. _____

Solve each nonlinear system.

16. $2x - y = 9$
    $xy = 5$

16. _____

17. _____

17. $x - 4 = 3y$
$$x^2 + y^2 = 8$$

18. _____

18. $x^2 + y^2 = 25$
$$x^2 - 2y^2 = 16$$

19.

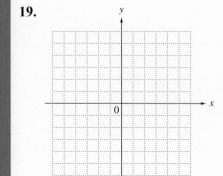

19. Graph the inequality $y < x^2 - 2$.

20.

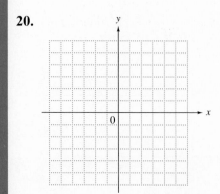

20. Graph the system $\begin{aligned} x^2 + 25y^2 &\le 25 \\ x^2 + y^2 &\le 9 \end{aligned}$.

Cumulative Review Exercises

CHAPTERS 1–12

1. Simplify $-10 + |-5| - |3| + 4$.

Solve.

2. $4 - (2x + 3) + x = 5x - 3$

3. $-4k + 7 \geq 6k + 1$

4. $|5m| - 6 = 14$

5. $|2p - 5| > 15$

6. Find the slope of the line through $(2, 5)$ and $(-4, 1)$.

7. Find the equation of the line through $(-3, -2)$ and perpendicular to the graph of $2x - 3y = 7$.

Solve each system.

8. $3x - y = 12$
$2x + 3y = -3$

9. $x + y - 2z = 9$
$2x + y + z = 7$
$3x - y - z = 13$

10. $xy = -5$
$2x + y = 3$

Solve each problem.

11. Al and Bev traveled from their apartment to a picnic 20 mi away. Al traveled on his bike while Bev, who left later, took her car. Al's average speed was half of Bev's average speed. The trip took Al $\frac{1}{2}$ hr longer than Bev. What was Bev's average speed?

12. The president of InstaTune, a chain of franchised automobile tune-up shops, reports that people who buy a franchise and open a shop pay a weekly fee (in dollars) to company headquarters, according to the linear function defined by

$$f(x) = .07x + 135,$$

where $f(x)$ is the fee and x is the total amount of money taken in during the week by the shop. Find the weekly fee if $2000 is taken in for the week. (*Source: Business Week.*)

Perform the indicated operations.

13. $(5y - 3)^2$

14. $(2r + 7)(6r - 1)$

15. $\dfrac{8x^4 - 4x^3 + 2x^2 + 13x + 8}{2x + 1}$

Factor.

16. $12x^2 - 7x - 10$

17. $2y^4 + 5y^2 - 3$

18. $z^4 - 1$

19. $a^3 - 27b^3$

Perform each operation.

20. $\dfrac{5x - 15}{24} \cdot \dfrac{64}{3x - 9}$

21. $\dfrac{y^2 - 4}{y^2 - y - 6} \div \dfrac{y^2 - 2y}{y - 1}$

22. $\dfrac{5}{c + 5} - \dfrac{2}{c + 3}$

23. $\dfrac{p}{p^2 + p} + \dfrac{1}{p^2 + p}$

Solve.

24. Kareem and Jamal want to clean their office. Kareem can do the job alone in 3 hr, while Jamal can do it alone in 2 hr. How long will it take them if they work together?

Simplify. Assume all variables represent positive real numbers.

25. $\left(\dfrac{4}{3}\right)^{-1}$

26. $\dfrac{(2a)^{-2}a^4}{a^{-3}}$

27. $4\sqrt[3]{16} - 2\sqrt[3]{54}$

28. $\dfrac{3\sqrt{5x}}{\sqrt{2x}}$

29. $\dfrac{5 + 3i}{2 - i}$

Solve.

30. $2\sqrt{k} = \sqrt{5k + 3}$

31. $10q^2 + 13q = 3$

32. $(4x - 1)^2 = 8$

33. $3k^2 - 3k - 2 = 0$

34. $2(x^2 - 3)^2 - 5(x^2 - 3) = 12$

35. $F = \dfrac{kwv^2}{r}$ for v

36. If $f(x) = x^3 + 4$, find $f^{-1}(x)$.

37. Evaluate $3^{\log_3 4}$.

38. Evaluate $e^{\ln 7}$.

39. Use properties of logarithms to write

$$2 \log(3x + 7) - \log 4$$

as a single logarithm.

40. Solve $\log(x + 2) + \log(x - 1) = 1$.

41. If $10,000 is invested at 5% for 4 yr, how much will there be in the account if interest is compounded

(a) quarterly,

(b) continuously?

The bar graph shows on-line retail sales (in billions of dollars) over the Internet. A reasonable model for sales y in billions of dollars is the exponential function defined by

$$y = 1.38(1.65)^x.$$

The years are coded such that x is the number of years since 1995.

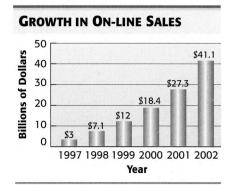

GROWTH IN ON-LINE SALES

Source: Jupiter Communications.

42. Use the model to estimate sales in the year 2000. (*Hint:* Let $x = 5$.)

43. Use the model to estimate sales in the year 2003.

44. If $f(x) = x^2 + 2x - 4$ and $g(x) = 3x + 2$, find
   (a) $(g \circ f)(1)$ **(b)** $(f \circ g)(x)$.

Graph.

45. $f(x) = -3x + 5$

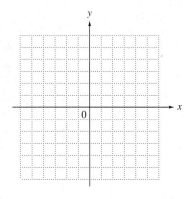

46. $f(x) = -2(x - 1)^2 + 3$

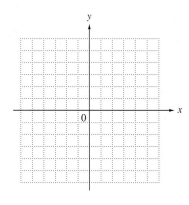

47. $\dfrac{x^2}{25} + \dfrac{y^2}{16} \leq 1$

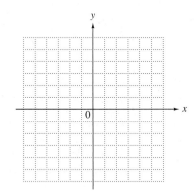

48. $f(x) = \sqrt{x - 2}$

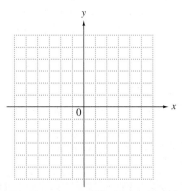

49. $\dfrac{x^2}{4} - \dfrac{y^2}{16} = 1$

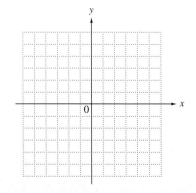

50. $f(x) = 3^x$

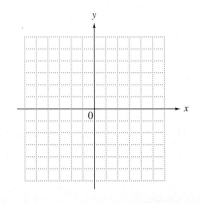

Appendix A
Strategies for Problem Solving

Appendix A Strategies for Problem Solving

OBJECTIVE 1 Learn additional problem-solving strategies. In **Section 2.3,** we introduce a six-step method for problem solving that we use throughout this text. This method is based on a problem-solving process developed by Hungarian native George Polya, among whose many publications is the modern classic *How to Solve It.*

Polya's Four-Step Process for Problem Solving

Step 1 **Understand the problem.** You cannot solve a problem if you do not understand what you are asked to find. The problem must be read and analyzed carefully. You may need to read it several times. After you have done so, ask yourself, "What must I find?"

Step 2 **Devise a plan.** There are many ways to attack a problem. Decide what plan is appropriate for the particular problem you are solving.

Step 3 **Carry out the plan.** Once you know how to approach the problem, carry out your plan. You may run into "dead ends" and unforeseen roadblocks, but be persistent. If you are able to solve a problem without a struggle, it isn't much of a problem, is it?

Step 4 **Look back and check.** Check your answer to see that it is reasonable. Does it satisfy the conditions of the problem? Have you answered all the questions the problem asks? Can you solve the problem a different way and come up with the same answer?

George Polya (1887–1985)

1 Compare the six-step problem-solving method given in **Section 2.3** with Polya's four steps.

Work Problem 1 at the Side. ▶▶▶

ANSWERS
1. Step 1 compares to Polya's first step, Steps 2 and 3 compare to his second step, Step 4 compares to his third step, and Step 6 compares to his fourth step.

In Step 2 of Polya's problem-solving process, we are told to devise a plan. The box on the next page lists some strategies that may prove useful.

Problem-Solving Strategies

Make a table or a chart.

Look for a pattern.

Solve a similar simpler problem.

Draw a sketch.

Write an equation and solve it.

If a formula applies, use it.

Work backward.

Guess and check.

Use trial and error.

Use common sense.

Look for a "catch" if an answer seems too obvious or impossible.

A particular problem solution may involve one or more of the strategies listed here, and you should try to be creative in your problem-solving techniques. The examples that follow illustrate some of these strategies.

The problem in Example 1 is a famous one in the history of mathematics and first appeared in *Liber Abaci*, a book written by the Italian mathematician Leonardo Pisano (also known as Fibonacci) in the year 1202. We apply Polya's four-step process to solve it.

EXAMPLE 1 Using a Table or a Chart

A man put a pair of rabbits in a cage. During the first month the rabbits produced no offspring, but each month thereafter produced one new pair of rabbits. If each new pair thus produced reproduces in the same manner, how many pairs of rabbits will there be at the end of one year?

Step 1 **Understand the problem.** After several readings, we can reword the problem as follows:

How many pairs of rabbits will the man have at the end of one year if he starts with one pair, and they reproduce this way: During the first month of life, each pair produces no new rabbits, but each month thereafter each pair produces one new pair?

Step 2 **Devise a plan.** Since there is a definite pattern to how the rabbits will reproduce, we can construct a table as shown below. Once the table is completed, the final entry in the final column is our answer.

Month	Number of Pairs at Start	Number of New Pairs Produced	Number of Pairs at End of Month
1st			
2nd			
3rd			
4th			
5th			
6th			
7th			
8th			
9th			
10th			
11th			
12th			

Continued on Next Page

Step 3   **Carry out the plan.** At the start of the first month there is only one pair of rabbits. No new pairs are produced during the first month, so there is $1 + 0 = 1$ pair present at the end of the first month. This pattern continues throughout the table. We add the number in the first column of numbers to the number in the second column to get the number in the third.

Month	Number of Pairs at Start	+	Number of New Pairs Produced	=	Number of Pairs at End of Month	
1st	1		0		1	$1 + 0 = 1$
2nd	1		1		2	$1 + 1 = 2$
3rd	2		1		3	$2 + 1 = 3$
4th	3		2		5	.
5th	5		3		8	.
6th	8		5		13	.
7th	13		8		21	.
8th	21		13		34	.
9th	34		21		55	.
10th	55		34		89	.
11th	89		55		144	.
12th	144		89		233	$144 + 89 = \mathbf{233}$

There will be 233 pairs of rabbits at the end of one year.

Step 4   **Look back and check.** This problem can be checked by going back and making sure that we have interpreted it correctly, which we have. Double-check the arithmetic. We have answered the question posed by the problem, so the problem is solved.

> **NOTE**
> The sequence shown in color in the table in Example 1 is called the *Fibonacci sequence,* and many of its interesting properties are investigated in other mathematics courses.

Work Problem 2 at the Side. ▶▶▶

In the remaining examples of this section, we use Polya's process but we do not list the steps specifically as we did in Example 1.

EXAMPLE 2   **Working Backward**

Rob Zwettler goes to the racetrack with his buddies on a weekly basis. One week he tripled his money, but then lost $12. He took his money back the next week, doubled it, but then lost $40. The following week he tried again, taking his money back with him. He quadrupled it, and then played well enough to take that much home with him, a total of $224. How much did he start with the first week?

This problem asks us to find Rob's starting amount, given information about his winnings and losses. We also know his final amount. The method of working backward can be applied quite easily.

Continued on Next Page

---

2 Refer to Example 1, and observe the sequence of numbers in color, the Fibonacci sequence. Choose any four consecutive terms. Multiply the first one chosen by the fourth, and then write the product. Now multiply the two middle terms and write the product. Repeat this process a few more times. What do you notice when the two products are compared?

ANSWERS
2. The products will always differ by 1.

❸ Solve each problem.

(a) Phyllis Crittenden bought a book for $10 and then spent half her remaining money on a train ticket. She then bought lunch for $4 and spent half her remaining money at a bazaar. She left the bazaar with $20. How much money did she start with?

(b) If a, b, and c are digits for which

$$
\begin{array}{r}
7\ a\ 2 \\
-\ 4\ 8\ b \\
\hline
c\ 7\ 3,
\end{array}
$$

then $a + b + c = $ _____.

A. 14 **B.** 15 **C.** 16
D. 17 **E.** 18.
(*Source: Mathematics Teacher* calendar, September 22, 1999.)

❹ Solve each problem.

(a) Assuming that he lives that long, one of the authors of this book will be 76 yr old in the year x^2, where x is a counting number. In what year was he born?

(b) Place each of the digits 1, 2, 3, 4, 5, 6, 7, and 8 in separate boxes so that boxes that share common corners do not contain successive digits.
(*Source: Mathematics Teacher* calendar, November 29, 1997.)

ANSWERS
3. (a) $98 **(b)** D
4. (a) 1949 **(b)** Here is one possible solution.

Since his final amount was $224 and this represents four times the amount he started with on the third week, we *divide* $224 by 4 to find that he started the third week with $56. Before he lost $40 the second week, he had this $56 plus the $40 he lost, giving him $96. This represented double what he started with, so he started with $96 *divided by* 2, or $48, the second week. Repeating this process once more for the first week, before his $12 loss he had

$$\$48 + \$12 = \$60,$$

which represents triple what he started with. Therefore, he started with

$$\$60 \div 3, \quad \text{or} \quad \mathbf{\$20.}$$

To check our answer, $20, observe the following equations that depict the winnings and losses:

First week: $(3 \times \$20) - \$12 = \$60 - \$12 = \mathbf{\$48}$
Second week: $(2 \times \$48) - \$40 = \$96 - \$40 = \mathbf{\$56}$
Third week: $(4 \times \$56) = \$224.$ His final amount

◀◀◀ Work Problem 3 at the Side.

Recall that $5^2 = 5 \cdot 5 = 25$, that is, 5 squared is 25. Thus, 25 is called a **perfect square.** Other perfect squares include

$$4, \quad 9, \quad 16, \quad 36, \quad \text{and so on.} \quad \text{Perfect squares}$$

EXAMPLE 3 **Using Trial and Error**

The mathematician Augustus De Morgan lived in the nineteenth century. He once made the following statement: "I was x years old in the year x^2." In what year was De Morgan born?

We must find the year of De Morgan's birth. The problem tells us that he lived in the nineteenth century, which is another way of saying that he lived during the 1800s. One year of his life was a perfect square, so we must find a number between 1800 and 1900 that is a perfect square. Use trial and error.

$$
\begin{aligned}
42^2 &= 42 \cdot 42 = 1764 \\
43^2 &= 43 \cdot 43 = \mathbf{1849} \qquad \text{1849 is between 1800 and 1900.} \\
44^2 &= 44 \cdot 44 = 1936
\end{aligned}
$$

The only natural number whose square is between 1800 and 1900 is 43, since $43^2 = 1849$. Therefore, De Morgan was 43 yr old in 1849. The final step in solving the problem is to subtract 43 from 1849 to find the year of his birth:

$$1849 - 43 = \mathbf{1806.} \qquad \text{He was born in 1806.}$$

While the following check may seem unorthodox, it works: Look up De Morgan's birth date in a book on mathematics history, such as *An Introduction to the History of Mathematics*, Sixth Edition, by Howard W. Eves.

◀◀◀ Work Problem 4 at the Side.

As mentioned above, $5^2 = 25$. The inverse (opposite) of squaring a number is called taking the **square root.** (See **Section 1.3.**) We indicate the positive square root using a **radical sign** $\sqrt{}$. Thus, $\sqrt{25} = 5$. Also,

$$\sqrt{4} = 2, \quad \sqrt{9} = 3, \quad \sqrt{16} = 4, \quad \text{and so on.} \quad \text{Square roots}$$

The next problem dates back to Hindu mathematics, circa 850.

EXAMPLE 4 **Guessing and Checking**

One-fourth of a herd of camels was seen in the forest. Twice the square root of that herd had gone to the mountain slopes, and 3 times 5 camels remained on the riverbank. What is the numerical measure of that herd of camels?

The numerical measure of the herd of camels must be a natural number. Since the problem mentions "one-fourth of a herd" and "the square root of that herd," the number of camels must be both a multiple of 4 and a perfect square, so no fractions will be encountered. The smallest natural number that satisfies both conditions is 4. We write an equation where x represents the numerical measure of the herd, and then substitute 4 for x to see if it is a solution.

$$\underset{\substack{\text{"one-fourth of}\\\text{the herd"}}}{\frac{1}{4}x} + \underset{\substack{\text{"twice the square}\\\text{root of that herd"}}}{2\sqrt{x}} + \underset{\substack{\text{"3 times}\\\text{5 camels"}}}{3 \cdot 5} = \underset{\substack{\text{"the numerical measure}\\\text{of the herd"}}}{x}$$

$$\frac{1}{4}(4) + 2\sqrt{4} + 3 \cdot 5 = 4 \qquad \text{Let } x = 4.$$
$$1 + 4 + 15 = 4 \qquad ? \ \sqrt{4} = 2$$
$$20 \neq 4$$

Since 4 is not the solution, try 16, the next perfect square that is a multiple of 4.

$$\frac{1}{4}(16) + 2\sqrt{16} + 3 \cdot 5 = 16 \qquad \text{Let } x = 16.$$
$$4 + 8 + 15 = 16 \qquad ? \ \sqrt{16} = 4$$
$$27 \neq 16$$

Since 16 is not a solution, try 36.

$$\frac{1}{4}(36) + 2\sqrt{36} + 3 \cdot 5 = 36 \qquad \text{Let } x = 36.$$
$$9 + 12 + 15 = 36 \qquad ? \ \sqrt{36} = 6$$
$$36 = 36$$

We see that 36 is the numerical measure of the herd. Check in the words of the problem: "One-fourth of 36, plus twice the square root of 36, plus 3 times 5" gives 9 plus 12 plus 15, which equals 36.

Work Problem 5 at the Side. ▶▶▶

EXAMPLE 5 **Considering a Similar Simpler Problem and Looking for a Pattern**

The digit farthest to the right in a natural number is called the *ones* or *units* digit, since it tells how many ones are contained in the number when grouping by tens is considered. What is the ones (or units) digit in 2^{4000}?

Recall that 2^{4000} means that 2 is used as a factor 4000 times:

$$2^{4000} = \underbrace{2 \times 2 \times 2 \times \cdots \times 2.}_{4000 \text{ factors}}$$

Certainly, we are not expected to evaluate this number. To answer the question, we examine some smaller powers of 2 and then look for a pattern. We start with the exponent 1 and look at the first twelve powers of 2.

Continued on Next Page

❺ Solve each problem.

(a) I am thinking of a positive number. If I square it, double the result, take half of that result, and then add 12, I get 21. What is my number?

(b) The same author mentioned in Margin Problem 4(a) graduated from high school in the year that satisfies these conditions:

(1) The sum of the digits is 23;

(2) The hundreds digit is 3 more than the tens digit;

(3) No digit is an 8.

In what year did he graduate?

ANSWERS
5. (a) 3   **(b)** 1967

6 Solve each problem.

(a) What is the units digit in 7^{491}?

(b) What is the 103rd digit in the decimal representation for $\frac{1}{11}$?

7 Solve each problem.

(a) What is the maximum number of small squares in which we may place a cross ($\times$) and not have any row, column, or diagonal completely filled with crosses? Illustrate your answer.

(b) By drawing two straight lines, divide the face of a clock into three regions such that the numbers in the regions have the same total. (*Source: Mathematics Teacher* calendar, October 28, 1998.)

ANSWERS
6. **(a)** 3   **(b)** 0
7. **(a)** 6

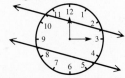

One of several possibilities

(b) Each region has a sum of 26.

$$2^1 = 2 \qquad 2^5 = 32 \qquad 2^9 = 512$$
$$2^2 = 4 \qquad 2^6 = 64 \qquad 2^{10} = 1024$$
$$2^3 = 8 \qquad 2^7 = 128 \qquad 2^{11} = 2048$$
$$2^4 = 16 \qquad 2^8 = 256 \qquad 2^{12} = 4096$$

Notice that in each of the four rows above, the ones digit is the same. The final row, which contains the exponents 4, 8, and 12, has the ones digit 6. Each of these exponents is divisible by 4, and since 4000 is divisible by 4, we observe the pattern to predict that the units digit in 2^{4000} is **6**.

The units digit for any other power of 2 can be found if we divide the exponent by 4 and compare the remainder to the preceding list of powers. For example, to find the units digit of 2^{543}, we divide 543 by 4 to get a quotient of 135 and a remainder of 3. The units digit is the same as that of 2^3, which is 8.

Work Problem 6 at the Side.

EXAMPLE 6 Drawing a Sketch

An array of nine dots is arranged in a 3×3 square, as shown in Figure 1. Is it possible to join the dots with exactly four straight lines if you are not allowed to pick up your pencil from the paper and may not trace over a line that has already been drawn? If so, show how.

Figure 1

Figure 2 shows three attempts. In each case, something is wrong. In the first sketch, one dot is not joined. In the second, the figure cannot be drawn without picking up your pencil from the paper or tracing over a line that has already been drawn. In the third figure, all dots have been joined, but you have used five lines as well as retraced over the figure.

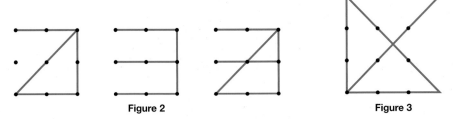

Figure 2   **Figure 3**

However, the conditions of the problem can be satisfied, as shown in Figure 3. We "went outside of the box," which was not prohibited by the conditions of the problem. This is an example of creative thinking—we used a strategy that is usually not considered at first, since our initial attempts involved "staying within the confines" of the figure.

Work Problem 7 at the Side.

The final example falls into a category of problems that involve a "catch." Some of these problems seem too easy or perhaps impossible at first, because we tend to overlook an obvious situation. We must look carefully at the use of language in such problems. And, of course, we should never forget to use common sense.

EXAMPLE 7 Using Common Sense

Two currently minted U.S. coins together have a total value of $1.05. One is not a dollar. What are the two coins?

Our initial reaction might be, "The only way to have two such coins with a total of $1.05 is to have a nickel and a dollar, but the problem says that one of them is not a dollar." This statement is indeed true. What we must realize here is that the one that is not a dollar is the nickel, and the *other* coin is a dollar! So the two coins are a dollar and a nickel.

Work Problem 8 at the Side.

⑧ Solve each problem.

(a) Which is correct? Three cubed *is* nine or three cubed *are* nine?

(b) If you take 7 bowling pins from 10 bowling pins, what do you have?

(c) If it takes $7\frac{1}{2}$ min to boil an egg, how long does it take to boil 5 eggs?

ANSWERS
8. (a) Neither is correct, since $3^3 = 27$.
   (b) 7 bowling pins
   (c) $7\frac{1}{2}$ min (Boil them all at the same time.)

Appendix A Exercises

FOR EXTRA HELP

Addison-Wesley Math Tutor Center

MathXL

Student's Solutions Manual

MyMathLab

Interactmath.com

Exercises 1–13 are from the popular monthly calendar feature in the journal Mathematics Teacher. *The authors wish to thank the many journal contributors for permission to use these problems. Original calendar dates are included.*

Use the various problem-solving strategies to solve each problem. In many cases, there is more than one possible approach, so be creative.

1. You are working in a store that has been very careless with the stock. Three boxes of socks are each incorrectly labeled. The labels say *red socks, green socks,* and *red and green socks.* How can you relabel the boxes correctly by taking only one sock out of one box, without looking inside the boxes? (October 22, 2001)

2. Three dice with faces numbered 1 through 6 are stacked as shown. Seven of the eighteen faces are visible, leaving eleven faces hidden on the back, on the bottom, and between faces. The total number of dots not visible in this view is _____.

   A. 21
   B. 22
   C. 31
   D. 41
   E. 53
   (September 17, 2001)

3. At his birthday party, Mr. Green would not directly tell how old he was. He said, "If you add the year of my birth to this year, subtract the year of my tenth birthday and the year of my fiftieth birthday, and then add my present age, the result is eighty." How old was Mr. Green? (December 14, 1997)

4. Today is your first day driving a city bus. When you leave downtown, you have 23 passengers. At the first stop, 3 people exit and 5 people get on the bus. At the second stop, 11 people exit and 8 people get on the bus. At the third stop, 5 people exit and 10 people get on. How old is the bus driver? (April 1, 2002)

5. You and a friend are playing tick-tack-toe, where three in a row *loses.* You are O. If you want to win, what must your next move be? (October 21, 2001)

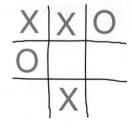

6. How can you connect each square with the triangle that has the same number? Lines cannot cross, enter a square or triangle, or go outside the diagram. (October 15, 1999)

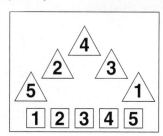

7. Pat and Chris have the same birthday. Pat is twice as old as Chris was when Pat was as old as Chris is now. If Pat is now 24 yr old, how old is Chris? (December 3, 2001)

8. Balls numbered 1 through 6 are arranged in a *difference triangle*. Note that in any row, the difference between the larger and the smaller of two successive balls is the number of the ball that appears below them. Arrange balls numbered 1 through 10 in a difference triangle. (May 6, 1998)

9. Only one of these numbers is a perfect square. Which one is it? (October 8, 1997)

329,476   389,372   964,328
326,047   724,203

10. While traveling to his grandmother's for Christmas, George fell asleep halfway through the journey. When he awoke, he still had to travel half the distance that he had traveled while sleeping. For what part of the entire journey had he been asleep? (December 25, 1998)

11. You have brought two unmarked buckets to a stream. The buckets hold 7 gal and 3 gal of water, respectively. How can you obtain exactly 5 gal of water to take home? (October 19, 1997)

12. Chip and Dale collected 32 acorns on Monday and stored them with their acorn supply. After Chip fell asleep, Dale ate half the acorns. This pattern continued through Friday night, with 32 acorns being added and half the supply being eaten. On Saturday morning, Chip counted the acorns and found that they had only 35. How many acorns had they started with on Monday morning? (March 12, 1997)

13. What are the final two digits of 7^{1997}? (November 29, 1997)

14. If you raise 3 to the 324th power, what is the units digit of the result?

15. A frog is at the bottom of a 20-ft well. Each day it crawls up 4 ft but each night it slips back 3 ft. After how many days will the frog reach the top of the well?

16. A lily pad grows so that each day it doubles its size. On the twentieth day of its life, it completely covers a pond. On what day was the pond half covered?

17. Some children are standing in a circular arrangement. They are evenly spaced and arranged in numerical order. The fourth child is standing directly opposite the twelfth child. How many children are there in the circle?

18. A *perfect number* is a natural number that is equal to the sum of all its counting number divisors except itself. For example, 28 is a perfect number, since its divisors other than itself are 1, 2, 4, 7, and 14, and $1 + 2 + 4 + 7 + 14 = 28$. What is the least perfect number?

19. Draw a diagram that satisfies the following description, using the minimum number of birds: "Two birds above a bird, two birds below a bird, and a bird between two birds."

20. Donna is taller than David but shorter than Bill. Dan is shorter than Bob. What is the first letter in the name of the tallest person?

21. A *magic square* is a square array of numbers that has the property that the sum of the numbers in any row, column, or diagonal is the same. Fill in the square below so that it becomes a magic square, and all digits 1, 2, 3, . . . , 9 are used exactly once.

6		8
	5	
		4

22. Refer to Exercise 21. Complete the magic square below so that all counting numbers 1, 2, 3, . . . , 16 are used exactly once, and the sum in each row, column, or diagonal is 34.

6			9
	15		14
11		10	
16		13	

23. What is the minimum number of pitches that a baseball player who pitches a complete game can make in a regulation 9-inning baseball game?

24. What is the least natural number whose written name in the English language has its letters in alphabetical order?

25. You have eight coins. Seven are genuine and one is a fake, which weighs a little less than the other seven. You have a balance scale, which you may use only three times. Tell how to locate the bad coin in three weighings. Then show how to detect the bad coin in only *two* weighings.

26. A person must take a wolf, a goat, and some cabbage across a river. The rowboat to be used has room for the person plus either the wolf, the goat, or the cabbage. If the person takes the cabbage in the boat, the wolf will eat the goat. While the wolf crosses in the boat, the cabbage will be eaten by the goat. The goat and cabbage are safe only when the person is present. Even so, the person gets everything across the river. Explain how. (This problem dates back to around the year 750.)

27. When the diagram shown is folded to form a cube, what letter is opposite the face marked Z?

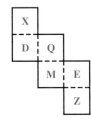

28. Draw a square in the following figure so that no two cats share the same region.

29. (This is an ancient Hindu problem.) Beautiful maiden with beaming eyes, tell me . . . which is the number that when multiplied by 3, then increased by $\frac{3}{4}$ the product, then divided by 7, diminished by $\frac{1}{3}$ of the quotient, multiplied by itself, diminished by 52, by the extraction of the square root, addition of 8, and division by 10 gives the number 2?

30. A teenager's age increased by 2 gives a perfect square. Her age decreased by 10 gives the square root of that perfect square. She is 5 yr older than her brother. How old is her brother?

31. Draw the following figure without picking up your pencil from the paper and without tracing over a line you have already drawn.

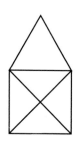

32. Repeat Exercise 31 for the figure shown here.

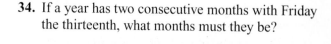

33. James, Dan, Jessica, and Cathy form a pair of married couples. Their ages are 36, 31, 30, and 29. Jessica is married to the oldest person in the group. James is older than Jessica but younger than Cathy. Who is married to whom, and what are their ages?

34. If a year has two consecutive months with Friday the thirteenth, what months must they be?

35. The brother of the chief executive officer (CEO) of a major industrial firm died. The man who died had no brother. How is this possible?

36. Some months have 30 days and some have 31 days. How many months have 28 days?

37. How much dirt is there in a cubical hole, 6 ft on each side?

38. Becky's mother has three daughters. She named her first daughter Penny and her second daughter Nichole. What did she name her third daughter?

39. Place one of the arithmetic operations $+$, $-$, $\times$, or $\div$ between each pair of successive numbers on the left side of this equation to make it true. Any operation may be used more than once or not at all.

$$1 \quad 2 \quad 3 \quad 4 \quad 5 \quad 6 \quad 7 \quad 8 \quad 9 = 100$$

40. In the addition problem below, some digits are missing as indicated by the blanks. If the problem is done correctly, what is the sum of the missing digits?

$$
\begin{array}{r}
\_\ 3\ 5 \\
8\ \_\ 6 \\
+\ 1\ 4\ \_ \\
\hline
\_\ 4\ 0\ 8
\end{array}
$$

41. Fill in the blanks so that the multiplication problem below uses all digits 0, 1, 2, 3, . . . , 9 exactly once, and is correctly worked.

$$
\begin{array}{r}
\_\ 0\ 2 \\
\times \qquad 3\_ \\
\hline
\_\ 5,\ \_\ \_\ \_
\end{array}
$$

42. Based on your knowledge of elementary arithmetic, describe the pattern that can be observed when the following operations are performed:

$$9 \times 1, \quad 9 \times 2, \quad 9 \times 3, \quad \ldots, \quad 9 \times 9.$$

(*Hint:* Add the digits in the answers. What do you notice?)

43. How many triangles are in the following figure?

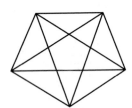

44. How many squares are in the following figure?

45. Volumes 1 and 2 of *The Complete Works of Wally Smart* are standing in numerical order from left to right on your bookshelf. Volume 1 has 450 pages and Volume 2 has 475 pages. Excluding the covers, how many pages are between page 1 of Volume 1 and page 475 of Volume 2?

46. At a hardware store, I can buy 1 for $.75 and I can buy 68356 for $3.75. What am I buying?

47. Eve said to Adam, "If you give me one dollar, then we will have the same amount of money." Adam then replied, "Eve, if you give me one dollar, I will have double the amount of money you are left with." How much does each have?

48. A drawer contains 20 black socks and 20 white socks. If the light is off and you reach into the drawer to get your socks, what is the minimum number of socks you must pull out in order to be sure that you have a matching pair?

Appendix B
Review of Fractions

Appendix B Review of Fractions

The numbers used most often in everyday life are the **whole numbers,**

$$0, 1, 2, 3, 4, 5, \ldots$$

and **fractions,** such as

$$\frac{1}{3}, \quad \frac{5}{4}, \quad \text{and} \quad \frac{11}{12}.$$

The parts of a fraction are named as follows.

$$\text{Fraction bar} \longrightarrow \frac{4}{7} \begin{array}{l} \longleftarrow \text{Numerator} \\ \longleftarrow \text{Denominator} \end{array}$$

OBJECTIVES

1. Identify prime numbers.
2. Write numbers in prime factored form.
3. Write fractions in lowest terms.
4. Multiply and divide fractions.
5. Add and subtract fractions.

If the numerator of a fraction is smaller than the denominator, we call it a **proper fraction.** A proper fraction has a value less than 1. If the numerator is greater than or equal to the denominator, the fraction is an **improper fraction.** An improper fraction that has a value greater than 1 is often written as a **mixed number.** For example,

$$\underset{\text{Improper fraction}}{\underset{\uparrow}{\frac{12}{5}}} \quad \text{may be written as} \quad \underset{\text{Mixed number}}{\underset{\uparrow}{2\frac{2}{5}}}.$$

OBJECTIVE 1 Identify prime numbers.   In work with fractions, we will need to write the numerators and denominators as products. A **product** is the answer to a multiplication problem. When 12 is written as the product $2 \cdot 6$, for example, 2 and 6 are called **factors** of 12. Other factors of 12 are 1, 3, 4, and 12. A whole number is **prime** if it has exactly two different factors (itself and 1). The first dozen primes are listed here.

$$2, 3, 5, 7, 11, 13, 17, 19, 23, 29, 31, 37 \qquad \text{Prime numbers}$$

A whole number greater than 1 that is not prime is called a **composite number.** Some examples follow.

$$4, 6, 8, 9, 10, 12 \qquad \text{Composite numbers}$$

The number 1 is neither prime nor composite.

1 Tell whether each number is *prime* or *composite*.

(a) 12

(b) 13

(c) 27

(d) 59

(e) 1806

2 Write each number in prime factored form.

(a) 70

(b) 72

(c) 693

(d) 97

EXAMPLE 1 Distinguishing between Prime and Composite Numbers

Decide whether each number is *prime* or *composite*.

(a) 33
33 has factors of 3 and 11 as well as 1 and 33, so it is composite.

(b) 43
Since there are no numbers other than 1 and 43 itself that divide *evenly* into 43, the number 43 is prime.

(c) 9832
9832 can be divided by 2, giving 2 · 4916, so it is composite.

◀◀◀ Work Problem 1 at the Side.

OBJECTIVE 2 Write numbers in prime factored form. To factor a number means to write it as the product of two or more numbers. Factoring is the reverse of multiplying two numbers to get the product.

Multiplication	Factoring
$6 \cdot 3 = 18$	$18 = 6 \cdot 3$
↑ ↑ ↑	↑ ↑ ↑
Factors Product	Product Factors

In algebra, a dot · is used instead of the × symbol to indicate multiplication because × may be confused with the letter *x*. A composite number written using factors that are all prime numbers is in **prime factored form.**

EXAMPLE 2 Writing Numbers in Prime Factored Form

Write each number in prime factored form.

(a) 35
Factor 35 as the product of the prime factors 5 and 7, or as $35 = 5 \cdot 7$.

(b) 24
We use a factor tree, as shown below. The prime factors are circled.

Divide by the smallest prime, 2, to get $24 = 2 \cdot 12$.

Now divide 12 by 2 to find factors of 12. $24 = 2 \cdot 2 \cdot 6$

Since 6 can be factored as $2 \cdot 3$, $24 = 2 \cdot 2 \cdot 2 \cdot 3$, where all factors are prime.

◀◀◀ Work Problem 2 at the Side.

OBJECTIVE 3 Write fractions in lowest terms. A fraction is in **lowest terms** when the numerator and denominator have no factors in common (other than 1). The following properties are useful.

Properties of 1

Any nonzero number divided by itself is equal to 1; for example, $\frac{3}{3} = 1$.

Any number multiplied by 1 remains the same; for example, $7 \cdot 1 = 7$.

ANSWERS
1. (a) composite (b) prime (c) composite
   (d) prime (e) composite
2. (a) $2 \cdot 5 \cdot 7$ (b) $2 \cdot 2 \cdot 2 \cdot 3 \cdot 3$
   (c) $3 \cdot 3 \cdot 7 \cdot 11$ (d) 97 is prime.

Writing a Fraction in Lowest Terms

Step 1 Write the numerator and denominator in prime factored form.

Step 2 Replace each pair of factors common to the numerator and denominator with 1.

Step 3 Multiply the remaining factors in the numerator and in the denominator.

(This procedure is sometimes called "simplifying the fraction.")

EXAMPLE 3 **Writing Fractions in Lowest Terms**

Write each fraction in lowest terms.

(a) $\dfrac{10}{15} = \dfrac{2 \cdot 5}{3 \cdot 5} = \dfrac{2}{3} \cdot \dfrac{5}{5} = \dfrac{2}{3} \cdot 1 = \dfrac{2}{3}$

Since 5 is a common factor of 10 and 15, we use the first property of 1 to replace $\frac{5}{5}$ with 1.

(b) $\dfrac{15}{45} = \dfrac{3 \cdot 5}{3 \cdot 3 \cdot 5} = \dfrac{1 \cdot 3 \cdot 5}{3 \cdot 3 \cdot 5} = \dfrac{1}{3} \cdot \dfrac{3}{3} \cdot \dfrac{5}{5} = \dfrac{1}{3} \cdot 1 \cdot 1 = \dfrac{1}{3}$

Multiplying by 1 in the numerator does not change the value of the numerator and makes it possible to rewrite the expression as the product of three fractions in the next step.

(c) $\dfrac{150}{200}$

It is not always necessary to factor into *prime* factors in Step 1. Here, if you see that 50 is a common factor of the numerator and the denominator, factor as follows:

$$\frac{150}{200} = \frac{3 \cdot 50}{4 \cdot 50} = \frac{3}{4} \cdot 1 = \frac{3}{4}.$$

NOTE

When writing a fraction in lowest terms, look for the largest common factor in the numerator and the denominator. If none is obvious, factor the numerator and the denominator into prime factors. *Any* common factor can be used and the fraction can be simplified in stages.

For example, $\dfrac{150}{200} = \dfrac{15 \cdot 10}{20 \cdot 10} = \dfrac{3 \cdot 5 \cdot 10}{4 \cdot 5 \cdot 10} = \dfrac{3}{4}.$

Work Problem 3 at the Side. ▶▶▶

OBJECTIVE 4 **Multiply and divide fractions.**

Multiplying Fractions

To multiply two fractions, multiply the numerators to get the numerator of the product, and multiply the denominators to get the denominator of the product. The product must be written in lowest terms.

3 Write each fraction in lowest terms.

(a) $\dfrac{8}{14}$

(b) $\dfrac{35}{42}$

(c) $\dfrac{120}{72}$

ANSWERS

3. (a) $\dfrac{4}{7}$ (b) $\dfrac{5}{6}$ (c) $\dfrac{5}{3}$

4 Find each product, and write it in lowest terms.

(a) $\dfrac{5}{8} \cdot \dfrac{2}{10}$

(b) $\dfrac{1}{10} \cdot \dfrac{12}{5}$

(c) $\dfrac{7}{9} \cdot \dfrac{12}{14}$

(d) $3\dfrac{1}{3} \cdot 1\dfrac{3}{4}$

EXAMPLE 4 Multiplying Fractions

Find each product, and write it in lowest terms.

(a) $\dfrac{3}{8} \cdot \dfrac{4}{9} = \dfrac{3 \cdot 4}{8 \cdot 9}$ — Multiply numerators. Multiply denominators.

$= \dfrac{3 \cdot 4}{2 \cdot 4 \cdot 3 \cdot 3}$ — Factor.

$= \dfrac{1}{2 \cdot 3} = \dfrac{1}{6}$ — Write in lowest terms.

(b) $2\dfrac{1}{3} \cdot 5\dfrac{1}{2} = \dfrac{7}{3} \cdot \dfrac{11}{2}$ — Write as improper fractions.

$= \dfrac{77}{6}$ or $12\dfrac{5}{6}$ — Multiply numerators and denominators; write as a mixed number.

Work Problem 4 at the Side.

Two fractions are **reciprocals** of each other if their product is 1. For example, $\dfrac{3}{4}$ and $\dfrac{4}{3}$ are reciprocals because

$$\dfrac{3}{4} \cdot \dfrac{4}{3} = 1.$$

The numbers $\dfrac{7}{11}$ and $\dfrac{11}{7}$ are reciprocals also. Other examples are $\dfrac{1}{5}$ and 5, $\dfrac{4}{9}$ and $\dfrac{9}{4}$, and 16 and $\dfrac{1}{16}$.

Because division is the opposite or inverse of multiplication, we use reciprocals to divide fractions.

Dividing Fractions

To divide two fractions, multiply the first fraction by the reciprocal of the second. The result, called the **quotient,** must be written in lowest terms.

The reason this method works will be explained in **Section 1.6.** However, as an example, we know that $20 \div 10 = 2$, and $20 \cdot \dfrac{1}{10} = 2$.

EXAMPLE 5 Dividing Fractions

Find each quotient, and write it in lowest terms.

(a) $\dfrac{3}{4} \div \dfrac{8}{5} = \dfrac{3}{4} \cdot \dfrac{5}{8} = \dfrac{3 \cdot 5}{4 \cdot 8} = \dfrac{15}{32}$

Multiply by the reciprocal of the second fraction.

(b) $\dfrac{3}{4} \div \dfrac{5}{8} = \dfrac{3}{4} \cdot \dfrac{8}{5} = \dfrac{3 \cdot 8}{4 \cdot 5} = \dfrac{3 \cdot 4 \cdot 2}{4 \cdot 5} = \dfrac{6}{5}$ or $1\dfrac{1}{5}$

(c) $\dfrac{5}{8} \div 10 = \dfrac{5}{8} \div \dfrac{10}{1} = \dfrac{5}{8} \cdot \dfrac{1}{10} = \dfrac{5 \cdot 1}{8 \cdot 10} = \dfrac{5 \cdot 1}{8 \cdot 2 \cdot 5} = \dfrac{1}{16}$

Write 10 as $\dfrac{10}{1}$.

Continued on Next Page

ANSWERS

4. (a) $\dfrac{1}{8}$ (b) $\dfrac{6}{25}$ (c) $\dfrac{2}{3}$ (d) $\dfrac{35}{6}$ or $5\dfrac{5}{6}$

(d) $1\frac{2}{3} \div 4\frac{1}{2} = \frac{5}{3} \div \frac{9}{2}$ Write as improper fractions.

$= \frac{5}{3} \cdot \frac{2}{9}$ Multiply by the reciprocal of the second fraction.

$= \frac{10}{27}$ Multiply numerators and denominators.

CAUTION
Notice that *only* the second fraction (the divisor) is replaced by its reciprocal in the multiplication.

Work Problem 5 at the Side. ▶▶▶

OBJECTIVE 5 Add and subtract fractions.   The result of adding two numbers is called the **sum** of the numbers. For example, since $2 + 3 = 5$, the sum of 2 and 3 is 5.

Adding Fractions
To find the sum of two fractions with the *same* denominator, add their numerators and keep the *same* denominator.

EXAMPLE 6 Adding Fractions with the Same Denominator

Add. Write sums in lowest terms.

(a) $\frac{3}{7} + \frac{2}{7} = \frac{3 + 2}{7} = \frac{5}{7}$ Add numerators; denominator does not change.

(b) $\frac{2}{10} + \frac{3}{10} = \frac{2 + 3}{10} = \frac{5}{10} = \frac{1}{2}$ Write in lowest terms.

Work Problem 6 at the Side. ▶▶▶

If the fractions to be added do not have the same denominator, the procedure above can still be used, but only *after* the fractions are rewritten with a common denominator. For example, to rewrite $\frac{3}{4}$ as a fraction with a denominator of 32,

$$\frac{3}{4} = \frac{?}{32},$$

we must find the number that can be multiplied by 4 to give 32. Since $4 \cdot 8 = 32$, we use the number 8. By the second property of 1, we can multiply the numerator and the denominator by 8.

$$\frac{3}{4} = \frac{3}{4} \cdot 1 = \frac{3}{4} \cdot \frac{8}{8} = \frac{3 \cdot 8}{4 \cdot 8} = \frac{24}{32}$$

5 Find each quotient, and write it in lowest terms.

(a) $\frac{3}{10} \div \frac{2}{7}$

(b) $\frac{3}{4} \div \frac{7}{16}$

(c) $\frac{4}{3} \div 6$

(d) $3\frac{1}{4} \div 1\frac{2}{5}$

6 Add. Write sums in lowest terms.

(a) $\frac{3}{5} + \frac{4}{5}$

(b) $\frac{5}{14} + \frac{3}{14}$

ANSWERS

5. **(a)** $\frac{21}{20}$ or $1\frac{1}{20}$ **(b)** $\frac{12}{7}$ or $1\frac{5}{7}$

(c) $\frac{2}{9}$ **(d)** $\frac{65}{28}$ or $2\frac{9}{28}$

6. **(a)** $\frac{7}{5}$ or $1\frac{2}{5}$ **(b)** $\frac{4}{7}$

7 Add. Write sums in lowest terms.

(a) $\dfrac{7}{30} + \dfrac{2}{45}$

(b) $\dfrac{17}{10} + \dfrac{8}{27}$

(c) $2\dfrac{1}{8} + 1\dfrac{2}{3}$

(d) $132\dfrac{4}{5} + 28\dfrac{3}{4}$

Finding the Least Common Denominator (LCD)

Step 1 Factor all denominators to prime factored form.

Step 2 The LCD is the product of every (different) factor that appears in any of the factored denominators. If a factor is repeated, use the largest number of repeats as factors of the LCD.

Step 3 Write each fraction with the LCD as the denominator, using the second property of 1.

EXAMPLE 7 **Adding Fractions with Different Denominators**

Add. Write sums in lowest terms.

(a) $\dfrac{4}{15} + \dfrac{5}{9}$

Step 1 To find the LCD, factor the denominators to prime factored form.

$$15 = 5 \cdot 3 \quad \text{and} \quad 9 = 3 \cdot 3$$

3 is a factor of both denominators.

Step 2 $\qquad \qquad \text{LCD} = 5 \cdot 3 \cdot 3 = 45$

In this example, the LCD needs one factor of 5 and two factors of 3 because the second denominator has two factors of 3.

Step 3 Now we can use the second property of 1 to write each fraction with 45 as the denominator.

$$\dfrac{4}{15} = \dfrac{4}{15} \cdot \dfrac{3}{3} = \dfrac{12}{45} \quad \text{and} \quad \dfrac{5}{9} = \dfrac{5}{9} \cdot \dfrac{5}{5} = \dfrac{25}{45}$$

Now add the two equivalent fractions to get the required sum.

$$\dfrac{4}{15} + \dfrac{5}{9} = \dfrac{12}{45} + \dfrac{25}{45} = \dfrac{37}{45}$$

(b) $3\dfrac{1}{2} + 2\dfrac{3}{4} = \dfrac{7}{2} + \dfrac{11}{4}$ Change to improper fractions.

$\qquad \qquad = \dfrac{14}{4} + \dfrac{11}{4}$ Get a common denominator.

$\qquad \qquad = \dfrac{25}{4} \quad \text{or} \quad 6\dfrac{1}{4}$ Add; write as a mixed number.

(c) $45\dfrac{2}{3} + 73\dfrac{1}{2}$

We use a vertical method here.

$$45\dfrac{2}{3} = 45\dfrac{4}{6}$$
$$+ 73\dfrac{1}{2} = 73\dfrac{3}{6}$$

Add the whole numbers and the fractions separately.

$$118\dfrac{7}{6} = 118 + \left(1 + \dfrac{1}{6}\right) = 119\dfrac{1}{6}$$

ANSWERS

7. (a) $\dfrac{5}{18}$ (b) $\dfrac{539}{270}$ or $1\dfrac{269}{270}$

(c) $\dfrac{91}{24}$ or $3\dfrac{19}{24}$ (d) $161\dfrac{11}{20}$

◀◀◀ Work Problem 7 at the Side.

The **difference** between two numbers is found by subtracting the numbers. For example, $9 - 5 = 4$, so the difference between 9 and 5 is 4. We find the difference between two fractions as follows.

Subtracting Fractions

To find the difference between two fractions with the *same* denominator, subtract their numerators and keep the *same* denominator.

If the fractions have *different* denominators, write them with a common denominator first.

EXAMPLE 8 **Subtracting Fractions**

Subtract. Write differences in lowest terms.

(a) $\dfrac{15}{8} - \dfrac{3}{8} = \dfrac{15 - 3}{8}$ Subtract numerators; denominator does not change.

$= \dfrac{12}{8} = \dfrac{3}{2}$ Lowest terms

(b) $\dfrac{15}{16} - \dfrac{4}{9}$

Since $16 = 2 \cdot 2 \cdot 2 \cdot 2$ and $9 = 3 \cdot 3$ have no common factors, the LCD is $16 \cdot 9 = 144$.

$\dfrac{15}{16} - \dfrac{4}{9} = \dfrac{15 \cdot 9}{16 \cdot 9} - \dfrac{4 \cdot 16}{9 \cdot 16}$ Get a common denominator.

$= \dfrac{135}{144} - \dfrac{64}{144}$

$= \dfrac{71}{144}$ Subtract numerators; keep the same denominator.

(c) $2\dfrac{1}{2} - 1\dfrac{3}{4} = \dfrac{5}{2} - \dfrac{7}{4}$ Write as improper fractions.

$= \dfrac{10}{4} - \dfrac{7}{4}$ Get a common denominator.

$= \dfrac{3}{4}$ Subtract.

Alternatively, we could use a vertical method.

$$2\dfrac{1}{2} = 2\dfrac{2}{4} = 1\dfrac{6}{4}$$
$$- 1\dfrac{3}{4} = 1\dfrac{3}{4} = 1\dfrac{3}{4}$$
$$\overline{\phantom{-1\dfrac{3}{4} = 1\dfrac{3}{4} = 1}\dfrac{3}{4}}$$

Work Problem 8 at the Side. ⟩⟩⟩

We often see mixed numbers used in applications of mathematics, as shown in Examples 9 and 10 on the next page.

8 Subtract.

(a) $\dfrac{9}{11} - \dfrac{3}{11}$

(b) $\dfrac{13}{15} - \dfrac{5}{6}$

(c) $2\dfrac{3}{8} - 1\dfrac{1}{2}$

(d) $50\dfrac{1}{4} - 32\dfrac{2}{3}$

ANSWERS

8. (a) $\dfrac{6}{11}$ (b) $\dfrac{1}{30}$ (c) $\dfrac{7}{8}$ (d) $17\dfrac{7}{12}$

9 Solve the problem.

To make a three-piece outfit from the same fabric, Wei Jen needs $1\frac{1}{4}$ yd for the blouse, $1\frac{2}{3}$ yd for the skirt, and $2\frac{1}{2}$ yd for the jacket. How much fabric does she need?

10 Solve the problem.

A gallon of paint covers 500 ft^2. (ft^2 means square feet.) To paint his house, Tram needs enough paint to cover 4200 ft^2. How many gallons of paint should he buy?

ANSWERS

9. $5\frac{5}{12}$ yd

10. $8\frac{2}{5}$ gal are needed, so he must buy 9 gal.

EXAMPLE 9   **Solving an Applied Problem Requiring Addition of Fractions**

The diagram in Figure 1 appears in the book *Woodworker's 39 Sure-Fire Projects.* It is a view of a bookcase/desk. Add the fractions in the diagram to find the height of the bookcase/desk to the top of the writing surface.

We must add the following measures ($''$ means inches):

$$\frac{3}{4}, \quad 4\frac{1}{2}, \quad 9\frac{1}{2}, \quad \frac{3}{4}, \quad 9\frac{1}{2}, \quad \frac{3}{4}, \quad 4\frac{1}{2}.$$

We begin by changing $4\frac{1}{2}$ to $4\frac{2}{4}$ and $9\frac{1}{2}$ to $9\frac{2}{4}$, since the common denominator is 4. Then we use the method of Example 7(c).

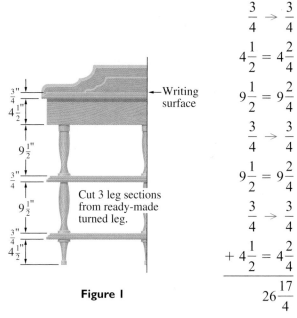

Figure 1

$$\begin{aligned}
\frac{3}{4} &\rightarrow \frac{3}{4} \\
4\frac{1}{2} &= 4\frac{2}{4} \\
9\frac{1}{2} &= 9\frac{2}{4} \\
\frac{3}{4} &\rightarrow \frac{3}{4} \\
9\frac{1}{2} &= 9\frac{2}{4} \\
\frac{3}{4} &\rightarrow \frac{3}{4} \\
+ 4\frac{1}{2} &= 4\frac{2}{4} \\
\hline
& \quad 26\frac{17}{4}
\end{aligned}$$

Since $\frac{17}{4} = 4\frac{1}{4}$, $26\frac{17}{4} = 26 + 4\frac{1}{4} = 30\frac{1}{4}$. The height is $30\frac{1}{4}$ in.

Work Problem 9 at the Side.

EXAMPLE 10   **Solving an Applied Problem Requiring Division of Fractions**

An upholsterer needs $2\frac{1}{4}$ yd of fabric to cover a chair. How many chairs can be covered with $23\frac{2}{3}$ yd of fabric?

To better understand the problem, we replace the fractions with whole numbers. Suppose each chair requires 2 yd, and we have 24 yd of fabric. Dividing 24 by 2 gives the number of chairs (12) that can be covered. To solve the original problem, we must divide $23\frac{2}{3}$ by $2\frac{1}{4}$.

$$23\frac{2}{3} \div 2\frac{1}{4} = \frac{71}{3} \div \frac{9}{4}$$

$$= \frac{71}{3} \cdot \frac{4}{9}$$

$$= \frac{284}{27} \quad \text{or} \quad 10\frac{14}{27}$$

Thus, 10 chairs can be covered with some fabric left over.

Work Problem 10 at the Side.

Appendix B Exercises

FOR EXTRA HELP

 Addison-Wesley Math Tutor Center

Math XL MathXL

Student's Solutions Manual

MyMathLab MyMathLab

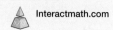 Interactmath.com

Decide whether each statement is true *or* false. *If it is* false, *say why.*

1. In the fraction $\frac{3}{7}$, 3 is the numerator and 7 is the denominator.

2. The mixed number equivalent of $\frac{41}{5}$ is $8\frac{1}{5}$.

3. The fraction $\frac{17}{51}$ is in lowest terms.

4. The reciprocal of $\frac{8}{2}$ is $\frac{4}{1}$.

5. The product of 8 and 2 is 10.

6. The difference between 12 and 2 is 6.

Identify each number as prime, composite, *or* neither. *See Example 1.*

7. 19

8. 99

9. 52

10. 61

11. 2468

12. 3125

13. 1

14. 83

Write each number in prime factored form. See Example 2.

15. 30

16. 40

17. 252

18. 168

19. 124

20. 165

21. 29

22. 31

Write each fraction in lowest terms. See Example 3.

23. $\frac{8}{16}$

24. $\frac{4}{12}$

25. $\frac{15}{18}$

26. $\frac{16}{20}$

27. $\frac{15}{75}$

28. $\frac{24}{64}$

29. $\frac{144}{120}$

30. $\frac{132}{77}$

31. For the fractions $\frac{p}{q}$ and $\frac{r}{s}$, which can serve as a common denominator?

    A. $q \cdot s$

    B. $q + s$

    C. $p \cdot r$

    D. $p + r$

32. Which is the correct way to write $\frac{16}{24}$ in lowest terms?

    A. $\dfrac{16}{24} = \dfrac{8 + 8}{8 + 16} = \dfrac{8}{16} = \dfrac{1}{2}$

    B. $\dfrac{16}{24} = \dfrac{4 \cdot 4}{4 \cdot 6} = \dfrac{4}{6}$

    C. $\dfrac{16}{24} = \dfrac{8 \cdot 2}{8 \cdot 3} = \dfrac{2}{3}$

    D. $\dfrac{16}{24} = \dfrac{14 + 2}{21 + 3} = \dfrac{2}{3} + \dfrac{2}{3} = \dfrac{4}{3}$

Find each product or quotient, and write it in lowest terms. See Examples 4 and 5.

33. $\dfrac{4}{5} \cdot \dfrac{6}{7}$

34. $\dfrac{5}{9} \cdot \dfrac{10}{7}$

35. $\dfrac{1}{10} \cdot \dfrac{12}{5}$

36. $\dfrac{6}{11} \cdot \dfrac{2}{3}$

37. $\dfrac{15}{4} \cdot \dfrac{8}{25}$

38. $\dfrac{4}{7} \cdot \dfrac{21}{8}$

39. $2\dfrac{2}{3} \cdot 5\dfrac{4}{5}$

40. $3\dfrac{3}{5} \cdot 7\dfrac{1}{6}$

41. $\dfrac{5}{4} \div \dfrac{3}{8}$

42. $\dfrac{7}{6} \div \dfrac{9}{10}$

43. $\dfrac{32}{5} \div \dfrac{8}{15}$

44. $\dfrac{24}{7} \div \dfrac{6}{21}$

45. $\dfrac{3}{4} \div 12$

46. $\dfrac{2}{5} \div 30$

47. $2\dfrac{5}{8} \div 1\dfrac{15}{32}$

48. $2\dfrac{3}{10} \div 7\dfrac{4}{5}$

49. In your own words, explain how to divide two fractions.

50. In your own words, explain how to add two fractions that have different denominators.

Find each sum or difference, and write it in lowest terms. See Examples 6–8.

51. $\dfrac{7}{12} + \dfrac{1}{12}$ **52.** $\dfrac{3}{16} + \dfrac{5}{16}$ **53.** $\dfrac{5}{9} + \dfrac{1}{3}$ **54.** $\dfrac{4}{15} + \dfrac{1}{5}$

55. $3\dfrac{1}{8} + \dfrac{1}{4}$ **56.** $5\dfrac{3}{4} + \dfrac{2}{3}$ **57.** $\dfrac{7}{12} - \dfrac{1}{9}$ **58.** $\dfrac{11}{16} - \dfrac{1}{12}$

59. $6\dfrac{1}{4} - 5\dfrac{1}{3}$ **60.** $8\dfrac{4}{5} - 7\dfrac{4}{9}$ **61.** $\dfrac{5}{3} + \dfrac{1}{6} - \dfrac{1}{2}$ **62.** $\dfrac{7}{15} + \dfrac{1}{6} - \dfrac{1}{10}$

Use the chart, which appears on a package of Quaker Quick Grits, to answer the questions in Exercises 63 and 64.

63. How many cups of water would be needed for eight microwave servings?

64. How many teaspoons of salt would be needed for five stove top servings? (*Hint:* 5 is halfway between 4 and 6.)

	Microwave		Stove Top		
Servings	1	1	4	6	
Water	$\dfrac{3}{4}$ cup	1 cup	3 cups	4 cups	
Grits	3 Tbsp	3 Tbsp	$\dfrac{3}{4}$ cup	1 cup	
Salt (optional)	Dash	Dash	$\dfrac{1}{4}$ tsp	$\dfrac{1}{2}$ tsp	

Solve each applied problem. See Examples 9 and 10.

65. A motel owner has decided to expand his business by buying a piece of property next to the motel. The property has an irregular shape, with five sides as shown in the figure. Find the total distance around the piece of property. This is called the **perimeter** of the figure.

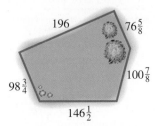

196    $76\dfrac{5}{8}$ $100\dfrac{7}{8}$ $98\dfrac{3}{4}$ $146\dfrac{1}{2}$

Measurements in feet

66. A triangle has sides of lengths $5\dfrac{1}{4}$ ft, $7\dfrac{1}{2}$ ft, and $10\dfrac{1}{8}$ ft . Find the perimeter of the triangle. See Exercise 65.

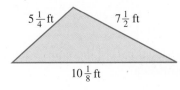

$5\dfrac{1}{4}$ ft $7\dfrac{1}{2}$ ft $10\dfrac{1}{8}$ ft

67. A hardware store sells a 40-piece socket wrench set. The measure of the largest socket is $\frac{3}{4}$ in., while the measure of the smallest socket is $\frac{3}{16}$ in. What is the difference between these measures?

68. Two sockets in a socket wrench set have measures of $\frac{9}{16}$ in. and $\frac{3}{8}$ in. What is the difference between these two measures?

69. Under existing standards, most of the holes in Swiss cheese must have diameters between $\frac{11}{16}$ and $\frac{13}{16}$ in. To accommodate new high-speed slicing machines, the USDA wants to reduce the minimum size to $\frac{3}{8}$ in. How much smaller is $\frac{3}{8}$ in. than $\frac{11}{16}$ in.? (*Source:* U.S. Department of Agriculture.)

70. Tex's favorite recipe for barbecue sauce calls for $2\frac{1}{3}$ cups of tomato sauce. The recipe makes enough barbecue sauce to serve 7 people. How much tomato sauce is needed for 1 serving?

71. It takes $2\frac{3}{8}$ yd of fabric to make a costume for a school play. How much fabric would be needed for 7 costumes?

72. A cake recipe calls for $1\frac{3}{4}$ cups of sugar. A caterer has $15\frac{1}{2}$ cups of sugar on hand. How many cakes can he make?

More than 8 million immigrants were admitted to the United States between 1990 and 1998. The pie chart gives the fractional number from each region of birth for these immigrants. Use the chart to answer the following questions.

73. What fractional part of the immigrants were from other regions?

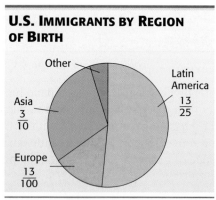

U.S. IMMIGRANTS BY REGION OF BIRTH

Source: U.S. Bureau of the Census.

74. What fractional part of the immigrants were from Latin America or Asia?

75. How many (in millions) were from Europe?

76. How many more immigrants were from Latin America than all of the other regions combined?

Appendix C
Determinants and Cramer's Rule

Appendix C Determinants and Cramer's Rule

Recall from **Section 5.4** that an ordered array of numbers within square brackets is called a *matrix* (plural *matrices*). Matrices are named according to the number of rows and columns they contain. A *square matrix* has the same number of rows as columns.

$$\text{Rows} \begin{bmatrix} 2 & 3 & 5 \\ 7 & 1 & 2 \end{bmatrix} \quad \begin{matrix} 2 \times 3 \\ \text{matrix} \end{matrix} \qquad \begin{bmatrix} -1 & 0 \\ 1 & -2 \end{bmatrix} \quad \begin{matrix} 2 \times 2 \\ \text{square matrix} \end{matrix}$$

Columns

Associated with every *square matrix* is a real number called the **determinant** of the matrix. A determinant is symbolized by the entries of the matrix placed between two vertical lines, such as

$$\begin{vmatrix} 2 & 3 \\ 7 & 1 \end{vmatrix} \quad \begin{matrix} 2 \times 2 \\ \text{determinant} \end{matrix} \qquad \begin{vmatrix} 7 & 4 & 3 \\ 0 & 1 & 5 \\ 6 & 0 & 1 \end{vmatrix}. \quad \begin{matrix} 3 \times 3 \\ \text{determinant} \end{matrix}$$

Like matrices, determinants are named according to the number of rows and columns they contain.

OBJECTIVE 1 Evaluate 2 × 2 determinants. As mentioned above, the value of a determinant is a *real number*. The value of the 2 × 2 determinant

$$\begin{vmatrix} a & b \\ c & d \end{vmatrix}$$

is defined as follows.

Value of a 2 × 2 Determinant

$$\begin{vmatrix} a & b \\ c & d \end{vmatrix} = ad - bc$$

1 Evaluate each determinant.

(a) $\begin{vmatrix} -4 & 6 \\ 2 & 3 \end{vmatrix}$

EXAMPLE 1 Evaluating a 2 × 2 Determinant

Evaluate the determinant.

$$\begin{vmatrix} -1 & -3 \\ 4 & -2 \end{vmatrix}$$

Here $a = -1$, $b = -3$, $c = 4$, and $d = -2$, so

$$\begin{vmatrix} -1 & -3 \\ 4 & -2 \end{vmatrix} = -1(-2) - (-3)4 = 2 + 12 = 14.$$

◀◀◀ **Work Problem 1 at the Side.**

A 3 × 3 determinant can be evaluated in a similar way.

Value of a 3 × 3 Determinant

$$\begin{vmatrix} a_1 & b_1 & c_1 \\ a_2 & b_2 & c_2 \\ a_3 & b_3 & c_3 \end{vmatrix} = (a_1b_2c_3 + b_1c_2a_3 + c_1a_2b_3) - (a_3b_2c_1 + b_3c_2a_1 + c_3a_2b_1)$$

(b) $\begin{vmatrix} 3 & -1 \\ 0 & 2 \end{vmatrix}$

Because this rule for evaluating a 3 × 3 determinant is hard to remember, we use a method based on the rule that is easier to apply. Rearranging terms and using the distributive property gives

$$\begin{vmatrix} a_1 & b_1 & c_1 \\ a_2 & b_2 & c_2 \\ a_3 & b_3 & c_3 \end{vmatrix} = a_1(b_2c_3 - b_3c_2) - a_2(b_1c_3 - b_3c_1) + a_3(b_1c_2 - b_2c_1). \quad (1)$$

Each of the quantities in parentheses represents a 2 × 2 determinant that is the part of the 3 × 3 determinant remaining when the row and column of the multiplier are eliminated, as shown below.

$$a_1(b_2c_3 - b_3c_2) \qquad \begin{vmatrix} a_1 & b_1 & c_1 \\ a_2 & b_2 & c_2 \\ a_3 & b_3 & c_3 \end{vmatrix}$$

$$a_2(b_1c_3 - b_3c_1) \qquad \begin{vmatrix} a_1 & b_1 & c_1 \\ a_2 & b_2 & c_2 \\ a_3 & b_3 & c_3 \end{vmatrix}$$

(c) $\begin{vmatrix} -2 & 5 \\ 1 & 5 \end{vmatrix}$

$$a_3(b_1c_2 - b_2c_1) \qquad \begin{vmatrix} a_1 & b_1 & c_1 \\ a_2 & b_2 & c_2 \\ a_3 & b_3 & c_3 \end{vmatrix}$$

These 2 × 2 determinants are called **minors** of the elements in the 3 × 3 determinant. In the determinant above, the minors of a_1, a_2, and a_3 are, respectively,

$$\begin{vmatrix} b_2 & c_2 \\ b_3 & c_3 \end{vmatrix}, \quad \begin{vmatrix} b_1 & c_1 \\ b_3 & c_3 \end{vmatrix}, \quad \text{and} \quad \begin{vmatrix} b_1 & c_1 \\ b_2 & c_2 \end{vmatrix}. \qquad \text{Minors}$$

OBJECTIVE **2** **Use expansion by minors to evaluate 3 × 3 determinants.** We evaluate a 3 × 3 determinant by multiplying each element in the first column by its minor and combining the products as indicated in equation (1). This procedure is called **expansion of the determinant by minors** about the first column.

ANSWERS
1. (a) -24 (b) 6 (c) -15

EXAMPLE 2 **Evaluating a 3 × 3 Determinant**

Evaluate the determinant using expansion by minors about the first column.

$$\begin{vmatrix} 1 & 3 & -2 \\ -1 & -2 & -3 \\ 1 & 1 & 2 \end{vmatrix}$$

In this determinant, $a_1 = 1$, $a_2 = -1$, and $a_3 = 1$. Multiply each of these numbers by its minor, and combine the three terms using the definition. Notice that the second term in the definition is *subtracted*.

$$\begin{vmatrix} 1 & 3 & -2 \\ -1 & -2 & -3 \\ 1 & 1 & 2 \end{vmatrix} = 1 \begin{vmatrix} -2 & -3 \\ 1 & 2 \end{vmatrix} - (-1) \begin{vmatrix} 3 & -2 \\ 1 & 2 \end{vmatrix} + 1 \begin{vmatrix} 3 & -2 \\ -2 & -3 \end{vmatrix}$$

$$= 1[-2(2) - (-3)1] + 1[3(2) - (-2)1]$$
$$\quad + 1[3(-3) - (-2)(-2)]$$
$$= 1(-1) + 1(8) + 1(-13)$$
$$= -1 + 8 - 13$$
$$= -6$$

Work Problem 2 at the Side. ▶▶▶

To get equation (1) on the preceding page, we could have rearranged terms in the definition of the determinant and used the distributive property to factor out the three elements of the second or third column or of any of the three rows. Therefore, expanding by minors about any row or any column results in the same value for a 3 × 3 determinant. To determine the correct signs for the terms of other expansions, the following **array of signs** is helpful.

Array of Signs for a 3 × 3 Determinant

$$\begin{matrix} + & - & + \\ - & + & - \\ + & - & + \end{matrix}$$

The signs alternate for each row and column beginning with a + in the first row, first column position. For example, if the expansion is to be about the second column, the first term would have a minus sign associated with it, the second term a plus sign, and the third term a minus sign.

EXAMPLE 3 **Evaluating a 3 × 3 Determinant**

Evaluate the determinant of Example 2 using expansion by minors about the second column.

$$\begin{vmatrix} 1 & 3 & -2 \\ -1 & -2 & -3 \\ 1 & 1 & 2 \end{vmatrix} = -3 \begin{vmatrix} -1 & -3 \\ 1 & 2 \end{vmatrix} + (-2) \begin{vmatrix} 1 & -2 \\ 1 & 2 \end{vmatrix} - 1 \begin{vmatrix} 1 & -2 \\ -1 & -3 \end{vmatrix}$$

$$= -3(1) - 2(4) - 1(-5)$$
$$= -3 - 8 + 5$$
$$= -6 \quad \text{The result is the same as in Example 2.}$$

Work Problem 3 at the Side. ▶▶▶

2 Evaluate each determinant using expansion by minors about the first column.

(a) $\begin{vmatrix} 0 & -1 & 0 \\ 2 & 4 & 2 \\ 3 & 1 & 5 \end{vmatrix}$

(b) $\begin{vmatrix} 2 & 1 & 4 \\ -3 & 0 & 2 \\ -2 & 1 & 5 \end{vmatrix}$

3 Evaluate each determinant using expansion by minors about the second column.

(a) $\begin{vmatrix} 2 & 1 & 3 \\ -1 & 0 & 4 \\ 2 & 4 & 3 \end{vmatrix}$

(b) $\begin{vmatrix} 5 & -1 & 2 \\ 0 & 4 & 3 \\ -1 & 2 & 0 \end{vmatrix}$

ANSWERS
2. (a) 4 (b) −5
3. (a) −33 (b) −19

▦ **Calculator Tip**   The graphing calculator function det(A) assigns to each square matrix A one and only one real number, the determinant of A. For example, Figure 1 shows how a graphing calculator displays the correct value for the determinant in Example 1. Similarly, Figure 2 supports the results of Examples 2 and 3.

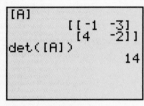

Figure 1   **Figure 2**

O B J E C T I V E   3   Understand the derivation of Cramer's rule.   We can use determinants to solve a system of equations of the form

$$a_1 x + b_1 y = c_1 \quad (1)$$
$$a_2 x + b_2 y = c_2. \quad (2)$$

The result will be a formula that can be used to solve any system of two equations with two variables. To get this general solution, we eliminate y and solve for x by first multiplying each side of equation (1) by b_2 and each side of equation (2) by $-b_1$. Then we add these results and solve for x.

$$a_1 b_2 x + b_1 b_2 y = c_1 b_2 \qquad \text{Multiply equation (1) by } b_2.$$
$$\underline{-a_2 b_1 x - b_1 b_2 y = -c_2 b_1} \qquad \text{Multiply equation (2) by } -b_1.$$
$$(a_1 b_2 - a_2 b_1)x = c_1 b_2 - c_2 b_1$$
$$x = \frac{c_1 b_2 - c_2 b_1}{a_1 b_2 - a_2 b_1} \qquad (\text{if } a_1 b_2 - a_2 b_1 \neq 0)$$

To solve for y, we multiply each side of equation (1) by $-a_2$ and each side of equation (2) by a_1 and add.

$$-a_1 a_2 x - a_2 b_1 y = -a_2 c_1 \qquad \text{Multiply equation (1) by } -a_2.$$
$$\underline{a_1 a_2 x + a_1 b_2 y = a_1 c_2} \qquad \text{Multiply equation (2) by } a_1.$$
$$(a_1 b_2 - a_2 b_1)y = a_1 c_2 - a_2 c_1$$
$$y = \frac{a_1 c_2 - a_2 c_1}{a_1 b_2 - a_2 b_1} \qquad (\text{if } a_1 b_2 - a_2 b_1 \neq 0)$$

We can write both numerators and the common denominator of these values for x and y as determinants because

$$a_1 c_2 - a_2 c_1 = \begin{vmatrix} a_1 & c_1 \\ a_2 & c_2 \end{vmatrix},$$

$$c_1 b_2 - c_2 b_1 = \begin{vmatrix} c_1 & b_1 \\ c_2 & b_2 \end{vmatrix},$$

and $$a_1 b_2 - a_2 b_1 = \begin{vmatrix} a_1 & b_1 \\ a_2 & b_2 \end{vmatrix}.$$

Using these results, the solutions for x and y become

$$x = \frac{\begin{vmatrix} c_1 & b_1 \\ c_2 & b_2 \end{vmatrix}}{\begin{vmatrix} a_1 & b_1 \\ a_2 & b_2 \end{vmatrix}} \quad \text{and} \quad y = \frac{\begin{vmatrix} a_1 & c_1 \\ a_2 & c_2 \end{vmatrix}}{\begin{vmatrix} a_1 & b_1 \\ a_2 & b_2 \end{vmatrix}}, \quad \begin{vmatrix} a_1 & b_1 \\ a_2 & b_2 \end{vmatrix} \neq 0.$$

For convenience, we denote the three determinants in the solution as

$$\begin{vmatrix} a_1 & b_1 \\ a_2 & b_2 \end{vmatrix} = D, \quad \begin{vmatrix} c_1 & b_1 \\ c_2 & b_2 \end{vmatrix} = D_x, \quad \text{and} \quad \begin{vmatrix} a_1 & c_1 \\ a_2 & c_2 \end{vmatrix} = D_y.$$

Notice that the elements of D are the four coefficients of the variables in the given system; the elements of D_x are obtained by replacing the coefficients of x by the respective constants; the elements of D_y are obtained by replacing the coefficients of y by the respective constants.

These results are summarized as **Cramer's rule.**

Cramer's Rule for 2 × 2 Systems

Given the system

$$a_1 x + b_1 y = c_1$$
$$a_2 x + b_2 y = c_2 \quad \text{with} \quad a_1 b_2 - a_2 b_1 = D \neq 0,$$

then

$$x = \frac{\begin{vmatrix} c_1 & b_1 \\ c_2 & b_2 \end{vmatrix}}{\begin{vmatrix} a_1 & b_1 \\ a_2 & b_2 \end{vmatrix}} = \frac{D_x}{D} \quad \text{and} \quad y = \frac{\begin{vmatrix} a_1 & c_1 \\ a_2 & c_2 \end{vmatrix}}{\begin{vmatrix} a_1 & b_1 \\ a_2 & b_2 \end{vmatrix}} = \frac{D_y}{D}.$$

NOTE

Swiss mathematician and physicist Gabriel Cramer (1704–1752) was looking for a method to determine the equation of a curve given several points on the curve. In 1750, he wrote down the general equation for a curve and then substituted each point for which he had two coordinates into the equation. For the resulting system of equations, he gave "a rule very convenient and general to solve any number of equations and unknowns which are of no more than first degree." This is the rule that now bears his name.

OBJECTIVE 4 Apply Cramer's rule to solve linear systems.   To use Cramer's rule to solve a system of equations, find the three determinants, D, D_x, and D_y, and then write the necessary quotients for x and y.

CAUTION

As indicated in the box, *Cramer's rule does not apply if $D = a_1 b_2 - a_2 b_1$ is 0.* When $D = 0$, the system is inconsistent or has dependent equations. For this reason, it is a good idea to evaluate D first.

4 Solve each system using Cramer's rule.

(a) $x + y = 5$
   $x - y = 1$

EXAMPLE 4 **Using Cramer's Rule to Solve a 2 × 2 System**

Use Cramer's rule to solve the system.

$$5x + 7y = -1$$
$$6x + 8y = 1$$

By Cramer's rule, $x = \dfrac{D_x}{D}$ and $y = \dfrac{D_y}{D}$. As previously mentioned, it is a good idea to find D first since if $D = 0$, Cramer's rule does not apply. If $D \neq 0$, then find D_x and D_y.

$$D = \begin{vmatrix} 5 & 7 \\ 6 & 8 \end{vmatrix} = 5(8) - 7(6) = -2$$

$$D_x = \begin{vmatrix} -1 & 7 \\ 1 & 8 \end{vmatrix} = -1(8) - 7(1) = -15$$

$$D_y = \begin{vmatrix} 5 & -1 \\ 6 & 1 \end{vmatrix} = 5(1) - (-1)6 = 11$$

From Cramer's rule,

$$x = \frac{D_x}{D} = \frac{-15}{-2} = \frac{15}{2} \quad \text{and} \quad y = \frac{D_y}{D} = \frac{11}{-2} = -\frac{11}{2}.$$

(b) $2x - 3y = -26$
   $3x + 4y = 12$

The solution set is $\{(\frac{15}{2}, -\frac{11}{2})\}$, as we can verify by checking in the given system.

◀◀◀ Work Problem 4 at the Side.

In a similar manner, we can apply Cramer's rule to systems of three equations with three variables.

Cramer's Rule for 3 × 3 Systems

Given the system

$$a_1x + b_1y + c_1z = d_1$$
$$a_2x + b_2y + c_2z = d_2$$
$$a_3x + b_3y + c_3z = d_3$$

(c) $4x - 5y = -8$
   $3x + 7y = -6$

with

$$D_x = \begin{vmatrix} d_1 & b_1 & c_1 \\ d_2 & b_2 & c_2 \\ d_3 & b_3 & c_3 \end{vmatrix}, \quad D_y = \begin{vmatrix} a_1 & d_1 & c_1 \\ a_2 & d_2 & c_2 \\ a_3 & d_3 & c_3 \end{vmatrix},$$

$$D_z = \begin{vmatrix} a_1 & b_1 & d_1 \\ a_2 & b_2 & d_2 \\ a_3 & b_3 & d_3 \end{vmatrix}, \quad D = \begin{vmatrix} a_1 & b_1 & c_1 \\ a_2 & b_2 & c_2 \\ a_3 & b_3 & c_3 \end{vmatrix} \neq 0,$$

then

$$x = \frac{D_x}{D}, \quad y = \frac{D_y}{D}, \quad \text{and} \quad z = \frac{D_z}{D}.$$

ANSWERS
4. (a) $\{(3, 2)\}$ (b) $\{(-4, 6)\}$ (c) $\{(-2, 0)\}$

EXAMPLE 5 **Using Cramer's Rule to Solve a 3 × 3 System**

Use Cramer's rule to solve the system.

$$x + y - z + 2 = 0$$
$$2x - y + z + 5 = 0$$
$$x - 2y + 3z - 4 = 0$$

To use Cramer's rule, first rewrite the system in the form

$$x + y - z = -2$$
$$2x - y + z = -5$$
$$x - 2y + 3z = 4.$$

Expand by minors about row 1 to find D.

$$D = \begin{vmatrix} 1 & 1 & -1 \\ 2 & -1 & 1 \\ 1 & -2 & 3 \end{vmatrix}$$

$$= 1 \begin{vmatrix} -1 & 1 \\ -2 & 3 \end{vmatrix} - 1 \begin{vmatrix} 2 & 1 \\ 1 & 3 \end{vmatrix} + (-1) \begin{vmatrix} 2 & -1 \\ 1 & -2 \end{vmatrix}$$

$$= 1(-1) - 1(5) - 1(-3)$$

$$= -3$$

Expanding D_x by minors about row 1 gives

$$D_x = \begin{vmatrix} -2 & 1 & -1 \\ -5 & -1 & 1 \\ 4 & -2 & 3 \end{vmatrix}$$

$$= -2 \begin{vmatrix} -1 & 1 \\ -2 & 3 \end{vmatrix} - 1 \begin{vmatrix} -5 & 1 \\ 4 & 3 \end{vmatrix} + (-1) \begin{vmatrix} -5 & -1 \\ 4 & -2 \end{vmatrix}$$

$$= -2(-1) - 1(-19) - 1(14)$$

$$= 7.$$

Expanding D_y by minors about row 1 gives

$$D_y = \begin{vmatrix} 1 & -2 & -1 \\ 2 & -5 & 1 \\ 1 & 4 & 3 \end{vmatrix}$$

$$= 1 \begin{vmatrix} -5 & 1 \\ 4 & 3 \end{vmatrix} - (-2) \begin{vmatrix} 2 & 1 \\ 1 & 3 \end{vmatrix} + (-1) \begin{vmatrix} 2 & -5 \\ 1 & 4 \end{vmatrix}$$

$$= 1(-19) + 2(5) - 1(13)$$

$$= -22.$$

Work Problem 5 at the Side. ▶▶▶

Using the results for D, D_x, D_y, and the result from Problem 5 at the side for D_z, apply Cramer's rule to get

$$x = \frac{D_x}{D} = \frac{7}{-3} = -\frac{7}{3}, \quad y = \frac{D_y}{D} = \frac{-22}{-3} = \frac{22}{3}, \quad z = \frac{D_z}{D} = \frac{-21}{-3} = 7.$$

Check that the solution set is $\{(-\frac{7}{3}, \frac{22}{3}, 7)\}$.

Work Problem 6 at the Side. ▶▶▶

5 Find D_z for Example 5.

6 Solve each system using Cramer's rule.

(a) $x + y + z = 2$
$2x \quad - z = -3$
$y + 2z = 4$

(b) $3x - 2y + 4z = 5$
$4x + y + z = 14$
$x - y - z = 1$

ANSWERS
5. $D_z = -21$
6. (a) $\{(-1, 2, 1)\}$ (b) $\{(3, 2, 0)\}$

7 Solve by Cramer's rule (if applicable).

$$x - y + z = 6$$
$$3x + 2y + z = 4$$
$$2x - 2y + 2z = 14$$

As mentioned earlier, Cramer's rule does not apply when $D = 0$. The next example illustrates this case.

EXAMPLE 6 Determining When Cramer's Rule Does Not Apply

Use Cramer's rule, if possible, to solve the system.

$$2x - 3y + 4z = 8$$
$$6x - 9y + 12z = 24$$
$$x + 2y - 3z = 5$$

First, find D.

$$D = \begin{vmatrix} 2 & -3 & 4 \\ 6 & -9 & 12 \\ 1 & 2 & -3 \end{vmatrix}$$

$$= 2\begin{vmatrix} -9 & 12 \\ 2 & -3 \end{vmatrix} - 6\begin{vmatrix} -3 & 4 \\ 2 & -3 \end{vmatrix} + 1\begin{vmatrix} -3 & 4 \\ -9 & 12 \end{vmatrix}$$

$$= 2(3) - 6(1) + 1(0)$$

$$= 0$$

Since $D = 0$ here, Cramer's rule does not apply and we must use another method to solve the system. Multiplying each side of the first equation by 3 shows that the first two equations have the same solution set, so this system has dependent equations and an infinite solution set.

Work Problem 7 at the Side.

ANSWERS
7. Cramer's rule does not apply.

Appendix C Exercises

FOR EXTRA HELP Addison-Wesley Math Tutor Center MathXL Student's Solutions Manual MyMathLab Interactmath.com

Decide whether each statement is true *or* false.

1. A matrix is an array of numbers, while a determinant is a single number.

2. A square matrix has the same number of rows as columns.

3. The determinant $\begin{vmatrix} a & b \\ c & d \end{vmatrix}$ is equal to $ad + bc$.

4. The value of $\begin{vmatrix} 0 & 0 \\ x & y \end{vmatrix}$ is 0 for any replacements for x and y.

Evaluate each determinant. See Example 1.

5. $\begin{vmatrix} -2 & 5 \\ -1 & 4 \end{vmatrix}$

6. $\begin{vmatrix} 3 & -6 \\ 2 & -2 \end{vmatrix}$

7. $\begin{vmatrix} 1 & -2 \\ 7 & 0 \end{vmatrix}$

8. $\begin{vmatrix} -5 & -1 \\ 1 & 0 \end{vmatrix}$

9. $\begin{vmatrix} 0 & 4 \\ 0 & 4 \end{vmatrix}$

10. $\begin{vmatrix} 8 & -3 \\ 0 & 0 \end{vmatrix}$

Evaluate each determinant using expansion by minors about the first column. See Example 2.

11. $\begin{vmatrix} -1 & 2 & 4 \\ -3 & -2 & -3 \\ 2 & -1 & 5 \end{vmatrix}$

12. $\begin{vmatrix} 2 & -3 & -5 \\ 1 & 2 & 2 \\ 5 & 3 & -1 \end{vmatrix}$

13. $\begin{vmatrix} 1 & 0 & -2 \\ 0 & 2 & 3 \\ 1 & 0 & 5 \end{vmatrix}$

14. $\begin{vmatrix} 2 & -1 & 0 \\ 0 & -1 & 1 \\ 1 & 2 & 0 \end{vmatrix}$

15. $\begin{vmatrix} 1 & 0 & 0 \\ 0 & 1 & 0 \\ 0 & 0 & 1 \end{vmatrix}$

16. $\begin{vmatrix} 0 & 0 & 1 \\ 0 & 1 & 0 \\ 1 & 0 & 0 \end{vmatrix}$

Evaluate each determinant by expansion about any row or column. (Hint: If possible, choose a row or column with 0s.) See Example 3.

17. $\begin{vmatrix} 4 & 4 & 2 \\ 1 & -1 & -2 \\ 1 & 0 & 2 \end{vmatrix}$

18. $\begin{vmatrix} 3 & -1 & 2 \\ 1 & 5 & -2 \\ 0 & 2 & 0 \end{vmatrix}$

19. $\begin{vmatrix} 2 & 0 & 1 \\ -1 & 0 & 2 \\ 5 & 0 & 4 \end{vmatrix}$

20. $\begin{vmatrix} 2 & -4 & 0 \\ 3 & -5 & 0 \\ 6 & -7 & 0 \end{vmatrix}$

21. $\begin{vmatrix} -6 & 3 & 5 \\ -3 & 2 & 2 \\ 0 & 0 & 0 \end{vmatrix}$

22. $\begin{vmatrix} 0 & 0 & 0 \\ 4 & 0 & -2 \\ 2 & -1 & 3 \end{vmatrix}$

23. $\begin{vmatrix} 3 & 5 & -2 \\ 1 & -4 & 1 \\ 3 & 1 & -2 \end{vmatrix}$

24. $\begin{vmatrix} 1 & 3 & 2 \\ 3 & -1 & -2 \\ 1 & 10 & 20 \end{vmatrix}$

25. For the system

$$8x - 4y = 8$$
$$x + 3y = 22,$$

$D_x = 112, D_y = 168,$ and $D = 28.$ What is the solution set of the system?

26. For the system

$$x + 3y - 6z = 7$$
$$2x - y + z = 1$$
$$x + 2y + 2z = -1,$$

the solution set is $\{(1, 0, -1)\}$ and $D = -43.$ Find the values of $D_x, D_y,$ and $D_z.$

Use Cramer's rule to solve each system. See Example 4.

27. $\quad 3x + 5y = -5$
$\quad -2x + 3y = 16$

28. $5x + 2y = -3$
$\quad 4x - 3y = -30$

29. $3x + 2y = 3$
$\quad 2x - 4y = 2$

30. $7x - 2y = 6$
$\quad 4x - 5y = 15$

31. $8x + 3y = 1$
$\quad 6x - 5y = 2$

32. $3x - y = 9$
$\quad 2x + 5y = 8$

Use Cramer's rule (where applicable) to solve each system. If Cramer's rule does not apply, say so. See Examples 5 and 6.

33. $2x + 3y + 2z = 15$
$\quad x - y + 2z = 5$
$\quad x + 2y - 6z = -26$

34. $\quad x - y + 6z = 19$
$\quad 3x + 3y - z = 1$
$\quad x + 9y + 2z = -19$

35. $2x + 2y + z = 10$
$\quad 4x - y + z = 20$
$\quad -x + y - 2z = -5$

36. $\quad x + 3y - 4z = -12$
$\quad 3x + y - z = -5$
$\quad 5x - y + z = -3$

37. $\quad 2x - 3y + 4z = 8$
$\quad 6x - 9y + 12z = 24$
$\quad -4x + 6y - 8z = -16$

38. $\quad 7x + y - z = 4$
$\quad 2x - 3y + z = 2$
$\quad -6x + 9y - 3z = -6$

39. $3x + 5z = 0$
$\quad 2x + 3y = 1$
$\quad -y + 2z = -11$

40. $-x + 2y = 4$
$\quad 3x + y = -5$
$\quad 2x + z = -1$

41. $\quad x - 3y = 13$
$\quad 2y + z = 5$
$\quad -x + z = -7$

Appendix D
Synthetic Division

Appendix D Synthetic Division

We begin by reviewing the terminology for the parts of a division problem. The *divisor* is the quantity we are dividing by, the *dividend* is the quantity we are dividing into, and the *quotient* is the result of the division.

$$\text{Divisor} \longrightarrow 247 \overline{)385{,}814} \quad \begin{matrix} 1\ 562 & \longleftarrow \text{Quotient} \\[6pt] & \longleftarrow \text{Dividend} \end{matrix}$$

OBJECTIVES

1 Use synthetic division to divide by a polynomial of the form $x - k$.

2 Use the remainder theorem to evaluate a polynomial.

3 Decide whether a given number is a solution of an equation.

OBJECTIVE 1 Use synthetic division to divide by a polynomial of the form $x - k$. When one polynomial is divided by a second, if the divisor has the form $x - k$, where the coefficient of the x-term is 1, there is a shortcut method for performing the division. Look at the division of $3x^3 - 2x + 5$ by $x - 3$ on the left below. Notice that we inserted 0 for the missing x^2-term.

$$
\begin{array}{r}
3x^2 + 9x + 25 \\
x - 3 \overline{)3x^3 + 0x^2 - 2x + 5} \\
\underline{3x^3 - 9x^2} \\
9x^2 - 2x \\
\underline{9x^2 - 27x} \\
25x + 5 \\
\underline{25x - 75} \\
80
\end{array}
\qquad
\begin{array}{r}
3 \quad 9 \quad 25 \\
1 - 3 \overline{)3 \quad 0 \quad -2 \quad 5} \\
\underline{3 \quad -9} \\
9 \quad -2 \\
\underline{9 \quad -27} \\
25 \quad 5 \\
\underline{25 \quad -75} \\
80
\end{array}
$$

On the right, the same division is shown written without the variables. This is why it is *essential* to use 0 as a placeholder in synthetic division. All the numbers in color on the right are repetitions of the numbers directly above them, so they are omitted to condense the work, as shown on the left below.

$$
\begin{array}{r}
3 \quad 9 \quad 25 \\
1 - 3 \overline{)3 \quad 0 \quad -2 \quad 5} \\
-9 \\
9 \quad -2 \\
-27 \\
25 \quad 5 \\
-75 \\
80
\end{array}
\qquad
\begin{array}{r}
3 \quad 9 \quad 25 \\
1 - 3 \overline{)3 \quad 0 \quad -2 \quad 5} \\
-9 \\
9 \\
-27 \\
25 \\
-75 \\
80
\end{array}
$$

The numbers in color on the left are again repetitions of the numbers directly above them; they too are omitted, as shown on the right above. If the 3 in the dividend is brought down to the beginning of the bottom row, the top row can be omitted since it duplicates the bottom row.

1 Divide, using synthetic division.

(a) $\dfrac{3z^2 + 10z - 8}{z + 4}$

$$
1 - 3\overline{)\begin{array}{rrrr} 3 & 0 & -2 & 5 \\ & -9 & -27 & -75 \\ \hline 3 & 9 & 25 & 80 \end{array}}
$$

Now we can condense the problem. We omit the 1 at the upper left, since it represents $1x$, which will *always* be the first term in the divisor. Also, to simplify the arithmetic, we replace subtraction in the second row by addition. We compensate for this by changing the -3 at the upper left to its additive inverse, 3. The result of doing all this is shown below.

Additive inverse of $-3 \longrightarrow$ $3\overline{)\begin{array}{rrrr} 3 & 0 & -2 & 5 \\ & 9 & 27 & 75 \\ \hline 3 & 9 & 25 & 80 \end{array}}$ $\longleftarrow$ Change signs.
$\longleftarrow$ Remainder

The quotient is read from the bottom row.

$$3x^2 + 9x + 25 + \dfrac{80}{x - 3}$$

The first three numbers in the bottom row are the coefficients of the quotient polynomial with degree 1 less than the degree of the dividend. The last number gives the remainder.

Synthetic Division

This shortcut method is called **synthetic division.** It is used *only* when dividing a polynomial by a binomial of the form $x - k$.

(b) $(2x^2 + 3x - 5) \div (x + 1)$

EXAMPLE 1 **Using Synthetic Division**

Use synthetic division to divide $5x^2 + 16x + 15$ by $x + 2$.

As mentioned above, we use synthetic division only when dividing by a polynomial of the form $x - k$. We change $x + 2$ to this form by writing it as

$$x + 2 = x - (-2),$$

where $k = -2$. Then we write the coefficients of $5x^2 + 16x + 15$.

$x + 2$ leads to -2. $\longrightarrow$ $-2\overline{)\begin{array}{rrr} 5 & 16 & 15 \end{array}}$ $\longleftarrow$ Coefficients

We bring down the 5, and multiply: $-2 \cdot 5 = -10$.

$$
-2\overline{)\begin{array}{rrr} 5 & 16 & 15 \\ & -10 & \\ \hline 5 & & \end{array}}
$$

We add 16 and -10, getting 6. Multiply 6 and -2 to get -12.

$$
-2\overline{)\begin{array}{rrr} 5 & 16 & 15 \\ & -10 & -12 \\ \hline 5 & 6 & \end{array}}
$$

We add 15 and -12, getting 3.

$$
-2\overline{)\begin{array}{rrr} 5 & 16 & 15 \\ & -10 & -12 \\ \hline 5 & 6 & 3 \end{array}}
$$ $\longleftarrow$ Remainder

We read the result from the bottom row.

$$\dfrac{5x^2 + 16x + 15}{x + 2} = 5x + 6 + \dfrac{3}{x + 2}$$

Work Problem 1 at the Side.

ANSWERS

1. (a) $3z - 2$ (b) $2x + 1 + \dfrac{-6}{x + 1}$

EXAMPLE 2 Using Synthetic Division with a Missing Term

Use synthetic division to find $(-4x^5 + x^4 + 6x^3 + 2x^2 + 50) \div (x - 2)$.
Use the steps given above, inserting a 0 for the missing x-term.

$$
\begin{array}{r|rrrrrr}
2) & -4 & 1 & 6 & 2 & 0 & 50 \\
   &   & -8 & -14 & -16 & -28 & -56 \\
\hline
   & -4 & -7 & -8 & -14 & -28 & -6
\end{array}
$$

Read the result from the bottom row.

$$\frac{-4x^5 + x^4 + 6x^3 + 2x^2 + 50}{x - 2} = -4x^4 - 7x^3 - 8x^2 - 14x - 28 + \frac{-6}{x - 2}$$

Work Problem 2 at the Side.

OBJECTIVE 2 Use the remainder theorem to evaluate a polynomial.
We can use synthetic division to evaluate polynomials. For example, in the synthetic division of Example 2, where the polynomial was divided by $x - 2$, the remainder was -6.

Replacing x in the polynomial with 2 gives

$$
\begin{aligned}
-4x^5 + x^4 + 6x^3 + 2x^2 + 50 &= -4 \cdot 2^5 + 2^4 + 6 \cdot 2^3 + 2 \cdot 2^2 + 50 \\
&= -4 \cdot 32 + 16 + 6 \cdot 8 + 2 \cdot 4 + 50 \\
&= -128 + 16 + 48 + 8 + 50 \\
&= -6,
\end{aligned}
$$

the same number as the remainder. Thus, dividing by $x - 2$ produced a remainder equal to the result when x is replaced with 2. This always happens, as the following remainder theorem states.

Remainder Theorem
If the polynomial $P(x)$ is divided by $x - k$, then the remainder is equal to $P(k)$.

This result is proved in more advanced courses.

EXAMPLE 3 Using the Remainder Theorem

Let $P(x) = 2x^3 - 5x^2 - 3x + 11$. Find $P(-2)$.
Use the remainder theorem; divide $P(x)$ by $x - (-2)$.

$$
\begin{array}{r|rrrr}
\text{Value of } x \rightarrow -2) & 2 & -5 & -3 & 11 \\
   &   & -4 & 18 & -30 \\
\hline
   & 2 & -9 & 15 & -19 \leftarrow \text{Remainder}
\end{array}
$$

By this result, $P(-2) = -19$.

Work Problem 3 at the Side.

2 Divide, using synthetic division.

(a) $\dfrac{3a^3 - 2a + 21}{a + 2}$

(b) $(-4x^4 + 3x^3 + 18x + 2) \div (x - 2)$

3 Let $P(x) = x^3 - 5x^2 + 7x - 3$. Use synthetic division to find each value.

(a) $P(1)$ (Divide by $x - 1$.)

(b) $P(-2)$

ANSWERS

2. (a) $3a^2 - 6a + 10 + \dfrac{1}{a + 2}$

   (b) $-4x^3 - 5x^2 - 10x - 2 + \dfrac{-2}{x - 2}$

3. (a) 0 (b) -45

4 Use synthetic division to decide whether 2 is a solution of each equation.

(a) $3x^3 - 11x^2 + 17x - 14 = 0$

OBJECTIVE 3 **Decide whether a given number is a solution of an equation.** The remainder theorem can also be used to show that a given number is a solution of an equation.

EXAMPLE 4 **Using the Remainder Theorem**

Show that -5 is a solution of the equation

$$2x^4 + 12x^3 + 6x^2 - 5x + 75 = 0.$$

One way to show that -5 is a solution is to substitute -5 for x in the equation. However, an easier way is to use synthetic division and the remainder theorem.

Proposed solution $\rightarrow -5)\overline{\begin{array}{ccccc} 2 & 12 & 6 & -5 & 75 \\ & -10 & -10 & 20 & -75 \\ \hline 2 & 2 & -4 & 15 & 0 \end{array}} \leftarrow$ Remainder

Since the remainder is 0, the polynomial has a value of 0 when $x = -5$, so -5 is a solution of the given equation.

Work Problem 4 at the Side.

The synthetic division in Example 4 shows that $x - (-5)$ divides the polynomial with 0 remainder. Thus $x - (-5) = x + 5$ is a *factor* of the polynomial and

$$2x^4 + 12x^3 + 6x^2 - 5x + 75 = (x + 5)(2x^3 + 2x^2 - 4x + 15).$$

The second factor is the quotient polynomial found in the last row of the synthetic division.

(b) $4x^5 - 7x^4 - 11x^2 + 2x + 6 = 0$

ANSWERS
4. (a) yes (b) no

Appendix D Exercises

FOR EXTRA HELP · Tutor Center Addison-Wesley Math Tutor Center · MathXL MathXL · Student's Solutions Manual · MyMathLab MyMathLab · Interactmath.com

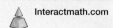

Choose the letter of the correct setup to perform synthetic division on the indicated quotient.

1. $\dfrac{x^2 + 3x - 6}{x - 2}$

   A. $-2\overline{)1 \quad 3 \quad -6}$ **B.** $-2\overline{)-1 \quad -3 \quad 6}$

   C. $2\overline{)1 \quad 3 \quad -6}$ **D.** $2\overline{)-1 \quad -3 \quad 6}$

2. $\dfrac{x^3 - 3x^2 + 2}{x - 1}$

   A. $1\overline{)1 \quad -3 \quad 2}$ **B.** $-1\overline{)1 \quad -3 \quad 2}$

   C. $1\overline{)1 \quad -3 \quad 0 \quad 2}$ **D.** $1\overline{)-1 \quad 3 \quad 0 \quad -2}$

Use synthetic division to find each quotient. See Examples 1 and 2.

3. $\dfrac{x^2 - 6x + 5}{x - 1}$

4. $\dfrac{x^2 - 4x - 21}{x + 3}$

5. $\dfrac{4m^2 + 19m - 5}{m + 5}$

6. $\dfrac{3k^2 - 5k - 12}{k - 3}$

7. $\dfrac{2a^2 + 8a + 13}{a + 2}$

8. $\dfrac{4y^2 - 5y - 20}{y - 4}$

9. $(p^2 - 3p + 5) \div (p + 1)$

10. $(z^2 + 4z - 6) \div (z - 5)$

11. $\dfrac{4a^3 - 3a^2 + 2a - 3}{a - 1}$

12. $\dfrac{5p^3 - 6p^2 + 3p + 14}{p + 1}$

13. $(x^5 - 2x^3 + 3x^2 - 4x - 2) \div (x - 2)$

14. $(2y^5 - 5y^4 - 3y^2 - 6y - 23) \div (y - 3)$

15. $(-4r^6 - 3r^5 - 3r^4 + 5r^3 - 6r^2 + 3r) \div (r - 1)$

16. $(-3t^5 + 2t^4 - 5t^3 + 6t^2 - 3t - 2) \div (t - 2)$

17. $(-3y^5 + 2y^4 - 5y^3 - 6y^2 - 1) \div (y + 2)$

18. $(m^6 + 2m^4 - 5m + 11) \div (m - 2)$

19. $\dfrac{y^3 + 1}{y - 1}$

20. $\dfrac{z^4 + 81}{z - 3}$

Use the remainder theorem to find P(k). See Example 3.

21. $P(x) = 2x^3 - 4x^2 + 5x - 3;\ k = 2$

22. $P(x) = x^3 + 3x^2 - x + 5;\ k = -1$

23. $P(r) = -r^3 - 5r^2 - 4r - 2;\ k = -4$

24. $P(z) = -z^3 + 5z^2 - 3z + 4;\ k = 3$

25. $P(x) = 2x^3 - 4x^2 + 5x - 33;\ k = 3$

26. $P(x) = x^3 - 3x^2 + 4x - 4;\ k = 2$

Use synthetic division to decide whether the given number is a solution of each equation. See Example 4.

27. $x^3 - 2x^2 - 3x + 10 = 0;\ x = -2$

28. $x^3 - 3x^2 - x + 10 = 0;\ x = -2$

29. $m^4 + 2m^3 - 3m^2 + 8m - 8 = 0;\ m = -2$

30. $r^4 - r^3 - 6r^2 + 5r + 10 = 0;\ r = -2$

31. $3x^3 + 2x^2 - 2x + 11 = 0;\ x = -2$

32. $3z^3 + 10z^2 + 3z - 9 = 0;\ z = -2$

33. Explain why it is important to insert 0s as placeholders for missing terms before performing synthetic division.

34. Explain why a 0 remainder in synthetic division of $P(x)$ by k indicates that k is a solution of the equation $P(x) = 0$.

Answers to Selected Exercises

In this section we provide the answers that we think most students will obtain when they work the exercises using the methods explained in the text. If your answer does not look exactly like the one given here, it is not necessarily wrong. In many cases there are equivalent forms of the answer that are correct. For example, if the answer section shows $\frac{3}{4}$ and your answer is .75, you have obtained the correct answer but written it in a different (yet equivalent) form. Unless the directions specify otherwise, .75 is just as valid an answer as $\frac{3}{4}$.

In general, if your answer does not agree with the one given in the text, see whether it can be transformed into the other form. If it can, then it is the correct answer. If you still have doubts, talk with your instructor.

DIAGNOSTIC PRETEST

(page xxix)
1. -3 2. 32,243 ft   3. -9 4. $\{-5\}$ 5. faster train: 67 mph; slower train: 54 mph   6. $24°, 31°, 125°$

7. $\left(-\infty, \frac{1}{2}\right]$

8. $(-4, 5]$

9. $\left\{-\frac{5}{8}, \frac{13}{2}\right\}$ 10. $-\frac{11}{7}$

11. x-intercept: $(-5, 0)$; y-intercept: $(0, 3)$

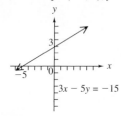

12.

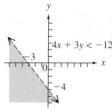

13. B; The set A includes three ordered pairs with the same first component (25.1) and different second components, so it is not a function.   14. $\{(-8, -1)\}$ 15. $\{(5, -3)\}$ 16. $\{(1, -2, 3)\}$

17. 20 lb of nuts, 12 lb of raisins   18. $\dfrac{49t^2}{s^8}$

19. $3k^3 - 5k^2 + 8k + 3$ 20. $16x^2 - 72xy + 81y^2$

21. $x^2 + 3x - 5 + \dfrac{2}{2x - 1}$ 22. (a) $(6q + 5p)(q - 4p)$

(b) $(y + 3)(y + 2)(y - 2)$ 23. $\left\{-5, \dfrac{3}{4}\right\}$ 24. $\dfrac{z + 4}{z + 1}$

25. $\dfrac{5x^2 + 11x + 12}{(x + 3)(x - 3)}$ 26. $4x + 1$ 27. $\dfrac{16m^{10}}{n^6}$ 28. $5y^2z^3\sqrt[3]{2yz^2}$

29. $\{7\}$ 30. 89   31. $\left\{-5, -\dfrac{1}{3}\right\}$

32. $\left\{\dfrac{-5 + \sqrt{37}}{6}, \dfrac{-5 - \sqrt{37}}{6}\right\}$ 33. east: 60 mi; south: 45 mi

34. vertex: (3, 5)
   domain: $(-\infty, \infty)$
   range: $(-\infty, 5]$

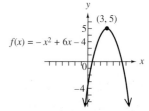

35. $f^{-1}(x) = \sqrt[3]{x + 8}$ 36. $\{2\}$ 37. $\{27\}$ 38. $\{8\}$
39. (a) 12   (b) 13   (c) $4x^2 - 4x + 4$ (d) $2x^2 + 5$
40.

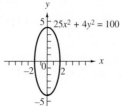

41.

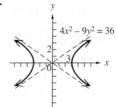

42. $\{(-1, -3), (3, 5)\}$

CHAPTER 1 REVIEW OF THE REAL NUMBER SYSTEM

Section 1.1 (page 11)
1. $\{1, 2, 3, 4, 5\}$ 3. $\{5, 6, 7, 8, \dots\}$ 5. $\{10, 12, 14, 16, \dots\}$
7. $\varnothing$ 9. $\{-4, 4\}$ 11. $\{x \mid x$ is an even natural number less than or equal to 8$\}$ 13. $\{x \mid x$ is a multiple of 4 greater than 0$\}$

15.   17.

19. (a) $5, 17, \dfrac{40}{2}$ (or 20)   (b) $0, 5, 17, \dfrac{40}{2}$ (c) $-8, 0, 5, 17, \dfrac{40}{2}$

(d) $-8, -.6, 0, \dfrac{3}{4}, 5, \dfrac{13}{2}, 17, \dfrac{40}{2}$ (e) $-\sqrt{5}, \sqrt{3}, \pi$ (f) All are

real numbers.   21. false; Some are integers, but others, like $\dfrac{3}{4}$,

are not.   23. false; No irrational number is an integer.   25. true
27. true   29. true   31. (a) -6 (b) 6   33. (a) 12   (b) 12

35. (a) $-\dfrac{6}{5}$ (b) $\dfrac{6}{5}$ 37. 8   39. $\dfrac{3}{2}$ 41. -5 43. -2 45. -4.5

47. 5 **49.** 6 **51.** 0 **53. (a)** Mumbai (Bombay); It increased 2.80%.
(b) Beijing; It increased .02%. **55.** Pacific Ocean, Indian Ocean,
Caribbean Sea, South China Sea, Gulf of California **57.** true
59. true **61.** false **63.** true **65.** true **67.** $7 > y$ **69.** $5 \geq 5$
71. $3t - 4 \leq 10$ **73.** $5x + 3 \neq 0$ **75.** $-6 < 10$; true
77. $10 \geq 10$; true **79.** $-3 \geq -3$; true **81.** $-8 > -6$; false
83. greater than **85.** Iowa (IA), Ohio (OH), California (CA),
Texas (TX), Minnesota (MN) **87.** $x > y$

Section 1.2 (page 21)

1. the numbers are additive inverses; $4 + (-4) = 0$ **3.** negative;
$-7 + (-21) = -28$ **5.** the positive number has larger absolute
value; $15 + (-2) = 13$ **7.** the number with smaller absolute value
is subtracted from the one with larger absolute value; $-15 - (-3) =$
-12 **9.** negative; $-5(15) = -75$ **11.** 9 **13.** -19
15. $-\frac{19}{12}$ **17.** -1.85 **19.** -11 **21.** 21 **23.** -13 **25.** -10.18
27. $\frac{67}{30}$ **29.** -6 **31.** -35 **33.** 40 **35.** 2 **37.** -12 **39.** $\frac{6}{5}$
41. 1 **43.** 5.88 **45.** -10.676 **47.** $\frac{1}{6}$ **49.** $-\frac{1}{7}$ **51.** $-\frac{3}{2}$
53. 5 **55.** 50 **57.** -1000 **59.** -7 **61.** 6 **63.** -4 **65.** 0
67. undefined **69.** $\frac{25}{102}$ **71.** $-\frac{9}{13}$ **73.** -2.1 **75.** 10,000
77. -11 **79.** 16 **81.** -4 **83.** -19 **85.** 112°F
87. $-\$37.5$ million **89.** 2000: $129 billion; 2010: $206 billion;
2020: $74 billion; 2030: $-\$501$ billion **91.** $-\$16,743$ million
93. C

Section 1.3 (page 31)

1. false; $-4^6 = -(4^6)$ **3.** true **5.** true **7.** true **9.** false; The
base is 3. **11. (a)** 64 **(b)** -64 **(c)** 64 **(d)** -64 **13.** 8^3
15. $\left(\frac{1}{2}\right)^2$ **17.** $(-4)^4$ **19.** z^7 **21.** 16 **23.** .021952 **25.** $\frac{1}{125}$
27. $\frac{343}{1000}$ **29.** -125 **31.** 256 **33.** -729 **35.** -4096
37. exponent: 7; base: -4.1 **39.** exponent: 7; base: 4.1 **41.** 9
43. 13 **45.** -20 **47.** $\frac{10}{11}$ **49.** $-.7$ **51.** not a real number
53. (a) B **(b)** C **(c)** A **55.** not a real number **57.** 24 **59.** 4
61. 14 **63.** 15 **65.** 55 **67.** -91 **69.** -8 **71.** -48 **73.** 8
75. -2 **77.** -2 **79.** undefined **81.** -1 **83.** 17 **85.** -96
87. $-\frac{15}{238}$ **89.** $1572 **91.** $3296 **93.** .035 **95.** Decreased
weight will result in higher BACs; .040; .053

Section 1.4 (page 41)

1. B **3.** A **5.** product; 0 **7.** grouping **9.** like **11.** $8k$
13. $-2r$ **15.** cannot be simplified **17.** $6a$ **19.** $2m + 2p$
21. $-10d + 5f$ **23.** 1900 **25.** 75 **27.** 431 **29.** $-6y + 3$
31. $p + 11$ **33.** $-2k + 15$ **35.** $m - 14$ **37.** -1 **39.** $2p + 7$
41. $-6z - 39$ **43.** $(5 + 8)x = 13x$ **45.** $(5 \cdot 9)r = 45r$
47. $9y + 5x$ **49.** 7 **51.** $8(-4) + 8x = -32 + 8x$
53. Answers will vary. One example is washing your face and brush-
ing your teeth. **55.** associative property **56.** associative property
57. commutative property **58.** associative property
59. distributive property **60.** add

Chapter 1 Review Exercises (page 47)

1. [number line marked $\frac{9}{4}$; $-4, -2, 0, 2, 4$] **2.** [number line marked $-\frac{11}{4}$, $-.5$, $\frac{13}{3}$; $-5, -3, -1, 0, 1, 3, 5$]

3. 16 **4.** 23 **5.** -4 **6.** 5 **7.** $0, \frac{12}{3}$ (or 4)

8. $-9, -\sqrt{4}$ (or -2), $0, \frac{12}{3}$ (or 4) **9.** $-9, -\frac{4}{3}, -\sqrt{4}$ (or -2),
$-.25, 0, .\overline{35}, \frac{5}{3}, \frac{12}{3}$ (or 4) **10.** All are real numbers except $\sqrt{-9}$.
11. $\{4, 5, 6, 7, 8\}$ **12.** $\{0, 1, 2, 3\}$ **13.** true **14.** false **15.** true
16. Hyundai; 50% **17.** General Motors; -5% **18.** false
19. true **20.** $\frac{41}{24}$ **21.** $-\frac{1}{2}$ **22.** -3 **23.** -17.09 **24.** -39
25. -1 **26.** $\frac{23}{20}$ **27.** $-\frac{5}{18}$ **28.** -35 **29.** 11,331 ft **30.** -90
31. $\frac{2}{3}$ **32.** -11.408 **33.** -15 **34.** 3.21 **35.** $\frac{5}{7 - 7}$
36. 10,000 **37.** $\frac{27}{343}$ **38.** -125 **39.** -125 **40.** 2.89 **41.** 20
42. -14 **43.** $\frac{8}{11}$ **44.** $-.9$ **45.** not a real number **46.** -4
47. 44 **48.** -2 **49.** -30 **50.** -30 **51.** $-\frac{8}{51}$ **52. (a)** 25
(b) Answers will vary. **53.** $21q$ **54.** $-4z$ **55.** $5m$ **56.** $4p$
57. $-2k - 6$ **58.** $6r + 18$ **59.** $18m + 27n$ **60.** $-p - 3q$
61. $y + 1$ **62.** 0 **63.** $-18m$ **64.** $(2 + 3)x = 5x$ **65.** -4
66. $(2 \cdot 4)x = 8x$ **67.** $13 + (-3) = 10$ **68.** 0 **69.** $5x + 5z$
70. 7 **71.** 1 **72.** $(3 + 5 + 6)a = 14a$ **73.** 0 **74.** 51,897 million
dollars; negative **75.** 52,844 million dollars; negative
76. 48,165 million dollars; negative **77.** $\frac{256}{625}$ **78.** 25 **79.** 31
80. 9 **81.** 0 **82.** -5 **83.** $\frac{4}{3}$ **84.** -6.16 **85.** -9 **86.** 2
87. 2 **88.** not a real number **89.** -116 **90.** Work inside the
parentheses first.

Chapter 1 Test (page 51)

1. [number line marked .75, $\frac{5}{3}$, 6.3; $-2, 0, 2, 4, 6$] **2.** $0, 3, \sqrt{25}$ (or 5), $\frac{24}{2}$ (or 12)
3. $-1, 0, 3, \sqrt{25}$ (or 5), $\frac{24}{2}$ (or 12) **4.** $-1, -.5, 0, 3, \sqrt{25}$ (or 5),
$7.5, \frac{24}{2}$ (or 12) **5.** All are real numbers except $\sqrt{-4}$. **6.** 0
7. -26 **8.** 19 **9.** 1 **10.** $\frac{16}{7}$ **11.** $\frac{11}{23}$ **12.** 50,395 ft
13. 37,486 ft **14.** 1345 ft **15.** 14 **16.** -15 **17.** not a real
number **18. (a)** a must be positive. **(b)** a must be negative.
(c) a must be 0. **19.** 2 **20.** $-\frac{6}{23}$ **21.** $10k - 10$ **22.** Both terms
change sign and are added to $3r + 8$; $7r + 2$. **23.** B **24.** E **25.** D
26. A **27.** F **28.** C **29.** C **30.** E

CHAPTER 2 LINEAR EQUATIONS AND APPLICATIONS

Section 2.1 (page 61)

1. A and C **3.** Both sides are evaluated as 30, so 6 is a solution.
5. Any number is a solution. For example, if the last name is
Lincoln, $x = 7$. Both sides are evaluated as -48. **7. (a)** equation
(b) expression **(c)** equation **(d)** expression **9.** The solution
contains a sign error when the distributive property was applied. The
left side of the second line of the solution should be $8x - 4x + 6$.
The correct solution is 1. **11.** $\{-1\}$ **13.** $\{-4\}$ **15.** $\{-7\}$
17. $\{0\}$ **19.** $\{4\}$ **21.** $\left\{-\frac{7}{8}\right\}$ **23.** $\left\{-\frac{5}{3}\right\}$ **25.** $\left\{-\frac{1}{2}\right\}$
27. $\{2\}$ **29.** $\{-2\}$ **31.** $\{-1\}$ **33.** $\{7\}$ **35.** $\{2\}$
37. $\{-8\}$ **39.** 12 **41. (a)** 10^2 or 100 **(b)** 10^3 or 1000

43. {6} **45.** {4} **47.** {−30} **49.** {0} **51.** {3} **53.** {0} **55.** {2000} **57.** {25} **59.** {40} **61.** {3} **63.** A conditional equation has at least one solution, an identity has infinitely many solutions, and a contradiction has no solution. **65. (a)** B **(b)** A **(c)** C **67.** contradiction; $\emptyset$ **69.** conditional; {0} **71.** identity; {all real numbers} **73.** identity; {all real numbers}

Section 2.2 (page 71)

1. (a) $3x = 5x + 8$ **(b)** $ct = bt + k$ **2. (a)** $3x - 5x = 8$
(b) $ct - bt = k$ **3. (a)** $-2x = 8$; distributive property
(b) $t(c - b) = k$; distributive property **4. (a)** $x = -4$
(b) $t = \dfrac{k}{c - b}$ **5.** $c \neq b$; If $c = b$, the denominator is 0.

6. To solve an equation for a particular variable, such as solving the second equation for t, go through the same steps as you would in solving for x in the first equation. Treat all other variables as constants.

7. $W = \dfrac{A}{L}$ **9.** $L = \dfrac{P - 2W}{2}$ or $L = \dfrac{P}{2} - W$ **11. (a)** $W = \dfrac{V}{LH}$

(b) $H = \dfrac{V}{LW}$ **13.** $r = \dfrac{C}{2\pi}$ **15. (a)** $h = \dfrac{2A}{b + B}$

(b) $B = \dfrac{2A}{h} - b$ or $B = \dfrac{2A - bh}{h}$ **17.** $C = \dfrac{5}{9}(F - 32)$

19. D **21.** $r = \dfrac{-2k - 3y}{a - 1}$ or $r = \dfrac{2k + 3y}{1 - a}$

23. $y = \dfrac{-x}{w - 3}$ or $y = \dfrac{x}{3 - w}$ **25.** 3.735 hr **27.** 104°F

29. 230 m **31.** radius: 240 in.; diameter: 480 in. **33.** 8 ft
35. 75% water, 25% alcohol **37.** 3% **39.** $10.51 **41.** $45.66
43. (a) .567 **(b)** .567 **(c)** .484 **(d)** .419 **45.** 41%
47. 1500 **49.** 14,350 **51.** $53,272 **53.** 18%; yes

Section 2.3 (page 85)

1. (a) $x + 12$ **(b)** $12 > x$ **3. (a)** $x - 4$ **(b)** $4 < x$ **5.** D

7. $2x - 13$ **9.** $12 + 3x$ **11.** $8(x - 12)$ **13.** $\dfrac{3x}{7}$

15. $x + 6 = -31; -37$ **17.** $x - (-4x) = x + 9; \dfrac{9}{4}$

19. $12 - \dfrac{2}{3}x = 10; 3$ **21.** expression **23.** equation

25. expression **27.** *Step 1:* We are asked to find the number of patents each university secured; *Step 2:* the number of patents Stanford secured; *Step 3:* x, $x - 38$; *Step 4:* 134; *Step 5:* 134, 96; *Step 6:* 38, MIT patents, 96, 230 **29.** width: 165 ft; length: 265 ft
31. 850 mi, 925 mi, 1300 mi **33.** Eiffel Tower: 984 ft; Leaning Tower: 180 ft **35.** Yankees: $149.7 million; Mets: $116.9 million
37. 1.9% **39.** 374,787 **41.** $225 **43.** $4000 at 3%; $8000 at 4% **45.** $10,000 at 4.5%; $19,000 at 3% **47.** $58,000 **49.** 5 L
51. 4 L **53.** 1 gal **55.** 150 lb **57.** We cannot expect the final mixture to be worth more than either of the ingredients.
59. (a) $800 - x$ **(b)** $800 - y$ **60. (a)** $.05x$; $.10(800 - x)$
(b) $.05y$; $.10(800 - y)$ **61. (a)** $.05x + .10(800 - x) = 800(.0875)$ **(b)** $.05y + .10(800 - y) = 800(.0875)$
62. (a) $200 at 5%; $600 at 10% **(b)** 200 L of 5% acid; 600 L of 10% acid **63.** The processes are the same. The amounts of money in Problem A correspond to the amounts of solution in Problem B.

Section 2.4 (page 95)

1. $4.50 **3.** 52 mph **5.** The problem asks for the *distance* to the workplace. To find this distance, we must multiply the rate, 10 mph,

by the time, $\dfrac{3}{4}$ hr. **7.** 17 pennies, 17 dimes, 10 quarters

9. 26 quarters, 21 half-dollars **11.** 28 $10 coins, 13 $20 coins
13. 450 floor tickets, 100 balcony tickets **15.** 7.91 m per sec

17. 8.42 m per sec **19.** $2\dfrac{1}{2}$ hr **21.** 7:50 P.M. **23.** 15 mph

25. $\dfrac{1}{2}$ hr **27.** 60°, 60°, 60° **29.** 40°, 45°, 95° **31.** 40°, 80°

32. 120° **33.** The sum is equal to the measure of the angle found in Exercise 32. **34.** The sum of the measures of angles ① and ② is equal to the measure of angle ③. **35.** Both measure 122°.
37. 64°, 26° **39.** 19, 20, 21 **41.** 61 yr old

Summary Exercises on Solving Applied Problems (page 99)

1. length: 8 in.; width: 5 in. **2.** length: 60 m; width: 30 m
3. $30.96 **4.** $425 **5.** $800 at 4%; $1600 at 5% **6.** $12,000 at 3%; $14,000 at 4% **7.** *Frasier:* 31; *The Simpsons:* 20 **8.** *Titanic:* $600.8 million; *Star Wars:* $461 million **9.** Cheruiyot: 2.17 hr;

Zakharova: 2.42 hr **10.** 10 ft **11.** $1\dfrac{1}{2}$ cm **12.** 6 in., 12 in., 16 in.

13. 4-yr university: $3746; 2-yr community college: $1379
14. 31, 32, 33 **15.** 20°, 30°, 130°

Chapter 2 Review Exercises (page 105)

1. $\left\{-\dfrac{9}{5}\right\}$ **2.** $\left\{\dfrac{1}{3}\right\}$ **3.** {10} **4.** $\left\{-\dfrac{7}{5}\right\}$ **5.** $\emptyset$ **6.** {0} **7.** {16}

8. {300} **9.** B **10.** Begin by subtracting 5 from each side. Then divide each side by −2. **11.** identity; {all real numbers}

12. contradiction; $\emptyset$ **13.** conditional; {0} **14.** $L = \dfrac{V}{WH}$

15. $b = \dfrac{2A}{h} - B$ or $b = \dfrac{2A - Bh}{h}$ **16.** $d = \dfrac{C}{\pi}$ **17.** 6 ft

18. 9.6% **19.** 6.5% **20.** 25° **21.** approximately 17,415,099

22. 100 mm **23.** $9 - \dfrac{1}{3}x$ **24.** $\dfrac{4x}{x + 9}$ **25.** length: 13 m;

width: 8 m **26.** 17 in., 17 in., 19 in. **27.** 12 kg **28.** 30 L
29. 10 L **30.** $10,000 at 6%; $6000 at 4% **31.** A
32. (a) 530 mi **(b)** 328 mi **33.** 2.2 hr **34.** 50 km per hr;

65 km per hr **35.** 1 hr **36.** 46 mph **37.** $\left\{\dfrac{7}{6}\right\}$ **38.** {0}

39. $\emptyset$ **40.** 12 in., 24 in., 32 in. **41.** $k = \dfrac{6t - bt}{a + s}$ or

$k = \dfrac{bt - 6t}{-a - s}$ **42.** {all real numbers} **43.** 6 in. **44.** Kerry: 252;

Bush: 286 **45.** eastbound car: 3 hr; westbound car: 2 hr

46. $1300 at 4%; $1800 at 5% **47.** {1500} **48.** $x = \dfrac{C - By}{A}$

Chapter 2 Test (page 109)

1. {−19} **2.** {5} **3.** {4} **4. (a)** contradiction; $\emptyset$
(b) identity; {all real numbers} **(c)** conditional equation; {0}

5. $v = \dfrac{S + 16t^2}{t}$ or $v = \dfrac{S}{t} + 16t$ **6.** $r = \dfrac{-2 - 6t}{a - 3}$ or

$r = \dfrac{2 + 6t}{3 - a}$ **7.** 3.497 hr **8.** 6.25% **9.** 73.1% **10.** $8000 at

3%; $20,000 at 5% **11.** faster car: 60 mph; slower car: 45 mph
12. 40°, 40°, 100° **13.** 10% **14.** 13.33% **15.** 1050

Cumulative Review Exercises: Chapters 1–2 (page 111)

1. $9, \sqrt{36}$ (or 6) **2.** $0, 9, \sqrt{36}$ (or 6) **3.** $-8, 0, 9, \sqrt{36}$ (or 6)

4. $-8, -\dfrac{2}{3}, 0, \dfrac{4}{5}, 9, \sqrt{36}$ (or 6) **5.** $-\sqrt{6}$ **6.** All are real numbers.

7. $-\dfrac{22}{21}$ **8.** 7.9 **9.** 8 **10.** 0 **11.** -243 **12.** $\dfrac{216}{343}$ **13.** 4096

14. -4096 **15.** $\sqrt{-36}$ **16.** $\dfrac{4+4}{4-4}$ **17.** -16 **18.** -34

19. 184 **20.** $\dfrac{27}{16}$ **21.** $-20r + 17$ **22.** $13k + 42$

23. commutative property **24.** distributive property **25.** inverse property **26.** $\{5\}$ **27.** $\{30\}$ **28.** $\{15\}$ **29.** $c = P - a - b$
30. $\emptyset$ **31.** {all real numbers} **32.** 2 L **33.** 9 pennies, 12 nickels, 8 quarters **34.** \$5000 at 5%; \$7000 at 6% **35.** $\dfrac{1}{8}$ hr **36.** 44 mg

37. 25.7 **38.** **(a)** 276 **(b)** 15.7% **39.** $107°, 73°$

CHAPTER 3 LINEAR INEQUALITIES AND ABSOLUTE VALUE

Section 3.1 (page 123)
1. D **3.** B **5.** F **7.** Use a parenthesis when an endpoint is not included; use a bracket when it is included.

9. $[5, \infty)$

11. $(7, \infty)$

13. $(-4, \infty)$

15. $(-\infty, -40]$

17. $(-\infty, 4]$

19. $\left(-\infty, -\dfrac{15}{2}\right)$

21. $\left[\dfrac{1}{2}, \infty\right)$

23. $(3, \infty)$

25. $(-\infty, 4)$

27. $\left(-\infty, \dfrac{23}{6}\right]$

29. $\left(-\infty, \dfrac{76}{11}\right)$

31. $\{-9\}$

32. $(-9, \infty)$

33. $(-\infty, -9)$

34. We obtain the set of all real numbers.

35. $(-\infty, -3)$

37. $(1, 11)$

39. $[-14, 10]$

41. $[-5, 6]$

43. $(-6, -4)$

45. $\left[-\dfrac{13}{3}, \dfrac{11}{3}\right]$

47. from about 8:00 A.M to 10:15 A.M. and after about 9:00 P.M.
49. about $65°F - 67°F$ **51.** at least 82 **53.** 628.6 mi
55. 921 deliveries **57.** **(a)** 130 to 157 beats per min **(b)** Answers will vary. **59.** all numbers between -6 and 4, that is, $(-6, 4)$
61. all numbers greater than or equal to 13, that is, $[13, \infty)$ **63.** all numbers less than or equal to 4, that is, $(-\infty, 4]$

Section 3.2 (page 135)
1. true **3.** false; The union is $(-\infty, 6) \cup (6, \infty)$. **5.** $\{4\}$ or D
7. $\emptyset$ **9.** $\{1, 2, 3, 4, 5, 6\}$ or A **11.** $\{1, 3, 5, 6\}$

13.

15.

17.

19. Answers will vary. One example is: The intersection of two streets is the region common to *both* streets.

21. $(-3, 2)$

23. $(-\infty, 2]$

25. $\emptyset$

27. $[5, 9]$

29. $(-\infty, 4]$

31. $(-\infty, 8]$

33. $[-2, \infty)$

35. $(-\infty, \infty)$

37. $(-\infty, -5) \cup (5, \infty)$

39. $(-\infty, 2) \cup (2, \infty)$

41. $[-4, -1]$ **43.** $[-9, -6]$ **45.** $(-\infty, 3)$ **47.** $[3, 9)$

49. intersection; $(-5, -1)$

51. union; $(-\infty, 4)$

53. intersection; $[4, 12]$

55. union; $(-\infty, 0] \cup [2, \infty)$

57. Mario, Joe   **58.** none of them   **59.** none of them   **60.** Luigi, Than   **61.** {Tuition and fees}   **63.** {Tuition and fees, Board rates}

Section 3.3 (page 145)

1. E; C; D; B; A   **3.** Use *or* for the $=$ statement and the $>$ statement. Use *and* for the $<$ statement.   **5.** $\{-12, 12\}$

7. $\{-5, 5\}$ **9.** $\{-6, 12\}$ **11.** $\{-5, 4\}$ **13.** $\left\{-3, \dfrac{11}{2}\right\}$

15. $\left\{-\dfrac{19}{2}, \dfrac{9}{2}\right\}$ **17.** $\{-10, -2\}$ **19.** $\left\{-8, \dfrac{32}{3}\right\}$

21. $(-\infty, -3) \cup (3, \infty)$

23. $(-\infty, -4] \cup [4, \infty)$

25. $(-\infty, -12) \cup (8, \infty)$

27. $\left(-\infty, -\dfrac{7}{3}\right] \cup [3, \infty)$

29. $(-\infty, -2) \cup (8, \infty)$

31. (a)

(b)

33. $[-3, 3]$

35. $(-4, 4)$

37. $[-12, 8]$

39. $\left(-\dfrac{7}{3}, 3\right)$

41. $[-2, 8]$

43. $(-\infty, -5) \cup (13, \infty)$

45. $\{-6, -1\}$

47. $\left[-\dfrac{10}{3}, 4\right]$

49. $\left[-\dfrac{7}{6}, -\dfrac{5}{6}\right]$

51. $\{-5, 5\}$ **53.** $\{-5, -3\}$ **55.** $(-\infty, -3) \cup (2, \infty)$

57. $[-10, 0]$ **59.** $\{-1, 3\}$ **61.** $\left\{-3, \dfrac{5}{3}\right\}$ **63.** $\left\{-\dfrac{1}{3}, -\dfrac{1}{15}\right\}$

65. $\left\{-\dfrac{5}{4}\right\}$ **67.** $\emptyset$ **69.** $\left\{-\dfrac{1}{4}\right\}$ **71.** $\emptyset$ **73.** $(-\infty, \infty)$

75. $\left\{-\dfrac{3}{7}\right\}$ **77.** $(-\infty, \infty)$ **79.** $\left(-\infty, -\dfrac{7}{10}\right) \cup \left(-\dfrac{7}{10}, \infty\right)$

81. $|x - 1000| \leq 100$; $900 \leq x \leq 1100$ **83.** 535.9 ft
84. Hyatt Regency   **85.** Transamerica Tower, Hyatt Regency, Power and Light Building   **86. (a)** $|x - 535.9| \geq 75$ **(b)** $x \geq 610.9$ or $x \leq 460.9$ **(c)** KCTV Tower, One Kansas City Place, City Hall, Federal Office Building, 1201 Walnut Street, Commerce Tower, City Center Square   **(d)** It makes sense because it includes all buildings *not* listed earlier.

Summary Exercises on Solving Linear and Absolute Value Equations and Inequalities (page 151)

1. $\{12\}$ **2.** $\{-5, 7\}$ **3.** $\{7\}$ **4.** $\left\{-\dfrac{2}{5}\right\}$ **5.** $\emptyset$ **6.** $(-\infty, -1]$

7. $\left[-\dfrac{2}{3}, \infty\right)$ **8.** $\{-1\}$ **9.** $\{-3\}$ **10.** $\left\{1, \dfrac{11}{3}\right\}$ **11.** $(-\infty, 5]$

12. $(-\infty, \infty)$ **13.** $\{2\}$ **14.** $(-\infty, -8] \cup [8, \infty)$ **15.** $\emptyset$

16. $(-\infty, \infty)$ **17.** $(-5.5, 5.5)$ **18.** $\left\{\dfrac{13}{3}\right\}$ **19.** $\left\{-\dfrac{96}{5}\right\}$

20. $(-\infty, 32]$ **21.** $(-\infty, -24)$ **22.** $\left\{\dfrac{3}{8}\right\}$ **23.** $\left\{\dfrac{7}{2}\right\}$

24. $(-6, 8)$ **25.** $(-\infty, \infty)$ **26.** $(-\infty, 5)$ **27.** $(-\infty, -4) \cup (7, \infty)$

28. $\{24\}$ **29.** $\left\{-\dfrac{1}{5}\right\}$ **30.** $\left(-\infty, -\dfrac{5}{2}\right]$ **31.** $\left[-\dfrac{1}{3}, 3\right]$

32. $[1, 7]$ **33.** $\left\{-\dfrac{1}{6}, 2\right\}$ **34.** $\{-3\}$

35. $(-\infty, -1] \cup \left[\dfrac{5}{3}, \infty\right)$ **36.** $\left[\dfrac{3}{4}, \dfrac{15}{8}\right]$ **37.** $\left\{-\dfrac{5}{2}\right\}$

38. $\{60\}$ **39.** $\left[-\dfrac{9}{2}, \dfrac{15}{2}\right]$ **40.** $(1, 9)$ **41.** $(-\infty, \infty)$ **42.** $\left\{\dfrac{1}{3}, 9\right\}$

43. $(-\infty, \infty)$ **44.** $\left\{-\dfrac{10}{9}\right\}$ **45.** $\{-2\}$ **46.** $\emptyset$

47. $(-\infty, -1) \cup (2, \infty)$ **48.** $[-3, -2]$

Chapter 3 Review Exercises (page 157)

1. $(-9, \infty)$

2. $(-\infty, -3]$

3. $\left(\dfrac{3}{2}, \infty\right)$

4. $\left(-\infty, -\dfrac{14}{9}\right)$

5. $[-3, \infty)$

6. $[-3, 12]$ [number line: bracket at -3 to bracket at 12]

7. $[3, 5)$ [number line: bracket at 3 to parenthesis at 5]

8. $\left(-3, \dfrac{7}{2}\right)$ [number line: parenthesis at -3 to parenthesis at $\frac{7}{2}$]

9. 38 m or less   **10.** 480 mi or less   **11.** any grade greater than or equal to 61%   **12.** Because the statement $-8 < -13$ is *false*, the inequality has no solution.   **13.** $\{a, c\}$ **14.** $\{a\}$
15. $\{a, c, e, f, g\}$ **16.** $\{a, b, c, d, e, f, g\}$

17. $(6, 9)$ [number line: parenthesis at 6 to parenthesis at 9]

18. $(8, 14)$ [number line: parenthesis at 8 to parenthesis at 14]

19. $(-\infty, -3] \cup (5, \infty)$ [number line: bracket at -3, parenthesis at 5]

20. $(-\infty, \infty)$ [number line, all shaded, at 0]

21. $\emptyset$ [number line, empty]

22. $(-\infty, -2] \cup [7, \infty)$ [number line: bracket at -2, bracket at 7]

23. $(-3, 4)$ **24.** $(-\infty, 2)$ **25.** $(4, \infty)$ **26.** $(1, \infty)$
27. (a) production, transportation, and material moving

(b) all occupations   **28.** $\{-7, 7\}$ **29.** $\{-11, 7\}$ **30.** $\left\{-\dfrac{1}{3}, 5\right\}$

31. $\emptyset$ **32.** $\{0, 7\}$ **33.** $\left\{-\dfrac{3}{2}, \dfrac{1}{2}\right\}$ **34.** $\left\{-\dfrac{3}{4}, \dfrac{1}{2}\right\}$ **35.** $\left\{-\dfrac{1}{2}\right\}$

36. $(-14, 14)$ [number line: parenthesis at -14 to parenthesis at 14]

37. $[-1, 13]$ [number line: bracket at -1 to bracket at 13]

38. $[-3, -2]$ [number line: bracket at -3 to bracket at -2]

39. $(-\infty, \infty)$ [number line, all shaded, at 0]

40. $\left(-\infty, -\dfrac{8}{5}\right) \cup (2, \infty)$ [number line: parenthesis at $-\frac{8}{5}$, parenthesis at 2]

41. $(-\infty, \infty)$ [number line, all shaded, at 0]

42. $\left(-\infty, \dfrac{7}{6}\right]$ **43.** $[-4, 5)$ **44.** $\left(-\infty, \dfrac{14}{17}\right)$ **45.** any amount greater than or equal to \$1100   **46.** $(-\infty, 2]$

47. $(-\infty, -1) \cup \left(\dfrac{11}{7}, \infty\right)$ **48.** $\{-5, 15\}$ **49.** $[-16, 10]$

50. $(-\infty, \infty)$ **51.** $\left\{-4, -\dfrac{2}{3}\right\}$ **52.** [number line: parenthesis at 6 to parenthesis at 8]

53. [number line: bracket at -2, bracket at 7, at 0]   **54. (a)** $\emptyset$ **(b)** $(-\infty, \infty)$ **(c)** $\emptyset$

Chapter 3 Test (page 161)

1. Reverse the direction of the inequality symbol.

2. $[1, \infty)$ [number line: bracket at 1]

3. $(-\infty, 28)$ [number line: parenthesis at 28]

4. $[-3, 3]$ [number line: bracket at -3 to bracket at 3]

5. C   **6. (a)** 1997, 1998, 1999, 2000   **(b)** 1994, 1995, 1996
(c) 1997, 1998, 1999   **7.** 82%   **8.** $[500, \infty)$

9. (a) $\{1, 5\}$ **(b)** $\{1, 2, 5, 7, 9, 12\}$ **10.** $\{2\}$

11. $[2, 9)$ [number line: bracket at 2 to parenthesis at 9]

12. $(-\infty, 3) \cup [6, \infty)$ [number line: parenthesis at 3, bracket at 6]

13. $\left[-\dfrac{5}{2}, 1\right]$ [number line: bracket at $-\frac{5}{2}$ to bracket at 1]

14. $\left(-\infty, -\dfrac{7}{6}\right) \cup \left(\dfrac{17}{6}, \infty\right)$ [number line: parenthesis at $-\frac{7}{6}$, parenthesis at $\frac{17}{6}$]

15. $\emptyset$ **16.** $\left\{-\dfrac{5}{3}, 3\right\}$ **17.** $\left\{-\dfrac{5}{7}, \dfrac{11}{3}\right\}$ **18.** $\emptyset$

Cumulative Review Exercises: Chapters 1–3 (page 163)

1. $\dfrac{3}{4}$ **2.** true   **3.** $\dfrac{37}{60}$ **4.** $\dfrac{48}{5}$ **5.** 11   **6.** -8 **7.** -36

8. -125 **9.** $\dfrac{81}{16}$ **10.** -34 **11.** $\dfrac{3}{16}$ **12.** distributive property

13. commutative property   **14.** $2k - 11$ **15.** $\{-1\}$ **16.** $\{-12\}$

17. $\{26\}$ **18.** $\left\{\dfrac{3}{4}, \dfrac{7}{2}\right\}$ **19.** $y = \dfrac{24 - 3x}{4}$ **20.** $n = \dfrac{A - P}{iP}$

21. $[-14, \infty)$ [number line: bracket at -14]

22. $\left[\dfrac{5}{3}, 3\right)$ [number line: bracket at $\frac{5}{3}$ to parenthesis at 3]

23. $(-\infty, 0) \cup (2, \infty)$ [number line: parenthesis at 0, parenthesis at 2]

24. $\left(-\infty, -\dfrac{1}{7}\right] \cup [1, \infty)$ [number line: bracket at $-\frac{1}{7}$, bracket at 1]

25. \$5000   **26.** $6\dfrac{1}{3}$ g   **27.** 74% or greater   **28.** 40 mph; 60 mph

29. (a) 143   **(b)** 8.9%   **30.** 4 cm; 9 cm; 27 cm

CHAPTER 4   GRAPHS, LINEAR EQUATIONS, AND FUNCTIONS

Section 4.1 (page 173)
1. (a) x represents the year; y represents the percent of women in math or computer science professions.   **(b)** 1990–2000   **(c)** 1990
(d) 1980   **3.** You should choose a bar graph.   **5.** origin   **7.** y; x
9. two   **11. (a)** I   **(b)** III   **(c)** II   **(d)** IV   **(e)** no quadrant
13. (a) I or III   **(b)** II or IV   **(c)** II or IV   **(d)** I or III

15–24.

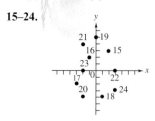

25. -3; 3; 2; -1

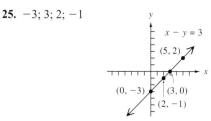

27. $\dfrac{5}{2}$; 5; $\dfrac{3}{2}$; 1

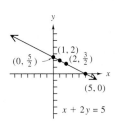

29. -4; 5; $-\dfrac{12}{5}$; $\dfrac{5}{4}$

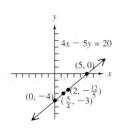

31. $(6, 0)$; $(0, 4)$

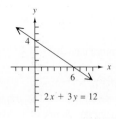

33. $(6, 0)$; $(0, -2)$

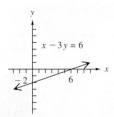

35. $(3, 0)$; $\left(0, -\dfrac{9}{7}\right)$

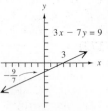

37. none; $(0, 5)$

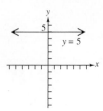

39. $(5, 0)$; none

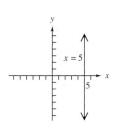

41. $(0, 0)$; $(0, 0)$

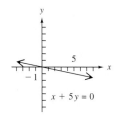

43. $(0, 0)$; $(0, 0)$

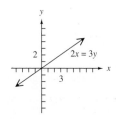

45. $(6, -2)$ **46.** $(5, -2)$ **47.** $(6, 0)$ **48.** $(5, 0)$
49. 5; 0   **50.** The x-coordinate of M is the average of the
x-coordinates of P and Q. The y-coordinate of M is the average
of the y-coordinates of P and Q.   **51.** $(-5, -1)$ **52.** $(2, 5)$

53. $\left(\dfrac{9}{2}, -\dfrac{3}{2}\right)$ **54.** $\left(-\dfrac{3}{2}, \dfrac{5}{2}\right)$ **55.** $\left(0, \dfrac{11}{2}\right)$ **56.** $\left(\dfrac{3}{2}, 0\right)$

Section 4.2 (page 185)

1. A, B, D   **3.** 2   **5.** undefined   **7.** 2   **9.** $\dfrac{5}{2}$ **11.** 0   **13.** 8

15. $\dfrac{5}{6}$ **17.** 0   **19.** $-\dfrac{5}{2}$ **21.** undefined   **23.** B   **25.** A

27. $-\dfrac{1}{2}$

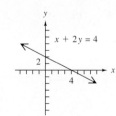

29. 1

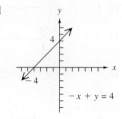

31. $-\dfrac{6}{5}$

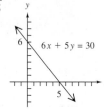

33. undefined

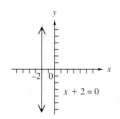

35. 4

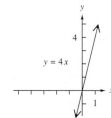

37. 0

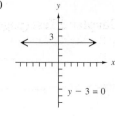

39.

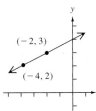

41.

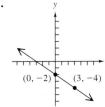

43.

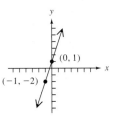

45. parallel **47.** perpendicular **49.** neither **51.** parallel
53. neither **55.** perpendicular **57.** $\dfrac{7}{10}$ **59.** $-\$4000$ per yr;

The value of the machine is decreasing $4000 each year during these years. **61.** 0% per yr (or no change); The percent of pay raise is not changing—it is 3% each year during these years. **63. (a)** 3.8% per yr **(b)** The positive slope means the percent of tax returns filed electronically *increased* an average of 3.8% each year. **65.** $-\$69$ per yr; The price decreased an average of $69 each year from 1997 to 2002. **66.** $\dfrac{1}{3}$ **67.** $\dfrac{1}{3}$ **68.** $\dfrac{1}{3}$

69. $\dfrac{1}{3}=\dfrac{1}{3}=\dfrac{1}{3}$ is true. **70.** They are collinear. **71.** They are not collinear.

Section 4.3 (page 199)
1. A **3.** A **5.** $3x+y=10$ **7.** A **9.** C **11.** H **13.** B
15. $y=5x+15$ **17.** $y=-\dfrac{2}{3}x+\dfrac{4}{5}$ **19.** $y=\dfrac{2}{5}x+5$
21. (a) $y=x+2$ **(b)** 1 **(c)** $(0,2)$
(d)

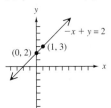

23. (a) $y=\dfrac{4}{5}x-4$ **(b)** $\dfrac{4}{5}$ **(c)** $(0,-4)$
(d)

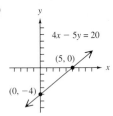

25. (a) $y=-\dfrac{1}{2}x-2$ **(b)** $-\dfrac{1}{2}$ **(c)** $(0,-2)$
(d)

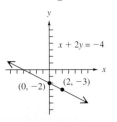

27. $3x+4y=10$ **29.** $2x+y=18$ **31.** $x-2y=-13$
33. $y=12$ **35.** $x=9$ **37.** $y=.2$ **39.** $2x-y=2$
41. $x+2y=8$ **43.** $2x-13y=-6$ **45.** $y=5$ **47.** $x=7$
49. $y=3x-19$ **51.** $y=\dfrac{1}{2}x-1$ **53.** $y=-\dfrac{1}{2}x+9$
55. $y=7$ **57.** $y=45x$; $(0,0),(5,225),(10,450)$ **59.** $y=2.00x$; $(0,0),(5,10.00),(10,20.00)$ **61. (a)** $y=39x+99$ **(b)** $(5,294)$; The cost of a 5-month membership is $294. **(c)** \$567
63. (a) $y=50x+25$ **(b)** $(5,275)$; The cost of the plan for 5 months is $275. **(c)** \$1225 **65. (a)** $y=.20x+50$ **(b)** $(5,51)$; The charge for driving 5 mi is $51. **(c)** 173 mi **67. (a)** $y=5.6x+9$; The percent of households accessing the Internet by broadband is increasing 5.6% per yr. **(b)** 43% **69. (a)** $y=-103.2x+28{,}908$ **(b)** 28,082; The result using the model is a little high. **71.** 32; 212
72. $(0,32)$ and $(100,212)$ **73.** $\dfrac{9}{5}$ **74.** $F=\dfrac{9}{5}C+32$
75. $C=\dfrac{5}{9}(F-32)$ **76.** $-40°$ **77.** $60°$ **78.** $59°$; They differ by 1°. **79.** $90°$; $86°$; They differ by 4°. **80.** Since $\dfrac{9}{5}$ is a little less than 2, and 32 is a little more than 30, $\dfrac{9}{5}C+32\approx 2C+30$.

Section 4.4 (page 209)
1. solid; below **3.** dashed; above **5.** The graph of $Ax+By=C$ divides the plane into two regions. In one of these regions, the ordered pairs satisfy $Ax+By<C$; in the other, they satisfy $Ax+By>C$.
7.

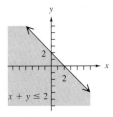

9.

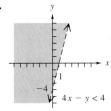

11.

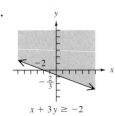

13.

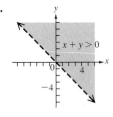

15.

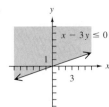

17.

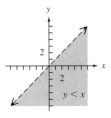

19.

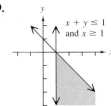

21.

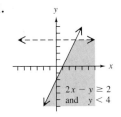

23.

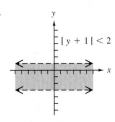

25.

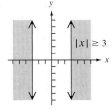

27.

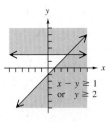

29.

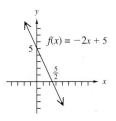

31.

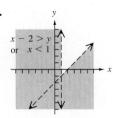

33.

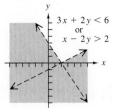

Section 4.5 (page 223)

1. independent variable **3. (a)** A relation is a set of ordered pairs.
(b) The domain is the set of all first components (x-values).
(c) The range is the set of all second components (y-values).
(d) A function is a relation in which each domain element is paired
with one and only one range element. **5.** function; domain:
$\{5, 3, 4, 7\}$; range: $\{1, 2, 9, 3\}$ **7.** not a function; domain: $\{2, 0\}$;
range: $\{4, 2, 6\}$ **9.** function; domain: $\{-3, 4, -2\}$; range: $\{1, 7\}$
11. not a function; domain: $\{1, 0, 2\}$; range: $\{1, -1, 0, 4, -4\}$
13. function; domain: $\{2, 5, 11, 17, 3\}$; range: $\{1, 7, 20\}$ **15.** not a
function; domain: $\{1\}$; range: $\{5, 2, -1, -4\}$ **17.** function;
domain: $(-\infty, \infty)$; range: $(-\infty, \infty)$ **19.** function; domain: $(-\infty, \infty)$;
range: $(-\infty, 4]$ **21.** not a function; domain: $[3, \infty)$; range: $(-\infty, \infty)$
23. function; domain: $(-\infty, \infty)$ **25.** not a function; domain: $[0, \infty)$
27. function; domain: $(-\infty, \infty)$ **29.** not a function; domain: $(-\infty, \infty)$
31. function; domain: $[0, \infty)$ **33.** function; domain: $(-\infty, 0) \cup (0, \infty)$
35. function; domain: $\left[-\dfrac{1}{2}, \infty\right)$ **37.** function; domain:
$(-\infty, 9) \cup (9, \infty)$ **39. (a)** $[0, 3000]$ **(b)** 25 hr; 25 hr **(c)** 2000 gal
(d) $g(0) = 0$; The pool is empty at time 0. **41.** Here is one example.
The cost of gasoline; number of gallons purchased; cost; number
of gallons **43.** 4 **45.** -11 **47.** $-3p + 4$ **49.** $3x + 4$
51. $-3x - 2$ **53.** $-\dfrac{p^2}{9} + \dfrac{4p}{3} + 1$ **55. (a)** 2 **(b)** 3 **57. (a)** 15
(b) 10 **59. (a)** 3 **(b)** -3 **61.** line; -2; linear; $-2x + 4$; -2;
$3; -2$ **63. (a)** $f(x) = \dfrac{12 - x}{3}$ **(b)** 3 **65. (a)** $f(x) = 3 - 2x^2$
(b) -15 **67. (a)** $f(x) = \dfrac{8 - 4x}{-3}$ **(b)** $\dfrac{4}{3}$

69. domain: $(-\infty, \infty)$; range: $(-\infty, \infty)$

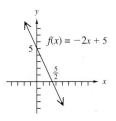

71. domain: $(-\infty, \infty)$; range: $(-\infty, \infty)$

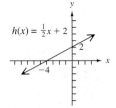

73. domain: $(-\infty, \infty)$; range: $\{-4\}$

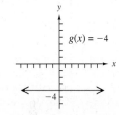

75. (a) \$0; \$1.50; \$3.00; \$4.50 **(b)** $1.50x$
(c) **77.** 194.53 cm **79.** 177.41 cm

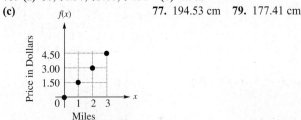

Section 4.6 (page 235)

1. inverse **3.** direct **5.** joint **7.** combined **9.** 36 **11.** .625
13. $222\dfrac{2}{9}$ **15.** increases; decreases **17.** If y varies inversely as
x, x is in the denominator; however, if y varies directly as x, x is
in the numerator. Also, for $k > 0$, with inverse variation, as
x increases, y decreases. With direct variation, y increases as
x increases. **19.** $\$1.69\dfrac{9}{10}$ **21.** about 450 cm^3 **23.** 256 ft
25. $13\dfrac{1}{3}$ amperes **27.** $21\dfrac{1}{3}$ foot-candles **29.** \$420 **31.** 448.1 lb
33. approximately 68,600 calls **35.** 11.8 lb **37.** $(0, 0), (1, 1.75)$
38. 1.75 **39.** $y = 1.75x + 0$ or $y = 1.75x$ **40.** $a = 1.75, b = 0$
41. It is the price per gallon and the slope of the line. **42.** It can
be written in the form $y = kx$ (where $k = a$). The value of a is called
the constant of variation.

Chapter 4 Review Exercises (page 243)

1. $3; 2; \dfrac{10}{3}$

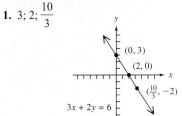

2. $-4, 3; -5; 4$

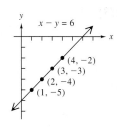

3. $(3, 0); (0, 4)$

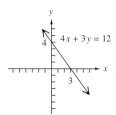

4. $(3, 0); \left(0, \dfrac{15}{7}\right)$

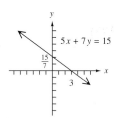

5. $-\dfrac{8}{5}$ **6.** 2   **7.** $\dfrac{3}{4}$ **8.** 0   **9.** $-\dfrac{2}{3}$ **10.** $-\dfrac{1}{3}$ **11.** positive slope

12. negative slope   **13.** 0 slope   **14.** undefined slope   **15.** 12 ft

16. \$1367 per yr   **17.** $y = \dfrac{3}{5}x - 8$ **18.** $y = -\dfrac{1}{3}x + 5$

19. $y = 12$ **20.** $x = 2$ **21.** $y = 4$ **22.** $x = .3$ **23.** $y = -9x + 13$

24. $y = \dfrac{7}{5}x + \dfrac{16}{5}$ **25.** $y = 4x - 26$ **26.** $y = -\dfrac{5}{2}x + 1$

27. (a) $y = 57x + 159$; \$843   **(b)** $y = 47x + 159$; \$723

28.

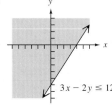

$3x - 2y \le 12$

29.

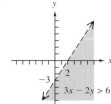

$5x - y > 6$

30.

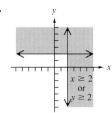

$x \ge 2$ or $y \ge 2$

31.

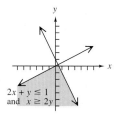

$2x + y \le 1$ and $x \ge 2y$

32. domain: $\{-4, 1\}$; range: $\{2, -2, 5, -5\}$; not a function
33. domain: $\{9, 11, 4, 17, 25\}$; range: $\{32, 47, 69, 14\}$; function
34. domain: $[-4, 4]$; range: $[0, 2]$; function   **35.** function; linear function; domain: $(-\infty, \infty)$ **36.** not a function; domain: $(-\infty, \infty)$
37. function; domain: $(-\infty, \infty)$ **38.** function; domain: $\left[-\dfrac{7}{4}, \infty\right)$
39. not a function; domain: $[0, \infty)$ **40.** function; domain: $(-\infty, 36) \cup (36, \infty)$ **41. (a)** yes   **(b)** domain: $\{1994, 1995, 1996, 1997, 1998, 1999\}$; range: $\{40, 60, 80, 130, 180, 200\}$ **(c)** Answers will vary. Two possible answers are $(1994, 40)$ and $(1995, 60)$.   **(d)** 40; In 1994, CNBC profits were \$40 million.

(e) 1998   **42.** -6 **43.** -15 **44.** $-2p^2 + 3p - 6$

45. $-2k^2 - 3k - 6$ **46.** $f(x) = 2x^2$; 18   **47.** C   **48.** C   **49.** $\dfrac{16}{5}$

50. 430 mm   **51.** 36 ft^3

Chapter 4 Test (page 247)

1. $\dfrac{1}{2}$ **2.** $\dfrac{3}{2}$; $\left(\dfrac{13}{3}, 0\right)$; $\left(0, -\dfrac{13}{2}\right)$ **3.** 0; none; $(0, 5)$
4. The graph is a vertical line.
5. $(-3, 0); (0, 4)$

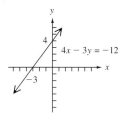

$4x - 3y = -12$

6. none; $(0, 2)$

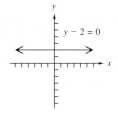

$y - 2 = 0$

7. $(0, 0); (0, 0)$

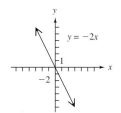

$y = -2x$

8.

$3x - 2y > 6$

9. $y = 14$ **10.** $y = -5x + 19$ **11. (a)** $y = -\dfrac{3}{5}x - \dfrac{11}{5}$

(b) $y = -\dfrac{1}{2}x - \dfrac{3}{2}$ **12.** D; domain: $(-\infty, \infty)$; range: $[0, \infty)$

13. D; domain: $\{0, 3, 6\}$; range: $\{1, 2, 3\}$ **14.** 0; $-a^2 + 2a - 1$
15. 200 amps   **16.** .8 lb

Cumulative Review Exercises: Chapters 1–4 (page 249)

1. always true   **2.** always true   **3.** never true   **4.** sometimes true; for example, $3 + (-3) = 0$, but $3 + (-1) = 2 \ne 0$ **5.** 4   **6.** .64

7. not a real number   **8.** $\dfrac{8}{5}$ **9.** $4m - 3$ **10.** $2x^2 + 5x + 4$

11. $-\dfrac{19}{2}$ **12.** $(-3, 5]$ **13.** no   **14.** -24 **15.** 56

16. undefined   **17.** $\left\{\dfrac{7}{6}\right\}$ **18.** $\{-1\}$ **19.** $h = \dfrac{3V}{\pi r^2}$ **20.** 2 hr

21. 4 white pills   **22.** 6 in.   **23.** The union of the three solution sets is $(-\infty, \infty)$.   **24.** $\left(-\dfrac{1}{2}, \infty\right)$ **25.** $(2, 3)$ **26.** $(-\infty, 2) \cup (3, \infty)$

27. $\left\{-\dfrac{16}{5}, 2\right\}$ **28.** $(-11, 7)$ **29.** $(-\infty, -2] \cup [7, \infty)$

30. $(0, -3), (4, 0), \left(2, -\dfrac{3}{2}\right)$

31. x-intercept: $(-2, 0)$; y-intercept: $(0, 4)$

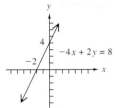

32. $-\dfrac{3}{2}$ **33.** $-\dfrac{1}{2}$ **34.** $-\dfrac{3}{4}$ **35.** $y = -\dfrac{3}{4}x - 1$ **36.** $y = -2$

37. $y = -\dfrac{4}{3}x + \dfrac{7}{3}$ **38. (a)** $(-\infty, \infty)$ **(b)** 22   **39.** 10.5

40. the segment for 1992 through 2000

CHAPTER 5 SYSTEMS OF LINEAR EQUATIONS

Section 5.1 (page 261)
1. $3; -6$ **3.** $\emptyset$ **5.** 0   **7.** D; The ordered pair solution must be in quadrant IV, since that is where the graphs of the equations intersect.   **9. (a)** B   **(b)** C   **(c)** A   **(d)** D
11. $\{(-2, -3)\}$

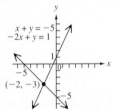

13. yes   **15.** no   **17.** $\{(1, 2)\}$ **19.** $\left\{\left(\dfrac{22}{9}, \dfrac{22}{3}\right)\right\}$ **21.** $\{(2, 3)\}$

23. $\{(5, 4)\}$ **25.** $\left\{\left(-5, -\dfrac{10}{3}\right)\right\}$ **27.** $\{(2, 6)\}$ **29.** $\{(3, -1)\}$

31. $\{(2, -3)\}$ **33.** $\left\{\left(\dfrac{3}{2}, -\dfrac{3}{2}\right)\right\}$ **35.** $\{(x, y) \mid 7x + 2y = 6\}$;
dependent equations   **37.** $\{(2, -4)\}$ **39.** $\emptyset$; inconsistent system
41. $y = -\dfrac{3}{7}x + \dfrac{4}{7}$; $y = -\dfrac{3}{7}x + \dfrac{3}{14}$; 0

43. Both are $y = -\dfrac{2}{3}x + \dfrac{1}{3}$; infinitely many

45. (a) Use substitution since the second equation is solved for y.
(b) Use elimination since the coefficients of the y-terms are opposites.   **(c)** Use elimination since the equations are in standard form with no coefficients of 1 or -1. Solving by substitution would involve fractions.   **47.** $\{(-3, 2)\}$ **49.** $\{(-4, 6)\}$

51. $\{(x, y) \mid 4x - y = -2\}$ **53.** $\left\{\left(1, \dfrac{1}{2}\right)\right\}$ **55. (a)** \$4

(b) 300 half-gallons   **(c)** supply: 200 half-gallons; demand: 400 half-gallons   **57. (a)** Houston, Phoenix, Dallas
(b) Philadelphia   **(c)** Dallas, Phoenix, Philadelphia, Houston
(d) 2010; 1.45 million   **(e)** $(2025, 2.8)$ **59.** 2000, 2001, 2002, first half of 2003   **61.** $\left(\dfrac{25}{7}, \dfrac{733}{70}\right)$ (million)   **63.** $\{(1, 3)\}$

64. $f(x) = -3x + 6$; linear   **65.** $g(x) = \dfrac{2}{3}x + \dfrac{7}{3}$; linear

66. one; 1; 3; 1; 3; 1; 3

Section 5.2 (page 273)
1. The statement means that when -1 is substituted for x, 2 is substituted for y, and 3 is substituted for z in the three equations, the resulting three statements are true.   **3.** $\{(3, 2, 1)\}$ **5.** $\{(1, 4, -3)\}$

7. $\left\{\left(1, \dfrac{3}{10}, \dfrac{2}{5}\right)\right\}$ **9.** $\{(0, 2, -5)\}$ **11.** $\left\{\left(-\dfrac{7}{3}, \dfrac{22}{3}, 7\right)\right\}$

13. $\{(4, 5, 3)\}$ **15.** $\{(2, 2, 2)\}$ **17.** $\left\{\left(\dfrac{8}{3}, \dfrac{2}{3}, 3\right)\right\}$ **19.** Answers

will vary. Some possible answers are   **(a)** two perpendicular walls and the ceiling in a normal room,   **(b)** the floors of three different levels of an office building, and   **(c)** three pages of this book (since they intersect in the spine).   **21.** $\emptyset$ **23.** $\{(x, y, z) \mid x - y + 4z = 8\}$
25. $\{(x, y, z) \mid 2x + y - z = 6\}$ **27.** $\{(0, 0, 0)\}$ **29.** $\{(3, 0, 2)\}$
31. $128 = a + b + c$ **32.** $140 = 2.25a + 1.5b + c$
33. $80 = 9a + 3b + c$ **34.** $a + b + c = 128$;
$2.25a + 1.5b + c = 140$; $9a + 3b + c = 80$; $\{(-32, 104, 56)\}$
35. $f(x) = -32x^2 + 104x + 56$ **36.** height; time   **37.** 56 ft
38. 140.5 ft

Section 5.3 (page 285)
1. wins: 90; losses: 72   **3.** length: 78 ft; width: 36 ft
5. Wal-Mart: \$259 billion; ExxonMobil: \$213 billion   **7.** $x = 40$
and $y = 50$, so the angles measure $40°$ and $50°$.   **9.** NHL: \$253.65;
NBA: \$261.26   **11.** Junior Roast Beef: \$1.49; Big Montana: \$4.19
13. (a) 6 oz   **(b)** 15 oz   **(c)** 24 oz   **(d)** 30 oz   **15.** \$.99x
17. 6 gal of 25%; 14 gal of 35%   **19.** 6 L of pure acid; 48 L of
10% acid   **21.** 14 kg of nuts; 16 kg of cereal   **23.** \$1000 at 2%;
\$2000 at 4%   **25. (a)** $10 - x$ mph   **(b)** $10 + x$ mph
27. freight train: 50 km per hr; express train: 80 km per hr
28. boat: 21 mph; current: 3 mph   **31.** McCartney: \$103.3 million;
Rolling Stones: \$87.9 million   **33.** 76 general admission;
108 with student ID   **35.** 8 for a citron; 5 for a wood apple
37. $x + y + z = 180$; angle measures: $70°, 30°, 80°$ **39.** first:
$20°$; second: $70°$; third: $90°$ **41.** shortest: 12 cm; middle: 25 cm;
longest: 33 cm   **43.** wins: 10; losses: 5; ties: 1   **45.** jersey: \$22;
bobblehead set: \$120; basketball: \$40   **47.** type A: 80;
type B: 160; type C: 250

Section 5.4 (page 297)
1. (a) $0, 5, -3$ **(b)** $1, -3, 8$ **(c)** yes; The number of rows is the
same as the number of columns (three).   **(d)** $\begin{bmatrix} 1 & 4 & 8 \\ 0 & 5 & -3 \\ -2 & 3 & 1 \end{bmatrix}$

(e) $\begin{bmatrix} 1 & -\dfrac{3}{2} & -\dfrac{1}{2} \\ 0 & 5 & -3 \\ 1 & 4 & 8 \end{bmatrix}$ **(f)** $\begin{bmatrix} 1 & 15 & 25 \\ 0 & 5 & -3 \\ 1 & 4 & 8 \end{bmatrix}$

3. $\begin{bmatrix} 1 & 2 & | & 11 \\ 2 & -1 & | & -3 \end{bmatrix}$; $\begin{bmatrix} 1 & 2 & | & 11 \\ 0 & -5 & | & -25 \end{bmatrix}$; $\begin{bmatrix} 1 & 2 & | & 11 \\ 0 & 1 & | & 5 \end{bmatrix}$; $x + 2y = 11$;

$y = 5$; $\{(1, 5)\}$ **5.** $\{(4, 1)\}$ **7.** $\{(1, 1)\}$ **9.** $\{(-1, 4)\}$ **11.** $\emptyset$
13. $\{(x, y) \mid 2x + y = 4\}$ **15.** $\{(0, 0)\}$

17. $\begin{bmatrix} 1 & 1 & -1 & | & -3 \\ 0 & -1 & 3 & | & 10 \\ 0 & -6 & 7 & | & 38 \end{bmatrix}$; $\begin{bmatrix} 1 & 1 & -1 & | & -3 \\ 0 & 1 & -3 & | & -10 \\ 0 & -6 & 7 & | & 38 \end{bmatrix}$;

$\begin{bmatrix} 1 & 1 & -1 & | & -3 \\ 0 & 1 & -3 & | & -10 \\ 0 & 0 & -11 & | & -22 \end{bmatrix}$; $\begin{bmatrix} 1 & 1 & -1 & | & -3 \\ 0 & 1 & -3 & | & -10 \\ 0 & 0 & 1 & | & 2 \end{bmatrix}$; $x + y - z = -3$;

$y - 3z = -10$; $z = 2$; $\{(3, -4, 2)\}$ **19.** $\{(4, 0, 1)\}$
21. $\{(-1, 23, 16)\}$ **23.** $\{(3, 2, -4)\}$ **25.** $\emptyset$
27. $\{(x, y, z) \mid x - 2y + z = 4\}$ **29.** $\{(0, 0, 0)\}$

Chapter 5 Review Exercises (page 303)
1. $\{(2, 2)\}$ 2. (a) 1978 and 1982   (b) just less than 500,000
3. $\left\{\left(-\dfrac{8}{9}, -\dfrac{4}{3}\right)\right\}$ 4. $\{(0, 4)\}$ 5. $\{(2, 4)\}$ 6. $\{(-1, 2)\}$
7. $\{(-6, 3)\}$ 8. $\left\{\left(\dfrac{68}{13}, -\dfrac{31}{13}\right)\right\}$ 9. $\{(x, y) \mid 3x - y = -6\}$
10. $\emptyset$ 11. Answers will vary.

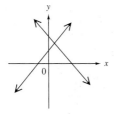

12. Answers will vary.

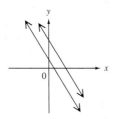

13. Answers will vary.

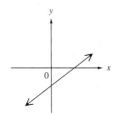

14. Because the lines have the same slope (3) but different y-intercepts ($(0, 2)$ and $(0, -4)$), the lines do not intersect. Thus, the system has no solution.   15. $\{(1, -5, 3)\}$ 16. $\emptyset$
17. $\{(1, 2, 3)\}$ 18. length: 200 ft; width: 85 ft   19. 3 weekend days; 3 weekdays   20. plane: 300 mph; wind: 20 mph   21. 30 lb of $2 per lb nuts; 70 lb of $1 per lb candy   22. 4 vats of green algae; 7 vats of brown algae   23. 85°, 60°, 35°   24. 5 L of 8%; 3 L of 20%; none of 10%   25. Mantle: 54; Maris: 61; Blanchard: 21
26. $\{(3, -2)\}$ 27. $\{(-1, 5)\}$ 28. $\{(0, 0, -1)\}$ 29. $\{(12, 9)\}$
30. $\left\{\left(\dfrac{82}{23}, -\dfrac{4}{23}\right)\right\}$ 31. $\{(3, -1)\}$ 32. $\{(5, 3)\}$ 33. $\{(0, 4)\}$
34. $\emptyset$ 35. 20 L   36. U.S.: 103; Russia: 92; China: 63
37. $2a + b + c = -5$ 38. $-a + c = -1$ 39. $3a + 3b + c = -18$
40. $a = 1, b = -7, c = 0; x^2 + y^2 + x - 7y = 0$
41. The relation is not a function because a vertical line intersects its graph more than once.

Chapter 5 Test (page 307)
1. 1997   2. about 110 thousand

3. $\{(6, 1)\}$

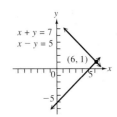

4. $\{(6, -4)\}$ 5. $\{(x, y) \mid 12x - 5y = 8\}$ 6. $\left\{\left(-\dfrac{9}{4}, \dfrac{5}{4}\right)\right\}$
7. $\{(3, 3)\}$ 8. $\{(0, -2)\}$ 9. $\emptyset$ 10. $\left\{\left(-\dfrac{2}{3}, \dfrac{4}{5}, 0\right)\right\}$

11. $\{(3, -2, 1)\}$ 12. *Ocean's Eleven:* $183.4 million; *Runaway Bride:* $152.1 million   13. 45 mph, 75 mph   14. 4 L of 20%; 8 L of 50%   15. AC adaptor: $8; rechargeable flashlight: $15
16. 60 oz of Orange Pekoe; 30 oz of Irish Breakfast; 10 oz of Earl Grey   17. $\left\{\left(\dfrac{2}{5}, \dfrac{7}{5}\right)\right\}$ 18. $\{(-1, 2, 3)\}$

Cumulative Review Exercises: Chapters 1–5 (page 309)
1. 81   2. -81 3. -81 4. .7   5. $-.7$ 6. not a real number
7. -199 8. 455   9. 14   10. $\left\{-\dfrac{15}{4}\right\}$ 11. $\{11\}$
12. $x = \dfrac{d - by}{a - c}$ or $x = \dfrac{by - d}{c - a}$ 13. $\left\{\dfrac{2}{3}, 2\right\}$ 14. $\left(-\infty, \dfrac{240}{13}\right]$
15. $\left[-2, \dfrac{2}{3}\right]$ 16. $(-\infty, \infty)$ 17. 2010; 1813; 62.8%; 57.2%
18. pennies: 35; nickels: 29; dimes: 30   19. 46°, 46°, 88°
20. $y = 6$ 21. $x = 4$ 22. $-\dfrac{4}{3}$ 23. $\dfrac{3}{4}$ 24. $4x + 3y = 10$
25. $f(x) = -\dfrac{4}{3}x + \dfrac{10}{3}$

26.

27.

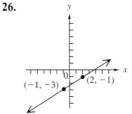

$-3x - 2y \le 6$

28. $\{(3, -3)\}$ 29. $\{(5, 3, 2)\}$ 30. 50 lb of $1.20 candy; 30 lb of $2.40 candy   31. $10,000 at 8%; $7000 at 10%; $8000 at 9%
32. $x = 8$ or 800 parts; $3000   33. about $400

CHAPTER 6 EXPONENTS, POLYNOMIALS, AND POLYNOMIAL FUNCTIONS

Section 6.1 (page 323)
1. incorrect; $(ab)^2 = a^2b^2$ 3. incorrect; $\left(\dfrac{4}{a}\right)^3 = \dfrac{4^3}{a^3}$
5. correct   7. 13^{12} 9. x^{17} 11. $-27w^8$ 13. $18x^3y^8$
15. The product rule does not apply.   17. (a) B   (b) C   (c) B   (d) C   19. 1   21. -1 23. 1   25. -2
27. (a) B   (b) D   (c) B   (d) D
29. $\dfrac{1}{5^4}$ or $\dfrac{1}{625}$ 31. $\dfrac{1}{8}$ 33. $\dfrac{1}{16x^2}$ 35. $\dfrac{4}{x^2}$ 37. $-\dfrac{1}{a^3}$
39. $\dfrac{1}{a^4}$ 41. $\dfrac{11}{30}$ 43. $-\dfrac{5}{24}$ 45. 16   47. $\dfrac{27}{4}$
49. $\dfrac{27}{8}$ 51. $\dfrac{25}{16}$ 53. (a) B   (b) D   (c) D   (d) B
55. 4^2 or 16   57. x^4 59. $\dfrac{1}{r^3}$ 61. 6^6 63. $\dfrac{1}{6^{10}}$
65. 7^2 or 49   67. r^3 69. The quotient rule does not apply.
71. x^{18} 73. $\dfrac{27}{125}$ 75. $64t^3$ 77. $-216x^6$
79. $-\dfrac{64m^6}{t^3}$ 81. $\dfrac{1}{3}$ 83. $\dfrac{1}{a^5}$ 85. $\dfrac{1}{k^2}$ 87. $-4r^6$
89. $\dfrac{625}{a^{10}}$ 91. $\dfrac{z^4}{x^3}$ 93. $\dfrac{1}{5p^{10}}$ 95. $\dfrac{4}{a^2}$ 97. $\dfrac{1}{6y^{13}}$

99. $\dfrac{2^2k^5}{m^2}$ or $\dfrac{4k^5}{m^2}$ **101.** $\dfrac{2k^5}{3}$ **103.** $\dfrac{8}{3pq^{10}}$ **105.** 5.3×10^2
107. 8.3×10^{-1} **109.** 6.92×10^{-6} **111.** -3.85×10^4
113. 72,000 **115.** .00254 **117.** $-60,000$ **119.** .000012
121. .0000025 **123.** 200,000 **125.** 1×10^9; 1×10^{12};
2.128×10^{12}; 1.44419×10^5 **127. (a)** 2.814×10^8
(b) 1×10^{12} **(c)** $3554 **129.** $40,045 **131.** approximately
$3.2 \times 10^4 = 32,000$ hr (about 3.7 yr) **133. (a)** 20,000 hr
(b) 833 days **135.** $-a^n = (-a)^n$ when n is an odd number. When
n is even, $-a^n \neq (-a)^n$. **137.** Write the fraction as its reciprocal
raised to the opposite of the negative power.

Section 6.2 (page 333)
1. neither **3.** ascending **5.** descending **7.** 7; 1
9. $-15; 2$ **11.** 1; 4 **13.** $-1; 6$ **15.** monomial; 0
17. binomial; 1 **19.** trinomial; 3 **21.** none of these; 5
23. $8z^4$ **25.** $7m^3$ **27.** $5x$ **29.** already simplified
31. $-3y^2 + 7y$ **33.** $8k^2 + 2k - 7$ **35.** $-2n^4 - n^3 + n^2$
37. $-9p^2 + 11p - 9$ **39.** $5a + 18$
41. $14m^2 - 13m + 6$ **43.** $13z^2 + 10z - 3$
45. $10y^3 - 7y^2 + 5y + 8$ **47.** $-5a^4 - 6a^3 + 9a^2 - 11$
49. $r + 13$ **51.** $8x^2 + x - 2$ **53.** $-2a^2 - 2a - 7$
55. $-3z^5 + z^2 + 7z$

Section 6.3 (page 341)
1. (a) -10 **(b)** 8 **3. (a)** 8 **(b)** 2 **5. (a)** 8 **(b)** 74
7. (a) -11 **(b)** 4 **9. (a)** 15,191 **(b)** 17,335 **(c)** 18,891
11. (a) 9 million **(b)** 61 million **(c)** 118 million
13. (a) $8x - 3$ **(b)** $2x - 17$ **15. (a)** $-x^2 + 12x - 12$
(b) $9x^2 + 4x + 6$ **17.** $x^2 + 2x - 9$ **19.** 6 **21.** $x^2 - x - 6$
23. 6 **25.** -33 **27.** 0
29. domain: $(-\infty, \infty)$; range: $(-\infty, \infty)$

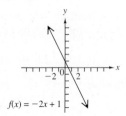

$f(x) = -2x + 1$

31. domain: $(-\infty, \infty)$; range: $(-\infty, 0]$

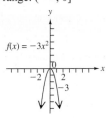

$f(x) = -3x^2$

33. domain: $(-\infty, \infty)$; range: $(-\infty, \infty)$

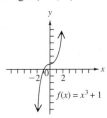

$f(x) = x^3 + 1$

Section 6.4 (page 349)
1. $-24m^5$ **3.** $-6x^2 + 15x$ **5.** $-2q^3 - 3q^4$
7. $18k^4 + 12k^3 + 6k^2$ **9.** $6m^3 + m^2 - 14m - 3$
11. $4x^5 - 4x^4 - 24x^3$ **13.** $6y^2 + y - 12$

15. $25m^2 - 9n^2$ **17.** $-2b^3 + 2b^2 + 18b + 12$
19. $8z^4 - 14z^3 + 17z^2 + 20z - 3$ **21.** $6p^4 + p^3 + 4p^2 - 27p - 6$
23. $m^2 - 3m - 40$ **25.** $12k^2 + k - 6$ **27.** $3z^2 + zw - 4w^2$
29. $12c^2 + 16cd - 3d^2$ **31.** $.1x^2 + .63x - .13$
33. $3r^2 - \dfrac{23}{4}ry - \dfrac{1}{2}y^2$ **35.** The product of two binomials is the
sum of the product of the first terms, the product of the outer terms,
the product of the inner terms, and the product of the last terms.
37. $4p^2 - 9$ **39.** $25m^2 - 1$ **41.** $9a^2 - 4c^2$ **43.** $16x^2 - \dfrac{4}{9}$
45. $16m^2 - 49n^4$ **47.** $5y^5 - 20y^3$ **49.** $y^2 - 10y + 25$
51. $4p^2 + 28p + 49$ **53.** $16n^2 - 24nm + 9m^2$
55. $k^2 - \dfrac{10}{7}kp + \dfrac{25}{49}p^2$ **57.** $(x + y)^2 = x^2 + 2xy + y^2$, because it
is a perfect square trinomial. Thus, it differs from $x^2 + y^2$ by $2xy$.
59. $25x^2 + 10x + 1 + 60xy + 12y + 36y^2$
61. $4a^2 + 4ab + b^2 - 9$ **63.** $4h^2 - 4hk + k^2 - j^2$
65. $x^3 + 6x^2 + 12x + 8$ **67.** $125r^3 - 75r^2s + 15rs^2 - s^3$
69. $q^4 - 8q^3 + 24q^2 - 32q + 16$ **71.** $a - b$
72. $A = s^2; (a - b)^2$ **73.** $(a - b)b$ or $ab - b^2; 2ab - 2b^2$
74. b^2 **75.** $a^2; a$ **76.** $a^2 - (2ab - 2b^2) - b^2 = a^2 - 2ab + b^2$
77. They must be equal to each other.
78. $(a - b)^2 = a^2 - 2ab + b^2$; This reinforces the special product
for the square of a binomial difference. **79.** $10x^2 - 2x$
81. $2x^2 - x - 3$ **83.** $8x^3 - 27$ **85.** -20 **87.** 32 **89.** 20
91. $(2 + 3)^3 \neq 2^3 + 3^3$ because $125 \neq 35$;
$(x + y)^3 = x^3 + 3x^2y + 3xy^2 + y^3$

Section 6.5 (page 357)
1. $3x^3 - 2x^2 + 1$ **3.** $3y + 4 - \dfrac{5}{y}$ **5.** $3m + 5 + \dfrac{6}{m}$
7. $n - \dfrac{3n^2}{2m} + 2$ **9.** $y - 3$ **11.** $t + 5$ **13.** $p - 4 + \dfrac{44}{p + 6}$
15. $z^2 + 3$ **17.** $x^2 + 2x - 3 + \dfrac{6}{4x + 1}$
19. $2x - 5 + \dfrac{-4x + 5}{3x^2 - 2x + 4}$ **21.** $3x^2 + 6x + 11 + \dfrac{26}{x - 2}$
23. $2x^2 - x - 5 + \dfrac{3}{x - 5}$ **25.** $2k^2 + 3k - 1$
27. $9z^2 - 4z + 1 + \dfrac{-z + 6}{z^2 - z + 2}$ **29.** $\dfrac{2}{3}x - 1$
31. $\dfrac{3}{4}a - 2 + \dfrac{1}{4a + 3}$ **33.** $5x - 1; 0$ **35.** $2x - 3; -1$
37. $4x^2 + 6x + 9; \dfrac{3}{2}$ **39.** $\dfrac{x^2 - 9}{2x}, x \neq 0$ **41.** $-\dfrac{5}{4}$
43. $\dfrac{x - 3}{2x}, x \neq 0$ **45.** 0 **47.** $2p + 7$ ft

Chapter 6 Review Exercises (page 363)
1. 64 **2.** $\dfrac{1}{81}$ **3.** -125 **4.** 18 **5.** $\dfrac{81}{16}$ **6.** $\dfrac{16}{25}$ **7.** $\dfrac{11}{30}$ **8.** 0
9. $-12x^2y^8$ **10.** $-\dfrac{2n}{m^5}$ **11.** $\dfrac{10p^8}{q^7}$ **12.** $\dfrac{x^2}{y^2}$ **13.** $\dfrac{1}{3^8}$ **14.** x^8
15. $\dfrac{y^6}{x^2}$ **16.** $\dfrac{1}{z^{15}}$ **17.** $\dfrac{25}{m^{18}}$ **18.** $\dfrac{r^{17}}{9}$ **19.** $\dfrac{25}{z^4}$ **20.** $\dfrac{1}{96m^7}$
21. $\dfrac{2025}{8r^4}$ **22.** 1.345×10^4 **23.** 7.65×10^{-8} **24.** 1.38×10^{-1}
25. 2.814×10^8; 5.0454×10^4; 1×10^2 **26.** 1,210,000
27. .0058 **28.** 2×10^{-4}; .0002 **29.** 1.5×10^3; 1500
30. 4.1×10^{-5}; .000041 **31.** 2.7×10^{-2}; .027 **32.** 14
33. -1 **34. (a)** $11k^3 - 3k^2 + 9k$ **(b)** trinomial **(c)** 3

35. (a) $9m^7 + 14m^6$ **(b)** binomial **(c)** 7
36. (a) $-7q^5r^3$ **(b)** monomial **(c)** 8
37. Answers will vary. An example is $x^5 + 2x^4 - x^2 + x + 2$.
38. $-x^2 - 3x + 1$ **39.** $-5y^3 - 4y^2 + 6y - 12$
40. $6a^3 - 4a^2 - 16a + 15$ **41.** $8y^2 - 9y + 5$
42. $12x^2 + 8x + 5$ **43. (a)** -11 **(b)** 4
44. (a) $5x^2 - x + 5$ **(b)** $-5x^2 + 5x + 1$
(c) 11 **(d)** -9 **45. (a)** 30 million **(b)** 52.325 million
(c) 86.7 million

46.

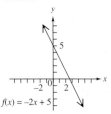

47.

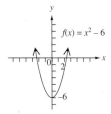

48.

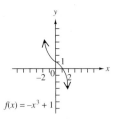

49. $-12k^3 - 42k$ **50.** $14y^2 + 5y - 24$ **51.** $6w^2 - 13wt + 6t^2$
52. $10p^4 + 30p^3 - 8p^2 - 24p$ **53.** $9z^4 - 12z^3 + 16z^2 - 11z + 2$
54. $36r^4 - 1$ **55.** $z^2 - \dfrac{9}{25}$ **56.** $16m^2 + 24m + 9$
57. $8x^3 + 60x^2 + 150x + 125$ **58.** $y^2 - 3y + \dfrac{5}{4}$
59. $p^2 + 6p + 9 + \dfrac{54}{2p - 3}$ **60.** $p^2 + 3p - 6$ **61. (a)** A
(b) G **(c)** C **(d)** C **(e)** A **(f)** E **(g)** B **(h)** H **(i)** F
62. 980 mi^2 **63.** $8x^2 - 10x - 3$ **64.** $\dfrac{y^4}{36}$ **65.** $\dfrac{1}{16y^{18}}$
66. $4x^2 - 36x + 81$ **67.** $2y^2x + \dfrac{3y^3}{2x} + \dfrac{5x^2}{2}$
68. $21p^9 + 7p^8 + 14p^7$ **69.** $-\dfrac{1}{5z^9}$ **70.** $x^2 + 2x - 3$
71. $-14 + 16w - 8w^2$ **72.** $-3k^2 + 4k - 7$

Chapter 6 Test (page 367)

1. (a) C **(b)** A **(c)** D **(d)** A **(e)** E **(f)** F **(g)** B **(h)** G
(i) C **2.** $\dfrac{4x^7}{9y^{10}}$ **3.** $\dfrac{6}{r^{14}}$ **4.** $\dfrac{16}{9p^{10}q^{28}}$ **5.** $\dfrac{16}{x^6 y^{16}}$ **6.** .00000091
7. 3×10^{-4}; .0003 **8. (a)** -18 **(b)** $-2x^2 + 12x - 9$
(c) $-2x^2 - 2x - 3$ **(d)** -7
9.

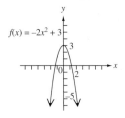

10. (a) 615 thousand **(b)** 738 thousand **(c)** 812 thousand
11. $x^3 - 2x^2 - 10x - 13$ **12.** $10x^2 - x - 3$
13. $6m^3 - 7m^2 - 30m + 25$ **14.** $36x^2 - y^2$

15. $9k^2 + 6kq + q^2$ **16.** $4y^2 - 9z^2 + 6zx - x^2$
17. $4p - 8 + \dfrac{6}{p}$ **18.** $x^2 + 4x + 4$ **19. (a)** $x^3 + 4x^2 + 5x + 2$
(b) 0 **20. (a)** $x + 2, x \neq -1$ **(b)** 0

Cumulative Review Exercises: Chapters 1–6 (page 369)

1. A, B, C, D, F **2.** B, C, D, F **3.** D, F **4.** C, D, F **5.** E, F
6. D, F **7.** 32 **8.** 0 **9.** $\{-65\}$ **10.** $(-\infty, \infty)$ **11.** $(-\infty, 6)$
12. $\left\{ -\dfrac{1}{3}, 1 \right\}$ **13.** $\left(-\infty, -\dfrac{8}{3} \right] \cup [2, \infty)$
14. 32%; 390; 270; 10% **15.** 15°, 35°, 130°
16. $-\dfrac{4}{3}$; $4x + 3y = -1$
17.

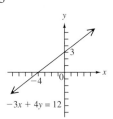

18.

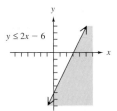

19.

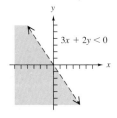

20. (a) 2505.1 per yr; The number of twin births increased
an average of 2505.1 per yr. **(b)** $y = 2505.1x + 93,865$
(c) about 123,926 **21.** domain: $\{-4, -1, 2, 5\}$; range:
$\{-2, 0, 2\}$; function **22.** $\{(3, 2)\}$ **23.** $\emptyset$ **24.** $\{(1, 0, -1)\}$
25. length: 42 ft; width: 30 ft **26.** $\dfrac{8m^9n^3}{p^6}$ **27.** $\dfrac{y^7}{x^{13}z^2}$ **28.** $\dfrac{m^6}{8n^9}$
29. $2x^2 - 4x + 38$ **30.** $15x^2 + 7xy - 2y^2$ **31.** $64m^2 - 25n^2$
32. $m^2 - 2m + 3$

CHAPTER 7 FACTORING

Section 7.1 (page 377)
1. To factor a polynomial means to write it as the product of
two or more polynomials. **3.** $3m$ **5.** $3(r + t)^2$ **7.** A
9. $10(x - 3)$ **11.** $8(s + 2t)$ **13.** $6(1 + 2r)$ **15.** $8k(k^2 + 3)$
17. $xy(3 - 5y)$ **19.** $-2p^2q^4(2p + q)$ **21.** $7x^3(3x^2 + 5x - 2)$
23. $5ac(3ac^2 - 5c + a)$ **25.** cannot be factored
27. $(m - 4)(2m + 5)$ **29.** $11(2z - 1)$
31. $(2 - x)^2(10 - x - x^2)$
33. $r(-r^2 + 3r + 5)$; $-r(r^2 - 3r - 5)$
35. $12s^4(-s + 4)$; $-12s^4(s - 4)$
37. $2x^2(-x^3 + 3x + 2)$; $-2x^2(x^3 - 3x - 2)$ **39.** $(m + 3q)(x + y)$
41. $(5m + n)(2 + k)$ **43.** $(2 - q)(2 - 3p)$ **45.** $(p + q)(p - 4z)$
47. $(2y + 3)(3y + 2x)$ **49.** $(m + 4)(m^2 - 6)$
51. $(a^2 + b^2)(-3a + 2b)$ **53.** $(y - 2)(x - 2)$
55. $(3y - 2)(3y^3 - 4)$ **57.** $m^{-5}(3 + m^2)$ **59.** $p^{-3}(3 + 2p)$

Section 7.2 (page 385)
1. D **3.** C **5.** $(y - 3)(y + 10)$ **7.** $(p - 8)(p + 7)$
9. $-(m - 10)(m - 6)$ **11.** $(a + 5b)(a - 7b)$ **13.** prime
15. $(xy + 9)(xy + 2)$ **17.** $-(6m - 5)(m + 3)$
19. $(5x - 6)(2x + 3)$ **21.** $(4k + 3)(5k + 8)$

23. $(3a - 2b)(5a - 4b)$ **25.** $(6m - 5)^2$ **27.** prime
29. $(2xz - 1)(3xz + 4)$ **31.** $3(4x + 5)(2x + 1)$
33. $-5(a + 6)(3a - 4)$ **35.** $-11x(x - 6)(x - 4)$
37. $2xy^3(x - 12y)(x - 12y)$ **39.** $(5k + 4)(2k + 1)$
41. $(3m + 3p + 5)(m + p - 4)$ **43.** $(a + b)^2(a - 3b)(a + 2b)$
45. $(2x^2 + 3)(x^2 - 6)$ **47.** $(4x^2 + 3)(4x^2 + 1)$
49. $(6p^3 - r)(2p^3 - 5r)$ **51.** no **52.** 1, 3, 5, 9, 15, 45; no
53. no **54.** $(5x + 2)(2x + 5)$; no **55.** Since k is odd, 2 is not a factor of $2x^2 + kx + 8$, and because 2 is a factor of $2x + 4$, the binomial $2x + 4$ cannot be a factor. **56.** $3y + 15$ cannot be a factor of $12y^2 - 11y - 15$ because 3 is a factor of $3y + 15$, but 3 is not a factor of $12y^2 - 11y - 15$.

Section 7.3 (page 391)

1. A, D **3.** B, C **5.** The sum of two squares can be factored only if the terms of the binomial have a common factor. **7.** $(p + 4)(p - 4)$
9. $(5x + 2)(5x - 2)$ **11.** $2(3a + 7b)(3a - 7b)$
13. $4(4m^2 + y^2)(2m + y)(2m - y)$ **15.** $(y + z + 9)(y + z - 9)$
17. $(4 + x + 3y)(4 - x - 3y)$ **19.** $4pq$ **21.** $(k - 3)^2$
23. $(2z + w)^2$ **25.** $(4m - 1 + n)(4m - 1 - n)$
27. $(2r - 3 + s)(2r - 3 - s)$ **29.** $(x + y - 1)(x - y + 1)$
31. $2(7m + 3n)^2$ **33.** $(p + q + 1)^2$ **35.** $(a - b + 4)^2$
37. $(2x - y)(4x^2 + 2xy + y^2)$ **39.** $(4g + 3h)(16g^2 - 12gh + 9h^2)$
41. $3(2n + 3p)(4n^2 - 6np + 9p^2)$
43. $(y + z - 4)(y^2 + 2yz + z^2 + 4y + 4z + 16)$
45. $(m^2 - 5)(m^4 + 5m^2 + 25)$
47. $(k^2 + k + 3)(k^4 - k^3 - 2k^2 + 6k + 9)$
49. $(x^3 - y^3)(x^3 + y^3)$; $(x - y)(x^2 + xy + y^2)(x + y)(x^2 - xy + y^2)$
50. $(x^2 + xy + y^2)(x^2 - xy + y^2)$
51. $(x^2 - y^2)(x^4 + x^2y^2 + y^4)$; $(x - y)(x + y)(x^4 + x^2y^2 + y^4)$
52. $x^4 + x^2y^2 + y^4$ **53.** The product must equal $x^4 + x^2y^2 + y^4$. Multiply $(x^2 + xy + y^2)(x^2 - xy + y^2)$ to verify this. **54.** Start by factoring as a difference of squares.

Summary Exercises on Factoring (page 393)

1. $(10a + 3b)(10a - 3b)$ **2.** $(5r - 1)(2r + 3)$
3. $6p^3(3p^2 - 4 + 2p^3)$ **4.** $5x(3x - 4)$ **5.** $(x + 7)(x - 5)$
6. $(3 + a - b)(3 - a + b)$ **7.** prime
8. $(x - 10)(x^2 + 10x + 100)$ **9.** $(6b + 1)(b - 3)$ **10.** prime
11. $3mn(3m + 2n)(2m - n)$ **12.** $(3t - 7u)(2t + 11u)$
13. $(2p + 5q)(p + 3q)$ **14.** $9m(m - 5 + 2m^2)$ **15.** $(2k + 7r)^2$
16. $2(3m - 10)(9m^2 + 30m + 100)$ **17.** $(m - 2)(n + 5)$
18. $(3m - 5n + p)(3m - 5n - p)$ **19.** $(x + 3)^2(x - 3)$
20. $7(2k - 5)(4k^2 + 10k + 25)$ **21.** prime
22. $(2p - 5)(4p^2 + 10p + 25)$ **23.** $(3k + 1)(2k - 1)$
24. $3(3m + 8n)^2$ **25.** $(x^2 + 25)(x + 5)(x - 5)$
26. $(5m^2 + 6)(25m^4 - 30m^2 + 36)$ **27.** $(a + 6)(b + c)$
28. $(p + 4)(p^2 - 4p + 16)$ **29.** $4y(y - 2)$
30. $(3a^2 + 2)(2a^2 - 5)$ **31.** $(7z + 2k)(2z - k)$
32. $6z(2z^2 - z + 3)$ **33.** $16(4b + 5c)(4b - 5c)$ **34.** prime
35. $8(5z + 4)(25z^2 - 20z + 16)$ **36.** $(8m + 5n)(8m - 5n)$
37. $(5r - s)(2r + 5s)$ **38.** $(3k - 5q)(4k + q)$
39. $8x^2(4 + 2x - 3x^3)$ **40.** $3(4k^2 + 9)(2k + 3)(2k - 3)$
41. $(7x + 5q)(2x - 5q)$ **42.** $5p(p - 2)$ **43.** $(y + 5)(y - 2)$
44. $(b - 9a)(b + 2a)$ **45.** $2a(a^2 + 3a - 2)$
46. $4rx(3m^2 + mn + 10n^2)$ **47.** $(9p - 5r)(2p + 7r)$
48. $(3a + b)(7a - 4b)$ **49.** $(x - 2y + 2)(x - 2y - 2)$
50. $(3m - n + 5)(3m - n - 5)$ **51.** $(5r + 2s - 3)^2$
52. $(p + 8q - 5)^2$ **53.** $(z + 2)(z - 2)(z^2 - 5)$
54. $(3m^2 - 5)(7m^2 + 1)$

Section 7.4 (page 401)

1. First rewrite the equation so that one side is 0. Factor the other side and set each factor equal to 0. The solutions of these linear equations are solutions of the quadratic equation. **3.** $\{-10, 5\}$

5. $\left\{-\dfrac{8}{3}, \dfrac{5}{2}\right\}$ **7.** $\{-2, 5\}$ **9.** $\{-6, -3\}$ **11.** $\left\{-\dfrac{1}{2}, 4\right\}$
13. $\left\{-\dfrac{1}{3}, \dfrac{4}{5}\right\}$ **15.** $\left\{-\dfrac{3}{4}\right\}$ **17.** $\left\{-5, -\dfrac{1}{5}\right\}$
19. $\{0, 6\}$ **21.** $\{-3, 3\}$ **23.** $\{-2, 2\}$ **25.** $\{-4, 2\}$
27. $\left\{-\dfrac{1}{2}, 6\right\}$ **29.** $\{1, 6\}$ **31.** $\left\{-\dfrac{1}{2}, 0, 5\right\}$ **33.** $\left\{-\dfrac{4}{3}, 0, \dfrac{4}{3}\right\}$
35. $\left\{-\dfrac{5}{2}, -1, 1\right\}$ **37.** By dividing each side by a variable expression, she "lost" the solution 0. **39.** width: 16 ft; length: 20 ft **41.** base: 8 m; height: 11 m **43.** 50 ft by 100 ft
45. length: 15 in.; width: 9 in. **47.** 3 sec and 5 sec; 1 sec and 7 sec **49.** 6 sec

Chapter 7 Review Exercises (page 407)

1. $7y(3y + 5)$ **2.** $4qb(3q + 2b - 5q^2b)$ **3.** $(x + 3)(x - 3)$
4. $(z + 1)(3z - 1)$ **5.** $(m + q)(4 + n)$ **6.** $(x + y)(x + 5)$
7. $(m + 3)(2 - a)$ **8.** $(a - b)(2m - p)$ **9.** $(3p - 4)(p + 1)$
10. $(3r + 1)(4r - 3)$ **11.** $(2m + 5)(5m + 6)$
12. $(2k - h)(5k - 3h)$ **13.** prime **14.** $2x(4 + x)(3 - x)$
15. $(2k^2 + 1)(k^2 - 3)$ **16.** $(p + 2)^2(p + 3)(p - 2)$
17. $(4x + 5)(4x - 5)$ **18.** $(3t + 7)(3t - 7)$ **19.** $(x + 7)^2$
20. $(3k - 2)^2$ **21.** $(r + 3)(r^2 - 3r + 9)$
22. $(5x - 1)(25x^2 + 5x + 1)$
23. $(m + 1)(m^2 - m + 1)(m - 1)(m^2 + m + 1)$
24. $(x^4 + 1)(x^2 + 1)(x + 1)(x - 1)$ **25.** $(x + 3 + 5y)(x + 3 - 5y)$
26. $\left\{-1, -\dfrac{2}{5}\right\}$ **27.** $\{2, 3\}$ **28.** $\left\{-\dfrac{5}{2}, \dfrac{10}{3}\right\}$ **29.** $\left\{-\dfrac{3}{2}, \dfrac{1}{3}\right\}$
30. $\{-3, 3\}$ **31.** $\left\{-\dfrac{3}{2}, 0\right\}$ **32.** $\left\{\dfrac{1}{2}, 1\right\}$ **33.** $\{4\}$
34. $\left\{-\dfrac{7}{2}, 0, 4\right\}$ **35.** 3 ft **36.** length: 60 ft; width: 40 ft
37. after 16 sec **38.** after 1 sec and after 15 sec
39. The rock reaches a height of 240 ft once on its way up and once on its way down. **40.** $a(6 - m)(5 + m)$
41. $(2 - a)(4 + 2a + a^2)$ **42.** prime **43.** $5y^2(3y + 4)$
44. $\left\{-\dfrac{3}{5}, 4\right\}$ **45.** $\{-1, 0, 1\}$ **46.** width: 25 ft; length: 110 ft

Chapter 7 Test (page 409)

1. $11z(z - 4)$ **2.** $5x^2y^3(2y^2 - 1 - 5x^3)$ **3.** $(x + y)(3 + b)$
4. $-(2x + 9)(x - 4)$ **5.** $(3x - 5)(2x + 7)$ **6.** $(4p - q)(p + q)$
7. $(4a + 5b)^2$ **8.** $(x + 1 + 2z)(x + 1 - 2z)$
9. $(a + b)(a - b)(a + 2)$ **10.** $(3k + 11j)(3k - 11j)$
11. $(y - 6)(y^2 + 6y + 36)$ **12.** $(2k^2 - 5)(3k^2 + 7)$
13. $(3x^2 + 1)(9x^4 - 3x^2 + 1)$ **14.** $-(x + 5)(x - 6)$
15. $(t^2 + 8)(t^2 + 2)$ **16.** It is not in factored form because there are two terms: $(x^2 + 2y)p$ and $3(x^2 + 2y)$. The common factor is $x^2 + 2y$, and the factored form is $(x^2 + 2y)(p + 3)$.
17. D **18.** $\left\{-2, -\dfrac{2}{3}\right\}$ **19.** $\left\{0, \dfrac{5}{3}\right\}$ **20.** $\left\{-\dfrac{2}{5}, 1\right\}$
21. length: 8 in.; width: 5 in. **22.** 2 sec and 4 sec

Cumulative Review Exercises: Chapters 1–7 (page 411)

1. $-2m + 6$ **2.** $4m - 3$ **3.** $2x^2 + 5x + 4$ **4.** -24 **5.** 204
6. undefined **7.** 10 **8.** $\left\{\dfrac{7}{6}\right\}$ **9.** $\{-1\}$ **10.** $\left(-\infty, \dfrac{15}{4}\right]$
11. $\left(-\dfrac{1}{2}, \infty\right)$ **12.** $(2, 3)$ **13.** $(-\infty, 2) \cup (3, \infty)$
14. $\left\{-\dfrac{16}{5}, 2\right\}$ **15.** $(-11, 7)$ **16.** $(-\infty, -2] \cup [7, \infty)$

17. $h = \dfrac{V}{lw}$ **18.** 2 hr

19.

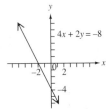

$4x + 2y = -8$

20. -1 **21.** 0   **22.** -1 **23.** $\left(-\dfrac{7}{2}, 0\right)$ **24.** $(0, 7)$

25. $\{(1, 5)\}$ **26.** $\{(1, 1, 0)\}$ **27.** $\dfrac{y}{18x}$ **28.** $\dfrac{5my^4}{3}$

29. $x^3 + 12x^2 - 3x - 7$ **30.** $49x^2 + 42xy + 9y^2$

31. $10p^3 + 7p^2 - 28p - 24$ **32.** $(2w + 7z)(8w - 3z)$

33. $(2x - 1 + y)(2x - 1 - y)$ **34.** $(2y - 9)^2$

35. $(10x^2 + 9)(10x^2 - 9)$ **36.** $(2p + 3)(4p^2 - 6p + 9)$

37. $\left\{-4, -\dfrac{3}{2}, 1\right\}$ **38.** $\left\{\dfrac{1}{3}\right\}$ **39.** 4 ft

40. longer sides: 18 in.; distance between: 16 in.

CHAPTER 8   RATIONAL EXPRESSIONS AND FUNCTIONS

Section 8.1 (page 421)
1. C   **3.** D   **5.** E   **7.** Replacing x with 2 makes the denominator 0 and the value of the expression undefined. To find the values excluded from the domain, set the denominator equal to 0 and solve the equation. All solutions of the equation are excluded from the domain.   **9.** 7; $\{x | x \neq 7\}$ **11.** $-\dfrac{1}{7}$; $\left\{x | x \neq -\dfrac{1}{7}\right\}$ **13.** 0; $\{x | x \neq 0\}$

15. $-2, \dfrac{3}{2}$; $\left\{x | x \neq -2, \dfrac{3}{2}\right\}$ **17.** none; $(-\infty, \infty)$ **19.** none; $(-\infty, \infty)$

21. (a) numerator: $x^2, 4x$; denominator: $x, 4$ **(b)** First factor the numerator, getting $x(x + 4)$, then divide the numerator and denominator by the common factor of $x + 4$ to get $\dfrac{x}{1}$ or x.   **23.** B   **25.** x

27. $\dfrac{x - 3}{x + 5}$ **29.** $\dfrac{x + 3}{2x(x - 3)}$ **31.** already in lowest terms   **33.** $\dfrac{6}{7}$

35. $\dfrac{z}{6}$ **37.** $\dfrac{2}{t - 3}$ **39.** $\dfrac{x - 3}{x + 1}$ **41.** $\dfrac{4x + 1}{4x + 3}$ **43.** $a^2 - ab + b^2$

45. $\dfrac{c + 6d}{c - d}$ **47.** $\dfrac{a + b}{a - b}$ **49.** -1 *In Exercises 51–55, there are other acceptable ways to express each answer.*   **51.** $-(x + y)$

53. $-\dfrac{x + y}{x - y}$ **55.** $-\dfrac{1}{2}$ **57.** already in lowest terms   **59.** $\dfrac{x + 4}{x - 2}$

61. $\dfrac{2x + 3}{x + 2}$ **63.** $-\dfrac{35}{8}$ **65.** $\dfrac{7x}{6}$ **67.** $-\dfrac{p + 5}{2p}$ (There are other ways.)   **69.** $\dfrac{-m(m + 7)}{m + 1}$ (There are other ways.)   **71.** -2

73. $\dfrac{x + 4}{x - 4}$ **75.** $\dfrac{2x + 3y}{2x - 3y}$ **77.** $\dfrac{k + 5p}{2k + 5p}$ **79.** $(k - 1)(k - 2)$

Section 8.2 (page 431)
1. To add or subtract rational expressions that have a common denominator, first add or subtract the numerators. Then place the result over the common denominator. Write the answer in lowest terms.   **3.** $\dfrac{9}{t}$ **5.** $\dfrac{2}{x}$ **7.** 1   **9.** $x - 5$ **11.** $\dfrac{1}{p + 3}$ **13.** $a - b$

15. $72x^4y^5$ **17.** $z(z - 2)$ **19.** $2(y + 4)$ **21.** $(x + 9)^2(x - 9)$ **23.** $(m + n)(m - n)$ **25.** $x(x - 4)(x + 1)$ **27.** $(t + 5)(t - 2)(2t - 3)$ **29.** $2y(y + 3)(y - 3)$ **31.** Yes, they could both be correct because the expressions are equivalent.

Multiplying $\dfrac{3}{5 - y}$ by 1 in the form $\dfrac{-1}{-1}$ gives $\dfrac{-3}{y - 5}$.   **33.** $\dfrac{31}{3t}$

35. $\dfrac{5 - 22x}{12x^2y}$ **37.** $\dfrac{1}{x(x - 1)}$ **39.** $\dfrac{5a^2 - 7a}{(a + 1)(a - 3)}$ **41.** 3

43. $\dfrac{3}{x - 4}$ or $\dfrac{-3}{4 - x}$ **45.** $\dfrac{w + z}{w - z}$ or $\dfrac{-w - z}{z - w}$ **47.** $\dfrac{-13}{12(3 + x)}$

49. $\dfrac{2(2x - 1)}{x - 1}$ **51.** $\dfrac{7}{y}$ **53.** $\dfrac{6}{x - 2}$ **55.** $\dfrac{3x - 2}{x - 1}$ **57.** $\dfrac{4x - 7}{x^2 - x + 1}$

59. $\dfrac{2x + 1}{x}$ **61.** $\dfrac{x}{(x - 2)^2(x - 3)}$ **63.** $\dfrac{10x + 23}{(x + 2)^2(x + 3)}$

65. $\dfrac{2x(x + 12y)}{(x + 2y)(x - y)(x + 6y)}$ **67.** $c(x) = \dfrac{10x}{49(101 - x)}$

69. $\dfrac{8}{9}$ **70.** $\dfrac{3}{7} + \dfrac{5}{9} - \dfrac{6}{63}$; They are the same.   **71.** $\dfrac{8}{9}$; yes

72. Answers will vary. Suppose the name is Sosa, so that $x = 4$. The problem is $\dfrac{3}{2} + \dfrac{5}{4} - \dfrac{6}{8}$. The predicted answer is $\dfrac{8}{4} = 2$, which is correct.   **73.** It causes $\dfrac{3}{x - 2}$ and $\dfrac{6}{x^2 - 2x}$ to be undefined, since 0 appears in the denominators.   **74.** 0

Section 8.3 (page 439)
1. Begin by simplifying the numerator. Then simplify the denominator. Write as a division problem, and proceed.   **3.** $\dfrac{2x}{x - 1}$

5. $\dfrac{2(k + 1)}{3k - 1}$ **7.** $\dfrac{5x^2}{9z^3}$ **9.** $\dfrac{1 + x}{-1 + x}$ **11.** $\dfrac{y + x}{y - x}$ **13.** $4x$

15. $x + 4y$ **17.** $\dfrac{3y}{2}$ **19.** $\dfrac{x^2 + 5x + 4}{x^2 + 5x + 10}$ **21.** $\dfrac{m^2 + 6m - 4}{m(m - 1)}$

22. $\dfrac{m^2 - m - 2}{m(m - 1)}$ **23.** $\dfrac{m^2 + 6m - 4}{m^2 - m - 2}$ **24.** $m(m - 1)$

25. $\dfrac{m^2 + 6m - 4}{m^2 - m - 2}$ **26.** Method 1 involves simplifying the numerator and the denominator separately and then performing a division. Method 2 involves multiplying the fraction by a form of 1, the identity element for multiplication. (Preferences will vary.)

27. $\dfrac{x^2y^2}{y^2 + x^2}$ **29.** $\dfrac{y^2 + x^2}{xy^2 + x^2y}$ or $\dfrac{y^2 + x^2}{xy(y + x)}$ **31.** $\dfrac{1}{2xy}$

Section 8.4 (page 445)

1. (a) $-1, 2$ **(b)** $\{x | x \neq -1, 2\}$ **3. (a)** $-\dfrac{5}{3}, 0, -\dfrac{3}{2}$

(b) $\left\{x | x \neq -\dfrac{5}{3}, 0, -\dfrac{3}{2}\right\}$ **5. (a)** 0   **(b)** $\{x | x \neq 0\}$ **7. (a)** $4, \dfrac{7}{2}$

(b) $\left\{x | x \neq 4, \dfrac{7}{2}\right\}$ **9. (a)** $0, 1, -3, 2$ **(b)** $\{x | x \neq 0, 1, -3, 2\}$

11. $\{1\}$ **13.** $\{-6, 4\}$ **15.** $\left\{-\dfrac{7}{12}\right\}$ **17.** $\emptyset$ **19.** $\{-3\}$

21. $\{5\}$ **23.** $\{5\}$ **25.** $\emptyset$ **27.** $\left\{\dfrac{27}{56}\right\}$ **29.** $\emptyset$ **31.** $\{-10\}$

33. $\emptyset$ **35.** $\{0\}$ **37.** $\left\{x | x \neq -\dfrac{3}{2}, \dfrac{3}{2}\right\}$

39. $x = 0$; $y = 0$

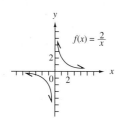

41. $x = 2$; $y = 0$

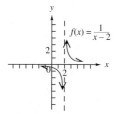

43. (a) 0 **(b)** 1.6 **(c)** 4.1 **(d)** The waiting time also increases.

45. Substituting -1 for x gives a true statement, $\dfrac{4}{3} = \dfrac{4}{3}$.

Substituting -2 for x leads to 0 in the first and third denominators.
46. $C = -4$; $\{-2\}$; -1 is rejected. **47.** $C = 24$; $\{-4\}$; 3 is rejected. **48.** Answers will vary. However, in every case, $-B$ will be the rejected solution, and $\{-A\}$ will be the solution set.

Summary Exercises on Rational Expressions and Equations (page 449)

1. equation; $\{20\}$ **2.** expression; $\dfrac{2(x + 5)}{5}$ **3.** expression; $-\dfrac{22}{7x}$

4. expression; $\dfrac{y + x}{y - x}$ **5.** equation; $\left\{\dfrac{1}{2}\right\}$ **6.** equation; $\{7\}$

7. expression; $\dfrac{43}{24x}$ **8.** equation; $\{1\}$ **9.** expression; $\dfrac{5x - 1}{-2x + 2}$ or

$\dfrac{5x - 1}{-2(x - 1)}$ **10.** expression; $\dfrac{25}{4(r + 2)}$ **11.** expression;

$\dfrac{x^2 + xy + 2y^2}{(x + y)(x - y)}$ **12.** expression; $\dfrac{24p}{p + 2}$ **13.** expression; $-\dfrac{5}{36}$

14. equation; $\{0\}$ **15.** expression; $\dfrac{b + 3}{3}$ **16.** expression; $\dfrac{5}{3z}$

17. expression; $\dfrac{2x + 10}{x(x - 2)(x + 2)}$ **18.** equation; $\{2\}$

19. expression; $\dfrac{-x}{3x + 5y}$ **20.** equation; $\{-13\}$ **21.** expression;

$\dfrac{3y + 2}{y + 3}$ **22.** equation; $\left\{\dfrac{5}{4}\right\}$ **23.** equation; $\emptyset$ **24.** expression;

$\dfrac{2z - 3}{2z + 3}$ **25.** expression; $\dfrac{-1}{x - 3}$ or $\dfrac{1}{3 - x}$ **26.** expression; $\dfrac{t - 2}{8}$

27. equation; $\{-10\}$ **28.** expression; $\dfrac{13x + 28}{2x(x + 4)(x - 4)}$

29. equation; $\emptyset$ **30.** expression; $\dfrac{k(2k^2 - 2k + 5)}{(k - 1)(3k^2 - 2)}$

Section 8.5 (page 459)

1. A **3.** D **5.** 65.625 **7.** $\dfrac{25}{4}$ **9.** $G = \dfrac{Fd^2}{Mm}$ **11.** $a = \dfrac{bc}{c + b}$

13. $v = \dfrac{PVt}{pT}$ **15.** $r = \dfrac{nE - IR}{In}$ **17.** $b = \dfrac{2A}{h} - B$ or

$b = \dfrac{2A - Bh}{h}$ **19.** $r = \dfrac{eR}{E - e}$ **21.** Multiply each side by $a - b$.

23. 15 girls, 5 boys **25.** $\dfrac{1}{2}$ job per hr **27.** 2000
29. 1996 **31.** 168,800 fast food restaurants **33.** 25,000 fish
35. \$95.75 **37.** 2.4 mL **39.** $x = \dfrac{7}{2}$; $AC = 8$; $DF = 12$
41. 3 mph **43.** 900 mi **45.** 480 mi **47.** 190 mi **49.** $6\dfrac{2}{3}$ min
51. 12 hr **53.** 20 hr **55.** $2\dfrac{4}{5}$ hr

Chapter 8 Review Exercises (page 469)

1. (a) -6 **(b)** $\{x \mid x \neq -6\}$ **2. (a)** $2, 5$ **(b)** $\{x \mid x \neq 2, 5\}$

3. (a) 9 **(b)** $\{x \mid x \neq 9\}$ **4.** $\dfrac{x}{2}$ **5.** $\dfrac{5m + n}{5m - n}$ **6.** $\dfrac{-1}{2 + r}$

7. The reciprocal of a rational expression is another rational expression such that the two rational expressions have a product of 1.

8. $\dfrac{3y^2(2y + 3)}{2y - 3}$ **9.** $\dfrac{-3(w + 4)}{w}$ **10.** $\dfrac{z(z + 2)}{z + 5}$ **11.** 1 **12.** $96b^5$

13. $9r^2(3r + 1)$ **14.** $(3x - 1)(2x + 5)(3x + 4)$ **15.** $\dfrac{16z - 3}{2z^2}$

16. 12 **17.** $\dfrac{71}{30(a + 2)}$ **18.** $\dfrac{13r^2 + 5rs}{(5r + s)(2r - s)(r + s)}$

19. $\dfrac{3 + 2t}{4 - 7t}$ **20.** -2 **21.** $\dfrac{1}{3q + 2p}$ **22.** $\dfrac{y + x}{xy}$ **23.** $\{-3\}$

24. $\{-2\}$ **25.** $\{0\}$ **26.** $\emptyset$ **27.** Although her algebra was correct, 3 is not a solution because it is not in the domain of the equation. Thus, $\emptyset$ is correct. **28.** In simplifying the expression, we are combining terms to get a single fraction with a denominator of $6x$, while in solving the equation, we are finding a value for x that makes the equation true. **29.** C; $x = 0$; $y = 0$ **30.** $\dfrac{15}{2}$ **31.** $m = \dfrac{Fd^2}{GM}$

32. $M = \dfrac{m\mu}{v - \mu}$ **33.** 6000 passenger-km per day **34.** 16 km per hr

35. $4\dfrac{4}{5}$ min **36.** $3\dfrac{3}{5}$ hr **37.** $\dfrac{1}{x - 2y}$ **38.** $\dfrac{x + 5}{x + 2}$ **39.** $\dfrac{6m + 5}{3m^2}$

40. $\dfrac{k - 3}{36k^2 + 6k + 1}$ **41.** $\dfrac{x^2 - 6}{2(2x + 1)}$ **42.** $\dfrac{x(9x + 1)}{3x + 1}$

43. $\dfrac{3 - 5x}{6x + 1}$ **44.** $\dfrac{11}{3 - x}$ or $\dfrac{-11}{x - 3}$ **45.** $\dfrac{1}{3}$ **46.** $\dfrac{s^2 + t^2}{st(s - t)}$

47. $\dfrac{5a^2 + 4ab + 12b^2}{(a + 3b)(a - 2b)(a + b)}$ **48.** $\dfrac{acd + b^2d + bc^2}{bcd}$

49. $\left\{\dfrac{1}{3}\right\}$ **50.** $r = \dfrac{AR}{R - A}$ or $r = \dfrac{-AR}{A - R}$ **51.** $\{1, 4\}$

52. $\left\{-\dfrac{14}{3}\right\}$ **53. (a)** 8.32 **(b)** 44.9 **54.** $8\dfrac{4}{7}$ min

Chapter 8 Test (page 473)

1. $-2, \dfrac{4}{3}$; $\left\{x \mid x \neq -2, \dfrac{4}{3}\right\}$ **2.** $\dfrac{2x - 5}{x(3x - 1)}$ **3.** $\dfrac{3(x + 3)}{4}$

4. $\dfrac{y + 4}{y - 5}$ **5.** $\dfrac{x + 5}{x}$ **6.** $t^2(t + 3)(t - 2)$ **7.** $\dfrac{7 - 2t}{6t^2}$

8. $\dfrac{13x + 35}{(x - 7)(x + 7)}$ **9.** $\dfrac{4}{x + 2}$ **10.** $\dfrac{72}{11}$ **11.** $-\dfrac{1}{a + b}$

12. $\dfrac{2y^2 + x^2}{xy(y - x)}$ **13. (a)** expression; $\dfrac{11(x - 6)}{12}$ **(b)** equation; $\{6\}$

14. $\left\{\dfrac{1}{2}\right\}$ **15.** $\{5\}$ **16.** A solution cannot make a denominator 0.

17. $x = -1; y = 0$

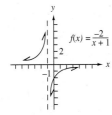

18. $\ell = \dfrac{2S}{n} - a$ or $\ell = \dfrac{2S - na}{n}$ **19.** $3\dfrac{3}{14}$ hr **20.** 15 mph

21. 48,000 fish **22. (a)** 3 units **(b)** 0

Cumulative Review Exercises: Chapters 1–8 (page 475)

1. -199 **2.** 12 **3.** $\left\{-\dfrac{15}{4}\right\}$ **4.** $\left\{\dfrac{2}{3}, 2\right\}$

5. $x = \dfrac{d - by}{a - c}$ or $x = \dfrac{by - d}{c - a}$ **6.** $\left(-\infty, \dfrac{240}{13}\right]$

7. $(-\infty, -2] \cup \left[\dfrac{2}{3}, \infty\right)$ **8.** \$4000 at 4%; \$8000 at 3% **9.** 6 m

10. x-intercept: $(-2, 0)$; y-intercept: $(0, 4)$

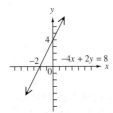

11. $-\dfrac{3}{2}$ **12.** $-\dfrac{3}{4}$ **13.** $y = -\dfrac{3}{2}x + \dfrac{1}{2}$

14. **15.**

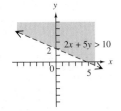

16. function; domain: {1990, 1992, 1994, 1996, 1998, 2000}; range:
{1.25, 1.61, 1.80, 1.21, 1.94, 2.26} **17.** not a function; domain:
$[-2, \infty)$; range: $(-\infty, \infty)$ **18.** function; domain: $[-2, \infty)$; range:
$(-\infty, 0]$ **19. (a)** $f(x) = \dfrac{5x - 8}{3}$ or $f(x) = \dfrac{5}{3}x - \dfrac{8}{3}$ **(b)** -1

20. $3x + 15$ **21.** $\{(-1, 3)\}$ **22.** $\{(-2, 3, 1)\}$ **23.** $\emptyset$

24. Lisa Leslie: 3768; Sheryl Swoopes: 2969 **25.** $\dfrac{a^{10}}{b^{10}}$ **26.** $\dfrac{m}{n}$

27. $4y^2 - 7y - 6$ **28.** $-6x^6 + 18x^5 - 12x^4$ **29.** $12f^2 + 5f - 3$

30. $49t^6 - 64$ **31.** $\dfrac{1}{16}x^2 + \dfrac{5}{2}x + 25$ **32.** $x^2 + 4x - 7$

33. (a) $2x^3 - 2x^2 + 6x - 4$ **(b)** $2x^3 - 4x^2 + 2x + 2$ **(c)** -14
34. $(2x + 5)(x - 9)$ **35.** $25(2t^2 + 1)(2t^2 - 1)$

36. $(2p + 5)(4p^2 - 10p + 25)$ **37.** $\left\{-\dfrac{7}{3}, 1\right\}$ **38.** $\dfrac{y + 4}{y - 4}$

39. $\dfrac{2x - 3}{2(x - 1)}$ **40.** $\dfrac{a(a - b)}{2(a + b)}$ **41.** 3 **42.** $\dfrac{2(x + 2)}{2x - 1}$

43. $\{-4\}$ **44.** $q = \dfrac{fp}{p - f}$ or $q = \dfrac{-fp}{f - p}$ **45.** 150 mph **46.** $1\dfrac{1}{5}$ hr

---

CHAPTER 9 ROOTS, RADICALS, AND ROOT FUNCTIONS

Section 9.1 (page 485)

1. E **3.** D **5.** A **7.** C **9.** C **11. (a)** not a real number
(b) negative **(c)** 0 **13.** -9 **15.** 6 **17.** -4 **19.** -8 **21.** 6

23. -3 **25.** not a real number **27.** 2 **29.** -9 **31.** $\dfrac{8}{9}$ **33.** $\dfrac{2}{3}$

35. $\dfrac{1}{2}$ **37.** $[-3, \infty)$; $[0, \infty)$

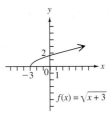

39. $[0, \infty)$; $[-2, \infty)$

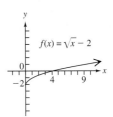

41. $(-\infty, \infty)$; $(-\infty, \infty)$

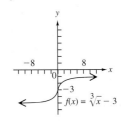

43. $|x|$ **45.** x **47.** $|x^5|$ **49.** $|x^3|$ **51.** 97.381 **53.** 16.863
55. 2.1 **57.** 1.5 **59.** -4 and 4 **60.** 4 **61.** 4, -4
62. $\{-4, 4\}$ **63.** $\pm\sqrt{16}$ represents the two numbers $\sqrt{16} = 4$
and $-\sqrt{16} = -4$. **64.** $\sqrt{x^2}$ is always nonnegative, so it must be
simplified as $|x|$ because x may be negative. **65.** 1,183,000 cycles
per sec **67.** 10 mi **69.** 392,000 mi^2 **71.** 1.732 amps

Section 9.2 (page 495)

1. C **3.** A **5.** H **7.** B **9.** D **11.** 13 **13.** 9

15. 2 **17.** $\dfrac{8}{9}$ **19.** -3 **21.** 1000 **23.** -1024

25. not a real number **27.** $\dfrac{1}{512}$ **29.** $\dfrac{9}{4}$ **31.** $(-64)^{1/2}$ is an even
root of a negative number. No real number squared will give -64.
On the other hand, $-64^{1/2} = -\sqrt{64} = -8$, which is a
real number. ($-64^{1/2}$ is the opposite of $64^{1/2}$.) **33.** $\sqrt{12}$

35. $(\sqrt[4]{8})^3$ **37.** $(\sqrt[8]{9q})^5 - (\sqrt[3]{2x})^2$ **39.** $\dfrac{1}{(\sqrt{2m})^3}$

41. $(\sqrt[3]{2y + x})^2$ **43.** $\dfrac{1}{(\sqrt[3]{3m^4 + 2k^2})^2}$

45. $\sqrt{a^2 + b^2} = \sqrt{3^2 + 4^2} = 5; a + b = 3 + 4 = 7; 5 \neq 7$
47. 64 **49.** 64 **51.** x^{10} **53.** $\sqrt[6]{x^5}$ **55.** $\sqrt[15]{t^8}$ **57.** 9 **59.** 4

61. y **63.** $k^{2/3}$ **65.** x^3y^8 **67.** $\dfrac{1}{x^{10/3}}$ **69.** $\dfrac{1}{m^{1/4}n^{3/4}}$ **71.** p^2

73. $\dfrac{c^{11/3}}{b^{11/4}}$ **75.** $\dfrac{q^{5/3}}{9p^{7/2}}$ **77.** $p + 2p^2$ **79.** $k^{7/4} - k^{3/4}$ **81.** $6 + 18a$

83. $\dfrac{5}{m^3}$ **85.** $y^{3/2}$ **87.** $\dfrac{1}{k^{2/3}}$ **89.** $x^{1/3}z^{5/6}$ **91.** $k^{1/6}$ **93.** $y^{1/30}$

95. $x^{5/27}$ **97.** 72 in.; 6.0 ft **99.** $m^{5/2}(m^{1/2} - 3)$

101. $x^{-1/3}(8x + 5)$ **103.** $m^{1/8}(2 - m^{1/2})$

Section 9.3 (page 505)

1. true; Both are equal to $4\sqrt{3}$ and approximately 6.92820323.

3. true; Both are equal to $6\sqrt{2}$ and approximately 8.485281374.

5. Because there are only two factors of $\sqrt[3]{x}$, $\sqrt[3]{x} \cdot \sqrt[3]{x} = (\sqrt[3]{x})^2$ or $\sqrt[3]{x^2}$. **7.** D **9.** $\sqrt{30}$ **11.** $\sqrt{14x}$ **13.** $\sqrt{42pqr}$ **15.** $\sqrt[3]{14xy}$

17. $\sqrt[4]{33}$ **19.** $\sqrt[4]{6xy^2}$ **21.** This product cannot be simplified

using the product rule. **23.** $\dfrac{8}{11}$ **25.** $\dfrac{\sqrt{3}}{5}$ **27.** $\dfrac{\sqrt{x}}{5}$ **29.** $\dfrac{p^3}{9}$

31. $\dfrac{3}{4}$ **33.** $-\dfrac{\sqrt[3]{r^2}}{2}$ **35.** $-\dfrac{3}{x}$ **37.** $\dfrac{1}{x^3}$ **39.** $2\sqrt{3}$ **41.** $12\sqrt{2}$

43. $-4\sqrt{2}$ **45.** $-2\sqrt{7}$ **47.** not a real number **49.** $4\sqrt[3]{2}$

51. $-2\sqrt[3]{2}$ **53.** $2\sqrt[3]{5}$ **55.** $-4\sqrt[4]{2}$ **57.** $2\sqrt[5]{2}$

59. His reasoning was incorrect. Here 8 is a term, not a factor.

61. $6k\sqrt{2}$ **63.** $\dfrac{3\sqrt[3]{3}}{4}$ **65.** $11x^3$ **67.** $-3t^4$ **69.** $-10m^4z^2$

71. $5a^2b^3c^4$ **73.** $\dfrac{1}{2}r^2t^5$ **75.** $5x\sqrt{2x}$ **77.** $-10r^5\sqrt{5r}$

79. $x^3y^4\sqrt{13x}$ **81.** $2z^2w^3$ **83.** $-2zt^2\sqrt[3]{2z^2t}$ **85.** $3x^3y^4$

87. $-3r^3s^2\sqrt[4]{2r^3s^2}$ **89.** $\dfrac{y^5\sqrt{y}}{6}$ **91.** $\dfrac{x^5\sqrt[3]{x}}{3}$ **93.** $4\sqrt{3}$

95. $\sqrt{5}$ **97.** $x^2\sqrt{x}$ **99.** $\sqrt[6]{432}$ **101.** $\sqrt[12]{6912}$ **103.** $\sqrt[6]{x^5}$

105. 5 **107.** $8\sqrt{2}$ **109.** 13 **111.** $9\sqrt{2}$ **113.** $\sqrt{17}$ **115.** 5

117. $6\sqrt{2}$ **119.** $\sqrt{5y^2 - 2xy + x^2}$ **121.** 27.0 in. **123.** .003

125. 15.3 mi

Section 9.4 (page 513)

1. B **3.** 15; Each radical expression simplifies to a whole number.

5. -4 **7.** $7\sqrt{3}$ **9.** $14\sqrt[3]{2}$ **11.** $5\sqrt[4]{2}$ **13.** $24\sqrt{2}$

15. cannot be simplified further **17.** $20\sqrt{5}$ **19.** $12\sqrt{2x}$

21. $-2m\sqrt{2}$ **23.** $\sqrt[3]{2}$ **25.** $2\sqrt[3]{x}$ **27.** $-\sqrt[3]{x^2y}$

29. $-x\sqrt[3]{xy^2}$ **31.** $19\sqrt[4]{2}$ **33.** $x\sqrt[4]{xy}$ **35.** $9\sqrt[4]{2a^3}$

37. $\dfrac{5\sqrt{5}}{6}$ **39.** $\dfrac{7\sqrt{2}}{6}$ **41.** $\dfrac{5\sqrt{2}}{3}$ **43.** $5\sqrt{2} + 4$

45. $\dfrac{5 - 3x}{x^4}$ **47.** $\dfrac{m\sqrt[3]{m^2}}{2}$ **49.** $\dfrac{3x\sqrt[3]{2} - 4\sqrt[3]{5}}{x^3}$

51. $12\sqrt{5} + 5\sqrt{3}$ in. **53.** $24\sqrt{2} + 12\sqrt{3}$ in.

Section 9.5 (page 521)

1. E **3.** A **5.** D **7.** $6 - 4\sqrt{3}$ **9.** $6 - \sqrt{6}$ **11.** 2 **13.** 9

15. $3\sqrt{2} - 5\sqrt{3} + 2\sqrt{6} - 10$ **17.** $3x - 4$ **19.** $4x - y$

21. $16x + 24\sqrt{x} + 9$ **23.** $81 - \sqrt[3]{4}$ **25.** $6 - 4\sqrt{3}$ is not

equal to $2\sqrt{3}$ because 6 and $4\sqrt{3}$ are not like terms, so they

cannot be combined. **27.** $\sqrt{7}$ **29.** $5\sqrt{3}$ **31.** $\dfrac{\sqrt{6}}{2}$

33. $\dfrac{9\sqrt{15}}{5}$ **35.** $-\sqrt{2}$ **37.** $\dfrac{-8\sqrt{3k}}{k}$ **39.** $\dfrac{6\sqrt{3}}{y}$

41. Both methods lead to the same result, $\dfrac{6\sqrt{3}}{y}$, but multiplying the

numerator and denominator by $\sqrt{y}$ produces this result more directly,

with less simplification required. **43.** $\dfrac{\sqrt{14}}{2}$ **45.** $-\dfrac{\sqrt{14}}{10}$

47. $\dfrac{2\sqrt{6x}}{x}$ **49.** $-\dfrac{7r\sqrt{2rs}}{s}$ **51.** $\dfrac{12x^3\sqrt{2xy}}{y^5}$ **53.** $\dfrac{\sqrt[3]{18}}{3}$

55. $\dfrac{\sqrt[3]{12}}{3}$ **57.** $-\dfrac{\sqrt[3]{2pr}}{r}$ **59.** $\dfrac{2\sqrt[4]{x^3}}{x}$ **61.** Multiply the

numerator and denominator by $4 - \sqrt{3}$, so the denominator

becomes $(4 + \sqrt{3})(4 - \sqrt{3}) = 16 - 3 = 13$, a rational number.

63. $\dfrac{2(4 - \sqrt{3})}{13}$ **65.** $3(\sqrt{5} - \sqrt{3})$ **67.** $\sqrt{3} + \sqrt{7}$

69. $\sqrt{7} - \sqrt{6} - \sqrt{14} + 2\sqrt{3}$ **71.** $\dfrac{4\sqrt{x}(\sqrt{x} + 2\sqrt{y})}{x - 4y}$

73. $\dfrac{x\sqrt{2} - \sqrt{3xy} - \sqrt{2xy} + y\sqrt{3}}{2x - 3y}$ **75.** Square each side to

show that each square is equal to $\dfrac{2 - \sqrt{3}}{4}$. **77.** $\dfrac{5 + 2\sqrt{6}}{4}$

79. $\dfrac{4 + 2\sqrt{2}}{3}$ **81.** $\dfrac{6 + 2\sqrt{6x}}{3}$ **83.** $\dfrac{319}{6(8\sqrt{5} + 1)}$

84. $\dfrac{9a - b}{(\sqrt{b} - \sqrt{a})(3\sqrt{a} - \sqrt{b})}$

85. $\dfrac{(3\sqrt{a} + \sqrt{b})(\sqrt{b} + \sqrt{a})}{b - a}$ **86.** In Exercise 84, we

multiplied the numerator and denominator by the conjugate of the

numerator, while in Exercise 85 we multiplied by the conjugate of the

denominator.

Summary Exercises on Operations with Radicals and Rational Exponents (page 525)

1. $-6\sqrt{10}$ **2.** $7 - \sqrt{14}$ **3.** $2 + \sqrt{6} - 2\sqrt{3} - 3\sqrt{2}$

4. $4\sqrt{2}$ **5.** $73 + 12\sqrt{35}$ **6.** $\dfrac{-\sqrt{6}}{2}$ **7.** $4(\sqrt{7} - \sqrt{5})$

8. $3\sqrt[3]{2x^2}$ **9.** $-3 + 2\sqrt{2}$ **10.** -2 **11.** -44 **12.** $\dfrac{\sqrt{x} + \sqrt{5}}{x - 5}$

13. $2abc^3\sqrt[3]{b^2}$ **14.** $5\sqrt[3]{3}$ **15.** $3(\sqrt{5} - 2)$ **16.** $\dfrac{\sqrt{15x}}{5x}$

17. $\dfrac{8}{5}$ **18.** $\dfrac{\sqrt{2}}{8}$ **19.** $-\sqrt[3]{100}$ **20.** $11 + 2\sqrt{30}$ **21.** $-3\sqrt{3x}$

22. $52 - 30\sqrt{3}$ **23.** 1 **24.** $\dfrac{\sqrt[3]{117}}{9}$ **25.** $t^2\sqrt[4]{t}$

26. $3\sqrt{2} + \sqrt{15} + \sqrt{42} + \sqrt{35}$ **27.** $2\sqrt[4]{27}$

28. $\dfrac{1 + \sqrt[3]{3} + \sqrt[3]{9}}{-2}$ **29.** $\dfrac{x\sqrt[3]{x^2}}{y}$ **30.** $-4\sqrt{3} - 3$ **31.** $xy^{6/5}$

32. $x^{10}y$ **33.** $\dfrac{1}{25x^2}$ **34.** $\dfrac{-6y^{1/6}}{x^{1/24}}$ **35.** $7 + 4 \cdot 3^{1/2}$ or $7 + 4\sqrt{3}$

36. 1

Section 9.6 (page 531)

1. (a) yes **(b)** no **3. (a)** yes **(b)** no **5.** no; There is no

solution. The radical expression, which is positive, cannot equal a

negative number. **7.** $\{11\}$ **9.** $\left\{\dfrac{1}{3}\right\}$ **11.** $\emptyset$ **13.** $\{5\}$ **15.** $\{18\}$

17. $\{5\}$ **19.** $\{4\}$ **21.** $\{17\}$ **23.** $\{5\}$ **25.** $\emptyset$ **27.** $\{0\}$ **29.** $\{0\}$

31. $\left\{-\dfrac{1}{3}\right\}$ **33.** $\emptyset$ **35.** You cannot just square each term. The

right side should be $(8 - x)^2 = 64 - 16x + x^2$. The correct first step

is $3x + 4 = 64 - 16x + x^2$, and the solution set is $\{4\}$. **37.** $\{1\}$

39. $\{-1\}$ **41.** $\{14\}$ **43.** $\{8\}$ **45.** $\{0\}$ **47.** $\emptyset$ **49.** $\{7\}$

51. $\{7\}$ **53.** $\{4, 20\}$ **55.** $\emptyset$ **57.** $\left\{\dfrac{5}{4}\right\}$ **59.** $\{9, 17\}$

61. $\left\{\dfrac{1}{4}, 1\right\}$ **63.** $K = \dfrac{V^2 m}{2}$ **65.** $L = \dfrac{1}{4\pi^2 f^2 C}$

67. $r = \dfrac{a}{4\pi^2 N^2}$

Section 9.7 (page 541)

1. i **3.** $-i$ **5.** $a + bi$ is a complex number if a and b are real numbers and i is the imaginary unit. Therefore, for every real number a, if $b = 0$, $a = a + 0i$ is a complex number. **7.** $13i$ **9.** $-12i$ **11.** $i\sqrt{5}$ **13.** $4i\sqrt{3}$ **15.** -15 **17.** -10 **19.** $\sqrt{3}$ **21.** $5i$ **23.** $-1 + 7i$ **25.** 0 **27.** $7 + 3i$ **29.** -2 **31.** $1 + 13i$ **33.** $6 + 6i$ **35.** $4 + 2i$ **37.** -81 **39.** -16 **41.** $-10 - 30i$ **43.** $10 - 5i$ **45.** $-9 + 40i$ **47.** 153 **49.** (a) $a - bi$ (b) $a^2; b^2$ **51.** $1 + i$ **53.** $-1 + 2i$ **55.** $2 + 2i$ **57.** $-\dfrac{5}{13} - \dfrac{12}{13}i$ **59.** (a) $4x + 1$ (b) $4 + i$ **60.** (a) $-2x + 3$ (b) $-2 + 3i$ **61.** (a) $3x^2 + 5x - 2$ (b) $5 + 5i$ **62.** (a) $-\sqrt{3} + \sqrt{6} + 1 - \sqrt{2}$ (b) $\dfrac{1}{5} - \dfrac{7}{5}i$ **63.** Because $i^2 = -1$, two pairs of like terms can be combined in Exercise 61(b). **64.** Because $i^2 = -1$, additional terms can be combined in the numerator and denominator. **65.** $\dfrac{5}{41} + \dfrac{4}{41}i$ **67.** -1 **69.** i **71.** 1 **73.** $-i$ **75.** Since $i^{20} = (i^4)^5 = 1^5 = 1$, the student multiplied by 1, which is justified by the identity property for multiplication. **77.** $\dfrac{1}{2} + \dfrac{1}{2}i$ **79.** $(1 + 5i)^2 - 2(1 + 5i) + 26$ will simplify to 0 when the operations are applied.

Chapter 9 Review Exercises (page 549)

1. 42 **2.** -17 **3.** not a real number **4.** 6 **5.** -2 **6.** $|x|$ **7.** x **8.** $|x^5|$ **9.** domain: $[1, \infty)$; range: $[0, \infty)$

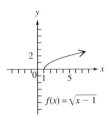

10. domain: $(-\infty, \infty)$; range: $(-\infty, \infty)$

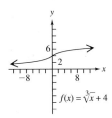

11. n must be even, and a must be negative. **12.** It is not a real number. **13.** 6.325 **14.** 8.775 **15.** 17.607 **16.** 1.9 sec **17.** 66 in.^{2} **18.** 7 **19.** -2 **20.** not a real number **21.** By a power rule for exponents and the definition of $x^{1/n}$, $a^{m/n} = (a^m)^{1/n} = \sqrt[n]{a^m}$. **22.** 32 **23.** -4 **24.** $-\dfrac{216}{125}$ **25.** -32 **26.** $\dfrac{1000}{27}$ **27.** 49 **28.** 96 **29.** $\dfrac{k^{17/12}}{2}$

30. $\sqrt[5]{2^4}$ or $\sqrt[5]{16}$ **31.** 3^9 **32.** $7^4\sqrt{7}$ **33.** $m^4\sqrt[3]{m}$ **34.** $k^2\sqrt[4]{k}$ **35.** $\sqrt[6]{m}$ **36.** $2y\sqrt[4]{y}$ **37.** $\sqrt[15]{y^8}$ **38.** $\sqrt[12]{y^5}$ **39.** $\sqrt{66}$ **40.** $\sqrt{5r}$ **41.** $\sqrt[3]{30}$ **42.** $\sqrt[4]{21}$ **43.** $2\sqrt{5}$ **44.** $-5\sqrt{5}$ **45.** $-3x\sqrt[3]{4xy}$ **46.** $4pq^2\sqrt[3]{p}$ **47.** $\dfrac{7}{9}$ **48.** $\dfrac{y\sqrt{y}}{12}$ **49.** $\dfrac{m^5}{3}$ **50.** $\dfrac{\sqrt[3]{r^2}}{2}$ **51.** $\sqrt[12]{2}$ **52.** $\sqrt[10]{x^3}$ **53.** $\sqrt{130}$ **54.** $\sqrt{53}$ **55.** $-11\sqrt{2}$ **56.** $23\sqrt{5}$ **57.** $7\sqrt{3y}$ **58.** $26m\sqrt{6m}$ **59.** $19\sqrt[3]{2}$ **60.** $-8\sqrt[4]{2}$ **61.** $1 - \sqrt{3}$ **62.** 2 **63.** $9 - 7\sqrt{2}$ **64.** $86 + 8\sqrt{55}$ **65.** $15 - 2\sqrt{26}$ **66.** $12 - 2\sqrt{35}$ **67.** $-3\sqrt{6}$ **68.** $\dfrac{3\sqrt{7py}}{y}$ **69.** $-\dfrac{\sqrt[3]{45}}{5}$ **70.** $\dfrac{3m\sqrt[3]{4n}}{n^2}$ **71.** $\dfrac{\sqrt{2} - \sqrt{7}}{-5}$ **72.** $\dfrac{-5(\sqrt{6} + \sqrt{3})}{3}$ **73.** $\{2\}$ **74.** $\{6\}$ **75.** $\varnothing$ **76.** $\{0, 5\}$ **77.** $\{9\}$ **78.** $\{3\}$ **79.** $\{7\}$ **80.** $\left\{-\dfrac{1}{2}\right\}$ **81.** $\{6\}$ **82.** $5i$ **83.** $10i\sqrt{2}$ **84.** $4i\sqrt{10}$ **85.** $-10 - 2i$ **86.** $14 + 7i$ **87.** $-\sqrt{35}$ **88.** -45 **89.** 3 **90.** $5 + i$ **91.** $32 - 24i$ **92.** $1 - i$ **93.** $4 + i$ **94.** $-i$ **95.** 1 **96.** $-i$ **97.** $-13ab^2$ **98.** $\dfrac{1}{100}$ **99.** $\dfrac{1}{y^{1/2}}$ **100.** $\dfrac{x^{3/4}}{z^{3/4}}$ **101.** k^6 **102.** $3z^3 t^2\sqrt[3]{2t^2}$ **103.** $57\sqrt{2}$ **104.** $6x\sqrt[3]{y^2}$ **105.** $\sqrt{35} + \sqrt{15} - \sqrt{21} - 3$ **106.** $-\dfrac{\sqrt{3}}{6}$ **107.** $\dfrac{\sqrt[3]{60}}{5}$ **108.** $\dfrac{2\sqrt{z}(\sqrt{z} + 2)}{z - 4}$ **109.** $7i$ **110.** $3 - 7i$ **111.** $-5i$ **112.** $\{5\}$ **113.** $\left\{\dfrac{3}{2}\right\}$ **114.** 7.9 ft **115.** $(12\sqrt{3} + 5\sqrt{2})$ ft

Chapter 9 Test (page 553)

1. -29 **2.** 5 **3.** C **4.** 12.09 **5.** domain: $[-6, \infty)$; range: $[0, \infty)$

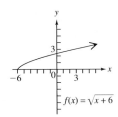

6. $\dfrac{1}{256}$ **7.** $\dfrac{9y^{3/10}}{x^2}$ **8.** $3x^2 y^3\sqrt{6x}$ **9.** $2ab^3\sqrt[4]{2a^3 b}$ **10.** $\sqrt[6]{200}$ **11.** $26\sqrt{5}$ **12.** $66 + \sqrt{5}$ **13.** $-2(\sqrt{7} - \sqrt{5})$ **14.** $\dfrac{-\sqrt{10}}{4}$ **15.** $\dfrac{2\sqrt[3]{25}}{5}$ **16.** $\sqrt{26}$ **17.** $\sqrt{145}$ **18.** $\{-1\}$ **19.** $\{6\}$ **20.** $\{-3\}$ **21.** $-5 - 8i$ **22.** $-10 + 10i$ **23.** $3 + 4i$ **24.** $-i$

Cumulative Review Exercises: Chapters 1–9 (page 555)

1. $\left\{\dfrac{4}{5}\right\}$ **2.** $\left\{\dfrac{11}{10}, \dfrac{7}{2}\right\}$ **3.** $(-6, \infty)$ **4.** $(1, 3)$ **5.** $(-2, 1)$ **6.** $12x + 11y = 18$ **7.** C **8.** (a) $(0, 6)$ (b) $(2, 0)$ **9.** $\$120$

10.

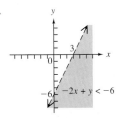

11. Both angles measure 80°.   **12.** {(7, −2)}   **13.** ∅   **14.** infinite number of solutions   **15.** 2-oz letter: $.60; 3-oz letter: $.83

16. $-k^3 - 3k^2 - 8k - 9$ **17.** $8x^2 + 17x - 21$ **18.** $z - 2 + \dfrac{3}{z}$

19. $3y^3 - 3y^2 + 4y + 1 + \dfrac{-10}{2y+1}$ **20.** $(2p - 3q)(p - q)$

21. $(3k^2 + 4)(6k^2 - 5)$ **22.** $(x + 8)(x^2 - 8x + 64)$ **23.** $\dfrac{y}{y+5}$

24. $\dfrac{4x + 2y}{(x+y)(x-y)}$ **25.** $-\dfrac{9}{4}$ **26.** $-\dfrac{1}{a+b}$ **27.** $\left\{-3, -\dfrac{5}{2}\right\}$

28. $\left\{-\dfrac{2}{5}, 1\right\}$ **29.** $\dfrac{1}{243}$ **30.** $x^{1/12}$ **31.** $8\sqrt{5}$ **32.** $\dfrac{-9\sqrt{5}}{20}$

33. $4(\sqrt{6} + \sqrt{5})$ **34.** $6\sqrt[3]{4}$ **35.** $\sqrt{29}$ **36.** {6}

37. 15 mph   **38.** $\dfrac{80}{39}$ or $2\dfrac{2}{39}$ L   **39.** 17 dimes and 12 quarters

40. Brenda: 8 mph; Chuck: 4 mph

CHAPTER 10 QUADRATIC EQUATIONS, INEQUALITIES, AND FUNCTIONS

Section 10.1 (page 565)

1. The equation is also true for $x = -4$.   **3. (a)** A quadratic equation in standard form has a second-degree polynomial in decreasing powers equal to 0.   **(b)** The zero-factor property states that if a product equals 0, then at least one of the factors equals 0.   **(c)** The square root property states that if the square of a quantity equals a number, then the quantity equals the positive or negative square root of the number.   **5.** {9, −9}   **7.** $\{\sqrt{17}, -\sqrt{17}\}$

9. $\{4\sqrt{2}, -4\sqrt{2}\}$ **11.** $\{2\sqrt{5}, -2\sqrt{5}\}$ **13.** $\{2\sqrt{6}, -2\sqrt{6}\}$

15. {−7, 3}   **17.** $\{4 + \sqrt{3}, 4 - \sqrt{3}\}$

19. $\{-5 + 4\sqrt{3}, -5 - 4\sqrt{3}\}$ **21.** $\left\{\dfrac{1 + \sqrt{7}}{3}, \dfrac{1 - \sqrt{7}}{3}\right\}$

23. $\left\{\dfrac{-1 + 2\sqrt{6}}{4}, \dfrac{-1 - 2\sqrt{6}}{4}\right\}$ **25.** 5.6 sec   **27.** square root

property for $(2x + 1)^2 = 5$; completing the square for $x^2 + 4x = 12$

29. (a) 9   **(b)** 49   **(c)** 36   **(d)** $\dfrac{9}{4}$ **(e)** $\dfrac{81}{4}$ **(f)** $\dfrac{1}{16}$ **31.** 4

33. 25   **35.** $\dfrac{1}{36}$ **37.** {−4, 6}   **39.** $\{-2 + \sqrt{6}, -2 - \sqrt{6}\}$

41. $\{-5 + \sqrt{7}, -5 - \sqrt{7}\}$ **43.** $\left\{-\dfrac{8}{3}, 3\right\}$

45. $\left\{\dfrac{-5 + \sqrt{41}}{4}, \dfrac{-5 - \sqrt{41}}{4}\right\}$ **47.** $\left\{\dfrac{5 + \sqrt{15}}{5}, \dfrac{5 - \sqrt{15}}{5}\right\}$

49. $\left\{\dfrac{4 + \sqrt{3}}{3}, \dfrac{4 - \sqrt{3}}{3}\right\}$ **51.** $\left\{\dfrac{2 + \sqrt{3}}{3}, \dfrac{2 - \sqrt{3}}{3}\right\}$

53. $\{1 + \sqrt{2}, 1 - \sqrt{2}\}$ **55.** $\{2i\sqrt{3}, -2i\sqrt{3}\}$

57. $\{5 + i\sqrt{3}, 5 - i\sqrt{3}\}$ **59.** $\left\{\dfrac{1 + 2i\sqrt{2}}{6}, \dfrac{1 - 2i\sqrt{2}}{6}\right\}$

61. $\{-2 + 3i, -2 - 3i\}$ **63.** $\left\{\dfrac{-2 + 2i\sqrt{2}}{3}, \dfrac{-2 - 2i\sqrt{2}}{3}\right\}$

65. $\{-3 + i\sqrt{3}, -3 - i\sqrt{3}\}$ **67.** x^2 **68.** x **69.** $6x$ **70.** 1

71. 9   **72.** $(x + 3)^2$ or $x^2 + 6x + 9$

Section 10.2 (page 575)

1. The student was incorrect, since the fraction bar should extend

under the term $-b$.   **3.** {3, 5}   **5.** $\left\{\dfrac{-2 + \sqrt{2}}{2}, \dfrac{-2 - \sqrt{2}}{2}\right\}$

7. $\left\{\dfrac{1 + \sqrt{3}}{2}, \dfrac{1 - \sqrt{3}}{2}\right\}$ **9.** $\{5 + \sqrt{7}, 5 - \sqrt{7}\}$

11. $\left\{\dfrac{-1 + \sqrt{2}}{2}, \dfrac{-1 - \sqrt{2}}{2}\right\}$ **13.** $\left\{\dfrac{-1 + \sqrt{7}}{3}, \dfrac{-1 - \sqrt{7}}{3}\right\}$

15. $\{1 + \sqrt{5}, 1 - \sqrt{5}\}$ **17.** $\left\{\dfrac{-2 + \sqrt{10}}{2}, \dfrac{-2 - \sqrt{10}}{2}\right\}$

19. $\{-1 + 3\sqrt{2}, -1 - 3\sqrt{2}\}$ **21.** $\left\{\dfrac{3 + i\sqrt{59}}{2}, \dfrac{3 - i\sqrt{59}}{2}\right\}$

23. $\{3 + i\sqrt{5}, 3 - i\sqrt{5}\}$ **25.** $\left\{\dfrac{1 + i\sqrt{6}}{2}, \dfrac{1 - i\sqrt{6}}{2}\right\}$

27. $\left\{\dfrac{-2 + i\sqrt{2}}{3}, \dfrac{-2 - i\sqrt{2}}{3}\right\}$ **29.** B   **31.** C   **33.** A   **35.** D

37. The equations in Exercises 29, 30, 33, and 34 can be solved by factoring.

Section 10.3 (page 585)

1. Multiply by the LCD, x.   **3.** Substitute a variable for $r^2 + r$.

5. The potential solution −1 does not check. The solution set is {4}.

7. {−4, 7}   **9.** $\left\{-\dfrac{2}{3}, 1\right\}$ **11.** $\left\{-\dfrac{14}{17}, 5\right\}$ **13.** $\left\{-\dfrac{11}{7}, 0\right\}$

15. $\left\{\dfrac{-1 + \sqrt{13}}{2}, \dfrac{-1 - \sqrt{13}}{2}\right\}$ **17.** $\left\{\dfrac{2 + \sqrt{22}}{3}, \dfrac{2 - \sqrt{22}}{3}\right\}$

19. (a) $(20 - t)$ mph   **(b)** $(20 + t)$ mph   **21.** 25 mph

23. 80 km per hr   **25.** 3.6 hr   **27.** 9 min   **29.** {1, 4}

31. {3}   **33.** $\left\{\dfrac{8}{9}\right\}$ **35.** {16}   **37.** $\left\{\dfrac{2}{5}\right\}$ **39.** {−3, 3}

41. $\left\{-\dfrac{3}{2}, -1, 1, \dfrac{3}{2}\right\}$ **43.** $\{-2\sqrt{3}, -2, 2, 2\sqrt{3}\}$

45. $\left\{\dfrac{\sqrt{9 + \sqrt{65}}}{2}, -\dfrac{\sqrt{9 + \sqrt{65}}}{2}, \dfrac{\sqrt{9 - \sqrt{65}}}{2}, -\dfrac{\sqrt{9 - \sqrt{65}}}{2}\right\}$

47. {−6, −5}   **49.** {−4, 1}   **51.** $\left\{-\dfrac{1}{3}, \dfrac{1}{6}\right\}$ **53.** {−8, 1}

55. {−64, 27}   **57.** {25}   **59.** It would cause both denominators

to be 0, and division by 0 is undefined.   **60.** $\dfrac{12}{5}$

61. $\left(\dfrac{x}{x-3}\right)^2 + 3\left(\dfrac{x}{x-3}\right) - 4 = 0$ **62.** The numerator can

never equal the denominator, since the denominator is 3 less than

the numerator.   **63.** $\left\{\dfrac{12}{5}\right\}$; The values for t are −4 and 1. The

value 1 is impossible because it leads to a contradiction

$\left(\text{since } \dfrac{x}{x-3} \text{ is never equal to } 1\right)$.   **64.** $\left\{\dfrac{12}{5}\right\}$; The values for s are

$\dfrac{1}{x}$ and $\dfrac{-4}{x}$. The value $\dfrac{1}{x}$ is impossible, since $\dfrac{1}{x} \neq \dfrac{1}{x-3}$ for all x.

Summary Exercises on Solving Quadratic Equations (page 589)

1. square root property   **2.** factoring   **3.** quadratic formula
4. quadratic formula   **5.** factoring   **6.** square root property

7. $\{\sqrt{47}, -\sqrt{47}\}$ **8.** $\left\{-\dfrac{3}{2}, \dfrac{5}{3}\right\}$ **9.** $\{-4 + \sqrt{10}, -4 - \sqrt{10}\}$

10. $\{-3, 11\}$ **11.** $\left\{-\dfrac{1}{2}, 5\right\}$ **12.** $\left\{-3, \dfrac{1}{3}\right\}$

13. $\left\{\dfrac{9 + \sqrt{33}}{6}, \dfrac{9 - \sqrt{33}}{6}\right\}$ **14.** $\{2i\sqrt{3}, -2i\sqrt{3}\}$

15. $\left\{\dfrac{1}{2}, 2\right\}$ **16.** $\left\{-\dfrac{\sqrt{6}}{3}, -\dfrac{1}{2}, \dfrac{1}{2}, \dfrac{\sqrt{6}}{3}\right\}$

17. $\left\{\dfrac{-5 + 2\sqrt{3}}{2}, \dfrac{-5 - 2\sqrt{3}}{2}\right\}$ **18.** $\left\{\dfrac{4}{5}, 3\right\}$

19. $\{-\sqrt{7}, -\sqrt{2}, \sqrt{2}, \sqrt{7}\}$ **20.** $\left\{\dfrac{-2 + \sqrt{14}}{2}, \dfrac{-2 - \sqrt{14}}{2}\right\}$

21. $\left\{\dfrac{-1 + i\sqrt{7}}{2}, \dfrac{-1 - i\sqrt{7}}{2}\right\}$

22. $\left\{\sqrt{4 + \sqrt{15}}, -\sqrt{4 + \sqrt{15}}, \sqrt{4 - \sqrt{15}}, -\sqrt{4 - \sqrt{15}}\right\}$

23. $\left\{\dfrac{3}{2}\right\}$ **24.** $\left\{\dfrac{2}{3}\right\}$ **25.** $\{6\sqrt{2}, -6\sqrt{2}\}$ **26.** $\left\{-\dfrac{2}{3}, 2\right\}$

27. $\{-4, 9\}$ **28.** $\{13, -13\}$ **29.** $\left\{\dfrac{3 + i\sqrt{3}}{3}, \dfrac{3 - i\sqrt{3}}{3}\right\}$

30. $\{3\}$ **31.** $\left\{-\dfrac{1}{3}, \dfrac{1}{6}\right\}$ **32.** $\left\{\dfrac{1 + i\sqrt{47}}{6}, \dfrac{1 - i\sqrt{47}}{6}\right\}$

33. $\{3\}$ **34.** $\left\{-\dfrac{8}{3}, -1\right\}$ **35.** $\left\{-i, i, -\dfrac{1}{2}i, \dfrac{1}{2}i\right\}$ **36.** $\{-2, 7\}$

Section 10.4 (page 595)

1. Solve for w^2 by dividing each side by g.   **3.** $m = \sqrt{p^2 - n^2}$

5. $t = \dfrac{\pm\sqrt{dk}}{k}$ **7.** $d = \dfrac{\pm\sqrt{skI}}{I}$ **9.** $v = \dfrac{\pm\sqrt{kAF}}{F}$

11. $r = \dfrac{\pm\sqrt{3\pi Vh}}{\pi h}$ **13.** $t = \dfrac{-B \pm \sqrt{B^2 - 4AC}}{2A}$ **15.** $h = \dfrac{D^2}{k}$

17. $\ell = \dfrac{p^2 g}{k}$ **19.** 2.3, 5.3, 5.8   **21.** eastbound ship: 80 mi;

southbound ship: 150 mi   **23.** 1 ft   **25.** 20 in. by 12 in.
27. 2.4 sec and 5.6 sec   **29.** 9.2 sec   **31.** It reaches its *maximum*
height at 5 sec because this is the only time it reaches 400 ft.
33. \$.80   **35. (a)** 2.4 million   **(b)** 2.4 million; They are the same.
37. 1995; The graph indicates that sales reached 2 million in 1996.
39. 5.5 m per sec   **41.** 5 or 14

Section 10.5 (page 605)

1. (a) B   **(b)** C   **(c)** A   **(d)** D   **3.** $(0, 0)$ **5.** $(0, 4)$ **7.** $(1, 0)$
9. $(-3, -4)$ **11.** In Exercise 9, the parabola is shifted 3 units to
the left and 4 units down. The parabola in Exercise 10 is shifted
5 units to the right and 8 units down.   **13.** down; wider
15. up; narrower   **17. (a)** I   **(b)** IV   **(c)** II   **(d)** III
19.

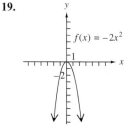

21.

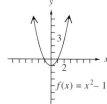

$f(x) = x^2 - 1$

23.

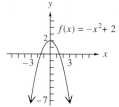

$f(x) = -x^2 + 2$

25. vertex: $(4, 0)$; axis: $x = 4$; domain: $(-\infty, \infty)$; range: $[0, \infty)$

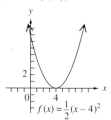

$f(x) = \dfrac{1}{2}(x - 4)^2$

27. vertex: $(-2, -1)$; axis: $x = -2$; domain: $(-\infty, \infty)$; range: $[-1, \infty)$

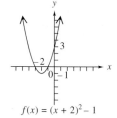

$f(x) = (x + 2)^2 - 1$

29. vertex: $(-3, 4)$; axis: $x = -3$; domain: $(-\infty, \infty)$; range: $(-\infty, 4]$

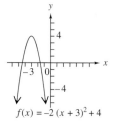

$f(x) = -2(x + 3)^2 + 4$

31. vertex: $(-2, 1)$; axis: $x = -2$; domain: $(-\infty, \infty)$; range: $(-\infty, 1]$

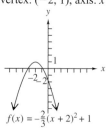

$f(x) = -\dfrac{2}{3}(x + 2)^2 + 1$

33. It is shifted 6 units up.
34.

$G(x) = x + 6$

35. It is shifted 6 units up. **36.** It is shifted 6 units to the right.

37.

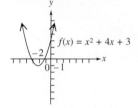
$G(x) = x - 6$

38. It is shifted 6 units to the right. **39.** quadratic; positive
41. quadratic; negative **43.** linear; positive
45. (a) **COMPANY BANKRUPTCY FILINGS**

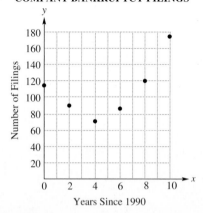

Years Since 1990

(b) quadratic; positive **(c)** $y = 2.969x^2 - 23.125x + 115$
(d) 265 **(e)** No. About 16 companies filed for bankruptcy each
month, so at this rate, filings for 2002 would be about 192. The
approximation using the model seems high. **47. (a)** 222.7
(per 100,000) **(b)** The approximation using the model is high.

Section 10.6 (page 619)

1. If x is squared, it has a vertical axis; if y is squared, it has a
horizontal axis. **3.** Use the discriminant of the corresponding
quadratic equation. If it is positive, there are two x-intercepts. If it
is 0, there is just one x-intercept (the vertex), and if it is negative,
there are no x-intercepts. **5.** $(-1, 3)$; up; narrower; no x-intercepts
7. $\left(\dfrac{5}{2}, \dfrac{37}{4}\right)$; down; same; two x-intercepts
9. $(-3, -9)$; to the right; wider

11. vertex: $(-2, -1)$; axis: $x = -2$; domain: $(-\infty, \infty)$; range: $[-1, \infty)$

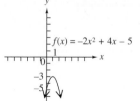
$f(x) = x^2 + 4x + 3$

13. vertex: $(1, -3)$; axis: $x = 1$; domain: $(-\infty, \infty)$; range: $(-\infty, -3]$

$f(x) = -2x^2 + 4x - 5$

15. vertex: $(1, 5)$; axis: $y = 5$; domain: $(-\infty, 1]$; range: $(-\infty, \infty)$

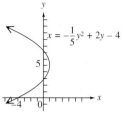
$x = -\dfrac{1}{5}y^2 + 2y - 4$

17. vertex: $(-7, -2)$; axis: $y = -2$; domain: $[-7, \infty)$; range: $(-\infty, \infty)$

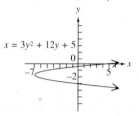
$x = 3y^2 + 12y + 5$

19. F **21.** C **23.** D **25.** 30 and 30 **27.** 160 ft by 320 ft
29. 16 ft; 2 sec **31. (a)** $R(x) = 20,000 + 200x - 4x^2$ **(b)** 25
(c) \$22,500 **33. (a)** minimum **(b)** 1995; 1.7% **35. (a)** The
coefficient of x^2 is negative because the parabola opens down.
(b) $(18.45, 3860)$ **(c)** In 2018 Social Security assets will reach
their maximum value of \$3860 billion.

Section 10.7 (page 629)

1. (a) $\{1, 3\}$ **(b)** $(-\infty, 1) \cup (3, \infty)$ **(c)** $(1, 3)$
3. (a) $\left\{-3, \dfrac{5}{2}\right\}$ **(b)** $\left[-3, \dfrac{5}{2}\right]$ **(c)** $(-\infty, -3] \cup \left[\dfrac{5}{2}, \infty\right)$
5. Include the endpoints if the symbol is $\geq$ or $\leq$. Exclude the end-
points if the symbol is $>$ or $<$.
7. $(-\infty, -1) \cup (5, \infty)$

9. $(-4, 6)$

11. $(-\infty, 1] \cup [3, \infty)$

13. $\left(-\infty, -\dfrac{3}{2}\right] \cup \left[\dfrac{3}{5}, \infty\right)$

15. $\left(-\dfrac{2}{3}, \dfrac{1}{3}\right)$

17. $\left(-\infty, -\dfrac{1}{2}\right] \cup \left[\dfrac{1}{3}, \infty\right)$

19. $(-\infty, 3 - \sqrt{3}] \cup [3 + \sqrt{3}, \infty)$

21. $(-\infty, \infty)$ **23.** $\varnothing$

25. $(-\infty, 1) \cup (2, 4)$

27. $\left[-\dfrac{3}{2}, \dfrac{1}{3}\right] \cup [4, \infty)$

29. $(-\infty, 1) \cup (4, \infty)$

31. $\left[-\frac{3}{2}, 5\right)$

33. $(2, 6]$

35. $\left(-\infty, \frac{1}{2}\right) \cup \left(\frac{5}{4}, \infty\right)$

37. $[-4, -2)$

39. $\left(0, \frac{1}{2}\right) \cup \left(\frac{5}{2}, \infty\right)$

41. $(-\infty, 2] \cup (4, \infty)$

43. 3 sec and 13 sec **44.** between 3 sec and 13 sec **45.** at 0 sec (the time when it is initially projected) and at 16 sec (the time when it hits the ground) **46.** between 0 and 3 sec and between 13 and 16 sec

Chapter 10 Review Exercises (page 637)

1. $\{11, -11\}$ **2.** $\{\sqrt{3}, -\sqrt{3}\}$ **3.** $\left\{-\frac{15}{2}, \frac{5}{2}\right\}$

4. $\left\{\frac{2 + 5i}{3}, \frac{2 - 5i}{3}\right\}$ **5.** $\{-2 + \sqrt{19}, -2 - \sqrt{19}\}$

6. $\left\{\frac{1}{2}, 1\right\}$ **7.** By the square root property, $x = \sqrt{12}$ or $x = -\sqrt{12}$. **8.** 3.1 sec **9.** $\left\{-\frac{7}{2}, 3\right\}$

10. $\left\{\frac{-5 + \sqrt{53}}{2}, \frac{-5 - \sqrt{53}}{2}\right\}$ **11.** $\left\{\frac{1 + \sqrt{41}}{2}, \frac{1 - \sqrt{41}}{2}\right\}$

12. $\left\{\frac{-3 + i\sqrt{23}}{4}, \frac{-3 - i\sqrt{23}}{4}\right\}$

13. $\left\{\frac{2 + i\sqrt{2}}{3}, \frac{2 - i\sqrt{2}}{3}\right\}$

14. $\left\{\frac{-7 + \sqrt{37}}{2}, \frac{-7 - \sqrt{37}}{2}\right\}$ **15.** C **16.** A **17.** D

18. B **19.** $\left\{-\frac{5}{2}, 3\right\}$ **20.** $\left\{-\frac{1}{2}, 1\right\}$ **21.** $\{-4\}$

22. $\left\{-\frac{11}{6}, -\frac{19}{12}\right\}$ **23.** $\left\{-\frac{343}{8}, 64\right\}$ **24.** $\{-2, -1, 1, 2\}$

25. 7 mph **26.** 40 mph **27.** 4.6 hr **28.** Greg: 2.6 hr; Carter: 3.6 hr **29.** $v = \frac{\pm\sqrt{rFkw}}{kw}$ **30.** $y = \frac{6p^2}{z}$

31. $t = \frac{3m \pm \sqrt{9m^2 + 24m}}{2m}$ **32.** 9 ft, 12 ft, 15 ft

33. 12 cm by 20 cm **34.** 1 in. **35.** 3 min **36.** 5.2 sec
37. .7 sec and 4.0 sec **38.** $.50 **39.** 4.5%
40. (a) 305; It is close to the number shown in the graph.
(b) $x \approx 9$, which represents 1999; Based on the graph, the number of e-mail boxes did not quite reach 200 million in 1999.

41. $(1, 0)$ **42.** $(3, 7)$ **43.** $\left(\frac{2}{3}, -\frac{2}{3}\right)$ **44.** $(-4, 3)$

45. vertex: $(2, -3)$; axis: $x = 2$; domain: $(-\infty, \infty)$; range: $[-3, \infty)$

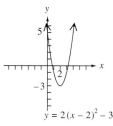

$y = 2(x - 2)^2 - 3$

46. vertex: $(2, 3)$; axis: $x = 2$; domain: $(-\infty, \infty)$; range: $(-\infty, 3]$

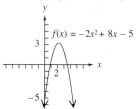

$f(x) = -2x^2 + 8x - 5$

47. vertex: $(-4, -3)$; axis: $y = -3$; domain: $[-4, \infty)$; range: $(-\infty, \infty)$

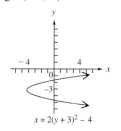

$x = 2(y + 3)^2 - 4$

48. vertex: $(4, 6)$; axis: $y = 6$; domain: $(-\infty, 4]$; range: $(-\infty, \infty)$

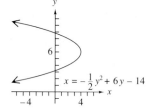

$x = -\frac{1}{2}y^2 + 6y - 14$

49. (a) $c = 10.54$; $9a + 3b + c = 18.49$; $36a + 6b + c = 27.69$
(b) $f(x) = .0694x^2 + 2.442x + 10.54$ **(c)** $24.49; The result using the model is close, but slightly low.
50. 5 sec; 400 ft **51.** length: 50 m; width: 50 m

52. $\left(-\infty, -\frac{3}{2}\right) \cup (4, \infty)$

53. $[-4, 3]$

54. $(-\infty, -5] \cup [-2, 3]$

55. $\emptyset$ **56.** $\left(-\infty, \frac{1}{2}\right) \cup (2, \infty)$

57. $[-3, 2)$

58. $R = \frac{\pm\sqrt{Vh - r^2h}}{h}$ **59.** $\left\{\frac{3 + i\sqrt{3}}{3}, \frac{3 - i\sqrt{3}}{3}\right\}$

60. $\{-2, -1, 3, 4\}$ **61.** $(-\infty, -6) \cup \left(-\frac{3}{2}, 1\right)$

62. $\left\{ \dfrac{-11 + \sqrt{7}}{3}, \dfrac{-11 - \sqrt{7}}{3} \right\}$

63. $d = \dfrac{\pm\sqrt{SkI}}{I}$ **64.** $\{4\}$ **65.** $\left\{ -\dfrac{5}{3}, -\dfrac{3}{2} \right\}$ **66.** $\left(-5, -\dfrac{23}{5} \right]$

67. vertex: $\left(-\dfrac{1}{2}, -3 \right)$; axis: $x = -\dfrac{1}{2}$; domain: $(-\infty, \infty)$; range: $[-3, \infty)$

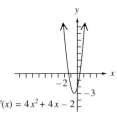

$f(x) = 4x^2 + 4x - 2$

68. (a) 21.92 trillion ft^3 **(b)** The result using the model is a little low.   **(c)** 2005

Chapter 10 Test (page 643)

1. $\{3\sqrt{6}, -3\sqrt{6}\}$ **2.** $\left\{ -\dfrac{8}{7}, \dfrac{2}{7} \right\}$ **3.** $\{-1 + \sqrt{2}, -1 - \sqrt{2}\}$

4. $\left\{ \dfrac{3 + \sqrt{17}}{4}, \dfrac{3 - \sqrt{17}}{4} \right\}$ **5.** $\left\{ \dfrac{2 + i\sqrt{11}}{3}, \dfrac{2 - i\sqrt{11}}{3} \right\}$

6. $\left\{ \dfrac{2}{3} \right\}$ **7.** A   **8.** discriminant: 88; two irrational solutions

9. $\left\{ -\dfrac{2}{3}, 6 \right\}$ **10.** $\left\{ \dfrac{-7 + \sqrt{97}}{8}, \dfrac{-7 - \sqrt{97}}{8} \right\}$

11. $\left\{ -2, -\dfrac{1}{3}, \dfrac{1}{3}, 2 \right\}$ **12.** $\left\{ -\dfrac{5}{2}, 1 \right\}$ **13.** $r = \dfrac{\pm\sqrt{\pi S}}{2\pi}$

14. Andrew: 11.1 hr; Kent: 9.1 hr   **15.** 7 mph   **16.** 2 ft   **17.** 16 m

18. (a) 6.5%   **(b)** 1996; 3.5%   **19.** A

20. vertex: $(0, -2)$; axis: $x = 0$; domain: $(-\infty, \infty)$; range: $[-2, \infty)$

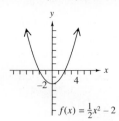

$f(x) = \dfrac{1}{2}x^2 - 2$

21. vertex: $(2, 3)$; axis: $x = 2$; domain: $(-\infty, \infty)$; range: $(-\infty, 3]$

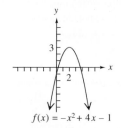

$f(x) = -x^2 + 4x - 1$

22. vertex: $(-5, -2)$; axis: $y = -2$; domain: $[-5, \infty)$; range: $(-\infty, \infty)$

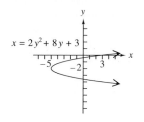

$x = 2y^2 + 8y + 3$

23. 140 ft by 70 ft; 9800 ft^2

24. $(-\infty, -5) \cup \left(\dfrac{3}{2}, \infty \right)$

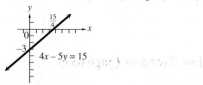

25. $(-\infty, 4) \cup [9, \infty)$

Cumulative Review Exercises: Chapters 1–10 (page 647)

1. (a) $-2, 0, 7$ **(b)** $-\dfrac{7}{3}, -2, 0, .7, 7, \dfrac{32}{3}$ **(c)** all except $\sqrt{-8}$

(d) All are complex numbers.   **2.** 6   **3.** 41   **4.** 930   **5.** 720

6. 990   **7.** 930   **8.** $\{1\}$ **9.** $[1, \infty)$ **10.** $\left[2, \dfrac{8}{3} \right]$

11. slope: $\dfrac{1}{2}$; y-intercept: $\left(0, -\dfrac{7}{4} \right)$ **12.** $x + 3y = -1$

13. function; domain: $(-\infty, \infty)$; range: $(-\infty, \infty)$

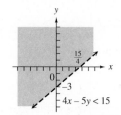

$4x - 5y = 15$

14. not a function

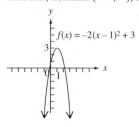

$4x - 5y < 15$

15. function; domain: $(-\infty, \infty)$; range: $(-\infty, 3]$

$f(x) = -2(x - 1)^2 + 3$

16. (a) $y = 123x + 600$ **(b)** \$1461 million; The result using the model is a little high.   **17.** No. Every ordered pair has first component 5. Such a relation is not a function.   **18.** $\{(1, -2)\}$

19. $\{(3, -4, 2)\}$ **20. (a)** $x + y = 34.2; x = 4y - .3$

(b) AOL: \$27.3 billion; Time Warner: \$6.9 billion   **21.** $\dfrac{x^8}{y^4}$

22. $\dfrac{4}{xy^2}$ **23.** $\dfrac{4}{9}t^2 + 12t + 81$ **24.** $-3t^3 + 5t^2 - 12t + 15$

25. $4x^2 - 6x + 11 + \dfrac{4}{x + 2}$ **26.** 4,600,000,000

27. $x(4 + x)(4 - x)$ **28.** $(4m - 3)(6m + 5)$ **29.** $(3x - 5y)^2$

30. $-\dfrac{5}{18}$ **31.** $-\dfrac{8}{k}$ **32.** $\dfrac{r - s}{r}$ **33.** $\dfrac{3\sqrt[3]{4}}{4}$ **34.** $\sqrt{7} + \sqrt{5}$

35. $\left\{ \dfrac{2}{3} \right\}$ **36.** $\left\{ \dfrac{2 + \sqrt{10}}{2}, \dfrac{2 - \sqrt{10}}{2} \right\}$ **37.** $\{-3, 5\}$ **38.** $\emptyset$

39. $\{-3, -1, 1, 3\}$ **40.** southbound car: 57 mi; eastbound car: 76 mi

CHAPTER 11 EXPONENTIAL AND LOGARITHMIC FUNCTIONS

Section 11.1 (page 657)

1. It is one-to-one. 3. Two or more siblings might be in the class. They would be paired with the same mother. 5. B 7. A
9. $\{(6, 3), (10, 2), (12, 5)\}$ 11. not one-to-one
13. $f^{-1}(x) = \dfrac{x - 4}{2}$ 15. $g^{-1}(x) = x^2 + 3, x \geq 0$
17. not one-to-one 19. $f^{-1}(x) = \sqrt[3]{x + 4}$ 21. (a) 8 (b) 3
23. (a) 1 (b) 0
25. (a) one-to-one 27. (a) not one-to-one

29. (a) one-to-one 31.

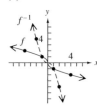

33. 35.

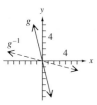
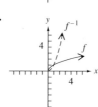

37.

Section 11.2 (page 667)

1. C 3. A
5. 7.

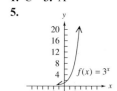

9.

11. $\{2\}$ 13. $\left\{\dfrac{3}{2}\right\}$ 15. $\{7\}$ 17. $\{-3\}$ 19. $\{-1\}$
21. $\{-3\}$ 23. (a) rises; falls (b) It is one-to-one and thus has an inverse. 25. (a) 1.0°C (b) .4°C 27. (a) 3.0°C (b) .7°C

Section 11.3 (page 675)

1. (a) C (b) F (c) B (d) A (e) E (f) D 3. $\log_4 1024 = 5$
5. $\log_{1/2} 8 = -3$ 7. $\log_{10} .001 = -3$ 9. $\log_{625} 5 = \dfrac{1}{4}$
11. $4^3 = 64$ 13. $10^{-4} = \dfrac{1}{10,000}$ 15. $6^0 = 1$ 17. $9^{1/2} = 3$
19. Since the radical $\sqrt{9} = 9^{1/2} = 3$, the exponent to which 9 must be raised is 1/2. 21. $\left\{\dfrac{1}{3}\right\}$ 23. $\{81\}$ 25. $\left\{\dfrac{1}{5}\right\}$ 27. $\{1\}$
29. $\{x|x > 0, x \neq 1\}$ 31. $\{5\}$ 33. $\left\{\dfrac{5}{3}\right\}$ 35. $\{4\}$ 37. $\left\{\dfrac{3}{2}\right\}$
39. 41.

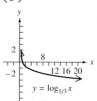

43. Answers will vary. 45. 8 47. 24 49. Since every real number power of 1 equals 1, if $y = \log_1 x$, then $x = 1^y$ and so $x = 1$ for every y. This contradicts the definition of a function.
51. $f(x) = 3^x$ 53. (a) 4385 billion ft³ (b) 5555 billion ft³
(c) 6140 billion ft³ 55. about 4 times as powerful

Section 11.4 (page 685)

1. false; $\log_b x + \log_b y = \log_b xy$ 3. true 5. $\log_7 4 - \log_7 5$
7. $\dfrac{1}{4} \log_2 8$ or $\dfrac{3}{4}$ 9. $\log_4 3 + \dfrac{1}{2} \log_4 x - \log_4 y$
11. $\dfrac{1}{3} \log_3 4 - 2 \log_3 x - \log_3 y$
13. $\dfrac{1}{2} \log_3 x + \dfrac{1}{2} \log_3 y - \dfrac{1}{2} \log_3 5$
15. $\dfrac{1}{3} \log_2 x + \dfrac{1}{5} \log_2 y - 2 \log_2 r$ 17. The distributive property tells us that the *product* $a(x + y)$ equals the sum $ax + ay$. In the notation $\log_a(x + y)$, the parentheses do not indicate multiplication. They indicate that $x + y$ is the result of raising a to some power.
19. $\log_b xy$ 21. $\log_a \dfrac{m^3}{n}$ 23. $\log_a \dfrac{rt^3}{s}$ 25. $\log_a \dfrac{125}{81}$
27. $\log_{10}(x^2 - 9)$ 29. $\log_p \dfrac{x^3 y^{1/2}}{z^{3/2} a^3}$ 31. For the power rule $\log_b x^r = r \log_b x$ to be true, x must be in the domain of $g(x) = \log_b x$, so $x > 0$. 33. 4 34. It is the exponent to which 3 must be raised in order to obtain 81. 35. 81 36. It is the exponent to which 2 must be raised in order to obtain 19. 37. 19 38. m

Section 11.5 (page 691)

1. C 3. C 5. 19.2 7. 1.6335 9. 2.5164 11. -1.4868
13. 9.6776 15. 2.0592 17. -2.8896 19. 5.9613 21. 4.1506
23. 2.3026 25. Answers will vary. Suppose the name is Paul Bunyan, with $m = 4$ and $n = 6$. (a) $\log_4 6$ is the exponent to which 4 must be raised in order to obtain 6. (b) 1.29248125
(c) 6 (the value of n) 27. An error message appears because we cannot find the common logarithm of a negative number.
29. bog 31. 5.9 33. 1.0×10^{-2} 35. 2.5×10^{-5} 37. (a) 35.0 yr
(b) 14.2 yr (c) 9.0 yr 39. (a) 107 dB (b) 100 dB
(c) 98 dB 41. (a) 55% (b) 90% 43. (a) 5.6 yr
(b) $t > 0$ and $\dfrac{87 - L}{63}$ is positive and in the domain of the function only if $24 < L < 87$.

Section 11.6 (page 703)

1. $\log 5^x = \log 125$ 2. $x \log 5 = \log 125$ 3. $x = \dfrac{\log 125}{\log 5}$

4. $\dfrac{\log 125}{\log 5} = 3; \{3\}$ 5. $\{.827\}$ 7. $\{.833\}$ 9. $\{1.201\}$

11. $\{2.269\}$ 13. $\{15.967\}$ 15. $\{261.291\}$ 17. $\{-10.718\}$
19. $\{3\}$ 21. $\{5.879\}$ 23. Natural logarithms are a better choice

because e is the base.   25. $\left\{\dfrac{2}{3}\right\}$ 27. $\left\{\dfrac{33}{2}\right\}$ 29. $\{-1 + \sqrt[3]{49}\}$

31. 2 cannot be a solution because $\log(2 - 3) = \log(-1)$, and -1

is not in the domain of $\log x$.   33. $\left\{\dfrac{1}{3}\right\}$ 35. $\{2\}$ 37. $\emptyset$

39. $\{8\}$ 41. $\left\{\dfrac{4}{3}\right\}$ 43. $\{8\}$ 45. (a) $2539.47   (b) 10.2 yr

47. $4934.71   49. 15.4 yr   51. about 180 g   53. 1.4315
55. 3.3030   57. $-.0947$ 59. -2.3219 61. -2.0639 63. .673

Chapter 11 Review Exercises (page 711)

1. not one-to-one   2. one-to-one   3. This function is not one-to-one
because two sodas in the list have 41 mg of caffeine.

4. $f^{-1}(x) = \dfrac{x - 7}{-3}$ or $\dfrac{7 - x}{3}$ 5. $f^{-1}(x) = \dfrac{x^3 + 4}{6}$

6. not one-to-one

7.

8.
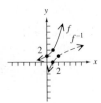

9.
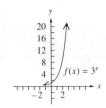

10.
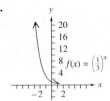

11.

12. $\{4\}$ 13. $\left\{\dfrac{3}{7}\right\}$ 14. $\{0\}$ 15. (a) 37.7 million

(b) 42.3 million   16. (a) $5^4 = 625$ (b) $\log_5 .04 = -2$
17. (a) $\log_b a$ represents the exponent to which b must be raised
to obtain a.   (b) a

18.
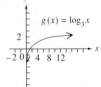

19.

20. $\{2\}$ 21. $\{-2\}$ 22. $\{8\}$ 23. $\{b \mid b > 0, b \neq 1\}$

24. $\log_4 3 + 2 \log_4 x$ 25. $2 \log_2 p + \log_2 r - \dfrac{1}{2} \log_2 z$

26. $\log_b \dfrac{3x}{y^2}$ 27. $\log_3 \dfrac{x + 7}{4x + 6}$ 28. 1.4609   29. $-.5901$
30. 3.3638   31. -1.3587 32. 6.4   33. 8.4   34. 2.5×10^{-5}
35. (a) 18 yr   (b) 12 yr   (c) 7 yr   (d) 6 yr   (e) Each comparison shows approximately the same number. For example, in part (a)

the doubling time is 18 yr (rounded) and $\dfrac{72}{4} = 18$. Thus, the formula

$t = \dfrac{72}{100r}$ (called the *Rule of 72*) is an excellent approximation of the

doubling time formula. (See the Focus on Real Data Application
following Section 11.6 exposition.)   36. (a) $g(92) = -3.4$;
$g(99) = 7.1$ (b) $g(92)$ agrees closely with the y-value from the
graph of about -3; $g(99)$ also is reasonably close to the y-value from
the graph of about 7.5.   37. $\{2.042\}$ 38. $\{4.907\}$ 39. $\{18.310\}$

40. $\left\{\dfrac{1}{9}\right\}$ 41. $\{-6 + \sqrt[3]{25}\}$ 42. $\{2\}$ 43. $\left\{\dfrac{3}{8}\right\}$ 44. $\{4\}$

45. $\{1\}$ 46. $7112.11   47. Plan A; it would pay $2.92 more.
48. $4267   49. about 11%   50. .9251   51. 1.7925   52. 1.4315

53. $\{72\}$ 54. $\{5\}$ 55. $\left\{\dfrac{1}{9}\right\}$ 56. $\left\{\dfrac{4}{3}\right\}$ 57. $\{3\}$ 58. $\left\{\dfrac{1}{8}\right\}$

59. $\left\{\dfrac{11}{3}\right\}$ 60. 6.8 yr

Chapter 11 Test (page 715)

1. (a) not one-to-one   (b) one-to-one
2. $f^{-1}(x) = x^3 - 7$

3.

4.
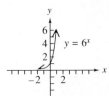

5.
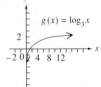

6. Interchange the x- and y-values of the ordered pairs, because the

functions are inverses.   7. $\{-4\}$ 8. $\left\{-\dfrac{13}{3}\right\}$ 9. (a) 775 millibars

(b) 265 millibars   10. $\log_4 .0625 = -2$ 11. $7^2 = 49$

12. $\{32\}$ 13. $\left\{\dfrac{1}{2}\right\}$ 14. $\{2\}$ 15. $2 \log_3 x + \log_3 y$

16. $\log_b \dfrac{r^{1/4} s^2}{t^{2/3}}$ 17. (a) 1.3284   (b) $-.8440$ (c) 2.1245

18. $\{3.9656\}$ 19. $\{3\}$ 20. (a) $12,507.51   (b) 15.5 yr

Cumulative Review Exercises: Chapters 1–11 (page 717)

1. $-2, 0, 6, \dfrac{30}{3}$ (or 10)   2. $-\dfrac{9}{4}, -2, 0, .6, 6, \dfrac{30}{3}$ (or 10)

3. $-\sqrt{2}, \sqrt{11}$ 4. 16   5. -39 6. $\left\{-\dfrac{2}{3}\right\}$ 7. $[1, \infty)$

8. $\{-2, 7\}$ 9. $\left\{-\dfrac{16}{3}, \dfrac{16}{3}\right\}$ 10. $\left[\dfrac{7}{3}, 3\right]$

11. $(-\infty, -3) \cup (2, \infty)$

12.

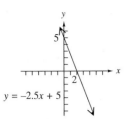

$y = -2.5x + 5$

13.

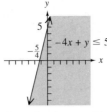

$-4x + y \le 5$

$-\frac{5}{4}$

11. $(-\infty, 0) \cup (0, \infty); (-\infty, 1) \cup (1, \infty)$

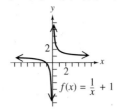

$f(x) = \frac{1}{x} + 1$

14. (a) yes **(b)** 1619.3; The number of travelers increased by an average of 1619.3 thousand per year during the period 1990–2000.

15. $y = \frac{3}{4}x - \frac{19}{4}$ **16.** $\{(4, 2)\}$ **17.** $\{(1, -1, 4)\}$ **18.** 6 lb of $1.00 candy and 10 lb of $1.96 candy **19.** $6p^2 + 7p - 3$
20. $16k^2 - 24k + 9$ **21.** $-5m^3 + 2m^2 - 7m + 4$
22. $2t^3 + 5t^2 - 3t + 4$ **23.** $x(8 + x^2)$ **24.** $(3y - 2)(8y + 3)$
25. $z(5z + 1)(z - 4)$ **26.** $(4a + 5b^2)(4a - 5b^2)$

27. $(2c + d)(4c^2 - 2cd + d^2)$ **28.** $(4r + 7q)^2$ **29.** $-\dfrac{1875p^{13}}{8}$

30. $\dfrac{x + 5}{x + 4}$ **31.** $\dfrac{-3k - 19}{(k + 3)(k - 2)}$ **32.** $\dfrac{22 - p}{p(p - 4)(p + 2)}$

33. $\left\{ -\dfrac{1}{3} \right\}$ **34.** $12\sqrt{2}$ **35.** $-\dfrac{1}{4}$ **36.** $-27\sqrt{2}$ **37.** $\{0, 4\}$

38. 41 **39.** $\left\{ \dfrac{1 + \sqrt{13}}{6}, \dfrac{1 - \sqrt{13}}{6} \right\}$ **40.** $(-\infty, -4) \cup (2, \infty)$

41. $\{-2, -1, 1, 2\}$

13. $[2, \infty); [0, \infty)$

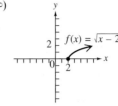

$f(x) = \sqrt{x - 2}$

42.

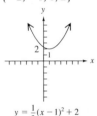

$y = \frac{1}{3}(x - 1)^2 + 2$

43.

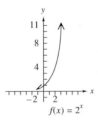

$f(x) = 2^x$

15. $(-\infty, 2) \cup (2, \infty); (-\infty, 0) \cup (0, \infty)$

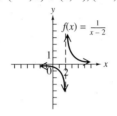

$f(x) = \frac{1}{x - 2}$

44.

$f(x) = \log_3 x$

17. $[-3, \infty); [-3, \infty)$

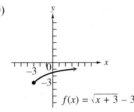

$f(x) = \sqrt{x + 3} - 3$

19.

$f(x) = [\![x - 3]\!]$

21.

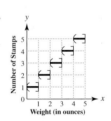

Number of Stamps

Weight (in ounces)

45. $\{-1\}$ **46.** $\{1\}$ **47.** $5^3 = 125$ **48.** $3 \log x + \dfrac{1}{2} \log y - \log z$

49. .110 or 11.0% **50.** $\{30\}$

23. 16 **25.** 83 **27.** 13 **29.** $4x^2 + 12x + 13$
31. $x^2 + 10x + 29$ **33.** $2x + 8$ **35.** $(f \circ g)(x) = 63{,}360x$; It computes the number of inches in x miles. **37.** $(A \circ r)(t) = 4\pi t^2$; This is the area of the circular layer as a function of time.

CHAPTER 12 NONLINEAR FUNCTIONS, CONIC SECTIONS, AND NONLINEAR SYSTEMS

Section 12.1 (page 729)
1. 0 **3.** 0; 0 **5.** B **7.** A
9. $(-\infty, \infty); [0, \infty)$

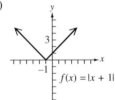

$f(x) = |x + 1|$

Section 12.2 (page 739)
1. (a) $(0, 0)$ **(b)** 5 **(c)**

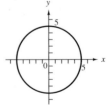

3. B **5.** D **7.** $(x + 4)^2 + (y - 3)^2 = 4$
9. $(x + 8)^2 + (y + 5)^2 = 5$ **11.** $(-2, -3); r = 2$
13. $(-5, 7); r = 9$ **15.** $(2, 4); r = 4$ **17.** The thumbtack acts as the center and the length of the string acts as the radius.

19.

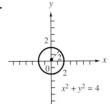

$x^2 + y^2 = 4$

21.

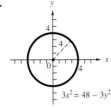

$3x^2 = 48 - 3y^2$

11.

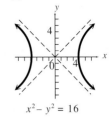

$\frac{x^2}{25} - \frac{y^2}{36} = 1$

13. hyperbola

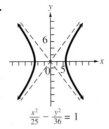

$x^2 - y^2 = 16$

23.

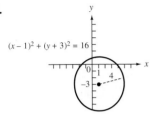

$(x - 1)^2 + (y + 3)^2 = 16$

15. ellipse

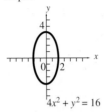

$4x^2 + y^2 = 16$

17. circle

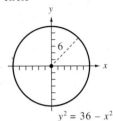

$y^2 = 36 - x^2$

25.

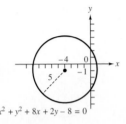

$x^2 + y^2 + 8x + 2y - 8 = 0$

19. hyperbola

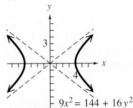

$9x^2 = 144 + 16y^2$

21. ellipse

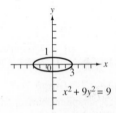

$x^2 + 9y^2 = 9$

27.

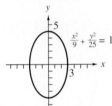

$\frac{x^2}{9} + \frac{y^2}{25} = 1$

29.

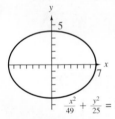

$\frac{x^2}{36} + \frac{y^2}{16} = 1$

23. $[-3, 3]$; $[0, 3]$

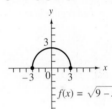

$f(x) = \sqrt{9 - x^2}$

25. $[-5, 5]$; $[-5, 0]$

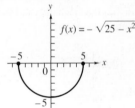

$f(x) = -\sqrt{25 - x^2}$

31.

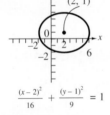

$\frac{x^2}{49} + \frac{y^2}{25} = 1$

33.

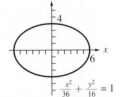

$(2, 1)$

$\frac{(x - 2)^2}{16} + \frac{(y - 1)^2}{9} = 1$

27. $[-4, \infty)$; $[0, \infty)$

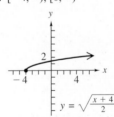

$y = \sqrt{\dfrac{x + 4}{2}}$

35. (a) 10 m **(b)** 36 m **37. (a)** 154.7 million mi
(b) 128.7 million mi (Answers are rounded.)

Section 12.3 (page 749)

1. C **3.** D **5.** When written in one of the forms given in the box titled "Equations of Hyperbolas" in this section, it will open up and down if the $-$ sign precedes the x^2-term; it will open left and right if the $-$ sign precedes the y^2-term.

7.

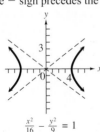

$\frac{x^2}{16} - \frac{y^2}{9} = 1$

9.

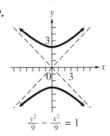

$\frac{y^2}{9} - \frac{x^2}{9} = 1$

29. (a) 50 m **(b)** 69.3 m
31. for V greater than 4325.68 m per sec

Section 12.4 (page 759)

1. Substitute $x - 1$ for y in the first equation. Then solve for x. Find the corresponding y-values by substituting back into $y = x - 1$. In the first equation, both variables are squared and in the second, both variables are to the first power, so the elimination method is not appropriate. **3.** one **5.** none

7.

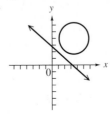

9.

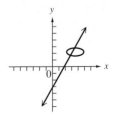

11.

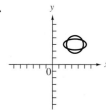

13. $\left\{(0,0), \left(\dfrac{1}{2}, \dfrac{1}{2}\right)\right\}$ **15.** $\{(-6, 9), (-1, 4)\}$

17. $\left\{\left(-\dfrac{1}{5}, \dfrac{7}{5}\right), (1, -1)\right\}$ **19.** $\left\{(-2, -2), \left(-\dfrac{4}{3}, -3\right)\right\}$

21. $\{(-3, 1), (1, -3)\}$ **23.** $\left\{\left(-\dfrac{3}{2}, -\dfrac{9}{4}\right), (-2, 0)\right\}$

25. $\{(-\sqrt{3}, 0), (\sqrt{3}, 0), (-\sqrt{5}, 2), (\sqrt{5}, 2)\}$

27. $\{(-2, 0), (2, 0)\}$ **29.** $\{(i\sqrt{2}, -3i\sqrt{2}), (-i\sqrt{2}, 3i\sqrt{2}),$
$(-\sqrt{6}, -\sqrt{6}), (\sqrt{6}, \sqrt{6})\}$ **31.** $\{(-2i\sqrt{2}, -2\sqrt{3}),$
$(-2i\sqrt{2}, 2\sqrt{3}), (2i\sqrt{2}, -2\sqrt{3}), (2i\sqrt{2}, 2\sqrt{3})\}$

33. $\{(-\sqrt{5}, -\sqrt{5}), (\sqrt{5}, \sqrt{5})\}$ **35.** $\{(i, 2i), (-i, -2i),$
$(2, -1), (-2, 1)\}$ **37.** length: 12 ft; width: 7 ft **39.** \$20;
800 calculators **41.** 1981; 470 thousand

Section 12.5 (page 767)

1. C **3.** Graph the corresponding equation as a solid curve
if the inequality is $\leq$ or $\geq$, or as a dashed curve if the inequality
is $<$ or $>$. Use a test point to decide which side of the boundary
satisfies the inequality, and shade it. The shaded region is the
solution set.

5.

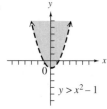

$y > x^2 - 1$

7.

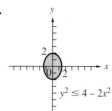

$y^2 \leq 4 - 2x^2$

9.

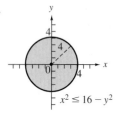

$x^2 \leq 16 - y^2$

11.

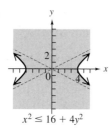

$x^2 \leq 16 + 4y^2$

13.

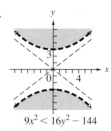

$9x^2 < 16y^2 - 144$

15.

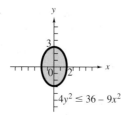

$4y^2 \leq 36 - 9x^2$

17.

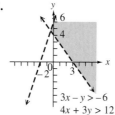

$x \geq y^2 - 8y + 14$

19.

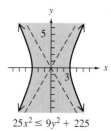

$25x^2 \leq 9y^2 + 225$

21.

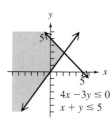

$3x - y > -6$
$4x + 3y > 12$

23.

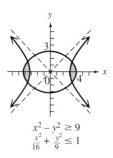

$4x - 3y \leq 0$
$x + y \leq 5$

25.

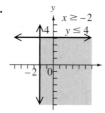

$x \geq -2$
$y \leq 4$

27.

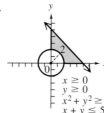

$x^2 - y^2 \geq 9$
$\dfrac{x^2}{16} + \dfrac{y^2}{9} \leq 1$

29.

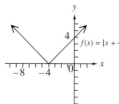

$x \geq 0$
$y \geq 0$
$x^2 + y^2 \geq 4$
$x + y \leq 5$

31.

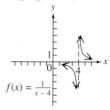

$y < x^2$
$y > -2$
$x + y < 3$
$3x - 2y > -6$

Chapter 12 Review Exercises (page 775)

1.

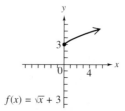

$f(x) = |x + 4|$

2.

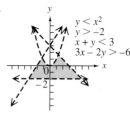

$f(x) = \dfrac{1}{x - 4}$

3.

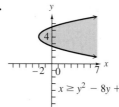

$f(x) = \sqrt{x} + 3$

4. 12 **5.** 2 **6.** -5 **7. (a)** 167 **(b)** 1495 **8. (a)** 20
(b) 42 **9. (a)** $75x^2 + 220x + 160$ **(b)** $15x^2 + 10x + 2$
10. No, composition of functions is not a commutative operation.
For example, the results of Exercise 9 show that $(f \circ g)(x) \neq$
$(g \circ f)(x)$ in this case. **11.** $(x + 2)^2 + (y - 4)^2 = 9$
12. $(x + 1)^2 + (y + 3)^2 = 25$ **13.** $(x - 4)^2 + (y - 2)^2 = 36$
14. $(-3, 2); r = 4$ **15.** $(4, 1); r = 2$ **16.** $(-1, -5); r = 3$

17. $(3, -2)$; $r = 5$

18.

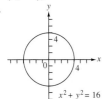

$x^2 + y^2 = 16$

19.

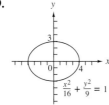

$\frac{x^2}{16} + \frac{y^2}{9} = 1$

20.

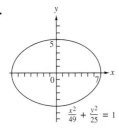

$\frac{x^2}{49} + \frac{y^2}{25} = 1$

21. $\dfrac{x^2}{65,286,400} + \dfrac{y^2}{2,560,000} = 1$ **22. (a)** 348.2 ft   **(b)** 1787.6 ft

23.

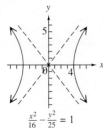

$\frac{x^2}{16} - \frac{y^2}{25} = 1$

24.

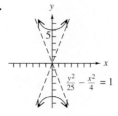

$\frac{y^2}{25} - \frac{x^2}{4} = 1$

25.

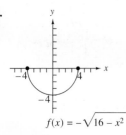

$f(x) = -\sqrt{16 - x^2}$

26. circle   **27.** parabola   **28.** hyperbola   **29.** ellipse

30. parabola   **31.** hyperbola   **32.** $\dfrac{x^2}{625} - \dfrac{y^2}{1875} = 1$

33. $\{(6, -9), (-2, -5)\}$ **34.** $\{(1, 2), (-5, 14)\}$

35. $\{(4, 2), (-1, -3)\}$ **36.** $\{(-2, -4), (8, 1)\}$

37. $\{(-\sqrt{2}, 2), (-\sqrt{2}, -2), (\sqrt{2}, -2), (\sqrt{2}, 2)\}$

38. $\{(-\sqrt{6}, -\sqrt{3}), (-\sqrt{6}, \sqrt{3}), (\sqrt{6}, -\sqrt{3}), (\sqrt{6}, \sqrt{3})\}$

39. 0, 1, or 2   **40.** 0, 1, 2, 3, or 4

41.

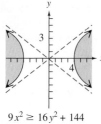

$9x^2 \geq 16y^2 + 144$

42.

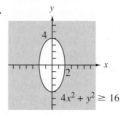

$4x^2 + y^2 \geq 16$

43.

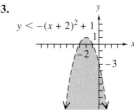

$y < -(x + 2)^2 + 1$

44.

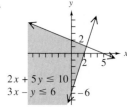

$2x + 5y \leq 10$
$3x - y \leq 6$

45.

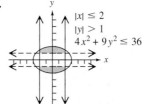

$|x| \leq 2$
$|y| > 1$
$4x^2 + 9y^2 \leq 36$

46.

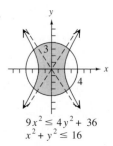

$9x^2 \leq 4y^2 + 36$
$x^2 + y^2 \leq 16$

47.

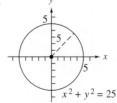

$x^2 + y^2 = 25$

48.

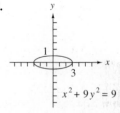

$x^2 + 9y^2 = 9$

49.

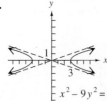

$x^2 - 9y^2 = 9$

50.

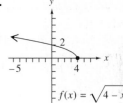

$f(x) = \sqrt{4 - x}$

51.

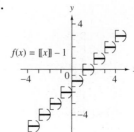

$f(x) = [\![x]\!] - 1$

52.

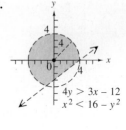

$4y > 3x - 12$
$x^2 < 16 - y^2$

53. There are cases where one x-value will yield two y-values. In a function, every x yields one and only one y.

Chapter 12 Test (page 779)

1. C   **2.** A   **3.** D   **4.** B

5.

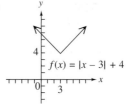

$f(x) = |x - 3| + 4$

6. (a) 23   **(b)** $3x^2 + 11$ **(c)** $9x^2 + 30x + 27$

7. center: $(2, -3)$; radius: 4

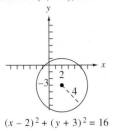

$(x - 2)^2 + (y + 3)^2 = 16$

8. center: $(-4, 1)$; radius: 5

9.

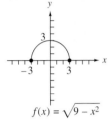

$f(x) = \sqrt{9 - x^2}$

10.

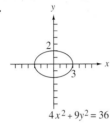

$4x^2 + 9y^2 = 36$

11.

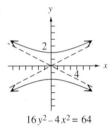

$16y^2 - 4x^2 = 64$

12.

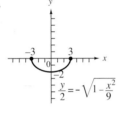

$\frac{y}{2} = -\sqrt{1 - \frac{x^2}{9}}$

13. ellipse   **14.** hyperbola   **15.** parabola

16. $\left\{\left(-\dfrac{1}{2}, -10\right), (5, 1)\right\}$ **17.** $\left\{(-2, -2), \left(\dfrac{14}{5}, -\dfrac{2}{5}\right)\right\}$

18. $\{(-\sqrt{22}, -\sqrt{3}), (-\sqrt{22}, \sqrt{3}), (\sqrt{22}, -\sqrt{3}), (\sqrt{22}, \sqrt{3})\}$

19.

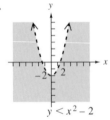

$y < x^2 - 2$

20.

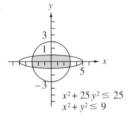

$x^2 + 25y^2 \le 25$
$x^2 + y^2 \le 9$

Cumulative Review Exercises: Chapters 1–12
(page 783)

1. -4 **2.** $\left\{\dfrac{2}{3}\right\}$ **3.** $\left(-\infty, \dfrac{3}{5}\right]$ **4.** $\{-4, 4\}$

5. $(-\infty, -5) \cup (10, \infty)$ **6.** $\dfrac{2}{3}$ **7.** $3x + 2y = -13$

8. $\{(3, -3)\}$ **9.** $\{(4, 1, -2)\}$ **10.** $\left\{(-1, 5), \left(\dfrac{5}{2}, -2\right)\right\}$

11. 40 mph   **12.** \$275   **13.** $25y^2 - 30y + 9$ **14.** $12r^2 + 40r - 7$

15. $4x^3 - 4x^2 + 3x + 5 + \dfrac{3}{2x + 1}$ **16.** $(3x + 2)(4x - 5)$

17. $(2y^2 - 1)(y^2 + 3)$ **18.** $(z^2 + 1)(z + 1)(z - 1)$

19. $(a - 3b)(a^2 + 3ab + 9b^2)$

20. $\dfrac{40}{9}$ **21.** $\dfrac{y - 1}{y(y - 3)}$ **22.** $\dfrac{3c + 5}{(c + 5)(c + 3)}$ **23.** $\dfrac{1}{p}$

24. $1\dfrac{1}{5}$ hr   **25.** $\dfrac{3}{4}$ **26.** $\dfrac{a^5}{4}$ **27.** $2\sqrt[3]{2}$ **28.** $\dfrac{3\sqrt{10}}{2}$

29. $\dfrac{7}{5} + \dfrac{11}{5}i$ **30.** $\emptyset$ **31.** $\left\{\dfrac{1}{5}, -\dfrac{3}{2}\right\}$

32. $\left\{\dfrac{1 + 2\sqrt{2}}{4}, \dfrac{1 - 2\sqrt{2}}{4}\right\}$ **33.** $\left\{\dfrac{3 + \sqrt{33}}{6}, \dfrac{3 - \sqrt{33}}{6}\right\}$

34. $\left\{-\dfrac{\sqrt{6}}{2}, \dfrac{\sqrt{6}}{2}, -\sqrt{7}, \sqrt{7}\right\}$ **35.** $v = \dfrac{\pm\sqrt{rFkw}}{kw}$

36. $f^{-1}(x) = \sqrt[3]{x - 4}$ **37.** 4   **38.** 7   **39.** $\log \dfrac{(3x + 7)^2}{4}$

40. $\{3\}$ **41. (a)** \$12,198.90   **(b)** \$12,214.03   **42.** \$16.9 billion

43. \$75.8 billion   **44. (a)** -1 **(b)** $9x^2 + 18x + 4$

45.

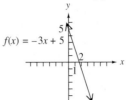

$f(x) = -3x + 5$

46.

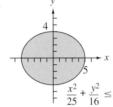

$f(x) = -2(x - 1)^2 + 3$

47.

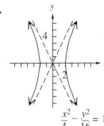

$\dfrac{x^2}{25} + \dfrac{y^2}{16} \le 1$

48.

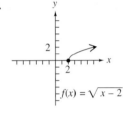

$f(x) = \sqrt{x - 2}$

49.

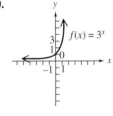

$\dfrac{x^2}{4} - \dfrac{y^2}{16} = 1$

50.

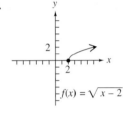

$f(x) = 3^x$

APPENDIX A STRATEGIES FOR PROBLEM SOLVING
(page 795)

1. You should choose a sock from the box labeled *red and green socks*. Since it is mislabeled, it contains only red socks or only green socks, determined by the sock you choose. If the sock is green, relabel this box *green socks*. Since the other two boxes were mislabeled, switch the remaining label to the other box and place the label that says *red and green socks* on the unlabeled box. No other choice guarantees a correct relabeling, since you can remove only one sock. **3.** 70   **5.** You must place the O in the bottom-left square. No other choice guarantees you a win.   **7.** 18   **9.** 329,476   **11.** One possible sequence is shown here. The numbers represent the number of gallons in each bucket in each successive step.

Big	7	4	4	1	1	0	7	5	5
Small	0	3	0	3	0	1	1	3	0

13. 07

15. 17 days   **17.** 16   **19.**   **21.**

6	1	8
7	5	3
2	9	4

23. 25 pitches (The visiting team's pitcher retires 24 consecutive batters through the first eight innings, using only one pitch per batter. His team does not score either. Going into the bottom of the ninth inning tied 0–0, the first batter for the home team hits his first pitch for a home run. The pitcher threw 25 pitches and loses the game by a score of 1–0.)   **25.** For three weighings, first balance four against four. Of the lighter four, balance two against the other two. Finally, of the lighter two, balance them one against the other. To find the bad coin in two weighings, divide the eight coins into groups of 3, 3, 2. Weigh the groups of three against each other on the scale. If the groups weigh the same, the fake is in the two left out and can be found in one additional weighing. If the two groups of three do not weigh the same, pick the lighter group. Choose any two of the coins and weigh them. If one of these is lighter, it is the fake; if they weigh the same, then the third coin is the fake.   **27.** Q   **29.** 28

31.

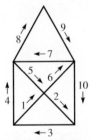

33. Dan (36) is married to Jessica (29); James (30) is married to Cathy (31).   **35.** The CEO is a woman.   **37.** None, since there is no dirt in a hole.   **39.** One solution is $1 + 2 + 3 + 4 + 5 + 6 + 7 + 8 \times 9 = 100$.   **41.** The correct problem follows.

$$\begin{array}{r} 402 \\ \times \quad 39 \\ \hline 15{,}678 \end{array}$$

43. 35   **45.** none   **47.** Eve has \$5 and Adam has \$7.

APPENDIX B   REVIEW OF FRACTIONS

(page 809)

1. true   **3.** false; The fraction $\dfrac{17}{51}$ can be simplified to $\dfrac{1}{3}$.

5. false; *Product* indicates multiplication, so the product of 8 and 2 is 16.   **7.** prime   **9.** composite   **11.** composite   **13.** neither

15. $2 \cdot 3 \cdot 5$ **17.** $2 \cdot 2 \cdot 3 \cdot 3 \cdot 7$ **19.** $2 \cdot 2 \cdot 31$ **21.** 29

23. $\dfrac{1}{2}$ **25.** $\dfrac{5}{6}$ **27.** $\dfrac{1}{5}$ **29.** $\dfrac{6}{5}$ **31.** A   **33.** $\dfrac{24}{35}$ **35.** $\dfrac{6}{25}$

37. $\dfrac{6}{5}$ or $1\dfrac{1}{5}$ **39.** $\dfrac{232}{15}$ or $15\dfrac{7}{15}$ **41.** $\dfrac{10}{3}$ or $3\dfrac{1}{3}$ **43.** 12   **45.** $\dfrac{1}{16}$

47. $\dfrac{84}{47}$ or $1\dfrac{37}{47}$ **49.** Multiply the first fraction (the dividend) by the reciprocal of the second fraction (the divisor) to divide two fractions.

51. $\dfrac{2}{3}$ **53.** $\dfrac{8}{9}$ **55.** $\dfrac{27}{8}$ or $3\dfrac{3}{8}$ **57.** $\dfrac{17}{36}$ **59.** $\dfrac{11}{12}$ **61.** $\dfrac{4}{3}$ or $1\dfrac{1}{3}$

63. 6 cups   **65.** $618\dfrac{3}{4}$ ft   **67.** $\dfrac{9}{16}$ in.   **69.** $\dfrac{5}{16}$ in.   **71.** $16\dfrac{5}{8}$ yd

73. $\dfrac{1}{20}$ **75.** more than $1\dfrac{1}{25}$ million

APPENDIX C   DETERMINANTS AND CRAMER'S RULE

(page 821)

1. true   **3.** false   **5.** -3 **7.** 14   **9.** 0   **11.** 59   **13.** 14
15. 1   **17.** -22 **19.** 0   **21.** 0   **23.** 20   **25.** $\{(4, 6)\}$

27. $\{(-5, 2)\}$ **29.** $\{(1, 0)\}$ **31.** $\left\{ \left(\dfrac{11}{58}, -\dfrac{5}{29} \right) \right\}$

33. $\{(-2, 3, 5)\}$ **35.** $\{(5, 0, 0)\}$ **37.** Cramer's rule does not apply.

39. $\{(20, -13, -12)\}$ **41.** $\left\{ \left(\dfrac{62}{5}, -\dfrac{1}{5}, \dfrac{27}{5} \right) \right\}$

APPENDIX D   SYNTHETIC DIVISION

(page 827)

1. C   **3.** $x - 5$ **5.** $4m - 1$ **7.** $2a + 4 + \dfrac{5}{a + 2}$

9. $p - 4 + \dfrac{9}{p + 1}$ **11.** $4a^2 + a + 3$

13. $x^4 + 2x^3 + 2x^2 + 7x + 10 + \dfrac{18}{x - 2}$

15. $-4r^5 - 7r^4 - 10r^3 - 5r^2 - 11r - 8 + \dfrac{-8}{r - 1}$

17. $-3y^4 + 8y^3 - 21y^2 + 36y - 72 + \dfrac{143}{y + 2}$

19. $y^2 + y + 1 + \dfrac{2}{y - 1}$ **21.** 7   **23.** -2 **25.** 0   **27.** yes

29. no   **31.** no   **33.** Since the variables are not present, a missing term will not be noticed in synthetic division, so the quotient will be wrong if placeholders are not inserted.

Index

A

Absolute value
  evaluating, 6–7
  explanation of, 43, 45
  simplifying roots using, 482–483
Absolute value equations
  explanation of, 139–140, 153
  solving, 140–143, 155
Absolute value functions, 722
Absolute value inequalities
  explanation of, 139, 153
  solving, 141–142, 144
Addition
  associative property of, 38, 39
  commutative property of, 38, 39
  of complex numbers, 537, 548
  explanation of, 45
  of fractions, 805–808
  identity property for, 37
  of polynomial functions, 336–337
  of polynomials, 330–331, 360
  of radical expressions, 511–512, 547
  of rational expressions, 425–430, 466
  of real numbers, 15–18
Addition property
  of equality, 54–56
  of inequality, 116–117
Additive identity, 37
Additive inverse, 5–6, 19, 43
Agreement on domain, 216–217
Algebraic expressions
  evaluating, 29–30
  examples of, 54
  explanation of, 43, 359
  translating from words to, 77–78
Algebraic fractions. See Rational
  expressions
Alternate forms of logarithms, 683–684
Angles
  complementary, 98, 101
  supplementary, 98, 101
  of triangle, 94
  vertical, 98, 101
Apogee, 742
Applied problems. See also Problem-
  solving strategies
  with distance, rate, and time, 454–456
  with fractions, 807–808

linear inequalities to solve, 121–122
methods to solve, 78–80, 102–103,
  615
modeled by quadratic functions,
  593–594
Pythagorean formula to solve, 592
requiring zero-factor property,
  398–399
with work rate, 456–457
"Approximately equal to" symbol, 44,
  545
Approximations, 484
Area
  formulas for, 593
  of triangle, 488
Arithmetic mean, 458
Array of signs, 815
Ascending powers, 329
Associative properties, 38–39, 46
Asymptotes
  explanation of, 414, 771
  horizontal, 444, 465
  of hyperbola, 744–745, 771
  of reciprocal function, 722
  vertical, 444, 465
Augmented matrix, 291–292, 299
Average, 458
Average rate of change, 183–184
Average speed, 68
Average test scores, 122, 156
Axis
  of parabola, 599, 633
  x, 167, 239
  y, 167, 239

B

Bar graph, 166
Bases
  explanation of, 25, 43
  of exponents, 26–27, 312
Best buys, 104
Binomial(s)
  explanation of, 330, 359
  factoring, 393
  greatest common factor, 373
  multiplication of, 344–345, 347, 515,
    516
  square of, 529–530

Boundary lines, 239
Braces, 2
Break-even point, 762
British thermal units (Btu), 166

C

Calculator tips
  for complex numbers, 539
  for determinants, 816
  for exponents, 26
  for logarithms, 687–690, 709
  for matrices, 291
  for order of operations, 29
  for quadratic regression, 604
  for reciprocals, 18
  for roots, 483–484
  for scientific notation, 321
  for systems of equations, 253
Cartesian coordinate system. See
  Rectangular coordinate system
Center
  of circle, 733–735, 771
  of ellipse, 736
Center-radius form of circle,
  734
Change-of-base rule, 700–701
Circle graphs, 166
Circle(s)
  center of, 733–735, 771
  center-radius form of, 734
  circumference of, 229
  equation of, 733–734, 773
  explanation of, 771
  graphs of, 733, 734, 746
  radius of, 229, 733, 735, 771
Circumference of circle, 229
Closed intervals, 115
Coefficients
  explanation of, 43
  numerical, 329, 359
Columns, of matrix, 291, 299
Combined variation, 234
Combining like terms, 38, 43, 330,
  331
Common logarithms
  applications of, 687–688
  evaluating, 687
  explanation of, 707, 709

Commutative properties, 38–39, 46
Complementary angles, 98, 101
Completing the square
    explanation of, 561, 634
    to find vertex, 611–612
    to graph horizontal parabolas, 617
    solving quadratic equations by,
        561–564, 589
Complex conjugates, 545
Complex fractions
    explanation of, 465
    simplifying, 435–438, 467
Complex numbers
    addition of, 537, 548
    on calculator, 539
    conjugates of, 538
    division of, 538–539, 548
    explanation of, 536–537, 545, 548
    imaginary part of, 537
    multiplication of, 538, 548
    powers of i and, 540
    real part of, 537, 545
    simplifying, 535–536
    standard form of, 545
    subsets of, 536–537
    subtraction of, 537–538, 548
Components, of ordered pair, 167, 239
Composite functions
    explanation of, 726–727, 771, 772
    method for evaluating, 727–728
Composite numbers, 801–802
Composition. See Composite functions
Compound inequalities
    explanation of, 129–131, 153
    solving with and, 129–131, 154
    solving with or, 132–133
Compound interest formula, 698
Compound interest problems, 81,
    698–699
Computer network problem, 362
Conditional equations, 101
Conic sections
    examples of, 599, 733
    explanation of, 733, 771
    summary of, 746
Conjugates, 524, 545
Consecutive integers, 98, 101
Consistent systems, 254
Constant functions, 222, 240
Constant of variation, 229–230
Continuously compounded interest, 699
Contradictions, 58, 59, 101
Coordinates
    explanation of, 43
    on line, 3
    of corresponding point, 167, 239
    of vertex, 615
Coordinate system, rectangular. See
    Rectangular coordinate system

Counting numbers. See Natural numbers
Cramer, Gabriel, 817
Cramer's rule
    applications of, 817–820
    derivation of, 816–817
Cube root function, 481
Cubes
    difference of, 389, 406
    sum of, 389–390, 406
Cubing function, 338, 339, 359
Currency exchange, 104

D

Decimals, 58
Degree, 335, 359
De Morgan, Augustus, 790
Denominators
    of complex fractions, 435–437
    of fractions, 801
    least common, 426, 465, 806
    operations with, 805–807
    of rational expressions, 425, 427–430
    rationalizing, 516–520, 545
Dependent equations
    explanation of, 299
    solving system of, 258–259
    in three variables, 272
Dependent variable, 239
Descartes, René, 166
Descending powers, 329, 359
Determinants
    array of signs for, 815
    evaluation of, 814–816
    of matrices, 813–816
    minors of, 814–816
Difference
    of cubes, 389, 406
    explanation of, 16, 43, 807
    of squares, 406
Direct variation, 229–231
Discriminant
    explanation of, 633, 634
    use of, 572–573, 614
Distance, rate, and time problems,
    281–282, 454–456
Distance formula, 92, 232, 503–504,
    547, 743
Distributive property
    explanation of, 35
    to solve linear equations, 56
    to solve linear inequalities, 119–120
    to subtract rational expressions, 428
    use of, 35–36, 67
Dividend, 823
Dividend, stock, 10
Division
    of complex numbers, 538–539, 548
    explanation of, 45
    of fractions, 804–805, 808

of polynomial functions, 356
of polynomials, 353–356, 361,
    823–825
of radical expressions, 515–520, 547
of rational expressions, 420, 466
of real numbers, 19–20
synthetic, 823–826
by zero, 19
Divisor, 823
Domain
    agreement on, 216–217
    explanation of, 239
    of functions, 215, 216
    of rational equations, 441, 465
    of rational expressions, 414–415, 465
    of rational functions, 414, 415

E

e, 688
Earthquake intensity, 322
Elements
    of matrix, 291, 299
    of set, 2, 43, 45
Elimination method
    to solve linear systems, 256–258,
        268–270, 300, 301
    to solve nonlinear systems, 755
Ellipse(s)
    center of, 736
    equation of, 736–737, 773
    explanation of, 733, 771
    foci of, 736
    graphs of, 737–738, 746, 748
    intercepts of, 736
Empty set
    explanation of, 2, 43
    symbol for, 2, 44
Endpoints of line segment, 176
Equality
    addition property of, 54–56
    multiplication property of, 54–56
Equation(s)
    absolute value, 139–140, 153, 155
    Celsius to Fahrenheit, 72, 204
    of circle, 733–734, 773
    conditional, 58, 59, 101
    contradiction, 58, 59, 101
    dependent, 258–259, 299
    of ellipse, 736–737, 773
    equivalent, 54, 101
    explanation of, 7, 43
    exponential, 663–665, 695–696,
        699–701, 707, 709, 710
    expressions vs., 78, 449
    factoring to solve, 395–399
    first-degree, 169, 239
    graphs of, 169, 239, 747
    of horizontal lines, 171–172, 195
    of hyperbola, 743–744, 773

identifying functions from, 218
identity, 58, 59, 101
independent, 299
of inverse function, 655
linear in one variable, 54–59, 101–103
linear in two variables, 169, 191–198, 239, 241
linear systems of, 252, 299
logarithmic, 696–698710
nonlinear, 54, 754
nonlinear systems of, 753–757, 771, 774
of parallel lines, 195
of perpendicular lines, 195
power rule for, 527–528
quadratic, 396–397, 633
quadratic in form, 577–583, 633, 634
radical, 545
with rational expressions, 441–444, 467
second-degree, 558, 746, 756
translating from words to, 78
variation, 230–233
of vertical lines, 171–172, 195
Equilibrium price, 762
Equity, 10
Equivalent equations, 54, 101
Equivalent forms of fraction, 20
Equivalent inequalities, 153
Euclid, 94
Euclidean geometry, 94
Euler, Leonhard, 688
Expansion of determinant by minors, 814, 815
Exponential equations
    explanation of, 663, 707
    with logarithms, 696–698
    properties of, 663, 695, 709–710
    solving, 663–665, 695–696
Exponential expressions
    evaluating, 26
    explanation of, 25–26, 43
    simplifying, 317–318
Exponential form to logarithmic form, 670
Exponential functions
    applications of, 665–666
    explanation of, 661, 708
    graphs of, 661–663, 708
Exponential growth and decay problems, 665–666, 674, 700
Exponent(s)
    base of, 26–27, 312
    explanation of, 25, 43, 45
    identification of, 26–27
    integer, 360

negative, 313–315, 317, 360
power rules for, 316, 317, 360
product rule for, 312–313, 317, 360, 679
quotient rule for, 315, 317, 360
rational, 489–493, 546
zero, 313, 360
Expressions
    algebraic, 29–30, 43, 54, 77–78, 359
    equations vs., 78, 449
    evaluating, 6–7, 26, 29–30
    exponential, 25, 26, 43
    mathematical, 77–78
    radical, 545
    rational, 414–420, 425–430, 465, 468
Extraneous solutions, 527, 545

F
Factoring
    difference of cubes, 389, 390, 406
    difference of squares, 387, 390, 406
    by grouping, 374–376, 406
    perfect square trinomials, 387–388, 390, 406
    polynomials, 372, 393
    to solve equations, 395–399, 406, 589
    substitution method for, 384
    sum of cubes, 389–390, 406
    trinomials, 379–384, 406
Factors
    explanation of, 25, 43, 67, 801
    greatest common, 372–374, 405
Fibonacci (Leonardo Pisano), 788
Fibonacci sequence, 789
Finite set, 2
First-degree equations, 169, 239. See also Linear equations in one variable; Linear equations in two variables
Foci
    of ellipse, 736
    of hyperbola, 743
FOIL method, 345, 361, 379, 515, 537, 538, 543
Formula(s)
    area, 593
    for circumference of circle, 229
    compound interest, 698
    distance, 92, 232, 503–504, 547
    explanation of, 65, 101
    finding unknown variables in, 451
    Galileo's, 560
    involving squares and square roots, 591–592
    midpoint, 176
    Pythagorean, 503, 547, 592
    quadratic, 569–573, 633, 634
    slope, 177
    to solve applied problems, 68

to solve percent problems, 68–69
solving for specified variable, 65–67, 102, 451–452
vertex, 612
Fraction(s)
    addition of, 805–808
    applied problems with, 807–808
    complex, 435–438, 465, 467
    denominator of, 801–803, 806
    division of, 804–805, 808
    equivalent forms of, 20
    explanation of, 801
    improper, 801
    linear equations with, 57
    linear inequalities with, 120
    in lowest terms, 802–803
    multiplication of, 803–804
    negative exponents with, 317
    numerator of, 524, 801–803, 807
    proper, 801
    reciprocal of, 804
    subtraction of, 807
Froude, William, 598
Froude number, 598
Function notation
    explanation of, 239
    use of, 219–221
Function(s)
    absolute value, 722
    composite, 726–727, 771, 772
    constant, 222, 240
    cube root, 481
    cubing, 338, 339, 359
    defined by radical expressions, 481–482, 546
    definitions of, 213, 219, 239, 242
    domain of, 215, 216
    equation of the inverse, 654–655
    exponential, 661–666, 708
    exponential decay, 666
    exponential growth, 665–666
    greatest integer, 725, 771
    identification of, 218
    identity, 338, 339, 359
    inverse, 652–656, 707, 708
    linear, 222, 240
    logarithmic, 671–674, 708
    one-to-one, 707, 708
    polynomial, 335–339
    quadratic, 633, 635
    range of, 215, 216
    rational, 232, 414, 415, 465
    reciprocal, 722
    relations as, 214–215
    square root, 481
    squaring, 338, 339, 359, 722
    step, 771
    translation of, 722, 723
    vertical line test for, 217

Fundamental property of rational numbers, 415–418
Fundamental rectangle of hyperbola, 771
$f(x)$ notation, 219, 240

G

Galilei, Galileo, 560
Galileo's formula, 560
Geometry
    Euclidean, 94
    solving problems in, 79
    of systems of linear equations, 254, 267–268
Golden ratio, 584
Graph(s)
    of absolute value function, 722
    bar, 166
    circle, 166
    of circles, 733, 734, 746
    of cube root functions, 481
    of elementary functions, 722–724
    of ellipses, 737–738, 746
    of equations, 169, 239, 747, 748
    explanation of, 3, 43
    of exponential functions, 661–663, 708
    of horizontal lines, 171
    of hyperbolas, 744–746
    of inverse functions, 656
    line, 166
    of linear equations, 191, 192, 252–253, 267–268
    of linear inequalities, 114–115, 242
    of linear systems, 252–253
    of lines, 191–193
    of logarithmic functions, 671–672
    of numbers, 3
    of ordered pairs, 167
    of parabolas, 599–601, 616–618, 636, 746
    of polynomial functions, 338–339, 360
    of quadratic functions, 599–604, 613, 635
    of radical functions, 481–482
    of rational functions, 444, 467
    of reciprocal functions, 722
    of relations, 216
    of second-degree inequalities, 763–766
    of semicircles, 748
    to solve quadratic inequalities, 623–624
    of square root functions, 481, 722–723, 747–748, 773
    of systems of inequalities, 774
    of vertical lines, 171

"Greater than or equal to" symbol, 8, 44
"Greater than" symbol, 8, 44
Greatest common factor (GCF)
    explanation of, 372, 405
    factoring out, 373–374
Greatest integer functions
    explanation of, 771
    graphs of, 725
    method for applying, 725
Greenwich Mean Time (GMT), 60
Grouping, factoring by, 374–376, 405
Growth and decay applications, 665–666, 674
Guessing-and-checking strategy, 791

H

Half-open interval, 115
Harmonic mean, 458
Heron's formula, 488
Horizontal asymptotes, 444, 465
Horizontal lines
    equations of, 171–172, 196, 241
    graphs of, 171
    slope of, 179–180
Horizontal line test, 654
Horizontal parabolas, 616–617
Horizontal shift
    of ellipse, 738
    method for applying, 723–724
    of parabola, 599–602
Hyperbola(s)
    asymptotes of, 744–745, 771
    equations of, 743–744, 773
    explanation of, 733, 771
    foci of, 743
    fundamental rectangle of, 771
    graphs of, 744–746
    intercepts of, 743
Hypotenuse, 503

I

i, 540, 545
Identities, 58, 59, 101
Identity function, 338, 339, 359
Identity properties, 37–38, 46
Idle prime time, 400
Imaginary numbers, 545
Imaginary parts, 537, 545
Imaginary unit, 535
Improper fractions, 801
Incidence rate, 448
Inconsistent systems
    explanation of, 254, 295–296
    solving, 259–260
Independent equations, 99
Independent variable, 239
Index of radical, 480, 502, 512, 530, 545
Inequality(ies)
    absolute value, 139–144, 153
    addition property of, 116–117

compound, 129–131, 153, 154
    explanation of, 7, 43, 114, 115
    interval notation for, 114
    linear in one variable, 114–122
    linear in two variables, 205–208, 242
    multiplication property of, 117–118
    on number line, 8
    polynomial, 626
    quadratic, 623–626, 633, 636
    rational, 627–628, 633, 636
    second-degree, 763–766, 771, 774
    symbols for, 7–9, 44
    systems of, 774
    three-part, 120–121
Infinite set, 2
Infinity, 153
Integer exponents, 360
Integers
    consecutive, 98, 101
    explanation of, 3, 4, 45
Intercepts
    of ellipse, 736
    finding x- and y-, 169–170
    of hyperbola, 743
Interest
    compound, 698, 699
    simple, 81
International Date Line, 60
International time zones, 60, 70
Intersection
    of linear inequalities, 133, 207, 208
    of sets, 129, 133, 153
Interval notation, 114, 153
Intervals
    on number line, 114, 153
    types of, 115
Inverse
    additive, 5–6, 19, 43
    multiplicative, 18
Inverse functions
    equations of, 654–655
    explanation of, 652–653, 707, 708
    graphs of, 656
    horizontal line test and, 654
    notation for, 707
Inverse properties, 36–37, 46
Inverse variation, 232–234
Investments, 81–82, 702
Irrational numbers, 3, 4, 45

J

Joint variation, 233–234

L

Least common denominator (LCD)
    explanation of, 465
    method to find, 426, 806
Legs of right triangle, 503
"Less than or equal to" symbol, 8, 44

"Less than" symbol, 8, 44
Like terms
    combining, 38, 43, 330, 331
    explanation of, 43, 330
Linear equations in one variable
    applications of, 77–84, 91–94, 103
    with decimals, 58
    explanation of, 54, 101, 102
    with fractions, 57
    solution of, 54–59, 102
    solution set of, 54
    types of, 58–59
Linear equations in two variables
    explanation of, 169, 239, 241
    graphs of, 191, 192
    intercepts of, 191
    point-slope form of, 193, 241
    real data described by, 197, 198
    slope-intercept form of, 191, 192, 241
    standard form of, 241
    systems of, 252–265, 279–282
Linear functions, 222, 240
Linear inequalities in one variable
    addition property to solve, 116–117
    applications of, 121–122
    distributive property to solve, 119–120
    explanation of, 114–115, 153, 154
    graphs of, 114–115
    multiplication property to solve, 117–118
    steps to solve, 119
    with three parts, 120–121
Linear inequalities in two variables
    explanation of, 205, 239
    graphs of, 205–208, 242
    intersection of, 207, 208
    union of, 208
Linear systems of equations, 252, 299, 817. See also Systems of linear equations; Systems of linear equations in three variables; Systems of linear equations in two variables
Line graphs, 166
Lines. See also Number lines
    boundary, 239
    horizontal, 179–180, 241
    parallel, 182, 195–196
    perpendicular, 182–183, 195–196
    slopes of, 177–184
    vertical, 179–180, 241
Line segments, 176
Literal equation, 65
Logarithmic equations
    explanation of, 670, 707
    properties of, 695, 709–710
    solving, 670–671, 696–698
Logarithmic form to exponential form, 670

Logarithmic functions
    applications of, 672–673
    with base a, 671
    explanation of, 671, 708
    graphs of, 671–672, 708
Logarithms
    alternate forms of, 683–684
    change-of-base rule for, 700–701
    common, 687–688, 707, 709
    explanation of, 669, 707
    natural, 688–689, 707, 709
    properties of, 671, 679–684, 709
Long-distance costs, 134
Lowest terms
    of fractions, 802–803
    of rational expressions, 415–418, 466, 520

M
Mathematical expressions, 77–78
Mathematical models
    comparing, 340
    explanation of, 65, 101
Matrices
    augmented, 291–292, 299
    columns of, 291, 299
    determinants of, 813–816
    elements of, 299
    explanation of, 291, 299, 813
    row echelon form of, 292
    row operations on, 292–296
    rows of, 291, 299
    square, 291, 299, 813
Matrix method for solving systems, 291–296, 302
Maximum value problems, 614–616
Mean, 458
Members of set, 2, 43
Midpoint formula, 176
Midpoint of line segment, 176
Minimum value problems, 614–616
Minors of determinants, 814–816
Mixed numbers, 801
Mixture problems, 83–84, 279–280
Money denomination problems, 91–92
Monomials
    dividing polynomials by, 353
    explanation of, 330, 359
    multiplication of, 343
Motion problems, 92–94, 577–578
Multiplication
    associative property of, 38, 39
    of binomials, 344–345, 347, 515, 516
    commutative property of, 38, 39
    of complex numbers, 538, 548
    explanation of, 45
    FOIL method of, 345, 361, 379
    of fractions, 803–804

identity property for, 37
of monomials, 343
of polynomials, 343–348, 361
of radical expressions, 515–520, 547
of rational expressions, 418–419
of real numbers, 18–19
of sum and difference of two terms, 345–346
Multiplication property
    of equality, 54–56
    of inequality, 117–118
    of zero, 40, 46
Multiplicative identity, 37
Multiplicative inverse, 18

N
Natural logarithms
    applications of, 689–690
    evaluation of, 688–689
    explanation of, 707, 709
Natural numbers, 2, 4, 45
Negative
    of number, 5
    of polynomial, 331–332, 359
Negative exponents
    explanation of, 313–315, 360
    special rules for, 317
Negative infinity, 153
Negative numbers
    explanation of, 3
    operations with, 15–18
    square roots of, 535
Negative slope, 181
Nonlinear equations, 54, 753, 771
Nonlinear systems of equations
    explanation of, 753, 771
    solving, 753–757, 774
Notation
    function, 219–221, 239
    interval, 114, 153
    scientific, 318–321, 360
    set-builder, 2–3, 43, 44
"Not equal to" symbol, 8, 44
nth roots, 482–483
Null set. See Empty set
Number lines
    addition and subtraction of real numbers on, 15
    explanation of, 3–4, 43
    inequalities on, 8, 114, 115
    intervals on, 153
Number(s)
    absolute value of, 45
    additive inverse of, 5–6, 43
    complex, 535–540, 545, 548
    composite, 801–802
    counting, 2
    factors of, 25, 43, 67, 801
    Froude, 598

Number(s) (continued)
graphs of, 3
imaginary, 545
integers, 3, 4, 45
irrational, 3, 4, 45
mixed, 801
multiplicative inverse of, 18
natural, 2, 4, 45
negative, 3, 5, 15–18, 535
opposites of, 5
positive, 3
prime, 400, 801–802
prime factored form of, 802
rational, 3, 4, 45
real, 3, 4, 45, 46
reciprocal of, 18–19
sets of, 4–5, 45
signed, 6, 43
whole, 2, 4, 45
Numerators
of fractions, 801–803, 807
rationalizing, 524
Numerical coefficients, 329, 359

O

Ohm's law, 544
One-to-one functions
explanation of, 652–653, 707, 708
horizontal line test for, 654
inverse of, 653–655
Open intervals, 115
Operations
order of, 28–29, 45
on real numbers, 15–20, 45
set, 129, 131, 133
Opposites of numbers, 5
Order of radical. See Index of radical
Ordered pairs
explanation of, 167–168, 239
plotting, 167
as solution of linear systems, 253–254
subscripts with, 177
symbol for, 240
Ordered triple, 267, 299
Order of operations
on calculators, 29
explanation of, 45
use of, 28–29, 574
Origin, 167, 239

P

Pairs, ordered. See Ordered pairs
Parabola(s)
applications of, 614–616
axis of, 602, 633

explanation of, 599, 633
graphs of, 599–601, 616–618, 636, 746
horizontal, 616–617
horizontal shift of, 599–602
symmetry of, 599
vertex of, 602, 611–612, 633
vertical, 611–612, 614, 618
vertical shift of, 599–602
x-intercepts, number of, 614
Parallel lines
equations of, 195–196
slopes of, 182
Parentheses
representing ordered pairs, 167
solving formula with, 67
Percent
explanation of, 68, 101
formula for, 68
problems with, 68–69, 81
reading graphs, 69
Perfect squares, 790
Perfect square trinomials
explanation of, 387–388, 406
factoring, 388–389
Perigee, 742
Perimeter, 811
Perpendicular lines
equations of, 195–196
slopes of, 182–183
pH, 687–688
Pie charts, 166
Pi (π), 688
Pisano, Leonardo (Fibonacci), 788
Plane, 167
Plot, 167, 239
Point-slope form, 193, 196, 241
Polya, George, 787
Polynomial functions
addition of, 336–337
of degree n, 335, 359
division of, 356
evaluating, 335
graphs of, 338–339, 360
modeling data with, 336
subtraction of, 336–337
Polynomial inequalities, 626
Polynomial(s)
addition of, 330–331, 360
in ascending powers, 333
degree of, 330, 359
in descending powers, 329
division of, 353–356, 361, 823–825
explanation of, 329–330, 359

factoring, 372, 393
greatest common factor of, 372–374, 405
multiplication of, 343–348, 361
negative of, 331–332, 359
prime, 379, 380, 405
subtraction of, 330, 332, 360
terms of, 329
types of, 330
in x, 359
Positive numbers, 3
Positive or negative symbol, 545
Positive slope, 181
Power rules
for exponents, 317, 360
for logarithms, 681–683, 709
for radical equations, 527–530, 547
Powers. See also Exponents
ascending, 329
descending, 359
explanation of, 25
of i, 540
Prime factored form, 802
Prime meridian, 60
Prime numbers, 400, 801–802
Prime polynomials, 379, 380, 405
Principal nth roots, 480–481, 545
Principal square roots, 27
Problem-solving strategies. See also Applied problems
distinguishing between expressions and equations as, 78
drawing sketches as, 792
examples of, 788
guessing and checking as, 791
for investment problems, 81–82
looking for patterns as, 791–792
for mixture problems, 83–84
for percent problems, 81
Polya's four-step method as, 787
six-step method for solving applied problems, 78–80
translating from words as, 77–78
trial and error as, 790–791
using common sense as, 792
using tables or charts as, 788–789
working backward as, 789–790
Product rule
for exponents, 312–313, 317, 360, 679
for logarithms, 679–680, 683, 709
for radicals, 499, 536, 547
Product
explanation of, 43, 801
of sum and difference of terms, 345–347

Proper fractions, 801
Proportions
    explanation of, 452, 465
    solving applications using, 452–453
Pure imaginary numbers, 537, 545
$P(x)$ notation, 336
Pythagorean formula, 503, 547

Q

Quadrants, 167, 239
Quadratic equation(s)
    applications of, 398–399, 577–580
    completing the square to solve,
        561–564
    discriminant of, 572–573, 633, 634
    explanation of, 396, 633
    factoring to solve, 396–397
    methods for solving, 589
    nonreal complex solutions of, 564
    quadratic formula to solve, 570–572
    solving equation that leads to, 443, 577
    square root property to solve,
        560–561
    standard form of, 396, 405
Quadratic formula
    derivation of, 569–570
    explanation of, 633, 634
    solving quadratic equations using,
        570–571, 589
    use of discriminant and, 572–573
Quadratic functions
    applications of, 399
    applied problems modeled by,
        593–594
    discriminant of, 614
    explanation of, 399, 599, 633
    graphs of, 599–604, 613, 635
Quadratic inequalities
    explanation of, 623, 633
    graphing to solve, 623–624
    special cases of, 626
    steps to solve, 625
    test numbers to solve, 624–625
Quadratic regression, 604
Quotient rule
    for exponents, 315, 317, 360
    for logarithms, 680, 683, 709
    for radicals, 500, 547
Quotient
    explanation of, 19, 43, 823
    radical, 520

R

Rabbit problem, 788–789
Radical equations
    explanation of, 545
    power rule for, 527–530, 547

Radical expressions
    addition of, 511–512, 547
    division of, 515–520, 547
    explanation of, 481, 545, 546
    functions defined by, 481–482, 546
    multiplication of, 515–520, 547
    simplifying, 499–504, 511–512, 547
    subtraction of, 511–512, 547
Radicals
    converting between rational
        exponents and, 492
    equations with, 773
    explanation of, 480, 545
    index of, 480
    order of, 480
    product rule for, 499, 536, 547
    quotient rule for, 500, 547
    simplifying, 500–504, 547
    solving equations with, 527–530, 548,
        580–581
Radical symbol, 27, 44, 545, 790
Radicand, 480, 545
Radius, 229, 733, 735, 771
Range
    explanation of, 239
    of relations, 215, 216
Rates of change
    comparison of, 7
    interpreting slope as average,
        183–184
Rational equations, 441–443
Rational exponents
    converting between radicals and, 492
    evaluating, 489–490
    explanation of, 546
    negative, 490–491
    rules for, 492–493
Rational expressions
    addition of, 425–430, 466
    applications of, 451–458, 468
    division of, 420, 466
    domain of, 414–415, 465
    equations with, 441–444, 467
    explanation of, 414, 465
    lowest terms of, 415–418, 466, 520
    multiplication of, 418–419, 466
    reciprocals of, 419
    solving equations with, 441–443
    subtraction of, 425–430, 466
Rational functions
    domains of, 414, 415
    explanation of, 232, 414, –415, 465
    graphs of, 444, 467
Rational inequalities
    explanation of, 627, 633
    solving, 627–628, 636

Rationalizing the denominator
    with binomials involving radicals,
        519–520
    explanation of, 545
    with one radical term, 516–518,
        545
Rationalizing the numerator, 524
Rational numbers
    explanation of, 3, 4, 45
    fundamental property of, 415–418
Ratios, 465
Real-data applications
    average rate, 458
    average test scores, 156
    best buy, 104
    comparing mathematical models,
        340
    computer network problem, 362
    currency exchange, 104
    distance, 758
    earthquake intensity, 322
    exponential decay, 674
    golden ratio, 584
    idle prime time, 400
    international time zones, 60, 70
    investment evaluation, 702
    long-distance costs, 134
    order of operations, 574
    stock splits, 10
    windchill factor, 494
Real numbers
    basic concepts of, 2–9
    explanation of, 3, 4, 45
    exponents and, 25–27
    operations on, 15–20, 45
    order of operations and,
        28–30
    properties of, 35–40, 46
    reciprocals of, 18–19
    sets of, 153
    square roots and, 27–28
Real parts, 537, 545
Reciprocal function, 722
Reciprocals
    on calculator, 18
    explanation of, 43
    of fractions, 804
    of rational expressions, 419
    of real numbers, 18–19
Rectangular coordinate system
    explanation of, 167, 239, 241
    graphing lines in, 169–172
    ordered pairs in, 167–168
Relation(s)
    domain of, 215–216
    explanation of, 213–215, 239

Relation(s) (continued)
    as functions, 214–215
    graphs of, 216
    range of, 215–216
Remainder theorem, 825
Resonant frequency, 484, 487
Richter scale, 322, 678
Rise, 177, 239
Roots
    approximations for, 484
    on calculator, 483–484
    cube, 481
    explanation of, 45, 480
    nth, 482–483
    principal, 480–481, 545
    simplifying, 480, 500–501
    square, 27–28
Rotational rate N of space station
        formula, 534
Row echelon form of matrix, 292,
    299
Row operations on matrix, 292–296
Rows, of matrix, 291, 299
Rule of 72, 702
Run, 177, 239

S

Scientific notation
    applications of, 318–321
    converting to and from, 319, 320
    explanation of, 318, 360
Second-degree equations, 558, 746,
    756
Second-degree inequalities
    explanation of, 763, 771
    graphs of, 673–766, 774
Semicircles, 748
Semiperimeter, 488
Set-builder notation
    explanation of, 2, 43
    symbol for, 3, 44
Set operations, 129, 131, 133
Set(s)
    elements of, 2, 43, 45
    empty, 2, 43, 44
    explanation of, 2, 43
    finite, 2
    infinite, 2
    intersection of, 129, 153
    of numbers, 4–5, 45
    of real numbers, 153
    review of, 45
    solution, 54, 101, 252, 299, 764–766
    union of, 131, 153
Signed numbers, 6, 15, 43
Similar triangles, 598
Simple interest, 81

Slope(s)
    as average rate of change, 183–184
    from equations of lines, 178–179
    explanation of, 177, 239, 241
    formula, 177
    of horizontal lines, 179–180
    of lines, 177–184
    negative, 181
    of parallel lines, 182
    of perpendicular lines, 182–183
    positive, 181
    symbol for, 240
    of vertical lines, 179–180
    zero, 181
Slope-intercept form, 191, 196, 241
Solutions
    of equations, 54, 101, 826
    extraneous, 527, 545
Solution set
    explanation of, 54, 101
    of system of equations, 252, 299
    of system of inequalities, 764–766
Specified variables, solving for, 65–67,
    102, 451–452
Square matrix
    determinant of, 813
    explanation of, 291, 299, 813
Square root functions
    explanation of, 481, 747
    graphs of, 481, 722–723, 747–748, 773
Square root property
    applications of, 560
    explanation of, 558–560, 634
    solving quadratic equations with,
        560–561, 589
Square roots
    on calculator, 28
    explanation of, 27, 43
    method for finding, 27–28
    of negative numbers, 535
    principal, 27
    simplifying, 482–483, 535
    solving for variables involving, 571
    symbol for, 44
Squares
    of binomials, 529–530
    difference of, 387, 406
    perfect, 790
    solving for variables involving, 571
Squaring function, 338, 339, 359, 722
Standard form
    of complex numbers, 545
    of linear equations, 169, 196, 241
    of quadratic equations, 396, 405,
        558
Step functions, 771
Stocks, 100

Stock splits, 10
Subscripts, 177
Substitution method
    for factoring, 384
    for linear systems, 254–256
    to solve equation quadratic in form,
        581–582
    to solve nonlinear systems, 753–754
Subtraction
    of complex numbers, 537–538, 548
    explanation of, 45
    of fractions, 807
    of polynomial functions, 336–337
    of polynomials, 330, 332, 360
    of radical expressions, 511–512
    of rational expressions, 425–430,
        466
    of real numbers, 16–18
Sum
    of the angles of triangle, 94
    of cubes, 389–390
    explanation of, 43, 805
Supplementary angles, 98, 101
Symmetry, of parabola, 599
Synthetic division
    to determine solutions of equations,
        826
    to divide by polynomial, 823–825
    to evaluate polynomials, 825
    explanation of, 824
Systems of equations. See also Systems
        of linear equations
    consistent, 299
    explanation of, 252, 299
    inconsistent, 299
    nonlinear, 771
    special cases of, 258–260
Systems of inequalities
    explanation of, 764
    graphs of, 774
    solution set of, 764–766
Systems of linear equations
    applications of, 277–284, 301
    Cramer's rule to solve, 817–820
    inconsistent, 295–296
    matrix method to solve, 291–296, 302
Systems of linear equations in three
        variables
    elimination method to solve,
        268–271, 301
    explanation of, 267, 282
    geometry of, 267–268
    graphs of, 267–268
    inconsistent, 271
    matrix method to solve, 294–295
    with missing terms, 270–271
    special cases of, 271–272

Systems of linear equations in two variables
  consistent, 254
  elimination method to solve, 256–258, 300
  graphs of, 252–253
  inconsistent, 254
  matrix method to solve, 293–294
  ordered pair as solution of, 253–254
  special cases of, 258–260
  steps to solve, 279–282, 300
  substitution method to solve, 254–265
Systems of nonlinear equations, 753, 771

T

Terms
  degree of, 330, 359
  explanation of, 43, 359
  like, 38, 43, 330, 331
  of polynomial, 329
  unlike, 38
Three-part inequalities, 120–121
Threshold weight, 487
Time zones, 60, 70
Traffic intensity, 448
Translations
  of functions, 723, 722
  of parabola, 599
Trial-and-error strategy, 790
Triangle(s)
  angles of, 94
  area of, 488
  problems involving angles of, 94
  similar, 598
Trinomials
  explanation of, 330, 359
  factoring, 379–384, 393, 406
  perfect square, 387–389
Triple, ordered, 267, 299

U

Union
  of linear inequalities, 208
  of sets, 131, 133, 153
Universal constant, 688

V

Variable(s)
  dependent, 213, 239
  explanation of, 2, 43
  independent, 213, 239
  simplifying radicals involving, 501–502
  solving for specified, 65–67, 102, 451–452
  solving for squared, 591–592
Variation
  combined, 234
  constant of, 229–230
  direct, 229
  inverse, 232–233
  joint, 233–234
Variation equation, 230–233
Vertex
  coordinates of, 615
  explanation of, 633
  formula, 612
  of parabolas, 602, 611–612
Vertical angles, 98, 101
Vertical asymptotes, 444, 465
Vertical lines
  equations of, 171–172, 196, 241
  graphs of, 171
  slope of, 179–180
Vertical line test, 217
Vertical parabolas
  explanation of, 618
  vertex of, 611–612
  x-intercepts of, 614
Vertical shift
  of ellipse, 738
  explanation of, 600
  method for applying, 724
  of parabola, 599–602

W

Whole numbers, 2, 4, 45, 801
Windchill factor, 494
Word expressions
  translated to equations, 78
  translated to mathematical expressions, 77–78
Working backward strategy, 789–790
Work rates, 456–457, 579–580

X

x-axis, 167, 239
x-intercepts
  explanation of, 169, 239
  finding, 169–170
  of parabola, number of, 614

Y

y-axis, 167, 239
y-intercepts
  explanation of, 169, 239
  finding, 169–170
  slope and, 191

Z

Zero
  division by, 19
  multiplication property of, 40, 46
Zero exponent, 313, 360
Zero-factor property
  applications of, 398–399
  explanation of, 395–398, 558